As students work through this text, they can supplement their understanding of each chapter's main themes, ideas and objectives using the attached CD-Rom and the Dynamic Earth 5e website:

www.wiley.com/college/skinner

GeoDiscoveries CD-Rom

This easy to use and student-focused CD-Rom helps reinforce and illustrate key concepts from the text. It also provides interactive media content that helps students prepare for tests and improve their grades. Each **GeoDiscoveries Module** has a three-part structure:

- Presentations that use videos, animations, and other visual resources to further explore key concepts from the chapter;

- Interactivities that challenge and engage students in concept-based exercises; and

- Assessment self-tests that allow students to measure their comprehension of the concept being explored.

The CD also features a **Virtual Petroscope** containing a library of images, tools and resources designed to simulate the real-world process of identifying rock specimens. Students are first introduced to criteria and tools geologists use to identify rocks and minerals, and are then encouraged to apply these resources to a collection of unidentified specimens.

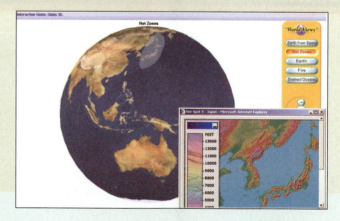

The CD-based **Interactive 3-D Globe** allows students to explore and understand the world by changing its face, using 5 distinct textures.

The Dynamic Earth Student Companion Site

www.wiley.com/college/skinner

This web site provides additional resources that compliment the textbook in ways that help teachers teach and students learn. Enhance your understanding of Geology and improve your grade by using the following resources:

- **Chapter Review Quizzes** provide immediate feedback to true/false, multiple choice, and short answer questions.

- **Interactive Flashcards** help students review and quiz themselves on the concepts, ideas and terms discussed in each chapter.

- **Interactive Drag-and-Drop Exercises** challenge students to correctly label important illustrations from the textbook.

- **The Audio Glossary** helps students learn how to pronounce difficult words.

- **Annotated Web Links** put useful electronic resources into context.

- **The Dynamic Earth Notes Outlines** correlate with the PowerPoint presentations created for instructors, providing an illustrated framework for their classroom notes.

Dynamic Earth

An Introduction to Physical Geology

Fifth Edition

BRIAN J. SKINNER
Yale University

STEPHEN C. PORTER
University of Washington

JEFFREY PARK
Yale University

WILEY
John Wiley & Sons. Inc.

Senior Acquisitions Editor   Ryan Flahive
Marketing Manager   Kevin Molloy
Senior Developmental Editor   Marian Provenzano
Assistant Editor   Denise Powell
Editorial Assistants   Alijah Vinikoor and Lili Kalish
Senior Production Editor   Patricia McFadden
Senior Designer   Harold Nolan
Interior Design   Lisa Delgado/Delgado Design
Cover Design   Howard Grossman
Illustration Coordinator   Anna Melhorn
Photo Editor   Jennifer MacMillan
Photo Researchers   Alexandra Truitt and Jerry Marshall
Cover Photos   Galen Rowell/Mountain Light Photography Inc.

Freelance Production Services   Ingrao Associates

This book was set in 10/12 Times Roman by LCI Design and printed and bound
by Von Hoffmann, Inc. The cover was printed by Von Hoffmann, Inc.

This book is printed on acid-free paper. ∞

L.C. Call no.   Dewey Classification No.   L.C. Card No.
ISBN 0-471-15228-5
WIE ISBN: 0-471-45157-6

Printed in the United States of America

10 9 8 7 6 5 4 3 2 1

Preface

This is a book about change. We are all aware that weather changes from day to day, and if we live near the shore, we know that tides and storms are ever changing the sea. We are not so likely to be aware that the solid Earth is also changing. The changes are mostly slow compared to scales of human activities, but over time mountains rise and are eroded away, new mountains rise, continents shift position, ocean basins grow and shrink and sometimes even disappear. Although these slow changes were apparent to geologists, until recently the reasons for them were not understood. Then, about 40 years ago, a new theory emerged, a theory that provided answers to how changes happened, and that also provided an all-encompassing framework within which to investigate and understand the way Earth works. That theory is the plate tectonic theory. For 40 years scientists have been hard at work revising and refining their understanding of Earth and its processes in light of the plate tectonic theory. It has been—and still is—a time of extraordinary excitement. There is still much left to do and to be investigated. We want to share some of that excitement with students, because it is they who will make the investigations and take the next steps in understanding Earth.

As a result of plate tectonics Earth's surface slowly changes. When mountains rise as a result of collisions between plates, wind patterns change. When an ocean appears because a plate splits apart, the climate changes. Careful study shows that all parts of Earth are interdependent. The atmosphere, the ocean, the soils and rocks, and the abundant life forms, interact in innumerable ways to maintain Earth's present state. Our planet, with all its parts, is a system—the Earth system, and the study of our planet as a system is called Earth system science. Today, with satellites in the sky, with measuring devices in remote places like the ocean floor and the great Antarctic ice sheet, it is possible to observe and measure movement of materials and energy, and to detect changes as they happen.

Plate tectonics and the Earth system are two parts of the basic framework around which this book is written. The two themes are introduced in the first two chapters. Thereafter, in order to stress the importance of the themes and their interconnectedness, each chapter closes with a section titled *Revisiting Plate Tectonics and the Earth System*.

WHAT'S NEW IN THE FIFTH EDITION?

This book is designed for a one-semester introduction to the way Earth works. We have done our best to make the book current by including the latest advances in science, and do so in a way that is accessible to students for whom this is a first introduction to physical geology.

We have emphasized plate tectonics early in the book. The concept is introduced in the first chapter, and the second chapter discusses the plate tectonic theory in detail. Most students will have been introduced to the concept of drifting continents and other manifestations of plate tectonics in grade school. The traditional way to discuss plate tectonics has been to do so on the basis of evidence from paleomagnetism and the random reversal of Earth's magnetic field. Modern evidence for plate tectonics involves Global Positioning System (GPS) measurements. GPS is an acronym that will be familiar to most college students. Airplanes, ships, automobiles, and geologists in the field use GPS to tell precisely where they are on Earth. Similarly, GPS can be used to detect motion of a continental landmass (or an island in the ocean) as it moves slowly over Earth's vast interior. Such measurements were unavailable in the 1960s and 1970s when the plate tectonic theory was first advanced. Today it is the way contemporary plate motions are studied. We use GPS measurements as the way to introduce plate tectonics in Chapter 2. Historic evidence, based on paleomagnetism, is discussed much later in the book, in Chapter 20, in which we discuss *Earth Through Geologic Time*.

Responding to insightful reviews from colleagues, we have changed the order of chapters from previous editions. Following an introduction to physical geology in Chapter 1, which includes the Earth system, and an introduction to plate tectonics in Chapter 2, we proceed to minerals in Chapter 3. Igneous processes are covered in two chapters rather than one, as in previous editions. Chapter 4 discusses *Igneous Rocks: Products of Earth's Internal Fire*, while Chapter 5 discusses *Magmas and Volcanoes*. Chapters 6, 7, and 8 cover weathering, sediments and sedimentary rocks, and metamorphism and metamorphic rocks, respectively. We have moved the chapters on structure and earthquakes to follow metamorphism; they are now Chapters 9 and 10, respectively. This order of topics places Chapter 11 on relative and numerical dating later in the sequence than in previous editions.

With Chapter 12 we have again introduced a topic change. This chapter, *The Changing Face of the Land*, has been moved from late in the book to become the introduction to several chapters on surficial processes. We believe that this order of topics is a more effective way to stress the importance of Earth as a system. The remaining chapters in the book introduce topics in the sequence they were introduced in previous editions.

Revisions, updates, and corrections are present in all chapters. About 25% of the illustrations in this edition are new, and many of the new illustrations are based on current and ongoing research. The new edition features four exciting two-page images called Changing Landscapes, which illustrate the role of geologic processes in shaping our planet. These new Changing Landscapes show how plate tectonics expresses itself in the south Pacific Ocean, how the eruption of Mayon volcano in the Philippines influenced surrounding communities, how earthquakes are expressed within patterns of topographic relief of western North America, and how glacier ice moves within the Antarctic ice sheet. The Changing Landscapes images follow pivotal chapters in the text, so that the reader can view a colorful synthesis of material developed in previous chapters. Each chapter in this fifth edition opens with a brief essay by one of us based on work we have done during our professional careers. Most of the essays concern field-based studies in various parts of the world. Following each of the major topic sections in a chapter, there is a short section titled *Before You Go On*, in which two to five brief questions are asked in order to stress the important items covered.

To remind students that the two main themes of the text are plate tectonics and the Earth system, beginning with Chapter 3 each chapter closes with a section containing two brief essays, one on plate tectonics one on the Earth system, in which we point out how those themes are involved in some aspect or aspects of the topics that were discussed in the chapter.

Finally, because we live on a planet with a population of 6 billion and counting, we integrate discussions of human effects on the environment throughout the book. The human race is now a force so far as changes to Earth are concerned. Chapter 19, *The Climate and Our Changing Planet*, ties earlier presented evidence together and presents the argument that humans have changed, and continue to change, the Earth system, and the climate of the world is changing in response.

THE TEACHING AND LEARNING PACKAGE

This 5e of Dynamic Earth is supported by a comprehensive supplements package that includes an extensive selection of print, visual, and electronic materials.

SUPPLEMENTS TO HELP TEACHERS TEACH

DYNAMIC EARTH 5E INSTRUCTORS' SITE

This comprehensive website includes numerous resources to help you enhance your current presentations, create new presentations and employ our pre-made PowerPoint presentations. These resources include:

- A complete collection of PowerPoint presentations. One for each chapter.
- All of the line art from the text.
- Access to hundreds of photographs.
- A comprehensive collection of animations.
- Research materials such as additional, brief articles or handouts in PDF.
- A comprehensive test bank.
- Advisory website with teaching and other career related suggestions.

INSTRUCTOR'S MANUAL AND TEST BANK

This manual includes a test bank, chapter overviews and outlines, and lecture suggestions. The Test Bank is also available on the password protected Instructor's website.

OVERHEAD TRANSPARENCY SET

Includes full-color textbook illustrations, resizes and edited for maximum effectiveness in large lecture halls.

SUPPLEMENTS TO HELP STUDENTS LEARN

GEODISCOVERIES CD-ROM

This easy-to-use and student-focused CD-Rom helps reinforce and illustrate key concepts from the text. It also provides interactive media content that helps students prepare for tests and improve their grades. The CD also features a **Virtual Petroscope** containing a library of images, tools, and resources designed to simulate the real-world process of identifying rock specimens. Students are first introduced to criteria and tools geologists use to identify rocks and minerals, and are then encouraged to apply these resources to a collection of unidentified specimens. The CD-based **Interactive 3-D Globe** allows students to explore and understand the world by changing its face, using five distinct textures.

DYNAMIC EARTH STUDENT COMPANION SITE: WWW.WILEY.COM/COLLEGE/SKINNER

This website provides additional resources that compliment the textbook. Enhance your understanding of Geology and improve your grade by using the following resources:

- **Chapter Review Quizzes** provide immediate feedback to true/false, multiple choice, and short answer questions.

- **Interactive Flashcards** help students review and quiz themselves on the concepts, ideas, and terms discussed in each chapter.

- **Interactive Drag-and-Drop Exercises** challenge students to correctly label important illustrations from the textbook.

- **Annotated Web Links** put useful electronic resources into context.

- **Dynamic Earth Notes Outlines** correlate with the PowerPoint presentations created for instructors, providing an illustrated framework for their classroom notes.

STUDY GUIDE TO ACCOMPANY DYNAMIC EARTH, FIFTH EDITION

This internet-based workbook by Denyse Lemaire of Rowan University presents on-line labs founded in Columbia University's Earthscape—an on-line resource on the global environment.

Geoscience Laboratory: The Discovery Edition 3e

The interactive format of this lab manual stimulates the spirit of inquiry and discovery, challenging students to tackle problems by thinking critically about geologic processes. Eighteen chapters in a three-ring, loose-leaf binder allow bringing maps and answer pages into alignment with the text. This format allows instructors to collect answer pages for scoring. An American Optical hand-help stereoscope is included for FREE, and a comprehensive Instructor's Guide is available.

GeoSciences Today

This casebook and interactive website provide students with eight cases from around the world in which to see and explore the interaction of people and their environment. The casebook was authored and developed by Robert Ford, Westminster University and James Hipple, University of Missouri.

Dynastic Earth: A Study Guide to Accompany the Dynamic Earth 5e

This manual integrates the use of independent multiple choice questioning throughout the course. It encourages students to actively participate in the study of physical geology. This guide is authored and developed by Michael and Susan Kimberly of North Carolina State University.

ACKNOWLEDGMENTS

Plans for the fifth edition of Dynamic Earth were commenced under the direction of Clifford Mills, then Earth Science editor at John Wiley and Sons. With Cliff's retirement, responsibility for the earth sciences, and for completion of the fifth edition passed to his successor Ryan Flahive, and publisher Anne Smith. This dynamic duo set tight deadlines and challenged us to produce. We authors have been extremely fortunate, both with the energy and enthusiasm of the earth-science team, and with the high level of expertise provided by all the staff of John Wiley and Sons; they always responded when we needed assistance. We have been especially fortunate to be able to work with developmental editor Marian Provenzano. Her unfailing good humor, patient chiding when we missed deadlines, and careful attention to detail have made the experience of completing this revision an easier and more enjoyable one that it would otherwise have been. We are very grateful to her.

The professional skills and competence of all the staff at John Wiley and Sons who worked on this edition, and of all the freelance professionals who were involved have been outstanding. Trish McFadden oversaw production; Jennifer MacMillan sought exactly the photos we needed; Anna Melhorn managed the illustration program; Harry Nolan designed the text; Denise Powell and Aliyah Vinikoor obtained the text's supplements; Martin Batey, Tom Kulesa, and Bridget O'Lavin oversaw multimedia development; Kevin Molloy developed the marketing program; and Lili Kalish and Yolanda Pagan provided editorial assistance. They and everyone else at Wiley were always cordial and unfailingly helpful, no matter how badly our travel schedules squeezed the carefully planned production schedules. It is only fair that we acknowledge the critical contributions of three professional colleagues; J. Marion Wampler of the Georgia Inst. of Technology provided a careful and detailed review of the fourth edition of the text, and his advice has stood us in good stead in the present edition. Theodore J. Bornhorst of the Michigan Technological University has been especially helpful with critical reviews for the clarity and content of figures, and with a sharp-eyed reading of the page proof. Richard J. Stewart critically read Chapter 7 on Sediments and Sedimentary rocks. Two expert freelance editors helped lighten the writing style; Frederick Cree Schroyer brought a whimsical turn of phrase and a sharpened pencil to various parts of the text, and Beverly Peavler untangled the syntax and lightened turgid prose in several chapters. Working with these skilled professionals has been an educational pleasure and a privilege.

We are especially indebted to our professional friends and colleagues who have responded when asked for advice and help on illustrations arising from the latest research. David Evans, Yale University, provided figures arising from his research for Chapter 20. For the Changing Landscapes section that follows Chapter 19, Ian Loughlin, Jet Propulsion Lab, California Institute of Technology, provided a spectacular image of flow in the Antarctic ice sheet; and Michael Studinger and Robin Bell, Lamont-Doherty Earth Observatory, provided Lake Vostok graphics. Ross Stein, U.S. Geological Survey, Gabi Laske, University of California, San Diego; Vadim Levin, Rutgers University; Nikolai Shapiro and Michael Ritzwoller of the University of Colorado provided earthquake and seismic images for Chapter 10. Peter Reiners, Yale University and Jonathan Tomkin, Louisiana State University, provided images of fission tracks and denudation simulations, respectively, for Chapter 12. Peter Bunge, Princeton University; Mark Richards and Michael Manga, University of California, Berkeley; John Baumgartner, Los Alamos National Laboratory; Uwe Walzer, University of Jena; and Dorothy Koch, NASA Goddard Institute of Space Sciences, provided images of mantle convection for Chapter 2.

We are indebted to the many people who provided the elegant photographs that appear in the book. Many of the photographers are geologists, and their discerning eyes and understanding of their topics can be sensed through the beautiful photographs they took. The names of the photographers are listed in the photo credits at the back of the book, but being so placed is no reflection on their importance.

We especially thank the thoughtful and dedicated teachers who commented on the previous texts or reviewed the present text. These fine people not only helped us keep a reasonable balance to the book, they also helped us keep the volume as up-to-date as possible without downplaying the great geological discoveries of the past. In particular, we thank the reviewers who have helped us prepare the fifth edition:

Raymond Beiersdorfer
Youngstown State University

Theodore J. Bornhorst
Michigan Technological University

Lynn A. Brant
University of Northern Iowa

Wang-Ping Chen
University of Illinois

Kevin Cole
Grand Valley State University

René DeHon
University of Louisiana at Monroe

Craig Dietsch
University of Cincinatti

James Evans
Bowling Green State University

Mark Feigenson
Rutgers University

Lisa Gilbert
Highline Community College

Julian W. Green
University of South Carolina-Spartanburg

Robert L. Hooper
University of Wisconsin-Eau Claire

Richard L. Josephs
University of North Dakota

John Madsen
University of Delaware

Charlie Onasch
Bowling Green State University

Todd A. Radenbaugh
University of Regina

Brady Rhodes
California State University-Fullerton

David R. Schwimmer
Columbus State University

Kevin Stewart
University of North Carolina-Chapel Hill

Jennifer A. Thomson
Eastern Washington University

Paul Tomascak
University of Maryland

Jan Tullis
Brown University

Gene C. Ulmer
Temple University

Mari A. Vice
University of Wisconsin-Platteville

Michael Wizevich
Southern Connecticut State University

Thomas H. Wolosz
State University of New York at Plattsburgh

Earlier editions were reviewed by:

Gary C. Allen
University of New Orleans

J.C. Allen
Bucknell University

Thomas B. Anderson
Sonoma State University

N.L. Archbold
Western Illinois University

Charles W. Barnes
Northern Arizona University

Philip Brown
University of Wisconsin

Collete D. Burke
Wichita State University

J. Allan Cain
University of Rhode Island

Robert Christman
Western Washington University

George Clark
Kansas State University

Nicholas K. Coch
Queens College

Kristine Crossen
University of Alaska

John Diemer
University of North Carolina at Charlotte

Grenville Draper
Florida International University

M. Ira Dubins
State University of New York at Oneonta

Timothy W. Duex
University of Southwestern Louisiana

William R. Dupré
University of Houston

John Ernisee
Clarion University of Pennsylvania

Stewart Farrar
Eastern Kentucky University

Mike Follo
University of North Carolina at Chapel Hill

Katherine Giles
New Mexico State University

Julian W. Green
University of South Carolina at Spartanburg

C.B. Gregor
Wright State University

Ann G. Harris
Youngstown State University

Richard A Heimlich
Kent State University

Robert L. Hopper
University of Wisconsin

Robert Horodyski
Tulane University

Albert Hsui
University of Illinois at Urbana-Champaign

Michael M. Kimberley
North Carolina State University

Allan Kolker
University of Nebraska-Lincoln

Peter L. Kresan
University of Arizona

Albert M. Kudo
University of New Mexico

Judith Kusnick
California State University

Peter B. Leavens
University of Delaware

Nancy Lindsley-Griffin
University of Nebraska

William W. Locke
Montana State University

Tim K. Lowenstein
State University of New York at Binghamton

David N. Lumsden
Memphis State University

Bart S. Martin
Ohio Wesleyan University

Gerald Matisoff
Case Western University

Robert McConnell
Mary Washington College

Stephen A. Nelson
Tulane University

Bruce Nocita
University of South Florida

Anne Pasch
University of Alaska-Anchorage

Gary Peters
California State University, Long Beach

Loren A. Raymond
Appalachian State University

John Renton
West Virginia University

Jack Rice
University of Oregon

Mary Jo Richardson
Texas A&M University

Robert W. Ridky
University of Maryland at College Park

Donald Ringe
Central Washington University

Len Saroka
St. Cloud State University

Paul A. Schroeder
University of Georgia

Brief Table of Contents

1   **Meet Planet Earth** 2

2   **Global Tectonics: Our Dynamic Planet** 30

3   **Atoms, Elements, Minerals, Rocks: Earth's Building Materials** 66

4   **Igneous Rocks: Products of Earth's Internal Fire** 98

5   **Magmas and Volcanoes** 120

6   **Weathering and Soils** 147

7   **Sediments and Sedimentary Rocks: Archives of Earth History** 172

8   **Metamorphism and Metamorphic Rocks: New Rocks from Old** 202

9   **How Rock Bends, Buckles, and Breaks** 224

10   **Earthquakes and Earth's Interior** 246

11   **Geologic Time and the Rock Record** 276

12   **The Changing Face of the Land** 302

13   **Mass Wasting** 324

14   **Streams and Drainage Systems** 352

15   **Groundwater** 382

16   **Glaciers and Glaciation** 410

17   **Atmosphere, Winds and Deserts** 446

18   **The Oceans and Their Margins** 474

19   **Climate and Our Changing Planet** 508

20   **Earth Through Geologic Time** 530

21   **Resources of Minerals and Energy** 560

Appendix A
Units and Their Conversions A1

Appendix B
Tables of the Chemical Elements and Naturally Occurring Isotopes A4

Appendix C
Table of the Properties of Common Minerals A7

Glossary G1
Selected References R1
Credits C1
Index I1

Contents

1 Meet Planet Earth 2
Opening Essay:
Discovering Geology, 3
Introduction, 4
About Geology and Geologists, 5
    Where We Work and What We Do, 5
    Physical Versus Historical Geology, 7
The Scientific Method, 8
How Rapid Are Geologic Processes?, 8
    James Hutton Applies the Scientific Method, 9
    Uniformitarianism, 9
    Catastrophism, 10
    Geologic Time and Earth's Age, 12
Earth's Neighborhood: The Solar System, 12
    The Planets—Terrestrial and Jovian, 12
    Origin of the Solar System—and Earth, 12
    Planetary Accretion Continues Today, 16
Earth's Internal Structure: Energy, Heat, Gravity, and Buoyancy, 16
    Layers of Differing Composition, Sorted by Density, 17
    Layers of Differing Physical Properties, 18
    Plate Tectonics, 18
The System Concept, 19
    The Earth System, 20
    Our Planet's "Four Spheres," 21
Interactions among Earth's Open Systems, 21
    Cyclical Movements, 21
    The Hydrologic Cycle, 23
    The Rock Cycle, 23
    The Tectonic Cycle, 26
    Uniformitarianism and Rates of Geologic Processes, 26
Changes Caused By Human Activities, 27
What's Ahead?, 27
Chapter Summary, 28
The Language of Geology, 29
Questions For Review, 29
GeoDiscoveries: The Earth System, 29
Understanding Our Environment Box 1.1 *Epitaph for the Aral Sea*, 7
The Science of Geology Box 1.2 *Cycling of Biogeochemical Elements Important for Life*, 24

2 Global Tectonics: Our Dynamic Planet 30
Opening Essay:
Earth is a Noisy Planet, If You Keep Your Ear to the Ground, 31

Introduction, 32
    Why Don't We Live on Venus?, 33
    Plate Tectonics: From Hypothesis to Theory, 33
What Earth's Surface Features Tell Us, 34
    Earth Is Solid But Not Rigid, 34
    Isostasy: Why Some Rocks Float Higher than Others, 34
    Earth's Surface: Land versus Water, 35
What Earth's Internal Phenomena Tell Us, 37
    Geothermal Gradients, 37
    Mantle Convection, 38
    Adiabatic Expansion of Rock, 40
Plates and Mantle Convection, 41
    GPS Proves That Continents Move, 41
Four Types of Plate Margins and How They Move, 42
    Seismology and Plate Margins, 45
Type I: Divergent Margin, 46
    Midocean Ridges, 46
    Birth of the Atlantic Ocean, 46
    Characteristics of Spreading Centers, 48
    Role of Seawater at Spreading Centers, 48
    The CO_2 Connection, 48
Type II: Convergent Margin/Subduction Zone, 48
    The CO_2 Connection, Again, 49
    Volcanoes Above Subduction Zones, 49
    Earthquakes in Subduction Zones, 50
Type III: Convergent Margin/Collision Zone, 51
Type IV: Transform Fault Margin, 52
Topography of the Ocean Floor, 53
    Comparing Venusian Topography, 53
    An Icy Analog to Earth Tectonics, 54
Hot Spots and Absolute Motion, 54
    Relative versus Absolute Motion, 54
    Mantle Plumes as Fixed Reference Points, 54
    Volcanic Domes and Coronae on Venus, 57
    Plume Volcanism on Mars, 57
What Causes Plate Tectonics?, 59
    How Does Plate Tectonics Work?, 59
    Why Does Plate Tectonics Work?, 59
What's Ahead?, 61
Chapter Summary, 61
The Language of Geology, 62
Questions For Review, 62
GeoDiscoveries: Plate Tectonics, 63
The Science of Geology Box 2.1 *Rock Deformation: Strain and Stress*, 32
The Science of Geology Box 2.2 *Thermal Convection*, 39

3 Atoms, Elements, Minerals, Rocks: Earth's Building Materials 66
Opening Essay:
Mineral Exploration: The Story of Sudbury, 67
Introduction: What Is A Mineral?, 68
    Mineraloids—Not Quite Minerals, 69
    Key Characteristics of Minerals, 69
Composition of Minerals, 69
    Chemical Elements, 69
    Energy-Level Shells, 70
    Ions, 71
    Compounds, 71
    Bonds of Four Major Types, 71
    Periodic Table of Chemical Elements, 73
Crystal Structure of Minerals, 74
    Ionic Substitution, 74
Properties of Minerals, 77
    Crystal Form and Growth Habit, 77
    Cleavage, 79
    Luster, 80
    Color and Streak, 80
    Hardness and the Mohs Scale, 81
    Density and Specific Gravity, 81
    Mineral Properties and Bond Types, 82
Common Minerals, 83
Silicates: The Largest Mineral Group, 84
    The Sturdy Silicate Tetrahedron, 84
    Isolated Tetrahedra: Olivines and Garnets, 85
    Chains: Pyroxenes and Amphiboles, 86
    Sheets: Clays, Micas, Chlorites, and
    Serpentines, 86
    Quartz—Pure SiO_2, 88
    The Feldspar Group—Commonest Minerals in
    Earth's Crust, 88
The Carbonate, Phosphate, and Sulfate Mineral
Groups, 89
    Carbonates, 89
    Phosphates, 89
    Sulfates, 90
The Ore Mineral Groups—Our Source for
Metals, 90
    Sulfides, 90
    Oxides, 92
Minerals Give Clues to Their Environment of
Formation, 92
Rocks: Mixtures of Minerals, 92
    Distinguishing The Three Rock Types, 92
Revisiting Plate Tectonics and the Earth
System, 93
    Plate Tectonics and Ore Minerals, 93
    Rock Types and the Rock Cycle, 93
What's Ahead?, 95
Chapter Summary, 95
The Language of Geology, 96
Mineral Names to Remember, 96

Questions For Review, 97
GeoDiscoveries: Atoms, Elements, Minerals, and
Rocks, 97
Petroscope For Minerals, 97
The Science of Geology Box 3.1 The Three States
of Matter, 76
Understanding Our Environment Box 3.2 Long-
term Danger of Abandoned Mines, 82
Understanding Our Environment Box 3.3 Asbestos:
How Risky Is It?, 91

4 Igneous Rocks: Products of Earth's
Internal Fire 98
Opening Essay:
Watching Magma Solidify, 99
Introduction: What Is an Igneous Rock?, 100
    Igneous Rocks—Intrusive and Extrusive, 101
Texture in Igneous Rocks, 101
    Sizes of Mineral Grains, 101
    How Mineral Grains Pack in Igneous
    Rocks, 102
Mineral Assemblage in Igneous Rocks, 102
    Intrusive (Course-Grained) Igneous Rocks, 103
    Extrusive (Fine-Grained) Igneous Rocks, 103
Pyroclasts, Tephra, and Tuffs, 104
Plutons, 106
    Minor Plutons, 107
    Major Plutons, 107
Whence Magma?, 110
    Distribution of Volcanoes, 110
    Origin of Balsaltic Magma, 111
    Origin of Andesitic Magma, 113
    Orgin of Rhyolitic Magma, 113
Solidification of Magma, 114
    Bowen's Reaction Series, 114
    Valuable Magmatic Mineral Deposits, 116
Revisiting Plate Tectonics and the Earth
System, 117
    Plate Tectonics and the Ocean Floor, 117
    Igneous Rock and Life on Earth, 117
What's Ahead?, 118
Chapter Summary, 118
The Language of Geology, 119
Important Rock Names to Remember, 119
Questions For Review, 119
GeoDiscoveries: Volcanoes and Igneous Rocks, 119
Petroscope to Igneous Rocks, 119
The Science of Geology Box 4.1. How Rock
Melts, 112

5 Magmas and Volcanoes 120
Opening Essay:
Witnessing an Eruption at Close Hand, 121
Introduction: Earth's Internal Thermal Engine, 122

Magma: Molten Rock of Different Types, 122
    Composition of Magmas and Lavas, 123
    Temperature of Magmas and Lavas, 123
    Viscosity of Magmas and Lavas, 124
How Bouyant Magma Erupts on the Surface, 125
Eruption Style—Nonexplosive or Explosive?, 125
    Nonexplosive Eruptions (Hawaiian-Type), 125
    Explosive Eruptions, 126
Volcanoes, 129
    Central Eruptions, 129
    Other Features of Central Eruptions, 132
    Fissure Eruptions, 136
Posteruption Effects, 137
Volcanic Hazards, 138
Revisiting Plate Tectonics and the Earth
System, 140
    Plates and Volcanoes, 140
    Submarine Volcanism and the Composition of
    Seawater, 141
What's Ahead?, 142
Chapter Summary, 142
The Language of Geology, 143
Questions For Review, 143
GeoDiscoveries: Geohazards and Volcanoes and
Igneous Rocks, 143
Understanding Our Environment Box 5.1 *Life
Returns to Mount St. Helens*, 130
The Science of Geology Box 5.2 *Harnessing the
Heat Within*, 138

6   Weathering and Soils 146

Opening Essay:
Chinese Soils and Monsoon Climates, 147
Introduction: Weathering—The Breakdown of
Rock, 148
Physical Weathering, 149
    Development of Joints, 149
    Crystal Growth, 150
    Frost Wedging, 151
    Daily Heating and Cooling, 151
    Spalling Due to Fire, 151
    Wedging by Plant Roots, 152
Chemical Weathering, 152
    Weathering Pathways, 153
    Effects of Chemical Weathering on Common
    Minerals and Rocks, 155
    Exfoliation and Spheroidal Weathering, 156
Factors That Influence Weathering, 157
    Mineralogy, 158
    Rock Type and Structure, 158
    Slope Angle, 158
    Climate, 158
    Burrowing Animals, 159
    Time, 159
Soil: Origin and Classification, 160

    Origin of Soil, 160
    Soil Profiles, 160
    Soil Types, 162
    Rate of Soil Formation, 164
    Paleosols, 165
Soil Erosion, 166
    Erosion Due to Human Activity, 166
    Control of Soil Erosion, 167
Revisiting Plate Tectonics and the Earth
System, 168
    Global Weathering Rates and High
    Mountains, 168
    Ancient Laterites and the Northward Drift of
    Australia, 169
What's Ahead?, 170
Chapter Summary, 170
The Language of Geology, 171
Questions For Review, 171
GeoDiscoveries: Sedimentary Rock, 171
Understanding Our Environment Box 6.1 *Decaying
Buildings and Monuments*, 154
The Science of Geology Box 6.2 *Using Weathering
Rinds to Measure Relative Age*, 161
Understanding Our Environment Box 6.3 *The Soil
Erosion Crisis*, 167

7   Sediments and Sedimentary Rocks: Archives of Earth History 172

Opening Essay:
Interpreting Sedimentary Rocks Before Plate
Tectonics, 173
Introduction: Sediments in the Rock Cycle, 174
Sedimentation, Stratification, and Bedding, 174
Sediment Types and Characteristics, 175
    Characteristics of Clastic Sediment, 175
    Characteristics of Chemical Sediments, 181
    Characteristics of Biogenic Sediments, 182
Sedimentary Environments and Facies Changes, 182
    Nonmarine Depositional Environments, 184
    Shoreline and Continental Shelf
    Environments, 185
    Continental Slope and Rise Environments, 186
    Deep-Sea Depositional Environments, 189
Diagenesis: How Sediment Becomes Rock, 190
    Formation of Peat: Diagenesis Under
    Anaerobic Conditions, 190
Sedimentary Rocks: Clastic, Chemical, and
Biogenic, 191
    Clastic Sedimentary Rocks, 191
    Chemical Sedimentary Rocks, 192
    Biogenic Sedimentary Rocks, 193
    Oil Shale—an Organic-Rich Clastic Rock, 195
Environmental Clues in Sedimentary Rocks, 195
    Clues on Bedding Planes, 196
    Clues from Fossils, 196

Clues from Rock Color, 196
Clues from Isotopes and Magnetic
Properties, 197
Revisiting Plate Tectonics and the Earth
System, 198
Sedimentation and Plate Tectonics, 198
Limestone, CO_2, and Global Climate
Change, 199
What's Ahead?, 199
Chapter Summary, 199
The Language of Geology, 200
Important Rock Names to Remember, 200
Questions For Review, 200
GeoDiscoveries: Sedimentary Rocks and Petroscope
for Sedimentary Rocks, 201
The Science of Geology Box 7.1 *Graphic Display
of Geologic Data*, 176

8 **Metamorphism and Metomorphic Rocks:
New Rocks from Old** 202
Opening Essay:
Estimating a Temperature and a Pressure, 203
Introduction: All Things Metamorphose, 204
What Is Metamorphism?, 204
Major Factors in Metamorphism, 205
Chemical Composition of Original Rock, 205
Temperature and Pressure, 205
Fluids Facilitate Metamorphism, 208
Role of Time in Metamorphism, 209
The Upper and Lower Limits of
Metamorphism, 209
Role of Water in Determining Limits, 209
How Rocks Respond to Temperature and Pressure
Change in Metamorphism, 210
Texture Change, 210
Mineral Assemblage Change, 210
Types of Metamorphic Rock, 211
Metamorphism of Shale and Mudstone, 211
Metamorphism of Basalt, 212
Metamorphism of Limestone and Sandstone,
213
Types of Metamorphism, 214
Cataclastic Metamorphism, 215
Contact Metamorphism, 216
Burial Metamorphism, 217
Regional Metamorphism—A Consequence of
Plate Tectonics, 217
Metamorphic Facies, 218
Metasomatism, 219
Revisiting Plate Tectonics and the Earth
System, 220
Plate Tectonics and Metamorphism, 220
Metamorphism and the Rock Cycle, 221
What's Ahead?, 222
Chapter Summary, 222

The Language of Geology, 223
Questions For Review, 223
GeoDiscoveries: Metamorphic Rocks, 223
Petroscope For Metamorphic Rocks, 223
The Science of Geology Box 8.1
*Pressure–Temperature–Time Paths: The Tectonic
History of Metamorphic Rock*, 206
Understanding Our Environment Box 8.2
*Concrete: The Artificial Metamorphic Rock that
Changed Our Environment*, 215

9 **How Rock Bends, Buckles, and
Breaks** 224
Opening Essay:
The Ore Body that Sank, 225
Introduction, 226
How Is Rock Deformed?, 226
Stress and Strain, 227
Stages of Deformation, 227
Ductile Deformation versus Fracture, 228
Brittle–Ductile Properties of the
Lithosphere, 230
Deformation in Progress, 231
Abrupt Movement, 231
Gradual Movement, 231
Evidence of Former Deformation, 232
Strike and Dip, 232
Deformation by Fracture, 233
Classification of Faults, 235
Evidence of Movement Along Faults, 239
Deformation by Bending, 240
Revisiting Plate Tectonics and the Earth
System, 244
Strike Slip Faults, 244
Tectonics and Its Effect on Climate, 244
What's Ahead?, 244
Chapter Summary, 245
The Language of Geology, 245
Questions For Review, 245
GeoDiscoveries: Plate Tectonics and Metamorphic
Rocks, 245
Understanding Our Environment Box 9.1 *Rock
Deformation and Oil Pools*, 241

10 **Earthquakes and Earth's Interior** 246
Opening Essay:
A Temblor Strikes Home, 247
Introduction, 248
How Earthquakes Are Studied, 248
Seismometers, 248
Earthquake Focus and Epicenter, 249
Seismic Waves, 250
Determining the Epicenter, 254
Earthquake Magnitude and Frequency, 254

Earthquake Hazard, 255
    Earthquake Disasters, 256
    Earthquake Damage, 257
World Distribution of Earthquakes, 259
First-Motion Studies of the Earthquake Source, 260
Earthquake Forecasting and Prediction, 261
Using Seismic Waves as Earth Probes, 266
    Layers of Different Composition, 266
    The Crust, 266
    The Mantle, 267
    The Core, 267
    Layers of Different Physical Properties in the
    Mantle, 269
Revisiting Plate Tectonics and the Earth
System, 270
What's Ahead?, 272
Chapter Summary, 272
The Language of Geology, 273
Questions For Review, 273
GeoDiscoveries: Geohazards, 273
The Science of Geology Box 10.1 *Calculating a
Richter Magnitude*, 251
Understanding Our Environment Box 10.2 *A Great
Earthquake in the Pacific Northwest?*, 262

11 **Geologic Time and the Rock Record** 276
Opening Essay:
Determining How Rapidly Carbonate Sediments
Accumulate, 277
Introduction, 278
Reading the Record of Layered Rocks, 278
    The Laws of Stratigraphy, 278
    Gaps in the Stratigraphic Record, 280
Stratigraphic Classification, 282
    The Rock-Stratigraphic Record, 282
    The Time-Stratigraphic Record, 282
    Geologic Time Intervals, 283
Bridging the Gaps: Correlation of Rock Units, 283
    What Is Correlation?, 283
    How Correlation Is Accomplished, 284
The Geologic Column and the Geologic Time
Scale, 285
    Eons, 286
    Eras, 287
    Periods, 287
    Epochs, 288
Measuring Geologic Time Numerically, 288
    Early Attempts to Measure Geologic Time
    Numerically, 288
    Radioactivity and the Measurement of
    Numerical Time, 289
    Radiometric Dating and the Geologic
    Column, 294
The Magnetic Polarity Time Scale, 297
    Magnetism in Rocks, 297

The Polarity-Reversal Time Scale, 298
Revisiting Plate Tectonics and the Earth
System, 300
    When Did the Rock Cycle Begin?, 300
What's Ahead?, 300
Chapter Summary, 300
The Language of Geology, 300
Questions For Review, 300
GeoDiscoveries: The Rock Record and Geologic
Time, 301
The Science of Geology Box 11.1 *Potassium-
Argon ($^{40}K/^{40}AR$) Dating*, 292
Understanding Our Environment Box 11.2 *K/Ar
Dating of Hawaiian Glaciations and African
Hominids*, 295

12 **The Changing Face of the Land** 302
Opening Essay:
Why Is Italy Uplifting?, 303
Introduction: Whence Earth's Varied
Landscapes?, 304
Uplift and Denudation: Competing Geologic
Forces, 305
    Factors Controlling Uplift, 305
    Factors Controlling Denudation, 309
    Tectonic and Climatic Control of Continental
    Divides, 310
Hypothetical Models For Landscape
Evolution, 311
    The Geographic Cycle of W. M. Davis, 311
    Steady-State Landscapes, 312
    Rapid Landscape Changes: Threshold
    Effects, 312
How Can We Calculate Rates of Uplift and
Denudation?, 313
    How Can We Calculate Uplift Rates?, 313
    How Can We Calculate Exhumation and
    Denudation Rates?, 314
Ancient Landscapes of Low Relief, 317
Revisiting Plate Tectonics and the Earth System—
Uplift, Weathering, and the Carbon Cycle, 318
    Controls Over the Carbon Cycle, 319
    Role of Uplift in CO_2 Levels, 320
    Recent Mountain Uplift and Isostatic
    Feedbacks, 321
What's Ahead?, 322
Chapter Summary, 322
The Language of Geology, 323
Questions For Review, 323
GeoDiscoveries: Plate Tectonics and
Geohazards, 323
Understanding Our Environment Box 12.1 *Human
Activity as an Agent of Denudation*, 306
The Science of Geology Box 12.2 *Estimating
Exhumation Rates with Radioactive Decay*, 316

13 Mass Wasting 324
Opening Essay:
The Night the Mountain Fell, 325
Introduction: What Is Mass Wasting?, 326
Downslope Movement of Rock Debris, 326
   Role of Gravity and Slope Angle, 327
   The Role of Water, 328
Mass-Wasting Processes, 329
   Slope Failures, 330
   Sediment Flows, 332
Mass Wasting in Cold Climates, 340
   Frost Heaving and Creep, 340
   Gelifluction, 340
   Rock Glaciers, 340
Mass Wasting Under Water, 340
   Mass Wasting on Marine Deltas, 341
   Mass Wasting in the Western North
   Atlantic, 342
   Hawaiian Submarine Landslides, 342
What Triggers Mass-Wasting Events?, 343
   Shocks, 343
   Slope Modification, 343
   Undercutting, 344
   Exceptional Precipitation, 344
   Volcanic Eruptions, 344
   Submarine Slope Failures, 345
Hazards to Life and Property, 345
   Assessments of Hazards, 345
   Mitigation of Hazards, 346
Revisiting Plate Tectonics and the Earth
System, 347
   From Alpine Peaks to the Deep Ocean, 347
   Landslides, Floods, and Plate Tectonics, 347
What's Ahead?, 349
Chapter Summary, 349
The Language of Geology, 350
Questions For Review, 350
GeoDiscoveries: Geohazards, 351
The Science of Geology Box 13.1 *Downslope
Movement and the Safety Factor*, 327
The Science of Geology Box 13.2 *The Beverage-
Can Experiment*, 328

14 Streams and Drainage Systems 352
Opening Essay:
A Chinese Emperor's Buried Army, 353
Introduction: Streams in the Landscape, 354
Stream Channels, 355
   Cross-Sectional Shape, 355
   Long Profile, 356
Dynamics of Streamflow, 356
   Factors of Streamflow, 356
   Changes Downstream, 356
   Floods, 357
   Base Level, 360

Natural and Artificial Dams, 361
Hydroelectric Power, 361
Channel Patterns, 362
   Straight Channels, 362
   Meandering Channels, 363
   Braided Channels, 364
Erosion by Running Water, 366
The Stream's Load, 366
   Bed Load, 366
   Placer Deposits, 366
   Suspended Load, 368
   Dissolved Load, 368
   Downstream Changes in Particle Size, 368
   Downstream Changes in Composition, 369
   Sediment Yield, 369
Stream Deposits, 370
   Floodplains and Levees, 371
   Terraces, 371
   Alluvial Fans, 372
   Deltas, 372
Drainage Systems, 375
   Drainage Basins and Divides, 375
   Stream Order, 375
   Evolution of Drainage, 376
   Drainage Patterns, Rock Structure, and Stream
   History, 377
Revisiting Plate Tectonics and the Earth
System, 377
   Drainage and Landscape Evolution Near Rifted
   Plate Margins, 377
   Tampering with the Nile, 379
What's Ahead?, 380
Chapter Summary, 380
The Language of Geology, 380
Questions For Review, 381
GeoDiscoveries: The Water Cycle, 381
Understanding Our Environment Box 14.1 *Taming
the Yangtze River*, 362

15 Groundwater 382
Opening Essay:
Ice-Age Cave Dwellers in the Pyrenees, 383
Introduction: The Importance of Water, 384
Water in the Ground, 384
   Origin of Groundwater, 384
   Depth of Groundwater, 385
   The Water Table, 385
How Groundwater Moves, 386
   Porosity and Permeability, 386
   Recharge and Discharge of Groundwater, 387
Springs and Wells, 389
   Springs, 389
   Wells, 389
Aquifers, 390
   Unconfined and Confined Aquifers, 390

Artesian Systems, 392
The Floridian Aquifer, 393
Mining Groundwater, and its Consequences, 394
Lowering of the Water Table, 394
Subsidence of the Land Surface, 395
Water Quality and Groundwater
Contamination, 396
Chemistry of Groundwater, 396
Pollution by Sewage, 396
Contamination by Seawater, 396
Toxic Wastes and Agricultural Poisons, 397
Underground Storage of Hazardous
Wastes, 399
Geologic Activity of Groundwater, 400
Dissolution, 401
Chemical Cementation and Replacement, 402
Carbonate Caves and Caverns, 402
Cave Deposits, 403
Sinkholes, 404
Karst Topography, 404
Revisiting Plate Tectonics and the Earth
System, 407
Plate Tectonics and the High Plains
Aquifer, 407
From Rainfall to Mineral Deposits, 407
What's Ahead?, 407
Chapter Summary, 408
The Language of Geology, 408
Questions For Review, 409
GeoDiscoveries: The Water Cycle, 409
The Science of Geology Box 15.1 *How Fast Does
Groundwater Flow?*, 389
Understanding Our Environment Box 15.2 *Toxic
Groundwater in the San Joaquin Valley*, 398

16 Glaciers and Glaciation 410

Opening Essay:
Modern Analogs of Ice-Age Glaciers, 411
Introduction: Earth's Changing Cover of Snow and
Ice Glaciers, 412
Mountain Glaciers and Icecaps, 414
Ice Sheets and Ice Shelves, 414
Temperate and Polar Glaciers, 416
Glaciers and the Snowline, 417
Conversion of Snow to Glacier Ice, 417
Why Glaciers Change in Size, 418
How Glaciers Move, 421
Glaciation, 423
Glacial Erosion and Sculpture, 426
Transport of Sediment by Glaciers, 426
Glacial Deposits, 427
Periglacial Landscapes and Permafrost, 430
The Glacial Ages, 431
Ice-Age Glaciers, 431
Drainage Diversions and Glacial Lakes, 432

Lowering of Sea Level, 432
Deformation of Earth's Crust, 433
Earlier Glaciations, 433
Little Ice Ages, 434
What Causes Glacial Ages?, 435
Glacial Eras and Shifting Continents, 435
Ice Ages and Astronomical Theory, 436
Ice-Core Archives of Changing Climate, 436
Atmospheric Composition, 437
Changes in Ocean Circulation, 439
Solar Variations, Volcanic Activity, and Little
Ice Ages, 440
Revisiting Plate Tectonics and the Earth
System, 440
What Will Happen to Glaciers in a Warmer
World?, 440
African Ice Sheets, 441
What's Ahead?, 442
Chapter Summary, 442
The Language of Geology, 443
Questions For Review, 443
GeoDiscoveries: Glaciers, Glaciation, and Ice
Sheets, 443
Understanding Our Environment Box 16.1 *Living
with Permafrost*, 432
The Science of Geology Box 16.2 *Understanding
Milankovitch*, 438

17 Atmosphere, Winds, and Deserts 446

Opening Essay:
Living on the Edge of the Desert, 447
Introduction: Wind as a Geologic Agent, 448
Planetary Wind System, 448
Circulation of the Atmosphere, 448
The Coriolis Effect, 448
Climate, 450
Movement of Sediment By Wind, 450
Windblown Sand, 450
Windblown Dust, 452
Detrimental Effects of Wind-Blown
Sediment, 454
Wind Erosion, 454
Deflation, 454
Abrasion, 455
Eolian Deposits, 457
Dunes, 457
Loess, 460
Dust in Ocean Sediments and Glacier Ice, 461
Volcanic Ash, 461
Deserts, 462
Types and Origins of Deserts, 462
Desert Climate, 464
Surface Processes and Landforms in Deserts, 464
Weathering and Mass-Wasting in Deserts, 464
Desert Streams and Associated Landforms, 465

Desertification, 468
    Desertification in North America, 469
    Countering Desertification, 469
Revisiting Plate Tectonics and the Earth
System, 469
    Plate Tectonics and Deserts of the Past, 469
    Air: A Consequence of the Earth System, 470
What's Ahead?, 472
Chapter Summary, 472
The Language of Geology, 472
Questions For Review, 472
GeoDiscoveries: Geohazards, 473
The Science of Geology Box 17.1 *Measuring
Earth's Hum*, 451
Understanding Our Environment Box 17.2 *Tectonic
Desertification*, 470

18 The Oceans and Their Margins 474

Opening Essay:
Changing Sea Level and Atoll Evolution, 475
Introduction: The World Ocean, 476
The Ocean's Characteristics, 476
Depth and Volume of the Oceans, 476
    Ocean Salinity, 477
    Temperature and Heat Capacity of the
    Ocean, 478
    Vertical Stratification, 480
Ocean Circulation, 480
    Surface Currents of the Open Ocean, 480
    Major Water Masses, 481
    The Global Ocean Conveyor System, 481
Ocean Tides, 484
    Tide-Raising Force, 486
    Tidal Power, 486
Ocean Waves, 486
    Wave Motion, 486
Coastal Erosion and Sediment Transport, 489
    Erosion by Waves, 489
    Sediment Transport by Waves and Currents, 489
Coastal Deposits and Landforms, 490
    Elements of the Shore Profile, 490
    Factors Affecting the Shore Profile, 492
    Major Coastal Deposits and Landforms, 493
How Coasts Evolve, 497
    Types of Coasts, 497
    Geographic Influences on Coastal
    Processes, 498
    Changing Sea Level, 498
Coastal Hazards, 501
    Storms, 501
    Tsunamis, 501
    Landslides, 502
Protection against Shoreline Erosion, 502
    Protection of Sea Cliffs, 502
    Protection of Beaches, 502

Effects of Human Interference, 504
Revisiting Plate Tectonics and the Earth
System, 505
    Ocean Circulation and the Carbon Cycle, 505
    Sediments at the Beginning and End of an
    Ocean's Life Cycle, 505
What's Ahead?, 505
Chapter Summary, 506
The Language of Geology, 506
Questions For Review, 507
GeoDiscoveries: Geohazards, 507
Understanding Our Environment Box 18.1 *How to
Modify an Estuary*, 500

19 Climate and Our Changing Planet 508

Opening Essay:
The Sensitivity of Earth's Climate to Small
Changes, 509
Introduction: The Changing Atmosphere, 510
    The Carbon Cycle, 511
    The Greenhouse Effect, 512
    Trends in Greenhouse Gas Concentrations, 514
Global Warming, 517
    Historical Temperature Trends, 517
    Climate Models, 518
    Environmental Effects of Global Warming, 520
The Past as a Key to the Future, 521
    Lessons from the Past, 521
    Why Was the Middle Cretaceous Climate So
    Warm?, 521
    Eocene Warmth and Lithosphere
    Degassing, 523
    Modeling Past Global Changes, 524
Revisiting Plate Tectonics and the Earth
System, 525
What's Ahead?, 528
Chapter Summary, 528
The Language of Geology, 528
Questions For Review, 528
GeoDiscoveries: Glaciers, Glaciation, and Ice
Sheets, 529
Understanding Our Environment Box 19.1 *The
Ozone Hole*, 515
The Science of Geology Box 19.2 *The Younger
Dryas Event and the End of the Last Ice Age*, 526

20 Earth Through Geologic Time 530

Opening Essay:
Watching a Continent Grow in Kamchatka, 531
Introduction: Tracking Past Plate Motions, 532
    Former Ideas About Continents, 532
    Pangaea, 533
    Apparent Polar Wandering, 535
    Seafloor Spreading, 535

Relict Plate Boundaries in the Geologic Record, 538
Supercontinents and Vanished Oceans, 541
   The Formation of Pangaea, 541
   Rodinia: A Supercontinent of the late Proterozoic, 542
   Ice Ages in Earth's History, 542
Regional Structures of Continents, 544
   Cratons, 544
   Orogens, 544
   Continental Shields, 544
Continental Margins, 544
   Passive Continental Margins, 545
   Continental Convergent Margins, 547
   Continental Collision Margins, 548
   Transform Fault Margins, 550
   Accreted Terrane Margins, 550
Mountain Building, 551
   The Appalachians, 552
   The Alps, 553
   The Canadian Rockies, 554
Revisiting Plate Tectonics and the Earth System, 557
   A Plate Collision and the Monsoons, 557
What's Ahead?, 558
Chapter Summary, 558
The Language of Geology, 559
Questions For Review, 559
GeoDiscoveries: The Rock Record and Geologic Time, 559
The Science of Geology Box 20.1 *Watching a Continent Splinter*, 547

21 Resources of Minerals and Energy 560
Opening Essay:
The Golden Puzzle of the Witwatersrand, 561
Introduction: Natural Resources and Human History, 562
Mineral Resources, 562
   Supplies of Minerals, 563
Origin of Mineral Deposits, 564
   Hydrothermal Mineral Deposits, 565
   Magmatic Mineral Deposits, 566
   Sedimentary Mineral Deposits, 566
   Placers, 570
   Residual Mineral Deposits, 570
Useful Mineral Substances, 572
   Geochemically Abundant Metals, 572

   Geochemically Scarce Metals, 572
Energy Resources, 573
   Supplies of Energy, 573
Fossil Fuels, 573
   Coal, 574
   Petroleum: Oil and Natural Gas, 575
   How Much Fossil Fuel?, 578
Other Sources of Energy, 578
   Biomass Energy, 578
   Hydroelectric Power, 579
   Nuclear Energy, 579
   Geothermal Power, 579
   Energy from Winds, Waves, Tides, and Sunlight, 580
Revisiting Plate Tectonics and the Earth System, 582
   Plate Tectonics and Mineral Deposits, 582
   Mineral Production and the Earth System, 582
What's Ahead?, 583
Chapter Summary, 583
The Language of Geology, 583
Questions For Review, 583
Understanding Our Environment Box 21.1 *Hydrothermal Mineral Deposits Forming Today*, 568
Understanding Our Environment Box 21.2 *The Waste Disposal Problem: Geology and Politics*, 580

Appendix A
Units and Their Conversions A1
Appendix B
Tables of the Chemical Elements and Naturally Occurring Isotopes A4
Appendix C
Table of the Properties of Common Minerals A7

Glossary G1
Selected References R1
Credits C1
Index I1

We have been privileged to work as geologists and geophysicists all of our professional lives, and the times of our lives have been some of the most exciting times in the history of science. Discoveries made over the past 50 years have transformed human understanding of the way Earth works. Extraordinary discoveries continue at such a pace that almost every week professional journals record some remarkable new observation, or a new measurement, that requires a revision or refining of our ideas about Earth. In writing this new edition of Dynamic Earth, we have attempted to share the excitement and wonder of this great scientific revolution. It is a revolution that continues.

Each of us has worked in many parts of the world, and in this revision of Dynamic Earth we have drawn on our experiences, as well as those of our many colleagues who have shared observations and data with us, in selecting examples to illustrate geologic processes.

Brian Skinner was born, raised, and attended college in Australia. His first geological experiences were there, first as a student, and subsequently as a mine geologist in Tasmania. Following a PhD at Harvard he returned to Australia where he worked at the University of Adelaide. A move to the U.S. Geological Survey brought him back to the United States, and he subsequently moved to Yale University where he has spent most of his professional life. His research has been focused on the genesis of mineral deposits, and on the chemical and physical properties of ore minerals. Research has led him, together with his students, to work in Asia, Africa, Australia, South and North America, and Europe.

Stephen Porter joined the geology faculty of the University of Washington after receiving graduate degrees from Yale University. His research has largely focused on alpine glaciation in many of the world's major mountain systems, and on the climatic changes revealed by their deposits. He also has mapped and studied Hawaiian and Cascade volcanoes, their eruptive products, and how volcanic eruptions have influenced climate. Over the past 20 years he has investigated the thick deposits of windblown dust in central China to reconstruct 7 million years of east-Asian monsoon history. With foreign colleagues, he also has studied rockfall hazards in the Italian Alps and full-glacial permafrost conditions on the Tibetan Plateau.

Jeffrey Park's career as a seismologist was inspired by two events in his freshman year of high school in Southern California. First, the 1971 Sylmar earthquake tossed him out of bed on a chilly winter morning. Second, the first television documentary on plate tectonics was broadcast on PBS, full of raw speculation on how Earth's history, natural resources, volcanoes and earthquakes could be explained together. After majoring in physics in college, his research focused on the free vibrational oscillations of Earth, whose ringing can be observed worldwide after large earthquakes. After joining the Yale faculty, Jeffrey also has investigated evidence for cycles in global climate records, and has explored the connection between earthquakes, volcanoes and mountain belts in Kamchatka, Russia, and central Italy.

Dynamic Earth

Chapter 1

The Earth System at Work. Intense erosion of an old volcano on Kauai, most northerly of the main Hawaiian Islands. This scene is on the Na Pali coast on the northwestern side of the island.

Glacially polished pavement, Hallet Cove, South Australia.

Discovering Geology

Sir Douglas Mawson, a distinguished Antarctic explorer, led my first geologic field trip. The year was 1947 and we were just 3 weeks into the introductory geology course at the University of Adelaide, South

Australia. The trip was to Hallet Cove, on the southern coast of Australia.

On top of a seacliff was a layer of mixed clay and rounded boulders, and beneath that again a smooth, polished surface of slate. Deep, roughly parallel scratches crossed the polished surface, and the scratches could be seen to continue beneath the clay-boulder layer.

The relative order of events was clear. The slate was oldest, the polished surface and scratches were next, deposition of the clay-boulder mix was last. The boulders were not slate; they were completely different kinds of rock. But how could we explain the sequence?

Based on evidence from elsewhere, Professor Mawson told us, the slate was very old—he described it as Precambrian in age (today we know the rocks are 600 million years old). From his work in Antarctica, he said the polished surface and scratches were typical of those caused by moving ice. The polish and scratches were caused by debris carried in the ice, and the direction of the scratches indicated that the ice had flowed from the south. The clay-boulder mix was typical of ice-transported debris. Looking south to see where the ice had come from, all we could see was the sea. But glaciers, we learned, only form on land. So where did the ice come from?

The nearest land to the south is Antarctica, about 2500 miles away. Could a glacier have crossed such an ocean? It seemed very unlikely, so a hypothesis was suggested. Could Australia and Antarctica once have been joined, covered by glacial ice, and later drifted apart? The notion that continents might drift seemed preposterous, but today we have evidence to prove that they do. We have now learned, too, that the glacially polished surface was formed, and the boulder and clay deposited, about 250 million years ago by a great continental ice mass centered on Antarctica.

In my student days, the hypothesis of drifting continents was very controversial. Evidence from Hallet Cove and other places was suggestive but it did not explain how continents could move. Lacking an explanation, most geologists sought other ways to explain the evidence, but only drift seemed to account for the evidence.

That first field trip provided a powerful lesson about the challenging hypotheses one must consider in order to explain geologic evidence. If any one event can be said to have determined the course of my professional life, it was that field trip. From that day on I was determined to be a geologist.

Brian J. Skinner

KEY QUESTIONS

1. **How do geologists study Earth?**
2. **What is the scientific method?**
3. **How fast are geologic processes and how long do geologic events take?**
4. **How did the solar system form?**
5. **What is Earth's internal structure?**
6. **What is plate tectonics?**
7. **What is the Earth system?**
8. **How do humans affect the Earth system?**

INTRODUCTION

Earth is our home, and home to millions of other kinds of life. No other planet in the solar system has the right chemical and physical mix needed to support life, and no evidence has yet been discovered of life existing elsewhere in the universe. So far as we know, Earth is unique.

Let us help you see your home planet with new eyes—eyes that peer deeply and understand the geology that surrounds you. The geology underneath your home or school may not be as spectacular as that shown in the photo on the cover of this text, but the *geologic phenomena* that created the scenery in the photo operate worldwide, creating the geology you see everywhere. When you finish your geology course, we want you to be able to look around you, wherever you are, and say, "I know how this area came to look like this, and I understand the geology beneath my feet."

In addition to appreciating more about the extraordinary planet on which we live, there are many very practical reasons to study geology. Here are a few of the reasons:

- We humans are influencing Earth's external geologic processes—in other words, our behavior is affecting our own environment (air, water, rocks, and life itself). We are now so populous (more than 6 billion and adding approximately three people each second) that our daily activities are having measurable effects on rainfall, climate, air and water quality, erosion, and more. For example, we use annually a volume of minerals that is greater than the mammoth volume of sediment transported to the oceans by the world's thousands of rivers. Thus, we have become critical components in the way Earth works—a remarkably influential species.

- Each year the average world citizen uses 10 tons of mineral resources (metals, building materials, fuels, fertilizers, etc.). In North America, the figure is closer to 20 tons each. Some nations have bountiful fossil fuels (coal, oil, natural gas), whereas others are rapidly exhausting their supplies. Some metals have been mined so extensively that resources are growing thin. Finding new resources to replace those that have been mined is a great challenge for geologists.

- At least 40 nations are currently experiencing a crisis of clean water supply, and public health is suffering in those countries as a result. In some countries supplies of water are so low they are no longer adequate for both human consumption and food growth. The study of both surface and underground water is an important aspect of geology.

- Almost every week we can pick up the newspaper and read of a natural disaster somewhere in the world—a volcanic eruption, a mudslide, an earthquake, or a flood. These are all geologic events, and the people who study them, and who try to warn residents in time to escape, are geologists.

ABOUT GEOLOGY AND GEOLOGISTS

Geology comes from two Greek words: *geo*, meaning Earth, and *logia*, meaning study or science. Geology, therefore, is the science of Earth. We scientists who study Earth are **geologists**.

Geologists study how Earth came into existence, how it evolved and developed into the planet of today, and how all the parts of the planet work together to maintain conditions right for life. Increasingly, geologists are also being called on to study how the human population is changing Earth, and to offer advice on what must be done to keep Earth a friendly, habitable planet.

Geologists are quite interdisciplinary. We must know about:

- *Earth's processes* (like volcanism, glaciation, streamflow, minerals, and how rocks form).

- *Chemistry* (to understand how minerals and rocks form, how groundwater and stream water dissolve minerals, and how mineral resources form).

- *Physics* (to understand plate tectonics, how rocks bend and break under stress, volcanism, earthquakes, and landslides).

- *Biology* (to understand how life processes integrate with other Earth systems and how life has evolved, as evidenced in fossils in the rocks).

- *Meteorology* (to understand how weather erodes rocks, affects streamflow and groundwater levels, and climate).

- *Oceanography* (to understand the seafloor's role in plate tectonics and our dynamic shorelines).

- *Astronomy* (to understand how geologic principles apply on other worlds).

- *Mathematics* (to compute quantitative relationships and develop statistical views of phenomena).

- *Computer science* (because we use numerous electronic measuring instruments and software for calculations).

- *Economics* (because geology profoundly affects every national economy—it's all about mineral and energy resources).

Some view geology as a derivative science because it builds on so many others, but it is really an all-encompassing science of our planet.

WHERE WE WORK AND WHAT WE DO

We geologists work everywhere—on every continent, in every country and state, on and under every ocean, and, given the opportunity, on the Moon and other planets (Figure 1.1). When "working in the field," it can mean glacial peaks, erupting volcanoes, earthquake zones, river basins, prairies, deserts, coastlines, offshore, or in space. We work in offices and labs, at mines and drilling rigs, running seismic lines, studying erosion and flood damage, exploring for resources, and in information and education. We work in research, industry, business, government, and education.

Motivated by curiosity and employment, we geologists seek to understand all processes that operate on and inside Earth and to discover and document our planet's long, convoluted history. We pay special attention to water bodies (rivers and lakes) that are being altered by human activities and to hazardous processes such as earthquakes, volcanic eruptions, floods, and landslides. (For example, see *Understanding our Environment*, Box 1.1, *Epitaph for the Aral Sea*, on the human-caused shrinking of Uzbekistan's Aral Sea from Earth's fourth-largest lake to a desert.)

How do we geologists study rocks? First, we do the obvious: examine rocks that are exposed in natural outcrops, in highway and railroad cuts, and in mines. To investigate rocks we cannot reach directly, we drill deep holes (often hundreds of meters), extract "drill cores" of rock for study, and lower instruments and cameras to see and measure the rocks the wells have penetrated.

Beyond the reach of drill holes, we geologists must rely on indirect observations. Like a doctor listens with a stethoscope to your body's sounds, we employ sensitive measuring devices to "listen" to the rumble of distant earthquakes and sense the pulse of activities deep inside Earth. We also use satellite imagery to spot surface patterns, ground-penetrating radar to see beneath desert sands, and sensors to find variations in Earth's gravity.

From these observations, we interpret an area's geologic history, the origin of its landscapes, where new oil and gas fields lie, where ore deposits may be concealed, where wells may strike good water, and where hazards may lie.

A.

B.

C.

D.

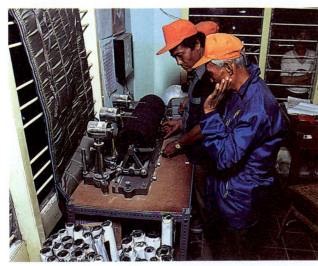

E.

F.

Figure 1.1 Geologists at Work A. Marine geologist studying coral growth. B. Planetary geologist Jack Schmidt sampling rocks on the Moon. C. Measuring the temperature of eruption and collecting gas and lava samples for analysis, Iceland. D. Geologist measuring the orientation of fractures in a rock outcropping. E. Studying a seismometer that is recording earthquake activity, Indonesia. F. Glaciologists measuring depth and flow in a glacial stream.

Box 1.1 U N D E R S T A N D I N G   O U R   E N V I R O N M E N T

EPITAPH FOR THE ARAL SEA

Just 40 years ago, the Aral Sea in Uzbekistan (Figure B1.1A) was the planet's fourth largest lake (after the Caspian Sea, Lake Superior, and Lake Victoria). It covered 68,000 km² (the area of West Virginia), averaged 16 m deep, and yielded 45,000 tons of fish each year. Fishing provided employment for 70,000 people.

Today the Aral Sea is the sixth largest lake, covers 40,000 km², averages 9 m deep, and has become so salty that the fishing industry is dead. Once-prosperous fishing villages are now 50 km from its shores. If the trend continues, the Aral Sea will be a waterless desert by about 2010 (Figure B1.1).

What happened? The story of the Aral Sea is a sobering lesson for all of us: when we alter the balance of nature, unforeseen side effects happen. Or, as Francis Bacon put it, "Nature, to be commanded, must be obeyed."

The Aral Sea is fed by two large rivers, the Amu Dar'ya and the Syr Dar'ya, which carry meltwater across the desert from the snowy mountains of northern Afghanistan. The sea has no outlet to the ocean, so water leaves only by evaporation. Thus the size of the sea is a balancing act between river inflow and evaporation. The sea is shrinking because the inflow has sharply declined.

A small part of the decline is climatic—there were several years of light snowfalls in the 1970s. But the big reason is irrigation, which has been practiced in the river valleys for millennia. Especially when Uzbekistan was part of the former Soviet Union, irrigation increased dramatically. By 1960, so much irrigation water was taken from the two rivers that inflow to the Aral Sea had dropped to a trickle. The sea has continued to shrink ever since.

The people who planned the irrigation expected the Aral Sea to shrink. But what they failed to anticipate were the side effects. We now know that the sea, because of its size, exerts a major influence on the area's climate. Because it is shrinking, less water evaporates into the air,

Figure B1.1 A New Dead Sea? Fishing boats stranded in the new desert that once was sea bottom in the Aral Sea in Uzbekistan. Are there lakes in the United States that could experience the same fate?

so local rainfall is declining, average temperature is rising, and wind velocities are increasing.

Most of the newly exposed sea bottom is covered with salt. The wind blows the salt around and has created withering salt storms. Potable water supplies have declined, and various diseases—especially intestinal—are afflicting the local population.

Simply reducing the amount of irrigation could reverse the situation. But the irrigated area has become one of central Asia's most prosperous, so the solution will probably have to be a compromise between restoring the Aral Sea to its original size and maintaining all the irrigated land. Had geologists been consulted and listened to, this disaster could have been prevented.

PHYSICAL VERSUS HISTORICAL GEOLOGY

We geologists divide our discipline into two broad areas—historical and physical:

Historical geology discovers the past, because it is the biggest clue to the present—for example, processes and events that formed a river valley over the past million years are very likely the same processes that operate today. Historical geology discovers the chronology of events, both physical and biological, that have occurred in the past.

Historical geologists try to answer questions like: When did the oceans form? When did dinosaurs first appear and last tread on Earth? Why are the Sierra Nevada Mountains so much younger than the Appalachian Mountains? When

did the first photosynthetic plants appear, and how did the oxygen they produced affect rocks worldwide?

We briefly touch on historical geology in Chapter 11, "Geologic Time and the Rock Record," but aspects of the long geological history of Earth are discussed at many places throughout the book.

Physical geology, the subject of this book, is concerned with understanding the *processes* that operate on and inside Earth, and the *materials* on which these processes operate. Processes include plate tectonics, volcanism, earthquakes, landslides, floods, formation of mineral deposits, mountain-building, shore erosion, landscape formation, and so on. Materials include soil, sand, rocks, minerals, air, and seawater.

Physical geology is our starting point for studying Earth. To understand our environment and make predictions about changes that lie ahead, we must understand how Earth works. To obtain that understanding, we must examine both processes and materials.

THE SCIENTIFIC METHOD

Like all scientists, geologists use a research strategy called the **scientific method**. Although it is not always a clear-cut process, the scientific method includes the following steps:

1. *Observe and measure.* Scientists focus on a phenomenon and acquire evidence about it that can be measured. *Figure 1.2 will be our example:* a geologist measures the thickness, extent, tilt, and content of a rock layer.

2. *Form a hypothesis.* Scientists attempt to explain their observations and measurements by developing a **hypothesis**—a plausible, but unproved, explanation for the way something happens. *Example:* a geologist hypothesizes that the horizontal layering of rocks in the figure is due to sediment being deposited in a body of water (lake, stream, ocean).

3. *Test the hypothesis.* Scientists use a hypothesis to make predictions about new observations. We *test* the hypothesis by comparing their predictions against the new observations. *Example:* if the hypothesis is that the horizontal layers in the figure result from sediment being deposited from water, then a prediction would be that fossil remains of aquatic plants and animals that once lived in the water might be present in the rocks.

4. *Formulate a theory.* Once a hypothesis withstands numerous tests, scientists become more confident in its validity, and the hypothesis is elevated to a **theory**. A theory is a generalization about natural phenomena. It is not the final word, however, and a theory is always open to further testing. (*Note:* In everyday language, people often misuse the term "theory" to mean "hypothesis," by saying, "Well, that's just a theory." What we mean is, "That's just a hypothesis." By the time a statement attains the status of a theory, it is very substantial and must be taken seriously.)

5. *Formulate a law or principle.* Eventually, a theory or group of theories may be formulated into a **law or principle**. Laws and principles are statements that some natural phenomenon is invariably observed to happen in the same way, and no deviations have ever been observed. *Example:* the **Law of Original Horizontality** states that water-deposited sediments are deposited in layers that are horizontal (or virtually so) and parallel to Earth's surface (or virtually so).

6. *Continually reexamine the law or principle.* The assumption that underlies all of science is that everything in the material world is governed by scientific laws. Because laws and principles, like hypotheses and theories, are always

Figure 1.2 Illustrating the Scientific Method A geologist observes and measures sedimentary rock layers in Deadman Wash, Wupatki National Monument, Arizona. The rocks are sandstones, shales, and mudstones. The geologist in this image hypothesizes that the rocks were deposited as sediment from water. If the geologist finds aquatic plant and animal fossils in the rock layers, would this support the hypothesis of deposition by water? Would the absence of such fossils indicate that the sediment was not deposited by water? This is how the scientific method works.

open to question when new evidence is found, hypotheses, theories, laws, and principles are continually examined. In fact, the key to the scientific method is *testability*. Any theory, for example, that cannot be tested is not scientific.

> ### Before you go on:
> **1.** What is geology?
> **2.** Can you explain the six steps of the scientific method?
> **3.** What is the difference between a hypothesis and a theory?

HOW RAPID ARE GEOLOGIC PROCESSES?

One of the great questions to confront geologists is: What is the rate of geologic processes? Are they catastrophically fast or very gradual? During the seventeenth and eighteenth centuries, before geology became the scientific discipline it is today, people believed that Earth's features—mountains, valleys, oceans, rivers—were permanent and had been produced by a few great catastrophes. These events were thought to be so huge that they could not be explained by natural processes and therefore were the result of supernatural intervention.

From this thinking was born the hypothesis of **catastrophism**. Because of the prescientific culture at the time, people did not consider catastrophism a hypothesis, but a belief. And because of the minimal geologic knowledge of the day, catastrophes were thought to have occurred within the past few thousand years and to fit a chronology of events recorded in the Judeo-Christian Bible. Non-Western cultures reflected similar beliefs.

JAMES HUTTON APPLIES THE SCIENTIFIC METHOD

During the late eighteenth century, early scientists reexamined catastrophism in light of slowly growing geologic evidence, and found the old hypothesis wanting. Scottish physician and gentleman farmer James Hutton (1726–1797) assembled much of the evidence and proposed a counterhypothesis called **gradualism**.

Hutton observed and recorded his evidence, then used logical arguments to draw conclusions from the evidence—in short, he used the scientific method. In 1795, he published *Theory of the Earth, with Proofs and Illustrations*; in those two volumes he introduced his counterhypothesis to catastrophism. Today, Hutton is regarded as the "father of modern scientific geology."

Hutton observed the slow but steady effects of erosion: the transport of rock particles by running water and their ultimate deposition in the ocean. Changes come about gradually—that is why his counter to catastrophism is called gradualism. He reasoned that, given enough time, mountains must slowly but surely be eroded to plains, that new rocks must somehow form from the debris of erosion, and that new rocks must be slowly thrust up to create new mountains. He didn't know the source of the energy that caused mountains to arise, but he argued that everything must move slowly in a repetitive, continuous cycle.

UNIFORMITARIANISM

Hutton's ideas evolved into what we now call the **Principle of Uniformitarianism**, which states that the same processes we observe today have been operating throughout Earth's history. This principle provided a first step in understanding Earth history, and was the key that unlocked our modern perceptions of geologic time.

The Principle of Uniformitarianism helps us understand how rocks form in each geologic environment. It tells us that we can examine any rock (however old), compare its characteristics with those of similar rocks that are forming today in a particular environment, and infer that the old rock very likely formed in the same kind of environment.

For example, travel to a sandy desert and you will observe sand dunes forming from windblown sand grains. Because of the way they form, dunes have a distinctive internal structure (Figure 1.3A). Using the Principle of Uniformitarianism, we can infer that any rock composed of cemented sand grains and having the same internal structure as modern dunes (Figure 1.3B) is the remains of

A.

B.

Figure 1.3 Principle of Uniformitarianism Seen in Sand Dunes, Modern (A) and Ancient (B) A. Layering is visible in a hole dug into a dune near Yuma, Arizona. B. A similar pattern exists in sandstone rocks in Zion National Park, Utah, that are hundreds of millions of years old. Using the Principle of Uniformitarianism, we can infer that these ancient rocks were once sand dunes—and that the environment was similar to that existing today where the modern sand dunes are.

Figure 1.4 Siccar Point, Scotland Here, in 1788, "father of geology" James Hutton first demonstrated that Earth processes of deposition-uplift-erosion repeat cyclically and endlessly. The vertical layers of 450-million-year-old sedimentary rock (A), originally horizontal, were uplifted and tilted to vertical. Exposed to weathering, these rocks eroded to form a new land surface (C) upon which layers of younger sediments (B) were laid down. These topmost layers, now gently sloping, are the Old Red Sandstone, aged 370 million years.

an ancient sand dune—and probably formed in a similar manner in a similar desert environment.

Hutton was especially impressed by evidence he observed at Siccar Point, not far from Edinburgh (Figure 1.4). There he saw the ancient sandstone layers, originally horizontal, but now standing vertically (A) and capped by layers of younger sandstone (B). He realized that the boundary between the layers was an ancient *erosion surface* (C).

The now-vertical layers are composed of debris eroded from an ancient landmass. Transported by streams, deposited on the seafloor, and formed into new rocks, these layers were uplifted (above sea level), tilted almost 90 degrees, and eroded in turn. Eventually, when erosion had formed a new, flat surface on the old, tilted sandstone layers, younger erosional debris was deposited on the fresh erosion surface. Eventually, the younger debris became rock, and uplift occurred again.

Hutton was immensely impressed by the cycle of uplift, erosion, transport, deposition, solidification into rock, and renewed uplift that he observed. There is, he wrote, "no vestige of a beginning, no prospect of an end" to Earth's geologic cycles.

Geologists who followed Hutton have been able to explain Earth's features logically with the Principle of Uniformitarianism. In doing so, they discovered the incredible age of our planet—about 4.55 billion years, quite beyond our ability to imagine. The big message is that most erosional processes are exceedingly slow.

It takes years, in the hundred-millions, to erode a mountain range down to vast volumes of sand and mud, to transport it all by streams to the ocean floor, to cement it into new rocks, and deform and uplift those new rocks through the actions of plate tectonics to form a new mountain range. Slow though it is, this cycle of erosion, formation of new rock, uplift, and more erosion has repeated many times during Earth's long history.

CATASTROPHISM

Uniformitarianism has defined modern geology, but does this mean that catastrophism is totally incorrect? Recently, a thin and very unusual rock layer has been discovered at many locations worldwide. It is rich in the rare metal iridium, which happens to be much more abundant in meteorites than in Earth's common rocks. This extraordinary iridium-rich layer has been discovered in Italy, Denmark, Arizona, and elsewhere (Figure 1.5). Its presence suggests that a catastrophic impact from a meteor may have occurred about 66 million years ago, and that many life-

Figure 1.5 Subtle Record of a Global Disaster? This thin, dark layer of rock (marked with coin) not only looks out-of-place in the thick sequence of pale limestones, but it also is rich in the rare element iridium ("rich" in the scientific sense: about 3 parts per billion). The layer, seen here in the Contessa Valley in Italy, has been identified at many places worldwide and may have been deposited from a globe-circling dust cloud that resulted from a large meteorite impact about 66 million years ago. This impact and its aftermath may have sent the dinosaurs into extinction.

forms, including the dinosaurs, may have become extinct as a result.

The hypothesis is that a gigantic meteorite impact threw so much debris into the atmosphere that most animals and many plants died. When the debris settled, it formed the thin, iridium-rich layer wherever sediments were being deposited around the world (Figure 1.6).

Even more dramatic extinctions have occurred quickly at other times in the past. The record in the rocks indicates that one event, about 245 million years ago, sent almost 90 percent of all living plants and creatures to extinction. We have not found the cause of that event, but it was clearly catastrophic.

The kind of catastrophe inferred from this evidence has led geologists to rethink the validity of catastrophism. These are different from the catastrophes perceived by seventeenth-century Biblical scholars, but they do seem to have caused massive changes. If we are interpreting these events correctly, we must conclude that catastrophism also plays a role in Earth's history.

Furthermore, other events such as earthquakes, volcanic eruptions, tsunami, floods, and landslides are certainly not gradual, because they occur in a geologic eye blink. These events are catastrophic, at least locally.

So, Earth's processes appear to occur along a spectrum from gradual to catastrophic, with the majority of processes being gradual, but a few being catastrophic in time-frames of moments to a few years.

Here is a final thought on catastrophism. A fascinating and frightening suggestion has been made that a catastrophe of a different sort may already be in motion: *our collective human activities may be changing Earth so rapidly, and so extensively, that plant and animal species may be going extinct at a rate similar in magnitude to the major extinctions in the geologic record, of which there have been at least 5, and possibly as many as 12, over the last 250 million years.* At present, this is a hypothesis, remaining to be extensively tested and thereby proved or dis-

Figure 1.6 Scar From an Ancient Impact Meteor Crater, near Flagstaff, Arizona, was formed by meteorite impact about 50,000 years ago. It is 1.2 km in diameter and 200 m deep. Note the raised rim and the wreath of broken rock debris thrown from the crater. Many impacts larger than this are believed to have occurred during Earth's history. (If this is true, why don't we all know about major craters in every country? The answer, as James Hutton observed for mountain ranges, is that erosion slowly grinds down Earth's surface, and evidence of ancient impacts is removed.)

proved. Nevertheless, the very fact that serious scientists are concerned that this hypothesis might advance to theory emphasizes an important fact: geology and the welfare of our species are inextricably linked.

GEOLOGIC TIME AND EARTH'S AGE

Several times in this chapter we have mentioned events that happened millions of years ago or that took millions of years to happen. Such vast stretches of time are difficult to comprehend because we humans measure time in units that are meaningful in terms of a human lifespan—days, weeks, and years. The historical record of the Roman Empire is very good, but it stretches our imagination to contemplate events that happened only 2000 years ago. How much more does it stretch our imagination to contemplate geologic events that happened hundreds of millions of years ago?

Long before geology emerged as a separate scientific discipline, early scientists recognized that a pile of sedimentary layers, such as that in Figure 1.2, records a sequence of past events—the layers at the bottom of the pile are the oldest, those at the top the youngest. The pile records a relative sequence of events, not a numerical measure in years.

Over many years geologists used such reasoning to work out the relative ages of many geologic events—they discovered, for example, that the Appalachians rose long before the modern Rockies were formed, and that the retreat of the most recent ice sheet form North America was a very recent event, geologically speaking. Slowly, geologists have been able to construct a sequential list of events around the world. But of course such a list does not tell us when things happened in terms of years.

One of the great scientific advances of the twentieth century was the discovery of natural radioactivity and the demonstration that radioactivity can be used to measure geologic events in terms of numerical time. The Appalachians rose about 300 million years ago, the modern Rockies about 70 million years, and the most recent great ice sheets retreated about 12,000 years ago.

It was through radioactive dating that we discovered that Earth was formed about 4.55 billion years ago. We will return to the issues of relative and numerical time, and radioactive dating in Chapter 11. But for the present we must try to grasp the immensity of geologic time. We cannot see a mountain eroded away over a human lifespan, but observation of geologic processes tells us that erosion is going on. Over millions of years the slow processes of erosion will eventually erode the mountain away.

Before you go on:

What is the difference between catastrophism and gradualism? Can you cite an example of each?

EARTH'S NEIGHBORHOOD: THE SOLAR SYSTEM

"Half a world to the left, half to the right; I can see it all. The Earth is so small," Russian Cosmonaut Vitali Sevastyanov reported to ground control. Cosmonauts and astronauts report that their strongest impressions in space are the aloneness, smallness, and seeming vulnerability of the little blue planet suspended in the vast emptiness of space.

To put Earth in perspective, we need to look at its neighborhood. Earth is not actually alone in space, of course, because it is part of the **solar system**. The solar system comprises all matter that is gravitationally retained by the Sun. The entire solar system consists of the Sun, nine planets, over five-dozen moons, a vast number of asteroids, millions of comets, and innumerable small fragments of rock and dust called meteoroids. All objects in the solar system move through space in orbits controlled by gravitational attraction. The planets, asteroids, and meteoroids all circle the Sun, while the moons circle their planets.

Distances between planets are so immense that it is difficult to comprehend them. To put the solar system in perspective, think of the Sun as basketball-sized. At this scale, the nearest planet, Mercury, would be a grain of sand about 12 m away. Earth would be a 2.5 mm granule about 30 m distant. Saturn would look like a 2.5 cm grape nearly 300 m away.

THE PLANETS—TERRESTRIAL AND JOVIAN

The solar system's nine planets can be divided into two groups, based on their proximity to the Sun and their density (Figure 1.7). The *terrestrial planets* are the four closest to the Sun, and all are similar to Earth in density (*terra* is the Latin word for Earth). All four terrestrial planets are small, rocky, and dense (3 g/cm^3 or greater).

The *Jovian planets* are those farther from the Sun than Mars. They are much larger than the terrestrial planets and much less dense (Pluto is an exception). Jupiter has 317 times Earth's mass but its density is only 1.3 g/cm^3. Saturn has 95 times Earth's mass but its density is only 0.7 g/cm^3. These planets take their group name from *Jove*, an alternative name for the Roman god, Jupiter.

All of the Jovian planets likely have solid cores that resemble terrestrial planets. But most of their mass is in thick shells that consist largely of the very light elements hydrogen and helium, keeping their densities low. Other, less volatile substances, like carbon dioxide, ammonia, and sulfuric acid, condense to form clouds we see in their atmospheres (Figure 1.8).

ORIGIN OF THE SOLAR SYSTEM— AND EARTH

How did the solar system form? We may never know the complete answer, but we have a solid theory based on

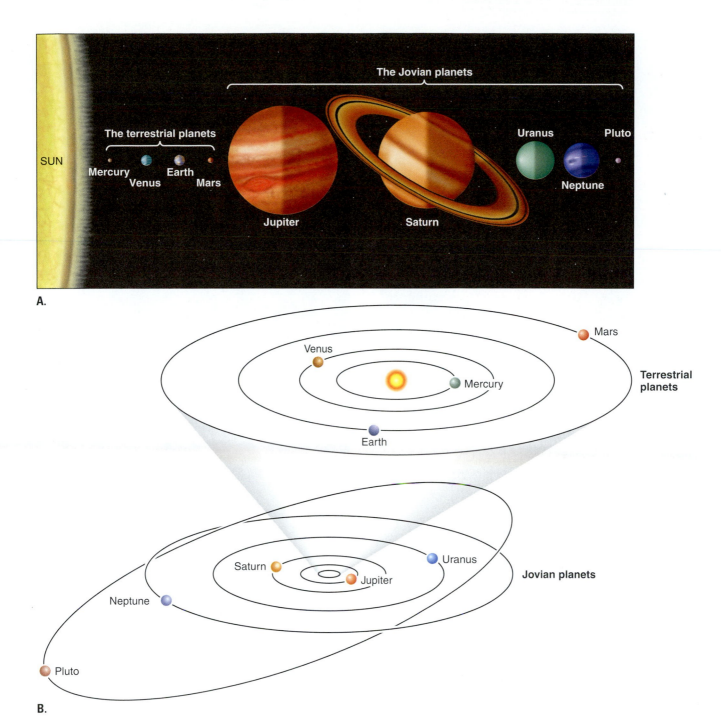

A.

B.

Figure 1.7 Family Portrait of the Solar System A. The solar system's nine planets, shown in proper size proportion to the Sun (but not in accurate distance, and they never line up this neatly—see B). The terrestrial planets are the four small, rocky ones nearest the Sun. The Jovian planets are the large, gas-rich bodies distant from the Sun. Images from space vehicles reveal ring systems around all four Jovian planets, but only Saturn's are large enough to depict at this scale. Pluto is the "odd planet out"—it is much smaller and lacks the gaseous envelope characteristic of the Jovian planets. B. The planets' orbits, approximately to scale. Note that all are in essentially the same plane except Pluto. Its inclined orbit may indicate that it is a wayward moon or a comet.

astronomical evidence, our knowledge of the solar system today, and the laws of physics and chemistry.

Our Sun and its planets were born in a way similar to that of other suns and planetary systems in the universe.

Birth began in a huge volume of space that was not entirely empty because earlier stars had exploded in *supernovas*, scattering atoms of various elements. Most of these atoms were hydrogen and helium, but all the other chemical ele-

Figure 1.8 Jupiter's Turbulent Atmosphere Jupiter, largest of the planets, is a gas-shrouded giant that probably conceals a solid interior. This *Voyager 2* image was taken in 1979 from 28 million km distance. The banded pattern is due to turbulence in Jupiter's atmosphere. A particularly violent storm is visible at lower left. Storms in Jupiter's atmosphere were first reported by Galileo in 1609, as he studied Jupiter through his early telescope. The largest of the storms has raged for centuries. The storm's genesis, size, and duration remain unexplained.

A.

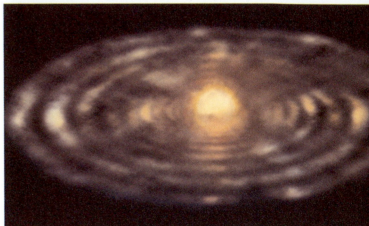

B.

Figure 1.9 Birth of the Solar System A. Over millions of years, gravity slowly gathered the thinly spread atoms into a thicker gas. B. The gathering of atoms created a rotating cloud of dense gas. Its center eventually became the Sun. The planets formed by condensation of the outer portions of the gas cloud.

ments were present in small percentages, too. Although thinly spread, these atoms formed a tenuous, turbulent, swirling cloud of cosmic gas.

Over millions of years, gravity slowly gathered the diffuse atoms into a denser gas. As the atoms moved closer together, the gas was squeezed, and grew hotter and denser as a result. Near the center of this gathering cloud, the temperature and density became so great that hydrogen atoms began to fuse to form helium atoms. When hydrogen fusion commenced, the Sun was born as a nuclear reactor. The time is estimated to have been about 4.6 billion years ago.

At some stage, the outer portions of the cosmic gas cloud cooled and became dense enough to allow solid objects to condense, just like water vapor condenses directly from the gaseous state to the solid state to form snow (Figure 1.9). These solid condensates eventually became the planets, moons, and the other solid objects of the solar system.

TEMPERATURE GRADIENT, DISTANCE FROM THE SUN, AND PLANETARY COMPOSITION

Refer to Figure 1.10. Planets and moons nearest the Sun, where the temperatures were highest, are built largely of compounds that condense only at high temperatures. These compounds contain iron, silicon, magnesium, and aluminum, mostly bound strongly with oxygen.

Planets and moons more distant from the Sun, where temperatures were lower, contain some of these compounds but also large quantities of volatile substances that condense at lower temperatures. For example, hydrogen and sulfur combine with oxygen to form compounds that are solid only at low temperatures (as illustrated by Jupiter's moon Io, rich in sulfur—Figure 1.11A). The farther from the Sun that condensation occurred, the lower was the temperature and the greater the fraction of these

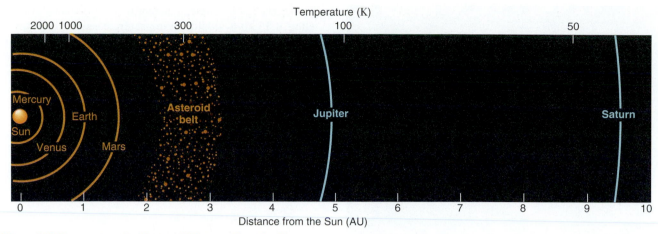

Temperature (K)

Distance from the Sun (AU)

Figure 1.10 Temperature Gradient and Distance from the Sun Close to the Sun, temperatures reached 2000 K (Kelvin), and the materials that condensed were largely oxides, silicates, and metallic iron and nickel. Farther from the Sun, in the Jupiter–Saturn region, temperatures were low enough to condense ices of water, ammonia, and methane. (See Appendix A for Kelvin temperature scale. Distances are measured in astronomical units or AU, the Earth-to-Sun distance.)

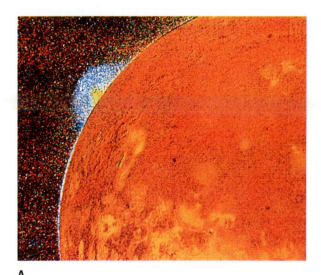

A.

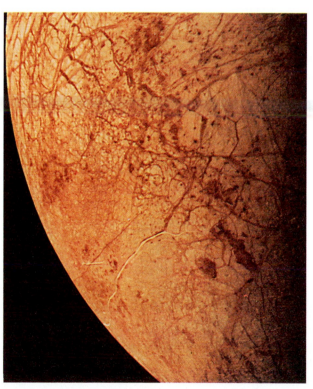

B.

Figure 1.11 Two Moons of Jupiter Are Rich in Volatile Substances A. Io is red because it is rich in sulfur. The plume from the volcanic eruption (bluish area) rises 100 km and is believed to be largely sulfur dioxide (SO_2). Several sites of active volcanism have been discovered on Io. B. Europa, smallest of Jupiter's four large moons, has low density, indicating it contains a substantial amount of ice. The surface is mantled by ice to 100 km depth. The fractures indicate an unknown internal process that is disturbing and renewing the moon's surface. The dark material in the fractures (appears red) apparently rises up from below. *Voyager 2* recorded this image in 1979.

PLANETARY ACCRETION AND METEORITES

For the terrestrial planets, condensation of solids from the gas cloud is only the first part of the birth story. Condensation formed innumerable small rocky fragments, but the fragments still had to be joined to form a terrestrial planet. This happened through impacts among fragments drawn together by gravitational attraction. The largest masses slowly swept up more and more of the con-

volatile substances. A striking demonstration of this is the large amount of frozen water in the moons of the Jovian planets (Figure 1.11B).

Figure 1.12 History from Space Meteorites carry some history of the earliest days of the solar system. This large meteorite is in the American Museum of Natural History, New York City.

densed rocky fragments, and grew larger to become the terrestrial planets.

Meteorites like the one in Figure 1.12 still are gravitationally attracted to Earth, proving that ancient, condensed rocky fragments still exist in space. Some meteorites resemble rocks formed on Earth, and these are believed to have been ejected into space when a meteorite struck a planet or a moon (like the impact that formed Meteor Crater—Figure 1.6).

Meteorites and the scars of ancient impacts provide evidence of the way terrestrial planets grew to their present sizes. This growth process—a gathering of more and more bits of solid matter from surrounding space—is called *planetary accretion*.

PLANETARY ACCRETION CONTINUES TODAY

Early in the lives of the terrestrial planets, planetary accretion happened much faster than it does today, and impacts of all sizes were common. Today, one sees an occasional meteor streaking across the night sky, and some do reach Earth, but large impacts are rare. A large (but not gigantic) impact formed Meteor Crater about 50,000 years ago. In 1908, a very large explosion occurred in a remote area of Siberia called Tunguska. Although we remain uncertain as to the cause, it is believed to have been a meteorite or a comet, making it the largest explosion due to impact in historical times. It is fortunate that it occurred in a sparsely inhabited area. What such impacts prove is that Earth and the other planets, even today, are hit by bits of ancient space debris.

Small meteorites enter Earth's atmosphere at 4–40 km/s (9000–90000 mi/h). If a large meteorite had such a velocity, the amount of energy released on impact would be enormous. For example, a meteorite 30 m in diameter traveling at 15 km/s would, on impact, release energy equivalent to 4 million tons of TNT. The resulting crater would be the size of Meteor Crater in Arizona—1200 m across and 200 m deep.

Cratering is a very rapid geologic process—the Meteor Crater event is estimated to have lasted about 1 minute.

> **Before you go on:**
> 1. How did the solar system form?
> 2. How did Earth form and evolve over time?

EARTH'S INTERNAL STRUCTURE: ENERGY, HEAT, GRAVITY, AND BUOYANCY

Earth formed by accretion, but is it a mass of gravitationally glued chunks of rock? Obviously not—the accreted pieces were melded together by subsequent events. These events were complex and involved energy transformation, radioactive decay, heat flow, gravity, buoyancy, convection, state (solid vs. liquid), and differentiation.

As accretion continued and the terrestrial planets grew larger, their temperatures must have risen. Why?

The reason for the rising temperature in the terrestrial planets is straightforward: energy can never be destroyed, but it can be transformed from one form to another—from electricity to heat, for example. Therefore, when a meteorite impacts a planet or moon, its energy of motion (called *kinetic energy*) is transformed to *heat energy*. As planet Earth grew larger and larger, the kinetic energy from continual impacts would have been transformed to heat energy, raising its temperature.

But impacts alone cannot account for the level of heat energy possessed by planet Earth. Heat must have been added continually from another source—*radioactive decay*. Among the many chemical elements in Earth, several are naturally radioactive, meaning that some of their atoms spontaneously transform to another element, simultaneously releasing energy. Examples are uranium and thorium, both of which transform to lead, and potassium, which transforms to argon and calcium.

Every time a radioactive transformation occurs, a tiny amount of heat energy is released. Thus, radioactivity continued to heat Earth internally, even as the frequency of impacts declined. Heat accumulated faster than it could be radiated into space, so Earth began to melt.

Here is where buoyancy enters the picture. Because Earth had become partly fluid, less-dense (more buoyant) molten materials were freed to migrate toward the surface—materials rich in the lighter elements silicon, aluminum, sodium, and potassium. Rocks at Earth's surface are still rich in these elements. This explains why we surface-dwellers have easy access to silicon for computer chips, aluminum for beverage cans, and salty seas.

Denser melted materials, such as molten iron, sank toward the center of the planet. The melting also released volatile substances, which escaped as gases through volcanoes. It was these escaped gases—mainly water vapor, but also smaller amounts of gaseous compounds of carbon dioxide and nitrogen—that formed Earth's atmosphere. From this same source also came the water we now find in Earth's oceans.

Thus, heat from impacts and radioactivity, combined with buoyancy, was responsible for creating the planet we call home. Partial melting changed Earth from its original homogeneous form to a compositionally layered one.

LAYERS OF DIFFERING COMPOSITION, SORTED BY DENSITY

Planet Earth has three main parts, like a hard-boiled egg—yolk, white, and shell. The three parts are distinguished by differences in composition and, consequently, density (Figure 1.13).

At the center is the densest part, the **core**. The spherical core is composed largely of metallic iron, with lesser amounts of nickel and other elements. Surrounding the core is the **mantle**, a thick shell of dense, rocky matter. Its density is between that of the core and the outermost layer. Surrounding the mantle lies the thinnest and outermost layer, the **crust**, which consists of the least-dense rocky matter.

The crust is far from uniform. It differs in thickness from place to place by a factor of about 10. The **oceanic crust** beneath the oceans averages about 8 km thick (as thin as 5 km in places), whereas **continental crust** averages about 45 km (ranges from 25 to 70 km).

We cannot directly see either the core or the mantle, so how do we know anything about their composition? The answer, of course, is indirect measurement. One method is to measure the time required for earthquake waves to travel through Earth by different paths, because the waves travel differently through materials that have different densities.

Earthquake wave speed depends on the density and elastic properties of the material the waves travel through. These properties, in turn, depend on the composition of the

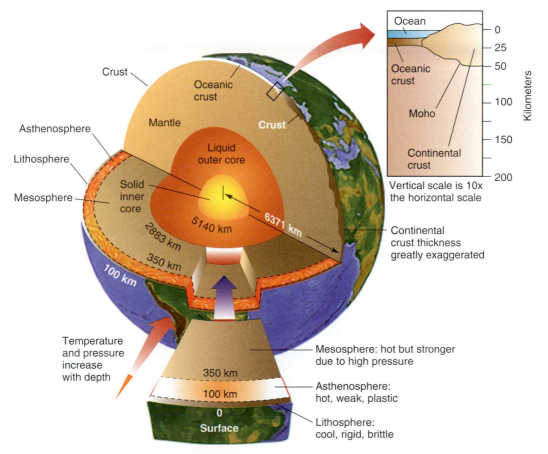

Figure 1.13 Inside Earth A sliced view of Earth reveals (1) layers of different composition and (2) zones of differing rock strength. The compositional layers, starting from the center, are the *core, mantle,* and *crust.* Note that the crust is thicker beneath the continents than under the oceans. Also note that boundaries between zones of different strength—lithosphere (outermost), asthenosphere, mesosphere—do not coincide with the compositional boundaries. This fact will become very important when we look at plate tectonics in Chapter 2.

material. At some depths, abrupt velocity changes in the waves indicate sharp changes in density and elastic properties, much like the refraction of light as it passes through one medium to another.

From these sharp changes, we can infer that Earth's interior lacks uniform composition and must instead consist of layers that have different characteristics. From what we know of these characteristics, we can estimate what the compositions of the different layers must be. (See more in Chapter 10, "Earthquakes and Earth's Interior.")

We cannot see and sample the mantle, so we have only hints of slight compositional variations that probably exist there. But we can see and sample the crust. This sampling shows that, even though the crust is quite varied, its overall composition and density are very different from those of the mantle, and the boundary between them is distinct.

The composition of the core presents the most difficulty. Temperatures and pressures in the core are so great that materials there probably have unusual properties. Some of the best evidence concerning core composition comes to us from *iron meteorites*. Such meteorites are believed to be fragments from the core of a small terrestrial planet that was shattered by a gigantic impact early in the history of the solar system. Scientists hypothesize that this now-shattered planet had compositional layers similar to those of Earth and the other terrestrial planets.

LAYERS OF DIFFERING PHYSICAL PROPERTIES

In addition to *compositional* layering, Earth has other zones with interesting *physical* properties. Most important are zones of differing rock strength and solid-versus-liquid. Physical property differences are largely controlled by temperature and pressure, rather than by rock composition. Importantly, the places where *physical* properties change do not coincide exactly with the *compositional* boundaries among the crust, mantle, and core (Figure 1.13).

INNER CORE (SOLID) AND OUTER CORE (LIQUID)

Within the core, an **inner core** exists where pressures are so great that iron is solid, despite its high temperature (Figure 1.13). Surrounding this inner core is the **outer core**, a zone where temperature and pressure are such that the iron is molten and exists as a liquid. Thus, the difference between the inner core and the outer core is not primarily composition (thought to be mostly elemental iron). Instead, the difference lies in the physical states of the two: the solid inner core, and the liquid outer core.

THE MESOSPHERE (HIGH STRENGTH)

The strength of a solid is controlled by both temperature and pressure. When a solid is heated, it loses strength. When it is compressed, it gains strength. Differences in temperature and pressure divide the mantle and crust into three *strength regions*. In the lower mantle, rock is so highly compressed that it has considerable strength, even though the temperature is very high. Nevertheless, rock in this zone is so hot it readily deforms under stress. Thus, a solid region of high temperature and relatively high strength exists within the mantle, from the core–mantle boundary (around 2900 km depth) to about 350 km depth, and is called the *mesosphere* (Greek: intermediate, or middle, sphere) (Figure 1.13).

THE ASTHENOSPHERE (WEAK AND PLASTIC)

Within the upper mantle, between 350 and 100–200 km below Earth's surface, is the **asthenosphere** ("weak sphere"). Here the balance between the effects of temperature and pressure is such that rocks are weak. Instead of being strong, like rock in the mesosphere, rock in the asthenosphere is very plastic and easily deformed, like butter or warm tar. The strength differences between the mesosphere and the asthenosphere are differences of degree, with the asthenosphere being notably weaker. As far as geologists can tell, the composition of the mesosphere and the asthenosphere is the same; the difference is one of physical properties. In this case, the property that changes is strength.

THE LITHOSPHERE (RIGID ROCKS)

Above the asthenosphere is the outermost strength zone, a region where rocks are cooler, stronger, and more rigid than those in the plastic asthenosphere. This hard outer region, which includes the uppermost mantle and all of the crust, is called the **lithosphere** ("rock sphere"). It is important to remember that, although the crust and mantle differ in composition, it is rock strength, not rock composition that differentiates the lithosphere from the asthenosphere.

Strength differences between lithospheric rock and asthenospheric rock are a function of temperature and pressure. At 1300°C and the pressure at 100 km depth, rocks of all kinds lose strength and become readily deformable (plastic). This is the base of the *oceanic lithosphere*. By contrast, the base of the *continental lithosphere* is at about 200 km and the temperature is about 1350°C. The reason for the difference between the two kinds of lithosphere is the way that temperature increases with depth, which we will look at in the next chapter when we discuss plate tectonics.

PLATE TECTONICS

Earth's interior is intensely hot—the temperature at the core–mantle boundary, for example, is about 5000°C, the same temperature as the surface of the Sun. Heat is still being added from present-day radioactivity. If Earth could not get rid of some of the heat the internal temperature would rise continually. The way Earth gets rid of heat and keeps a nearly constant internal temperature is through convection in the mesosphere and asthenosphere, a process we will discuss in detail in Chapter 2. We cannot observe Earth's internal convection directly, but we can see the effects of convection on the lithosphere and in particular on Earth's surface. The effect is called plate tectonics.

Plate tectonics is the "grand unifying theory" of geology, as fundamental to our science as evolution is to biology. **Plate tectonics theory** says that Earth's outermost 100 km "eggshell" (the lithosphere) is cracked, and *composed of about a dozen pieces* ("plates"). The plates float on the hot, deformable asthenosphere, moving ponderously (from a few mm to several cm a year) in various directions, spreading apart, slowly colliding, and grinding past one another. The locations of continents, oceans, mountains, volcanoes, and earthquakes are determined by plate tectonics. Throughout this book, we use plate tectonics as a unifying framework to integrate most geologic processes. Plate tectonics is the key link between Earth's internal and surface activities.

Half a century ago, most geologists assumed that Earth's oceans and continents had existed in their locations for much of Earth's 4.55-billion-year history. But in the 1960s, research by many geologists and oceanographers melded into the revolutionary hypothesis of plate tectonics, and today this is the overarching theory that appears to explain nearly all geologic phenomena.

Plate tectonics is a triumph for the scientific method, that methodology in which we constantly question, experiment, discard weak hypotheses, and continually seek new evidence to test our conclusions, even when we "feel certain." The remarkable story of plate tectonics theory, how it came to be, and why it explains most geologic phenomena are presented in Chapter 2.

Before you go on:

1. What are the three *compositional* layers of Earth that are sorted by density?

2. What are the key *physical* properties of the inner core, outer core, mesosphere, asthenosphere, and lithosphere?

3. What is plate tectonics?

THE SYSTEM CONCEPT

A **system** is any portion of the universe that can be isolated from the rest of the universe for observing and measuring change. The system concept is a powerful analytical tool because it allows you to break a large, complex problem into smaller, more easily studied pieces. A system can include whatever you select to observe—you choose its limits for the purpose of your study.

A system can be large or small, simple or complex (*examples:* the contents of a beaker, a flock of nesting birds, a lake, a small sample of rock, an entire ocean, a volcano, an entire mountain range, a continent, or even an entire planet). A leaf is a system, but it is also part of a larger system (a tree), which is part of a system (a forest), which is part of an even larger system, a biome.

When a scientist *isolates* a system from the rest of the universe, it means drawing a *boundary* to set the system apart from its surroundings. The nature of this boundary is one of the most important defining features of a system, and it leads us to the three basic kinds of systems—isolated, closed, and open. Figure 1.14 shows the three types.

The simplest kind of system to understand is an **isolated system**. In this case, the boundary completely prevents *the exchange of either matter or energy* between the system and its surroundings. The concept of an isolated system is an ideal because, although boundaries can prevent the passage of matter, no real-world boundary is so perfectly isolating that energy can neither enter nor escape.

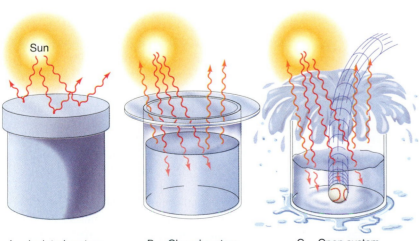

A. Isolated system   B. Closed system   C. Open system

Figure 1.14 Three Kinds of Systems A. An *isolated system* is "ideal"—it allows neither energy nor matter to cross its boundaries. Isolated systems cannot exist in nature. B. A *closed system* allows energy to cross its boundaries, but not matter. Closed systems are rare; the Earth is an example of a natural closed system—almost. Extremely light atoms of hydrogen and helium still escape gravity's pull and enter space, and meteors from space still enter the Earth system. But Earth is a close approximation of a closed system. C. An *open system* allows both energy and matter to freely cross its boundaries. Open systems are very common in nature—all of the Earth's subsystems are open systems.

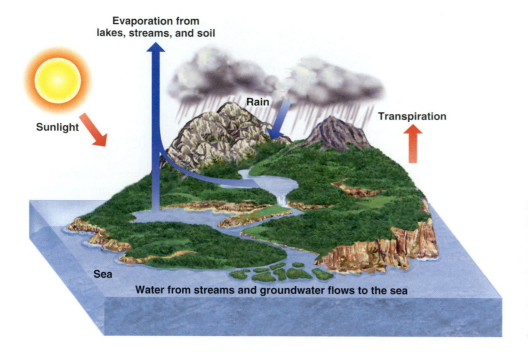

Evaporation from
lakes, streams, and soil

Rain

Transpiration

Sunlight

Sea

Water from streams and groundwater flows to the sea

Figure 1.15 Example of an Open System Energy (sunlight) and matter (rainfall) reach an island from external sources. The energy leaves the island both as reflected light and as heat. The water leaves the island by evaporating and draining into the sea.

The nearest thing to an isolated system in the real world is a **closed system**. Such a system has a boundary that permits *the exchange of energy with its surroundings, but not matter*. The Space Shuttle is an example of a nearly closed system. It allows the material inside to be heated and cooled but is designed to minimize the loss of any material.

An **open system** *can exchange both energy and matter across its boundary*. For example, consider an island on which rain is falling. Some of the water runs off via streams or seeps downward to become groundwater, while some is absorbed by plants, and some evaporates back into the atmosphere (Figure 1.15).

THE EARTH SYSTEM

Earth is an amazing collection of parts—the geosphere (rocks), the atmosphere (air), the hydrosphere (water), and the biosphere (life in all its forms)—each interacting dynamically with all the others. We call the assemblage of all of Earth's parts the **Earth system**. Each part is a system of smaller parts, and can be studied on its own, but we will study the interactions between the systems. Energy and materials (like water, carbon, and minerals) are transferred from one system to another. That's why we chose the opening photo to vividly display Earth's interacting systems.

To a close approximation, Earth is a closed system. It is only a close approximation, because meteorites do come in from space and fall on Earth, and a tiny trickle of gases leave the atmosphere and escape into space. Everything on and in Earth is part of an open system, meaning that matter and energy can move freely between them. Earth, a closed system, is an assemblage of open systems, and each open system interacts with all the other open systems. The

waters, the atmosphere, the plants and animals, and the rocks all interact with one another.

To illustrate this point, running water, acidic rain, freezing and thawing, oxidation, the activities of plants and microbes, and gravity all work together to break down solid rock into soil particles. From this soil, plants extract moisture and minerals. Through photosynthesis, plants use solar energy to power their conversion of water and carbon dioxide into their own food (carbohydrate). They also release a waste gas, oxygen, into the atmosphere—which animal species, including humans, inhale to "burn" their own food to release energy for life activities. Everything interacts.

In short, Earth is an assemblage of open systems, each dependent upon the others for the whole Earth "organism" to operate. It is these marvelous interactions—the Earth system—that create the unique environment in which life on Earth can prosper. No other world has this combination of systems.

A new branch of science focuses on these interacting systems: *Earth system science.* Geologists lead this science, but it is very interdisciplinary, with biologists, chemists, physicists, and others involved. Earth system science is abetted through sophisticated monitoring of Earth systems by satellite sensors, deep-diving submarines, air-monitoring and radiation-sensing devices, thermal and motion detectors, and so on. We now continually monitor interactions among parts of the Earth system.

Interestingly, any planet we might study will have its own "planet system science." We already have geologists who have become authorities on "Mars system science." But it is very different from Earth system science, because Mars has a radically different atmosphere (mostly carbon dioxide; no free oxygen), little water (all frozen), and no known life forms. It cannot match the magnificent complexity of Earth system science.

OUR PLANET'S "FOUR SPHERES"

Earth comprises four vast reservoirs or "spheres" of matter. Each is an open system, because both matter and energy flow back and forth among them (Figure 1.16). The four reservoirs are:

1. The **atmosphere**—the mixture of gases that surrounds Earth—predominantly nitrogen, oxygen, argon, carbon dioxide, and water vapor.

2. The **hydrosphere**—all of Earth's water, including oceans, lakes, streams, underground water, snow, and ice—excluding water vapor in the atmosphere (considered to be part of the atmosphere).

3. The **biosphere**—all of Earth's organisms, as well as any organic matter not yet decomposed (dead animals and plants).

4. The **geosphere**—the solid Earth from core to surface, composed principally of rock and **regolith** (the irregular blanket of loose, uncemented rock particles that covers Earth).

Each of the four systems can be further subdivided into smaller, more easily studied units. For example, we can divide the hydrosphere into the ocean system, glacial ice system, streams system, groundwater system. All of these and smaller systems within the Earth system are open systems.

Although this book focuses primarily on the geosphere, remember that the geosphere is a big subsystem of the Earth system and that many important processes involve the atmosphere, hydrosphere, and biosphere subsystems. For example, **erosion**, which is the breakdown of rock through the combined effects of rain, wind, snow, ice, plants, and animals, involves all four spheres.

INTERACTIONS AMONG EARTH'S OPEN SYSTEMS

We walk on the regolith, breathe the atmosphere, and feel the rain: the evidence is clear that Earth's external systems are places of intense and continual activity. Water and air penetrate the regolith and far into the crust. Chemical and physical disintegration of rock goes on because the atmosphere, biosphere, and hydrosphere combine to alter and break down the crust.

CYCLICAL MOVEMENTS

It is useful to envision the interactions between Earth's internal and external systems as processes that facilitate the movement of materials and energy among Earth's reservoirs. The movement of materials is cyclical, meaning it is continuous: this feature provides a convenient framework that helps us study how materials and energy are stored and how they are cycled among the principal reservoirs, the geosphere, hydrosphere, atmosphere, and biosphere, by the Earth system.

There are two key aspects to cycles: (1) the reservoirs in which the materials reside; and (2) the flows, or fluxes, of materials from reservoir to reservoir. For example, when raindrops form, they dissolve gases from the atmosphere (a reservoir) and carry them down to Earth's surface (part of the geosphere, and therefore another reservoir). The transport of water and dissolved gases from the atmosphere to Earth's surface is a flow. The water and dissolved gases react with minerals at the surface and convert some of the constituents of the minerals into new minerals, while other constituents dissolve and remain in solution. The dissolved products are transported by streams (a flow) to become concentrated in the ocean (a reservoir). This is the origin of many of the salts in seawater.

How long do materials reside in the reservoirs? The storage time can differ greatly. For example, an individual molecule of water may spend 200,000 years in a great ice sheet like the one that covers Antarctica, 40,000 years in the ocean, 1000 years in a deep underground reservoir, 10 years in a lake, 10 days in the atmosphere, and only 10 hours in an animals' body. Water is transferred between these and other reservoirs via innumerable pathways and at

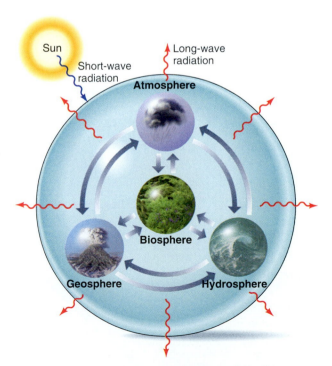

Figure 1.16 Earth Is Almost a Closed System—Except for Meteorites Energy reaches Earth from an external source and eventually returns to space as heat radiation. Smaller systems within Earth—the atmosphere, biosphere, hydrosphere, and geosphere—are all open systems.

Figure 1.17 Cycling Water Between Reservoirs Water in the environment. A. Example of the cycling of water. The Sun heats the ocean (a reservoir) and causes water to evaporate into the atmosphere (a reservoir). Water vapor in the atmosphere condenses to form clouds, rain, and snow. Compacted snow forms ice in a glacier (a reservoir). The glacier flows slowly down to the sea where huge masses of ice break off to form bergs that slowly melt, returning water to the ocean. In this photo, ice breaks off from Hubbard Glacier in Alaska. B. The hydrologic cycle for the conterminous United States showing the amounts of water exchanged between reservoirs, in millions of cubic meters a day.

A.

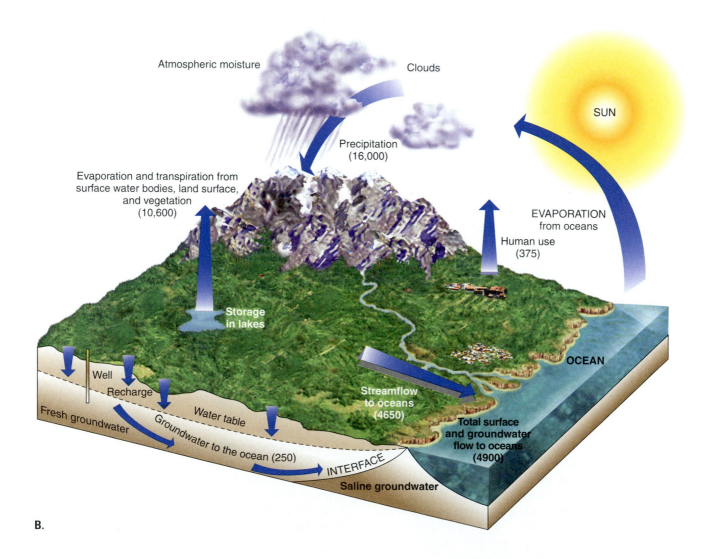

B.

greatly differing rates (Figure 1.17). A cycle can therefore include reservoirs of greatly differing size and involve processes that operate on many differing time scales.

The three most important cycles that involve the solid Earth, and therefore the most important cycles for the study of physical geology, are:

1. The **hydrologic cycle**, which is the day-to-day and long-term cyclical changes and movements of the water in Earth's hydrosphere, which are made evident by such elements as rain, snow, and running streams.

2. The **rock cycle**, which describes all of the various processes by which rock is formed, modified, decomposed, and reformed by the internal and external processes of Earth.

3. The **tectonic cycle**, which deals with the movements of plates of lithosphere, and the internal processes of Earth's deep interior that drive plate motions.

The rock cycle, tectonic cycle, and hydrologic cycle are closely linked through physical, chemical, and biological processes. The three cycles involve most of the topics discussed in this book. There are, of course, many other cycles. For example, a useful group of cycles is the **biogeochemical cycles**, which trace the movement of chemical elements that are essential for life, including carbon, oxygen, nitrogen, sulfur, hydrogen, and phosphorus (see *The Science of Geology*, Box 1.2, *Cycling of Biogeochemical Elements Important for Life*).

THE HYDROLOGIC CYCLE

The hydrologic cycle (Figure 1.17B) is powered by heat from the Sun, which evaporates water from the ocean and the land surface. The water vapor thus produced enters the atmosphere and moves with the flowing air. Some of the water vapor condenses and is precipitated as rain or snow back into the ocean, or on land. Rain falling on land may drain off into streams, percolate into the ground, or be evaporated back into the air, where it is further recycled. Part of the water in the ground is taken up by plants, which return water to the atmosphere through their leaves by a process called *transpiration*.

Snow may remain on the ground for one or more seasons until it melts and the meltwater flows away. Snow that nourishes glaciers remains locked up much longer, through many tens or even thousands of years, but eventually it too melts or evaporates and returns to the oceans. The many reservoirs and pathways of the hydrologic cycle are shown diagrammatically in Figure 1.18.

Bodies of water—especially the ocean—control the weather and influence the distribution of climatic zones. The varied landscapes we see around us are another important consequence of the hydrologic cycle. The erosional and depositional work of streams, waves, and glaciers, combined with tectonic movement, volcanism, and defor-

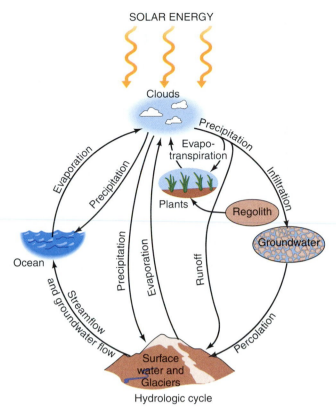

Figure 1.18 The Hydrologic Cycle The hydrologic cycle (or water cycle) describes the pathways by which water moves between the principal reservoirs, the ocean, ice, groundwater, and the atmosphere.

mation of crustal rocks, have produced a diversity of landscapes that makes Earth's surface unlike that of any planet in the solar system. In all these ways and more, the hydrologic cycle influences our everyday lives.

THE ROCK CYCLE

To discuss the rock cycle we must first define the term rock and then introduce the major kinds of rocks. **Rock** is any naturally formed, nonliving, firm and coherent aggregate of mineral matter that constitutes part of a planet. There are three large families of rocks, each defined by the processes that form them.

THE THREE ROCK FAMILIES

The first of the three major rock families is the **igneous rock** family (named from the Latin *igneus*, meaning fire). The cooling and solidification of magma forms igneous rocks. **Magma** is molten rock.

Some products of weathering are soluble and are carried away in solution by streams, but most weathering products are loose particles in the regolith. Particles of the regolith that are transported by water, wind, or ice will sooner or later settle to form deposits of **sediment** ("settling"). Sediment eventually becomes **sedimentary rock**, which is

BOX 1.2 **THE SCIENCE OF GEOLOGY**

CYCLING OF BIOGEOCHEMICAL ELEMENTS IMPORTANT FOR LIFE

Plants extract carbon dioxide from the atmosphere as well as the soil and, through the process of photosynthesis, combine the two to make carbohydrates and oxygen; this is the source of oxygen in the atmosphere (Figure B1.2). Photosynthesis is just one among many steps by which carbon, an element essential to all forms of life, cycles between the biosphere, atmosphere, hydrosphere, and geosphere. The natural cycle of describing the movements and interactions of a chemical element essential to life is called a biogeochemical cycle. The most important elements for life are carbon, oxygen, sulfur, hydrogen, nitrogen, and phosphorus. The biogeochemical cycle of carbon is discussed in detail later in the book. Here we discuss the cycle of nitrogen as an example of a biogeochemical cycle.

The Biogeochemical Cycle of Nitrogen

Amino acids are essential constituents of all living organisms. They are given the name *amino* because they contain amine groups (NH_2), and the key element in amines is nitrogen. As a consequence nitrogen is essential for all forms of life.

Nitrogen in nature exists in three forms: in its elemental form N_2, the most abundant gas in the atmosphere; in a reduced form such as ammonia, NH_3; and in oxidized forms such as nitrate, NO_3. It is only in the reduced form that nitrogen is incorporated into the newly formed organic matter of living organisms.

Most of Earth's nitrogen is present as N_2 in the atmosphere. The key to the nitrogen cycle is understanding how reduction (also called *fixation*) and oxidation (also called *denitrification*) take place and the way nitrogen moves between the five major reservoirs—the atmosphere, biosphere, ocean, soil, and sediment. As you read the following discussion, please study Figure B1.3 carefully; the figure is a diagram showing the reservoirs, the estimated number of grams of nitrogen in each, and the paths by which nitrogen moves between the reservoirs.

The nitrogen cycle is dominated by the atmosphere and by the fact that organisms cannot directly use N_2. Nitrogen is removed from the atmosphere in three ways:

1. By solution of N_2 in the ocean.
2. By combination of N_2 and O_2 (oxidation) to form NO_3 by lightning discharges, in which form it is rained out of the atmosphere and into the soil and the sea. Plants can reduce NO_3 to NH_3, thus making nitrogen assimilable to the biosphere.
3. By the reduction of N_2 to NH_3 through the actions of nitrogen-fixing bacteria in the soil or the sea. The reduced nitrogen is quickly assimilated by the biosphere.

The nitrogen cycle is interesting because of its complexity, but interesting too because parts of the cycle must

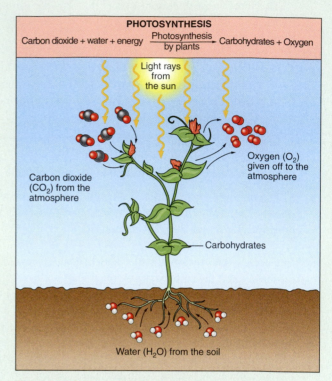

PHOTOSYNTHESIS

Carbon dioxide + water + energy $\xrightarrow[\text{by plants}]{\text{Photosynthesis}}$ Carbohydrates + Oxygen

Light rays from the sun

Carbon dioxide (CO_2) from the atmosphere

Oxygen (O_2) given off to the atmosphere

Carbohydrates

Water (H_2O) from the soil

Figure B1.2 Source of Earth's Oxygen Through photosynthesis, plants combine carbon dioxide (CO_2) and water (H_2O) to make carbohydrates, a basic food and energy resource. The process releases a waste gas, oxygen, to the atmosphere. Humans and animals require oxygen to survive.

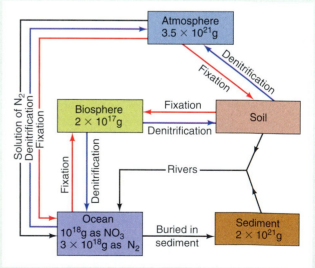

Atmosphere
$3.5 \times 10^{21}g$

Denitrification

Fixation

Solution of N_2
Denitrification
Fixation

Biosphere
$2 \times 10^{17}g$

Fixation

Soil

Denitrification

Fixation

Denitrification

Rivers

Ocean
$10^{18}g$ as NO_3
$3 \times 10^{18}g$ as N_2

Buried in sediment

Sediment
$2 \times 10^{21}g$

Figure B1.3 The Nitrogen Cycle A box diagram showing the biogeochemical cycle of nitrogen. The boxes are reservoirs, and the arrows represent the flows of material between the reservoirs. Numbers are the estimated amount of nitrogen in a given reservoir, in grams.

have evolved at different times in the past. The ancient atmosphere did not contain oxygen, so path 2 above must have evolved after the atmosphere became oxygenated. The atmosphere became oxygenated long after primitive life appeared on Earth. Because organisms cannot use N2 directly, either some reduced nitrogen must have been available when life arose or the earliest organisms had the ability to reduce N2. Anaerobic bacteria are those that live in oxygen-free places. Nitrogen-fixing anaerobic bacteria are very ancient, and the fixation chemistry that evolved with them will not work in the presence of oxygen. Such bacteria must therefore have evolved before the atmosphere contained oxygen (Figure B1.4).

Figure B1.4 Nitrogen Fixers Root nodules on a white clover plant produced by colonies of nitrogen-fixing bacteria.

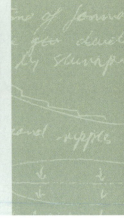

any rock formed by consolidation of sediment, including sediment formed by precipitation of materials that had been carried in solution to the sea by solution in stream water. Sedimentary rocks constitute the second rock family.

The final major rock family is the **metamorphic rock** family (from the Greek *meta*, meaning change, and *morphe*, meaning form: hence change of form). Metamorphic rocks are those whose original form has been changed as a result of high temperature, high pressure, or both. Metamorphism, the process that forms metamorphic rocks from sedimentary or igneous rocks, is analogous to the process that occurs when a potter fires a clay pot in an oven. The tiny mineral grains in the clay undergo a series of chemical reactions as a result of the increased temperature. New compounds form, and the formerly soft clay molded by the potter becomes hard and rigid.

THE CYCLE

When mountain ranges rise as a result of plate tectonic actions, or when lava and volcanic ash erupt from volcanoes, the newly exposed or newly formed rocks are quickly attacked by water, wind, and ice. These constantly modify Earth's surface, cutting away material here, depositing material there, and in the process create the landscapes we see around us. Through weathering and erosion, the atmosphere, hydrosphere, and biosphere continually react with and change the surface of the solid Earth and are thereby closely linked to the rock cycle. The rock cycle describes all the processes by which rock is formed, modified, transported, decomposed, and reformed as a result of Earth's internal and external processes.

Cycles are continuous and have no beginning or end. The rock cycle, which principally involves the continental crust, is like a merry-go-round, endlessly turning, powered by Earth's internal heat energy and by incoming energy from the Sun (Figure 1.19). In order to discuss the rock cycle we have to start somewhere, so let's start at the top of Figure 1.19 with uplift of continental crust and the exposure of crustal rocks.

Exposed rocks are vulnerable to weathering and are transformed into regolith by processes involving the atmosphere, hydrosphere, and biosphere. The combined processes of erosion then transport the regolith and even-

SOLAR ENERGY

Continental crust

Weathering

Regolith

Erosion and deposition

Sediment

Burial and cementation

Tectonic uplift

Tectonic uplift

Tectonic uplift

Sedimentary rocks

Metamorphis

Metamorphism

Metamorphic rocks

Melting

Magma

Intrusion and volcanism

Igneous rocks

Rock cycle

GEOTHERMAL ENERGY

Figure 1.19 The Rock Cycle The rock cycle traces the processes whereby materials within and on top of Earth's crust are weathered, transported, deposited, metamorphosed, and even melted. Note that rocks don't necessarily follow the path that leads around the outside of the circle; they can follow any of the short circuits from one stage to another.

tually deposit it as sediment. After deposition, the sediment is buried and compacted, eventually becoming sedimentary rock. Processes involving plate motions and crustal uplift lead to rock deformation. This paragraph describes the sequence of events first outlined by James Hutton, and illustrated in Figure 1.4.

Deeper burial turns sedimentary rock into metamorphic rock, and even deeper burial may cause some of the metamorphic rock to melt, forming magma from which new igneous rock will form. At any stage in the cycle, tectonic processes might elevate the crust and expose the rocks to weathering and erosion. As a consequence of the rock cycle, material of the continental crust is being constantly recycled.

Because erosion of land is generally very slow and the mass of the continental crust is very large, the average time rock spends in the continental crust is very long. Time estimates vary, and they are difficult to make, but the average age of all the rock in the continental crust seems to be about 650 million years.

THE TECTONIC CYCLE

The tectonic cycle involves plate tectonics. **Tectonics** is the study of the movement and deformation of the lithosphere. The word comes form the Greek, *tekton*, meaning carpenter.

The tectonic cycle and the rock cycle are closely linked. However, because the rock cycle is mainly a phenomenon of the continental crust and is dominated by interactions with the atmosphere, hydrosphere, and biosphere, it is driven in large part by solar energy. The tectonic cycle, by contrast, mainly involves oceanic crust and is dominated by processes deep in the Earth that are driven by Earth's internal heat energy.

The tectonic cycle is depicted in Figure 1.20. When magma rises from deep in the mantle, it forms new oceanic crust at midocean ridges. Old oceanic crust sinks back into the mantle at a rate equal to the rate at which new oceanic crust is created. The lifetime of oceanic crust is shorter than the lifetime of continental crust. The most ancient oceanic crust of the ocean basins is only about 180 million years, and the average of all oceanic crust is about 70 million years.

An important interaction between the oceanic crust and the continental crust involves volcanism. When old oceanic crust sinks back into the mantle, it carries some water with it. As the sinking crust heats up, the water is driven off and it causes melting in the mantle adjacent to the sinking crust. The magma so formed rises to form volcanoes. By this mechanism there is a continual addition of new mantle-derived material to the continental crust.

Another important interaction involving oceanic crust concerns the composition of seawater. The magma that rises to form new oceanic crust at midocean ridges forms hot igneous rocks that react with seawater. In the reaction, some constituents in the hot rock, such as calcium, are dissolved in the seawater and constituents already in the seawater, such as magnesium, are deposited in the igneous rock. Because the magma that forms the oceanic crust

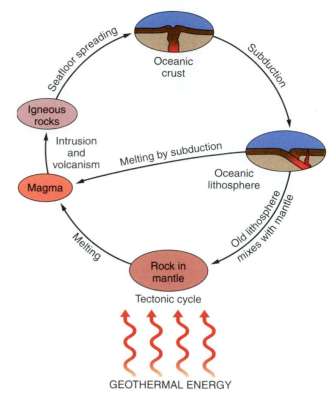

Figure 1.20 The Tectonic Cycle Processes driven by Earth's internal heat energy power the tectonic cycle. Rock in the mantle melts and the magma rises to make new oceanic crust. Eventually, the oceanic crust sinks back into the mantle, where water released from the old seafloor rocks causes partial melting of the mantle rocks in contact with the sinking oceanic crust. This magma, in turn, rises to create volcanoes.

comes from the mantle, the reactions between hot crust and seawater are one way in which the mantle plays a role in determining the composition of seawater, and a very important way that the materials and processes of the tectonic cycle interact with the those of the hydrologic cycle.

UNIFORMITARIANISM AND RATES OF GEOLOGIC PROCESSES

The great contributions James Hutton made to geology were mentioned earlier in the chapter. It was Hutton who recognized that the same external and internal processes we see in action today have been operating throughout Earth's long history, and it was Hutton's associates and successors who codified his observations into the Principle of Uniformitarianism.

During the nineteenth century, geologists tried to estimate how long the rock cycle had been operating by using the thickness of a pile of sedimentary rock as a way to estimate how long it took for the sediment from which it was formed to accumulate. They assumed that the Principle of Uniformitarianism applied to rates of geologic processes as well as to the processes themselves, and hence that the rates of deposition have always been constant and equal to today's rates. Thus, they thought, it should be possible to estimate the time needed to produce all the sedimentary rock still present on Earth. The figure so obtained, they argued, would be the minimum age for Earth. The results, we now know, were greatly in error. One reason for the error was the assumption that geologic rates were constant.

The more we learn of Earth history and the more accurately we determine the timing of past events through radiometric dating (Chapter 11), the clearer it becomes that the rates of geologic processes have not always been the same. The evidence is strongly against constancy; some rates were once more rapid, others much slower.

One reason the rates of processes involved in the rock cycle have changed through time is that Earth is very slowly cooling as its internal heat is lost. Earth's internal temperature is maintained, in part, by natural radioactivity. Early in Earth's history, more radioactive atoms were present than there are today, and more heat must have been produced than is produced today. Internal processes, which are all driven by Earth's internal heat, must have been more rapid than they are today. It is possible that a billion years ago oceanic crust was created at a faster rate than it is made today, and that continental crust was uplifted and eroded at a faster rate. Either or both actions would cause the rock cycle to speed up.

At the same time, the rates of external processes have also varied. Long-term changes in the rates have occurred because of slow increases in the heat output of the Sun and also because of the gradual slowing of the rate of rotation of Earth on its axis. For instance, scientists estimate that 600 million years ago there were 400 days in the year, and 2 billion years ago there were 450 days a year. Short-term effects on external processes have also arisen because of changes in the in the orientation of Earth's axis of rotation and changes in Earth's orbit. These changes, which are discussed more fully in Chapter 16, play an important role in the waxing and waning of the great ice sheets. It is clear, therefore, that even though the cycles have been continuous, the processes involved in the cycles have not maintained constant rates through time.

The conclusion that rates of geologic processes have differed in the past from today's rates is an important one. It means that the relative importance of different geologic processes has probably differed in the past. For example, just because glaciation is an important process today, we cannot assume it has been equally important throughout geologic time. But we can be confident that when glaciation did affect Earth in geologically remote times, the processes and effects of glaciation were the same as processes and effects of glaciation we observe in Antarctica today.

Before you go on:

1. What is a system?
2. What are the differences among open, closed, and isolated systems?
3. What are Earth's "four spheres"? How do they interact?
4. What are the three major cycles and how do they interact?

CHANGES CAUSED BY HUMAN ACTIVITIES

We have emphasized that Earth is a complex closed system made up of interacting open systems, and that the four major open systems of our planet are the four interactive reservoirs of matter involved in all geologic processes. When you see how change in one of the open systems can impact the other systems, you begin to appreciate how risky it can be to tamper with our home in an uncontrolled manner. This is why geologists and other scientists are deeply concerned about the potential for human activities to impact the environment, not only locally, but globally. Here are three of the ways we humans are affecting the Earth system:

1. By burning petroleum and coal, and by intensive farming activities, we are changing the composition of the atmosphere. The carbon dioxide content of the atmosphere is slowly increasing, and the changes are making the atmosphere an ever-more efficient heat blanket. This effect, colloquially known as **global change**, or **global warming**, is raising the temperature of the atmosphere.

2. For many years we all used, in an unquestioning way, refrigerators, pressure-spray cans, and other devices that used gases containing chlorine. When the gases escaped into the atmosphere, they interfered with natural reactions that produce ozone. The problem is that ozone high in the atmosphere protects us from damaging ultraviolet radiation and prevents it from reaching ground level. Ozone reduction puts us at risk.

3. Through the storage of water in giant reservoirs, we have changed the distribution of weight at Earth's surface. As a result we have changed, slightly but measurably, the rotation of the Earth on its axis.

What's Ahead?

The three great geologic cycles discussed in this chapter—the hydrologic, rock, and tectonic cycles—have been continuous since there was continental crust and oceans of seawater—at least 4 billion years. The rates of the processes may have changed but the processes themselves have been continuous. The interactions we observe between the many cyclical processes today must therefore have been going on for 4 billion years, too.

In Figure 1.21 you can see, in a diagrammatic manner, how the processes of the three main geolog-

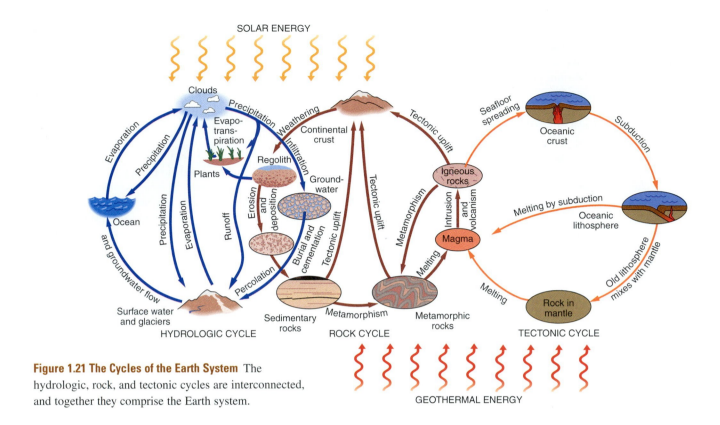

Figure 1.21 The Cycles of the Earth System The hydrologic, rock, and tectonic cycles are interconnected, and together they comprise the Earth system.

ic cycles interact. Figure 1.21 is a convenient way to think about the Earth system, and we will return to a discussion of the Earth system in each chapter. Figure 1.21 is also a convenient way to bring this chapter to a close because all of the subsequent chapters discuss one or more of the flows of energy and material between the reservoirs.

The Earth system is discussed in detail in this first chapter; plate tectonics is the subject of Chapter 2. Thereafter, starting with Chapter 3, we revisit the topics of the Earth system and plate tectonics at the end of each chapter. We do so to help you understand that the Earth system and plate tectonics are unifying themes that tie together all of Earth's processes and materials.

CHAPTER SUMMARY

1. Geology is the study of changes, past and present, that happen to Earth. Scientists who study Earth are called geologists.

2. Science is a system of learning and understanding that advances by application of the scientific method: observation, formation of a hypothesis, testing of the hypothesis, formation of a theory, more testing, and, in some instances, formation of a law.

3. The Principle of Uniformitarianism states that the internal and external processes operating today have been operating throughout Earth's history.

4. Random massive disasters, such as gigantic meteorite impacts, appear to have played important roles in Earth's history. These events cause catastrophic change in Earth's appearance, and to many forms of life, but are not attributed to supernatural forces the way the events of the outdated concept of catastrophism were.

5. There are nine planets in orbit around the Sun. The innermost four, Mercury, Venus, Earth, and Mars, are small, dense, rocky bodies. Four of the outer plants, Jupiter, Saturn, Uranus, and Neptune, are larger, less dense objects with thick atmospheres of hydrogen and helium. Pluto, the planet farthest from the Sun, is small, not gassy, and does not fit the mold of either terrestrial or Jovian planet.

6. The planets formed by a two-step process. First, small rocky fragments condensed from a disc-shaped envelope of gas that rotated around the Sun. Then the rocky fragments started accreting into ever larger masses. The largest, today's planets, had all formed by 4.6 billion years ago.

7. The heat released by decay of naturally occurring radioactive chemical elements was sufficient, early in Earth's history, to cause a fraction of Earth to melt. Heavy materials sank to the center, and lighter ones rose, giving Earth a compositionally three-layered structure: core, mantle, and crust.

8. The crust consists of two parts: oceanic crust with an average thickness of 8 km and continental crust with an average thickness of about 45 km.

9. Earth is also layered with respect to its physical properties, in

particular, strength. The lithosphere, the outer zone of the solid Earth, consists of rock that is strong and relatively rigid. Beneath the oceanic crust, the lithosphere is about 100 km thick; beneath the continental crust, it is 200 km thick. Below the lithosphere, down to a depth of 350 km, is the asthenosphere, a region where high temperatures make rock weak and easily deformed. Beneath the asthenosphere is the mesosphere, where rocks become gradually stronger. Within the core there are also two regions differing in physical properties but with the same composition: the inner core is solid, the outer core molten.

10. The lithosphere consists of six large and many small plates that slide slowly over the asthenosphere at rates up to 12 cm/yr, as a result of a process called plate tectonics.

11. Earth is a closed system—energy can enter and leave but the amount of matter present is constant.

12. All systems within the Earth—for example, geosphere, atmosphere, hydrosphere, and biosphere—are open systems.

13. Material and energy move back and forth between Earth's open systems in continuing cycles.

14. Energy to drive Earth's systems comes from the Sun and Earth's internal heat.

THE LANGUAGE OF GEOLOGY

asthenosphere (p. 18)
atmosphere (p. 21)

biogeochemical cycles (p. 23)
biosphere (p. 21)

catastrophism (p. 8)
closed system (p. 20)
continental crust (p. 17)
core (p. 17)
crust (p. 17)

Earth system (p. 20)
erosion (p. 21)

geologists (p. 5)
geology (p. 5)
geosphere (p. 21)
global change (p. 27)
global warming (p. 27)

gradualism (p. 9)

historical geology (p. 7)
hydrologic cycle (p. 23)
hydrosphere (p. 21)
hypothesis (p. 8)

igneous rock (p. 23)
inner core (p. 18)
isolated system (p. 19)

law (p. 8)
Law of Original Horizontality (p. 8)
lithosphere (p. 18)

magma (p. 23)
mantle (p. 17)
metamorphic rock (p. 25)

oceanic crust (p. 17)
open system (p. 20)

outer core (p. 18)

physical geology (p. 7)
plate tectonics theory (p. 19)
principle (p. 8)
Principle of Uniformitarianism (p. 9)

regolith (p. 21)
rock (p. 23)
rock cycle (p. 23)

scientific method (p. 8)
sediment (p. 23)
sedimentary rock (p. 23)
solar system (p. 12)
system (p. 19)

tectonics (p. 26)
tectonic cycle (p. 23)
theory (p. 8)

QUESTIONS FOR REVIEW

1. Suggest three human activities that affect Earth's external processes in a detectable way.

2. Identify three human activities in the area where you live that are causing big changes to the environment.

3. What is the scientific method? Illustrate your answer with an example of the scientific method in practice.

4. How does the Principle of Uniformitarianism help us understand the history of Earth?

5. Briefly describe the steps by which scientists believe planets form from a huge cloud of gaseous material.

6. Describe Earth's compositional layers. Discuss how the layers developed from an originally homogeneous Earth.

7. The Earth is layered with respect to its physical properties. Describe the major physical property layers and discuss how they arise.

8. What are the relationships between the crust, the mantle, and the lithosphere?

9. Why and how does the thickness of the lithosphere vary?

10. What are the differences between a closed and an open system?

11. What consequences arise from the fact that Earth is a closed system? Would Earth still be a closed system if we started a colony on Mars and started trading with the colony?

Click on *Presentation* and *Interactivity* in the **Earth Systems** module of your CD-ROM to further explore resources and activities presenting concepts from this chapter. Select *Assessment* in the same module to text your understanding of this chapter.

Chapter 2

Ariel view of the San Andreas Fault. The boundary between the Pacific and North American plates extends through California along the San Andreas Fault. Lateral motion of Earth's surface over millions of years has displaced streambeds and gullies that once crossed the fault in a straight line.

Global Tectonics: Our Dynamic Planet

Seismic waves from a California earthquake, as recorded "down under"

Earth Is a Noisy Planet, If You Keep Your Ear to the Ground

It is easy to be passionate about earthquakes. On average, every month our planet suffers a subterranean rock fracture that releases 10 times more energy than a thermonuclear bomb. One such "seismic event" in Los Angeles knocked me out of bed in the pre-dawn of a winter day in eighth grade. I have been fascinated by earthquakes ever since.

An earthquake briefly illuminates Earth's interior with vibrational, or seismic, waves, much as a camera flash brightens a dark room with a burst of light. Deep in our planet, seismic waves bend and bounce, focus and defocus, speed up and slow down, scatter and decay away. Seismologists reconstruct the passage of seismic waves through Earth to discover its distinctive layering with depth. The locations of earthquakes tell us where the boundaries of plates are. Within the plate interiors earthquakes are scarce, because the plate acts as a rigid shell. At their boundaries the plate edges crumple, bend, and slide past each other—this is where the earthquake action is. The largest earthquakes occur wherever two plates collide and one recycles into Earth's interior, in a process called subduction. Seismologists love these places. Only travel agents might know about the Kingdom of Tonga in the remote tropical Pacific, but to seismologists Tonga is the baddest subduction zone on the planet. A torn corner of the Pacific plate descends steeply under Tonga, through the asthenosphere and into the deeper mantle, crackling with earthquakes to a depth near 670 kilometers.

After any large earthquake, the Internet hums with automated computer requests for data from hundreds of seismic motion detectors worldwide. By the time I can reach my own computer, other computers have located the earthquake from the arrival times of its seismic waves. In June 1992 a large earthquake in southern California rattled my parents. A few hours later in Connecticut I looked at the earthquake's waves recorded in New Zealand and Australia. These waves had crossed the Pacific Ocean in 40 minutes, sideswiping Tonga along the way. In the data I found an unusual seismic wave that, like an echo in a shooting gallery, had deflected from some sort of barrier beneath Tonga. The barrier was the sinking Pacific Plate, which was stretching the surrounding asthenosphere like taffy.

Jeffrey Park

KEY QUESTIONS

1. **How do Earth's topography and bathymetry tell us that our planet's interior moves in a ductile manner?**

2. **How is plate tectonics related to the loss of Earth's internal heat to space?**

3. **How do we measure the drift of continents?**

4. **What happens when Earth's tectonic plates collide, split apart, or slide past each other?**

5. **How does plate tectonics influence Earth's climate?**

6. **Does plate tectonics operate on Venus, a planet with size and composition similar to Earth? If not, why not?**

INTRODUCTION

Each rocky body in the solar system, Earth included, has been losing its internal heat to space for approximately 4.5 billion years. Each rocky body, whether planet or moon, started with a hot interior, and each has been kept warm over time by energy released by the decay of radioactive isotopes. Despite radioactive heating, rocky bodies have cooled considerably since their formation, so that their outer layers have stiffened into thick motionless shells, or lithospheres. The interiors of Earth and Venus remain hot and geologically active. The stiff lithospheres of both Earth and Venus are thin compared to their planetary diameters, so that the lithospheres can readily bend, stretch, and break in response to the slow motion of hot interior rock.

The mantles of Earth and Venus lose internal heat by convection—slow flow of solid rock. Hot rock expands as its temperature increases and becomes buoyant. Cool rock contracts as its temperature decreases and loses its buoyancy. Hot rock rises upward to near the surface, releases its heat to the atmosphere and (eventually) to space, and sinks back into the planetary interior. Convection on Earth expresses itself in the form of plate tectonics. Earth's stiff lithosphere is broken into a collection of near-rigid plates. Most large-scale geologic events that originate within Earth's interior (e.g., earthquakes and volcanic eruptions) occur at the edges of plates, where large chunks of lithosphere collide, pull apart, or slide past each other. The plates participate in Earth's convection directly, by sinking into the interior once they have cooled, contracted, and become more dense. In fact, the cool sinking plates largely control the overall pattern of convection in Earth's mantle.

Plate tectonics is a large-scale process within the Earth system. Many other processes in the Earth system, such as the hydrologic and biogeochemical cycles, are profoundly affected by plate tectonics. The importance of plate tectonics to the world we live in becomes clear when we consider other rocky bodies in our solar system, particularly our sister planet, Venus. Each started with a chemical composition similar to Earth's and each has followed the same laws of physics. Despite these similarities, only Earth has plate tectonics, and only Earth is known to have sustained diverse communities of living organisms for billions of years. These two facts are connected.

WHY DON'T WE LIVE ON VENUS?

Why do we live on Earth, and not on our sister planet Venus? The orbit of Venus is closer to the blazing Sun, but not too close to preclude life. Venus has water vapor, and water is essential to life. Venus is a rocky planet like Earth, so it might be rich in minerals to support modern living organisms.

The Venusian surface is completely shrouded by clouds. Prior to the 1960s, scientists had considerable freedom to speculate on what might lay under them. However, early spacecraft flybys sensed infrared radiation from Venus— lots of it. This revealed a surface far hotter than a pizza oven, almost 500°C. The Venusian atmosphere, full of carbon dioxide and nitrogen, proved to be surprisingly dense. Atmospheric surface pressure on Venus is nearly 100 times that of Earth, comparable to the pressure one might feel at a depth of 1 km in the ocean.

In Earth's atmosphere, carbon dioxide, water, and several other trace gases cause the *greenhouse effect*—they create an invisible blanket around our planet that retains heat and thus maintains a climate hospitable to Earth's life forms. In the Venusian atmosphere, the greenhouse effect has run away to an extreme, leaving the planetary surface to roast beneath an overcast sky.

What geologic processes led Venus to become so inhospitable to life, and Earth to be so welcoming? Why is there so much carbon dioxide in the Venusian atmosphere? The answer may lie *beneath* the surface. The rocky mantles of both Earth and Venus transport heat upward by means of slow-motion convection, but they do so in different ways.

Earth's convection is characterized by *plate tectonics*— the creation, drifting, and eventual sinking of cool thick, stiff plates of lithosphere that cover the entire globe. Sections of plates that lie beneath the oceans collect sediment full of water and carbon compounds, and thereby extract greenhouse gases from the atmosphere. The plates eventually dive back into Earth's mantle. On their descent the chemical ingredients of greenhouse gases escape and are recycled bit by bit back to the atmosphere through volcanic eruptions. The ingredients for CO_2 reside within the lithosphere for 70 million years on average. Thus, plate tectonics helps maintain a vital part of the Earth system, a **carbon cycle** that moderates the greenhouse effect, and powerfully influences the climate that favors life on Earth.

But Venus is different. Radar-generated maps of the Venusian surface reveal many meteor craters. Their abundance suggests that Venus has not recycled its surface rocks in perhaps a half-billion years! Earth has recycled its entire ocean floor several times during the same interval. On Venus there is no plate tectonics, no carbon cycle, and

no life. The greenhouse gases in the Venusian atmosphere appear to be trapped in a geologic dead end. We don't live on Venus because plate tectonics does not operate there. On Earth, plate tectonics makes life, as we know it, possible.

Plate tectonics is not just an interesting scientific theory. The carbon atoms contained in the page that you are now reading were extracted by a tree from CO_2 in the atmosphere. Nearly all of these carbon atoms have spent a substantial part of their existence locked within the rocks and sediments of Earth's lithosphere, on a slow-motion trip from the seafloor to a volcano. The water that you drink has likewise spent considerable time chemically bound within rock-forming minerals, liberated for your refreshment by plate tectonic processes. Plate tectonics influences life on Earth in ways that are surprising, even to the original creators of the theory.

PLATE TECTONICS: FROM HYPOTHESIS TO THEORY

Plate tectonics is a scientific theory that explains two centuries of often-puzzling observations and hypotheses about our planet Earth. The basic plate tectonics concept is simple: Earth's surface is composed of stiff, interacting plates that slowly move, spread apart, dive beneath one another, or crumple together. Plate tectonics has helped explain how Earth works as a system, forming connections between the rock cycle, the carbon cycle, the rise and decline of mountains, the emplacement of economically important ore deposits, the history of Earth's climate, and much more.

An important feature of plate tectonics is that the continents on which we live are drifting very slowly across the face of our planet. **Continental drift** is a concept with a long history. A century ago, geologists puzzled over the uncanny fit of the shorelines of Africa and South America. They noted that fossils of extinct land-bound plants and animals, glacial deposits, and ancient lava flows could be matched together along coastlines that today are thousands of kilometers apart. Could the continents be reassembled like a jigsaw puzzle to form the picture of a former world? Such a world might have been far different from the present one. For instance, the first Antarctic explorers found coal deposits exposed in small gaps of an otherwise endless ice sheet. Coal is the fossilized remains of ancient swamp plants. The natural conclusion was that Antarctica had once had a warm climate, hard to explain at the frigid South Pole.

Faced with puzzling data, scientists develop hypotheses to explain them. A scientific hypothesis is like an Internet startup company; many appealing ideas will fail when faced with harsh reality. A startup company must eventually make a profit; a scientific hypothesis must explain data without glaring mismatches. Alfred Wegener proposed the most comprehensive early hypothesis for "continental drift" in 1912. Wegener's ideas were hotly debated in the following decades. They were widely rejected, and for good reason. Wegener proposed that continents plowed slowly through the solid rock of the ocean floor, drifting at a rate 10,000 times faster than we now know them to move. Skeptics of Wegener's hypothesis argued (correctly!) that the ocean floor was too strong to be plowed aside, and that Wegener had not proposed a plausible force that could induce the continents to drift.

Even an outlandish hypothesis can become an accepted scientific theory if the data support it. Unfortunately, attempts to test Wegener's hypothesis with observations had mixed success. Mounting contradictions led continental drift into disrepute by 1930. Another 30 years of puzzling scientific observations led a later generation of geologists to recast the *continental drift* hypothesis as the *plate tectonics* hypothesis. The evolution of thought was subtle, but far-reaching. Instead of continents drifting through the ocean floor, plate tectonics hypothesized that the ocean floor drifted as well. In fact, plate tectonics proposes that the drift of the oceanic lithosphere causes the drift of the continental lithosphere, and not vice versa.

A generation ago, complex arguments based on indirect geologic evidence were necessary to turn plate tectonics from a hypothesis into a theory that gained wide acceptance among geologists. Today the evidence for the theory is simple and direct—we can measure the slow drift of plates worldwide using satellite navigation systems. We also have better explanations for the causes of plate motion than Wegener had at his disposal. There are many challenging problems yet to be solved in the search for underlying causes (see later in this chapter). The basic premises of plate theory are secure, however, because they can be tested against a wide variety of observations.

Plate tectonics is the process by which Earth's hot interior loses heat. Earth's style of heat loss differs from that used by the other rocky planets and moons in our solar system. We don't have a complete understanding of why Earth is unique, but there are hypotheses. Throughout this chapter we compare Earth and its drifting plates with our sister planet Venus to show the profound implications of Earth's heat-loss strategy.

WHAT EARTH'S SURFACE FEATURES TELL US

The rocks beneath our feet are solid, but they are not rigid. Solid material can flow over time, usually at rates that are imperceptible to the naked eye. Long ago, geologists reached this conclusion from indirect evidence, largely from Earth's bumpy surface itself. Bumps in the solid surface of the planet are called *topography*. We quantify topography on land as a height above sea level. Topography on the ocean floor is called *bathymetry*, which we quantify as a depth below sea level. A trained eye can read patterns in Earth's seafloor bathymetry that reveal how plate tectonics operates. After you read this chapter, you can see this for yourself by turning to the Changing Landscapes display of bathymetry in the southern Pacific Ocean, which lies between Chapters 2 and 3.

EARTH IS SOLID BUT NOT RIGID

The largest bump in our planet's surface is a bulge around the equator. Seen from space, Earth seems an essentially perfect sphere. But careful measurement tells a different story. Earth bulges around its equator and is slightly flattened at the poles. The bulge is modest—the diameter at the equator is 12,756.4 km and at the poles it is 12,713.6 km, a difference of 42.8 km. The equatorial bulge represents a deformation of only one part in 300, but the difference in diameter is nearly 10 times larger than the average depth of the ocean floor.

What causes the bulge? All evidence points to centrifugal force caused by Earth's rotational spin. But isn't Earth a rigid solid and thus impervious to distortion by spinning? No! A French mathematician named Clairaut calculated in 1743 that a body the size and density of Earth, if entirely liquid and rotating at the same speed as Earth, would have the same equatorial bulge. This points to an important conclusion: the "solid" Earth has no permanent strength. When a stretching force (like the outward pull due to rotation) is applied for a long time—thousands or millions of years—Earth will deform.

Understanding that solid material can flow is extremely important in understanding the behavior of rocks, which can be bent, folded, and broken. Solid flow is how our mountains, plateaus, lowlands, and ocean basins came to be. Because the solid rock of Earth's asthenosphere can flow, the semirigid plates of Earth's lithosphere can drift over it.

ISOSTASY: WHY SOME ROCKS FLOAT HIGHER THAN OTHERS

Highs and lows on world topographic maps often reveal where lighter subsurface material has risen buoyantly and heavier material has subsided. If we could drain the oceans and view the dry Earth from space, the contrast between ocean basins and continents would be stark. The continents average about 4.5 km elevation above the ocean basin floor (Figure 2.1). They stand notably higher than the ocean basins because the thick continental crust is relatively light (average density 2.7 g/cm^3), whereas the thin oceanic crust is relatively heavy (average density 3.0 g/cm^3). Because the lithosphere floats on the asthenosphere, the portions of lithosphere capped by thick, light continental crust stand higher, whereas those capped by thin, heavy oceanic crust rest lower.

Recall our discussion of buoyancy in Chapter 1. If you jump into a pool, you quickly "find your level"—you float with a certain portion of yourself above and below the water surface, determined by the balance of forces acting to make you float instead of sink. Similarly, but much

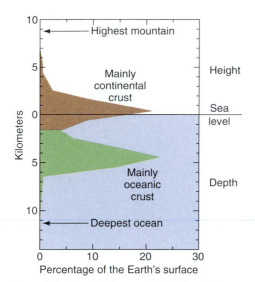

Figure 2.1   **Earth's Topography Is Bimodal**   Distribution of Earth's solid surface above and below sea level (by percent). Areas underlain by continental crust stand considerably higher than areas underlain by oceanic crust. This is a hypsometric curve, or one that shows elevations.

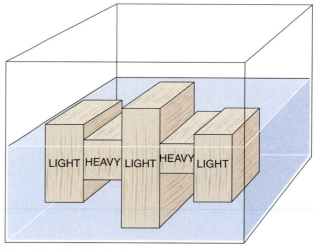

Figure 2.2   **Why Earth's Topography Is Bimodal**   How isostasy works. Heavy and thin wooden blocks float low in the water (equivalent to the oceanic crust, density 3.0 g/cm^3). Light and thick blocks float high (equivalent to continental crust, density 2.7 g/cm^3).

more slowly, portions of the lithosphere find their level, floating on the asthenosphere. They rise or subside until their mass is buoyantly balanced. This is the **Principle of Isostasy** (Greek, meaning "equal standing").

Isostasy explains many of the up-and-down motions of the lithosphere. For example, high mountain ranges and plateaus are typically underlain by rocks that are less dense than rocks beneath lowland areas. As erosion wears away a high mountain range, the thinning crust becomes lighter. It gradually adjusts isostatically by rising, just as an iceberg floating in the ocean rises as the top of the berg melts.

Because of isostasy, all parts of the lithosphere are in a floating balance, like so many blocks of wood of different densities floating in water. Low-density wood blocks float high and have deep underwater "roots," whereas high-density blocks float low and have shallow "roots" (Figure 2.2). Evidence for vertical continental motions associated with isostasy led geologists to deduce the presence of the asthenosphere beneath the lithosphere, many decades before most geologists accepted that continents could also move laterally.

EARTH'S SURFACE: LAND VERSUS WATER

The primary division of the world's surface is into water and land—the ocean covers 71 percent whereas land occupies only 29 percent (Figure 2.3). The *geographic* edge of the ocean is the shoreline. The *geologic* edge of the ocean basin—the tectonic boundary—is where oceanic crust meets continental crust. You might expect shorelines to

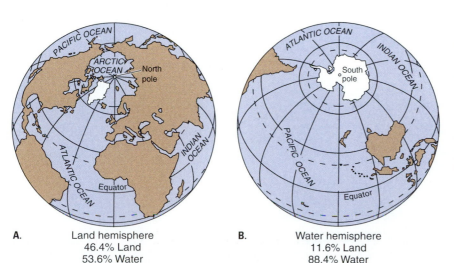

A.   Land hemisphere
46.4% Land
53.6% Water

B.   Water hemisphere
11.6% Land
88.4% Water

Figure 2.3   **Land and Sea**   The unequal distribution of land and sea is emphasized by these two views of Earth. A. View from directly above Spain; 65 percent of the world's landmass is in the Northern Hemisphere. B. View from directly above New Zealand; only 35 percent of the world's landmass is in the Southern Hemisphere, so nearly 90 percent of the hemisphere is ocean.

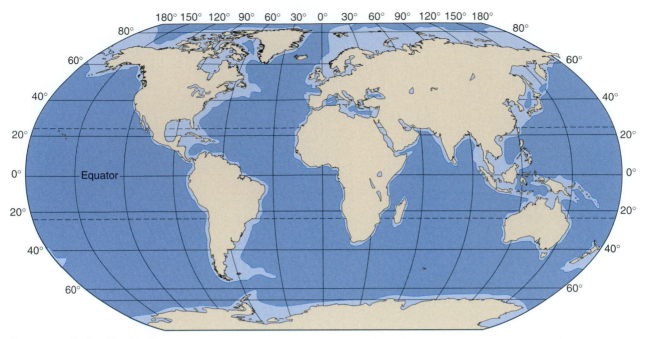

Figure 2.4 The Overflowing Ocean   The continental shelves and slopes (light blue) form about 25 percent of the mass of the continental crust, but they are water covered because there is more water in the ocean than is needed to fill the ocean basins.

coincide with the tectonic boundaries, but they don't, as you can see in Figure 2.4. This is partly because today's ocean basins are overflowing with water.

Sea level fluctuates over time spans of decades to millennia as water is frozen into or melted from the planet's great ice masses—the glaciers and ice caps. Fluctuating sea level makes shorelines mobile features. They move seaward or landward, depending on the volume of water in the oceans. When water becomes stored as ice, sea level falls and the shoreline retreats seaward, exposing former land surface. Conversely, when the ice melts and sea level rises, shorelines advance inland, drowning former real estate.

Seafloor bathymetry differs greatly from continental topography, a strong hint that oceanic and continental lithosphere were created in different ways. Large areas of the seafloor have little relief, and are much smoother than continental areas of similar size. The seafloor is divided by a spectacular network of undersea midocean ridges, which link together around the world to form a continuous feature more than 60,000 km long. According to the plate tectonics theory, midocean ridges mark where two oceanic plates spread apart and new lithosphere forms in the gap.

At the continental margins, plate tectonics identifies two types of continent–ocean transition: passive margins and active margins.

Passive margins have little earthquake or volcanic activity. Beyond the shoreline, the continent extends into a flooded margin known as the **continental shelf** (Figures 2.4, 2.5), which steepens slightly at 100–200 meter depth to become the **continental slope**. The tectonic boundary of the ocean basin is at the bottom of the continental slope.

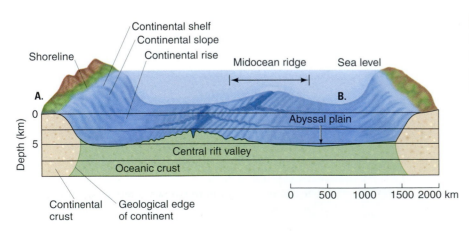

Figure 2.5 Topography of the Seafloor
Section across the northern part of the Atlantic Ocean showing the major topographic features. The Atlantic coastline here has passive continental margins.

A.   **B.**

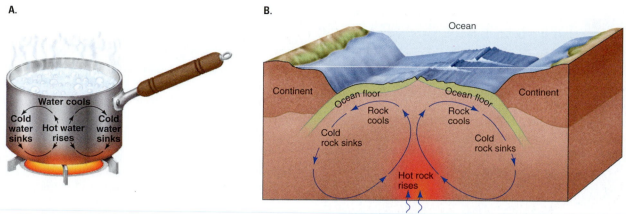

Figure 2.6 Convection in the Mantle   Convection shapes the Earth's surface. A. Convection in a saucepan full of water. Water that is heated expands and rises. As it rises, it starts to cool, flows sideways, and sinks, eventually to be reheated and pass again through the convection cell. B. Convection as it is thought to occur in Earth. Though much slower than convection in a saucepan, the principle is similar. Heat lost from the cooling lithosphere leads it to contract into a smaller volume, becoming

denser. Large slabs of lithospheric rock sink through the asthenosphere and remix with the deeper mantle. In Earth's mantle, cooling from the top, rather than heating from below, maintains convective motions. Once sinking starts, hot rock rises slowly from deep inside Earth to balance the downward flow. Eventually the upward-flowing rock flows sideways, cools, and sinks. The rising hot rock and sideways flow cause midocean ridge volcanism and the drift of ocean basins and continents.

The **continental rise** descends more gently from the base of the continental slope. It is underlain by oceanic crust, but is covered with erosional debris shed from the adjacent continent. The widths of passive margins range from a few hundred kilometers to more than 1000 km. The Atlantic Coast of North America is a good example of a passive margin. According to plate tectonics, passive margins mark where continental lithosphere has rifted apart to form a new ocean basin. Oceanic and continental lithosphere remain connected at the passive margin as the continent drifts and the ocean widens.

Earthquakes and volcanoes are common along **active margins**, and the continent–ocean transition is abrupt (Figure 2.6). The continental shelf steepens into an offshore ocean **trench**. Eroded material from the continent accumulates as sediment in the trench, which nevertheless is typically a kilometer or more deeper than the deep ocean floor beyond the margin. The Pacific Coast of South America is a good example of an active margin. According to plate tectonics, active margins mark where oceanic lithosphere and continental lithosphere converge at the boundary between two plates. Because oceanic lithosphere is the denser of the two, it descends under the active continental margin and sinks into the deeper mantle.

The large, flat **abyssal floors** of the open ocean lie 3 to 6 km below sea level adjacent to the continental rises of passive margins (Figure 2.5) and ocean trenches of the active margins. According to plate tectonics, oceanic lithosphere does not vary greatly in composition or thickness from place to place in an ocean basin, because all of it was produced by the same geologic processes at a single midocean ridge. It is no surprise that the abyssal ocean floors are flat!

WHAT EARTH'S INTERNAL PHENOMENA TELL US

If Earth's plates are constantly moving, something has to be driving them—there has to be an energy source. The energy source is Earth's internal heat engine. Recall from Chapter 1 that Earth's heat comes from two sources. Part is residual heat from the planet's formation 4.55 billion years ago (it's still cooling off) and part is from the continuing decay of radioactive uranium and other elements, which releases heat.

Rocks are poor conductors of heat, so Earth moves its internal heat by moving the rock itself. From as deep as the core–mantle boundary, hot mantle rock rises to the surface layers, releases its heat to the hydrosphere and atmosphere, and sinks back down to gain heat again. The circulation of hot rock is maintained by **mantle convection**. The rock that cools at Earth's surface forms lithosphere, and moves laterally as a collection of stiff plates. We expand this capsule description of plate tectonics in the following sections. In the essay *Rock Deformation: Strain and Stress*, we outline in more detail the ductile and brittle responses of solid rock to the forces associated with gravitational buoyancy and with the stretching and squeezing that accompany plate tectonics.

GEOTHERMAL GRADIENTS

A *gradient* is a progressive change in some physical or chemical property. Gradients can be steep or shallow, the same way that a hiking trail can ascend in varied terrain. Heat energy always moves down a temperature gradient, from a warmer place to a cooler one. Thus, heat from

THE SCIENCE OF GEOLOGY

ROCK DEFORMATION: STRAIN AND STRESS

Rocks are solid, but will respond when forces are applied to them. The response may involve tiny changes in the volume and shape of the rock, including slow flow. Such changes are called strain or deformation. Strain is a proportional change in dimension, such as "stretching by 1 part in 10." Rock deformation is discussed more fully in Chapter 9. Here we introduce some terminology to help you understand thermal convection and plate tectonics.

There are three main ways that rocks deform:

1. **Elastic deformation** is similar to the response of a bedspring, which returns to its original position after forces are removed.
2. **Ductile deformation** persists after forces are removed, like tooth marks that remain in chewing gum. *Ductile flow* occurs when steady forces cause ductile deformation to accumulate in a steady manner over time.
3. **Brittle deformation** occurs through internal breakage of the rock. Breakage can occur on large cracks or on a multitude of small fractures. Fractures whose faces slide past each other are known as **faults**.

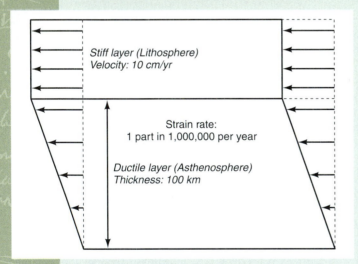

Strain rate:
1 part in 1,000,000 per year

Stiff layer (Lithosphere)
Velocity: 10 cm/yr

Ductile layer (Asthenosphere)
Thickness: 100 km

Figure B2.1   Viscous Rock, Ductile Flow.   Illustration of how shear strain in the asthenosphere allows the overlying stiff lithosphere to drift slowly over Earth's deep interior.

Strain is a response to *stress*, which is force per unit area. Stress within Earth is typically a combination of uniform stress and differential stress. *Uniform stress* is equal in all directions and can be directed either inward or outward on a surface. Uniform stress is also known as *pressure*, and is the kind of stress that induces volume changes in materials. *Differential stress* induces changes in shape rather than in volume. It occurs in different situations—shearing or squeezing (compressional) or stretching (tensional) of material along a single direction rather than in all directions equally.

Rocks are, of course, much stiffer than bedsprings. The pressure of 300–1000 Earth atmospheres* is required to compress a rock by 1 part in 1000, depending on the type of rock. Mantle rocks tend to be stiffer than crustal rocks, and igneous crustal rocks tend to be stiffer than sedimentary rocks. Earthquakes create elastic deformations that travel through rock as *seismic waves* in all directions, transmitting enormous forces over long distances.

Rocks also have very high *viscosity*, which measures the resistance of a solid or a fluid to ductile flow. Aside from flowing lava and creeping glaciers, ductile flow of rock at the surface is extremely slow and rarely significant. However, ductile deformation of rock is important tens and hundreds of kilometers beneath Earth's surface, where rocks are hot and less rigid. Geologists estimate that a force equivalent to 100 Earth atmospheres would cause the mantle rocks beneath the tectonic plates to deform at a steady rate of 1 part in 1,000,000 per year—in other words, 1 mm per kilometer.

If this strain rate were spread across the entire asthenosphere (roughly 100 km), the ductile flow would be 100 mm per year, or 10 cm per year. (Figure B2.1)

This flow velocity is roughly the speed at which human fingernails grow. It is also comparable to the maximum speed of tectonic plates. This slow accumulation of such tiny shear strains is the root cause of most earthquakes and volcanic activity on our planet.

*(Atmospheric pressure is roughly equal to 1 bar [1000 millibars], which is equal to 100,000 pascals in metric units; 1 bar is roughly equal to the pressure exerted by a 10-meter-tall column of water or, equivalently, the extra pressure felt by a scuba diver 10 meters below the sea surface.)

Earth's hot interior flows outward toward the cooler surface. The implications of this simple fact are great because this moving heat affects Earth's rocks from core to surface.

Careful temperature measurements have been made worldwide in underground mines and in wells drilled for petroleum exploration, to harvest rock samples, and to measure rock properties like temperature, radioactivity, and porosity. These measurements show that Earth's temperature increases with depth. This **geothermal gradient** varies widely with geography from 5°C/km to 75°C/km. Because the mantle is solid, not molten, these gradients must lessen with depth. Below 200 km the gradient is thought to be only 0.5°C/km.

MANTLE CONVECTION

Conduction is the process by which heat moves through solid rock, or any other solid body, without deforming it.

BOX 2.2 THE SCIENCE OF GEOLOGY

THERMAL CONVECTION

Thermal convection causes most differential stresses within Earth's mantle. Our planet has been losing heat to space since its formation 4.55 billion years ago. **Heat conduction** and **convection** compete for dominance at all depths beneath Earth's surface. Convection wins if the rocks actually start moving to transport their heat.

A balance of five competing factors must exist for convection to occur:

1. The temperature difference between warm and cool extremes must be sufficiently large.
2. The ability of mantle rock to conduct heat must be sufficiently feeble. The buoyancy of hot rock will weaken if its heat escapes.
3. Mantle rock must undergo significant *thermal expansion* to develop high-density (cooler) and low-density (warmer) regions.
4. The viscosity of mantle rock must not be too great.
5. The size of the convecting region must allow rising and sinking motions to spread over large distances so that internal strains are small.

Physicists discovered a way to quantify these five factors and combine them into a single number, named the **Rayleigh number** after Lord Rayleigh, a famous British physicist of the late nineteenth century. The Rayleigh number expresses the ratio of physical factors that encourage convection versus factors that inhibit convection. The likelihood of thermal convection in systems both large and small can be assessed by estimating the Rayleigh number of the system. If the Rayleigh number is greater than some critical value, convection will occur. We know already that the mantle convects because satellite geodesy can measure the motion of lithospheric plates. A careful translation of the five factors listed above into a Rayleigh number tells us more: the mantle should convect vigorously.

Vigorous convection implies that, though the mantle moves exceedingly slowly, its convection should be irregular and unpredictable over the long term. The factors that favor convection in Earth's mantle are so strong that the mantle could convect, in principle, simultaneously within its whole volume and within much smaller portions of its volume. Convective flows on larger and smaller scales combine to form changeable and chaotic patterns. If the past billion years of mantle convection could be accelerated to fit in a short video clip, its motions would resemble a boiling turmoil more than a lazy simmer.

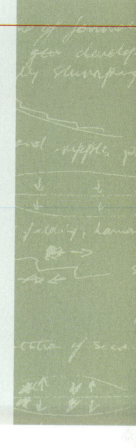

Heat conduction is familiar—it is the way heat moves along the handle of a skillet to burn your hand. Conduction is heat energy on the move, but in a manner that does not involve the movement of hot material from one place to another.

We know that molten rock moves within volcanoes, so Earth's heat clearly can move with its hot material in a second process called *convection* (Figure 2.6). The winds that accompany a summer thunderstorm are driven largely by convection. Convection can happen in gases, in liquids, or, given enough time, in ductile solids. A prerequisite condition for mantle convection is the **thermal expansion** of hot rock. Rock expands as its temperature increases, so the original mass of a heated rock occupies a slightly larger volume. Its density thereby decreases slightly, so that the hot rock finds itself buoyant relative to cooler rock in its immediate neighborhood. For typical mantle rocks, a 1 percent expansion requires an increase of 300–400°C, and leads to a 1 percent decrease in density.

The hot buoyant rock will tend to rise as fast as the stiffness of the surrounding rock allows. We express the propensity of rock to ductile flow as its *viscosity*. No portion of Earth's interior is completely rigid. The viscosity of the stiff lithosphere has been estimated to be at least 1000 times larger than the viscosity of the weak asthenosphere. This means that the lithosphere would flow at least 1000 times more slowly than the asthenosphere, if subjected to the same forces. We discuss the influence of rock viscosi-

ty and other material properties on convection in Earth's mantle in Box 2.2 *Thermal Convection*.

It is remarkable that such modest changes in density induce solid rock to move as much as we observe in nature, but the laws of physics are quite happy with the situation. Lab experiments demonstrate that rock does not need to melt before it can flow, although the flow rate for solid rock under mantle conditions is typically smaller than 1 part in 1,000,000 per year. This corresponds to 1 mm of motion in 1 km. Higher temperatures weaken rock, allowing it to flow more readily. Slow convection currents are possible deep inside Earth because the interior is very hot. The presence of H_2O or other volatile substances also encourages flow in solid rock. We know that CO_2, and other volatile substances are present in Earth's interior because volcanoes vent large quantities of them.

Convection currents bring hot rocks upward from Earth's interior. Hot rock near the surface cools and stiffens to form lithosphere. The rock in the lithosphere is too cool and too stiff for convection to continue, so heat moves through the lithosphere primarily by conduction. Conduction in rock is a slow process, however, so thermal gradients in the lithosphere are steep—that is, temperatures increase rapidly with depth in the lithosphere.

The temperature at the base of the lithosphere (the lithosphere–asthenosphere boundary) is 1300–1350°C, depending on depth. The average temperature at the top of

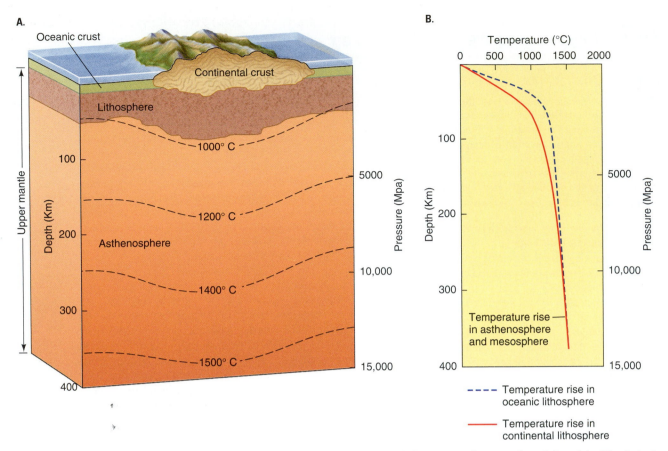

Figure 2.7 Deeper and Hotter A. Temperature increases with depth in Earth, as shown here. The dashed lines are isotherms, lines of equal temperature. Note that the temperature increases more slowly with depth under the continents than under the oceans, where it gets quite hot at a shallow depth. B. This shows the same information as in (A) but in graph form. Earth's surface is at the top, so depth (and corresponding pressure) increases downward. Temperature increases from left to right. The dashed curve shows how temperature increases with depth under the oceans. The solid curve shows how temperature increases more slowly with depth under the continents. The shallow geothermal gradients are large, consistent with heat conduction through the lithosphere. Deeper gradients are small, consistent with adiabatic expansion in the flowing asthenosphere and mesosphere.

the lithosphere—in other words, where we live—is close to 0°C. Because oceanic lithosphere is about 100 km thick, the average geothermal gradient in oceanic lithosphere is 1300°C/100 km, or 13°C/km.

By contrast, average continental lithosphere is 200 km thick, so its average geothermal gradient is about 1350°C/200 km, or 6.7°C/km (Figure 2.7). Within the asthenosphere and mesosphere, temperatures are high enough, and rocks weak enough, for convection to occur. Convection transports heat upward faster than conduction does. As a result, in the deeper portions of Earth, convection is a far more important process for the movement of heat than is conduction.

ADIABATIC EXPANSION OF ROCK

Temperature increases with depth beneath the lithosphere, reaching at least the melting point of iron at the top of the outer core. The temperature gradient remains positive because rock compresses into a smaller volume in response to higher pressure. You may have encountered this physical process when filling a bicycle tire. The energy required to stroke a bicycle pump converts partially to heat energy in the compressed air.

A descending rock mass experiences increasing pressures. These compress the descending rock into a smaller volume and raise its temperature. If a mass of hot rock, deep inside Earth, starts to rise convectively, the process operates in reverse. Pressure decreases on the way up, the rising mass expands somewhat, using heat energy to push outward, and its temperature decreases. If this process occurs without any change in the total energy of the rock (heat plus the energy associated with expansion and contraction) the process is called *adiabatic*. **Adiabatic expansion** means "expansion without loss or gain of energy."

The thermal gradient owing to adiabatic expansion is approximately 0.5°C/km. Earth's mantle convects very slowly, but beneath the lithosphere its convective heat transport outpaces its conductive heat flow by a wide margin. Below a depth of roughly 200 km, the oceanic and continental geothermal gradients merge and remain roughly constant at 0.5°C/km.

EARTH'S CONVECTION: DRIVEN FROM THE TOP

Below the lithosphere, rock masses in the deeper mantle rise and fall according to differences in temperature and buoyancy. It is easy to imagine Earth's convecting mantle by watching a pot of bubbling oatmeal or a lava lamp, but it would also be misleading. Unlike these household examples, Earth's convection is driven mainly by colder material sinking from the top, not hotter material rising from the bottom. Because the cool, dense lithosphere is broken into a collection of stiff plates, individual plates can slide down into the asthenosphere along their edges, like an iron raft slicing into a fishpond. Oceanic lithosphere leads the way down, owing to its greater density. Continental lithosphere drifts across Earth's surface in response to the motions of oceanic lithosphere, but is too light itself to sink back into the mantle.

In this section we have used isostasy to connect Earth's topography to the density of its lithosphere, and we have used evidence for ductile flow in Earth's interior to suggest how plate motions can be connected to Earth's loss of heat by convection. We can reason that the densest lithosphere, beneath the abyssal ocean floor, is most likely to sink back into the asthenosphere and the deeper mantle. This reasoning is supported by a variety of observations, which we present later in this chapter. Alfred Wegener did not discover this reasoning when he tried in the early twentieth century to explain the uncanny fit of the Atlantic coastlines. By focusing his hypotheses instead on continental drift, Wegener missed the underlying causes of lithospheric motion.

Before you go on:

1. What features in Earth's topography and bathymetry provide clues to the existence of moving lithospheric plates?

2. How do Wegener's continental drift hypothesis and modern plate tectonics theory differ?

3. What is the difference between heat conduction and convection?

4. Where in Earth's mantle does convection dominate? Why? Where does conduction dominate? Why?

PLATES AND MANTLE CONVECTION

As far as we can tell, all rocky bodies in the solar system as large or larger than our Moon either convect now or have convected in the past. A strong, stiff lithosphere and weak, ductile asthenosphere exist in all convecting rocky planets and moons. Plate tectonics as we now know it, however, is not the only way for a convecting mantle to lose heat at its surface. Plate tectonics was formulated and demonstrated with field observations, and not deduced from the laws of physics.

After Wegener's original continental drift hypothesis was found wanting, many geologists clung to parts of his ideas. Between Wegener's death in 1930 and the 1960s, geologists probed the seafloor with echosounders and collected enough other geologic data to formulate better hypotheses for the lateral movement of Earth's lithosphere. After decades of contentious debate, geologists arrived at a momentous conclusion—momentous not only to the science of geology but to everyone's understanding of why our planet is the way it is. They concluded that the ocean floor and the continents are slowly moving, sometimes together, sometimes in opposition, drifting laterally at up to 12 cm/yr.

Before we review evidence for plate tectonics, let's clarify how we think the plates behave. Continental and oceanic crust rides atop large fragments of cooling, and therefore stiff, mantle, together forming plates of mobile lithosphere. The lithospheric plates are not permanent. They have a life cycle that is tied to mantle convection (Figure 2.8). When continents split apart, a new ocean basin forms from rising hot mantle rock, which cools and hardens slowly to form a new plate roughly 100 km thick. The Red Sea is an example—a young ocean that started forming about 30 million years ago when a split developed between the Arabian Peninsula and Africa.

As time passes, the lithosphere beneath the seafloor cools and consequently grows denser and prone to sinking. However, because lithosphere is inflexible and broad, it does not sink easily—just like a sheet of plywood does not sink easily in water, even if weights are added. It is easier for lithosphere to move laterally than to sink vertically, so it does.

Once one end of old lithosphere sinks beneath the edge of an adjacent plate, gravity pulls the rest of the plate into the asthenosphere like a bank-deposit envelope is pulled into the slot of an automatic-teller machine. This process is called **subduction**. The line where the top of the sinking plate dives into the interior is marked by a deep ocean trench in the seafloor. Once plunged into the hot asthenosphere, most plates bend and descend more vertically. Subducting plates that are particularly old, cold, and dense, will sink more steeply.

GPS PROVES THAT CONTINENTS MOVE

Scientific debate over continental drift raged for half a century before anyone could measure lithospheric motion directly. In the 1960s, the U.S. Department of Defense established a network of satellites having orbits that could be used for reference in precisely determining location. This satellite network now has commercial use as well, and is the **Global Positioning System (GPS)**.

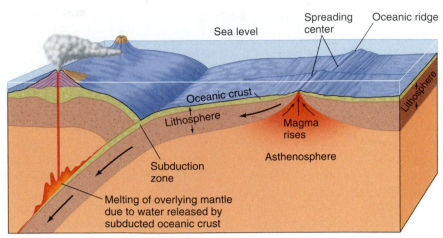

Figure 2.8   Anatomy of a Plate   Schematic diagram showing some major features of a plate. Near the spreading center, where the temperature is high because of rising magma, the lithosphere is thin. Away from the spreading center the lithosphere cools and becomes thicker so the lithosphere–asthenosphere boundary is deeper. As lithosphere sinks into the asthenosphere at a subduction zone, it reheats. At about 100 km depth, water and other volatile substances released from the oceanic crust cause the mantle rock overlying the sinking slab to start to melt. The resulting magma rises and creates a belt of volcanoes parallel to the subduction zone.

GPS has revolutionized **geodesy**, the science of precise position measurements, particularly since the late 1980s. GPS technology helps aircraft pilots, motorists, boaters, and wilderness hikers navigate in unfamiliar territory. GPS satellites feed positional data to navigation systems in some passenger vehicles. Geologists use GPS to detect small movements of Earth's surface, with precision of a few millimeters.

Two measurement methods are common. In a *GPS campaign*, researchers establish a network of fixed reference points on Earth's surface, often using monuments of concrete attached to bedrock. Using a portable GPS receiver to receive signals from the orbiting satellites, researchers remeasure the position of these monuments every few months or years.

In *continuous GPS measurement*, the receivers are attached permanently to monuments, and position is estimated at fixed intervals of a few seconds or minutes. Continuous GPS monitoring is not necessary to estimate the long-term motion of tectonic plates, but it has confirmed that the plates move at a steady rate (at least during the past 15 years!). Continuous GPS measurement has also enabled the discovery of short-term surface motions associated with earthquakes and volcanic eruptions.

GPS measurements confirm two essential features of plate tectonics (Figure 2.9). First, points on the Earth surface indeed are drifting at a slow, steady rate. Second, the plate interiors are approximately rigid. GPS observation points on different plates move relative to each other, but GPS observation points on a single plate tend to maintain their relative positions. This confirms that the lithosphere is stiff and that nearly all surface deformation occurs at plate margins. Within the Eurasian Plate, for example,

recent continuous GPS measurements indicate that the plate stretches and squeezes internally by less than 0.2 cm/yr over a lateral span of nearly 9000 km.

FOUR TYPES OF PLATE MARGINS AND HOW THEY MOVE

The lithosphere currently consists of 12 large plates and many smaller fragments (Figure 2.10). The 7 largest plates are the North American Plate, South American Plate, African Plate, Pacific Plate, Eurasian Plate, Australian–Indian Plate, and Antarctic Plate. All are moving at speeds ranging from 1 to 12 cm a year. As a plate moves, everything on it moves, too. If a plate is capped partly by oceanic crust and partly by continental crust, then both the ocean floor and the continent move with the same speed and in the same direction.

Relative motions occur at the boundaries between plates, and can lead to strains large enough to fracture the stiff rocks of the lithosphere. Not surprisingly, most earthquakes occur at plate boundaries. In fact, a map of worldwide earthquake activity over a decade gives a good snapshot of plate boundaries. (Figure 2.9)

Plates have four kinds of boundaries or margins, shown in Figure 2.11:

Type I. Divergent margin, also called a **spreading center** (Figure 2.11A). This is a break in the lithosphere where magma rises, to form new oceanic crust between the two pieces of the original plate (labeled Plate 1, Plate 2) as they drift apart.

Type II. Convergent margin/subduction zone, where two plates move toward each other and one sinks

A. Plate Model Velocities

5 cm/year ⟶   10 cm/year ⟶

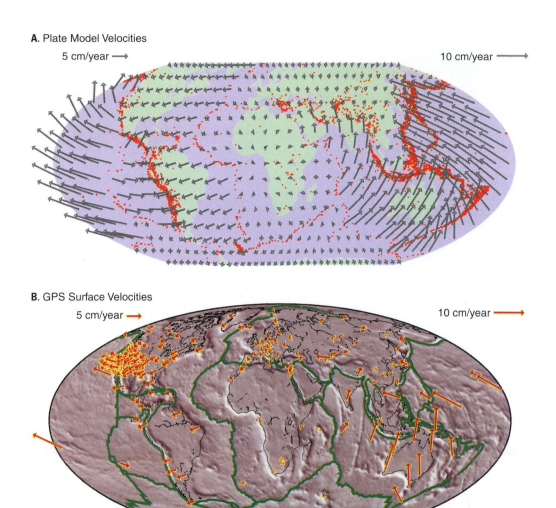

B. GPS Surface Velocities

5 cm/year ⟶   10 cm/year ⟶

C. GPS in Western USA

3 cm/year ⟶   6 cm/year ⟶

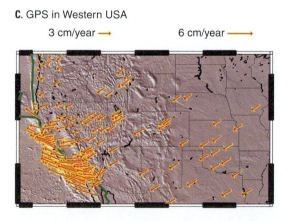

Figure 2.9 Predicted and Measured Plate Motions A. Predictions of present-day plate motion based on many sources of geologic data. The arrows show the surface velocity of each plate. Red dots mark the locations of significant earthquakes since 1965. These earthquakes trace out the major plate boundaries. They are numerous at convergent plate boundaries where older lithosphere is thicker, cooler, and more brittle, but sparse at divergent boundaries, where young lithosphere is thinner, warmer, and more ductile. B. The arrows show surface motions from one network of continuous GPS measurements. C. Expansion of part B to illustrate dense GPS coverage in North America. East of California, the United States drifts with the North American Plate. Coastal California drifts with the Pacific Plate.

beneath the other. (Figure 2.11B). A convergent margin is Type II if either of the colliding plates is oceanic.

Type III. Convergent margin/collision zone, where two continental plates move toward each other and crumple together without sinking, which creates a mountain range (Figure 2.11C).

Type IV. Transform fault margin, a fracture in the lithosphere where two plates slide past each other, grinding and abrading their edges (Figure 2.11D).

Convergent, divergent, and transform margins arise naturally from the motion of plates. If you built a model Earth and tried to push plates around its surface, plate divergence

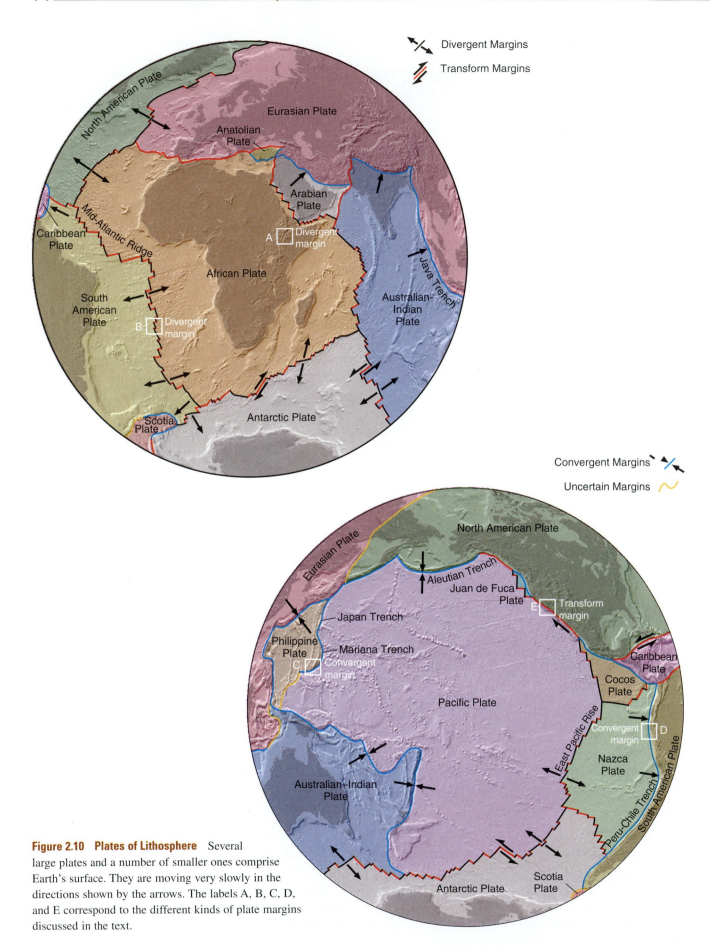

Figure 2.10 Plates of Lithosphere Several large plates and a number of smaller ones comprise Earth's surface. They are moving very slowly in the directions shown by the arrows. The labels A, B, C, D, and E correspond to the different kinds of plate margins discussed in the text.

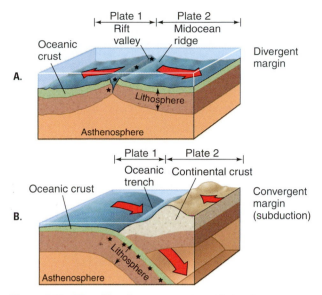

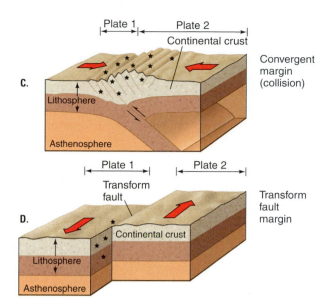

Figure 2.11 Plate Margins The various kinds of plate margins. Stars (★) indicate earthquake centers. A. Divergent margin or spreading center; its topographic expression is a midocean ridge. B. Convergent subduction margin; its topographic expression is an oceanic trench. C. Convergent collision margin; its topographic expression is a mountain range. D. Transform fault margin; it does not produce a consistent topographic expression but is often marked by a long, thin valley due to preferential erosion along the fault.

(spreading) in one location must be balanced by plate convergence (subduction or crumpling) someplace else. Transform motion tends to occur on plate margins that connect divergent and convergent boundaries together. As you might expect, each type of boundary has its own characteristic pattern of earthquakes, volcanic activity, and topography.

Distinct geologic processes that occur along each type of margin explain why each region of Earth is the way it is—why some places have volcanoes but others have quiet mountains, why some have killer earthquakes but others have small tremors, why some have mountains but others have plains, and so forth. In the following sections we will look at processes occurring at each type of boundary and the topography that results. After absorbing the following material, your understanding of the scenery wherever you go will be changed forever.

SEISMOLOGY AND PLATE MARGINS

Earthquakes occur in portions of the lithosphere that are not only stiff, but also *brittle* like a plate of glass. Brittle subsurface rock breaks repeatedly in successive earthquakes as lithosphere deforms. *Ductile* rock deforms without fracture like a ball of putty. Fracture surfaces within brittle rock persist as weak zones, subject to later fracture. For this reason, earthquakes usually occur on preexisting fracture surfaces, or *faults*. When an earthquake occurs, rocks on the two sides, or faces, of a fault scrape past each other. Using records of seismic waves that radiate from an earthquake, seismologists can deter-

mine the orientation of the fault and the relative motion between the fault faces.

We can classify earthquakes and their faults in three distinctive types that correlate nicely with motion at plate boundaries (Figure 2.12):

- **Strike-slip faults** are vertical or near-vertical fracture surfaces that accommodate lateral shear. At a plate boundary these are also known as *transform faults*. *Strike-slip earthquakes* occur when rocks on each side of the fault slide laterally with respect to each other. Motion is entirely horizontal.

- **Thrust faults** are fracture surfaces that dip at an angle between the horizontal and the vertical. Thrust faults accommodate convergent, or squeezing, motion within a volume of rock, or between different bodies of material. In a *thrust earthquake*, the upper fault face slides upward, and the lower fault face slides downward. Motion is partly horizontal, partly vertical.

- **Normal faults** are fracture surfaces that also dip, and accommodate divergent, or stretching, motion with and between bodies of rock. In a *normal earthquake*, the upper fault face slides downward, and the lower fault face slides upward. Motion is partly horizontal, partly vertical, but opposite to the motion on a thrust fault.

Rock deformation at many plate margins is complex, sometimes causing earthquakes of all three types. Nevertheless, earthquake patterns at plate boundaries confirm the predictions of plate theory well. For instance, thrust earthquakes are most prevalent at convergent margins.

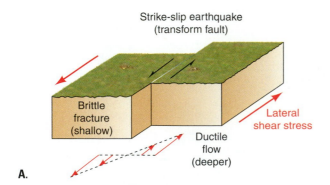

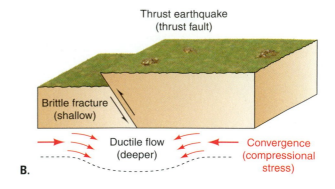

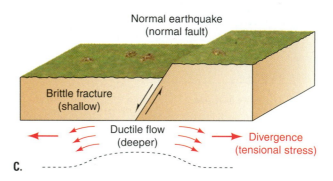

Figure 2.12 Earthquakes and Deformation Near Plate Boundaries
Earthquakes are the largest examples of brittle deformation in
Earth. Earthquakes in plate boundary zones are often accompanied
by ductile deformation in surrounding rock. (One important
exception: brittle deformation in subduction-zone thrust faults per-
sists to great depth at the top of the downgoing plate.) Shown here
are common geometries for earthquake faults at plate boundaries.
A. Strike-slip faults at transform boundaries under shear stress. B.
Thrust faults in convergent zones under compressional stress. C.
Normal faults in divergent zones under tensional stress.

TYPE I: DIVERGENT MARGIN

Where two plates spread apart at a divergent boundary, hot
asthenosphere rises to fill the gap. Although the rising
mantle rock is solid, as it ascends the rock experiences a
decrease in pressure and melts partially. The molten rock
is called *magma*.

MIDOCEAN RIDGES

When a divergent margin occurs in oceanic crust, it coincides with a **midocean ridge** (Figure 2.11). These midocean ridges are found in every ocean, and connect together to form a continuous chain that circles the globe. Thus,
it follows that the midocean ridges represent a common
process. We can't see into the mantle beneath the ridges,
but we can infer what must be happening: the magma that
forms in the asthenosphere beneath the midocean ridge
rises to the top of the lithosphere, where it emerges into the
cool seawater, quenches (quickly), and hardens to form
new oceanic crust (Figure 2.8). This process of melting
and formation of new oceanic crust is discussed in more
detail in Chapters 4 and 5.

Oceanic crust is about 8 km thick at many spreading
centers worldwide, suggesting that its formation is a
remarkably steady process. Oceanic crust forms the top
layer of newly formed lithosphere. The mantle layer of
new lithosphere forms from the asthenosphere that is left

behind after melting. This residual layer of unmelted rock
cools and stiffens slowly over time, as plate motion pulls it
away from the midocean ridge.

BIRTH OF THE ATLANTIC OCEAN

When a spreading center splits continental crust, an interesting sequence of events occurs. First, a great rift forms; a
famous modern example is the African Rift Valley (location
A in Figure 2.10). As the two pieces of continental crust
spread apart, the lithosphere thins, the underlying asthenosphere rises, and volcanism commences. Continued movement allows the rift to widen and deepen, eventually dropping below sea level, so the sea enters to form a long,
narrow water body (the Red Sea is a modern example).
Eventually, the fragments of continental crust move far
apart, new oceanic crust separates them, and a new ocean
has been formed. The edges of the continents are stretched
and thinned in the transition to oceanic crust, exactly the
pattern that we observe at many continental margins today.

The Atlantic Ocean did not exist 250 million years ago.
Instead, the continents that now border it were joined into
a single vast continent that Alfred Wegener named
Pangaea (Figure 2.13). The place that is now New York
City was then as distant from the ocean as central
Mongolia is today! Then, about 200 million years ago, new
spreading centers split the huge continent.

Abundant evidence marks where the torn margins formerly fitted together. If we reassemble the pieces, as in

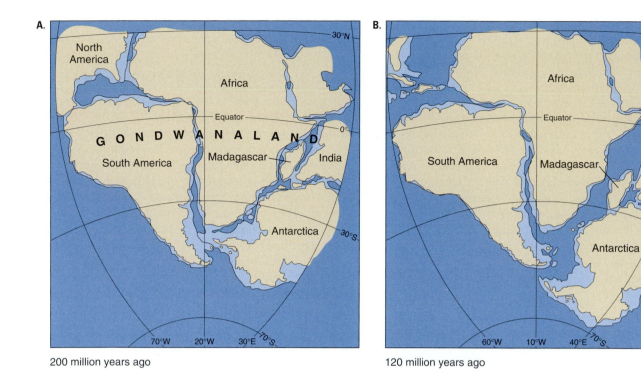

A.

200 million years ago

B.

120 million years ago

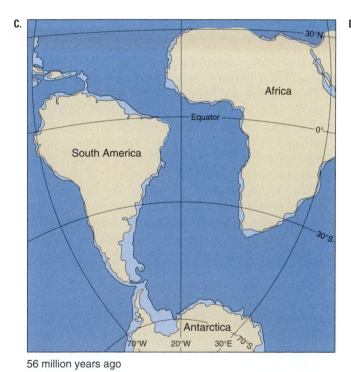

C.

56 million years ago

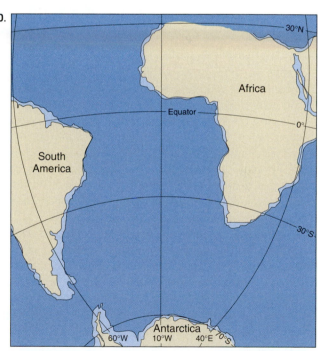

D.

Present

Figure 2.13 Breakup of a Supercontinent The continents of today's Southern Hemisphere were joined together 200 million years ago, as shown in A. They formed the southern half of a supercontinent called Pangaea. The southern half itself is called Gondwanaland. (The northern half, not shown, is Laurasia.) Magnetic data from the oceanic crust were used to plot how the southern part of the Atlantic Ocean opened as South America and Africa drifted apart (B, C, D). When we fit the continents back together along a line 2000 m below sea level (A), the match is very close. Notice how the continents have moved relative to the equator and how Antarctica slowly moved southward.

Figure 2.13, the continental slopes on each side of the ocean match like the pieces of a jigsaw puzzle. The line-of-match follows the original continental rift—which happens to match the shape of the present Mid-Atlantic Ridge (Figure 2.10–see location B).

We do not fully understand why continents split apart, but presumably it involves new convection currents in the asthenosphere and mesosphere. In the case of Pangaea, new spreading centers rifted the lithosphere, breaking the ancient single supercontinent into fragments—today's continents—that drifted slowly to their present positions.

At first, the Atlantic Ocean was a narrow strip of water that separated North America from Europe and North Africa. As spreading slowly continued, the ocean gradually widened and lengthened, splitting South America from Africa, and the ocean expanded to its present width of thousands of kilometers. The Atlantic continues to widen today at 2–4 cm/yr.

CHARACTERISTICS OF SPREADING CENTERS

Compared with the other types of plate boundaries, spreading centers have relatively few earthquakes and feature non-violent volcanic eruptions. The reason is that the rising asthenosphere is hot and tends to flow rather than break. Thus, earthquakes at midocean ridges occur only in the first 10 km beneath the seafloor and tend to be small. Normal faults form parallel lines along the rifted margin, a strong indication of extensional stretching. Volcanic activity at midocean ridges and continental rifts tends to occur along narrow parallel fissures, which open as the plates move apart.

The most obvious characteristic of a spreading center is its topography. The midocean ridges rise 2 km or more above surrounding seafloor because hot mantle asthenosphere lies close to the surface. Thermal expansion of hot mantle rock decreases its density. The principle of isostasy applies: lower-density rock rises to form a higher elevation (or in this case, shallower bathymetry).

As spreading continues, young lithosphere drifts away from the ridge and cools. As it cools, thermal expansion reverses, density increases, and isostasy predicts that the plate will subside. In fact, studies of ocean bathymetry have shown that steady cooling of new lithosphere, combined with isostasy, can predict the depth of seafloor around midocean ridges for nearly all oceanic lithosphere that is younger than 70 million years.

If the spreading rate is "fast," a larger amount of young, warm oceanic lithosphere will border the midocean ridge, and the ridge will be wider. Conversely, a "slow-spreading" ridge will be narrower. The Atlantic Ocean spreads slowly, growing wider at 2–4 cm/year, depending on latitude. The plates that border the Mid-Atlantic Ridge share this growth fairly evenly.

The Pacific spreading center is fast by comparison: 6–20 cm/year, and its location is asymmetrical, much closer to the Americas than to Asia. Yet, despite a faster spreading rate,

the Pacific Ocean basin is shrinking, not growing, because it is bordered by convergent plate boundaries, described later.

ROLE OF SEAWATER AT SPREADING CENTERS

As oceanic lithosphere cools, it gains water. To understand how this happens, we must look closely at heat transfer within the young plate. In most of the stiff plate, heat is transferred upward by conduction. But within 10 km of the seafloor, convection becomes important again, even though the rocks are too stiff to move. Instead, heated seawater convects. Seawater circulates through cracks beneath the ocean floor. Cold water percolates through these cracks, warms in contact with subsurface rock, and rises convectively to form undersea hot springs.

During this subterranean journey, seawater reacts chemically with lithospheric rock, leaching many metallic elements from it. A small fraction of the seawater remains in the rock, chemically bound within hydrous (water-bearing) minerals like serpentine and clays (Chapter 3). This water plays a pivotal role at the end of the lithospheric life cycle.

THE CO_2 CONNECTION

As oceanic lithosphere ages, it accumulates a thick layer of sediments. Seafloor sediments contain clays, which attach chemically bound water to the lithosphere. Also, the sediments add calcium carbonate ($CaCO_3$) from the shells and internal skeletons of countless marine organisms. The formation of calcium carbonate consumes carbon dioxide (CO_2) that is dissolved in seawater. Because most of this CO_2 enters the ocean from the atmosphere, the slow accumulation of seafloor sediments indirectly removes CO_2 from the atmosphere, exerting a long-term influence on the greenhouse effect and Earth's climate.

The Earth system forges links among atmosphere, hydrosphere, lithosphere, and biosphere to create and maintain the environment we all live in. The consumption of CO_2 is one link in the carbon cycle, an example of the dynamic cycles that bind the four "spheres" of our planet together.

TYPE II: CONVERGENT MARGIN/SUBDUCTION ZONE

Over 70 million years, oceanic lithosphere drifts 1500 to 3000 km from the spreading center. During this time interval, mantle rock beneath the ocean crust cools and stiffens to form a mature oceanic plate. As the plate cools, it grows denser. The Principle of Isostasy demands that the plate subsides as it grows denser. This process continues until heat conduction through the seafloor is balanced by heat flow from the convecting mantle at the base of the plate. When this heat balance is achieved, the lithosphere ceases to cool, stiffen, and subside. Oceanic lithosphere older than 70 million years lies beneath the broad flat expanses of the abyssal seafloor. In contrast to its balance of heat flow, the density of mature oceanic plate is seriously out of balance. Old, stiff lithosphere is denser than the hot, weak asthenosphere below it.

The process by which lithosphere sinks into the asthenosphere, and eventually into the mesosphere, is called *subduction*, Small patches of dense oceanic lithosphere cannot sink independently into the asthenosphere, because they are connected together into a broad stiff plate. The lithosphere must first break apart to form a new plate boundary. This allows one fractured plate edge to descend beneath the other. Geologists have proposed some places on Earth where such plate breakage might be occurring today, but most if not all subduction now occurs at preexisting convergent plate boundaries (Figure 2.11). The margins along which plates are subducted are called *subduction zones*. Subduction zones at which oceanic lithosphere sinks beneath a continental plate correspond to active continental margins.

As a subducting slab of oceanic lithosphere sinks slowly through the asthenosphere, it goes too deep for direct study. Consequently, we have deduced what happens next from indirect evidence.

The sinking slab warms, softens, and exchanges material with the surrounding mantle. The most important material exchange comes from the crust that lies atop the descending slab. Under elevated temperature and pressure, the crust expels a number of chemical compounds. The most important are water (H_2O), carbon dioxide (CO_2), and sulfur compounds. All of these are volatile, which means they exist as gases in Earth's atmosphere and can diffuse independently through the rock of the mantle above the slab.

As discussed in Chapter 4, a small addition of these volatile substances can lower the melting point of rock by several hundred degrees Celsius. The hot mantle rock immediately above the sinking slab starts to melt. Some of the magma rises to the surface to form volcanoes. As a result of volatile release from the slab, subduction zones are marked by an arc of volcanoes parallel to the edge of the plate.

The CO_2 Connection, Again

The same volatile substances that induce melting above the subducting slab also accompany the rising magma to the surface, where they are vented as gases from volcanoes. Water, carbon dioxide, and sulfuric gases like sulfur dioxide (SO_2) and hydrogen sulfide (H_2S) return to the atmosphere from their long imprisonment in the rocks and sediments of the seafloor. Thus, subduction zone volcanic activity raises the carbon dioxide level in the atmosphere, exerting a strong influence on the greenhouse effect and Earth climate.

Volcanism tends to replace the CO_2 that is lost from the atmosphere into the ocean and stored in the seafloor. These two processes in the global carbon cycle have not always been in perfect balance. By reconstructing past fluctuations in sediment burial and volcanism, geologists have argued that plate-tectonic activity must have induced fluctuations in atmospheric CO_2 in Earth's history. Warmer or cooler episodes in past Earth climate can be deduced from the fossils of ancient plants and animals, and many of these can be matched by CO_2 fluctuations predicted from past plate-tectonic activity.

Volcanoes Above Subduction Zones

Subduction zones have a distinctive topography and surface geology. At the plate boundary, the plunging plate draws the seafloor down into an ocean trench, some 10 km deep or more. A wedge of sediments accumulates along the trench from seafloor sediments that are scraped from the downgoing plate. On the plate overriding the subducting slab lies an arc of active volcanoes that parallels the trench.

- If the overriding plate is *oceanic lithosphere*, the volcanoes form a series of islands, a *volcanic island arc*. The Mariana Islands of the western Pacific (location C in Figure 2.10) and the Aleutian Islands of Alaska are examples.

- If the overriding plate is *continental lithosphere*, a *continental volcanic arc* forms on land (Figure 2.14). Sediment washed from the continent tends to fill the

Figure 2.14   Volcanoes Above a Subducting Plate   This chain of volcanoes in Oregon and Washington sits above the subducting Juan de Fuca Plate where it sinks below the western edge of the North American Plate. Six snow-capped volcanoes are visible in this aerial photograph. In the foreground is Bachelor Butte, behind are the Three Sisters, and in the distance are Mount Jefferson and Mount Hood.

offshore trench. The Cascade Range of the Pacific Northwest and the Andes of South America (location D in Figure 2.10) are examples.

Magma induced by volatile release from the slab supplies the volcanic arc. Chemical analyses of subduction-arc rocks disclose unusual concentrations of rare elements, such as boron, best explained by the expulsion of water and other volatile substances from the descending slab.

Interestingly, the chemistry of most subduction-arc magmas suggests that the slab itself melts very little. The exact location of mantle melting and the path and ascent rate of magmas to the surface remain unsolved mysteries. One important clue is that most subduction-arc volcanoes occur roughly where earthquakes indicate that the top of the slab is 100 km deep. It is not known, however, whether this depth is favorable for volatile release, magma transport, or some other subduction process.

EARTHQUAKES IN SUBDUCTION ZONES

The largest and the deepest earthquakes occur in subduction zones (Figure 2.15). This seems logical, considering that the subducting slab is cooler than the surrounding mantle, and that its rocks are more likely to break than flow as the slab bends and forces its way downward.

The locations of most earthquakes define the top surface of a slab as it slides into the mantle, extending from the surface to as deep as 670 km (Figure 2.16). Down to 100 km, large earthquakes occur on thrust faults as the descending plate scrapes its way beneath the overriding plate, but quakes deeper than 100 km are more likely associated with faults caused by stresses within the slab.

At all depths, earthquakes occur within the slab where it bends downward from the horizontal. Within the top half of the bending plate, extensional stresses lead to normal-fault earthquakes, not thrusts. In 2001, two such normal-

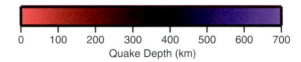

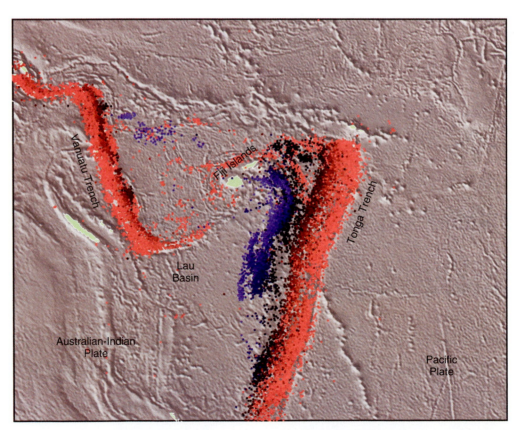

Figure 2.15 Earthquakes and Subduction Zones Deep earthquakes occur in many subduction zones as the slab sinks. Beneath the Tonga Islands, a torn edge of the Pacific Plate subducts beneath the Australian-Indian Plate. Earthquake depths are color-coded, and descend steeply from the trench, reaching depths near 670 km. Simultaneously, a torn piece of the Australian-Indian Plate subducts beneath the Pacific Plate beneath the Vanuatu Islands.

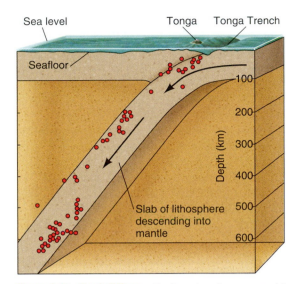

Figure 2.16 Benioff Zone Earthquakes that occurred beneath the Tonga Trench, Pacific Ocean, during several months in 1965. Each circle represents a single earthquake. The earthquakes define the Benioff zone and are generated by downward movement of a comparatively cold slab of lithosphere.

fault earthquakes rocked the cities of Seattle in the United States and San Salvador in the Central American country of El Salvador.

TYPE III: CONVERGENT MARGIN/COLLISION ZONE

Continental crust is not recycled into the mantle. Continental crust is lighter (less dense) and thicker than oceanic crust. Continental lithosphere, as a result, is less dense than oceanic lithosphere, so it resists subduction. When two fragments of continental lithosphere converge, the surface rocks crumple together to form a collision zone.

Continents collide only after there has been subduction of oceanic plate beneath one of the colliding fragments (Figure 2.17). The subduction zone becomes clogged with buoyant continental rock once the oceanic portion of the subducting plate is consumed. Although subduction of lithosphere halts, the actual collision process may last for tens of millions of years.

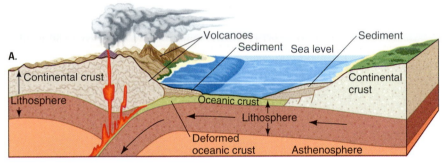

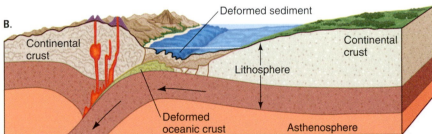

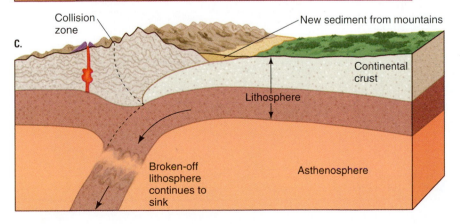

Figure 2.17 Collisions Form Mountains
Mountains formed by plate-tectonic collision between two masses of continental crust. A. Subducting oceanic lithosphere compresses and deforms sediments at the edge of continent (left). Sediments at the edge of approaching continent (right) are undeformed. B. Collision. Sediment at the edge of approaching continent (right) starts to become deformed and welded onto already deformed continental crust (left). C. After collision. The leading edge of the subducting plate breaks off and continues to sink. The two continental masses are welded together with a mountain range standing where once there was ocean.

Collision zones that mark the closure of a former ocean form spectacular mountain ranges. The Alps, the Himalayas, and the Appalachians are each the result of past continental collisions. Because continental crust is too buoyant to sink into the mantle, much of the evidence concerning ancient plates and their motions is recorded in the bumps and scars of past continental collisions.

Not all mountains are formed this way. The Rocky Mountains of the western United States, for instance, may have formed from a detachment of mantle and crust from the bottom of the North American Plate. We describe hypothetical models for such *intraplate* processes in Chapter 12.

TYPE IV: TRANSFORM FAULT MARGIN

Beside the spreading centers of divergent margins and the subduction and collision zones of convergent margins, there is a fourth kind of plate margin, the *transform fault margin*. Along a transform fault margin, two plates grind past each other in horizontal motion. These margins involve strike-slip faults in the shallow lithosphere and often a broader shear zone deeper in the lithosphere. Most transform faults occur underwater between oceanic plates, where they link up the endpoints of distinct spreading segments within the midocean ridge system (see Figure 2.10).

Two of Earth's most notorious and dangerous transform faults are on land: the North Anatolian Fault in Turkey and the San Andreas Fault in California (Figure 2.18). The North Anatolian Fault runs east–west across Turkey from near the Caucasus Mountains to the Aegean Sea. In 1999, a large strike-slip earthquake on this fault devastated the city of Izmit and killed an estimated 17,500 people.

The San Andreas Fault runs approximately north–south, separating the North American Plate on its east side from the Pacific Plate on its west side (Location E in Figure 2.10). Los Angeles sits on the Pacific Plate, and San Francisco on the North American Plate. Past earthquakes indicate that the Pacific Plate is moving northward along the fault. As the two plates grind and scrape past each other, Los Angeles is creeping toward San Francisco at roughly 4.5 cm/year.

Before you go on:

1. What is the life cycle of oceanic lithosphere?

2. How does seawater affect the temperature and chemistry of a newly formed oceanic plate?

3. Why are there two different types of convergent boundaries?

4. What two transform faults on land have significant earthquake hazard?

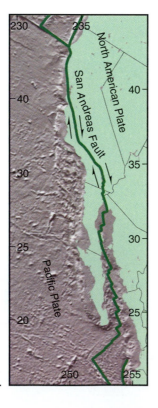

Figure 2.18 Transform Fault in Continents A. California's San Andreas Fault is a transform fault that separates the Pacific Plate from the North American Plate. Arrows show directions of motion. Los Angeles, on the Pacific Plate, is moving northward, while San Francisco, on the North American plate, is moving in the opposite direction, bringing the two cities ever closer together. B. Turkey's North Anatolian Fault extends from the Caucasus Mountains to the Aegean Sea. The bulk of Anatolia drifts from east to west with respect to the Eurasian Plate.

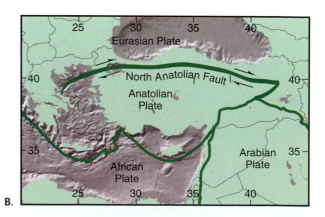

TOPOGRAPHY OF THE OCEAN FLOOR

Plate tectonics gives the ocean floor a distinctive topography. Broad, flat expanses of mature, cool oceanic lithosphere form the abyssal floors of the Pacific, Indian, and Atlantic ocean basins. Bordering the abyssal floor, in addition to the continental shelves, slopes, and rises, are distinctive underwater features: the midocean ridges and the oceanic trenches.

The midocean ridge system is by far the longest continuous chain of mountains on Earth, some 64,000 km in length, that twists and branches in a complex pattern through the ocean basins. This great mountain chain would be one of Earth's most impressive features viewed from space if the ocean basins were empty and dry.

The midocean ridges are, of course, the spreading centers that separate plates. A narrow valley, or rift, runs down the center of all oceanic ridges. The rifts are characterized by intense volcanic activity—in fact, midocean ridges are the most volcanically active features on Earth. At several places around the world, the oceanic ridge with its central rift reaches sea level and forms volcanic islands. The largest of these is Iceland, which lies on the center of the Mid-Atlantic Ridge.

Ocean trenches are the places on the seafloor where the lithosphere sinks into the mantle. The deepest places in the ocean are in trenches, but when trenches are close to land they tend to be filled with sediment.

COMPARING VENUSIAN TOPOGRAPHY

After plate tectonics became an accepted scientific theory in the 1970s, scientists sought to apply the same model to other planets. Venus was of special interest because it resembles Earth in size and chemical composition. Although thick clouds shroud its surface, Venusian topography was easily probed with radar. The Magellan Project orbited a satellite around Venus in 1990 and mapped its surface over several years. *Magellan*'s radar altimeter mapped surface topography with 100 m accuracy at a lateral spacing of only 50 km. Images of radar scattering could resolve a lateral spacing of 150 m and sense surface features with 30 m relief. With *Magellan*'s radar-scattering images, geologists could recognize individual volcanoes, extensional fissures, and other indicators of surface motion.

The Venusian surface clearly has tectonic activity. Many Venusian landforms, if they were on Earth, would imply surface divergence and convergence, including rift zones with long parallel surface cracks and compressional highlands. In addition, some elevated areas strongly suggest the presence of hot rising mantle beneath the Venusian surface.

Venusian tectonics is not *plate tectonics*, however. If the Venusian lithosphere were broken into a collection of rigid

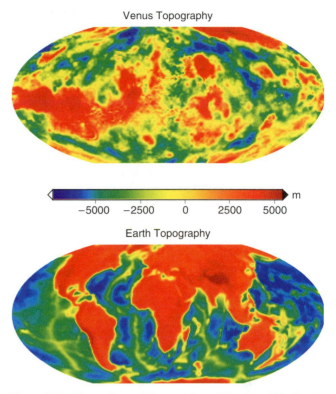

Venus Topography

Earth Topography

Figure 2.19   Comparison of Topography on Venus and Earth   If the water were removed from Earth's oceans and its topography were averaged to match our resolution of Venus, the total relief of its surface (roughly 12 km) would be comparable to the surface relief of Venus. However, Earth's topography is shaped by plate tectonics, which produces large expanses of abyssal ocean floor and narrow continental margins. By contrast, the high- and low-standing regions of Venus are more scattered geographically, with more gradual transitions. Note the clear expression of Earth's midocean ridge network, which has no counterpart in Venus topography.

plates, it would not be hard to identify the spreading centers where plates form and the subduction zones where plates sink back into the mantle. However, Venusian topography is not organized in a network of linear and arc-shaped features that resemble Earth's plate boundaries. Venus does not exhibit the long midocean ridges and subduction zones that dominate Earth's bathymetry.

Venusian topography resembles Earth's in some respects. The overall relief between high- and low-standing topography is similar (Figure 2.19). Venus has highlands that may, like Earth's continents, be composed of light (lower-density) rock. Beyond these attributes, the two planets differ greatly. Venus has no water ocean because of the extreme temperature of its surface (around 450–500°C; hot enough to melt the metal lead). More telling, however, is that Venus has no ocean floor. On Earth, broad expanses of mature oceanic lithosphere lie beneath the abyssal ocean floors, generated over time by plate divergence at the midocean ridges. On Venus such low-standing topography occurs only in small isolated patches, suggesting a different origin.

When compared on a common color scale (Figure 2.19), the geographic patterns of the two planets' topography differ greatly. Eroded continental landscapes and abyssal ocean floors form the two peaks in Earth's bimodal hypsometric curve (see Figure 2.1). Continental margins, both active and passive, delineate sharply the boundaries between topographic levels. Both active and passive continental margins on Earth are closely related to the plate-tectonic processes of subduction and continental rifting, respectively. Venusian hypsometry is unimodal, not bimodal, so that its high- and low-standing topographic regions are not delineated by narrow margins.

The final argument against Venusian plate tectonics comes from the roughly 900 meteor craters identified from *Magellan* satellite data. On Earth, subduction obliterates any crater in oceanic lithosphere within a few hundred million years, and craters on continents erode even faster. Without liquid surface water for erosion or plate tectonics for subduction, meteor craters on Venus persist until the surface is buried under flowing magma or…*something else.*

Some researchers speculate that Venus underwent a catastrophic "resurfacing" event roughly 500 million years ago. Venus has fewer craters per unit area than Mars and the Moon, which have surfaces many billions of years old. By comparing crater abundances, geologists estimate that the Venusian surface is roughly 500 million years old. On Earth the continents (continental crust) can be billions of years old, but no oceanic lithosphere is older than 200 million years old because of its life cycle.

The entire Venusian lithosphere could have subducted in a relatively short period of time (10 million years would be short in this context) and a new lithosphere could have cooled at the surface in its place. Whether it did or not, Venus is not resurfacing itself to any great extent now, by plate tectonics or by any other mechanism.

AN ICY ANALOGUE TO EARTH TECTONICS

The closest approximation to Earth tectonics in our solar system is found on a quite different type of heavenly body. Europa is one of Jupiter's four largest moons, 3138 km in diameter, large enough to be discovered in 1610 by Galileo with his early telescope. In the 1990s a spacecraft named after Galileo gave us a more detailed look at Europa, providing both photographs and radar images. Europa's interior has a rocky composition with density similar to Earth, but its surface layer consists mainly of water ice, perhaps more than 100 km of H_2O.

A network of linear features covers Europa's surface, suggesting deep cracks within its outer ice layer (Figure 1.11B). Large fragments of the icy surface appear to be rigid, and some of these are bounded by zones of complex

topography that suggest the freezing of upwelled water. Plates on Europa are much smaller than Earth plates, perhaps because ice is much weaker than mantle rock. Topography at Europa's plate margins suggests convergence, divergence, and transform-fault motion, just as with Earth's plate margins.

HOT SPOTS AND ABSOLUTE MOTION

When plate tectonics was first proposed in the 1960s, geologists recognized that their data did not tell the complete story of lithospheric motions and mantle convection. Geologic evidence gathered at plate margins supported the idea of *relative* motion—that plates move with respect to each other—but it did not reveal the *absolute motion* of the plates over the planetary interior. To determine absolute motion, a fixed point is required. Considering that all rocks on Earth's surface drift with the plates, where do we find a fixed reference point?

RELATIVE VERSUS ABSOLUTE MOTION

A familiar example of relative and absolute motion occurs when one automobile overtakes another. By looking only at each other's cars, and not at the surrounding terrain whizzing by, the drivers can judge only the *difference* in velocity between them. One car could be traveling 90 km/h and the overtaking car at 100 km/h, but only the *relative velocity* of 10 km/h could be determined. To judge absolute velocity, the drivers would have to look at the passing roadside, which is a fixed reference point. Another way is to use a speedometer. An automobile speedometer tells a driver the velocity with respect to the fixed road surface because it computes horizontal speed from the rotation rate of the car's tires.

Unfortunately for geologists, lithosphere does not roll across the asthenosphere on wheels, so its absolute speed must be measured in a different way.

A related gap in plate theory involved the dominant role of lithospheric motion in mantle convection. The volcanism associated with the life cycle of lithosphere is largely confined to the ridges and subduction zones, and appears to involve only the uppermost layers of the mantle, not much more than a few 100 km in nearly 3000 km of hot rock. If the most prominent heat transfers and rock motions occur near the top of the mantle, how does heat escape the deepest mantle, near the molten-iron core?

MANTLE PLUMES AS FIXED REFERENCE POINTS

An absolute reference frame was eventually provided by GPS navigation, because orbiting satellites respond to the

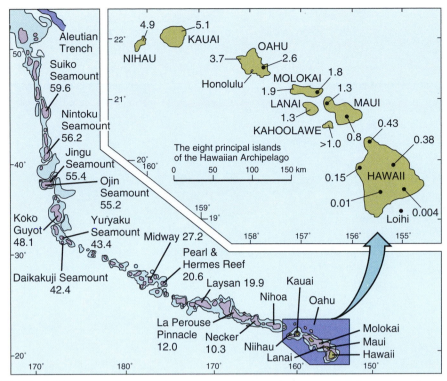

Figure 2.20   Chain of Hot Spot Volcanoes   The Hawaii-Emperor chain of volcanic islands and seamounts. The oldest reliable age of rocks associated with particular volcanoes is list-ed in millions of years (Myr). Note the steady progression of ages from recent to ancient volcanic activity as one progresses down the chain.

gravitational pull of Earth's entire mass, not to individual drifting surface plates. But long before GPS, geologists proposed a different measure of absolute plate motion that also helped to explain the missing heat transfer from the deep mantle. It is called the *hot spot hypothesis* or *mantle plume hypothesis*.

THE CURIOUS CASE OF THE HAWAIIAN ISLANDS

During the nineteenth century, American geologist James Dwight Dana (1813–1895) observed that the age of extinct volcanoes in the Hawaiian Island chain increases as one gets farther away from the active volcanoes on the "big island" (Figure 2.20). The age increases from the contemporary, active underwater volcano Loihi (on the map it is southeast of the big island of Hawaii) to nearly 60 million years at the Suiko Seamount 1300 km distant. The only active volcanoes are at the southeast end of the island chain, and the seamounts to the northwest are long extinct. Earthquakes occur only near the active volcanoes.

As the theory of plate tectonics took shape in the 1960s, J. Tuzo Wilson, the Canadian scientist who named transform faults, pointed out that Dana's observation about the progressively aged islands in the chain provided a possible test of the plate hypothesis. Wilson proposed that a long-lived hot spot lies anchored deep in the mantle beneath Hawaii. A hot, buoyant plume of mantle rock continually

rises from the hot spot, partially melting to form magma at the bottom of the lithosphere—magma that feeds Hawaii's active volcanoes.

If the seafloor were stationary over the deep mantle, an active volcano would just sit above the hot spot and erupt for as long as heat in the deep mantle could maintain the plume. But if the seafloor moves over the mantle plume, an active volcano could remain over the magma source only for about a million years. As the plate moved, the old volcano would pass beyond the plume and become dormant. A new volcano would sprout periodically through the plate above the hot spot, fed by plume magmas. Then the next volcano would emerge, and so on (Figure 2.21). Eventually a chain of volcanoes would form, lined up in order of increasing age away from the hot spot.

The Hawaiian Islands connect with a chain of **seamounts** to the northwest. These are underwater mountains formed from dormant seafloor volcanoes that have sunk below the sea surface by a combination of erosion and isostasy. By taking the ratio of distance (from an extinct volcano to the active volcanoes on the big island of Hawaii) and time (the age of an extinct volcano), you can compute the average speed of the Pacific Plate. Further, the orientation of the island/seamount chain, also known as a *hot spot track*, indicates the direction of plate motion.

In the case of Hawaii and the Emperor Seamounts, where the seamounts are about 40 million years old, the Hawaii hot

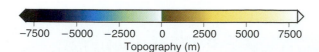

Topography (m)

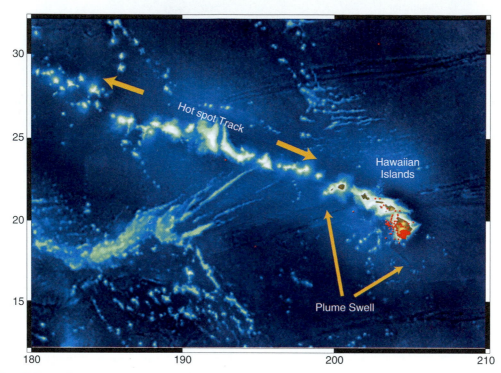

Figure 2.21 Hawaii Plume Causes an Upward Bulge in the Pacific Plate Hawaiian hot spot activity affects Pacific bathymetry in two ways. First, magma from a mantle plume rises through the lithosphere and accumulates on the seafloor to form the enormous volcanoes that comprise the Hawaiian Islands. Second, the isostatic buoyancy of the hot plume beneath the Pacific Plate causes a broad uplift in the seafloor that is known as a hot spot swell. In this color bathymetric relief map around Hawaii, the hot spot swell is visible as a bright area of seafloor around the main islands. The plume swell subsides slowly in the direction of past hot spot volcanism as the lithosphere drifts past the hot spot and cools. Volcanic islands associated with a midplate hot spot eventually subside beneath sea level to become seamounts.

spot track changes direction on the Pacific seafloor. The bend in the hot spot track suggests that the Pacific lithospheric plate changed its direction of motion at this time.

THE WORLD'S HOT SPOTS

Several dozen **hot spots** have been identified worldwide (Figure 2.22). Not all hot spots tap a steady source of hot rock from the deepest mantle, but most give geologists useful reference points for the motion of Earth's tectonic plates. For example, because hot spot volcanoes do not form tracks on the African Plate, geologists conclude that this plate must be very nearly stationary.

The African Plate is almost completely surrounded by spreading centers (Figure 2.10), and it is growing by adding oceanic lithosphere. It follows that the Mid-Atlantic Ridge must be moving westward and that the oceanic ridge that runs up the center of the Indian Ocean

must be moving eastward. If the African Plate is motionless or nearly so, then the *absolute* motion of the Mid-Atlantic Ridge in the southern Atlantic Ocean must be about 2 cm/yr, and the *absolute* velocity of the South American Plate must be 4 cm/yr.

By estimating the total heat output of volcanoes worldwide, and comparing this value with heat conduction through the lithosphere, it is estimated that hot spots transport roughly 10 percent of the total heat that escapes Earth. Although each plume tends to persist for tens of millions of years, there is evidence that plume activity has fluctuated over longer intervals of Earth's history.

Mantle plumes were probably more numerous 90–110 million years ago than today, because extinct seamount volcanoes of that age crowd together in the central Pacific Ocean over a broad region of elevated bathymetry. This concentration implies an ancient burst of volcanism on the open seafloor and a temporary acceleration of mantle con-

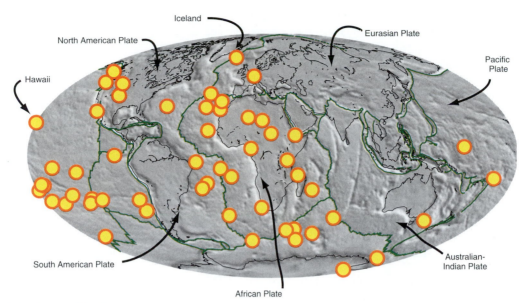

Figure 2.22 Volcanism and Hot Spots Most volcanism on Earth occurs underwater at the midocean ridges. Most volcanoes above sea level are associated with subduction zones. In addition, hot spot volcanism associated with rising mantle plumes (yellow circles) has been identified at several locations. Some hot spots, such as under Iceland, are associated with plate boundaries. Other hot spots, such as Hawaii, lie within plate interiors.

vection. Plate divergence at midocean ridges was also more rapid during this time. An abundance of hot mantle rock beneath the oceanic lithosphere would have lifted the seafloor like a blanket (remember isostasy!) and decreased the total volume of the world's ocean basins. The sea would have spilled over low-lying portions of continents. This prediction agrees well with what we know about sea level 100 million years ago. Many continental regions at that time were submerged beneath shallow seas, including much of what is now the Great Plains of the United States.

VOLCANIC DOMES AND CORONAE ON VENUS

Although evidence for plate tectonics is scarce on Venus, there is abundant evidence for hot spots. Numerous elevated domes and ring-shaped features called **coronae** have been detected by *Magellan*'s radar. Rough-surfaced lava flows on Venus should show up brightly on radar images, and many such bright spots seem to spill out from fissures in the domes and coronae.

Several hypotheses have been advanced for the origin of domes and coronae on Venus, but their remarkable circularity can be explained well by a variation of the hot spot model. As long as the Venusian lithosphere does not drift laterally, blobs of hot mantle (known as *diapirs*) that rise to the base of the lithosphere should spread evenly in all directions.

Mathematical flow models show that a diapir affects surface topography as it rises and spreads. It first forms a steep-sided dome, then a broad plateau, and finally the center collapses to leave a ring-like ridge (Figure 2.23).

In the diapir model, for Venus coronae, Venusian hot spots consist of a single rising diapir. Many of Earth's hot spot tracks are quite long, which seem to require longer-lived plumes. However, because the Venusian lithosphere does not drift, we cannot determine from topography how long a Venus hot spot persists. If a large number of domes and coronae are indeed related to hot spots, it follows that the Venusian mantle convects strongly beneath its unbroken lithosphere.

Earth and Venus can be compared to twin sisters separated at birth, yielding different but related outcomes from nearly the same starting point.

PLUME VOLCANISM ON MARS

Other rocky bodies in the solar system, such as Mars, Mercury, and our Moon, also have had volcanism in the past. However, their small size has limited their tectonic histories. Too small to retain internal heat for billions of years, the Moon, Mercury, and Mars now have thick immobile lithospheres—in fact, the lunar lithosphere may encompass the Moon's entire mantle, leaving no room for an asthenosphere. A smaller mantle also implies less vigorous convection (see Box 2.2). Volcanism was most persistent on Mars, which is larger than Mercury and the Moon and therefore started with more internal heat.

At least 20 huge volcanoes and many smaller cones have been identified on Mars. The largest is Olympus Mons (Figure 2.24), which stands 27 km above the surrounding plains and is capped by a complex caldera that is 80 km across. Mauna Loa in Hawaii, the largest volcano on

Figure 2.23 Venus Coronae A. One of several hundred ring-like topographic structures, ranging in diameter between 60 km and 1000 km, that have been identified as coronae in *Magellan* radar images. B. Physical model for the evolution of Venusian topography in response to an isolated hot rising diapir of mantle rock. The diapir starts as a sphere, but flattens and spreads laterally as it rises. An initial dome above the rising diapir widens as the diapir spreads. Eventually the center of the dome collapses to form the familiar corona ring structure.

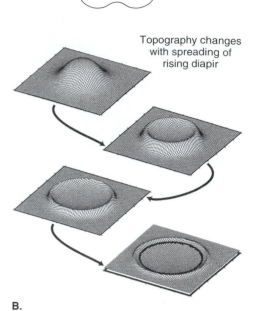

Topography changes with spreading of rising diapir

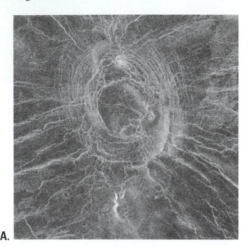

A.

B.

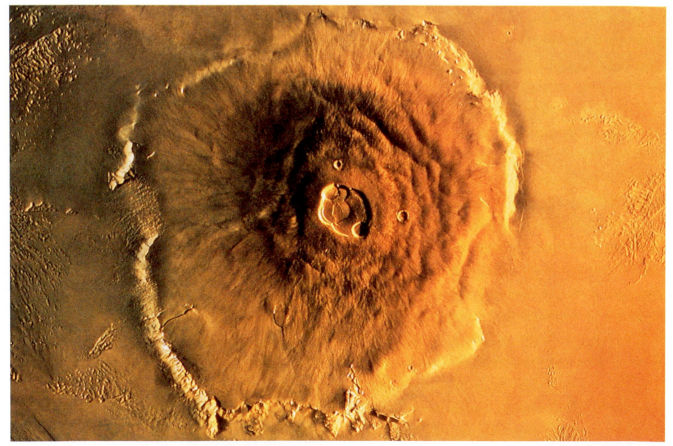

Figure 2.24 Largest Plume Volcano in the Solar System
Olympus Mons on Mars has a basal diameter of 600 km, roughly the distance from Boston to Washington, D.C. The image is a computer-enhanced composite of images taken by *Mariner* spacecraft. The size and symmetry of Olympus Mons suggest a long-lived mantle plume beneath an immobile lithosphere.

Earth, has a similar shape, but it is only 9 km high above the adjacent seafloor. It is estimated that the amount of volcanic rock in Olympus Mons exceeds all of the volcanic rock in the entire Hawaiian chain of volcanoes.

The presence of a huge volcanic edifice such as Olympus Mons implies several things. First, a long-lived mantle plume must once have been present in the Martian interior. Second, the plume must have remained connected to the volcanic vent for a very long time. This, in turn, means that the Martian lithosphere has been stationary, and plate tectonics has not been operating on Mars. A third implication is that the Martian lithosphere must be thick and strong. If it were not, isostasy would by now have caused it to subside from the weight of Olympus Mons.

Before you go on:

1. What topographic features are shaped by plate tectonics on Earth, but are not found on Venus?

2. How do hot spots support the hypothesis that Earth releases heat with plate tectonics? How can the same reasoning suggest the absence of plate motion on Venus and Mars?

WHAT CAUSES PLATE TECTONICS?

When Alfred Wegener proposed continental drift in 1912, he could not explain convincingly what made the continents move. Without a causative mechanism his hypothesis got into trouble with his colleagues. Nearly a century later we still are unable to say exactly how mantle convection makes plates of lithosphere move, although we obviously know much more about our planet's internal workings than Wegener did.

The situation is analogous to knowing the shape, color, size, and speed of an automobile, and knowing that gasoline supplies the energy needed for movement, but not knowing *how* the gasoline makes the engine work. The problem divides, somewhat imperfectly, into two parts, corresponding to "how" and "why" plate tectonics occurs on our planet.

The "how" question leads to a broad discussion of mechanical forces and energy. The "why" question is analogous to why one car starts easily on a cold morning while another car stumbles. What special factors have made plate tectonics common on Earth, and absent from our sister planet Venus?

HOW DOES PLATE TECTONICS WORK?

There is more agreement on *how* plate tectonics works than on *why* it works. The physical properties of hot rock and the temperature extremes of the mantle predict that vigorous convection must occur. Geophysicists have

attempted to reproduce mantle convection in computer simulations. These computations have a complexity that is similar to computer predictions of next week's weather.

Mantle-convection simulations are less reliable than weather simulations, however, because we have fewer observations of deep-Earth processes to calibrate the calculations. Despite the obstacles, computer simulations are steadily improving in their realism and in their ability to highlight critical physical processes.

Here are some results of recent computer simulations:

- Mantle convection occurs in a variety of patterns, large-scale and small-scale, with both persistent and transient flows. In some simulations the mantle flows steadily for long periods, then overturns suddenly like a capsized oil tanker.

- We know that hotter rock is less viscous than cooler rock, and computer simulations tell us that this property is critical to convective circulation. The simulations show hot, buoyant, low-viscosity material rising in narrow columns that resemble hot spot plumes. Cooler, stiffer material from the surface sinks into the mantle in sheets rather than columns, similar to subducting slabs (Figure 2.25).

- Computer simulations can readily generate a stiff surface layer that resembles a plate, but simple physical models that relate rock temperature to its viscosity cannot explain why the stiff surface layer should break apart and subduct. Something besides temperature must govern the response of the lithosphere to its surrounding forces.

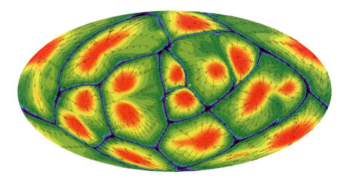

Figure 2.25 Computer Simulation of Convection in an Idealized Earth Mantle Many features of mantle convection can be simulated with computer models. This graphic illustrates temperature and flow for the case where rock viscosity decreases as temperature increases. Temperature increases by 800°C from the coolest (blue) to warmest (red) regions of the convecting system. Viscosity expresses the resistance of rock to ductile flow. Therefore, hot buoyant rock flows more readily, and rises upward in narrow plumes. By contrast, cooler rock is stiffer, and sinks in interconnected sheets. The arrows show the lateral velocity of ductile flow, diverging from the hot red plumes, and converging on the cooler blue sheets.

Simulation of mantle convection is both complex and imperfect, but we can also consider simpler ideas about the forces that move the plates. Mantle convection keeps the asthenosphere hot and weak by bringing up heat from the deep mantle and core. We can treat a lithospheric plate as an independent raft on the weak asthenosphere, and consider the forces that act on it.

Three forces seem likely to have a part in moving the lithosphere:

1. Ridge push. At the midocean ridge, the young lithosphere sits atop a topographic high. The downward pull of gravity will induce the lithosphere to slide down the gentle slopes of the ridge, pushing against the thicker lithosphere beneath the flat abyssal ocean floor.

2. Slab pull. At a subduction zone, the cold, dense slab is free to sink into the mantle, often sliding down a dipping trajectory beneath the overriding plate. As the slab descends, it pulls the rest of the lithosphere into the oceanic trench behind it.

3. Friction. This force is exerted by the viscous mantle rock that borders the lithosphere. We can distinguish between *slab friction*, which drags the top, the bottom, and the leading edge of descending lithosphere in the subduction zone, and *plate friction*, which drags elsewhere at the base of the plate.

These forces are estimated to be quite large relative to the masses of the plates themselves. Thus you might expect some rapid, dramatic movement in response to these forces. But plate movements are slow and steady, so the different forces must cancel one another almost exactly.

A search for a force balance that applies to all major Earth plates leads to the conclusion that slab pull, balanced by frictional resistance to the slab, is the most important determinant of plate motion. This balance helps to explain why plates that do not subduct along any boundary, or only a small part thereof, like the African Plate, tend to move slowly or not at all. A plate that subducts along much of its boundary, such as the Pacific Plate, tends to move more rapidly.

WHY DOES PLATE TECTONICS WORK?

Scientists consider a natural process understood if a scientific theory can explain all important observations about the process. Plate tectonics theory explains the motions of Earth's surface quite well, and much else besides. (We will prove this point in nearly every chapter of this book!) We therefore understand plate tectonics fairly well. Nevertheless, the theory does not explain *why* the plates exist. So there are still some gaps in our knowledge.

Any explanation of plate tectonics must eventually address why the Venusian mantle seems to convect without plates. One important factor is Venus's surface temperature, 450–500°C, very high compared to roughly 0°C at Earth's seafloor. Mantle rock softens and will flow con-

vectively at 1300°–1400°C, so Venus's surface temperature limits lithospheric cooling to little more than half that possible on Earth. A warmer Venusian lithosphere should be less stiff and less dense than Earth's lithosphere, and perhaps thinner too. Even if Venus's subduction could be initiated, would it be sustained? All the above factors imply much weaker slab pull than on Earth.

The laws of physics, represented in computer simulations of mantle convection, do not easily explain why cool rigid lithosphere breaks apart and subducts back into the mantle. In fact, we may understand convection on Venus better than on Earth. The Venusian lithosphere does not break up and subduct and this agrees with most computer simulations based on simple physical laws. To create a subduction zone, the computer simulations must add some kind of cheat factor, such as a preexisting weak zone in the lithosphere (Figure 2.26).

Make no mistake, however: the existence of plate tectonics on Earth does not contradict the laws of physics! Rather, it implies that the current recipe for mantle convection on Earth is missing an important ingredient. Finding this missing ingredient is one of the most important research tasks in geology.

At the present time, a number of scientific clues point to water as the missing ingredient in plate tectonics. Consider the circumstantial evidence:

- Both geologic field observations and laboratory experiments have shown that water molecules can diffuse slowly through solid rock.

- Water is known to weaken rock in several ways. We have already discussed how water released from the subduct-

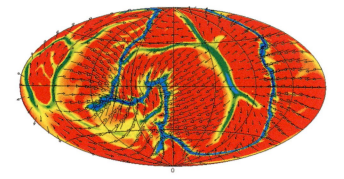

Figure 2.26 Computer Simulation of Mantle Convection With Stiff Plates Surface flow patterns for a convection model in which the shallow mantle is allowed to stiffen as it cools. In this simulation, rock viscosity is determined by a formula that maintains narrow weak zones at the plate boundaries, so that plates remain distinct and can move relative to each other. The viscosity varies by a factor of 30 between red (stiff) and blue (weak) regions. The arrows indicate surface velocity. Plate interiors are red. The arrows diverge from simulated midocean ridges, such as the blue strip at center right. The arrows converge at simulated subduction zones, such as the complex blue curve at lower center.

ing slab lowers the melting point of mantle rock, leading to subduction-arc volcanism. The melting point drops because water molecules weaken chemical bonds within the mineral grains that form rock. Weaker grains are easier to deform, so the presence of water in a subduction zone should also lower the viscosity of mantle asthenosphere, and lubricate the descent of the slab.

- Water has been implicated as an important factor in the rupture of earthquake faults, as discussed in Chapter 10. Therefore, water can facilitate the breakup and descent of lithosphere at all levels in a subduction zone, from brittle rocks in the crust to the viscous asthenosphere.

Water is not the only chemical compound that can diffuse through solid rock and weaken it, but it is arguably the most abundant. On Earth, water and lithosphere are closely connected, but on Venus they are not. On Earth, seawater interacts with oceanic lithosphere from the ridge to the subduction zone. On Venus, the lithosphere is dry, because all surface water has evaporated from the intense surface heat.

Water vapor, a greenhouse gas, contributes to the scorching surface temperature of Venus, ensuring that liquid water plays no role in its mantle convection. On Earth, the greenhouse effect never reached this extreme state, perhaps because the supply of water and carbon dioxide in the atmosphere and hydrosphere have been regulated, in a broad sense, by plate tectonics itself. Our hospitable environment owes a lot to the plates.

Before you go on:

1. What is the main force balance that seems to explain the motion of most plates on Earth?

2. How does water likely influence plate tectonics?

What's Ahead?

Plate tectonics is the unifying theory of how the solid Earth works, and how the tectonic cycle interacts with other components of the Earth system. This chapter has presented the broad outlines of how Earth's lithosphere moves and why it moves, and how the life cycle of a plate interacts with the other parts of the Earth system. The air we breathe and the water we drink have both traveled long distances with the drifting lithosphere.

Because continents drift with the plates, the map of our planet changes continually. By measuring small traces of magnetism in rocks that date from the time of their origin, geologists can infer their past location. Ancient configurations of land and sea can be reconstructed from such magnetic data, and will be discussed in Chapter 20.

Directly following this chapter is the first Changing landscapes graphic of this book. It focuses on how bathymetry in the soiuthern Pacific Ocean can be interpreted within the framework of plate tectonics.

In the next several chapters, we change focus from Earth's large-scale dynamics to the scale of rocks you can hold in your hand. Earth's large-scale processes depend on the physical and chemical behavior of the elemental substances that accreted to form our planet some 4.55 billion years ago. Most of these elements combine to form a variety of crystalline substances called minerals. Solid aggregates of one or more minerals are called rocks. The details of mineral composition and rock formation have many interesting features and practical applications, besides their influence on plate tectonics.

CHAPTER SUMMARY

1. Earth's mantle loses its internal heat with a combination of heat conduction and convection. As part of this process, the upper 100–200 km layer of Earth cools and stiffens to form the lithosphere. Internal heat conducts through the lithosphere to Earth's surface. Beneath the lithosphere, internal heat is transported with moving rock in a convective circulation.

2. Earth's lithosphere is currently composed of roughly 12 large distinct fragments that, owing to their stiffness, behave as rigid plates. The theory that describes the motions and interactions of these lithospheric fragments is called plate tectonics.

3. Since the 1990s, direct evidence for the slow drift of all continental landmasses has been available from GPS satellite geodesy data. GPS data also confirm that the plates are approximately rigid, with little relative motion in their interiors. In the 1960s, however, plate tectonics theory was proposed and verified using indirect geologic evidence.

4. The interaction between plates at their shared boundaries determines the overall pattern of mantle convection, and accounts for most of the earthquakes and volcanic activity on Earth.

5. Continental crust is thicker (20–70 km) and less dense, on average, than oceanic crust (8 km). This difference leads to greater buoyancy for continental lithosphere. The Principle of Isostasy predicts that continental lithosphere will float higher on the viscous asthenosphere, and the oceanic lithosphere lower.

6. Plates diverge at midocean ridges, and the underlying asthenosphere rises to fill the gap. Rising asthenosphere melts partially, and the resulting magma rises to the seafloor and solidifies to form oceanic crust with an average thickness of 8 km. The residual asthenosphere cools and stiffens to form the mantle portion of oceanic lithosphere, reaching a mature thickness near 100 km after roughly 70 Ma. As the young oceanic lithosphere cools, it grows denser and subsides to form the abyssal seafloor.

7. The cooling of a young oceanic plate is enhanced by seawater, which circulates convectively within crack networks in the upper 10 km of the lithosphere. Water leaches metals from the rock, but also is consumed to form hydrated minerals within the plate. Water and carbon compounds join the plate in sediments that accumulate on the seafloor. This accumulation reduces the supply of the important greenhouse gases CO_2 and water vapor to the atmosphere.

8. Where plates converge and at least one plate is oceanic, dense oceanic lithosphere on one side of the plate margin sinks into the asthenosphere in a subduction zone. The convergent boundary between two such plates is marked by an ocean trench. Large earthquakes on thrust faults occur at the boundary between the downgoing and overriding plates. Water, CO_2, and other volatile substances escape the descending lithosphere and lower the melting temperature of the surrounding asthenosphere. This mantle rock partially melts, and the magma rises to form an arc of volcanoes in the overriding plate, parallel to the trench. Water vapor and CO_2 return to the atmosphere by means of volcanic activity.

9. Continental lithosphere is too buoyant to subduct. When two continental plates converge, therefore, a collision zone forms, characterized by thickened crust and high mountains.

10. At transform boundaries, plates slide laterally past each other. Strike-slip earthquakes are common along such boundaries.

11. Mantle convection is induced by the thermal expansion of relatively warm rock, which leads to a lower density and a tendency to rise. Conversely, the thermal contraction of relatively cool rock leads to a higher density and a tendency to sink. Convective motions are resisted by rock viscosity.

12. In plate tectonics, the slow descent of relatively cold, dense oceanic lithosphere in subduction zones is the primary driver of convective flow. Most upward motion of hot buoyant mantle rock occurs passively, induced by plate divergence at midocean ridges.

13. Some upward motion of hot, buoyant mantle rock occurs in plumes that rise from the deep mantle. These plumes form persistent hot spots at Earth's surface. Volcanism from a mantle plume can form a chain of volcanoes in a plate, formed sequentially as it drifts over the hot spot.

14. Mantle plumes within other terrestrial planets appear to be fixed relative to their lithospheres, confirming the absence of plate motion on other planets.

15. Although physical considerations and direct observations suggest that all terrestrial bodies either convect now or have convected in the past, only Earth's convection is expressed in the form of plate tectonics. Without oceanic sediments to accumulate and retain carbon compounds and water, the greenhouse effect on Venus lacks an important regulatory mechanism.

THE LANGUAGE OF GEOLOGY

abyssal floor (p. 37)
active margin (p. 37)
adiabatic expansion (p. 40)

brittle deformation (p. 45)

carbon cycle (p. 33)
collision zone (p. 42)
continental drift (p. 33)
continental rise (p. 37)
continental shelf (p. 36)
continental slope (p. 36)
convergent margin (p. 42)
coronae (p. 57)

divergent margin (p. 42)
ductile deformation (p. 45)

elastic deformation (p. 38)

fault (p. 45)
friction (p. 60)

geodesy (p. 41)
geothermal gradient (p. 38)
Global Positioning System (GPS) (p. 41)

heat conduction (p. 39)
heat convection (p. 39)
hot spot (p. 56)

Isostasy, Principle of (p. 34)

mantle convection (p. 37)
midocean ridge (p. 46)

normal fault (p. 45)

passive margin (p. 36)
plate friction (p. 60)

Rayleigh number (p. 39)
ridge push (p. 60)

seamount (p. 55)
slab friction (p. 60)
slab pull (p. 60)
spreading centers (p. 42)
strike-slip fault (p. 45)
subduction (p. 41)
subduction zone (p. 42)

thermal expansion (p. 38)
thrust fault (p. 45)
transform fault margin (p. 42)
trenches (oceanic) (p. 37)

QUESTIONS FOR REVIEW

1. A creative geologist once called plate tectonics the "slow gear" of the geochemical cycle of carbon. Can you explain what she meant?

2. What reason can you give for the fact that shorelines all lie on the continental crust and that joins between oceanic crust and continental crust are everywhere covered by water?

3. Sketch a section through a margin of a continent, showing the shoreline continental shelf, continental slope, and continental rise. Also mark the approximate position of the join between continental and oceanic crust.

4. How are abyssal ocean floors thought to have formed? What does their absence from Venus tell us?

5. What is the Principle of Isostasy?

6. How would you distinguish a chain of islands caused by a hot spot from a chain of islands caused by a subduction zone?

7. Briefly describe the four kinds of plate margins.

8. Describe what happens when two plates topped by oceanic crust converge. Compare your description with what happens when the converging is between two plates capped by continental crust.

9. Identify the major topographic features of the ocean floor, and state how they are related to tectonic plates.

10. What features of computer simulations of mantle convection appear to match the behavior of Earth's mantle?

Click on *Presentation* and *Interactivity* in the **Plate Tectonics** module of your CD-ROM to further explore resources and activities presenting concepts from this chapter. Select *Assessment* in the same module to test your understanding of this chapter.

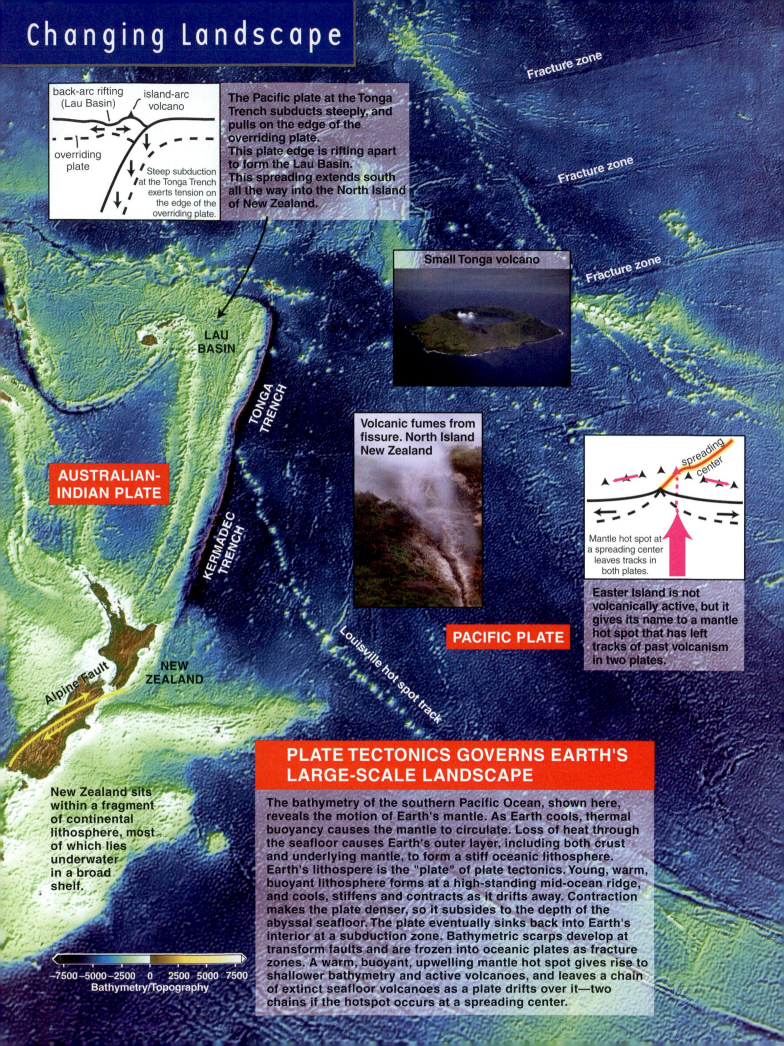

Changing Landscape

back-arc rifting (Lau Basin) | island-arc volcano

overriding plate

Steep subduction at the Tonga Trench exerts tension on the edge of the overriding plate.

The Pacific plate at the Tonga Trench subducts steeply, and pulls on the edge of the overriding plate.
This plate edge is rifting apart to form the Lau Basin.
This spreading extends south all the way into the North Island of New Zealand.

Fracture zone

Fracture zone

Fracture zone

Small Tonga volcano

LAU BASIN

TONGA TRENCH

AUSTRALIAN-INDIAN PLATE

KERMADEC TRENCH

Volcanic fumes from fissure. North Island New Zealand

spreading center

Mantle hot spot at a spreading center leaves tracks in both plates.

PACIFIC PLATE

Easter Island is not volcanically active, but it gives its name to a mantle hot spot that has left tracks of past volcanism in two plates.

Louisville hot spot track

NEW ZEALAND

Alpine Fault

New Zealand sits within a fragment of continental lithosphere, most of which lies underwater in a broad shelf.

PLATE TECTONICS GOVERNS EARTH'S LARGE-SCALE LANDSCAPE

The bathymetry of the southern Pacific Ocean, shown here, reveals the motion of Earth's mantle. As Earth cools, thermal buoyancy causes the mantle to circulate. Loss of heat through the seafloor causes Earth's outer layer, including both crust and underlying mantle, to form a stiff oceanic lithosphere. Earth's lithospere is the "plate" of plate tectonics. Young, warm, buoyant lithosphere forms at a high-standing mid-ocean ridge, and cools, stiffens and contracts as it drifts away. Contraction makes the plate denser, so it subsides to the depth of the abyssal seafloor. The plate eventually sinks back into Earth's interior at a subduction zone. Bathymetric scarps develop at transform faults and are frozen into oceanic plates as fracture zones. A warm, buoyant, upwelling mantle hot spot gives rise to shallower bathymetry and active volcanoes, and leaves a chain of extinct seafloor volcanoes as a plate drifts over it—two chains if the hotspot occurs at a spreading center.

−7500 −5000 −2500 0 2500 5000 7500
Bathymetry/Topography

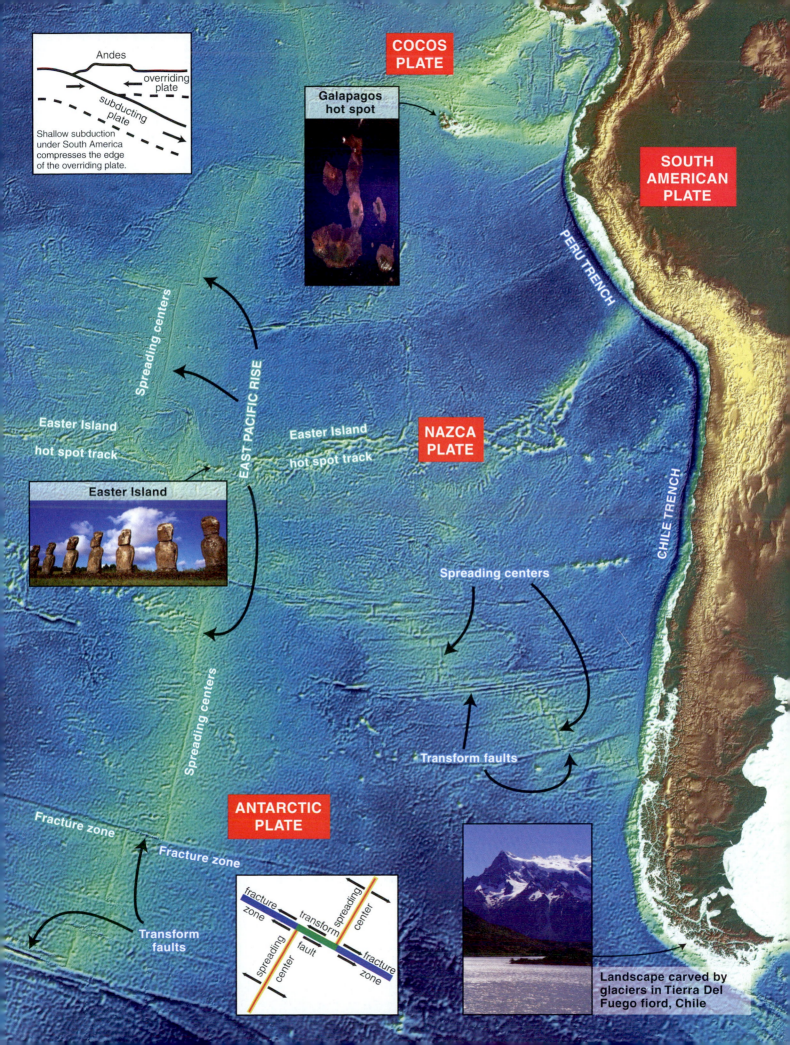

COCOS PLATE

Galapagos hot spot

SOUTH AMERICAN PLATE

Andes
overriding plate
subducting plate

Shallow subduction under South America compresses the edge of the overriding plate.

PERU TRENCH

Spreading centers

EAST PACIFIC RISE

Easter Island hot spot track

Easter Island hot spot track

NAZCA PLATE

CHILE TRENCH

Easter Island

Spreading centers

Spreading centers

Transform faults

ANTARCTIC PLATE

Fracture zone

Fracture zone

Transform faults

fracture zone
transform fault
spreading center
spreading center
fracture zone

Landscape carved by glaciers in Tierra Del Fuego fiord, Chile

Chapter 3

Uncommon Examples of Common Minerals. Feldspar and quartz are the two most common minerals in the continental crust. A. Quartz crystals from Mt. Ida, Arkansas. B. Iridescent plagioclase (variety, labradorite) from Ontario, Canada.

Atoms, Elements, Minerals, Rocks: Earth's Building Materials

Our home while we mapped the rocks SW of Sudbury, Ontario. 1952.

Mineral Exploration: The Story of Sudbury

In 1883, in the Ontario forests north of Lake Huron, workers on the new Canadian Pacific Railroad found traces of nickel and copper mineralization. By chance, they had discovered one of the world's richest deposits of nickel, copper, and platinum. Prospectors soon found similar rich deposits arrayed around the edge of an elliptical ring of igneous rocks 35 miles long and 17 miles wide. Sudbury is not just one deposit, it's an incredibly rich mineral district.

Early in my career I worked in the mining industry in Australia, but in 1952 I was back in school as a graduate student and that summer I was part of a geologic team sent by a mining company to study the west end of the ring. The company sought more deposits, and to help in the search they sought to understand the origin of the ring and thereby the origin of the deposits. Two hypotheses were considered—either ore was brought up from the mantle with the magma, or the igneous rocks served as a source of hot, watery solutions that deposited the ores. We camped on the edge of a lake, moved around by canoe and on foot, and were supplied with food by plane every two weeks.

Unfortunately, our summer's work was inconclusive because it failed to fully support either hypothesis.

In 1964, Robert Dietz proposed a new hypothesis. The ring, he suggested, is the scar of an ancient meteorite impact; the igneous rocks and the ores are the remains of the meteorite. Favorable evidence accumulated quickly and previously puzzling details began to fall into place. Rocks I had mapped that were neither clearly igneous nor clearly sedimentary turned out to be debris from the giant impact. Sedimentary features developed when rock fragments fell back to Earth. Igneous features developed because the fragments were so hot they approached melting. None of us working that summer had any knowledge of meteorite impacts. Today we know such impacts have been common through Earth's long history. One of those impacts, 1850 million years ago, created one of the world's greatest mineral treasures.

Brian J. Skinner

KEY QUESTIONS

1. **What is a mineral?**
2. **Why is the chemical composition of a mineral important?**
3. **Why is the crystal structure of minerals important?**
4. **What are the different properties of minerals and why are they important?**
5. **Which mineral group is most abundant in Earth's crust?**
6. **Why are silicates so abundant in Earth's crust?**
7. **What is the importance of the carbonate, phosphate, and sulfate groups to us?**
8. **What are the three main rock families and how does each form?**
9. **How does the rock cycle influence the percentages of different rocks we see in Earth's crust?**

INTRODUCTION: WHAT IS A MINERAL?

In this chapter, we'll look at the building components of our planet: atoms, elements, minerals, and rocks. First, we

need definitions of the four components, from simplest to most complex:

1. *Atom*—the smallest individual particle that retains the distinctive properties of a chemical element.

2. *Element*—any of the fundamental substances into which matter can be broken down chemically (for example, hydrogen, oxygen, carbon, silicon, lead). There are 92 naturally occurring elements.

3. *Mineral*—In common speech, mineral means any substance that is neither animal nor vegetable. But in geology, a mineral is any naturally formed inorganic solid that has a specific chemical composition and a distinct crystal structure. Examples are quartz, diamond, salt, and gold.

4. *Rock*—In common speech, a rock means a chunk or mass of stone. But in geology, a rock is a naturally formed, coherent mass of one or more minerals, sometimes including organic debris. Examples are granite (combining grains of several minerals), sandstone (commonly quartz grains cemented together), and coal (mineral matter and plant debris combined).

Let's look more closely at the definition of a mineral. To be a **mineral**, a substance must meet five requirements:

1. A mineral is *naturally formed*. This excludes the vast numbers of substances produced in factories and laboratories, like petroleum-derived plastics, concrete, and steel.

2. A mineral is *inorganic*, meaning it cannot be a hydrocarbon compound like a wax or a resin, made by some form of life. A mineral *can* contain just carbon—thus, diamond (pure carbon), graphite (pure carbon), and calcite ($CaCO_3$)

are minerals. But crystalline sugar ($C_{12}H_{22}O_{11}$) is a hydrocarbon compound made by plants, so it is not a mineral.

3. A mineral is *solid*. This excludes all liquids and gases. This requirement is based on the *state* of the material, not its composition. (A curiosity: ordinary *ice* is a mineral—but only if it stays frozen, as in a lake or a glacier!)

4. A mineral has a *specific chemical composition*. This means that minerals must be either chemical elements (gold, copper, diamond), or they must be chemical compounds containing atoms in specific ratio formulas (example: quartz, SiO_2, always has a silicon-to-oxygen ratio of 1 to 2).

Many minerals have more complicated formulas than quartz. A variety of mica called phlogopite combines six elements into the compound $KMg_3AlSi_3O_{10}(OH)_2$. Other minerals have even more complex formulas, but in every case the elements are combined in specific ratios.

Sometimes one element can substitute for another without changing the ratio requirement. For example, Fe can substitute for Mg in phlogopite, so that the formula becomes $K(Mg,Fe)_3AlSi_3O_{10}(OH)_2$, but the ratio of K to (Mg+Fe) remains 1 to 3.

5. A mineral has a *characteristic crystalline structure*. This excludes amorphous materials, such as glass and amber.

To summarize, using examples: quartz is a mineral—it is a naturally formed, inorganic solid that has a specific composition of silicon and oxygen atoms in a 1:2 ratio (SiO_2), and its atoms are in a regular geometric array (its crystal structure). In contrast, granite is also mostly silicon and oxygen, but it is not a mineral. It is a mixture of several different minerals that vary in proportion from sample to sample, so different granites vary in composition. Granite lacks a specific composition; it is a rock.

MINERALOIDS—NOT QUITE MINERALS

Some naturally occurring solid compounds do not meet the full definition of a mineral because they lack either a definite composition, a characteristic crystal structure, or both. Examples are natural glasses and resins, both of which have wide and variable composition ranges and are amorphous. Another example is opal, which has a more or less constant composition but is amorphous. The term **mineraloid** is used to describe such mineral-like materials.

KEY CHARACTERISTICS OF MINERALS

Minerals have two key characteristics:

1. Composition, which is the chemical elements that compose a mineral, and their proportions.

2. Crystal structure, which is the organized way in which the atoms of the elements are packed together in a mineral.

We will look first at composition.

> ### Before you go on:
>
> **1.** What are the relationships among an atom, an element, a mineral, and a rock?
>
> **2.** What is the five-part definition of a mineral?
>
> **3.** What are the two essential characteristics of minerals?

COMPOSITION OF MINERALS

A few minerals are composed of a single element; examples are diamond, graphite, gold, copper, and sulfur. But most minerals are compounds—they contain more than one chemical element. So, to help you understand minerals, let us briefly review how elements combine to form compounds.

CHEMICAL ELEMENTS

If you were a chemist analyzing a mineral or rock, you would record the kinds and amounts of the chemical elements the rock contains. *Chemical elements* are the most fundamental substances into which matter can be separated by chemical means. For example, table salt (NaCl) is not an element, because it can be separated into sodium and chlorine. Neither sodium (Na) nor chlorine (Cl) can be broken down further chemically, so each is an element.

Each element is identified by a symbol, such as H for hydrogen, Si for silicon, or Cl for chlorine. Some symbols, such as that for hydrogen, come from the element's name in English. Other symbols come from other languages—iron (Fe) is from the Latin word *ferrum*, copper (Cu) is from the Latin *cupru*, and sodium (Na) is from the Latin *natrium*. The 92 naturally occurring elements and their symbols are listed in Appendix B, Table B.2.

A piece of a pure element—even a tiny piece no bigger than a pin-head—consists of at least a billion trillion (10^{21}) particles called *atoms*. An **atom** is the smallest individual particle that retains the distinctive properties of a given chemical element. Atoms are so tiny that they can be seen only by using the most powerful microscopes; even then the image is imperfect because individual atoms are only about 10^{-10} m in diameter—they are so small it would take 10 million of them lined up side by side to equal a millimeter.

Atoms are built of *protons* (which have positive electrical charges), *neutrons* (no electrical charge), and *electrons* (negative electrical charges that balance exactly the positive charges of protons). Protons and neutrons are dense, very tiny particles that join to form the *nucleus (core)* of an atom. The protons give the nucleus a positive charge. Electrons are even tinier particles than protons or neutrons; they move, like a distant and diffuse cloud, in orbits around the nucleus (Figure 3.1A).

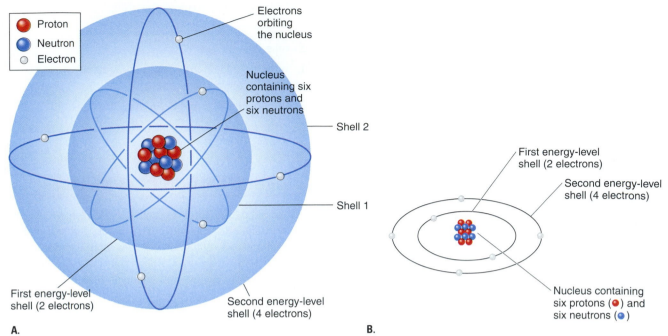

A.

B.

Figure 3.1 Structure of an Atom Diagram of an atom of carbon-12 (^{12}C). A. The nucleus contains six protons and six neutrons. Electrons circle the nucleus in complex paths called orbitals, so this diagram is schematic, not literal. Two electrons are in orbitals close to the nucleus, called energy-level shell 1. Four electrons are in more distant orbitals in energy-level shell 2. B. Two-dimensional representation of a ^{12}C atom.

The number of protons in the nucleus gives each atom its special chemical characteristics and determines what makes it a specific chemical element. The number of protons is the atomic number. Elements are catalogued by **atomic number**, beginning with hydrogen (atomic number 1, because hydrogen atoms contain one proton). Hydrogen atoms are followed by helium atoms (two protons), and so on (see Periodic Table, Figure 3.6).

Uranium, atomic number 92 (it has 92 protons in the nucleus), has the highest atomic number of the *naturally occurring* elements. Using nuclear reactors, scientists have synthesized unstable elements with higher atomic numbers; the highest, reported early in 1999, has an atomic number of 114.

The proton count in an atom's nucleus has interesting implications. For example, hydrogen and helium are so light that they can actually escape Earth's gravitational pull and migrate into space—which is one reason why we have little hydrogen and helium in Earth's atmosphere. At the other extreme, uranium's nucleus is so dense that uranium metal is heavier than lead, and for this reason it is used in anti-tank ammunition for greater impact.

What roles do neutrons play in the nucleus? Neutrons act like a glue that holds the nucleus together. The sum of the numbers of neutrons and protons in an atom is the **mass number**. All atoms of an element have the same *atomic number*—but these atoms can have different *mass numbers* because they can have different numbers of neutrons in their nuclei, yet still have the same chemical properties.

Atoms with the same atomic number but different mass numbers are called **isotopes**. For example, there are three natural isotopes of the element carbon: carbon-12, carbon-13, and carbon-14. Each of these carbon isotopes has six protons per atom and thus the same atomic number (six) and the same behavior in chemical reactions.

But the three isotopes contain different numbers of neutrons: six, seven, and eight per atom, respectively. Heavier isotopes react more slowly than lighter ones in a chemical reaction. *For this reason rapid reactions, like the ones that make the carbon compounds in our bodies, incorporate more of the six-neutron isotope of carbon than the seven-neutron isotope.* A chemist describes the situation by saying we are isotopically light.

Most of the common chemical elements are mixtures of two or more isotopes. Mass numbers of isotopes are written as superscripts; thus, the three isotopes of carbon are denoted ^{12}C, ^{13}C, and ^{14}C, respectively.

ENERGY-LEVEL SHELLS

Electrons move around the nucleus of an atom in complex three-dimensional patterns called *orbitals*. Figure 3.1 is a schematic diagram of an atom of carbon-12 (it is schematic because the paths of the electrons are too complex to show accurately).

Note that two electrons are in orbitals close to the nucleus, and four electrons are in more distant orbitals. The two groupings of orbitals are called **energy-level** shells. The maximum number of electrons that can have orbitals in a given energy-level shell is fixed. Shell 1, closest to the nucleus of an atom, can accommodate only 2 electrons; shell 2 can accommodate up to 8 electrons; shell 3 can accommodate 18; and shell 4 can accommodate 32.

IONS

When an energy-level shell is filled with electrons, it is very stable, like an evenly loaded boat. Atoms seek stability, so to attain it they share or transfer electrons among themselves. Normally, an atom is electrically neutral because it has the same number of protons and orbiting electrons. But when transfer of an electron occurs, the balance of electrical forces becomes upset.

An atom that loses an electron has lost a negative electrical charge and therefore is left with a net positive charge. On the other hand, an atom that gains an electron acquires a net negative charge. An atom that has such an excess positive or negative charge caused by electron transfer is called an **ion**. When the charge is positive (meaning that the atom gives up electrons), the ion is called a **cation**; when negative (meaning an atom adds electrons), it is called an **anion**. (To remember the difference, think: *anion* has two *n*'s, for *negative*.)

We show ionic charges like this: Li^{1+} is a lithium cation that has given up one electron, while F^{1-} is a fluorine anion that has accepted one electron.

Be careful not to confuse mass numbers and ionic charges. Both are shown as superscripts, but mass numbers are on the left while ionic charges are on the right. For example, sodium has a single electron in its second energy-level shell, and forms a cation with a charge of 1+ by giving up the electron. Sodium has a mass number of 23, so a sodium cation is written $^{23}Na^{1+}$. In practice, scientists only record the mass number when necessary.

COMPOUNDS

Chemical compounds form when atoms of different elements combine in a specific ratio. For example, lithium

and fluorine combine to form lithium fluoride (written LiF), a compound used in making ceramics and enamels. Writing the formula LiF indicates that for every Li atom there is one F atom. Similarly, the compound H_2O forms when hydrogen combines with oxygen in the ratio of two atoms of hydrogen to one of oxygen.

The formula of a compound is written by putting the element that tends to form cations (+) first and the element that tends to form anions (–) second. The relative number of atoms is indicated by subscripts, and for simplicity the charges of the ions are usually omitted. Thus, for water, we simply write H_2O rather than $H^{1+}_2O^{2-}$.

An example of how electron transfer leads to formation of a compound is shown in Figure 3.2 for lithium and fluorine. A lithium atom has energy-level shell 1 filled by two electrons, but has only one electron in shell 2, even though shell 2 can accommodate up to eight electrons.

The lone outer electron in shell 2 can easily be transferred to an element such as fluorine, which already has seven electrons in shell 2 and needs only one more to be completely filled. In this fashion, if the lithium and fluorine are in close proximity, both a lithium cation and a fluorine anion finish with filled shells, and the resulting positive charge on the lithium and the negative charge on the fluorine draw the two ions together.

Properties of compounds are quite different from the properties of their constituent elements. Ordinary table salt is a fine example: the element sodium (Na) is a violently reactive metal, and chlorine (Cl) is a highly reactive gas. But, combined into the compound sodium chloride (NaCl, table salt), the resulting substance is safe, nonreactive, and essential to human health.

BONDS OF FOUR MAJOR TYPES

The manner in which electrons are transferred or shared between atoms leads to four major types of bonds that bind atoms together.

A **molecule** is the smallest unit that has the distinctive chemical properties of a compound. Do not confuse a mol-

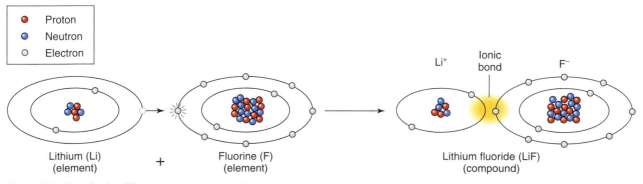

Legend:
- Proton
- Neutron
- Electron

Lithium (Li) (element)
+
Fluorine (F) (element)

Li⁺   Ionic bond   F⁻

Lithium fluoride (LiF) (compound)

Figure 3.2 Transferring Electrons An atom of lithium combines with an atom of fluorine to form lithium fluoride. The lithium atom transfers its lone electron in shell 2 in order to fill

the fluorine atom's shell 2, creating an Li^{1+} cation and a F^{1-} anion in the process. The electrostatic force that draws the lithium and fluorine ions together is an ionic bond.

ecule and an atom—the definitions are similar, *but a molecular compound always consists of two or more kinds of atoms held together*.

The force that holds the atoms together in a compound is called **bonding**. There are several different kinds. Because bonding determines the physical and chemical properties of a compound, let's review the types.

As we mentioned, an energy-level shell that is filled with its quota of electrons is very stable. Where shells are not filled, an atom will try to fill shells and reach a stable configuration. To achieve this, atoms transfer or share electrons, and it is this transferring and sharing of electrons that forms the strongest bonds between atoms. There are four important kinds of bonds:

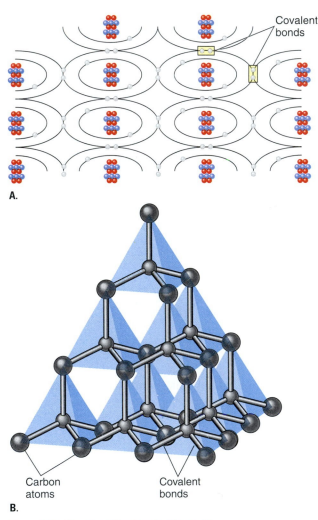

A.

B.

Figure 3.3 Covalent Bonding in Diamond   A. Schematic diagram showing how each carbon atom in diamond shares its four electrons in energy-level shell 2 with four other carbon atoms so that all atoms have eight electrons in shell 2. Each shared electron pair creates a covalent bond. B. The three-dimensional geometric arrangement of carbon atoms in diamond. Note that each atom is surrounded by four others. The actual covalent bonding in diamond is three dimensional, not two dimensional, as shown for simplification in A.

1. *Ionic bonding.* Electron transfers between atoms produce cations and anions that can exist as free entities, but an electrostatic attraction still exists between negatively and positively charged ions that draws them together, forming an **ionic bond**. An example is the bond between lithium and fluorine shown in Figure 3.2.

Compounds having ionic bonds tend to have moderate strength and moderate hardness. Table salt (NaCl) has ionic bonds. When you eat salt and it dissolves in your mouth, the NaCl separates into Na^{1+} and Cl^{1-} ions. The pure elements Na and Cl are toxic, but their ions in solution are not. It is the ions that create the familiar flavor of salt.

2. *Covalent bonding.* Some atoms *share* electrons rather than transferring them. Electron sharing doesn't form ions, but it does create a strong **covalent bond**. One important substance in which covalent bonding is present is *diamond*, a form of carbon. The second energy-level shell of carbon has four electrons, but requires eight for maximum stability. To reach stability, each carbon atom shares two electrons with four other carbon atoms, as shown in Figure 3.3.

Elements and compounds with covalent bonding tend to be strong and hard—like diamond. Covalent bonds also give diamond and other substances special optical properties. The sparkle that makes diamonds attractive gems is due to covalent bonding.

A very important substance in which covalent bonding is present is water (H_2O). The second energy-level shell of oxygen has six electrons but requires eight for maximum stability. A hydrogen atom has only one electron in the first energy-level shell but requires two for maximum stability. Thus, both kinds of atoms satisfy their electron needs by sharing, as shown in Figure 3.4.

3. *Metallic bonding.* Metals have a special kind of bonding that is found in a small group of minerals. In **metallic bonding**, the atoms are closely packed. Electrons in higher energy-level shells are shared among several atoms, and because they are loosely held, they can drift from one atom to another. The drifting electrons give metals their distinc-

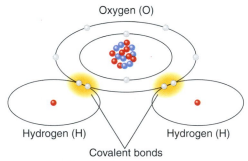

Figure 3.4 Covalent Bonding in Water   Two atoms of hydrogen form covalent bonds with an oxygen atom through sharing of electrons. The oxygen atom thereby has its most stable configuration with eight electrons in shell 2, and each of the hydrogen atoms fills shell 1 with two electrons, making the compound H_2O.

tive properties—for example, metals are opaque, malleable, and good conductors of heat and electricity. Gold, silver, iron, copper, mercury, and platinum are examples of naturally formed metals with metallic bonding.

4. *Van der Waals Bonding.* This bonding arises because of a weak secondary attraction between certain molecules formed by transferring electrons. Although much weaker than ionic, covalent, or metallic bonding, **van der Waals bonding** plays an important role in the structure of certain minerals.

Graphite is a good example. It contains only carbon atoms, and it has a sheetlike structure in which each carbon atom has three nearest neighbors at the corners of an equilateral triangle (Figure 3.5). Carbon atoms are bound covalently within the sheets, and as a result the sheets themselves are very strong. However, because they are sheets, they are flexible. The graphite used in a tennis racket or a golf club exploits the strong covalent bonds in the sheets.

Adjacent sheets of graphite are held together by van der Waals bonds. The bonds are so weak they are easily broken. This is why graphite feels slippery when you rub it between your fingers—the rubbing breaks the bonds, letting the sheets slide easily past each other. This is also why graphite is often used as a lubricant, and because it is stable to much higher temperatures than oils and greases, why it is especially used for very-high-temperature purposes. Talcum powder is another substance having van der Waals bonds. Talc, the mineral in talcum powder, has a sheet structure analogous to a graphite sheet, and just like graphite, the sheets are held together by van der Waals bonds. Talcum powder feels smooth and slippery because the bonds are so easily broken.

COMPLEX IONS

Sometimes two or more kinds of ions form such strong covalent bonds that the combined atoms act as if they were a single entity. Such a strongly bonded unit is called a **complex ion**. Complex ions act like single ions, forming compounds by ionic bonding with other atoms. For example, carbon and oxygen combine to form the complex carbonate ion $(CO_3)^{2-}$, which is the anion in calcite ($CaCO_3$), the mineral present in limestone and marble. Other examples of complex ions are the sulfate anion $(SO_4)^{2-}$ present in gypsum, $CaSO_4\cdot2H_2O$, which is the mineral present in wallboard and Plaster of Paris, and the silicate anion $(SiO_4)^{4-}$, present in quartz.

PERIODIC TABLE OF CHEMICAL ELEMENTS

More than two centuries ago, chemists started to group the chemical elements on the basis of similar chemical properties, such as their ability to combine with other chemical elements. Although early chemists did not know it, their groupings reflected the specific numbers of electrons in the different energy-level shells. The person most directly responsible for working out the way chemical elements

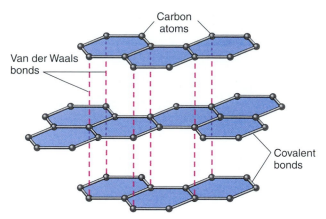

Figure 3.5 Why Graphite Feels Slippery Geometric arrangement of atoms in graphite. Bonding within the sheets of atoms is covalent (strong); bonding between sheets is van der Waals. The weak van der Waals bonds are easily broken, so when you rub graphite between your fingers, the covalently bonded sheets of carbon readily slide past one another.

can be grouped into a sequence based on properties was the Russian scientist Dmitri Mendeleev (1834–1907).

Figure 3.6 is a modern version of the *periodic table of chemical elements*. The table organizes the elements into rows and columns:

- Within rows, elements increase in atomic number from left to right.
- Elements within each column have the same number of electrons in their outermost energy-level shell.
- All elements in the first column readily give up the lone outer-shell electron and so form cations (H^+, Li^+, etc.).
- Elements in columns toward the right-hand side tend to gain electrons and therefore form anions (N^-, O^-, F^-, etc.).
- The farthest column to the right contains the six elements that have full energy-level shells. These six are called *noble gases* because they have no tendency to gain or lose electrons and thus no tendency to form compounds.

Before you go on:

1. What is an element's atomic number? its mass number?
2. What are energy-level shells? Why are they significant?
3. What are anions and cations?
4. What is the difference between an atom and a molecule?
5. What are the four kinds of chemical bonds? What is an example of each?

1 H Hydrogen																	2 He Helium
3 Li Lithium	4 Be Beryllium											5 B Boron	6 C Carbon	7 N Nitrogen	8 O Oxygen	9 F Fluorine	10 Ne Neon
11 Na Sodium	12 Mg Magnesium											13 Al Aluminum	14 Si Silicon	15 P Phosphorus	16 S Sulfur	17 Cl Chlorine	18 Ar Argon
19 K Potassium	20 Ca Calcium	21 Sc Scandium	22 Ti Titanium	23 V Vanadium	24 Cr Chromium	25 Mn Manganese	26 Fe Iron	27 Co Cobalt	28 Ni Nickel	29 Cu Copper	30 Zn Zinc	31 Ga Gallium	32 Ge Germanium	33 As Arsenic	34 Se Selenium	35 Br Bromine	36 Kr Krypton
37 Rb Rubidium	38 Sr Strontium	39 Y Yttrium	40 Zr Zirconium	41 Nb Nobelium	42 Mo Molybdenum	43 Tc Technetium	44 Ru Ruthenium	45 Rh Rhodium	46 Pd Palladium	47 Ag Silver	48 Cd Cadmium	49 In Indium	50 Sn Tin	51 Sb Antimony	52 Te Tellurium	53 I Iodine	54 Xe Xenon
55 Cs Cesium	56 Ba Barium	57* La Lanthanum	72 Hf Hafnium	73 Ta Tantalum	74 W Tungsten	75 Re Rhenium	76 Os Osmium	77 Ir Iridium	78 Pt Platinum	79 Au Gold	80 Hg Mercury	81 Tl Thallium	82 Pb Lead	83 Bi Bismuth	84 Po Polonium	85 At Astatine	86 Rn Radon
87 Fr Francium	88 Ra Radium	89† Ac Actinium															

* Lanthanides

58 Ce	59 Pr	60 Nd	61 Pm	62 Sm	63 Eu	64 Gd	65 Tb	66 Dy	67 Ho	68 Er	69 Tm	70 Yb	71 Lu

† Actinides

90 Th	91 Pa	92 U	93 Np	94 Pu	95 Am	96 Cm	97 Bk	98 Cf	99 Es	100 Fm	101 Md	102 No	103 Lr

Elements present in continental crust in amounts equal to or greater than 0.1% weight.

Figure 3.6 **Periodic Table of Chemical Elements** Chemical elements arranged in rows in order of increasing atomic number, from left to right, and in vertical columns so that all the elements in a column have the same number of electrons in their outermost energy-level shell, and therefore similar chemical properties. The most abundant and geologically important elements in Earth's continental crust are highlighted. Shown in separate rows are elements 58–71, the lanthanides or rare earths, and 90–103, the actinides. Lanthanides and actinides have the same number of electrons in their outermost shells but differ from each other in the electrons in their inner shells.

CRYSTAL STRUCTURE OF MINERALS

All minerals are solids. Thus, whether a given chemical element or compound can be called a mineral is controlled by its *state*—whether it occurs as a solid, liquid, or gas. As noted, ice in a glacier meets the definition of a mineral, but water in the ocean and water vapor in the atmosphere do not because they have no crystal structure. (See *The Science of Geology*, Box 3.1, *The Three States of Matter.*)

The molecules and atoms in gases and liquids are randomly jumbled, but the atoms in most solids are organized in regular, geometric patterns, like eggs in a carton, as shown in Figure 3.7A. The geometric pattern that atoms assume in a solid is called the *crystal structure*, and solids that have a crystal structure are said to be **crystalline**. All minerals are crystalline, and the crystal structure of a mineral is a unique property of that mineral. All specimens of a given mineral have identical crystal structure.

Solids that lack crystal structures are *amorphous* (Greek, "without form"). Glass and amber are examples of amorphous solids.

IONIC SUBSTITUTION

The bonding in most common minerals is ionic. This fact allows the important phenomenon of *ionic substitution* to occur. To understand this, consider three things: the crystal structure of a typical mineral, ion size, and ion electrical charge.

CRYSTAL STRUCTURE OF A TYPICAL MINERAL

The crystal structure of common lead mineral *galena* (PbS) is shown in Figure 3.7B. The bonding is ionic—lead is the cation (Pb^{2+}) and sulfur is the anion (S^{2-}). Note that the ions in Figure 3.7B are arranged in a cube-like grid in which each sulfur is surrounded by six leads and each lead by six sulfurs—bonds from one of the leads to the six adjacent sulfurs are highlighted for clarity.

ION SIZE

Note that the anions (S) in Figure 3.7B are larger than the cations (Pb). The size of ions is commonly expressed as the **ionic radius**, which is the distance from the center of the nucleus to the outermost shell of orbital electrons.

Figure 3.7   Seeing Atoms in a Mineral   The arrangement of atoms of lead (Pb) and sulfur (S) in galena (PbS), the most common lead mineral. Bonding between lead and sulfur is ionic; lead forms a cation Pb^{2+}, sulfur an anion, S^{2-}, so to maintain a charge balance there must be equal numbers of Pb atoms and S atoms in the structure. Atoms are so small that a cube of galena 1 cm on an edge contains about 10^{22} atoms each of Pb and S. A. Atoms at the surface of a galena crystal revealed with a scanning-tunneling microscope. S atoms are the large bumps; Pb atoms the smaller ones. B. The packing arrangement of atoms in a galena crystal. The atoms are shown pulled apart along the black lines to demonstrate how they fit together. Bonds from one Pb atom to the six nearest sulfur atoms are highlighted in red.

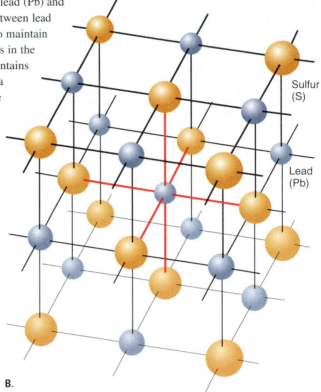

Sulfur (S)

Lead (Pb)

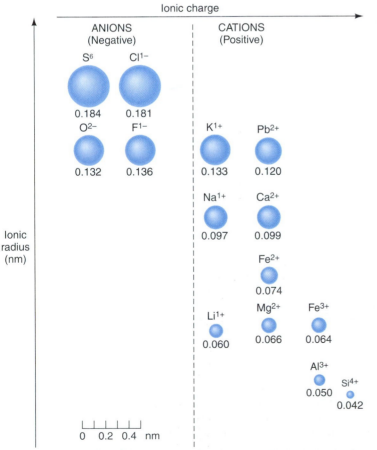

A.

B.

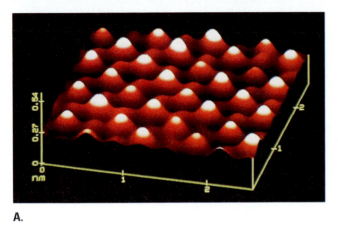

Ionic charge

ANIONS (Negative)

CATIONS (Positive)

S^{6} Cl^{1-}

0.184   0.181

O^{2-} F^{1-} K^{1+} Pb^{2+}

0.132   0.136   0.133   0.120

Na^{1+} Ca^{2+}

0.097   0.099

Fe^{2+}

0.074

Li^{1+} Mg^{2+} Fe^{3+}

0.060   0.066   0.064

Al^{3+}

0.050 Si^{4+}

0.042

Ionic radius (nm)

0   0.2   0.4   nm

Anions tend to have large radii because the addition of extra electrons to fill an energy-level shell means that the pull exerted on each orbital electron by the protons in the nucleus is slightly reduced. As a result, the electrons move a bit farther away from the nucleus. By contrast, cations tend to be small because they lose electrons and their remaining electrons are more tightly held.

Most of the volume of a crystal structure is taken up by the largest ions (the anions), and as a result the crystal structure of a mineral is determined largely by the packing arrangements of anions. The radii of some common ions are shown in Figure 3.8.

ION ELECTRICAL CHARGE

It is apparent from Figure 3.8 that certain ions have the same electrical charge and are nearly alike in size. For example, Fe^{2+} has a radius of 0.074 nanometers; Mg^{2+} is 0.066 nm.

Figure 3.8   Different-Sized Ions   Ionic radii of some geologically important ions range from Si^{4+} at lower right to S^{2-} at upper left. Ions are shown here in vertical groups based on charge, from $^{2-}$ at left to $^{4+}$ at right. Note that the anions tend to have larger ionic radii than cations. Ions in each of these pairs—Si^{4+} and Al^{3+}, Mg^{2+} and Fe^{2+}, and Na^{1+} and Ca^{2+}—are about the same size and commonly substitute for each other in crystal structures. Radii are expressed in nanometers (nm).

Box 3.1 THE SCIENCE OF GEOLOGY

THE THREE STATES OF MATTER

Solid, liquid, and gas are the three states in which compounds and chemical elements exist at Earth's surface. The state of a substance is determined by its *temperature* and *pressure*. For example, when ice gains heat, it will reach a temperature where the ice will melt to water—a change of state. If the temperature rises further, the water will evaporate and become water vapor—another change of state.

Conversely, if water vapor is cooled or compressed under pressure, it will condense to either a solid (ice or snow) or liquid (water or rain). Figure B3.1 shows the regions of temperature and pressure in which the different states of H_2O are stable. Similar diagrams can be drawn for most compounds and elements (their freezing and boiling points, are, of course, quite different from those of H_2O).

A change of state involves heat energy. The transition from ice to water, for example, requires the addition of heat energy. In the reverse case, when water freezes to ice, heat energy is released. The amount of heat per gram released or absorbed during a change of state is known as *latent heat* (from the Latin, *latens*, meaning hidden, hence hidden heat). The latent heat corresponding to changes of state in H_2O are shown in Figure B3.2.

Changes in the state of H_2O at Earth's surface play a great role in the distribution of heat energy in the atmosphere, hydrosphere, and lithosphere, and therefore in Earth's climate.

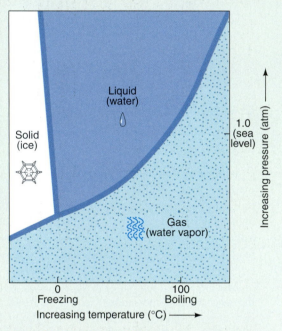

Figure B3.1 The States of Water The state in which the H_2O occurs is determined by the temperature and pressure at any given time and place. Along the boundary lines between any two states, both states can exist. Earth is the only body in the solar system in which H_2O exists in all three states at the surface.

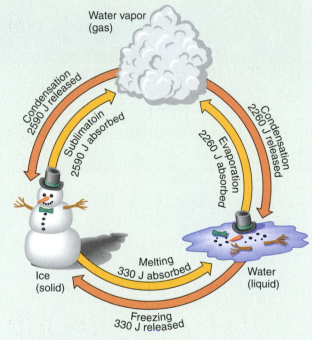

Figure B3.2 Latent Heat The amounts of heat absorbed by, or released from, one gram of H_2O during a change of state.

Now, putting together crystal structure, ion size, and electrical charge, you can see that some ions can substitute for one another in a mineral. For example, because of their similarity in size and charge, ions of Fe^{2+} can substitute for ions of Mg^{2+} in magnesium-bearing minerals. The crystal structure of the mineral is not changed by the substitution. This substitution of one ion for another in a random fashion throughout a crystal structure is **ionic substitution**.

Ionic substitutions are indicated in chemical formulas by showing the ions as (x,y). For example, *olivine* is actu-ally a *mineral group* comprising several minerals of variable chemical formulas. The general formula for the olivine group is $(Mg,Fe)_2SiO_4$. This means that Fe^{2+} and Mg^{2+} can substitute for one another (but not for any other atoms in the structure—Si or O).

The magnesium-dominant variety of olivine (Mg_2SiO_4) is called forsterite. The iron-dominant variety (Fe_2SiO_4) is called fayalite. But it is important to remember the mineral groups, because we use them repeatedly throughout this book and because their names are commonly used in science. It is less important to remember the specific mineral names.

Olivine is one of the most important mineral groups because olivine is a major mineral in rocks of the mantle. The properties of olivine largely determine the properties of the mantle, and the properties of the mantle, in turn, control plate tectonics.

Before you go on:

1. What is crystal structure?

2. Why is ionic substitution important in minerals? How does it affect their properties?

PROPERTIES OF MINERALS

Geology students usually find mineral properties interesting, because these properties include color, hardness, luster, and other characteristics that make individual minerals valuable or valueless, pretty or ugly, sparkly or dull, and so forth. The properties of minerals are determined by what you just studied: their composition and crystal structure.

Once you know which properties are characteristic of each mineral, you can use those properties to identify unknown minerals. Thus, you can discover a mineral's identity without analyzing it chemically or determining its crystal structure. We will show you the techniques that geologists use to identify minerals in the field.

The properties most often used to identify minerals are obvious ones, such as color, shape, and hardness—plus some less-obvious qualities, such as luster, cleavage, and specific gravity. Each property is briefly discussed below. A table showing these properties for individual minerals is in Appendix C.

CRYSTAL FORM AND GROWTH HABIT

When ancient Greeks saw glistening needles of ice covering the ground on a frosty morning, they were intrigued by the fact that the needles were six-sided and had smooth, planar surfaces. They had made many discoveries about geometry, but they could not explain how three-dimensional, geometric solids could apparently grow spontaneously.

They called ice *krystallos*, which the Romans latinized to *crystallum*. Eventually, the word **crystal** came to apply to any solid body that grows with planar surfaces. The planar surfaces that bound a crystal are **crystal faces**, and the geometric arrangement of crystal faces, the **crystal form**, became the subject of intense study during the seventeenth century.

CRYSTAL FORM

At that time, scientists discovered that crystal form can be used to identify minerals. But certain features were diffi-

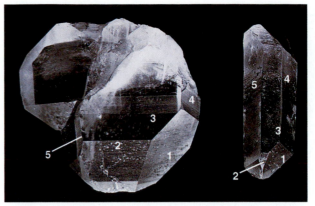

Figure 3.9 Crystal Faces and Angles in Quartz Two quartz crystals showing corresponding crystal faces. Although the sizes of the equivalent faces differ markedly between the two crystals, it is apparent that each numbered face on one crystal is parallel to an equivalent face on the other crystal. It is a fundamental property of crystals that, due to their internal crystal structure, the angles between adjacent faces are identical for all crystals of the same mineral.

cult for them to explain. Why did the sizes of crystal faces differ widely from sample to sample? Under some circumstances, a mineral may grow as a long, thin crystal, but under others, the same mineral may grow as a short, fat crystal, as in Figure 3.9. Superficially, the two quartz crystals in Figure 3.9 look very different. You can see that the overall size of crystals and their faces is not a unique property of a mineral.

The person who solved the mystery was a Danish physician, Nicolaus Steno. In 1669, he demonstrated that the unique property of a mineral's crystal is not its relative face size but *the angles between the faces*. The angle between any pair of crystal faces is constant, he wrote, and is the same across all specimens of a mineral, regardless of its overall shape or size.

Steno's discovery that *interfacial angles are constant* is made clear by the numbering in Figure 3.9. Corresponding faces occur on the two crystals. Each corresponding pair of faces is parallel; therefore, the angle between any two corresponding faces must be the same on each crystal.

Steno and other early scientists suspected that the capacity for crystals to form and for interfacial angles to be constant must depend on some kind of internal order. But they could not see the ordered particles—atoms—so they could only speculate. Proof that crystal form reflects internal order was finally achieved in 1912, when German scientist Max von Laue demonstrated with X-rays that crystals must be made up of atoms packed in fixed geometric arrays (Figure 3.7).

Crystals form only when mineral grains can grow freely in an open space. For this reason, crystals actually are uncommon in nature, because most minerals form in crowded, obstructed spaces.

Figure 3.10   Irregular Growth—The Usual Case   Quartz grains that grew in a cramped environment where other quartz grains prevented development of well-formed crystal faces. The amber-colored grains are iron carbonate ($FeCO_3$). Compare with Figure 3.9, which shows crystals that grew in open spaces, unhindered by adjacent grains.

Compare Figures 3.9 and 3.10. Figure 3.9 shows crystals of quartz that grew freely into an open space, so well-developed crystal faces were able to form. But in Figure 3.10, quartz grew as irregularly shaped grains in an environment crammed with other quartz grains.

Using von Laue's discovery, it is easy to show that, in both a quartz crystal and an irregularly shaped quartz grain, all the atoms present are packed in the same strict geometric arrangement—that is, both the quartz crystals and the irregular quartz grains are crystalline. This is why we use the word *crystalline*, rather than crystal, in the definition of a mineral.

Figure 3.11   Pyrite (aka Fool's Gold)   Distinctive crystal form of pyrite, FeS_2. One of the growth habits of pyrite is as cube-shaped crystals with pronounced striations on the cube face. The largest crystals in the photograph are 3 cm on an edge. The specimen is from Huaron, Peru.

GROWTH HABIT

Growth habit is the characteristic crystal form of each mineral—they have different habits. Some have such distinctive forms that you can use growth habit as an identification tool without having to measure angles between crystal faces. For example, the mineral *pyrite* (FeS_2) is commonly found as intergrown cubes with markedly striated faces (Figure 3.11).

A few minerals can develop distinctive growth habits even when they grow in restricted environments. These, too, can aid in identification. For example, Figure 3.12 shows chrysotile asbestos, a variety of the mineral serpentine that characteristically grows as fine, elongate threads.

Figure 3.12   Fibers of Chrysotile Asbestos
Some minerals have distinctive growth habits even though they do not develop well-formed crystal faces. The mineral chrysotile sometimes grows as fine, cotton-like threads that can be separated and woven into fireproof fabric, in which case it is referred to as asbestos. Chrysotile is one of several minerals that have asbestiform growth habits and are mined and commercially processed for asbestos. In commerce, chrysotile asbestos is commonly called "white asbestos."

POLYMORPHS

Some elements and compounds form two or more different minerals, because the atoms can be packed to form more than one kind of crystal structure. For example, the element carbon can have two dramatically different crystalline forms: diamond and graphite. As shown in Figures 3.3 and 3.5, these minerals have quite different internal structures. Different crystal structures are called **polymorphs** (meaning "many forms").

$CaCO_3$ is a compound that forms two different minerals. One is *calcite*, which composes limestone and marble. The other is *aragonite*, most commonly found in the shells of clams, oysters, and snails. Calcite and aragonite have identical compositions but entirely different crystal structures.

Some common mineral polymorphs are listed in Table 3.1.

TABLE 3.1 Examples of Polymorphs (Most Common Mineral Listed First)

Composition	Mineral Name
C	Graphite, Diamond
$CaCO_3$	Calcite, Aragonite
FeS_2	Pyrite, Marcasite
SiO_2	Quartz, Cristobalite

CLEAVAGE

A mineral's tendency to break in preferred directions along bright, reflective planar surfaces is called **cleavage**.

If you break a mineral with a hammer or drop a specimen on the floor so that it shatters, you will probably see that the broken fragments are bounded by surfaces that are smooth and planar, just like crystal faces. In exceptional cases, such as the *halite* (NaCl) fragments shown in Figure 3.13A, all of the breakage surfaces are smooth planar surfaces.

Don't confuse crystal faces and cleavage surfaces, however, even though the two often look alike. A cleavage surface is a *breakage* surface, whereas a crystal face is a *growth* surface.

The planar directions along which cleavage occurs are governed by the crystal structure (Figure 3.13B). They are planes along which the bonding between atoms is relatively weak. Because the cleavage planes are direct expressions of the crystal structure, the angles between cleavage planes are the same for all grains of a given mineral. Just as interfacial angles of crystals are constant, so are the angles between cleavage planes constant. Cleavage, therefore, is a valuable guide for the identification of minerals.

Many common minerals have distinctive cleavage planes. One of the most distinctive is found in mica (Figure 3.14). Clay minerals also have distinctive cleavage, and it is an easy cleavage direction that makes them feel smooth and slippery when rubbed between the fingers. Other minerals with distinctive cleavages are *fluorite* (CaF_2), which

A.

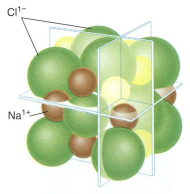

Cl^{1-}

Na^{1+}

B.

Figure 3.13 Mineral Cleavage Relation between crystal structure and cleavage. A. Halite, NaCl, has three well-defined cleavage directions and breaks into fragments bounded by perpendicular faces. B. The crystal structure in the same orientation as the cleavage fragments shows that the directions of breakage are planes in the crystal between equal numbers of sodium and chlorine atoms.

Figure 3.14 A Book of Mica Perfect cleavage of muscovite (a mica), shown by thin, planar flakes into which this specimen is being split. The cleavage flakes suggest leaves of a book, a resemblance embodied in the name "book of mica."

A.   **B.**

Figure 3.15 Cleavage in Two Common Minerals Two common minerals with distinctive cleavages. Fluorite (A) breaks along four planar directions. Potassium feldspar (B) breaks along two planar directions that are perpendicular.

breaks along four planar directions (Figure 3.15A), and *potassium feldspar* ($KAlSi_3O_8$), which breaks along two planar directions that are perpendicular, creating fragments that are approximately rectangular (Figure 3.15B).

LUSTER

The quality and intensity of light reflected from a mineral produce an effect known as **luster**. Two minerals with almost identical color can have quite different lusters. The most important lusters are:

Metallic, like that of a polished metal surface.

Vitreous, like that of glass.

Resinous, like that of resin.

Pearly, like that of pearl.

Greasy, as if the surface were covered by a film of oil (Figure 3.16).

COLOR AND STREAK

COLOR IS UNRELIABLE FOR IDENTIFICATION

The color of a mineral is often striking, but unfortunately color is an unreliable means for mineral identification.

Color is determined by several factors, but its main cause is chemical composition. Some elements can create strong color effects even when they are present only in very small amounts through ionic substitution.

For example, the mineral *corundum* (Al_2O_3) is commonly white or grayish, but when small amounts of Cr^{3+} replace Al^{3+} by ionic substitution, corundum becomes blood red, giving us the gem *ruby*. Similarly, when small amounts of Fe^{2+} and Ti^{4+} are present, corundum becomes deep blue, and we call it *sapphire* (Figure 3.17).

One reason color can be confusing is weathering. Water and oxygen will react with a mineral such as pyrite (FeS_2), forming an altered surface that differs in color from the unaltered mineral. Weathering reactions like this are common and, as discussed in *Understanding Our Environment*, Box 3.2, *Long-Term Danger of Abandoned Mines*, can have important consequences for the environment.

STREAK IS MUCH MORE RELIABLE FOR IDENTIFICATION

One way to reduce errors of judgment where color is concerned is to make a *streak* of the mineral specimen on an unglazed porcelain plate. The streak is a thin layer of powdered mineral made by rubbing the specimen on the plate. The powder gives a reliable color effect because all the

A.   **B.**   **C.**

Figure 3.16 Luster Three minerals with different lusters. A. Quartz has a vitreous luster. B. Sphalerite has a resinous luster—it resembles dried pine-tree resin. C. Talc has a pearly luster.

Figure 3.17   Same Mineral, Different Colors   Two specimens of the same mineral, corundum (Al_2O_3), with distinctly different colors. The red crystal, ruby, is from Tanzania and is about 2.5 cm across. The blue crystal, sapphire, comes from Newton, New Jersey, and is about 3 cm across.

Figure 3.18   Streak of a Mineral   Color contrast between hematite mineral specimen and a hematite streak. Massive hematite is opaque, has a metallic luster, and appears black. On a porcelain plate, hematite gives a red streak.

grains in a powder streak are very small and the effect of grain size is reduced.

A classic example is *hematite* (Fe_2O_3), a black, metallic mineral that has a striking red streak—not black (Figure 3.18). In general, color in opaque minerals with metallic lusters can be very confusing because the color is partly a property of the size of the individual mineral grains; for such minerals streak is a reliable property.

HARDNESS AND THE MOHS SCALE

The term **hardness** refers to a mineral's relative resistance to scratching. Like crystal form and cleavage, hardness is governed by crystal structure and by the strength of the bonds between atoms. The stronger the bond, the harder the mineral.

The *Mohs relative hardness scale* uses 10 minerals, each with its distinctive hardness (Table 3.2). At "1" on the scale is the softest mineral known, talc (the basic ingredient of many body powders or "talcum"). Minerals "2" through "9" are progressively harder. At "10," diamond is the hardest mineral known.

Values on the scale indicate *relative* hardness, not *absolute*—in other words, the numbers simply indicate that any mineral on the scale will scratch all those having a lower number. For example, diamond (10) will scratch every mineral below it. Quartz (7) will scratch every mineral below it, but none above it. Talc (1) can't scratch anything, but is scratched by everything above it.

For convenience, we often test relative hardness by using common objects, such as a copper penny (equivalent to fluorite's hardness of 4) or a steel knife blade or glass (equivalent to feldspar's hardness of 6).

The 10 hardness steps do not represent equal intervals of hardness—for example, diamond at 10 is *a lot* harder than corundum at 9. But, although this scale is "semiscientific," it is extremely useful in the field.

DENSITY AND SPECIFIC GRAVITY

Another obvious physical property of a mineral is its density—how heavy it feels. We know that two equal-sized baskets have different weights when one is filled with

TABLE 3.2   Mohs Scale of Relative Hardness*

Relative Number in the Scale	Mineral	Hardness of Some Common Objects
10	Diamond	
9	Corundum	
8	Topaz	
7	Quartz	
6	Potassium feldspar	
		Pocket knife; glass
5	Apatite	
4	Fluorite	
		Copper penny
3	Calcite	
		Fingernail
2	Gypsum	
1	Talc	

(Decreasing ↓)

* Named for Friedrich Mohs, an Austrian mineralogist, who chose the 10 minerals of the scale.

UNDERSTANDING OUR ENVIRONMENT

LONG-TERM DANGER OF ABANDONED MINES

Weathering is a general term for all of the reactions between minerals, air, and water. Most minerals are susceptible to weathering reactions, but the most susceptible family of minerals are the sulfide minerals. The two common iron sulfide minerals, pyrite (FeS_2)

and pyrrhotite (FeS), in particular, weather rapidly at Earth's surface. Both pyrite and pyrrhotite react and combine with oxygen in the air or dissolved in water. The reaction products are relatively insoluble iron hydroxide compounds and sulfuric acid. The acid thus produced will attack more resistant sulfide minerals such as sphalerite (ZnS), galena (PbS), and chalcopyrite ($CuFeS_2$), producing more sulfuric acid and solutions with high concentrations of heavy metals. The oxidation of sulfide minerals can proceed inorganically, but certain bacteria can speed up the process hundreds of times faster.

In nature, fresh rock is exposed relatively slowly, so the rate at which acid is released by oxidation is very slow. But when mining operations expose fresh rocks by blasting and digging, a lot of sulfide minerals can be exposed rapidly. The mining of sulfide ores usually proceeds so rapidly that acid solutions are not a problem. But when mining ceases and workings fill with water, bacteria start to proliferate and oxidation proceeds rapidly. The waters seeping from old mines can be so acidic that no plants or animals can live in them. Released into streams, acid mine waters can cause devastating damage (Figure B3.3).

Although no mine that contains sulfide minerals is entirely free of acid water, the problem is particularly serious in abandoned coal mines. All coal contains some iron sulfide minerals, and large coal mines can leave tens or even hundreds of miles of openings for water and air to enter. Of course, climate plays a role in acid mine drainage. In areas of plentiful rainfall, such as Pennsylvania, old mines fill quickly with water so that acid mine drainage is a continual problem.

So far, no one has found a way to prevent oxidation and the formation of acid mine waters. But forward planning for the design of mine openings, and the control of effluents once mining has ceased, can greatly ameliorate the problem.

Figure B3.3 An Abandoned Mining Mess Acid drainage seeping from an abandoned nineteenth-century silver mine, Silverton, Colorado. The acidity is due to the oxidation of pyrite.

---

feathers and the other with rocks. The property that causes this difference is *density*, or the average mass per unit volume. Minerals with a high density, such as gold, contain atoms with high mass numbers that are closely packed. Minerals with a low density, such as ice, have loosely packed atoms. The unit of density is grams per cubic centimeter (g/cm^3).

The relative density of minerals can be quickly sensed by hand. Simply hold one specimen in your left hand and the other in your right, and compare their "heft."

Density is easily measured using the property called specific gravity. **Specific gravity** *is the weight of a substance in air divided by the weight of an equal volume of pure water.* Since we know that the density of pure water is 1 g/cm^3, a mineral's specific gravity is directly equal to

its density. Specific gravity is a ratio of two weights, so it does not have any units.

Gold has a density of 19.3 g/cm^3 and feels very heavy. But many other minerals, such as galena (PbS) and magnetite (Fe_3O_4), which have densities of 7.5 g/cm^3 and 5.2 g/cm^3, respectively, also feel heavy by comparison with many common minerals that have densities of 2.5–3.0 g/cm^3. In general, metallic minerals (galena, pyrite) feel relatively heavy, whereas nearly all others feel relatively light.

MINERAL PROPERTIES AND BOND TYPES

Mineral properties depend strongly on the kinds of bonds present (Table 3.3). Strength and hardness are determined

TABLE 3.3 Examples of Mineral Properties That Depend on Bond Type

Bond Type	Mineral	Strength	Hardness	Electrical Conductance	Solubility in Water and Weak Acids
Ionic	Calcite ($CaCO_3$) Halite (NaCl)	High	Moderate to high	Very low	High
Covalent	Diamond (C) Sphalerite (ZnS)	Very high	High	Very low	Very low
Mixed ionic and covalent	Olivine (Mg_2SiO_4) Muscovite $KAl_2(Si_3Al)O_{10}(OH)_2$	Very high	Moderate to high	Very low	Low
Metallic	Gold (Au) Copper (Cu)	Moderate	Low	High	Very low
van der Waals	Graphite (C) Sulfur (S)	Very low	Very low	Low	Low

by the weakest bond type present because, just as the weakest link in a chain determines the strength of the chain, so the weakest bond type in a mineral determines its strength. Ionic and covalent bonds are strong, making minerals hard and strong. But metallic and van der Waals bonds are much weaker, so minerals containing them tend to be soft and easily deformed.

Before you go on:

1. What are crystal form and growth habit?

2. How does a cleavage direction differ from a crystal face?

3. What is the Mohs hardness scale? How is it useful for mineral identification?

4. What is the relationship between mineral properties and bond types?

COMMON MINERALS

Now that you have studied the chemical composition of minerals, their crystal structures, and their various properties, you are ready to explore the fascinating world of the minerals themselves—hopefully in a hands-on lab and in the field, as well as by reading this text.

Scientists have identified approximately 4000 minerals—(don't panic, you needn't learn them all—just learn the basic types and a few examples!). Most occur in the crust, but a few have been identified in meteorites, and two new ones were discovered in Moon rocks retrieved by astronauts. The number of minerals may seem large, but it is tiny by comparison to the astronomically large number of ways chemists can combine naturally occurring elements to form synthetic minerals.

The reason for the disparity between natural minerals and synthesized ones becomes apparent when you consider the relative abundances of the chemical elements in Earth's crust. As Figure 3.19 shows, only 12 elements occur in the continental crust in amounts greater than 0.1 percent by weight (left-hand pie).

Together, these 12 make up 99.23 percent of the crustal mass. The crust, therefore, is constructed mostly of a limited number of minerals—only about 30 are commonly encountered—in which one or more of the 12 abundant elements is an essential ingredient.

Many of the scarce elements do not form distinct minerals, but instead occur by ionic substitution. To repeat the example of the mineral group olivine, $(Mg,Fe)_2SiO_4$, these minerals can contain, in addition to Mg, Fe, Si, and O, trace amounts of Cu, Ni, Co, Mn, and many other elements as ionic substitutes for the Mg or Fe.

Minerals containing scarce elements certainly occur, but only in small amounts, and those small amounts form only under special and restricted circumstances. A few scarce elements, such as hafnium and rhenium, are not known to form minerals under any circumstances: they occur only by ionic substitution.

We will consider these three mineral groups:

1. Silicate minerals, $(SiO_4)^{4-}$, the most abundant in Earth's crust.

2. Carbonate $(CO_3)^{2-}$, phosphate $(PO_4)^{3-}$, and sulfate $(SO_4)^{2-}$ minerals.

3. Ore minerals—sulfides (S^{2-}) and oxides (O^{2-}) that contain valuable metals.

Before you go on:

1. Which two chemical elements are most abundant in Earth's crust?

2. Which mineral group is the most abundant in Earth's crust?

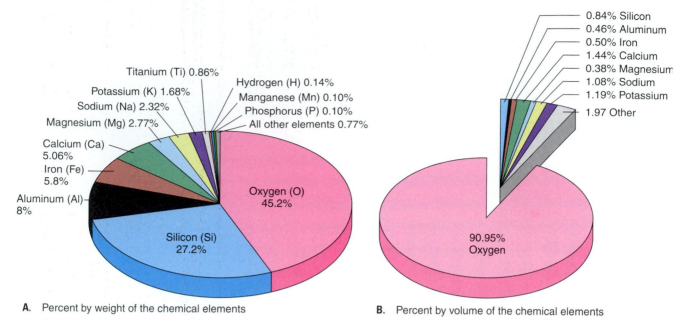

A. Percent by weight of the chemical elements

B. Percent by volume of the chemical elements

Figure 3.19 Average Composition of the Continental Crust A. Percent by weight of the chemical elements. B. Percent by volume of the chemical elements.

SILICATES: THE LARGEST MINERAL GROUP

As Figure 3.19A shows, two elements, oxygen and silicon, make up more than 70 percent of the weight of the continental crust. Oxygen forms a simple anion, O^{2-}, and compounds that contain the O^{2-} anion are called *oxides*. Silicon forms a simple cation, Si^{4+}.

Together, oxygen and silicon form an exceedingly strong complex ion, the **silicate anion** $(SiO_4)^{4-}$. Geologists distinguish between simple oxides (like MgO) and complex oxides (like Mg_2SiO_4) by calling the complex ones **silicates** or **silicate minerals**.

Silicate minerals are the most abundant of all naturally occurring inorganic compounds, and simple oxides are the second most abundant group. Silicates are known for their strength and variety.

THE STURDY SILICATE TETRAHEDRON

The four oxygen atoms in a silicate anion are tightly bound to the single silicon atom. The bond, which is very strong, is largely covalent. Oxygen has a large ionic radius, while silicon has a small ionic radius (Figure 3.8). In the silicate anion, the oxygen ions pack into the smallest space possible for four large spheres.

As you can see in Figure 3.20, the four oxygens sit at the corners of a tetrahedron, and the small silicon atom sits in the space between the oxygens at the center of the tetrahedron. Therefore, the shape of the silicate anion is a tetrahedron. The structures and properties of all silicate miner-

als are determined by the way the $(SiO_4)^{4-}$ silicate tetrahedra pack together in the crystal structure.

Each silicate tetrahedron has four negative charges. Silicon has four electrons in the outer shell, and oxygen has six. The silicon atom in a silicate tetrahedron shares one of its electrons with each of the four oxygens. Each of the four oxygens, in turn, shares one of their electrons with the silicon. This leaves the silicon with a stable outer shell of eight electrons, but each of the four oxygens still requires an additional electron for a stable octet.

The four oxygens can each attain a stable shell 2 of eight electrons in two ways:

1. They can each accept an electron from, or share an electron with, other atoms. An example of this is found in olivine (Mg_2SiO_4), in which Mg atoms transfer their two outer-shell electrons to the oxygens, forming ionic bonds in the process.

2. An oxygen atom can bond with two Si atoms at the same time. The shared oxygen is then covalently bonded to each of two silicons. The second electron shell of the shared oxygen is now filled because it shares an electron with each silicon, and the two tetrahedral-shaped silicate anions, now joined at a common apex, form an even larger complex anion with the formula $(Si_2O_7)^{6-}$, as shown in Figure 3.21.

POLYMERIZATION

The large $(Si_2O_7)^{6-}$ anion forms compounds by accepting or sharing electrons with other atoms in the same way that the smaller $(SiO_4)^{4-}$ anion does. However, just as more

A.

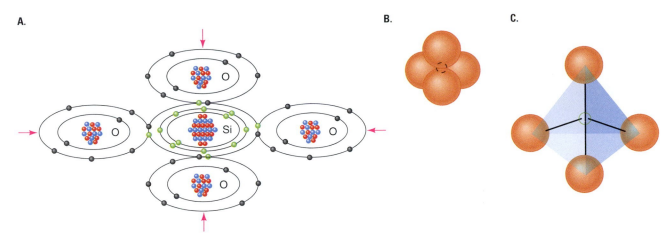

B.   C.

Figure 3.20 Silicate Tetrahedron Bonding and structure of the silicate anion. Electrons associated with the silicon atom are colored green; oxygen's electrons are colored black. A. Silicon has four electrons in the outer energy-level shell. Each of the four electrons is shared with one oxygen, and each oxygen in turn shares an electron with the silicon. The silicon finishes with eight electrons in the outer shell, each oxygen with seven electrons. Bonding between the silicate anion and other ions occurs when the four oxygens share or accept an additional electron, thereby filling their second energy-level shells. B. Tetrahedral-shaped silicate anion with oxygens touching each other in natural positions. Silicon (dashed circle) occupies central space. C. Expanded view showing large oxygens at the four corners, equidistant from a small silicon atom. The lines show bonds between silicon and oxygen; shading outlines the tetrahedron.

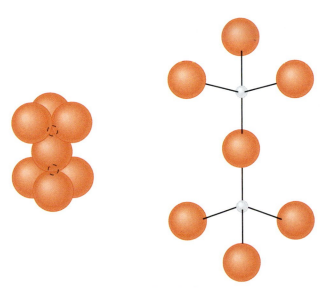

Figure 3.21 Linking Silicate Tetrahedra Two silicate tetrahedra share an oxygen, thereby satisfying some of the unbalanced electrical charges and, in the process, forming a larger and more complex anion group. A. The arrangement of oxygen and silicon atoms in a double tetrahedron, giving the complex anion $(Si_2O_7)^{6-}$. B. An expanded view of two tetrahedra sharing an oxygen.

and more beads can be strung on a necklace, so can **polymerization**, the process of linking silicate tetrahedra by oxygen sharing, be extended to form huge anions (Figure 3.22A).

Many common silicate minerals contain very large anions as a result of polymerization. As shown in Figure 3.22B, if a tetrahedron shares more than one oxygen with adjacent tetrahedra, the complex anionic structures so formed can have large circular shapes, endless chains, sheets, and even three-dimensional networks of tetrahedra. No matter how big an anion becomes by polymerization, any oxygen that does not have a stable outer shell of eight electrons can form bonds with cations by accepting or sharing electrons.

Observe in Figures 3.22A and B that an important restriction to the polymerization process is always met—two adjacent tetrahedra never share more than one oxygen. Stated another way, tetrahedra join only at their apexes, never along the edges or faces. The common polymerizations, together with the rock-forming minerals containing them, are illustrated in Figure 3.23 and discussed below in order of increasing complexity of polymerization.

ISOLATED TETRAHEDRA: OLIVINES AND GARNETS

Isolated tetrahedra are linked by the bonding of each oxygen to a cation such as Mg^{2+} or Fe^{2+}, and each cation bonds to six or eight oxygens. In this way each silicate tetrahedron is completely isolated by cations. Two very important rock-forming mineral groups, the *olivines* and the *garnets*, have crystal structures in which the silicate tetrahedra are isolated.

Olivine is among Earth's most abundant mineral groups, a very common constituent of igneous rocks in

A.

B.

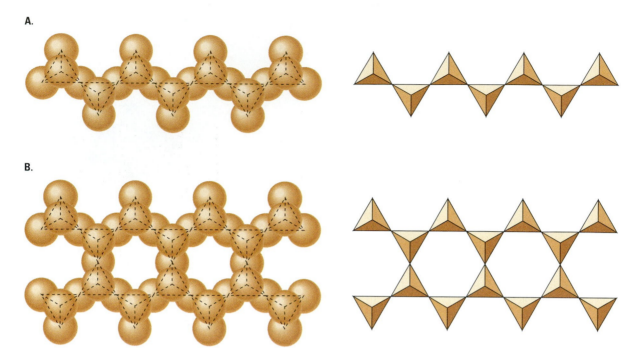

Figure 3.22 Large Anions Formed by Polymerization
Formation of complex silicate anions by polymerization. A. Polymerization of silicate anions to form a continuous chain in which each silicate anion shares two oxygens with adjacent

anions. A geometric representation of the chain is on the right. Formula of the complex anion is $(SiO_3)^{2n-}$. B. Double chain of polymerized silicate anions with formula $(Si_4O_{11})^{6n-}$.

oceanic crust and the upper part of the mantle. In rare cases, olivine occurs in such flawless and beautiful crystals that it is used as a gem, peridot.

CHAINS: PYROXENES AND AMPHIBOLES

Single chains of tetrahedra form when two oxygens in each tetrahedron bond with adjacent tetrahedra (Figure 3.23). The result is an open-ended chain in which each tetrahedron has two of the four oxygens with unsatisfied charges. The chains are linked together by the bonding of the unsatisfied oxygens to cations such as Ca^{2+}, Mg^{2+}, and Fe^{2+}. One of the most important mineral groups, the *pyroxenes*, contains single-chain linkages, and the most common pyroxene is called *augite*.

Two single-chain linkages may join together to form a double chain as shown in Figure 3.23. A double-chain linking still leaves 5 of every 16 oxygens with unsatisfied charges, and the double chains are linked together to make mineral structures by cations such as Ca^{2+}, Na^{1+}, Mg^{2+}, Al^{3+}.

A very common and important family of minerals, the *amphiboles*, contains double chains. The most common of the amphiboles is called *hornblende*. The pyroxenes and amphiboles look similar. Both mineral families tend to be dark green, brown, or black, and both commonly occur in igneous and metamorphic rocks, so they are hard to tell

apart. One reliable way to distinguish pyroxenes from amphiboles is by cleavage. Each mineral group has two good cleavages; the cleavages in pyroxene are at right angles (90°); those in amphibole are at 120°. (See Figure 3.24.)

SHEETS: CLAYS, MICAS, CHLORITES, AND SERPENTINES

When each tetrahedron shares three oxygens with adjacent tetrahedra, the result is a sheet of tetrahedra in which each tetrahedron has a single oxygen with an unsatisfied charge. Adjacent sheets are linked together by cations that bond to the single oxygens.

Several important mineral groups contain sheet linkages—the micas, clays, chlorites, and serpentines are examples. *Kaolinite*, $Al_4Si_4O_{10}(OH)_8$, is one of the most common *clays*, while *muscovite*, $KAl_2(Si_3Al)O_{10}(OH)_2$, is a common mica. *Chlorite*, which contains Mg^{2+} and Fe^{2+} cations, is usually greenish in color; its name derives from *chloros*, a Greek word for green. All sheet structure minerals have one well-developed cleavage parallel to the sheet (Figures 3.14 and 3.15).

The *serpentine group* consists of three polymorphs with the formula $Mg_6Si_4O_{10}(OH)_8$: chrysotile, antigorite, and lizardite. The three minerals commonly occur together as fine-grained greenish masses formed by the alteration of olivine or other magnesium silicates. *Chrysotile* is the

Silicate Structure		Mineral/Formula	Cleavage	Example of a Specimen
	 Single tetrahedron	Olivine Mg_2SiO_4	None	
	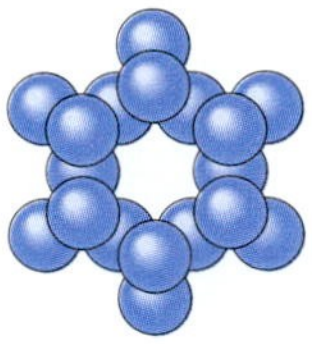 Hexagonal ring	Beryl (Gem form is emerald) $Be_3Al_2Si_6O_{18}$	One plane	
	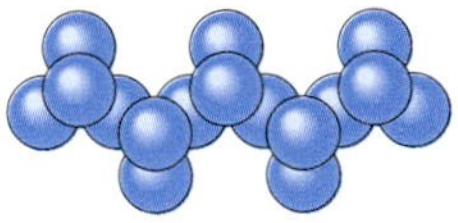 Single chain	Pyroxene group $CaMg(SiO_3)_2$ (variety: diopside)	Two planes at 90°	
	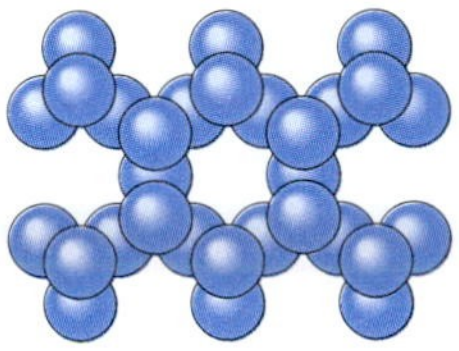 Double chain	Amphibole group $Ca_2Mg_5(Si_4O_{11})_2(OH)_2$ (variety: tremolite)	Two planes at 120°	
	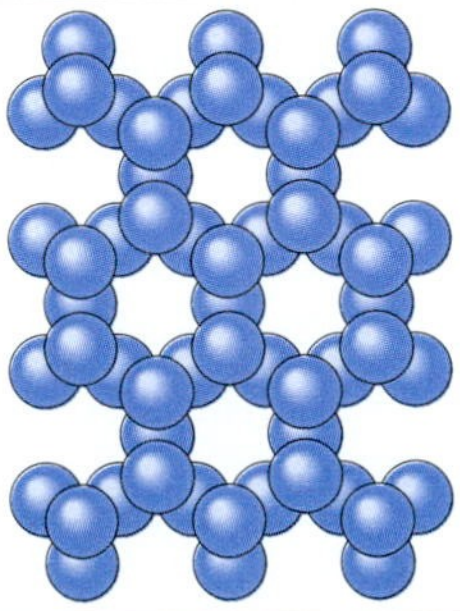 Sheet	Mica $KAl_2(AlSi_3)O_{10}(OH)_2$ (variety: muscovite) $K(Mg,Fe)_3(AlSi_3)O_{10}(OH)_2$ (variety: biotite)	One plane	
Too complex to draw.	Three-dimensional network	Feldspar $KAlSi_3O_8$ (variety: orthoclase)	Two directions at 90°	
		Quartz SiO_2	None	

Figure 3.23 Common Silicate Minerals Summary of the ways in which silicate anions can link together to form the common silicate minerals. Typical minerals of each type are listed and shown. Many other minerals exist in each polymerization cate-gory, far too many to show on a diagram like this. Silicate linkages other than those shown are also possible but do not occur in common minerals.

white asbestos of commerce (Figure 3.12). Chrysotile fibers form when the polymerized silicate sheets roll up, like a tightly wound carpet. For a discussion of the health effects of asbestos, *Understanding Our Environment*, Box 3.3, *Asbestos: How Risky Is It?*

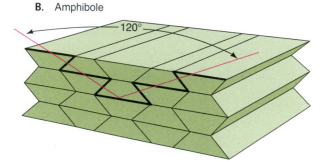

A. Pyroxene

90°

B. Amphibole

120°

Figure 3.24 Cleavage in the Chain Silicates Cleavage in pyroxene and amphibole. A. Stacking of pyroxene chains, viewed end-on. Bonds that hold the chains together break along the wide black line. The angle between the two cleavage directions (red line) is close to 90°. B. Stacking of amphibole double chains. The angle between cleavage directions is 120°.

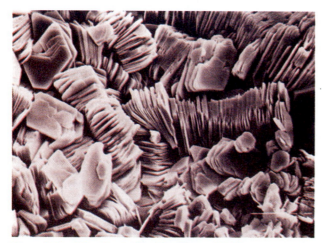

Figure 3.25 Why Clay Is Good for Pottery Making Minuscule crystals of the clay mineral kaolinite as seen through a scanning electron microscope. Pronounced cleavage parallel to the polymerized sheets of silicate anions is apparent. Clay absorbs water between the polymerized sheets, making the wet clay weak, slippery, and easy to mold into pots.

For the linguistically curious, the name *serpentine* is from the Latin word for serpent. No one knows how this mineral group acquired the name. Some suggest that a spotted serpentine-containing rock, *Verde antique*, fancifully resembles a spotted serpent. Others think the name comes from Greek writer Dioscorides, who, in A.D. 50, recommended serpentine to prevent snakebite. No evidence indicates that serpentine has curative powers for snakebites.

QUARTZ—PURE SiO$_2$

The only common mineral composed exclusively of silicon and oxygen is *quartz*, SiO$_2$. It provides an example of polymerization filling all the second energy-level shells and in the process forming a three-dimensional network of tetrahedra.

Quartz characteristically forms six-sided crystals (Figure 3.9) and is found in many beautiful colors, several of which are prized as gemstones (Figure 3.26). The colors come from minute amounts of iron, aluminum, titanium, and other elements present by ionic substitution. Quartz occurs in igneous, metamorphic, and sedimentary rocks and is one of the most widely used gem and ornamental minerals.

Certain specimens of quartz—those formed by precipitation from cool water solutions—are so fine-grained that they appear almost amorphous, and we can demonstrate the presence of an internal crystal structure only by using high-powered microscopes, X-ray machines, and other research tools. These fine-grained forms of quartz are called *chalcedony*. Varietal names are *agate* if it has color banding (Figure 3.27) or *flint* (gray) and *jasper* (red) if the color is uniform.

THE FELDSPAR GROUP—MOST COMMON MINERALS IN EARTH'S CRUST

The name *feldspar* derives from two Swedish words, *feld* (field) and *spar* (mineral). Early Swedish miners were familiar with feldspar in their mines and found the same

Figure 3.26 The Colors of Quartz Trace amounts of atoms other than silicon and oxygen give rise to a wide range of colors in quartz.

Figure 3.27 Banded Agate Agate, a color-banded microcrystalline variety of quartz formed by precipitation of SiO_2 in an open space from groundwater. Color banding is due to minute amounts of impurities. The sample is 10 cm across.

mineral group in the abundant rocks they had to clear from their fields before they could plant crops. They were so struck by the abundance of feldspar that they chose a name to indicate that their fields seemed to be growing an endless crop of the minerals.

Feldspar is indeed the most common mineral group in Earth's crust. It accounts for about 60 percent of all minerals in the continental crust, and together with quartz it con-

Figure 3.28 Green Feldspar Seen in this specimen are the two most common minerals in the continental crust, feldspar and quartz. This specimen from Crystal Peak, Colorado, is a group of green potassium feldspar crystals and dark gray quartz grains. Gem varieties of green feldspar are called amazonite.

stitutes about 75 percent of the volume of the continental crust. Feldspar is also abundant in rocks of the seafloor.

Feldspar, like quartz, has a structure formed by polymerization of all the oxygen atoms in the silicate tetrahedra. Unlike quartz, however, some of the tetrahedra contain Al^{3+} substituting for Si^{4+}, and so another atom must be added to the structure to donate an electron and balance the ionic charges. The cations that are present in minerals of the feldspar group are K^{1+}, Na^{1+}, or Ca^{2+}.

Potassium feldspar has several polymorphs, but their structural differences are subtle. They are sometimes pink or green due to ionic substitution of small amounts of Fe^{3+} for Al^{3+} (Figure 3.28).

Before you go on:

1. Why are silicates so abundant in Earth's crust?

2. What property allows silicate tetrahedra to link (polymerize) into sheets, chains, and networks?

3. What common mineral is pure silica?

4. Which mineral group accounts for about 60 percent of all minerals in the crust?

THE CARBONATE, PHOSPHATE, AND SULFATE MINERAL GROUPS

CARBONATES

The carbonate anion $(CO_3)^{2-}$ forms three common minerals: *calcite*, *aragonite*, and *dolomite*. We have already seen that calcite and aragonite have the same composition, $CaCO_3$, and are polymorphs. Calcite is much more abundant than aragonite. Dolomite has the formula $CaMg(CO_3)_2$.

Both calcite and dolomite are abundant minerals, and they look similar. They have the same vitreous luster and the same distinctive cleavage (Figure 3.29), both are relatively soft, and both are common in sedimentary rocks—calcite in limestone, dolomite in dolostone. One simple way to distinguish between them is by applying a few drops of dilute hydrochloric acid (HCl) to the mineral surface. Calcite reacts vigorously, bubbling and effervescing, whereas dolomite reacts very slowly with little or no effervescence.

PHOSPHATES

Apatite is by far the most important phosphate mineral. It is the substance from which our bones and teeth are made. It contains the complex anion $(PO_4)^{3-}$ and has the general formula $Ca_5(PO_4)_3(F,OH)$. It is also a common mineral in many varieties of igneous and sedimentary rocks and is the main source of the phosphorus used for making phosphate fertilizers.

Figure 3.29 Confusing Carbonates Calcite ($CaCO_3$) on the left and dolomite $CaMg(CO_3)_2$ on the right have similar crystal structures and, as a result, similar cleavages. Both cleave in three directions that are not perpendicular, yielding rhombohedral-shaped fragments. One quick way to distinguish between calcite and dolomite is to put a drop of diluted hydrochloric acid on the mineral—calcite will fizz and effervesce, while dolomite will slowly dissolve but not effervesce.

SULFATES

All sulfate minerals contain the sulfate anion, $(SO_4)^{2-}$. Although many sulfate minerals are known, only two are common, and both are calcium sulfate minerals: *anhydrite*

($CaSO_4$) and *gypsum* ($CaSO_4 \cdot 2H_2O$). Both form when seawater evaporates—anhydrite when temperatures are high, gypsum at lower temperatures.

Gypsum is the raw material used for making plaster. Plaster of Paris got its name from a quarry near Paris where a very desirable, pure-white form of gypsum was mined centuries ago.

THE ORE MINERAL GROUPS— OUR SOURCE FOR METALS

The term *ore mineral* is used for minerals that are sought and processed for their valuable metal contents. Such minerals tend to be pure elements (copper, gold), sulfides, or oxides. The important ore minerals are listed in Appendix C.

SULFIDES

The common sulfide minerals all have metallic lusters and high specific gravities. The two most common, *pyrite* (FeS_2) and *pyrrhotite* (FeS), are not actually mined for their iron content, but even so they are commonly referred to as ore minerals. Most of the world's lead is won from *galena* (PbS), most of the zinc from *sphalerite* (ZnS), and most of the copper from *chalcopyrite* ($CuFeS_2$) (Figure 3.30). Other familiar metals won from sulfide ore minerals are cobalt, mercury, molybdenum, and silver.

A.

B.

C.

Figure 3.30 Ore Minerals Example of three common ore minerals. A. Sphalerite (ZnS), the principal zinc mineral, from Joplin, Missouri. Sphalerite is the dark mineral. The clear, colorless mineral is calcite. B. Galena (PbS), the main lead mineral, from Joplin, Missouri. C. Chalcopyrite ($CuFeS_2$), an important copper mineral, from Ugo, Japan. Chalcopyrite is the gold-colored mineral. The white mineral is quartz.

Box 3.3 # UNDERSTANDING OUR ENVIRONMENT

ASBESTOS: HOW RISKY IS IT?

With every breath of air we inhale microscopic particles of suspended mineral matter. We call them *dust*. Over our lifespan, each of us inhales billions of mineral particles.

Most are expelled when we exhale, but a few lodge in our bronchial tubes or lungs. Our bodies have several ways of preventing these foreign bodies from causing harm. Our lungs envelop the foreign body in a sort of scar tissue, but they can only handle so much—if the load of particles becomes too great, our lungs' capacity starts to decline. In some cases, this can lead to lung cancer.

A contentious scientific debate surrounds the long-term effects of "asbestos"—fibrous mineral particles—in our lungs. Based on the experience of asbestos workers, some scientists argue that asbestos is a major hazard and that natural fibers pose severe health effects. Others insist that the issue has been blown out of proportion, and that it is far from clear that workplace concentrations of mineral fibers are relatable to the general living environment. The debate has arisen for several reasons.

The very term *asbestos* creates uncertainty. Ancient Greeks discovered that mineral fibers could be woven into a fireproof cloth, and they called such fibers *asbestos*, which means *unaffected by fire*. Many different minerals grow as long, flexible fibers and therefore are *asbestos*. Consequently, asbestos is not a specific mineral. *Asbestiform* is the correct textural term that describes any mineral found as a strong, flexible fiber. Today, commercial asbestos is made from six different minerals. The majority (95%) is a white mineral called chrysotile, which is a member of a group of minerals called *serpentines*. The five remaining commercial asbestos minerals all belong to a different mineral group, called the *amphiboles*.

Some of the most compelling evidence linking cancer and a certain type of asbestos involves a blue-colored amphibole called crocidolite (kro-SID-oh-lite), mined mainly in South Africa and Australia (Figure B3.4). Compositionally, the amphiboles and serpentines are quite different. So, can evidence derived from blue asbestos (crocidolite) be used to evaluate the effects of white asbestos (chrysotile)? Herein lies the source of some of the debate.

If it is the shape and size of the particles (asbestiform) that matters, then mineral differences may be unimportant. But if it is composition that matters (serpentine vs. amphibole), then we may be needlessly frightening people and denying ourselves use of a valuable material by saying that all asbestiform materials, regardless of composition, are dangerous.

Further, just how much asbestos is dangerous? Asbestos workers exposed to high levels of any kind of asbestiform dust can indeed develop scarred lungs and can eventually develop severe lung disorders. The rest of us may inhale a few fibers of chrysotile because it is such

Figure B3.4 Deadly Mineral   Waste pile of blue asbestos (crocidolite), Wittenoom Gorge, Hamersley Range, Western Australia. People who worked in the mining, processing, and use of blue asbestos have a high incidence of a form of cancer called mesothelioma.

a common mineral, but at some level the body is probably able to handle these fibers so that they don't pose a threat. What that level may be is a matter of research and therefore another source of debate.

The issue of asbestos and health has long been political. As a result, a large industry has emerged to remove asbestos from buildings, regardless of what kind of asbestos is involved. A legal specialty has also emerged—litigation over asbestos issues has become gargantuan.

The asbestos saga probably will be with us for a long time. And we may never hear the end of "dangerous mineral stories" because suspicions are being voiced about hazards from other common minerals such as talc and quartz. Your knowledge of geology and the "truth about minerals" may prove valuable to you in the workplace, home, and voting booth.

OXIDES

Because iron is one of the most abundant elements in Earth's crust, the iron oxides *magnetite* (Fe_3O_4) and *hematite* (Fe_2O_3) are the two most common oxide minerals.

Magnetite takes its name from the Greek word *Magnetis*, meaning stone of *Magnesia*, an ancient town in Asia Minor. Magnetis had the power to attract iron particles, and so the terms *magnet* and *magnetite* eventually joined our vocabulary.

Hematite refers to its red color when powdered, the Greek word for red blood being *haima*. Magnetite and hematite are the main ore minerals of iron.

Other oxide ore minerals are *rutile* (TiO_2), the principal source of titanium; *cassiterite* (SnO_2), the main ore mineral for tin; and *uraninite* (U_3O_8), the main source of uranium. Other metals won from oxide ore minerals are chromium, manganese, niobium, and tantalum.

MINERALS GIVE CLUES TO THEIR ENVIRONMENT OF FORMATION

Minerals are more than sources of valuable gems, metals, and beauty. Contained within them is historical evidence of the conditions under which they formed (and the rocks they are part of, as well). Thus, our study of minerals can provide insight into the chemical and physical conditions of long ago and beneath the surface where we cannot directly observe and measure. Here are a couple of examples; in later chapters, we will look further at how both rocks and minerals can be used to understand past environments.

• *Clues to temperature and pressure.* Our understanding of the environments where minerals grow has come largely through studying minerals in the laboratory. Through experiments, scientists have been able to determine the temperatures and pressures at which carbon will form a diamond or form graphite, its polymorph (Figure 3.31).

Because we can infer how temperature and pressure increase with depth in Earth, we can state with certainty that rocks in which diamonds are found were at one time subjected to pressures and temperatures equivalent to those in the mantle at least 150 km below Earth's surface.

• *Clues to climate. Regolith* is the blanket of loose rock particles that covers Earth. Some minerals form in regolith during the weathering process, so these are controlled by climate—cold/wet versus hot/dry, for example. Thus, we can decipher past climates from the kinds of minerals preserved in sedimentary rocks.

• *Clues to seawater composition.* We also can determine the composition of seawater in past ages from minerals

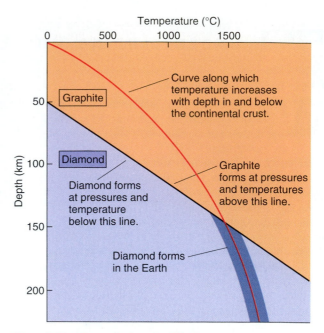

Figure 3.31   Diamonds from the Mantle   Line separating regions of temperature and pressure (here plotted as depth) in which the two polymorphs of carbon grow—graphite and diamond. At a pressure equal to that at 150 km depth, the line separating the graphite and diamond regions intersects the curve depicting the geothermal gradient in and below the continental crust. Thus, diamonds found at Earth's surface have formed at depths of 150 km or greater.

formed when the seawater evaporated and deposited its salts.

Before you go on:

1. What is the most common carbonate mineral?

2. Of what common mineral are our bones and teeth composed? To what mineral group does it belong?

3. What common mineral is used to make plaster?

4. What two mineral groups provide most of our metallic mineral resources?

5. How can minerals provide clues to past temperature, pressure, climate, and seawater compositions?

ROCKS: MIXTURES OF MINERALS

As we saw in Chapter 1, we group *rocks* into three large families, based on how they formed:

1. *Igneous rocks*, formed by solidification of magma.

2. *Sedimentary rocks*, formed by sedimentation of materials transported in solution or suspension.

3. *Metamorphic rocks*, formed by the alteration of preexisting sedimentary or igneous rocks in response to increased pressure and temperature.

Although we group rocks by their mode of formation, only rarely can we actually see a rock in the act of being formed. We may observe lavas solidifying to form igneous rocks, but we cannot see magma solidifying deep inside the Earth. Nor can we see metamorphism happening, nor sedimentary rock forming. In most cases, we have to use the evidence we can actually see and measure in order to deduce how a given rock is formed.

DISTINGUISHING THE THREE ROCK TYPES

Studying a large number of rock specimens soon makes it clear that, no matter what kind of rock is being examined—sedimentary, metamorphic, or igneous—the differences between samples can be described in terms of two features: texture and mineral assemblage.

TEXTURE

Texture is the overall appearance of a rock due to the size, shape, and arrangement of its constituent mineral grains. For example, the mineral grains may be flat and parallel to each other, giving the rock a texture like a deck of playing cards—layered or platy. In addition, the various minerals may be unevenly distributed and concentrated into specific layers. The rock texture is then both layered and platy.

Specific textural terms are used for each rock family, so we will introduce them in the chapters where the three rock families are discussed.

MINERAL ASSEMBLAGE

Mineral assemblage is the type and abundance of the minerals making up a rock. Although a few kinds of rock contain only one mineral, and natural glasses contain no minerals, most rocks contain two or more minerals. The mineral assemblage is an important indicator of how a rock formed.

A systematic description of a rock includes both texture and mineral assemblage. Two additional terms, megascopic and microscopic, are very useful. *Megascopic* refers to those textural features of rocks that we can see with the unaided eye or with the eye assisted by a simple lens that magnifies up to 10 times. *Microscopic* refers to those textural features of rocks that require high magnification to be viewed.

Examination of a microscopic texture commonly requires preparation of a special *thin section* of rock that must be viewed through a microscope. A thin section is prepared by first grinding a smooth, flat surface on a small piece of rock. The flat surface is glued to a glass slide, and then the rock is ground away until the fragment glued to

the glass is so thin that light passes through it easily. An example of a thin section is shown in Figure 3.32.

Before you go on:

1. What are the three rock families and how does each form?

2. What two features of rocks are used to describe them?

3. What is meant by the *texture* of a rock?

REVISITING PLATE TECTONICS AND THE EARTH SYSTEM

PLATE TECTONICS AND ORE MINERALS

Most ore minerals are rare. A few, such as magnetite, hematite, and rutile, can be found in small amounts in many igneous and metamorphic rocks, but ore minerals of metals such as lead, zinc, copper, gold, silver, and tin are rarely seen in common rocks. Such scarce chemical elements are certainly present in common rocks, but only in tiny amounts, and only by ionic substitution for more abundant elements such as iron, magnesium, and potassium, which are constituents of common minerals.

For an ore mineral to form, some process must first concentrate the scarce metal. The two most common processes of concentration are, first, by vapors released by a cooling body of magma. Such vapors can extract metals from the partly cooled magma. The second way is for a hot saline solution, such as heated seawater, to react with, and alter, a rock, and in the process extract the scarce metals. In both cases, the result is a solution that has concentrated scarce metals. As such a solution cools the metals are deposited in veins or other settings as ore minerals. In short, the concentration process forms an ore deposit.

How is plate tectonics involved in these complex processes? The answer becomes apparent when we look at the distribution of volcanoes around the world. Volcanoes are evidence that magma exists at depth, and magma can both release the vapors that carry scarce metals, or serve as the heat source that generates hot saline solutions. Volcanoes are concentrated around the edges of tectonic plates, and the magmas that cause volcanism are produced as a result of plate tectonic movements.

Without plate tectonics there would be fewer mineral deposits, and Earth would be a much less interesting place to prospect.

ROCK TYPES AND THE ROCK CYCLE

The concept of a rock cycle as an integral part of the Earth system was introduced in Chapter 1. Here we discuss how

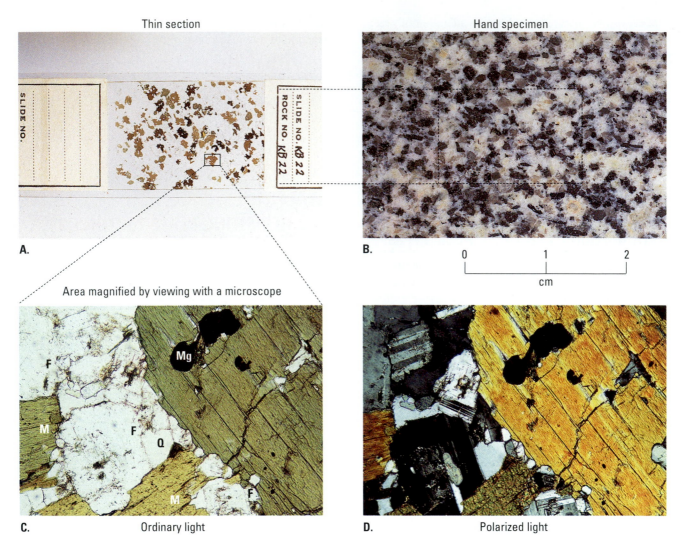

Thin section

Hand specimen

A.

B.

0   1   2
cm

Area magnified by viewing with a microscope

C.   Ordinary light

D.   Polarized light

Figure 3.32 Igneous Rock Seen Through a Microscope
Polished surfaces and thin slices reveal textures and mineral assemblages to great advantage. The specimen here is an igneous rock containing quartz (Q), feldspar (F), amphibole (A), mica (M), and magnetite (Mg). A. A thin slice mounted on glass. The slice is 0.03 mm thick, and light can pass through the minerals. B. A polished surface. The dashed rectangle indicates the area used to make the thin slice shown in part A. C. An area of the thin slice viewed under a microscope, magnified 253 times. D. The same view as in part C seen through polarizers, used to emphasize the shapes and orientations of individual grains.

the operation of the rock cycle influences the percentages of the different kinds of rocks we actually see in Earth's crust. The rock cycle illustrated in Figure 3.33 is the same as that shown in Figure 1.19.

The internal processes that form magma and that in turn lead to the formation of igneous rock interact with external processes through erosion. All internal processes are driven by Earth's internal heat energy. External processes such as weathering, erosion, and deposition are driven by the Sun's heat energy:

1. When rock erodes, the eroded particles form sediment.

2. The sediment may eventually become cemented, usu-ally by substances carried in water moving through the ground, and thereby converted into new sedimentary rock.

3. In places where such sedimentary rock forms, it can reach depths at which pressure and heat cause new com-pounds to form, so that the sedimentary rock becomes metamorphic rock.

4. Sometimes metamorphic rock settles so deep that the high temperatures melt it and magma is formed.

5. The new magma can then move upward through the crust, where it can cool and form another body of igneous rock.

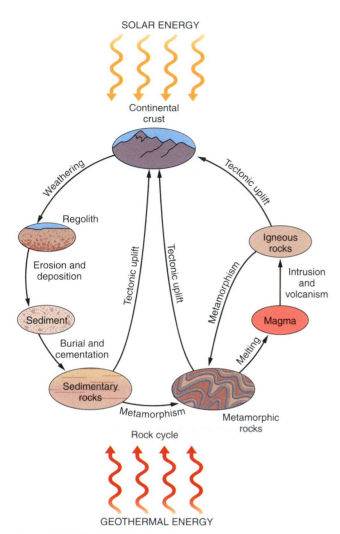

Figure 3.33 **The Rock Cycle** The rock cycle traces the processes whereby materials within and on top of Earth's crust are weathered, transported, deposited, metamorphosed, and even melted. Note that rocks don't necessarily follow the path that leads around the outside of the circle; they can follow any of the short circuits from one stage to another.

6. Eventually, the new body of igneous rock can be uncovered and subjected to erosion, the eroded particles start once more on their way to the sea, sediment is laid down, and the cycle repeats.

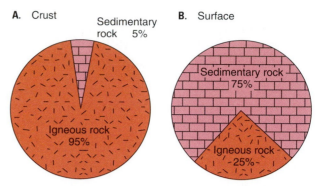

Figure 3.34 **Frequency of Rock Types** Relative amounts of sedimentary and igneous rock. (Metamorphic rocks are included in either sedimentary or igneous categories, depending on their origin.) A. The great bulk of the crust consists of igneous rock (95%), but sedimentary rock (5%) forms a thin covering at and near the surface. B. The extent of sedimentary rock cropping out at the surface is much larger than that of igneous rock, so 75 percent of all rock seen at the surface is sedimentary, and only 25 percent is igneous.

As James Hutton first recognized, the cycle has occurred again and again throughout the Earth's long history.

By studying deeply eroded parts of the crust, and from geophysical measurements, geologists have discovered that the crust is 95 percent igneous rock or metamorphic rock derived from igneous. However, as seen in Figure 3.34, most of the rock that we actually see at the Earth's surface is sedimentary. The difference arises because sediments are products of weathering, and as a result they are draped as a thin veneer over the largely igneous and metamorphic crust below. The distribution of rock types is an important consequence of the rock cycle.

What's Ahead?

Now that you have a background in minerals, you are ready to see how they assemble to form Earth's igneous, sedimentary, and metamorphic rocks. Because igneous rocks are the most abundant kind of rocks in both the continental and oceanic crust, they are the first rock family we will discuss in the next chapter.

CHAPTER SUMMARY

1. The nucleus of an atom contains protons (positive electrical charge) and neutrons (no electrical charge). Moving in orbitals around the nucleus are electrons (negative electrical charge). The positive electrical charge on a proton is equal but opposite to the negative charge on an electron. The number of protons in a nucleus is the atomic number of an element.

2. Orbital electrons have different energy levels, and the number of electrons that can occupy a specific energy-level shell is fixed: 2 in shell 1, 8 in shell 2, 18 in shell 3, 32 in shell 4.

3. The forces that hold atoms together in minerals are called bondings, and there are four different kinds of bondings. Ionic bondings arise when atoms transfer to, or accept orbital electrons from, other atoms. Covalent bondings involve the sharing of electrons between atoms. Metallic bondings involve the sharing of electrons in higher energy-level shells between several atoms. Van der Waals bondings are weak, secondary attractions that form as a result of the sharing or transfer of electrons.

4. Minerals are naturally formed inorganic solids that have a definite composition and a specific crystal structure. Minerals can be either chemical elements, such as sulfur and diamond, or compounds of two or more chemical elements.

5. The compositions of some minerals vary because of ionic substitution, whereby one ion in a crystal structure can be replaced by another ion having a like electrical charge and a like ionic radius.

6. Some compounds have the same composition but different crystal structures. Each structure is a separate mineral. Minerals that have the same compositions but different structures are called polymorphs.

7. The principal properties used to characterize and identify minerals are crystal form, growth habit, cleavage, luster, color and streak, hardness, and specific gravity.

8. Approximately 4000 minerals are known, but of these only about 30 are commonly encountered. The common 30 make up more than 95 percent of the Earth's crust.

9. Silicates are the most abundant minerals, followed by oxides, carbonates, sulfides, sulfates, and phosphates.

10. The basic building block of silicate minerals is the silicate tetrahedron, a complex anion in which a silicon atom is covalently bonded to four oxygen atoms. The four oxygen ions sit at the apexes of a tetrahedron, with the silicon at its center. Adjacent silicate tetrahedra can bond together to form larger complex anions by sharing one or more oxygens. The process is called polymerization.

11. The feldspars are the most abundant group of minerals in the continental crust, comprising approximately 60 percent of the volume. Quartz is the second most common mineral in the continental crust.

12. Rocks are grouped into three families according to the way they are formed. We rarely observe rocks being formed, however, so rocks are usually described in terms of their textures and mineral assemblages.

13. Igneous rocks, or metamorphic rocks derived from igneous rocks, account for 95 percent of all rocks in the Earth's crust; sedimentary rocks account for 5 percent.

THE LANGUAGE OF GEOLOGY

anion (p. 71)
atom (p. 69)
atomic number (p. 70)

bonding (p. 72)

cation (p. 71)
cleavage (p. 79)
complex ion (p. 73)
composition (p. 69)
covalent bonding (p. 72)
crystal (p. 77)
crystal faces (p. 77)
crystal form (p. 77)
crystalline (p. 74)
crystal structure (p. 69)

energy-level shells (p. 70)

hardness (p. 81)

ion (p. 71)
ionic bonding (p. 72)
ionic radius (p. 74)
ionic substitution (p. 76)
isotope (p. 70)

luster (p. 80)

mass number (p. 70)
metallic bonding (p. 72)
mineral (p. 68)
mineral assemblage (p. 93)

mineraloid (p. 69)
molecule (p. 71)

polymerization (p. 85)
polymorphs (p. 79)

silicate (p. 84)
silicate anion (p. 84)
silicate mineral (p. 84)
specific gravity (p. 82)

texture (of a rock) (p. 93)

van der Waals bonding (p. 73)

MINERAL NAMES TO REMEMBER

agate (p. 88)
amphibole group (p. 86)
anhydrite (p. 90)
apatite (p. 89)
aragonite (p. 89)
augite (p. 86)

calcite (p. 89)
cassiterite (p. 92)
chalcedony (p. 88)

chalcopyrite (p. 90)
chlorite group (p. 86)
chrysotile (p. 86)
clay group (p. 86)
corundum (p. 80)

diamond (p. 72)
dolomite (p. 89)

feldspar group (p. 88)

flint (p. 88)
fluorite (p. 79)

galena (p. 90)
garnet group (p. 85)
graphite (p. 73)
gypsum (p. 90)

halite (p. 79)
hematite (p. 81)

hornblende (p. 86)

ice (p. 69)

jasper (p. 88)

kaolinite (p. 86)

magnetite (p. 92)
muscovite (p. 86)

olivine group (p. 85)

potassium feldspar (p. 80)
pyrite (p. 78)
pyroxene group (p. 86)
pyrrhotite (p. 90)

quartz (p. 88)

ruby (p. 80)
rutile (p. 92)

sapphire (p. 80)
serpentine group (p. 86)
sphalerite (p. 90)

uraninite (p. 92)

QUESTIONS FOR REVIEW

1. Oxygen has an atomic number of 8, and there are three isotopes of oxygen—mass numbers 16, 17, and 18. For each isotope list the number of protons and neutrons in the nucleus.

2. How many electrons occupy energy-level shell 2 of an oxygen atom?

3. What is the difference between an atom and an ion? between an atom and a molecule?

4. Sketch the structure of a silicon atom and show how the silicate anion is formed by electron sharing with oxygen atoms.

5. What is a mineral? Give three reasons why the study of minerals is important.

6. Describe three ways chemical elements bond together to form compounds. Name two minerals that are examples of each bond type.

7. What properties besides composition and crystal structure can be used to identify minerals?

8. What are polymorphs? What are the mineral names of the two polymorphs of $CaCO_3$? of C?

9. What mineral has a crystal structure in which Pb and S atoms alternate at the corners of a three-dimensional, right-angled geometric grid?

10. What is ionic substitution? Illustrate your answer with two examples.

11. Approximately how many common minerals are there? What is the most abundant one?

12. Name five common minerals that are found in the area in which you live.

13. Describe how silicate anions join together to form silicate minerals.

14. Describe the polymerization of silicate anions in the pyroxenes, the micas, and the feldspars.

15. Name five common minerals that are not silicates, and name the anion each contains.

16. Are any minerals mined in the area in which you live? What are they and what are they mined for?

17. Why is it that igneous rocks are so much more abundant than sedimentary rocks, yet sedimentary rocks cover most of the continental crust and are therefore the most visible?

Click on *Presentation* and *Interactivity* in the **Atoms, Elements, Minerals, and Rocks** module of your CD-ROM to further explore resources and activities presenting concepts from this chapter. Select *Assessment* in the same module to test your understanding of this chapter.

Click on *Petroscope* for **Minerals** to learn more about the characteristics of minerals from this chapter as well as to learn how to identify unknown mineral samples.

Chapter 4

Volcanic Fury. Kilauea East, Hawaii, during an eruption in 1984. Trees in the foreground have been stripped of foliage by hot, volcanic ash falling on them.

Igneous Rocks: Products of Earth's Internal Fire

Drilling down to molten lava, 1963.

Watching Magma Solidify

Magma is dangerous stuff, but an eruption of Kilauea volcano in Hawaii, in 1963, provided an unusual chance to study the way it solidifies as it cools. As magma poured down the flank of Kilauea it filled a small crater, remnant of an ancient eruption. The result was a lake of magma 1000 by 800 feet and 50 feet deep. The temperature of the magma was 1140°C. A solid crust soon covered the lake and even though it was less than a foot thick, scientists could walk safely on the surface within two days. After 5 days a drill rig was erected in the middle of the lake by scientists of the Hawaiian Volcano Observatory, a hole was drilled through the crust to the magma below, and we started to study the solidification process. Drilling continued as the crust thickened at an average rate of 3 feet per month. I was not part of the drilling team—my job was to study samples collected at the base of the crust.

The interface between the magma and the solid crust was actually a zone a foot or so wide. The point at which there was 50 percent crystals and 50 percent liquid marked the point where liquid properties gave way to solid properties—the temperature at which this happened was 1067°C. A discovery in the material I was studying was quite unexpected. Magma samples collected at the solid–liquid interface contained tiny droplets of molten iron sulfide with traces of copper and nickel. Geologists had speculated for years that certain kinds of ore deposits might form when magma splits into two immiscible liquids, one a silicate liquid, the other a sulfide liquid. Here in Hawaii was evidence that the process really happened as hypothesized.

No one expects there to be an ore deposit at the bottom of the Hawaiian magma lake. The lessons we learned from watching magma solidify are lessons of process. Hawaii served as a natural laboratory that yielded results applicable around the world.

Brian J. Skinner

KEY QUESTIONS

1. **What distinguishes an igneous rock from other rocks?**
2. **How varied are igneous rocks?**
3. **What are pyroclastic rocks?**
4. **What are plutons and how varied are they?**
5. **Are all magmas the same?**
6. **What happens when magma cools and becomes an igneous rock?**

INTRODUCTION: WHAT IS AN IGNEOUS ROCK?

In this chapter we will look at the most common family of rocks, the igneous rocks. All igneous rocks are formed through the cooling and solidification of magma. Except for a thin covering of sediment, the oceanic crust is made entirely of igneous rocks, and those igneous rocks are created at plate-spreading centers. In the continental crust, as we discussed in Chapter 3, igneous rocks, or metamorphic rocks derived from igneous parents, make up an estimated 95 percent of the volume of the crust. Clearly we live off a planet that has been largely shaped by igneous activity.

Igneous rocks vary greatly; some contain large mineral grains, others contain grains so small they can barely be seen under a high-power microscope. Igneous rocks also vary greatly in color, and even a cursory examination reveals that color differences reflect differing mineral assemblages (Figure 4.1). Regardless of grain size or mineral composition, all igneous rocks have distinctive tex-

A.

B.

Figure 4.1 A Natural Jigsaw Puzzle A. Sample of granite from Vermont shows the way mineral grains interlock and produce an igneous texture. The specimen is 10 cm wide. B. Photograph of a granite, taken in polarized light through a microscope. Note how the mineral grains interlock to produce a tough structure. Minerals visible are muscovite (M, green), amphibole (A, yellow), quartz (Q, colorless), and feldspar (F, gray and pale blue). The specimen is 1 cm across.

tures that reflect their origin by crystallization of magma. That distinctive texture is a random interlocking of mineral grains. The specimens in Figure 4.1 both display the mosaic pattern of mineral grains that is typical of igneous rocks.

IGNEOUS ROCKS—INTRUSIVE AND EXTRUSIVE

All igneous rocks form through the cooling and solidification of magma. But they can form either underground or on the surface, and these different environments create two broad categories of igneous rocks:

1. Intrusive igneous rocks form when magma cools within existing rocks in Earth's crust, so they are "intruders" there.

2. Extrusive igneous rocks form when magma cools on Earth's surface, so they are "extruded" there.

Both types have similar chemistry, but their environment of formation controls their texture and their mineral assemblage—which are the bases for classifying and naming both intrusive and extrusive igneous rocks.

TEXTURE IN IGNEOUS ROCKS

The two most obvious textural features of an igneous rock are the *size* of its mineral grains and how the mineral grains are *packed* together.

SIZES OF MINERAL GRAINS

The key difference between intrusive and extrusive igneous rocks is that *intrusive rocks are coarse-grained* (because magma that solidifies in the crust cools slowly and has sufficient time to form large mineral grains), whereas *extrusive rocks are fine-grained* (because magma that solidifies on the surface usually cools rapidly, allowing insufficient time for large crystals to grow).

Coarse-grained igneous rock is called a **phanerite** (from a Greek word meaning visible). In other words, mineral grains are large enough to see without a magnifier or microscope. Practically, this means that the mineral grains are at least 2 mm in their largest dimension. Figure 4.2A shows an example of a coarse-grained igneous rock.

Igneous rock that contains unusually large mineral grains (2 cm or larger) is called a *pegmatite* (Figure 4.3). Individual mineral grains in pegmatites can be huge; single grains of muscovite and feldspar several meters across have been reported.

Fine-grained igneous rock is called an **aphanite** (from a Greek word meaning invisible). In other words, individual mineral grains can be seen clearly only with magnification (Figure 4.2B). This means that the mineral grains are less than 2 mm in their largest dimension.

A.

B.

Figure 4.2   Phanerite versus Aphanite   Two rocks with the same composition and containing the same minerals. A. Phanerite; a granite. B. Aphanite; a rhyolite. Both specimens are 10 cm wide.

A **porphyry** is an igneous rock in which 50% or more of the rock is coarse mineral grains scattered through a mixture of fine mineral grains, as shown in Figure 4.4. The isolated large grains are **phenocrysts**, and they form in the same way mineral grains in coarse-grained igneous rocks do—by slow cooling of magma in the crust. The fine-grained groundmass that encloses phenocrysts provides evidence that a partly solidified magma moved quickly upward. In its new setting, the magma cooled rapidly, so the later mineral grains forming the groundmass are tiny.

Figure 4.3   Very Large Grains   Vein of pegmatite cutting through a finer-grained phanerite.

Figure 4.4 Porphyritic Texture Porphyritic texture is characterized by a bimodal distribution in the sizes of the mineral grains. This photo of a porphyry from Oslo, Norway, shows large phenocrysts of feldspar in a fine-grained matrix. The rock is a porphyritic andesite, and the field of view is about 8 cm across.

GLASSY ROCKS

Lava can cool and solidify so rapidly that its atoms lack time to organize themselves into minerals. Glass, a mineraloid, forms instead. Extrusive igneous rocks that are largely or wholly glassy are called *obsidian*. When broken, such rocks display a distinctive conchoidal fracture: smooth, curved surfaces resembling a conch shell (Figure 4.5).

Another common variety of glassy igneous rock is pumice, a mass of glassy bubbles of volcanic origin. Volcanic ash is also mostly glassy because the fragments of magma that became ash particles cooled too quickly to crystallize following eruption.

HOW MINERAL GRAINS PACK IN IGNEOUS ROCK

Two centuries ago, while making field observations in Scotland, James Hutton noticed that some layered sedimentary rocks were intruded and cut across by coarse-

Figure 4.5 Glassy Rock Obsidian from the Jemez Mountains, New Mexico, is almost entirely glassy. The composition is rhyolitic. The curved ridges are typical of fracture patterns observed in glass that is broken by a sharp blow. The specimen is 17 cm across.

grained rocks of distinctive texture. The texture he observed was an interlocking of randomly oriented mineral grains, like tightly fitting jigsaw-puzzle pieces.

Hutton recognized that the interlocking texture was similar to the texture produced by slow crystallization of molten substances in the laboratory. So, he concluded—correctly—that the cross-cutting coarse-grained rocks must once have been molten and that the distinctive interlocking of their numerous grains resulted from slow solidification of magma.

Almost all igneous rocks, whether phaneritic or aphanitic, have the characteristic texture of interlocking mineral grains observed by Hutton (Figure 4.1).

Before you go on:

1. How do igneous rocks form?

2. What are intrusive igneous rocks?

3. What are the two main textural features that characterize igneous rocks?

4. How does an aphanite differ from a phanerite?

5. What is a pegmatite? a porphyry?

MINERAL ASSEMBLAGE IN IGNEOUS ROCKS

For magma of a given composition, the mineral assemblage that forms is the same for intrusive and extrusive rocks. The difference between intrusive and extrusive is purely textural. Thus, once you determine the texture of an igneous rock, its name will depend on its mineral assemblage. To name igneous rocks, geologists use the general model shown in Figure 4.6.

All common igneous rocks consist largely of the minerals or mineral groups named in the figure—*quartz, feldspar* (both potassium feldspar and plagioclase), *mica* (both muscovite and biotite), *amphibole, pyroxene*, and *olivine*.

The vertical axis in the figure indicates the percentage of each mineral. Once you estimate the percentage of each mineral in an igneous rock, you can draw a vertical line on the figure along those percentages. The vertical line will intersect two rock names in the top two rows. Then select fine-grained (aphanitic) or coarse-grained (phaneritic) texture, as appropriate.

For example, suppose you have a rock specimen that is 30 percent olivine, 30 percent pyroxene, and 40 percent plagioclase. To name this specimen, you estimate the point on the figure where the curve separating olivine from pyroxene cuts a horizontal line drawn through 30 percent. Draw a vertical line through the point. Your vertical line cuts the curve between pyroxene and plagioclase at a point close to 60 percent. Olivine (30 percent) plus pyroxene (30 percent) equals 60 percent, so plagioclase accounts for the remaining 40 percent. If your rock specimen is coarse grained, it is called gabbro; if fine grained, basalt. (Do not expect a perfect match—variation in mineral assemblage is too great to represent on this simple two-dimensional diagram.)

When a rock has a porphyritic texture, we use the name determined by the mineral assemblage as the noun and the term for the groundmass texture as the adjective. For example, if the groundmass is fine grained, we will call it a porphyritic rhyolite. If the groundmass is coarse grained, we will call it a porphyritic granite or a porphyritic granodiorite.

The overall lightness or darkness of a rock is valuable information. Geologists express this as "light-colored" or "dark-colored." Quartz, feldspar, and muscovite are light-colored minerals; biotite, amphibole, and pyroxene are dark-colored minerals. Olivine is not as dark as the others.

Note that rocks to the left in Figure 4.6 are light-colored and those to the right are increasingly dark-colored. The boundary line between diorite and gabbro is drawn where the dark-colored minerals reach 50 percent of the total and exceed the light-colored feldspars. The boundary is carried through between the fine-grained equivalents of diorite and gabbro—andesite and basalt, respectively—on the basis of color.

INTRUSIVE (COARSE-GRAINED) IGNEOUS ROCKS

GRANITE AND GRANODIORITE

Feldspar and quartz are the chief minerals in granite and granodiorite (Figures 4.6 and 4.7). A mica, either muscovite or biotite, is usually present, and many granites contain scattered grains of hornblende (an amphibole).

The name *granite* is applied only to quartz-bearing rocks in which potassium feldspar is at least 65 percent by volume of the total feldspar present. The name *granodiorite* applies to similar rocks in which plagioclase is 65 percent or more of the total feldspar present. Without special equipment the differences in feldspars are not always easily recognized, and in a general study the term *granitic* is extended to this whole group of rocks.

Granitic rocks are only found in the continental crust. This is so because granitic magma forms when continental crust is heated to its melting temperature. The most common place where such high temperatures are reached is in the deeper portions of mountain belts formed by the collision of two masses of continental crust.

DIORITE

The chief mineral in *diorite* is plagioclase. Quartz and mica are usually absent, but either or both amphibole and pyroxene are invariably present (Figures 4.6 and 4.7).

Diorite is a common igneous rock but not as common as granite and granodiorite. Diorite is another granitic rock, and forms in the same way as granite and granodiorite. It is found only in continental crust.

GABBRO AND PERIDOTITE

Dark-colored diorite grades into *gabbro* (Figures 4.6 and 4.7), in which the dark-colored minerals pyroxene and olivine exceed 50 percent of the volume of the rock. A coarse-grained igneous rock in which olivine is the most abundant mineral is called a *peridotite*. Gabbros and peridotites can be found in both the oceanic and the continental crust. This is so because the magma from which such rocks form is generated in the mantle, and can rise to intrude either kind of overlying crust.

EXTRUSIVE (FINE-GRAINED) IGNEOUS ROCKS

RHYOLITE AND DACITE

An aphanitic igneous rock with the composition of a granite is a *rhyolite* (Figures 4.6 and 4.7); if the composition is that of a granodiorite, the aphanitic rock is a *dacite*. Both rhyolites and dacites are quartz-bearing; the difference between them, as with granites and granodiorites, is in the predominance of potassium feldspar in rhyolites and plagioclase in dacite.

Because it is commonly impossible to identify fine-grained feldspars in aphanitic igneous rocks without special microscopes, dacites are very difficult to distinguish from rhyolites. Many geologists, when in doubt, simply call both rocks rhyolites, or even *rhyodacites*.

Luckily, many rhyolites and dacites have porphyritic textures in which phenocrysts of quartz, and sometimes of feldspar, are set in an aphanitic groundmass. When the phenocrysts are about 50 percent or more of the volume of the rock, the name porphyritic rhyolite or porphyritic dacite should be used, as appropriate.

The colors of rhyolites and dacites are always pale—ranging from nearly white to shades of gray, yellow, red, or purple. Many obsidians have the same compositions as rhyolites or dacites. Such obsidians may appear dark, even black, and seem to contradict the rule that igneous rocks with a high silica content are light-colored. But rhyolitic obsidian chipped to a thin edge appears transparent with very little color, like slightly smoky glass. The dark appearance results from a small amount of dark mineral matter distributed evenly in the glass.

ANDESITE

An igneous rock similar in appearance to a dacite, but lacking quartz, is an *andesite* (Figures 4.6 and 4.7). Named for the Andes, the major mountain system of western South America, andesite is equivalent in composition to a diorite and is a common rock. Common colors are shades of gray,

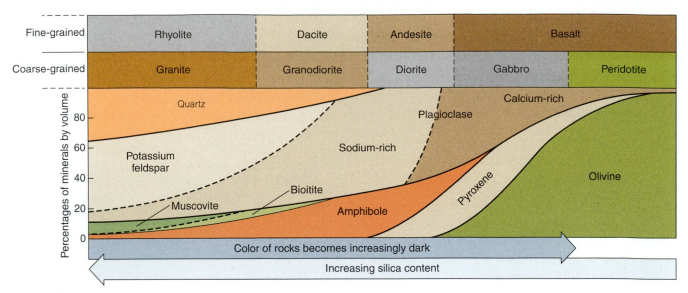

Figure 4.6 Naming Igneous Rocks The proportions of minerals in the common igneous rocks. Boundaries between rock types are not abrupt but gradational, as suggested by the broken lines between rock names. To determine the composition range for any rock type, project the broken lines vertically downward; then estimate the percentages of the minerals by means of the numbers at the edge of the diagram. Quartz- and feldspar-rich rocks are said to be felsic and are light in color (see Figure 4.7). Rocks rich in pyroxene, amphibole, and olivine are called mafic igneous rocks and are dark in color.

purple, and even dark green. Many andesites are porphyritic, with phenocrysts of amphibole, pyroxene, or plagioclase, but not quartz. When phenocrysts are abundant, the name andesite porphyry should be used.

BASALT

The dominant rock of the oceanic crust is *basalt*, a fine-grained igneous rock, sometimes porphyritic, that is always dark gray or black (Figures 4.6 and 4.7). Compositionally equivalent to gabbro, basalt is the most common kind of extrusive igneous rock. Phenocrysts, when present in a basalt porphyry, can be either plagioclase, pyroxene, or olivine.

> **Before you go on:**
> 1. What are the six mineral families that characterize all common igneous rocks?
> 2. How does granite differ from granodiorite? from rhyolite?

PYROCLASTS, TEPHRA, AND TUFFS

A fragment of rock ejected during a volcanic eruption is called a **pyroclast** (from the Greek words *pyro*, heat or fire, and *klastos*, broken; hence: hot, broken fragments). Rocks formed from pyroclasts are **pyroclastic rocks**. Geologists also commonly refer to a deposit of pyroclasts as **tephra**, a Greek term for ash.

Tephra is a collective term for all airborne pyroclasts, including fragments of newly solidified magma as well as fragments of older broken rock. It includes individual pyroclasts that fall directly to the ground during an eruption and those that move over the ground as part of a hot, moving flow. The terms used to describe tephra particles of different sizes—bombs, lapilli, and ash—are listed in Table 4.1 and illustrated in Figure 4.8.

There is a saying among geologists that "tephra is igneous when it goes up but sedimentary when it comes down." As a result, pyroclastic rocks are transitional between igneous and sedimentary rocks, and the names of pyroclastic rocks reflect this fact.

As you will see in Chapter 7, the names of one class of sedimentary rocks are determined by the sizes of the sediment fragments. So, too, with pyroclastic rocks: they are called *agglomerates* when tephra particles are bomb-sized,

TABLE 4.1 Names for Tephra and Pyroclastic Rock

Average Particle Diameter (mm)	Tephra (unconsolidated) material)	Pyroclastic Rock (consolidated material)
>64	Bombs	Agglomerate
2–64	Lapilli	Lapilli tuff
<2	Ash	Ash tuff

Silica Content of Magma	Approx. Melting Temperature of Rock	Viscosity of Magma	Color of Mineral Assemblage	Hand Samples of	
				Resulting Plutonic Rocks	Resulting Volcanic Rocks
High (≈70–75%)	Low (≈800°C)	High (not runny)	Felsic (light-colored)	Granite	Rhyolite
Intermediate (≈60%)	Medium (≈1000°C)	Medium	Intermediate	Diorite	Andesite
Low (≈45–50%)	High (≈1200°C)	Low (runny)	Mafic (dark-colored)	Gabbro	Basalt

Figure 4.7 Common Igneous Rocks Comparison of coarse-grained (phaneritic) and fine-grained (aphanatic) igneous rocks. Note that mafic igneous rocks are darker in color than felsic igneous rocks. Each specimen is about 7 cm in length.

C.

Figure 4.8 Tephra of Various Sizes   A. Large spindle-shaped bombs up to 50 cm in length cover the surface adjacent to a small volcanic cone on Haleakala volcano, Maui. B. Intermediate-sized tephra called lapilli cover the Kau Desert, Hawaii. The coin is 18 mm in diameter. C. Volcanic ash, the smallest-sized tephra, covers leaves in a garden in Anchorage, Alaska, following an eruption of Mount Spurr.

A.

B.

Figure 4.9 Pyroclastic Igneous rocks   Two kinds of tuff. A. Lapilli tuff formed by cementation of lapilli and ash, Isabela Island, Galapagos, Ecuador. The largest lapilli are about 1.5 cm in diameter. B. Welded tuff, Shoshone, California. Glassy shards of volcanic ash were so hot the shards welded to make solid rock. The bright band in the center is almost entirely glassy.

or *tuffs* when the particles are either lapilli or ash (Figure 4.9). As seen in Table 4.1, the appropriate rock names are lapilli tuff and ash tuff.

Tephra can be converted to pyroclastic rock in two ways. The most common is through the addition of a cementing agent, such as quartz or calcite, introduced by groundwater. Figure 4.9A is an example of a rhyolitic lapilli tuff formed by cementation. The second way tephra is transformed to pyroclastic rock is through welding of hot, glassy, ash particles. When ash is very hot and glassy, the individual particles can fuse together to form a glassy pyroclastic rock. Such a rock is called *welded tuff* (Figure 4.9B).

PLUTONS

Beneath every volcano lies a complex of chambers and channelways through which magma reaches the surface. We cannot study the magmatic channels of an active volcano, but we can look at ancient channelways that have been laid bare by erosion, as seen in Figure 4.10. What we

A.

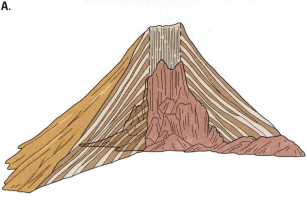

B.

Figure 4.10   Volcanic Neck, Shiprock, New Mexico   A. The conical tephra cone that once surrounded this volcanic neck has been removed by erosion. B. Diagram of the way the original volcano may have appeared prior to erosion.

find is that these ancient channelways are filled by intrusive igneous rock because they are the underground sites where magma solidified.

All bodies of intrusive igneous rock, regardless of shape or size, are called **plutons** after Pluto, the Greek god of the underworld. The magma that forms a pluton did not originate where we now find the pluton. Rather, the magma was intruded upward into the surrounding rock from the place where it formed.

Plutons are given special names depending on their shapes and sizes. The common plutons are illustrated in Figure 4.11, and each is briefly discussed below.

MINOR PLUTONS

DIKES, SILLS, AND LACCOLITHS

The most obvious and familiar evidence of past igneous activity is a **dike**, a tabular sheet-like (thin but laterally extensive) body of igneous rock that cuts across the layering or fabric of the rock it intrudes (Figure 4.12A). A dike forms when magma squeezes into a fracture. If you look ahead to Chapter 5 you can see in Figure 5.21 the eruption of lava from a fracture. The fracture is probably now filled by a dike of gabbro, formed as a result of the cooling of the basaltic magma that filled the opening at the end of the eruption.

Sills are tabular and sheet-like, the same as dikes, except they run parallel to the layering or fabric of the rocks they intrude, like a windowsill (Figure 4.12B).

Commonly, dikes and sills occur together as part of a network of plutons, as shown in Figure 4.11. Both dikes

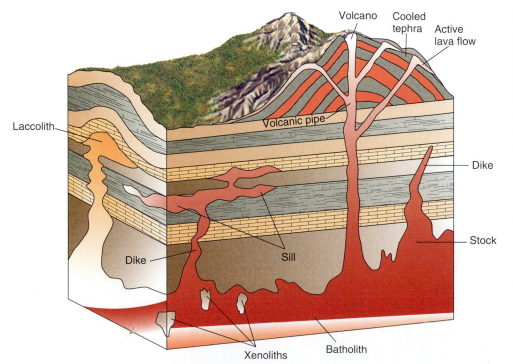

Figure 4.11   Plutons Diagrammatic section of part of the crust showing the various forms assumed by plutons. Many plutons were once connected with volcanoes, and a close relationship exists between intrusive and extrusive igneous rocks.

A.

B.

Figure 4.12   Dike and Sill: Sheet-Like Plutons   A. Three dikes of gabbro cut across the layering of sedimentary rocks, near Saglek, Torngat Mountains, Labrador, Canada. B. Sill of gabbro (black band) intruded parallel to the layering of sedimentary rocks, Glacier National Park, Montana.

and sills can be very large. For example, the Great Dike in Zimbabwe is a mass of gabbro nearly 500 km long and about 8 km wide, with parallel, vertical walls. The Great Dike fills what must once have been a huge fracture in the crust.

A large, well-known sill is the Palisades Sill. Cliffs of the sill line the Hudson River opposite New York City (Figure 4.13). The Palisades Sill reaches a thickness of about 300 m, and like the Great Dike is gabbro. The sill intruded between layers of ancient sedimentary rock about 200 million years ago. It is visible today because tectonic forces raised that portion of the crust, and erosion then removed the overlying sedimentary rocks.

A variation of a sill is a *laccolith*, an igneous body intruded parallel to the layering of the rocks it intrudes and above which the layers of the intruded rocks have been bent upward to form a dome (Figure 4.11).

VOLCANIC PIPES AND NECKS

A *volcanic pipe* is the roughly cylindrical conduit that once fed magma upward to a volcanic vent and later filled with igneous rock or fragments after eruption ended. If erosion strips away the surrounding rock, the body of rock that filled the pipe is called a *volcanic neck*. Figure 4.10 shows a famous example of a volcanic neck, together with associated dikes (the low ridge on the right), at Shiprock, New Mexico.

MAJOR PLUTONS

BATHOLITHS AND STOCKS

A **batholith** is the largest kind of pluton. It is an intrusive igneous body of irregular shape that cuts across the layering or other fabric of the rock it intrudes. Most batholiths

Figure 4.13   View from New York City   The Palisades Sill, New Jersey, forms the western bank of the Hudson River opposite New York City. The sill is a gabbro intrusion that was emplaced approximately 200 million years ago coincident with the breakup of Pangaea and the opening of the Atlantic Ocean.

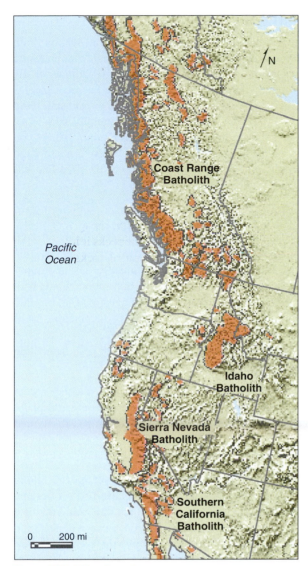

Figure 4.15   Xenoliths   Xenoliths (X) dislodged during intrusion of the East Range Granite. The outcrop is along Yakutat Bay, Tongass National Forest, Alaska.

Figure 4.14   Giant Intrusions   The Idaho, Sierra Nevada, and Southern California batholiths, largest in the conterminous United States, are dwarfed by the Coast Range Batholith in southern Alaska and British Columbia. Each of these giant batholiths formed from magma generated by the partial melting of continental crust, and each intrudes metamorphosed rocks.

(from *bathus*, Greek for deep, and *lithos*, rock; hence, deep rocks) are composite masses that comprise separate intrusive bodies of slightly differing composition. The differences reflect variations in composition of the magma from which the batholith formed. Some batholiths exceed 1000 km in length and 250 km in width; the largest in North America is the Coast Range Batholith of British Columbia and southern Alaska, about 1500 km long (Figure 4.14).

Figure 4.11 does not show how deep a batholith extends. Where it is possible to see them, the walls of batholiths are nearly vertical, leading to an old belief that they extend to great depths. But geophysical measurements and other studies now suggest that batholiths are at

most only 10 to 20 km thick, thin compared to their great lateral extents.

The magma from which a batholith forms intrudes upward from its source deep in the continental crust. Just how such a huge volume of magma rises and is emplaced remains a matter of controversy. In part the magma must rise through fractures, but other mechanisms are also thought to play a role. One important mechanism is known as *stoping*. The rising magma apparently dislodges fragments of the overlying rock, and the dislodged blocks, being cooler and more dense than the magma, sink. As they sink, those fragments may react with and be partly dissolved by the magma.

Some stoped fragments may dissolve completely, but more commonly such fragments retain their identity despite partial reaction with the magma. Some may sink to the floor of the magma chamber, but others sink so slowly through the sticky magma that they become suspended in the solidifying rock of the batholith. Any rock fragment still enclosed in a magmatic body when it solidifies is a xenolith (Figure 4.15) (from Greek *xenos*, meaning stranger, and *lithos*, stone).

Like batholiths, **stocks** are irregularly shaped intrusives no larger than 10 km in maximum dimension. As is apparent from Figure 4.11, a stock may merely be a companion body to a batholith or even the top of a partly eroded batholith.

Before you go on:

1. What is a pyroclast? tephra?

2. What is the difference between a tuff and an agglomerate?

3. What is a pluton?

4. How does a sill differ from a dike?

5. How are batholiths intruded?

WHENCE MAGMA?

How and where do magmas form? By studying magma erupting from volcanoes geologists have discovered that there are only three major kinds of *magma*—basaltic, andesitic, and rhyolitic. Many clues about how magma forms can be deduced from the geographic distribution of each volcano type. Here is a summary of current thinking (refer to Figure 4.16).

DISTRIBUTION OF VOLCANOES

RHYOLITIC MAGMA

Volcanoes that erupt rhyolitic magma are abundant on the continental crust. However, in oceanic crust, there are only a few places where volcanoes erupt some rhyolitic magma, and these are primarily andesitic eruptions that are building new continental-like crust.

These observations suggest that (1) the processes that form rhyolitic magma must be restricted to continental-type crust (including those places in the ocean where new crust of continental character is forming), and (2) the processes that form rhyolitic magma do not occur in oceanic crust.

Nor, presumably, do these processes occur in the mantle—if they did, the magma would rise to the surface, regardless of the kind of crust above, and therefore be just as common in the ocean as on land.

ANDESITIC MAGMA

Volcanoes that erupt andesitic magma occur on both oceanic and continental crust. This suggests that andesitic magma forms in the mantle and rises, regardless of the overlying crust. Further, a line around the Pacific separates andesitic volcanoes from those that erupt only basaltic lava (Figure 4.17). This *Andesite Line* is generally parallel to the plate subduction margins shown in Figure 2.10. As you shall see shortly, andesitic magma probably forms through the heating and dehydration of old oceanic crust that has been subducted back into the mantle.

BASALTIC MAGMA

Volcanoes that erupt basaltic magma also occur on both oceanic and continental crust. The source of basaltic magma, therefore, must also be the mantle. The geographic distribution of basaltic volcanoes does not coincide with a specific kind of crustal feature, suggesting that basaltic magma must form through melting of the mantle itself, regardless of the kind of overlying crust.

However, two other observations suggest something about the origin of basaltic magma:

1. Everywhere along the midocean ridges, volcanoes erupt basaltic magma. Midocean ridges coincide with plate-spreading centers, so we must consider the possibility that plate motion and the generation of spreading-center magma could be connected.

2. Some large basaltic volcanoes are not located along midocean ridges. The volcanoes of Hawaii sit on oceanic crust in the middle of the Pacific Plate, far from any plate edges. Those active today—Mauna Loa, Kilauea, and Loihi (a submarine volcano)—are just the youngest members of a long chain of mostly extinct volcanoes. The vol-

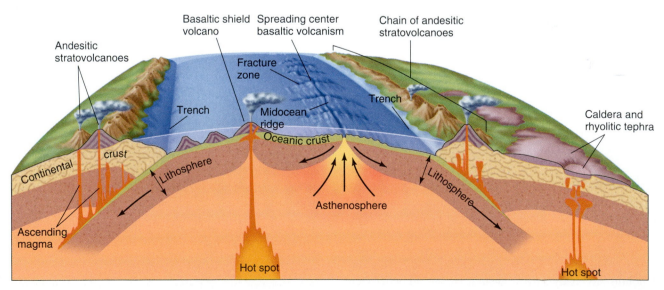

Figure 4.16   Locations of Volcanoes   Diagram illustrating the locations of the major kinds of volcanoes in a plate-tectonic setting.

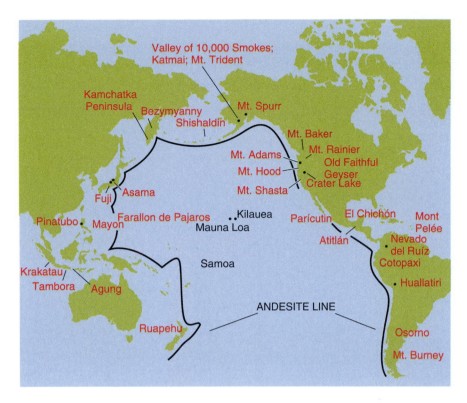

Figure 4.17   Where Andesites Are Found
The Andesite Line surrounds the Pacific Ocean basin and separates areas where andesitic magma is not found from areas where it is common. Volcanoes that are inside the Pacific basin, such as Mauna Loa, erupt basaltic magma but not andesitic magma. Those outside the line, such as Mount Shasta and Mount Fuji, may erupt basaltic magma too, but they also erupt andesitic magma.

canic rocks in the Hawaiian chain are progressively older to the northwest.

The Hawaiian volcanic chain is believed to have formed over the past 70 million years as the Pacific Plate moved slowly northwestward across a midplate hot spot above which frequent and voluminous eruptions built a succession of volcanoes. This implies that deep in the mantle there is a long-lived source of hot rock from which basaltic magma can form. The causes of hot-spot magmas are conjectural, but both spreading-center magmas and hot-spot magmas are believed to play important roles in plate tectonics and are caused by convection in the mantle.

ORIGIN OF BASALTIC MAGMA

When discussing the origin of magma, geologists first ask, "Was the rock that melted to form the magma wet or dry?" This is critical, because the presence of water lowers the temperature at which melting begins.

Then they ask, "What kind of rock melted?" The kind of rock that melts must obviously control the composition of the magma that forms.

Then they ask, "Did the rock melt completely or only partially?" Partial versus complete melting is not such an obvious question. A mineral melts at a specific temperature, but an assemblage of minerals (that is, a rock) melts over a temperature range of up to 200°C, and within that melting range, a mixture of magma and unmelted mineral grains coexist. The process of forming magma through the incomplete melting of rock is known as **chemical differ-**

entiation by partial melting. (For a more detailed discussion of melting, see *The Science of Geology*, Box 4.1, *How Rock Melts*.)

The dominant minerals in basalt and gabbro are olivine, pyroxene, and plagioclase. None of these contain water in their formulas. This fact suggests that basaltic magma is probably either a dry or a water-poor magma. Indeed, all evidence from observations of basaltic lava during eruption suggests that the water content of basaltic magma rarely exceeds 0.2 percent. Since there is not enough water to have much of an effect on the melting of silicate minerals, it may be concluded that basaltic magma originates by some sort of virtually dry partial-melting process, and, as we had earlier concluded, the process must occur in the mantle.

Much debate has centered on the chemical composition of the mantle, but it appears that the upper portion contains rocks called *garnet peridotites*, which are rich in olivine and garnet. Laboratory experiments on the dry partial-melting properties of garnet peridotite show that, at asthenospheric pressures and temperatures (100 km deep), a 5 to 10 percent partial melt has a basaltic composition. It is also important to note that 5 to 10 percent partial melting is the point where a melt is able to separate from the residual solid and move upward.

This leaves unexplained the heat source and why basaltic magma should develop in some parts of the asthenosphere but not others. Nevertheless, we can confidently conclude that basaltic magma forms by dry partial melting of rocks in the upper mantle. We will return to the

BOX 4.1 THE SCIENCE OF GEOLOGY

HOW ROCK MELTS

Making magma requires temperatures so high that the idea of a rock melting is difficult for many to grasp. An important turning point in the history of geology was the demonstration by a Scot, James Hall, 200 years ago, that common rocks can be melted and that such melts have the same properties as magma erupted from volcanoes.

No longer do we question whether a rock can melt. We now ask, "What kind of rock melted?" "Did all of the rock melt or only part of it?" "How does temperature increase with depth?" "How far below Earth's surface did melting occur?"

As can be seen in Figure 2.7, temperatures beneath the oceanic crust and the continental crust rise to about 1000°C at rather shallow depths (120 km or less) in the upper part of the mantle. Measurements made on lava prove that magma is fluid at 1000°C, so an immediate question is, "Why isn't Earth's mantle entirely molten?"

The answer is that pressure influences melting (Figure B4.1A). As pressure rises, the temperature at which a silicate mineral melts also rises. For example, the feldspar albite ($NaAlSi_3O_8$) melts at 1104°C at Earth's surface, where the pressure is 1 atmosphere. At a depth of 100 km, however, where the pressure is 35,000 times greater, the melting temperature is 1440°C. Therefore, whether a particular rock melts and forms a magma at a specified depth depends on both the temperature and pressure at that depth.

The Effect of Water on Melting

The effect of pressure on melting is straightforward if no other substances are present. However, if water is present, under sufficient pressure to contain the water vapor at high temperatures, a complication enters. At any given pressure of water vapor, a mineral will begin to melt at a lower temperature than it would in the absence of water. The effect is the same as that of salt and ice. Salt can melt the ice on an icy road. This happens because a mixture of ice and salt melts to a salt solution at a lower temperature than pure ice melts to water.

In the same way, a mineral and water mixture melts to magma at a lower temperature than the melting temperature of the pure mineral. Furthermore, as the pressure rises, the effect of water on the melting temperature increases. This is so because the higher the pressure, the greater the amount of water that will dissolve in the melt. Therefore, the temperature at which the mineral begins to melt in the presence of water decreases as pressure increases (exactly the opposite of what happens with a dry mineral), as Figure B4.1B shows.

A rock that is a composite of several minerals, as nearly all rocks are, does not melt at one specific temperature as a pure mineral would (at a fixed pressure). Instead, as shown in Figure B4.2, a rock melts over a temperature interval. Melting begins at a temperature lower than the temperature at which any of the minerals would melt by itself.

The early formed melt is a mixture of substances from different minerals, but it will contain relatively more of the substance of the mineral that has the lowest melting temperature. Thus, the melt and the mixture of unmelted minerals will almost always differ in composition, and both will differ from the starting composition of the parent rock.

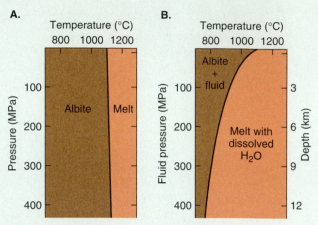

A.

B.

Figure B4.1 Effect of Pressure on Melting Influence of pressure on the melting temperature of albite ($NaAlSi_3O_8$). A. Dry-melting curve. Increasing pressure raises the melting temperature. B. Wet-melting curve. Water dissolves in the melt and decreases the melting temperature.

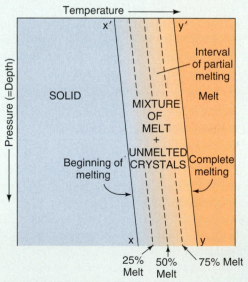

Figure B4.2 Fractional Melting Dry melting of rock containing several kinds of minerals. The pressure effects are similar to those shown in Figure B4.1A. Line x-x' marks the onset of melting, curve y-y' the completion of melting. Between the two lines is a region in which melt and a mixture of unmelted crystals coexist.

questions of a heat source and why melting occurs later in the chapter, when we examine plate tectonics and the ocean floor.

ORIGIN OF ANDESITIC MAGMA

Andesitic magma is close to the average composition of continental crust. Igneous rocks formed from andesitic magma commonly occur in the continental crust. From these two facts, we might suppose that andesitic magma forms by the complete melting of a portion of the continental crust. Some andesitic magma may indeed be generated in this way, but andesitic magma is also extruded from volcanoes that are far from the continental crust. In those cases, the magma must be developed either from the mantle or from the oceanic crust. Laboratory experiments provide a possible answer.

In the laboratory, wet partial melting of mantle rock under suitably high pressure yields a magma of andesitic composition. An interesting hypothesis suggests how this might happen in nature. When a moving plate of lithosphere plunges back into the asthenosphere, it carries with it a capping of basaltic oceanic crust saturated with seawater. The plate heats up and eventually the crust starts to release water, which is then thought to promote partial melting in the mantle above the downgoing plate. Wet partial melting that starts at a pressure that is equivalent to a depth of about 80 km produces a melt having the composition of andesitic magma. It is also possible that some of the downgoing oceanic crust may partially melt.

Some details concerning the wet partial melting of mantle rock and oceanic crust remain to be deciphered. But two pieces of evidence support the idea that much andesitic magma forms in this manner. First, the Andesite Line corresponds closely with plate subduction margins. Second, on the seafloor, a subduction zone is marked by the presence of a deep-sea trench. Beyond the trench, the lithosphere of one plate sinks into the asthenosphere, carrying with it its capping of wet oceanic crust. A horizontal distance of about 250 km, equivalent to a depth of about 80 km, marks the edge of a curved belt of volcanoes. The situation is nicely demonstrated by the andesitic volcanoes of Japan (Figure 4.18).

ORIGIN OF RHYOLITIC MAGMA

Two observations suggest a continental origin for rhyolitic magma:

1. Volcanoes that extrude rhyolitic magma are confined to the continental crust or to regions of andesitic volcanism.

2. Volcanoes that extrude rhyolitic magma give off a great deal of water vapor, and intrusive igneous rocks formed from rhyolitic magma (granite) contain significant quantities of OH-bearing (hydrous) minerals, such as mica and

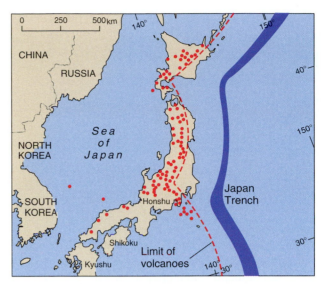

Figure 4.18 Volcanoes in Japan Relations between ocean trenches and arcs of volcanoes erupting andesitic magma. Arc-shaped Japanese islands are parallel to the Japan Trench. Andesitic volcanoes (dots) active during the last million years are also confined behind arc-shaped boundaries.

amphibole. The OH must have come from water dissolved in the magma.

These two points suggest that the generation of rhyolitic magma probably involves some sort of wet partial melting of rock having the composition of andesite—that is, with the average composition of the continental crust. Laboratory experiments bear out this suggestion. In the laboratory, when water-bearing rocks having the average composition of the continental crust have partially melted, the composition of the magma that forms is rhyolitic. The source of heat for such melting to occur in the crust is the mantle. The actual situation is thought to be that hot rock, or magma upwelling in the mantle, brings up the heat that causes crustal rock to partially melt.

Once a rhyolitic magma has formed, it starts to rise. However, the magma rises slowly because it is very viscous, with a high SiO_2 content (70 percent, as we will discuss in the next chapter). As it slowly rises, pressure on the magma decreases. The effectiveness of water in reducing the melting temperature is diminished by reduced pressure. Unless there is some way to heat it, therefore, a rising magma formed by wet partial melting will tend to solidify and form intrusive igneous rock underground.

But a rising magma traverses cool rock, and there is no source of heat to cause a temperature increase. As a result, the depth-temperature path of a rising body of rhyolitic magma brings it closer and closer to solidification. Therefore, most rhyolitic magma solidifies underground and forms *granitic batholiths* rather than reaching the surface and forming rhyolitic lava or tephra.

SOLIDIFICATION OF MAGMA

Literally hundreds of different kinds of igneous rock can be found. Most are rare, but the fact that they exist suggests an important point: a magma of a given composition can crystallize into many different kinds of igneous rock.

Solidifying magma forms several different minerals, and those minerals start to crystallize from the cooling magma at different temperatures. The process is just the opposite of partial melting, and the temperature interval across which solidification occurs is simply the reverse of the melting interval discussed in Box 4.1 and illustrated in Figure B4.2. If at any stage during crystallization the melt becomes separated from the crystals, a magma with a brand-new composition results, while the crystals left behind form an igneous rock with a composition that is quite different from the composition of the magma.

Crystal-melt separations can occur in a number of ways. For example, compression can squeeze melt out of a crystal-melt mixture. Another mechanism involves the sinking of the dense, early crystallized minerals to the bottom of a magma chamber, thereby forming a solid mineral layer covered by melt. However a separation occurs, the compositional changes it causes are called **magmatic differentiation by fractional crystallization**.

BOWEN'S REACTION SERIES

Canadian-born scientist N. L. Bowen (1887–1956) first recognized the importance of magmatic differentiation by fractional crystallization. Because basaltic magma is far more common than either rhyolitic or andesitic magma, he suggested that basaltic magma may be primary and the other magmas may be derived from it by magmatic differentiation.

At least in theory, Bowen argued, a single magma could crystallize into both basalt and rhyolite because of fractional crystallization. Such extreme differentiation rarely happens, we now know, but fractional crystallization is nevertheless an exceedingly important phenomenon in producing a wide range of rock compositions.

Bowen knew that plagioclases that crystallize from basaltic magma are usually calcium-rich (anorthitic), whereas those formed from rhyolitic magma are commonly sodium-rich (albitic). Andesitic magma, he observed, tends to crystallize plagioclases of intermediate composition.

Bowen's experiments provided an explanation for these observations. He discovered that, although the composition of the first plagioclase that crystallizes from a basaltic magma is anorthitic, the composition changes toward albitic as crystallization proceeds and the ratio of crystals-to-melt increases. This means that all the plagioclase crystals, even the ones formed earliest, must continually be changing their compositions as the magma cools. A chemical balance between crystals and melt, if maintained, is referred to as *chemical equilibrium*.

As explained in Chapter 3, the plagioclases involve a coupled ionic substitution in which $Ca^{2+} + Al^{3+}$ in the crystal are replaced by $Na^{1+} + Si^{4+}$. Bowen recognized that such substitution allows plagioclase to undergo a continuous change in composition in a cooling magma.

CONTINUOUS REACTION SERIES

Bowen called such a continuous reaction between crystals and melt a *continuous reaction series*, by which he meant that, although the composition changes continuously, the crystal structure remains unchanged. The speed at which the change occurs is controlled by the rates at which the four ions can diffuse through the plagioclase structure.

To maintain equilibrium, some Ca^{2+} and Al^{3+} must diffuse out of the early formed plagioclase and be replaced by Na^{1+} and Si^{4+}, which diffuse into the crystal structure from the melt. Such diffusion is exceedingly slow. Equilibrium is rarely attained, because crystals usually grow fast enough that diffusion doesn't maintain equilibrium between the surface and interior of a crystal. As a result, compositionally zoned crystals of anorthite are formed, with anorthite-rich inner zones and albite-rich outer zones. The anorthite-rich inner zones are out of chemical equilibrium with the albite-rich outer zones and with the residual magma.

Bowen also knew that plagioclase grains in many igneous rocks have concentric zones of differing compositions such that the innermost core (and therefore the earliest formed part of the crystal) is anorthitic and successive layers are increasingly albite rich (Figure 4.19A). He pointed out an important implication for the existence of zoned crystals of plagioclase. When anorthite-rich cores are present, the residual melt is necessarily richer in sodium and silica than it would have been if equilibrium had been maintained.

According to Bowen, the anorthite-rich cores are an example of magmatic differentiation by fractional crystallization. If, in a partially crystallized magma containing zoned crystals, the melt were somehow squeezed out of the crystal mush, the result would be an albite-rich magma and the residue would be an anorthite-rich rock.

DISCONTINUOUS REACTION SERIES

Bowen identified several sequences of reactions besides the continuous reaction series of the feldspars. One of the earliest minerals to form in a cooling basaltic magma is olivine. Olivine contains about 40 percent SiO_2 by weight,

A.

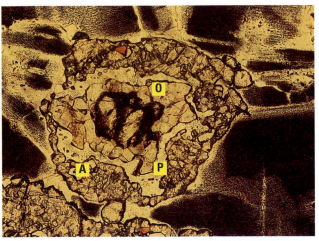

B.

Figure 4.19 Proof That Equilibrium Is Not Maintained
Textures illustrating Bowen's reaction series. A. Zoned plagioclase crystal in andesite, proof that a continuous reaction
occurred but that equilibrium was not maintained. Photographed
in polarized light to enhance the zoning. Bands near center are
anorthite-rich, progressing to albite-rich at the rim. The crystal
is 2 mm long. B. A grain of olivine (O) in a gabbro surrounded
by reaction rims of pyroxene (P) and amphibole (A) demonstrates discontinuous reaction. The diameter of the outer rim is
3 mm.

whereas a basaltic magma contains 50 percent SiO_2. Thus,
crystallization of olivine will leave the residual liquid a little richer in silica. Eventually, the solid olivine reacts with
silica in the melt to form a more silica-rich mineral, pyroxene (Figure 4.19B). The pyroxene in turn can react to form
amphibole, and then the amphibole can react to form
biotite. Such a series of reactions, where early formed minerals form entirely new minerals through reaction with the
remaining liquid, is called a *discontinuous reaction series*.

If a core of olivine is shielded from further reactions by
a rim of pyroxene, the remaining liquid will be more silica-rich than it would be if equilibrium were maintained
and all the olivine were converted to pyroxene. Bowen reasoned that if partial reactions occurred in both continuous
and discontinuous reaction series, magmatic differentiation by fractional crystallization in a basaltic magma could
eventually produce a residual magma with a rhyolitic
composition (Figure 4.20).

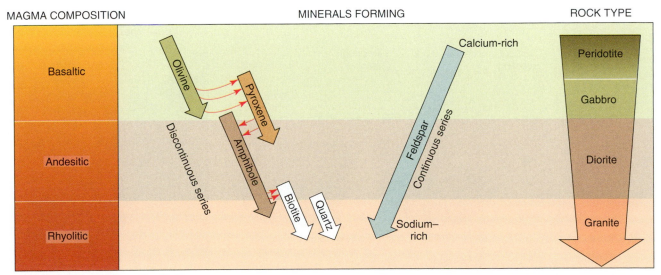

Figure 4.20 Bowen's Reaction Series Bowen demonstrated
how the cooling and crystallization of a primary magma of
basaltic composition, through reactions between mineral grains
and magma, followed by separation of mineral grains from the
remaining magma, can change from basaltic to andesitic to rhyolitic. Bowen identified two series of reactions: a continuous
series in which plagioclase changes from anorthitic to albitic in
composition, and a discontinuous series in which minerals
change abruptly from one kind to another, for example, from
olivine to pyroxene.

PROBLEMS WITH BOWEN'S CONCEPT

The answer to the question that Bowen posed—whether large volumes of rhyolitic magma can form from basaltic magma by fractional crystallization—is negative. The problems with Bowen's idea are:

1. If equilibrium is maintained, crystallization of a magma is complete long before a residual magma is siliceous enough to have a rhyolitic composition.

2. Calculations show that only a tiny percentage of the volume of a basaltic magma could ever be differentiated to rhyolitic magma.

3. Rhyolitic magma forms in the continental crust. If rhyolitic magma formed by fractional crystallization of basaltic magma, we would surely expect to find a lot of rhyolite in the oceanic crust, for it is there that basaltic magma is most common.

Although rhyolitic magma forms principally through partial melting of the continental crust, the importance of Bowen's reaction series has been demonstrated many times. Careful study of almost any igneous rock reveals evidence that fractional crystallization played a role in its formation.

VALUABLE MAGMATIC MINERAL DEPOSITS

The processes of partial melting and fractional crystallization in magmas sometimes lead to formation of large and potentially valuable mineral deposits. Because magma is involved in the formation process, such deposits are called *magmatic mineral deposits.*

When magma undergoes magmatic differentiation by fractional crystallization, the residual melt becomes progressively enriched in any chemical element that is not removed by the early crystallizing minerals. Separation and crystallization of the residual melt produces an igneous rock that contains the concentrated elements.

An important example of this kind of concentration process is provided by pegmatites, especially those formed through crystallization of rhyolitic magma. Some pegmatites form by magmatic differentiation during the formation of granitic stocks and batholiths. Pegmatites may contain significant enrichments of rare elements such as beryllium, tantalum, niobium, uranium, and lithium. If sufficiently concentrated, these can be profitably mined.

Another form of magmatic differentiation by fractional crystallization occurs when early formed dense minerals sink and accumulate on the floor of a magma chamber. The process is called *crystal settling*, and in some cases the segregated minerals make desirable ores.

Most of the world's chromium ores were formed in this manner by accumulation of the mineral chromite ($FeCr_2O_4$) (Figure 4.21). The largest known chromite deposits are in South Africa, Zimbabwe, and the former Soviet Union. Similarly, vast deposits of ilmenite ($FeTiO_3$), a source of titanium, were formed by magmatic differentiation. Large deposits occur in the Adirondack mountains of New York.

A form of concentration similar to crystal settling occurs when, for reasons not clearly understood, certain magmas separate, as oil and water do, into two immiscible liquids. The first observation of this happening in a magma was discussed in the essay at the beginning of this chapter. One of the liquids, a sulfide liquid rich in iron, copper, and nickel, sinks to the floor of the magma chamber because it is denser. After cooling and crystallization, the resulting igneous rock has a copper or nickel ore at the base. Many of the world's great nickel deposits, in Canada, Australia, Russia, and Zimbabwe, formed in this manner.

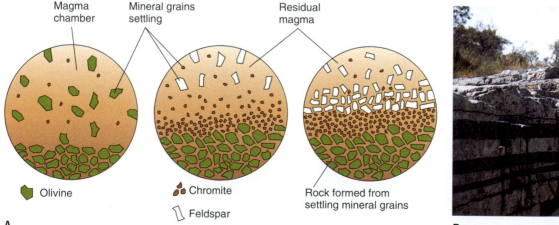

A.

B.

Figure 4.21 Magmatic Differentiation Fractional crystallization. A. Grains of three minerals—olivine, chromite (chromium iron-oxide), and feldspar—settle one after the other to the bottom of a magma chamber, producing three kinds of rocks whose compositions differ considerably from that of the parent magma. B. Layers of plagioclase (light gray) and chromite (black) formed by fractional crystallization in the Bushveld Igneous Complex, South Africa.

Before you go on:

1. What is magmatic differentiation by fractional crystallization?

2. What is a continuous reaction series? a discontinuous series?

3. What roles do partial melting and fractional crystallization play in the formation of magmatic mineral deposits?

REVISITING PLATE TECTONICS AND THE EARTH SYSTEM

PLATE TECTONICS AND THE OCEAN FLOOR

Underlying the ocean basins is a crust of basaltic composition. New crust is formed at spreading centers that coincide with the midocean ridges; these spreading centers are places where the lithosphere is splitting apart (see Chapter 2). Beneath a midocean ridge the boundary between the lithosphere and the asthenosphere rises close to the surface, as you can see in Figure 4.22. Hot rock of the asthenosphere moves upward, squeezing into the fracture zone. To understand what happens let us review the earlier discussion about rock melting (Box 4.1). The melting temperature of a rock increases with pressure. Therefore, if a hot mass of rock is under pressure and the pressure suddenly decreases, *decompression melting* can occur. That is exactly what happens in rock of the asthenosphere that rises up beneath the midocean ridge. The magma that forms by decompression melting in

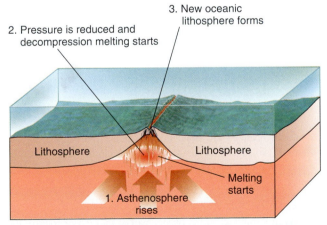

Figure 4.22 Making MORB The magma that makes MORB (midocean ridge basalt) is formed by decompression melting of rock in the asthenosphere. When the lithosphere splits to create a new spreading center, (1) the asthenosphere rises into the fracture; (2) pressure is reduced and melting begins; and (3) the emerging magma chills and solidifies to form new oceanic lithosphere.

the asthenosphere rises and eventually solidifies to form the basalt of the oceanic crust. The oceanic crust varies very little in composition around the world and is simply referred to as **MORB**, an acronym for *midocean ridge basalt*.

We know from laboratory melting experiments that decompression melting that produces MORB magmas starts in the mantle at a pressure equivalent to a depth of about 65 km. The process seems to happen when an upward-moving (convecting) mass of rock in the asthenosphere reaches a depth of 65 km below the surface. A 5 percent partial melt of mantle rock produces magma of MORB composition. Remember that the degree of partial melting influences the composition of the magma; if a larger or smaller fraction of the mantle rock melts, the resulting magma will not be MORB. The reason MORB is similar everywhere is the similarity of the tectonic environment in which these magmas are formed—the depth of decompression melting and the degree of partial melting vary little from one plate-spreading center to another.

The oceans cover 71 percent of Earth's surface. Consequently, MORB is the most common igneous rock at the surface. Common though it is, MORB is not easy to observe; the ridge and seafloor are everywhere covered by water except in a few places such as Iceland, where the midocean ridge stands above sea level. There is, however, one circumstance in which it is possible to see MORB above the sea: in places where a plate collision has caught up and crushed a fragment of oceanic crust between two colliding continental masses, like a nut between two sides of a nut cracker.

When oceanic crust is caught up and crushed in a plate collision, the minerals that are characteristic of basalt are transformed into an assemblage dominated by a green, fibrous mineral called *serpentine*. As we discussed in Chapter 3, serpentine is the most common asbestos mineral—it is from seafloor crushed during plate collisions that most of the world's asbestos is mined.

Serpentine-dominated fragments of oceanic crust found on continents are called **ophiolites** from the Greek word for serpent, *ophis*. Ophiolites are studied intensively because they provide evidence of ancient plate collisions and because they provide convenient samples of oceanic crust (Figure 4.23).

IGNEOUS ROCK AND LIFE ON EARTH

Life requires nutrients in order to survive and grow; among the nutrients are chemical elements such as potassium, sulfur, calcium, and phosphorus. Life on land, including human life, gets nutrients from the regolith, and the regolith is formed by chemical and physical breakdown of rock at the surface. The breakdown processes, which are all part of the Earth system, release soluble constituents, including the needed nutrients, and leave behind insoluble, barren residues. Long-continued breakdown and leaching can produce a land surface that can hardly support any life forms.

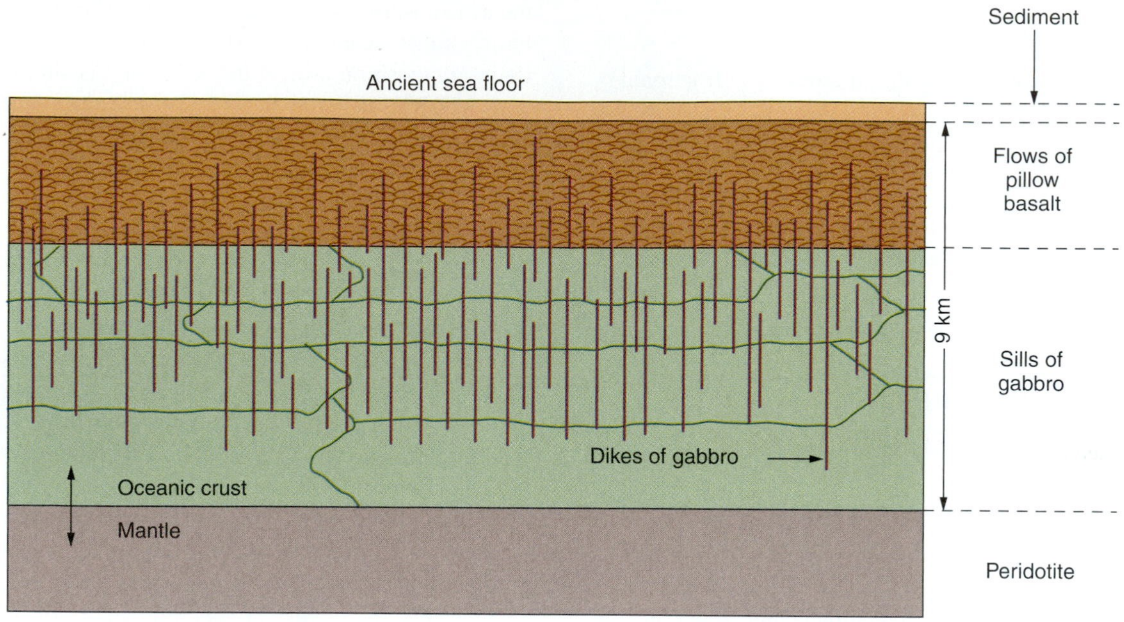

Figure 4.23 Ophiolite: Fragment of Oceanic Crust on Land
This is an idealized section of oceanic crust reconstructed from observations of an ophiolite. A thin layer of sediment overlies basalt that was extruded onto the seafloor, and then intruded by a thick pile of gabbro sills. Both the basalt and the gabbro sills are intruded by gabbro dikes. The composition of the basalt and the gabbro sills and dikes is the same: MORB.

A continent unaffected by any process of surface renewal, such as uplift or volcanic eruptions, but subjected to erosion for a hundred million years, would finish with low relief and almost barren soils. Without surface renewal, erosion can convert a continent to an essentially lifeless platform. What saves Earth from such a fate are two processes of surface renewal, the intrusion and extrusion of magma, and tectonic uplift. Both processes serve to bring new rock, and therefore new supplies of nutrients, to the surface.

We learned in Chapter 3 that the continental crust is more than 95 percent comprised of igneous rock. This vast reservoir contains all of the nutrients needed by life. Magma, which is less dense than the rock from which it forms by melting, rises buoyantly upward, bringing with it the nutrients on which life depends. Igneous rock and the rock cycle are essential for life support.

What's Ahead?

Now that you have a background in igneous rocks, Earth's most common and important kinds of rock, you are ready to explore how igneous processes influence Earth's surface. In the next chapter, we'll start with a discussion of the properties of magma and then explore volcanoes and volcanism.

CHAPTER SUMMARY

1. Igneous rock forms by the solidification and crystallization of magma.

2. Igneous rock may be intrusive (meaning it formed within the crust) or extrusive (meaning it formed on the surface). The grain sizes of igneous rocks indicate how and where the rocks formed.

3. Igneous rocks rich in quartz and feldspar, such as granite, granodiorite, and rhyolite, are characteristically found in the continental crust. Basalt, which is rich in pyroxene and olivine, is derived from magma formed in the mantle and is common in the oceanic crust.

4. All bodies of intrusive igneous rock are called plutons. Special names are given to plutons based on shape and size.

5. Modern volcanic activity is concentrated along plate margins. Andesitic volcanoes are found at subduction margins, and

basaltic volcanoes are concentrated along spreading margins. Rhyolitic volcanoes occur at collision margins and in places where rising basaltic magma causes the continental crust to melt.

6. Magma forms by the partial melting of rock. (Complete melting, if it happens at all, is rare.)

7. Basaltic magma forms by dry partial melting of rock in the mantle. Andesitic magma forms during subduction by wet partial melting of mantle rock as a result of water being released from heated oceanic crust. Rhyolitic magma forms by wet partial melting of rock in the continental crust.

8. Processes that separate remaining melt from already formed crystals in a cooling magma lead to the formation of a wide diversity of igneous rocks.

9. Magmatic mineral deposits, which form as a result of magmatic differentiation, are the world's major sources of nickel, chromium, vanadium, platinum, beryllium, and a number of other important industrial metals.

THE LANGUAGE OF GEOLOGY

aphanite (p. 101)

batholith (p. 108)

chemical differentiation by partial melting (p. 111)

dike (p. 107)

extrusive igneous rocks (p. 101)

intrusive igneous rocks (p. 101)

magma (p. 110)
magmatic differentiation by fractional crystallization (p. 114)
MORB (p. 117)

ophiolite (p. 117)

phanerite (p. 101)

phenocryst (p. 101)
pluton (p. 107)
porphyry (p. 101)
pyroclast (p. 104)
pyroclastic rocks (p. 104)

sill (p. 107)
stock (p. 109)

tephra (p. 104)

IMPORTANT ROCK NAMES TO REMEMBER

agglomerate (p. 104)
andesite (p. 103)

basalt (p. 104)

dacite (p. 103)
diorite (p. 103)

gabbro (p. 103)
granite (p. 103)
granodiorite (p. 103)

obsidian (p. 102)

pegmatite (p. 101)
peridotite (p. 103)

rhyodacite (p. 103)
rhyolite (p. 103)

tuff (p. 106)

welded tuff (p. 106)

QUESTIONS FOR REVIEW

1. What controls the grain size and texture in igneous rocks? Would you expect solidified lava flows to be basalt or gabbro?

2. What are the distinguishing features of pyroclastic rocks? How might you tell the difference between a rhyolite that flowed as a lava and a rhyolitic tuff?

3. Do porphyries always contain phenocrysts? What minerals might you find as phenocrysts in a porphyritic rhyolite? a porphyritic basalt?

4. What is the major difference between the mineral assemblage of a diorite and a granodiorite? between granite and granodiorite? between a gabbro and a peridotite?

5. How does a lapilli tuff differ from a welded tuff? Could both rocks form as a result of eruptions from the same volcano?

6. What does the term *partial melting* mean, and what role does it play in forming of basaltic magma? Where in Earth does basaltic magma form?

7. How does the magma-forming process of dry partial melting differ from wet partial melting? Can you suggest an example of a magma type formed by each kind of melting?

8. How might it be possible for fractional crystallization to produce more than one kind of igneous rock from a single magma? Comment on the role of fractional crystallization in the formation of mineral deposits.

9. Comment on the importance of Bowen's reaction series for understanding differentiation processes in cooling magmas. How does a continuous reaction series differ from a discontinuous series?

10. How does a dike, like the Great Dike of Zimbabwe, differ from a sill, such as the Palisades Sill?

11. How big can batholiths be, and by what mechanism are granitic magmas believed to move upward in the crust?

12. Briefly describe the plate-tectonic settings of andesitic and basaltic volcanoes. Where does Kilauea Volcano in Hawaii sit with respect to the Pacific Plate? Where does Mt. Rainier sit with respect to a plate boundary?

Click on *Presentation* and *Interactivity* in the **Volcanoes and Igneous Rocks** module of your CD-ROM to further explore resources and activities presenting concepts from this chapter. Select *Assessment* in the same module to test your understanding of this chapter.

Click on *Petroscope* for **Igneous Rocks** to learn more about the characteristics of igneous rocks from this chapter as well as to learn how to identify unknown igneous rock samples.

Nighttime Eruption. Galunggung, a stratovolcano in Indonesia during an eruption. Notice the lighting discharges (left) caused by the eruption of gas and volcanic ash.

Magmas and Volcanoes

Lava fountain, Kilauea, Hawaii, 1984.

Witnessing an Eruption at Close Hand

A telephone call from the Hawaiian Observatory jolted me awake before dawn one morning in 1984. A summit eruption of Kilauea volcano had begun, and if I would drive to the other side of the island in the next two hours, I could join several other geologists to observe the eruption at close hand. I was involved in studying the history of two much older volcanoes, Mauna Kea and Kohala, on the other side of the island, and here was a chance to see how the rocks I was studying had actually formed.

Near the top of Kilauea, we flew by helicopter across rivers of red-hot lava toward an impressive fountain of erupting lava that rose hundreds of meters above the summit caldera. The helicopter landed several hundred meters upwind from the lava fountain, and at this distance the heat and noise resembled the blast furnace of a steel mill. Approaching the lava fountain at a safe distance, a geologist measured the temperature of the eruption as more than 1200°C. Blobs of lava falling from the eruption column piled up to form a rampart around the vent, and larger ones cooled during their descent to form volcanic bombs. Nearby, a massive rubbly lava flow moved slowly through a forest of ohia trees. As the front advanced, incandescent lava was visible just beneath the surface rubble, and we approached close enough to obtain samples of the now-solid, but extremely hot new rock. Small fires along the flow margin burst forth as trees were overrun and rapidly converted to flammable gases.

Despite the spectacle it presents in full eruption, Kilauea is considered a comparatively safe volcano to approach and study close up. The eruption I witnessed allowed me to visualize how early lavas of the older nearby volcanoes must have formed many hundreds of thousands of years ago.

Stephen C. Porter

KEY QUESTIONS

1. **What is magma?**
2. **What causes eruptions?**
3. **What is the difference between violent and nonviolent eruptions?**
4. **What are the types of volcanoes?**
5. **What hazards do volcanoes present?**
6. **Are volcanoes and plate tectonics related?**

INTRODUCTION: EARTH'S INTERNAL THERMAL ENGINE

What do the following things have in common?

Rocks melting to form magma in Earth's crust and mantle.

Subduction zones.

Midocean spreading centers.

Hot spots.

Volcanoes.

Lava.

Igneous rocks.

Granite.

Basalt.

What they have in common is that all are components of Earth's internal thermal engine. Let's string them all together:

Melting occurs in the crust and mantle along subduction zones, at midocean spreading centers, and at hot spots; this melting forms magma; liquid magma is buoyant, so it slowly rises through the crust and may express itself as volcanoes that erupt lava; and this process forms igneous rocks, both underground and on the surface—rocks like granite and basalt.

This chapter ventures through this complex chain of events. We'll start with a look at magma and how it behaves, then examine volcanoes and the materials they erupt, and finally look at the dangers posed by volcanic eruptions.

MAGMA: MOLTEN ROCK OF DIFFERENT TYPES

When conditions are right in the crust and mantle, especially along subduction zones and spreading centers, rocks melt to form magma. We have used the term in previous chapters, and defined *magma* as molten rock beneath Earth's surface, but now we need to look a little closer. Magma also includes any unmelted mineral grains and dissolved gases. Because liquid magma is less dense than surrounding solid rock, and obviously more mobile, magma, once formed, rises toward the surface.

As we discussed in the previous chapter, some magma never reaches Earth's surface, because it cools and solidifies underground to form intrusive igneous rock. Magma that reaches the surface does so by erupting through vents we call **volcanoes**. (The term *volcano* comes from the name of the Roman god of fire, Vulcan.) Along with the magma, some solid rock debris and gases erupt, too. Once magma erupts through the surface, we give it a special name—**lava**.

The stereotype most people have of a volcano envisions a violently erupting mountain spewing streams of lava. But this stereotype is too simple. There are different types of volcanoes, different types of magma and therefore different kinds of lava, and eruptions vary from gentle flows (Hawaii and Iceland) to catastrophic explosions (Mount St. Helens, Mt. Pinatubo, Soufrière Hills). What's more, the majority of eruptions never make the news because they occur beneath the ocean, unobserved.

Because magma forms many miles underground, we cannot study it directly—except where volcanoes conveniently deliver samples to the surface. So, we will study magma by looking at volcanoes and the properties of the lava they erupt. In the previous chapter we examined the igneous rocks that form when magma cools. You should refer back to Chapter 4 to refresh your memory of the names of the common igneous rocks, as we will use them in this chapter.

By observing the eruption of lava, we can draw three important conclusions about magma:

1. Magma has a wide range of compositions, but silica (SiO_2) always dominates the mix.

2. Magma has high temperatures.

3. Magma is fluid—it has the ability to flow. This is true even though some magma flows so slowly that it seems almost stationary. Most magma actually is a mixture of liquid (often referred to as *melt*) and solid mineral grains.

COMPOSITION OF MAGMAS AND LAVAS

The composition of magmas and lavas is controlled by the most abundant elements in Earth—Si, Al, Fe, Ca, Mg, Na, K, H, and O. Because magmas consist almost entirely of elements combined chemically with oxygen, we usually express compositional variations of magmas in terms of oxides—SiO_2, Al_2O_3, CaO, and H_2O. The most abundant is SiO_2.

Three distinct types of magma are more common than all others: *basaltic* contains about 50 percent SiO_2, *andesitic* about 60 percent, and *rhyolitic* about 70 percent. These percentages have very important implications for the way each type of magma erupts and the rocks formed by each type. The common igneous rocks derived from each magma type are, respectively, basalt, andesite, and rhyolite (Figure 5.1).

The three magmas are not formed in equal abundance.

1. Basaltic magmas are erupted by approximately 80 percent of volcanoes worldwide, including those beneath the oceans (the seafloor worldwide is mostly basalt). Magma from Hawaiian volcanoes such as Kilauea and Mauna Loa is basaltic. The entire island of Iceland is basaltic.

2. Andesitic magmas are about 10 percent of the total. Magma from Mount St. Helens in Washington State and Krakatau in Indonesia is usually andesitic.

3. Rhyolitic magmas are about 10 percent of the total. Magmas erupted from volcanoes that once were active at Yellowstone Park are mostly rhyolitic.

As you'll see later in this chapter, the distribution of these different kinds of magmas and their associated volcanoes is closely related to plate tectonics.

GASES DISSOLVED IN MAGMA

Small amounts of gas (0.2 to 3% by weight) are dissolved in all magma. Although present in small amounts, these gases strongly influence a magma's properties. The principal gas is water vapor, which, together with carbon dioxide, accounts for more than 98 percent of all gases emitted from volcanoes. Other substances, which are rarely present in amounts exceeding 1 percent, include sulfur compounds, hydrogen chloride, nitrogen, and argon.

TEMPERATURE OF MAGMAS AND LAVAS

Magma temperature is difficult to measure, but sometimes it can be done during a volcanic eruption. Scientists who study magma are not eager to be roasted alive, so they make measurements from a safe distance, using optical

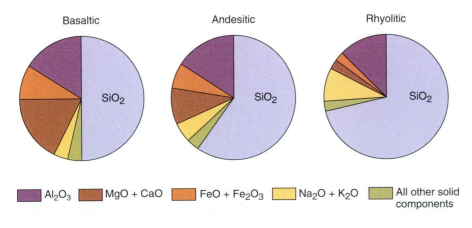

Basaltic   Andesitic   Rhyolitic

Figure 5.1   The Three Main Magmas
The average compositions (in weight percent) of the three principal kinds of magma. The analyses are of rocks that solidify from the magmas. In addition to the solid materials, the magmas contain dissolved gases, most of which escape during solidification. Basaltic magma has a low content of dissolved gas; andestic and rhyolitic magmas tend to be very gas-rich.

■ Al_2O_3 ■ $MgO + CaO$ ■ $FeO + Fe_2O_3$ ■ $Na_2O + K_2O$ ■ All other solid components

Figure 5.2   A House Is No Match for Lava   An advancing tongue of very fluid basaltic lava setting fire to a house in Kalapana, Hawaii, during an eruption of Kilauea Volcano in June 1989. Flames at the edge of the flow are burning lawn grass.

devices. Magma temperatures determined in this manner during eruptions, such as Kilauea in Hawaii and Mount Vesuvius in Italy, range from 1000° to 1200°C (hot enough to melt copper). Experiments on synthetic magmas in the laboratory suggest that, under some conditions, magma temperatures might reach 1400°C.

VISCOSITY OF MAGMAS AND LAVAS

Dramatic pictures of lava flowing rapidly down the side of a volcano prove that some magmas are very fluid. Basaltic lava moving down a steep slope on Mauna Loa in Hawaii has been clocked at 16 km/h. But such fluidity is rare, and flow rates are more commonly meters per hour or even meters per day. Figure 5.2 shows basaltic lava destroying a house in Hawaii. Although this lava type often flows quite rapidly, rates are usually slow enough that people can move out of harm's way. Magma containing 70 percent or more SiO_2 and very little dissolved gas flows so slowly that its movement can hardly be detected.

The internal property of a substance that offers resistance to flow is called **viscosity**. The more viscous a magma, the less easily it flows. Viscosity of a magma depends on temperature and composition (especially the silica and dissolved-gas contents).

TEMPERATURE AND VISCOSITY

The higher the temperature, the lower the viscosity, and the more readily a magma flows. A very hot magma erupted from a volcano may flow readily, but it soon begins to cool, becomes more viscous, and eventually slows to a

Figure 5.3   Same Lava, Different Flows   The way lava flows is controlled by viscosity. Shown are two strikingly different flows that have the same basaltic composition. The lower flow on which the geologist is standing is pahoehoe, formed from a low-viscosity lava like that shown in Figure 5.2. The upper flow being sampled by the geologist, which was very viscous and slow moving, is an aa flow erupted from Kilauea Volcano in 1989. The pahoehoe flow was erupted in 1959.

complete halt. In Figure 5.3, the smooth, ropy-surfaced lava, called *pahoehoe* (pah-hoy-hoy) in Hawaii, formed from a very hot, very fluid lava. By contrast, the rubbly, rough-looking lava formed from a cooler lava having a high viscosity. Hawaiians call this rough lava *aa* (ah-ah).

SILICA CONTENT AND VISCOSITY

The $(SiO_4)^{4-}$ anions that occur in silicate minerals (Chapter 3) are also present in magmas. Just as they do in minerals, these anions polymerize by sharing oxygens. However, unlike the anions in silicate minerals, those in magma form irregularly shaped groupings of silicate tetrahedra. The size of such a grouping depends on the number of silicate tetrahedra the group contains. As the average number of tetrahedra in the polymerized groups becomes larger, the magma grows more viscous—more resistant to flow.

The number of tetrahedra in the groups depends on the silica content of the magma. The greater the silica content, the larger are the polymerized groups. For this reason, rhyolitic magma (70% silica) is always more viscous than basaltic magma (50% silica), and andesitic magma has a viscosity that is intermediate between the two (60% silica).

HOW BUOYANT MAGMA ERUPTS ON THE SURFACE

Magma, like most liquids, is less dense than the solid rock from which it forms. Therefore, once formed, lower-density magma exerts an upward push on its enclosing higher-density solid rock. The magma attempts to force its way toward Earth's surface. Countering this is tremendous pressure on the rising magma from the mass of overlying rock. The pressure is proportional to depth (thickness of overlying rock). Therefore, as magma rises upward, the pressure on it decreases.

Pressure controls the amount of gas a magma can dissolve—more at high pressure, less at low. Gas dissolved in a rising magma acts the same way as gas dissolved in soda. When a container of soda is opened, the pressure inside the bottle abruptly drops, gas comes out of solution, and bubbles form.

Gas dissolved in an upward-moving magma also comes out of solution and forms bubbles. What happens to the bubbles once they are formed is determined by the viscosity of the magma.

> **Before you go on:**
>
> 1. What are the three characteristic properties of magma?
> 2. What are the three most common kinds of magma? What are their SiO_2 contents?
> 3. What controls the viscosity of magma?
> 4. Why does magma rise toward the surface?

ERUPTION STYLE— NONEXPLOSIVE OR EXPLOSIVE?

As mentioned, magma is far from uniform, and it varies widely in composition. This is the key to the difference between gentle eruptions that pose little threat to people and violent eruptions that can kill.

NONEXPLOSIVE ERUPTIONS (HAWAIIAN-TYPE)

People usually regard any volcanic eruption as hazardous. However, many volcanoes are comparatively safe and relatively easy to study. Nonexplosive eruptions occur notably in Hawaii, Iceland, and on the seafloor. They are relatively safe compared to violent, explosive events like the eruption of Mount St. Helens in Washington State (1980), Mt. Pinatubo in the Philippines (1991), or Soufrière Hills in the West Indies (2000). Each of these eruptions was explosive and caused substantial destruction or loss of life.

The difference between nonexplosive and explosive eruptions depends largely on magma viscosity and dissolved-gas content. Generally, low-viscosity magmas and low dissolved-gas contents produce nonexplosive eruptions.

However, even nonexplosive eruptions may appear violent during their initial stages. The reason is that gas bubbles in a low-viscosity basaltic magma will rise rapidly upward, like the gas bubbles in a glass of soda. If a basaltic magma rises rapidly, the pressure drop will be fast, allowing gas to bubble so rapidly from solution that spectacular *lava fountains* will occur (Figure 5.4). Bits of falling lava spatter when they strike the ground and can pile up to form a *spatter cone* or spatter rampart beside the vent (the opening from which the lava issues). In the photographs of tephra shown in Figure 4.8, the bombs were found around a spatter cone.

When the fountaining subsides, hot and fluid lava emerging from the vent flows rapidly downslope, sometimes forming streams (Figure 5.5). Because heat is lost quickly at the surface of the flowing lava, the surface solidifies onto a crust, beneath which the liquid lava continues to flow in well-defined channels called *lava tubes*. These tubes inhibit upward loss of heat and enable low-viscosity lava to continue flowing just below the surface for great distances (miles) from the vent.

As the lava cools and continues to lose dissolved gases, its viscosity increases and the character of flow changes. The very fluid lava initially forms thin pahoehoe flows, but with increasing viscosity the rate of movement slows and the stickier lava may be transformed into a rough-surfaced aa flow that moves very slowly. Thus, during a single nonexplosive Hawaiian-type eruption, spatter cones, pahoehoe, and aa may form from the same batch of magma.

Figure 5.5 Fast-Flowing Lava   This stream of low-viscosity basaltic lava moving smoothly away from an eruptive vent demonstrates how fluid and free-flowing lava can be. The initial temperature of the lava was about 1100°C. The eruption occurred in Hawaii in 1983.

Figure 5.4 Lava Fountain An eruption at Mauna Ulu, a vent on the flank of Kilauea Volcano, Hawaii, starts with spectacular fountaining as gases are released from the rising magma. Use of a telephoto lens foreshortens the field of view. The observer is several hundred meters away from the fountain, which reached as high as 300 meters.

As a basaltic lava cools and its viscosity increases, the gas bubbles find it increasingly difficult to escape. When the lava finally solidifies to rock, the last-formed bubbles become trapped and their bubble-shape is preserved. These bubble holes are called *vesicles*, and they produce an igneous rock that has *vesicular texture* (Figure 5.6).

Vesicular basalts are common. In many, the vesicles have been later filled with calcite, quartz, or some other mineral deposited by heated groundwater. Vesicles filled by secondary minerals are called *amygdules* (Figure 5.7).

EXPLOSIVE ERUPTIONS

In viscous andesitic or rhyolitic magmas, gas bubbles can rise only very slowly. When such a magma migrates toward the surface, the reduced confining pressure allows gas to form and expand within the magma, and this can lead to an explosive eruption. When confining pressure

Figure 5.6 Captured Bubbles   Vesicular basalt, Rangitoro Island, New Zealand. The specimen is 10 cm in diameter.

drops quickly, the gas in a magma can expand into a froth of innumerable glass-walled bubbles called *pumice*. Some pumice has such a low density from the bubbles that it will float on water. Beaches on mid-Pacific islands, for example, are often littered with pumice that has floated in on currents from distant volcanic eruptions.

In many instances, instead of forming pumice, small bubbles expanding within a huge mass of sufficiently gas-rich, viscous magma will shatter the magma into tiny fragments called volcanic ash (refer back to Figure 4.8). *Volcanic ash* is the most abundant product of explosive eruptions.

Figure 5.7 Filled Bubble Holes Amygdaloidal basalt formed when secondary minerals such as calcite fill vesicles in a basalt. The specimen is about 10 cm in diameter.

The word *ash* is misleading because it means, strictly, the solid matter left after something flammable, such as wood, has burned. However, the fine debris thrown out by volcanoes looks so much like true ash that it has become customary to use the word for this material also.

When little or no dissolved gas is present, a magma will erupt as a lava flow, regardless of composition. If dissolved gas is present, however, it must escape somehow, and the higher the viscosity, the greater the likelihood that the escaping gas will cause an explosive eruption.

ERUPTION COLUMNS AND TEPHRA FALLS

The largest and most violent explosive eruptions are associated with silica-rich magmas having a high dissolved-gas content. As the rising magma approaches the surface, rapid decompression causes the gases to form bubbles and expand. This, in turn, produces a violent eruption of densely mixed hot gas and tephra. This hot, turbulent mixture rises rapidly in the cooler air above the vent to form an *eruption column* that may tower as high as 45 km in the atmosphere (Figure 5.8). A violent eruption of this kind is called a Plinian eruption, so named after the A.D. 79 eruption of Mt. Vesuvius in Italy, in which the Roman author and statesman Pliny lost his life.

The rising, buoyant column of a Plinian eruption is driven by heat energy released from hot, newly formed pyroclasts. At a height where the density of the material in the column equals that of the surrounding atmosphere, the column begins to spread laterally to form an anvil-shaped cloud of the type so familiar in thunderstorm clouds.

As an eruption cloud begins to drift with the upper atmospheric winds, the particles of debris rain down in a *tephra fall* and eventually accumulate on the ground as tephra deposits. The eruptions of Mt. Pinatubo in the Philippines in 1991, and Soufrière Hills on Montserrat in the Caribbean in 2000, both led to such extensive tephra falls that local populations had to be evacuated (Figure 5.9). During exceptionally explosive eruptions, tephra can

Figure 5.8 Plinian Eruption Column An eruption column of hot gas and fine tephra rising above the summit of Mount St. Helens during the eruption event of May 1980. The cloud, a mixture of hot gas and tiny pyroclasts, rises, expands, and cools. When the column density equals that of the surrounding air, the column stops rising and starts spreading sideways.

Figure 5.9 Midday During an Eruption Volcanic ash from Mount Pinatubo, on the island of Luzon, Philippines, blocks out the midday sun. Residents of Tarlac, a town near the volcano, cover their faces and leave for safety. The eruption was in June 1991.

be spread 1000–2000 km from its source in amounts sufficient to form noticeable ash accumulations. Fine volcanic dust is carried much farther.

Some eruption columns reach such great heights that high-level winds transport the fine debris and associated sulfur-rich gas completely around the world. This natural atmospheric pollution blocks incoming solar energy, and this can lower average land-surface temperatures up to 1°C for a year or longer. It also can cause spectacular sunsets as the Sun's rays are refracted by the airborne particles.

PYROCLASTIC FLOWS—MONT PELÉE

When the mixture of hot gases and pyroclasts is more dense than the atmosphere, the turbulent mixture flows down the side of the volcano rather than forming an eruption column. A hot, highly mobile flow of tephra that rushes down the flank of a volcano during a major eruption is called a **pyroclastic flow**. These are among the most devastating and lethal forms of volcanic eruptions. Analysis of the worldwide geologic record of prehistoric pyroclastic flows shows that they can travel 100 km or more from source vents. Historic observations indicate that they can reach velocities of more than 700 km/h. They are also known by the French name nuée ardente—glowing cloud.

One of the most famous and destructive, on the Caribbean island of Martinique in 1902, rushed down the flanks of Mont Pelée Volcano at an estimated speed of 200 km/hr, and overwhelmed the city of St. Pierre, instantly killing some 29,000 people. Such a pyroclastic flow can be caused by the explosion of a hot magma mass near the top of the volcano. This produces a denser-than-air mixture of hot gases and ash, often with some larger pyroclastic materials as well. Geologists call the resulting poorly sorted deposit an *ignimbrite*.

Pyroclastic flows can also be generated by the partial or continuous collapse of an eruption column. During the 1980 eruption of Mount St. Helens, for example, a number of hot (850°C) pyroclastic flows, likely caused by column collapse, traveled up to 8 km down the north side of the mountain and covered an area of about 15 km².

LATERAL BLASTS—MOUNT ST. HELENS

In 1980, Mount St. Helens, a volcano in Washington, near the border with Oregon, erupted violently. Geologists had been monitoring the volcanic rumblings leading up to the eruption for many months, and the event, when it finally happened, displayed many features of a typical large explosive eruption. Nevertheless, the magnitude of the event caught geologists by surprise. Figure 5.10 shows the events leading to this eruption.

As magma rose under the volcano, the mountain's north flank began to bulge upward and outward. Finally, a strong earthquake dislodged a large mass of rock, which quickly moved downslope as a gigantic landslide. The landslide exposed the mass of hot magma in the core of the volcano.

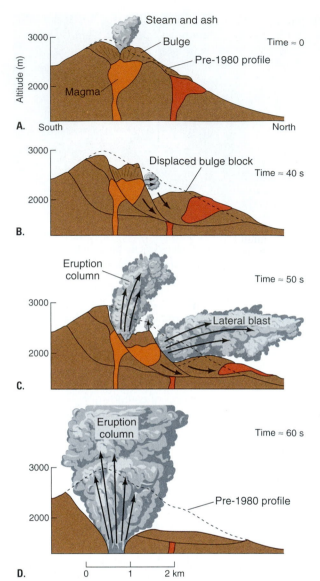

Figure 5.10 Eruption of Mount St. Helens Sequence of events leading to the eruption of Mount St. Helens in May 1980. A. Earthquakes and then puffs of steam over a period of several months indicated that magma was rising; a small crater formed, and the north face of the mountain bulged alarmingly. B. On the morning of May 18, an earthquake shook the mountain, the bulge broke loose, and a rock mass slid downward. This reduced the pressure on the magma and initiated the lateral blast that killed a geologist, David Johnston, who was monitoring the event for the U.S. Geological Survey. C. The violence of the eruption caused a second block to slide downward, exposing more of the magma, and initiated an eruption column. D. The eruption increased in intensity. The eruption column carried volcanic ash as high as 19 km into the atmosphere.

With the lid of rock removed, the gas-rich magma underwent such rapid decompression that a mighty blast resulted, blowing a mixture of magmatic gas, ash, and pulverized rock both sideways and upward. What was unusual about the Mount St. Helens eruption is that the initial blast was sideways rather than upward. Other lateral blasts have been recorded, but they are not common.

Within the devastated area that extends up to 30 km from the Mount St. Helens crater and covers some 600 km², trees in the once-dense forest were leveled and covered with hot debris.

Despite the trauma of the eruption, life returned quickly to the devastated area around Mount St. Helens (see *Understanding our Environment*, Box 5.1, *Life Returns to Mount St. Helens*).

> ### Before you go on:
> 1. Hawaiian-type eruptions tend to be nonexplosive. Why?
> 2. What is an eruption column? Why does it form?
> 3. How does a pyroclastic flow differ from an eruption column?

VOLCANOES

There are two broad families of volcanoes; those formed by eruptions from a central vent, and those that erupt through a long fissure. Central-vent eruptions build mounds of the kind most people associate with volcanoes. Fissure eruptions build plateaus. Within each family there are variations

caused by the kind of magma erupted. The landforms characteristic of each kind of volcano are distinctive. We will first discuss those formed by central eruptions.

CENTRAL ERUPTIONS

Based on their size and shape there are three broad classes of central-vent volcanoes—shield volcanoes, tephra cones, and stratovolcanoes. Each volcano type reflects the composition of the magma that creates it.

SHIELD VOLCANOES

A **shield volcano** is a broad, dome-shaped mountain with an average surface slope of only a few degrees. It somewhat resembles a warrior's shield lying on the ground, which gives the structure its name (Figure 5.11). Shield volcanoes typically form from the eruption of basaltic lava, in which the proportions of ash and other tephra are small. Low-viscosity basaltic lavas can flow for kilometers down gentle slopes and they cool and solidify into thin (centimeters thick) sheets of nearly uniform thickness. It is the accumulated lava from repeated eruptions of low-viscosity lava that builds a shield volcano.

In Figure 5.11, observe how the slope of Mauna Kea is very gentle at the summit but steepens on the flanks. This shape is a consequence of the low viscosity of the basaltic magma that built it. Near the summit, the slope is slight because the very fluid magma easily runs down even a

Figure 5.11 Shield Volcano Mauna Kea, a 4200-m-high shield volcano on Hawaii, as seen from Mauna Loa. Note the gentle slopes formed by highly fluid basaltic lava. The view is almost directly north. A pahoehoe flow is in the foreground on the northeast flank of Mauna Loa.

UNDERSTANDING OUR ENVIRONMENT

LIFE RETURNS TO MOUNT ST. HELENS

May 18, 1980, saw the explosive eruption of Mount St. Helens in Washington State. The volcano had been dormant for over a century, but on that morning a magnitude-5.2 earthquake triggered the eruption. The north face of the mountain collapsed, and the resulting avalanche buried Spirit Lake and the headwaters of the Toutle River. Hot pumice and ash flowed down the flanks of the mountain; mudflows swept along nearby rivers. Winds up to 1000 km/h and temperatures of 600°C leveled trees and scorched the soil. (See Figure B5.1).

More than 620 square kilometers of forest were destroyed, and the ground near the volcano was covered

A.

Figure B5.1   Devastation at Mount St. Helens   Life returns to Mount St. Helens. A. Trees snapped off and killed by the volcanic blast of May 18, 1980. B.

B.

Revegetation by Pacific silver fir and mountain hemlock trees, 19 years after the blast. Meta Lake Trail, Mount St. Helens National Volcanic Monument, Washington.

gentle slope, so the lava accumulates only in a thin layer. The farther lava flows down the flank, the cooler and more viscous it becomes, so the steeper the slope must be for it to flow. Slopes on young, growing shield volcanoes, such as Kilauea in Hawaii, typically range from less than 5° near the summit to 10° on the flanks.

The upper portions of several large shield volcanoes rise as islands in the ocean, including the Hawaiian Islands, Tahiti, Samoa, the Galapagos, and many others. The gentle slopes of the shield volcanoes can be confusing, making them appear smaller than they actually are. Mauna Loa volcano, for example, rises to a height of 4169 m above sea level, but if measured from the seafloor the height is 10,000 m, making Mauna Loa the highest mountain on Earth.

TEPHRA CONES

When gases continually bubble out of a magma that is exposed to the surface through a volcanic vent, a relatively gentle shower of pyroclastic debris may continue for many days. As this debris showers down, it builds a **tephra cone**

around the vent (Figure 5.12). The slopes of tephra cones are steep, typically about 30°, because the loose, coarse pyroclastic fragments are unstable on a steeper slope.

Fractures may split the flanks of a large volcano so that eruptions of lava and/or tephra occur. Small tephra cones of spattered lava or tephra may then develop above these fractures, like so many small pimples (Figure 5.13).

STRATOVOLCANOES

Some volcanoes, particularly those of andesitic composition, emit both viscous lava flows and tephra. The emissions tend to alternate, forming alternating strata of lava and tephra, building a **stratovolcano**. Stratovolcanoes are large, conical, and steep-sided. The lava flows are a major distinguishing feature between tephra cones and stratovolcanoes. The volume of tephra may equal or exceed the volume of the lava.

Stratovolcanoes may be thousands of meters high. Their slopes are steep, like those of tephra cones. Near the summit, a stratovolcano's slope may reach 40°; toward the base, the slope flattens to about 6°–10°. The steep slope

with more than three feet of ash. More than 60 people perished, including a geologist who was studying the mountain.

This was not the first time Mount St. Helens had devastated its surroundings. Over many millennia, the volcano has spewed millions of tons of lava, pumice, and ash, enough to build the mountain that erupted in 1980. Evidence of those ancient eruptions can be seen in ancient mudflow deposits along riverbanks and layers of ash and pumice along steep roadsides. After each eruption, the surrounding ecosystem slowly recovered. Today, in 2002, as these words are being written, it is still recovering form the 1980 eruption, but this time scientists are studying the process in great detail.

Perhaps the most significant conclusion of these studies is that natural disturbances like the Mount St. Helens eruption are essential to certain ecological processes: they initiate forest succession and create new habitats. The process begins with the plants and animals that survived the disturbance. At Mount St. Helens these included below-ground dwellers such as pocket gophers and deer mice, ground-dwelling insects and spiders, and animals that were hibernating at the time of the eruption. Other survivors included buried roots and bulbs, seedlings and shrubs covered by snowbanks, and the few trees that remained standing.

Recovery began when the snow melted and the plants that had been covered by snowbanks started to spread. Seeds sprouted in soil that had been buried in snow. Wind-blown seeds of plants like fireweed took root. As the plant life recovered, wildlife gradually returned to the blast zone.

An interesting effect of the eruption and its aftermath was the creation of a number of different habitats, which attracted a mix of species that had not previously lived in the area. This was especially true of birds, with high-country species joining the species common to eastern Washington, the dry lands of eastern Oregon and northern Nevada, and those seen in lowland meadows and wetlands. Ten years after the blast, willows, alders, and cottonwoods growing along streams created deciduous woodland in place of the former largely evergreen fir and pine forest, and more new bird species began to appear—in particular, neotropical migrating species that have been declining in recent decades.

In 1982, Congress created the Mount St. Helens Volcano Monument and placed it under the management of the Forest Service. With an ample food supply and little human disturbance, some species have thrived. Today a wide variety of plant and animal species can be seen in the 110,000-acre monument. Birds have been especially successful, but there are many kinds of mammals as well. Herds of elk and deer have expanded dramatically.

Small mammals such as mice and voles are multiplying, followed by their predators—weasels, hawks, and the like. While the blast was lethal to most insects, ant colonies survived and resumed their activities soon after the eruption. The most conspicuous plants are small fir and hemlock trees and weedy plant species such as fireweed and lupine.

Scientists view the recovery of vegetation at Mount St. Helens as the first step in a long-term sequence that will eventually transform an open, ash-covered blast zone to an old-growth forest over the next 200 to 500 years. As different habitats are created and filled with new or returning species, ecologists will refine their knowledge and use their findings to develop better strategies for maintaining natural diversity in managed areas.

A.

B.

Figure 5.12 Tephra Cones   A. Two small tephra cones forming as a result of an eruption in Kivu, Zaire. Arcs of lights are caused by the eruption of red-hot lapilli and bombs. B. Tephra cone in Arizona built from lapilli-sized tephra. Note the small basaltic lava flow coming from the base of the cone.

Figure 5.13 Volcanic Landscape Tephra cones on the flanks of Mauna Kea, Hawaii. The cone in the foreground is called Nohonaohae, and the large shield volcano in the background is Kohala. Nohonaohae was built about 20,000 years ago.

near the summit is due in part to the brief flows of viscous lava and in part to the tephra. As a stratovolcano develops, lava flows act as a cap to slow erosion of the loose tephra, and thus the volcano becomes much larger and steeper than a typical tephra cone.

Stratovolcanoes are among Earth's most picturesque sights. The snow-capped peak of Mount Fuji in Japan has inspired poets and writers for centuries (Figure 5.14). Mount Rainier and Mount Baker in Washington State and Mount Hood in Oregon are majestic examples in North America, and Mt. Mayon in the Philippines is both picturesque and dangerous (see Changing Landscapes No. 2).

OTHER FEATURES OF CENTRAL ERUPTIONS

A lot of additional interesting features are associated with eruptions from central vents. These include craters, lava domes, flank eruptions, calderas, and diatremes.

CRATERS

Near the summit of most volcanoes is the vent from which lava, gases, and tephra erupt. Around the vent, a **crater** commonly forms—a funnel-shaped depression, with steep-sided walls, that opens upward (Figure 5.15). Craters form in two ways: by the collapse of the steep sides of the vent,

Figure 5.14 Classic Stratovolcano Mount Fuji, Japan, a snow-clad giant that towers over the surrounding countryside, displays the classic profile of a stratovolcano.

Figure 5.15   Volcanic Crater   Crater at the summit of Sakurajima Volcano, Kyushu, Japan. The photograph was taken as an eruption was commencing.

or by an explosive eruption. As a lava eruption wanes, the last of the erupting magma drains back into the vent, and solidifies. With a subsequent eruption, pressure blasts open the vent, removing both the solidified magma from the previous eruption and part of the crater wall. In this way, a crater can grow slowly larger, eruption by eruption.

LAVA DOMES

Magma extruded at the close of a pyroclastic eruption of a stratovolcano or a tephra cone tends to have very little dissolved gas remaining, so instead of further pyroclastic activity, the gas-poor magma is extruded as a lava flow. If the magma is very viscous—as in a rhyolitic or andesitic magma, it squeezes out to form a *lava dome*. A lava dome more than 200 m high now fills the center of the Mount St. Helens crater (Figure 5.16).

CALDERAS

Many shield and stratovolcanoes are marked near their summit by a striking and much larger depression than a crater. This is a **caldera**, a roughly circular, steep-walled basin about a kilometer diameter or larger. (*Caldera* is from the Spanish word for cauldron.)

Calderas are created by collapse of the surface rock following an eruption and partial emptying of the underlying magma chamber. What happens is that rapid ejection of magma during a large eruption of tephra or lava can leave the magma chamber empty or partly so. The unsupported roof of the emptying chamber then sinks slowly under its own weight, like a snow-laden roof on an old barn, dropping downward on a ring of steep vertical fractures. Subsequent volcanic eruptions commonly occur along these fractures, creating roughly circular rings of small cones inside the caldera.

Crater Lake in Oregon occupies a circular caldera 8 km in diameter (Figure 5.17). The caldera formed about 6600

Figure 5.16   Lava Dome   A lava dome in the crater of Mount St. Helens, Washington, in May 1982. The plume rising above the dome is steam.

Figure 5.17 Crater Lake   Crater Lake in Oregon occupies a caldera 8 km in diameter that crowns the summit of what remains of a once-lofty stratovolcano, posthumously named Mount Mazama. Crater Lake is the deepest lake in North America.

Figure 5.18 Remains of a Long-ago Eruption   The Pinnacles, Crater Lake National Park. Striking erosional forms developed in the thick tephra blanket left by the eruption of Mount Mazama 6600 years ago.

years ago during a massive eruption that spread a thick tephra blanket over Crater Lake National Park and a vast area of the northwestern United States and Canada (Figure 5.18). The volcano that erupted has been posthumously called Mount Mazama. The tephra can be seen at many places in Crater Lake National Park and over a vast area of the northwestern United States. During this outpouring of about 75 km^3 of tephra—equal to a cube of tuff 4.25 km (about 2.6 miles) on edge—what remained of the volcano's roof collapsed into the magma chamber (Figure 5.19). Yellowstone National Park contains several giant overlapping calderas that formed during great tephra eruptions that occurred 2.0, 1.2, and 0.6 million years ago.

RESURGENT DOMES

A volcano does not necessarily go dormant following the formation of a caldera. Often, more magma enters the chamber and lifts the collapsed caldera floor to form a *resurgent dome*. Subsequently, small tephra cones and lava flows build up in the interior of the caldera. Wizard Island in Crater Lake is part of a resurgent dome.

DIATREMES

Diatremes are volcanic pipes filled with a rubble of broken rock. The walls of diatremes are vertical, or very nearly so, and the fragments of rubble tend to be angular and of many

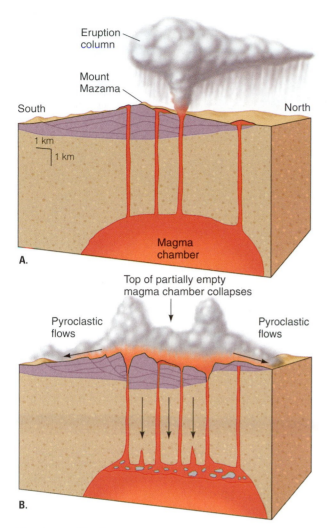

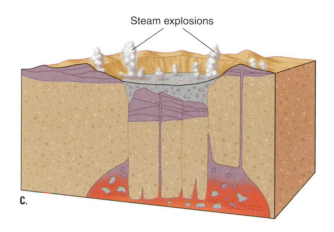

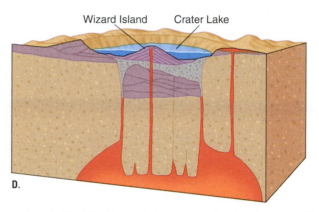

Figure 5.19 The Eruption of 6600 Years Ago Sequence of events that formed Crater Lake following the eruption of Mount Mazama 6600 years ago. A. Eruption column of tephra rises from the flank of Mount Mazama. B. The eruption reaches a climax. Dense clouds of ash fill the air, the hot pyroclastic flows sweep down the mountainside as the central part of the moun- tain subsides into the emptying magma chamber. It was at this stage that the deposits shown in Figure 5.18 were formed. C. The top of Mount Mazama collapses still further into the magma chamber, forming a caldera 8 km in diameter. D. During a final phase of eruption, Wizard Island is formed. The water-filled caldera is Crater Lake, shown in Figure 5.17.

different kinds of rock. Although no one in modern times has seen a diatreme erupting, there are many clues to suggest their origin. Diatremes range in diameter from a few meters to a kilometer or so, and among the rocks found in many is peridotite, a rock characteristic of the mantle. Observations suggest that diatremes form when gas-charged magma generated deep in the mantle literally blasts its way upward, punching a hole through to the surface.

A famous diatreme is the diamond mine in Kimberley, South Africa (Figure 5.20). The Kimberley pipe, as this diatreme is called, was the site of the world's first great

Figure 5.20 Famous Diamond Pipe The Kimberley diatreme, South Africa. It was here, in Kimberley, where the first diamond pipe was discovered. The pit is all that remains after mining. The vertical walls are the actual pipe walls.

underground diamond mine. Diamonds are mixed in with the rubble that fills the pipe. As discussed in Chapter 3, and shown in Figure 3.31, diamonds only form at pressures equal to or greater than those reached at depths of 150 km or more below the Earth's surface. The Kimberley pipe, therefore, was formed by magma that formed at least 150 km below the Earth's surface. Perhaps it is just as well that no one has observed a diatreme eruption, because it would surely be very violent.

FISSURE ERUPTIONS

Extrusion of lava along an elongate fracture in the crust is called a *fissure eruption* (Figure 5.21). Such eruptions, often very dramatic, are associated with basaltic magma. When fissure eruptions occur on land the low-viscosity basaltic lava tends to spread widely and to create flat lava plains; such lavas are called **plateau basalts**.

Although there have been a number of small fissure eruptions observed in Hawaii, only one eruption large enough to actually form a plateau has been observed in historic times. That eruption occurred in Iceland in 1783 along a fracture 32 km long. Known as the Laki eruption, lava from it flowed 64 km from one side of the fracture and nearly 48 km from the other, covering 588 km^2. The volume of the lava extruded was 12 km^3, enough to cover an area the size of Cape Cod, Massachusetts, to a depth of 20 m. The Laki eruption is the largest lava flow of any kind in historic times. It was also one of the most deadly. The flow destroyed homes and food supplies, killed livestock, and covered agricultural fields. Famine followed and more than 9000 died—20 percent of the Icelandic population.

There is good evidence that larger eruptions occurred prehistorically. The Roza flow, a great sheet of plateau basalt in eastern Washington State, can be traced over 22,000 km^2 (twice the area of Massachusetts) and shown to have a volume of 650 km^3 (about 1.3 times the volume of Lake Erie).

PILLOW BASALTS

The Laki eruption in Iceland occurred where a segment of the midocean ridge is exposed above sea level. Similar fissure eruptions also happen beneath the sea because Earth's most extensive volcanic system lies beneath the sea. Fissures that split the centers of the midocean ridges are channelways for rising basaltic magma. This magma spreads sideways away from the erupting fissure and forms sheets of basalt on the ocean floor.

When the basaltic magma erupts, seawater quenches (cools) it so rapidly that a very distinctive volcanic feature forms. Near submarine volcanic fissures, where lava temperature is highest, thin lava flows are rapidly quenched to form glassy surfaces. These flows build piles of basalt in which each sheet may have a thickness of only 20 cm or so. Farther from a vent, where the temperature of the flowing lava has decreased, "pillow structure" develops. The term **pillow basalt** describes discontinuous, pillow-shaped

Figure 5.21   Fissure Eruption   Lava fountains rise from an eruptive fissure on Kilauea Volcano, Hawaii, in 1983. The fissures cut through a forest and older lava flows. New lava is flowing toward the photographer.

masses of basalt, ranging from a few centimeters to a meter or more in greatest dimension (Figure 5.22).

Pillow structure forms when the basaltic lava surface is chilled quickly. The brittle, chilled surface cracks, making an opening for the still-molten magma inside to ooze out like a bead of toothpaste. In turn, the newly oozed lava chills, its surface cracks, and the process continues. The end result is a pile of lava pillows that resemble a jumbled pile of sandbags. Most of the lavas of the oceanic crust are pillow basalts, and because MORB (midocean ridge basalt; see Chapter 4) is the most common igneous rock on Earth, pillow basalt must be the most common kind of volcanic feature on the Earth.

ASH-FLOWS

Fissure eruptions of andesitic or rhyolitic magma are much less common than fissure eruptions of basaltic lava, but

Figure 5.22   Lava on the Seafloor   Tube-shaped pillows of basalt on the seafloor near Hawaii. The pillows were formed by an underwater eruption.

Figure 5.23 Ash-Flow Sheet An ash-flow sheet deposited as a result of the eruption of Mount St. Helens, Washington, in 1982.

when they happen they must be incredibly destructive because what is erupted is a great sheet of hot tephra. Sometimes the pyroclasts in the tephra are so hot that the fragments form welded tuff (Figure 4.9B). More commonly the pyroclasts in ash-flow deposits are tuffs in which the particles are held together by addition of a cement such as silica (Figure 5.23).

There have not been any large ash-flow eruptions in modern times, but there are many examples in the geolog-

ic record. Some 40 to 50 million years ago, huge ash-flow eruptions happened in Nevada, and the erupted products cover an area in excess of 200,000 km^2—equivalent to Arizona plus Nevada—and in places more than 2 km thick. If such an eruption were to happen today the results would be catastrophic.

POSTERUPTION EFFECTS

When active volcanism finally ceases, rock in and near an old magma chamber may remain hot for hundreds of thousands of years. Descending groundwater that comes into contact with the hot rock becomes heated and tends to rise again to the surface along rock fractures to form thermal springs. Thermal springs at many volcanic sites—for example, in Italy, Iceland, Japan, and New Zealand—have become famous health spas and also sources of energy (see *The Science of Geology*, Box 5.2, *Harnessing the Heat Within*).

A thermal spring that intermittently erupts water and steam is a *geyser*. The name comes from the Icelandic word *geysir*, meaning to gush, for Iceland is the home of many geysers (Figure 5.24). Most of the world's geysers outside Iceland are in New Zealand and in Yellowstone National Park.

Gases that bubble up from magma far below the surface may emerge either from the central volcanic vent or from small satellite vents. The emitted gases are mostly water vapor, but some evil-smelling sulfurous gas may be present, too. Such emerging gases can alter and discolor the rocks they contact (Figure 5.25).

Figure 5.24 Where the Geyser Got Its Name The great Geysir in Iceland, from which all geysers take their name. Geysers erupt steam and hot water intermittently. They do so when water at the bottom of the hot spring flashes to steam and starts an eruption.

Figure 5.25 Volcanic Gases Sulfurous gases stream from several vents following a 1965 eruption in Hawaii. Sulfur condensed from the acrid gases has killed the vegetation and covers the ground with a yellow coating of elemental sulfur.

THE SCIENCE OF GEOLOGY

HARNESSING THE HEAT WITHIN

Iceland is an unusual place. It lies even farther north than the northernmost part of Hudson Bay, and, as the name suggests, it's a very cold place, with numerous glaciers. But Iceland straddles the Mid-Atlantic Ridge, so it is also a place of active volcanism. The people of Iceland have found many clever ways to harness volcanic heat to combat the cold climate.

Using water warmed by hot volcanic rocks, they heat their houses, grow tomatoes in hothouses even though the temperature outside may be subfreezing, and swim year-round in naturally heated pools. Icelanders also generate most of the electricity they need using volcanically produced steam.

Everywhere, Earth's temperature increases with depth. Everyone, not just Icelanders, should be able to drill water-circulation holes deep enough so that the Earth's internal heat could be tapped—in theory. In practice, this is very difficult. To be used efficiently, geothermal steam should be 200°C or hotter. In most parts of the world, holes must be 5 to 7 km deep to reach rock that hot. We can drill that deep but it's very expensive to do so.

For the practical development of geothermal energy, as Earth's store of heat is called, we need to be able to reach rock temperatures of 200°C or higher within 3 kilometers of the surface. So far at least, it is also necessary for there to be a natural reservoir of underground water that is heated by the hot rocks, as shown in Figure B5.2.

Places on Earth where these conditions are met are places of recent volcanic activity. Most of the world's volcanic and magmatic activity is close to plate margins, and it is here that geothermal power is being used, in places such as New Zealand, the Philippines, Japan, Italy, Iceland, and the western United States.

Unfortunately, the total energy that can be recovered from natural geothermal fields is not very large. Today geothermal power is of local importance only. It is an interesting question whether the situation may change. Many

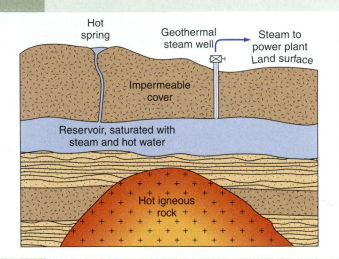

Figure B5.2 Geothermal Reservoir A typical geothermal steam reservoir. Water in a permeable aquifer, such as a tuff, is heated by magma or hot igneous rock. As steam and hot water are withdrawn through the well, cold water flows into the reservoir through the aquifer.

Before you go on:

1. What are three kinds of central-vent volcanoes?

2. What is the difference between a crater and a caldera?

3. What are plateau basalts? pillow basalts?

4. What is the origin of volcanic thermal springs?

VOLCANIC HAZARDS

Volcanic eruptions are not rare on land, and are essentially continuous on the seafloor. Every year about 50 volcanoes erupt on the continents, including one in Antarctica (Mt. Erebus). Most of the eruptions are basaltic. Those on the seafloor are benign to people and property. On land, eruptions of basaltic shield volcanoes often destroy property by burning it, but are seldom dangerous to people. Fissure eruptions of basaltic lava can be very dangerous, however, as we discussed earlier in the case of the Laki eruption in Iceland in 1783.

Unlike basaltic flows, tephra eruptions from andesitic or rhyolitic stratovolcanoes like Mount St. Helens and Krakatau can be disastrous, because millions live near them. Eruptions present five kinds of hazards:

1. *Hot, rapidly moving pyroclastic flows and laterally directed blasts can overwhelm people before they can evacuate.* The tragedies of Mont Pelée in 1902 and Mount St. Helens in 1980 are examples.

2. *Tephra and hot poisonous gases can bury or suffocate people.* Such a tragedy occurred in A.D. 79 when Mount Vesuvius, a supposedly dormant volcano in southern Italy, burst to life. Hot, poisonous gases killed many in the nearby Roman city of Pompeii and then tephra buried them (Figure 5.26).

3. *Tephra can be dangerous long after an eruption has*

volcanoes, like Mount St. Helens, are too active and too dangerous to be considered as geothermal energy sources. Some places, like Yellowstone National Park, could produce a large amount of energy, but if we drilled and pumped out the steam and hot water reservoirs beneath Yellowstone, the famous hot springs and geysers would soon be dry.

Interesting geothermal experiments have been conducted in New Mexico. The goal is to create our own geothermal fields. In the Jemez Mountains on the edge of an extinct (but still hot) volcano, scientists drilled two holes deep into the hot rock, as shown in Figure B5.3. Then they shattered the hot rock with explosives to create an artificial reservoir and pumped water through to absorb the heat and produce steam.

The first tests were only partly successful. A major difficulty was that water did not flow uniformly through the hot rock but instead flowed mostly through the wider flow paths, and the rocks that lined them soon cooled. Further tests also are proceeding in France, England, and other countries.

Hot, dry rocks are much more abundant than geothermal steam fields. If the water flow troubles encountered in the Jemez Mountains experiments can be overcome, geothermal energy may someday play a major role in meeting our energy needs.

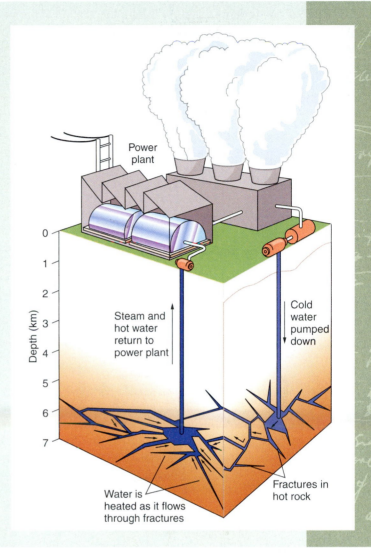

Figure B5.3   Power from Hot Rocks   Geothermal energy from hot dry rocks. Cold water is pumped through fractures at the bottom of a 6-km-deep well, becomes heated, then flows back up to geothermal power plant, where heat energy is converted to electricity.

ceased because of mudflows. Rain or meltwater from snow can loosen tephra piled on a steep volcanic slope and start a deadly mudflow that sweeps down the mountainside and can flow for miles from the base of the mountain. Such mudflows are called *lahars.* In 1985, the Colombian volcano Nevado del Ruíz experienced a small, nonthreatening eruption. But, when glaciers at the summit melted, massive mudflows of volcanic debris moved swiftly down the mountain, killing 20,000.

Figure 5.26   Ancient Volcanic Disaster   Evidence of an ancient disaster. Casts of bodies of five citizens of Pompeii, Italy, who were killed during the eruption of Mount Vesuvius in A.D. 79. Death was caused by poisonous gases, then the bodies were buried by lapilli. Over the centuries the bodies decayed away but the body shapes were imprinted in the tephra blanket. When excavators discovered the imprints, they carefully recorded them with plaster casts.

TABLE 5.1   Volcanic Disasters Since A.D. 1800 in Which a Thousand or More People Lost Their Lives

Volcano	Country	Year	Tephra Eruption	Mudflow	Tsunami	Famine
			Fatalities According to Primary Causes			
Mayon	Philippines	1814	1,200			
Tambora	Indonesia	1815	12,000			80,000
Galunggung	Indonesia	1822	1,500	4,000		
Mayon	Philippines	1825		1,500		
Awu	Indonesia	1826		3,000		
Cotopaxi	Ecuador	1877		1,000		
Krakatau	Indonesia	1883			36,417	
Soufrière	St. Vincent	1902	1,565			
Mont Pelée	Martinique	1902	29,000			
Santa Maria	Guatemala	1902	6,000			
Taal	Philippines	1911	1,332			
Kelud	Indonesia	1919		5,110		
Merapi	Indonesia	1930	1,300			
Lamington	Papua-New Guinea	1951	2,942			
Agung	Indonesia	1963	1,900			
El Chichón	Mexico	1982	1,700			
Nevado del Ruíz	Colombia	1985		25,000		

4. *Violent undersea eruptions can cause powerful sea waves called tsunami.* Tsunami triggered by the eruption of Krakatau in 1883 killed more than 36,000 coast dwellers on Java and nearby Indonesian islands. Krakatau was a volcanic island in the Sunda Strait, between Java and Sumatra. The eruption removed the island and part of the seafloor, leaving a large underwater depression.

5. *A tephra eruption can disrupt agriculture, both the growing of crops and raising of livestock, creating a famine.* Tephra eruptions from Mt. Mayon in the Philippines cause frequent disruptions to local agriculture—see Changing Landscapes No. 2.

Since A.D. 1800, there have been 18 volcanic eruptions in which a thousand or more people died (Table 5.1). As long as our planet remains tectonically active—which it should be for many millions of years to come—other violent and dangerous eruptions will occur. To some extent, volcanic hazards can be anticipated. Geologists who study volcanoes gather data before, during, and after eruptions so that they can better anticipate future eruptions. Geologists who are so prepared can advise civil authorities when to implement hazard warnings and when to evacuate endangered populations to areas of lower risk. Unfortunately, civil authorities do not always respond to the advice. Warnings were given, but ignored, before the dreadful lahar covered Nevado del Ruíz in 1985. In the case of the Mount Pinatubo eruption in the Philippines in 1994, however, civil authorities did respond and evacuated more than 250,000 people in time to avoid the worst of the eruption.

While there have been many cases where geologists have been able to help authorities avoid large death tolls, there have also been some terrible failures. The day before the great eruption of Mont Pelée, in 1902, for example, a Dr. Landes announced that Pelée presented no more danger to St. Pierre than Vesuvius does to Naples. Dr. Landes perished the following day with the other 29,000 inhabitants of St. Pierre. Landes had apparently forgotten the great eruption of A.D. 79 that obliterated the towns of Pompeii and Herculaneum. Naples still remains in danger from Vesuvius should a massive eruption happen.

Even volcanic experts lose their lives. The famous husband-and-wife team, Katia and Maurice Krafft, lost their lives during the pyroclastic eruption of Mount Unzen volcano, Japan, in 1991. And in Colombia, a group of volcanologists studying Galeras volcano during a professional field trip were overtaken by a sudden pyroclastic eruption in which six lost their lives.

Before you go on:

What are the five main volcanic hazards?

REVISITING PLATE TECTONICS AND THE EARTH SYSTEM

PLATES AND VOLCANOES

The distribution of active volcanoes around the world is strongly influenced by plates, and specifically plate margins (Figure 5.27). Most of the world's volcanism happens beneath the sea, along the 64,000-km midocean ridge. There, unseen by human eyes, basaltic volcanism continu-

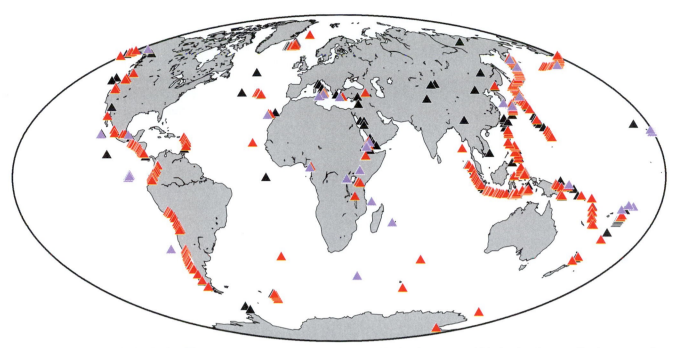

Figure 5.27 A Volcanically Active Planet Locations of volcanoes that have erupted in historic times and are considered to still be active. Red triangles are stratovolcanoes, blue triangles are shield volcanoes, and black triangles are all other types of eruptive centers, such as fissures and cinder cones.

ally creates new oceanic crust along plate spreading edges. The eruptions are mostly fissure eruptions and apparently of such small volume they are undetected. If the eruptions were much larger they would register on seismic and other recording devices.

On land, and above the sea where we can see volcanism in action, about 15 percent of all active volcanoes are located along spreading centers. For example, along the Mid-Atlantic Ridge there are volcanoes in Iceland and the Azores, and in the East African Rift Valley, there are many active volcanoes.

Most of the world's visible and active volcanoes are located where two plates collide and one is subducted beneath the other. There, water released from the subducted plate leads to the formation of andesitic magma by wet partial melting of mantle rock. The magma rises and forms great stratovolcanoes. The Pacific Ocean is ringed on three sides by subducting plate margins (Figure 2.10). As a result, a great necklace-like ring of volcanoes surrounds the ocean—a great ring of fire.

About 5 percent of all active volcanoes are located in the interiors of plates. The Hawaiian volcanoes are an example. Intraplate volcanoes are thought to be due to convective plumes of hot mantle rock rising from deep in the Earth and undergoing decompression melting deep in the asthenosphere. The resulting magma is basaltic, but it differs in composition from MORB, which is also formed by decompression melting of mantle rock. The difference arises because plume melting happens much deeper than MORB melting. Remember, as we discussed in Chapter 4,

that magma composition is influenced both by the degree of partial melting and by the pressure at which melting takes place.

SUBMARINE VOLCANISM AND THE COMPOSITION OF SEAWATER

The ocean is a wonderful example of an open system. It is the main reservoir of water for the hydrologic cycle, it is a sink for all of the sediment and dissolved salts brought down from the continents by the rivers of the world, and it dissolves gases from the atmosphere. One of those gases, carbon dioxide, is extracted by aquatic plants and animals to make their bodies and shells. When the plants and animals die, their remains sink to the bottom and get buried by sediment and thereby incorporated into the lithosphere.

Geologists have carefully measured the fluxes of material into and out of the ocean in order to try and understand why it has the composition it does, and why the composition does not seem to have varied through long geologic ages. The principal source of dissolved salts in seawater was long thought to be river water. But the balance of salts in river water is quite different from that of salts in seawater. The problem of the imbalance is not yet fully resolved, but with the discovery of the extent of volcanism along the midocean ridge, it is now clear that submarine volcanoes are also a major source of sea salts.

The midocean ridge is a place where rocks are being subjected to tensile forces that pull them apart. The tensile forces fracture the rocks and seawater penetrates deeply into the oceanic crust through the fractures. There, the seawater encounters hot magmatic rocks, reacts with them, and exchanges chemicals in solution with those in the hot rocks. The hot water, now modified by reaction, rises convectively to the seafloor, where it is discharged as submarine hot springs. The hot solutions mix with seawater and slowly change its composition. Magnesium and sulfur, for example, are removed from seawater by the hot rocks, and calcium and other chemical elements are added. The hot rocks of the midocean ridge, it now turns out, are a major factor in maintaining the composition of seawater. The system works because the flux of seawater through the rocks of the ridge is so great that a volume of water equal to the volume of the entire ocean passes through fractures in the midocean ridge in about 6 million years.

What's Ahead?

Igneous rocks are the most abundant rocks on Earth, and through continuing magmatic activity new igneous rocks are being created all the time. But igneous rocks are simply cooled and solidified magma, and magma is created deep in the Earth where chemical conditions are very different from those at the surface. When igneous rocks are exposed at the surface they are in contact with gases of the atmosphere and water of the hydrosphere and these react with, and break down, the rocks by a process called *weathering*. In the next chapter we will examine the processes of weathering and the formation of soils.

CHAPTER SUMMARY

1. Magma is molten rock beneath or on the Earth's surface. It also includes any unmelted mineral grains and dissolved gases.

2. Magma has a wide range of compositions but silica always dominates the mix. Magma is characterized by high temperatures and the ability to flow.

3. Three kinds of magma predominate: basaltic, andesitic, and rhyolitic. Approximately 80 percent of all magma erupted by volcanoes is basaltic.

4. Viscosity is the property of a substance that offers resistance to flow.

5. The principal controls on the physical properties of magma are temperature, SiO_2 content, and, to a lesser extent, dissolved-gas content. High temperature and low SiO_2 content result in fluid magma (basaltic). Lower temperature and high SiO_2 contents result in viscous magma (andesitic and rhyolitic magma). Dissolved gas reduces viscosity, but the main controls on viscosity are the SiO_2 content and temperature.

6. Once formed, magma rises buoyantly because it is less dense than the solid rocks from which it forms.

7. All magma contains dissolved gas—mainly water vapor, but also carbon dioxide and other gases. Pressure controls the amount of gas magma can dissolve.

8. Gas comes out of solution as magma rises and the pressure decreases.

9. Nonexplosive eruptions are characteristic of low-viscosity basaltic magmas. When gas comes out of solution, bubbles form and escape from basaltic magma, in some cases forming spectacular lava fountains.

10. Explosive eruptions are characteristic of viscous andesitic and rhyolitic magmas with high gas contents. Gas coming out of solution cannot escape the sticky magma and eventually the bubbles shatter the hot magma into volcanic ash.

11. Eruption columns of densely mixed hot volcanic gas and tephra can result from explosive eruptions. Such eruptions are called Plinian eruptions.

12. When a mixture of volcanic gas and pyroclasts in an explosive eruption is more dense than air, the mixture flows down the side of the volcano as a pyroclastic flow.

13. In some cases the first blast of an explosive eruption is sideways—a lateral blast. The eruption of Mount St. Helens in 1980 started with a devastating lateral blast.

14. Volcanoes erupt either from a central vent or from an elongate fissure.

15. The sizes and shapes of central-vent volcanoes depend on the kind of material erupted, viscosity of the lava, and explosiveness of the eruptions.

16. Viscous magmas erupt a lot of tephra and tend to build steep-sided tephra cones or stratovolcanoes.

17. Low-viscosity magmas low in SiO_2 tend to be erupted from central vents as fluid lavas that build gently sloping shield volcanoes. When such magma is erupted from a fissure the result is a lava plateau.

18. Near the summit of most central-vent volcanoes is a crater formed by a combination of eruption and collapse of the steep-sided walls.

19. Calderas are created by the collapse of the roof of a volcano into the underlying magma chamber following a major eruption.

20. Fissure eruptions form plateau basalts with low-viscosity magma, and ash-flows with high-viscosity magma. Submarine fissure eruptions form pillow basalts.

21. Low-viscosity eruptions are not dangerous to humans, though they can overrun buildings and landscapes. High-viscosity tephra eruptions are very dangerous.

THE LANGUAGE OF GEOLOGY

caldera (p. 133)
crater (p. 132)

diatreme (p. 134)

lava (p. 123)

pillow basalt (p. 136)
plateau basalt (p. 136)
pyroclastic flow (p. 128)

shield volcano (p. 129)
stratovolcano (p. 130)

tephra cone (p. 130)

viscosity (p. 124)
volcano (p. 124)

QUESTIONS FOR REVIEW

1. Is the major oxide component of magma SiO_2, MgO, or Al_2O_3? Briefly describe the effect of the SiO_2 content on the fluidity of magma. What effect does temperature have on viscosity?

2. What is the origin of andesitic magma such as that erupted by Mount St. Helens? With what kind of volcanoes are andesitic eruptions associated? What is the distribution of andesitic volcanoes with respect to today's tectonic plates?

3. Why does a shield volcano like Mauna Loa in Hawaii have a gentle surface slope, while a stratovolcano such as Mount Fuji in Japan has steep sides?

4. How does a tephra cone differ from a stratovolcano?

5. Describe the sequence of events leading to the formation of a caldera.

6. How does a lava dome such as the one that now lies in the crater of Mount St. Helens form? How, if at all, does a lava dome differ from a resurgent dome?

7. Describe the distinctive kind of basaltic lava flows formed by submarine fissure eruptions. How do such flows form?

8. How is the distribution of active volcanoes related to plate tectonics?

9. Briefly describe the plate-tectonic settings of andesitic and basaltic volcanoes. Where does Kilauea Volcano in Hawaii sit with respect to the Pacific Plate? Where does Mount St. Helens sit with respect to a plate boundary?

10. What role does submarine volcanism play in determining the composition of seawater?

Click on *Presentation* and *Interactivity* in the **GeoHazards** and the **Volcanoes and Igneous Rocks** modules of your CD-ROM to further explore resources and activities presenting concepts from this chapter. Select *Assessment* in the same modules to test your understanding of this chapter.

Changing Landscape

Mt. Mayon, with town of Legazpi at its foot

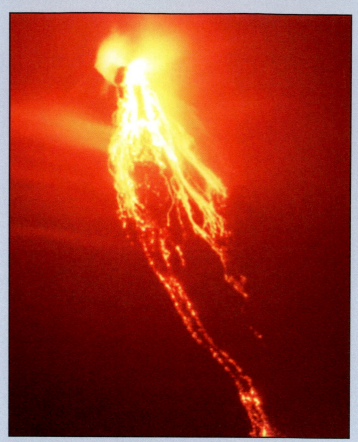

Mt. Mayon in eruption, July 24, 2001. A river of lava fragments cascades down the slope.

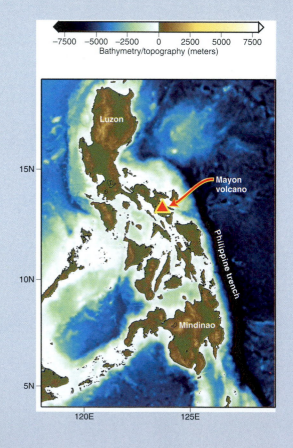

Mt. Mayon in eruption, July 26, 2001. Ash column 13km high.

Ash flow buries coconut farm, Mabinitu Village; eruption of March, 2000

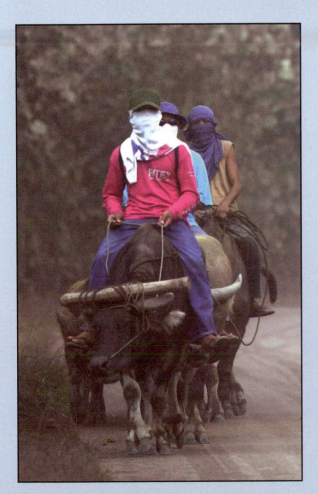

Residents of Ligao leave during eruption of
June 25, 2001

LIVING CLOSE TO DANGER

No landscape changes more rapidly than that around an active volcano. Mt. Mayon, located near the southern tip of the island of Luzon, Philippines, is the world's most active starto-volcanoe. Every eruption of a stratovolcano is potentially dangerous, and Mt. Mayon is no exception: some eruptions spill fluidized masses of red-hot lava fragments and volcanic ash which settles as a thick blanket across the countryside. In the rainy season, volcanic ash on the steep slopes of Mt. Mayon becomes a smothering mudflow that sweeps down the mountainside, covering everything in its path.

Despite the ever-present danger, a large population lives in the villages clustered around the volcano. The climate is pleasant, the soil fertile, and the people are apparently prepared to face the risks. The willingness of the local population to live so close to danger poses many problems for civil authorities who must decide where and when people have to be evacuated. The authorities rarely have much time to act, and the numbers to be relocated are huge. In the eruption of March, 2000, 83,000 people had to be evacuated; in June 2001, 46,000 had to flee to save their lives, and a month later, in July, 2001, another 10,000 had to be evacuated. Despite the devastation and the large numbers of displaced people, very few lives have been lost in recent years.

Chapter 6

Weathered boulders in the Alabama Hills near the eastern base of Mt. Whitney in California's Sierra Nevada.

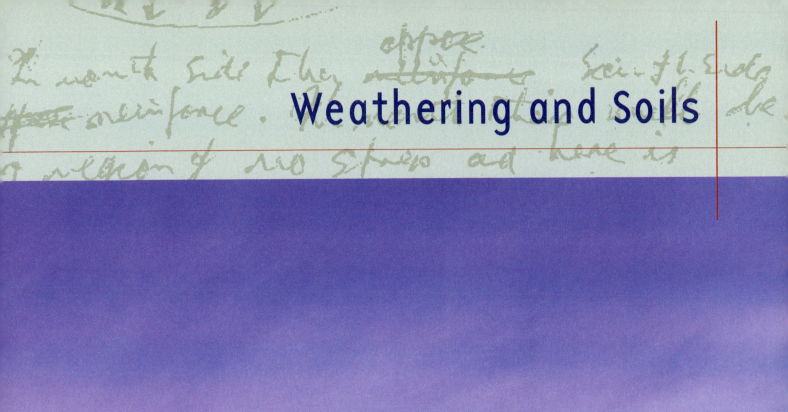

Weathering and Soils

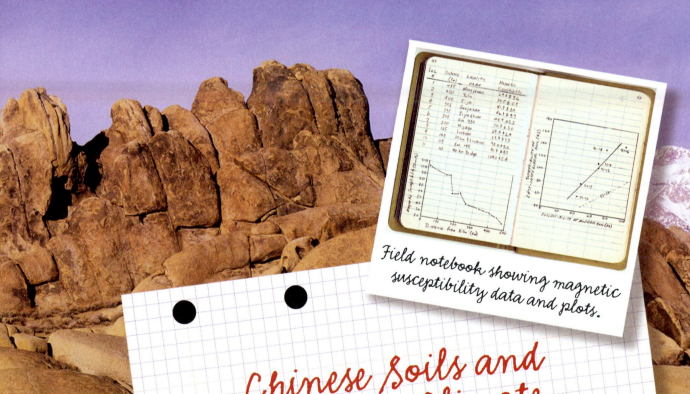

Field notebook showing magnetic susceptibility data and plots.

Chinese Soils and Monsoon Climate

A billion people in China rely on monsoon rainfall to sustain their agricultural productivity. What would happen if the monsoon rains weakened or failed to appear?

How will monsoon rainfall be affected if Earth's climate continues to warm in future decades? The answers may partly lie in the long geologic record of monsoon climate.

During the summer of 1991, I joined a team of scientists in central China to study profiles of yellowish wind-blown dust and interlayered reddish soils in order to reconstruct monsoon history. We carried a device to determine magnetic susceptibility, which is a measure of the concentration of magnetic minerals in sediment. Laboratory studies had shown that fine-grained magnetic minerals are produced by soil bacteria that thrive in a warm, moist climate. Earlier work had confirmed that high susceptibility values correspond to such warm-climate conditions, and low values to cold, dry climate when soil development is weak. Alternating high and low susceptibility values indicated that the monsoon climate has shifted between these extremes repeatedly during the last several million years.

Whenever we stopped our jeep to examine a sediment profile, we made susceptibility measurements at the top of the modern soil and recorded the average values. After a time, a clear pattern appeared. High values were measured in the southern, moister part of the region where summer monsoon rain is heaviest and modern soils are thick and rich in organic matter. The values decreased as we moved northwestward into drier desert regions of low annual rainfall and thin, poorly developed soils. We compared these results with measurements from a prominent buried soil in the same region. This well-developed ancient soil contains warm-climate fossils, and lies within part of today's dry zone, implying that when the soil formed about 120,000 years ago the climate was warmer and wetter than now. These results suggest that in a future, warmer Earth, the monsoon rains may strengthen, thereby enhancing, rather than hindering, Chinese agricultural productivity in the region we studied.

Stephen C. Porter

KEY QUESTIONS

1. **What is weathering, and what are the two types of weathering?**

2. **How does rock change as it weathers physically?**

3. **How does rock change as it weathers chemically?**

4. **What factors influence the intensity of weathering?**

5. **How are weathering and soil formation related?**

6. **Why is soil such an important natural resource?**

7. **How have human activities accelerated the natural rate of soil erosion?**

8. **What evidence can geologists use to infer a relationship between weathering and plate tectonics?**

INTRODUCTION: WEATHERING— THE BREAKDOWN OF ROCK

Whether it occurs rapidly or slowly, physical and chemical alteration of rock takes place throughout the zone where the lithosphere, hydrosphere, biosphere, and atmosphere mix. The zone extends downward into the ground to whatever depth air, water, and microscopic forms of life can penetrate. Within it, the rock constitutes a porous framework, full of fractures, cracks, and other openings. Some are very small, but all make the rock vulnerable. This open framework is continually being attacked, both chemically and physically, by water solutions and by microscopic life. Given sufficient time, conspicuous changes of the rock result.

When exposed at Earth's surface, no rock escapes the effects of **weathering**, the chemical alteration and mechanical breakdown of rock and rocky debris when exposed to air, moisture, and organic matter. In other words, weathering involves processes that convert rock to regolith. The results of weathering are often seen in landslide scars and large excavations, such as road cuts, that expose bedrock.

In Figure 6.1, loose, unorganized, earthy regolith in which the texture of the bedrock is no longer apparent grades downward into rock that has been altered but still retains its organized appearance, and eventually into practically unaltered bedrock. It is evident in such an exposure that alteration of the rock progresses from the surface downward.

A close look at the bedrock in Figure 6.1 would show that near the bottom of the exposure the cleavage surfaces of feldspar grains are bright and reflective, while higher up such surfaces are lusterless and stained. Soft, earthy material near the top of the outcrop no longer resembles the feldspar that was once present there and that now has largely decomposed. The changes that have occurred result mainly from **chemical weathering**, the decomposition of

Figure 6.1 Weathering Profile Long-continued chemical alteration has produced a deep, reddish weathering profile exposed in this road cut in Costa Rica. Plant roots are largely confined to the uppermost (brownish) soil near the top of the exposure. The profile grades downward into partly altered but distinctly layered rocks.

rocks and minerals as chemical reactions transform them into new chemical combinations that are stable at or near Earth's surface.

Some regolith consists of rock fragments identical to the adjacent bedrock. We often see piles of loose rock fragments at the base of bedrock cliffs. Compared with the bedrock, these coarse rock fragments show little or no evidence of chemical weathering. The mineral grains are fresh or only slightly altered. Clearly, then, bedrock can be broken down not only chemically but also by **physical weathering**, the disintegration (physical breakup) of rock.

Weathering is an integral part of the rock cycle, and, as we saw in Chapters 1 and 2, the rock cycle is linked to the tectonic cycle through plate tectonics. If fresh rocks were not continually brought to the surface by tectonic uplift and volcanism, erosion would lower the land, and a deep weathering profile would eventually develop on each con-

tinent. Therefore, the processes associated with plate tectonics provide the grist for the weathering mill.

In the following sections, we look first at physical weathering and then at chemical weathering. Although we treat them as being distinct from one another, in fact, the two processes generally work hand in hand, and their effects are inseparably blended. Finally, we will examine soils, which are the widespread result of weathering and are among our most important natural resources.

> **Before you go on:**
> 1. Where does weathering occur?
> 2. What is the relationship between physical and chemical weathering?

PHYSICAL WEATHERING

Physical weathering of rock may begin with the formation of fractures (joints), in which crystals of ice or salt can grow. It may also occur when rock is heated by fire or disrupted by the growth of plant roots.

DEVELOPMENT OF JOINTS

Most rocks are subject to a type of mechanical fracturing that plays an important role in weathering. Rocks in the upper crust are brittle and, like any other brittle material, they break at weak spots when they are twisted, squeezed, or stretched by tectonic forces. The timing and origin of such forces are not always obvious, but they leave their mark in the form of **joints**, which are fractures in a rock along which no appreciable movement has occurred.

Joints form as rocks buried deep in the crust are slowly uplifted and erode away. Removal of the weight of overlying rocks releases stress on the buried rock and causes joints to open slightly, thereby allowing water, air, and microscopic life to enter.

Rock adjusts to removal of overlying rock by expanding upward. As it does so, closely spaced fractures approximately parallel to the ground surface can develop, giving the rock a sheet-like appearance (Figure 6.2). Generally, such joints disappear below a depth of about 50 to 100 m, as the pressure of the overlying rock becomes too great for joints to form. More commonly joints occur as a widespread set or sets of parallel fractures (Figure 6.3). The pattern and spacing of joints strongly influence the way a rock breaks apart.

One class of joints is restricted to tabular bodies of igneous rock—such as dikes, sills, lava flows, and welded tuffs—that cooled rapidly at or close to the land surface. When such a body of igneous rock cools, it contracts and may fracture into pieces, in much the same way that a very hot glass bottle contracts and shatters when plunged into cold water. Unlike shattered glass, however, cooling frac-

Figure 6.2 Sheet Jointing   Sheet-like jointing in massive granite forms a stepped surface in Yosemite National Park, California.

Figure 6.3 Joint Sets   Two well-developed sets of vertical joints combine with horizontal bedding planes to cause this red sandstone in Brecon Beacons, Wales, to break into roughly rectangular blocks.

Figure 6.4 Columnar Joints   Columnar jointing in igneous rock near San Miguel Regla, Mexico, offers a challenge to a rock climber jambing his way up a crack between two adjacent columns.

tures in igneous rock tend to form regular patterns. Geologists use the term *columnar joints* for joints that split igneous rocks into long prisms or columns (Figure 6.4).

Crystal Growth

Water moving slowly through fractured rocks contains ions, which may precipitate out of solution to form salts. The force exerted by salt crystals growing either within rock cavities or along the boundaries between mineral grains can be very large and can result in the rupture or disaggregation of rocks.

The effects of such physical weathering can often be seen in deserts, where salt crystals grow as rising groundwater evaporates and its dissolved salts are precipitated. A quite different effect can be seen in the ice-free valleys of coastal Antarctica, where bizarre cavernous landforms have been produced by the physical weathering of granite boulders and bedrock outcrops (Figure 6.5). When salts crystallize out of solutions confined within the pores and fine cracks between mineral grains in the granite of these valleys, the rock slowly disintegrates. The resulting debris is carried away by strong winds blowing off the nearby ice

Figure 6.5 Cavernous Weathering Granitic bedrock on the flank of Gondola Ridge in Antarctica has been so strongly weathered that it resembles Swiss cheese. Such cavernous weathering results from granular disintegration as salts crystallize in small cavities and along grain boundaries.

Figure 6.6 Frost Wedging An extensive field of granite blocks, 1–2 m in diameter, covers the summit of Mount Whitney, the highest peak in California's Sierra Nevada. Frost wedging, the result of melting snow that percolates downward into cracks where it refreezes, has disrupted the granite bedrock, producing this vast litter of angular boulders.

sheet, leaving an unusual landscape in which the exposed rocks resemble Swiss cheese.

FROST WEDGING

Wherever temperatures fluctuate about the freezing point for part of the year, water in the ground periodically freezes and thaws. As water freezes to form ice, its volume increases by about 9 percent. When freezing occurs in the pore spaces of a rock, water is strongly attracted to the growing ice, thereby increasing the stresses against the rock. This process leads to a very effective type of physical weathering known as **frost wedging**, in which ice forms in a confined opening within a rock, and forces the rock apart. Frost wedging is strong enough to force apart tiny particles as well as huge blocks, some weighing many tons (Figure 6.6). Frost wedging probably is most effective at temperatures of −5° to −15°C. At higher temperatures ice pressures are too low to be very effective, and at lower temperatures the rate of ice growth decreases because the water necessary for crack growth is less mobile. Frost wedging likely is responsible for most of the rock debris seen on high mountain slopes.

DAILY HEATING AND COOLING

Some geologists have speculated that daily heating of rock in bright sunlight followed by cooling each night should cause physical breakdown of rocks because the common rock-forming minerals expand by different amounts when heated. Surface temperatures as high as 80°C have been measured on exposed desert rocks, and daily temperature variations of more than 40° have been recorded on rock surfaces. Highest temperatures are achieved by dark-colored rocks, like basalt, and by rocks that do not easily conduct heat inward. Despite a number of careful tests, no one has yet demonstrated that daily heating and cooling cycles have noticeable physical effects on rocks. These tests, however, have been carried out only over relatively brief time intervals. Perhaps thermal fracturing occurs only after repeated, extreme natural temperature fluctuations over long time intervals.

SPALLING DUE TO FIRE

Fire can be very effective in disrupting rocks, as anyone knows who has witnessed a rock beside a campfire become overheated and shatter explosively. Because rock is a relatively poor conductor of heat, an intense fire heats only a thin outer shell, which expands and breaks away as a *spall*. The heat of forest fires and brush fires can lead to the spalling of rock flakes from exposed bedrock or boulders (Figure 6.7). Studies of fire history in forested regions show that large natural fires, mostly started by lightning, may recur every several hundred years. Therefore, over

Figure 6.7 Spalling Due to Intense Heat A fast-moving forest fire in the pine forests of Yellowstone National Park, Wyoming, caused mechanical weathering of a boulder of igneous rock. Fresh, bright-colored spalls, which flaked off as heat caused the outermost part of the rock to expand rapidly, litter the fire-blackened ground.

Figure 6.8 Wedging by Roots A pine tree that began growing in a crack on this bedrock outcrop has caused a large flake of rock to break away, exposing the tree's expanding root system.

long intervals of geologic time, fires may contribute significantly to the physical breakdown of surface rocks.

WEDGING BY PLANT ROOTS

Seeds germinate in cracks in rocks, and as the plants grow, they extend their roots deeper into the cracks. The roots of trees growing in cracks can wedge apart adjoining blocks of bedrock (Figure 6.8). In much the same way, roots disrupt stone pavements, sidewalks, garden walls, and even buildings. Large trees swaying in the wind can cause cracks to widen, and, if blown over, can pry rock apart. Although it would be difficult to measure, the total amount of rock breakage done by plants must be very large. Much of it is obscured by chemical weathering, which takes advantage of the new openings as soon as they are created.

Before you go on:

1. What are joints, how do they form, and what role do they play in physical weathering?

2. How is physical weathering promoted by the growth of crystals in rocks or regolith?

3. What are several ways in which temperature affects physical weathering?

CHEMICAL WEATHERING

The principal agents of chemical weathering are water and substances dissolved in water—most importantly, carbon dioxide and oxygen. The effects of chemical weathering are greatly speeded up by high temperature, and because water is a major component in all aspects of chemical weathering, high rainfall is also important. In this section we first discuss the major ways in which chemical weathering occurs, then we consider some of the effects produced by chemical weathering. (See Box 6.1.)

When raindrops form in the atmosphere, they dissolve small amounts of carbon dioxide, producing **carbonic acid** (H_2CO_3)—all rainwater is slightly acidic as a result of this

TABLE 6.1 Common Chemical Weathering Reactions

1. Production of carbonic acid by solution of carbon dioxide, and the ionization of carbonic acid:

$$H_2O \ + \ CO_2 \ \rightleftarrows \ H_2CO_3 \ \rightleftarrows \ H^{1+} \ + \ HCO_3^{1-}$$

Water   Carbon   Carbonic   Hydrogen   Bicarbonate
           dioxide   acid   ion   ion

2. Hydrolysis of potassium feldspar:

$$4KAlSi_3O_8 \ + \ 4H^{1+} \ + \ 2H_2O \ \rightarrow \ 4K^{1+} \ + \ Al_4Si_4O_{10}(OH)_8 \ + \ 8SiO_2$$

Potassium   Hydrogen   Water   Potassium   Kaolinite   Silica
feldspar   ions   ions

3. Oxidation and hydrolysis of iron (Fe^{2+}) compounds to form ferric hydroxide:

$$4FeO \ + \ 6H_2O \ + \ O_2 \ \rightarrow \ 4Fe(OH)_3$$

Iron   Water   Oxygen   Ferric
oxide   hydroxide

4. Dehydration of ferric hydroxide to form goethite:

$$Fe(OH)_3 \ \rightarrow \ FeO \cdot (OH) \ + \ H_2O$$

Ferric   Geothite   Water
hydroxide

5. Dehydration of goethite to form hematite:

$$2FeO \cdot OH \ \rightarrow \ Fe_2O_3 \ + \ H_2O$$

Geothite   Hematite   Water

6. Dissolution and hydrolysis of carbonate minerals by carbonic acid:

$$CaCO_3 \ + \ H_2CO_3 \ \rightarrow \ Ca^{2+} \ + \ 2(HCO_3)^{1-}$$

Calcium   Carbonic   Calcium   Bicarbonate
carbonate   acid   ion   ions

reaction. As weakly acidic rainwater moves downward into the soil, additional carbon dioxide is dissolved from decaying vegetation. The carbonic acid ionizes to form hydrogen H^{1+} ions, and bicarbonate ions HCO_3^- (Table 6.1, reaction 1). The hydrogen ions are so small they can enter a mineral structure and replace other ions, thereby changing the composition.

WEATHERING PATHWAYS

There are four different chemical pathways by which chemical weathering proceeds. In many chemical-weathering reactions two or more different pathways proceed at the same time. The four pathways are dissolution, hydrolysis, leaching, and oxidation.

DISSOLUTION

The easiest reaction pathway to comprehend is **dissolution**, by which is meant the dissolving of a compound in water. (See *Understanding Our Environment*, Box 6.1.) There are a few soluble minerals; halite (NaCl) is an example of a mineral that can be removed completely from a rock by dissolution. Almost all water–mineral reactions involve dissolution, but in most cases water–mineral reactions also involve some other reaction.

HYDROLYSIS

Any reaction involving water that leads to the decomposition of a compound is a **hydrolysis** reaction. Where minerals are concerned it is hydrogen ions (H^{1+}) that most commonly cause hydrolysis. Consider the way in which hydrogen ions produced by the ionization of a carbonic acid solution decompose potassium feldspar (Table 6.1, reaction 2). Hydrogen ions and water interact with the aluminum-silicate crystal structure to release alumina and silica into solution slowly. Once in solution, the alumina and silica combine with water and precipitate out of solution as the clay mineral **kaolinite**. Kaolinite was not present in the original rock—it was created by the chemical reaction. Kaolinite is a common member of a large group of relatively insoluble minerals called clays that occur widely in the regolith.

Hydrolysis is one of the chief processes involved in the chemical breakdown of common rocks. Although we have discussed the case where hydrogen ions formed by ionization of a carbonic acid solution decompose feldspar, similar reactions can occur with other acids. For example, as water passes through the soil, organic acids released by plant roots, or by bacteria, can also be dissolved, then ionized, and decompose minerals.

UNDERSTANDING OUR ENVIRONMENT

DECAYING BUILDINGS AND MONUMENTS

In order to construct buildings, monuments, tombstones, and other structures, people have long sought stone that would be both attractive and durable. More than two millennia ago, the Greeks and Romans discovered that limestone and marble make ideal building materials, for they are solid, reasonably strong, and relatively easy to carve, and produce aesthetically pleasing structures. However, what they failed to appreciate was how vulnerable these carbonate rocks would be to decay.

In many large industrial cities, both new and ancient structures built of marble or limestone are slowly disappearing under the vigorous attack of airborne pollutants produced by the burning of fossil fuels. Especially damaging is sulfur dioxide gas, the main contributor to the acid rain that causes the calcite composing these rocks to dissolve. As a result, many European public statues and historic buildings, some dating to medieval times, have been extensively disfigured (Figure B6.1). Even in the United States, where most public buildings are less than two centuries old, widespread decay has begun to obliterate the surface of structures built of carbonate rocks.

Most of this damage has taken place since the beginning of the Industrial Revolution, when steam engines and coal fires came into widespread use. In the twentieth century, increasing combustion of fossil fuels has pumped vast amounts of pollutants into the air, thereby greatly accelerating the rate at which the perfectly natural geologic process of rock decay takes place. In the eastern United States, studies using freshly quarried stone have shown that the surfaces of limestone and marble slabs exposed at relatively nonpolluted sites are dissolving at an average rate of 0.001 to 0.002 cm per year. At highly polluted sites, rates are far higher. Clearing the air of harmful pollutants, an important challenge for all nations in the coming decades, will lead not only to healthier air to breathe, but to a healthier environment and longer life for buildings and monuments as well.

Figure B6.1 Corroded Statues Carved statues on the façade of Bayeux Cathedral in Normandy, France, show the disfiguring effects of chemical weathering caused by acid pollution.

LEACHING

A common process of chemical weathering is **leaching**, which is the continued removal, by water solution, of soluble matter from bedrock or regolith. The example of the solution of halite, discussed under dissolution, is a special example of leaching in that it refers to the total removal of the halite. More commonly we use the term leaching for the partial removal of material in solution. For example, when silica is released from rocks by chemical weathering, as just described in the decomposition of potassium feldspar by hydrolysis, some remains in the clay-rich regolith, and some is removed in solution by water moving through the ground. Potassium ions released by decomposition of feldspar also escape in solution and so are leached from the rock being altered. Soluble substances leached from rocks during weathering are present in all surface and groundwater. Sometimes their concentrations are high enough to give the water an unpleasant taste.

OXIDATION

The final pathway of chemical weathering is **oxidation**, a process by which an ion loses an electron and the oxidation state of the ion is said to increase. A common example is the oxidation of iron, a normal constituent of many rock-forming minerals. An iron atom that has given up two electrons forms a ferrous ion (Fe^{2+}) and is said to have been oxidized to the ferrous state. When a ferrous ion is oxidized further by giving up a third electron, the result is a ferric ion (Fe^{3+}).

When a mineral containing iron is chemically weathered, iron is released and, in the presence of oxygen, rapidly oxidized from Fe^{2+}, the most common state in minerals, to Fe^{3+}. Typically this results in the formation of a reddish-yellow precipitate of ferric hydroxide [$Fe(OH)_3$] (Table 6.1, reaction 3).

Note that in the oxidation reaction discussed above, water has been incorporated in the structure of the ferric hydroxide. The incorporation of water in a mineral structure is called **hydration**. The ferric hydroxide may **dehydrate**, meaning it will lose some water, in which case it will form goethite (FeO·OH) (Table 6.1, reaction 4). Goethite may dehydrate still further to form hematite (Fe_2O_3), a brick-red mineral (Table 6.1, reaction 5). The intensity of the colors of ferric hydroxide, goethite, and hematite, ranging from yellowish through brownish red to brick red, can provide clues to how much time has elapsed since weathering began, and to the degree or intensity of weathering.

COMBINED REACTIONS

Almost all instances of chemical weathering involve more than one of the reaction pathways. Dissolution, for example, plays a part in virtually all chemical weathering processes and is usually accompanied by hydrolysis and leaching. Consider the reaction of calcite in rainwater. Calcite is only slightly soluble in water. When it dissolves, Ca^{2+} and CO_3^{2-} ions are formed in the solution. If carbonic acid is present the process is speeded up, and the CO_3^{2-} ions react with H^+ ions in solution to form bicarbonate ions $(HCO_3)^-$. Dissolution and hydrolysis are combined in reaction 6 of Table 6.1, and leaching continually removes the dissolved matter. The effects of these processes are widely seen in the distinctive landscapes—including caves, caverns, and sinkholes—underlain by carbonate rocks (Chapter 15).

EFFECTS OF CHEMICAL WEATHERING ON COMMON MINERALS AND ROCKS

The minerals and soluble ions that result when an igneous rock weathers chemically depend on the original mineral composition of the rock. Granite has a higher silica content than basalt and a different mineralogical makeup. A typical granite contains quartz, which is relatively inactive chemically, potassium-bearing minerals such as muscovite and potassium feldspar, plagioclase feldspar, and minerals rich in iron and magnesium (called ferromagnesian minerals). When a granite decomposes, it does so by the combined effects of dissolution, hydrolysis, and oxidation. The feldspar, mica, and ferromagnesian minerals weather to clay minerals and soluble Na^{1+}, K^{1+}, and Mg^{2+} ions (Figure 6.9). The quartz grains, being relatively inactive chemically, remain essentially unaltered. Because basalt lacks quartz, mica, and potassium feldspar, quartz grains and K^{1+} ions are not among its chemical weathering products. Like granite, the plagioclase feldspar and ferromagnesian minerals in a basalt weather to clay minerals and soluble ions (Na^{1+}, Ca^{2+}, and Mg^{2+}), while the iron from ferromagnesian minerals, together with iron from magnetite, forms goethite.

When limestone, the most common sedimentary rock that contains calcium carbonate, is attacked by dissolution and hydrolysis, it is readily dissolved, leaving behind only the nearly insoluble impurities (chiefly clay and quartz) that are always present in small amounts in the rock. Therefore, as limestone weathers chemically, the residual regolith that develops from it consists mainly of clay and quartz.

Not only quartz but also a number of other minerals are so resistant to chemical attack that they can persist for long periods of time at the land surface. Minerals such as gold, platinum, and diamond, as previously mentioned, persist during weathering. Because some of these minerals have a higher specific gravity than common minerals such as

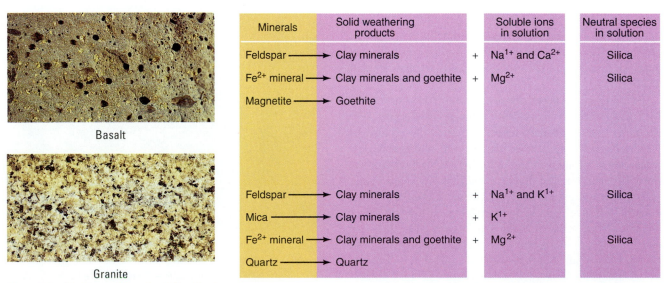

Figure 6.9 Products of Chemical Weathering When a basalt weathers chemically, its silicate minerals and magnetite are converted to clay minerals, goethite, soluble cations, and silica. When a granite weathers, these same products result from the breakdown of feldspar and ferromagnesian minerals, but they include grains of quartz that are resistant to chemical breakdown.

Figure 6.10 Exfoliation Exfoliating granite boulders (The Devil's Marbles) in central Australia. Thin, curved spalls flake off the rock as it weathers, gradually causing the boulders to increase in sphericity.

quartz, they concentrate at the beds of streams or along ocean beaches (Chapters 14 and 18). Some may end up sufficiently concentrated to form mineral deposits of economic value. Valuable mineral deposits may also form as a result of chemical weathering reactions other than concentration of residual minerals (see Chapter 21).

EXFOLIATION AND SPHEROIDAL WEATHERING

During weathering, concentric shells of rock may spall off from the outside of an outcrop or a boulder, a process known as **exfoliation** (Figure 6.10). Sometimes only a single exfoliation shell is present, but 10 or more shells may develop, thereby giving a rock the layered appearance of an onion.

Exfoliation is caused by differential stresses within a rock that result mainly from chemical weathering. For example, when feldspars weather to clay, the volume of weathered rock is greater than the volume of original rock. Stresses thus produced cause thin shells of rock to separate from the main mass of unweathered rock.

Beneath the ground surface, chemical weathering frequently will give rise to a halo of decayed rock around an unaltered rock core. Starting with a cube of solid, fresh rock, as water moves along joints and attacks the rock from all sides, the volume of unaltered rock will slowly decrease in size and become more spherical (Figure 6.11), a process known as **spheroidal weathering**. The results are often seen in fresh roadcuts where rounded boulders produced by such progressive decomposition often are found in rows running in several directions. The pattern results from intersecting joint sets that control the slow movement of water through the rock.

Figure 6.11 Spheroidal Weathering Spheroidal weathering of granitic bedrock in the northern Sierra Nevada of California produces a rounded core of solid granite surrounded by a halo of disintegrated rock. Although the boulders resemble rounded stream-transported boulders, their form is entirely the result of weathering.

At this point, two important relationships should be noted. First, the effectiveness of chemical weathering increases as the surface area exposed to weathering increases. Second, surface area increases simply from the subdivision of large blocks into smaller blocks. Subdividing a cube, while adding nothing to its volume, greatly increases the surface area (Figure 6.12). Repeated subdivision leads to a remarkable result: one cubic centimeter of rock has a surface area of 6 cm^2; when subdivided into particles the size of the smallest clay minerals, the total surface area increases to nearly 40 million cm^2. Chemical weathering therefore leads to a dramatic increase in surface area.

Before you go on:

1. How is water involved in chemical weathering?
2. What are the four pathways of chemical weathering?

3. Through what processes does a typical granite decompose, and what are the products of decomposition?
4. How is the rate of chemical weathering related to surface area of rock or detritus?

FACTORS THAT INFLUENCE WEATHERING

A visit to an early cemetery of a New England industrial town will quickly show us that weathering does not proceed at a uniform pace. The inscription on a mid-nineteenth-century marble tombstone may be degraded and nearly unreadable, whereas the writing on a nearby tombstone of equal age, but composed of granite or slate, might appear nearly as fresh as the day it was carved. From this observation we could guess that rock type must play a role

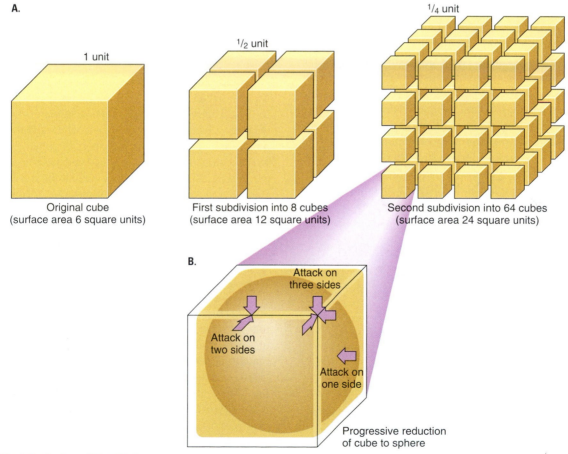

A.

1 unit

Original cube
(surface area 6 square units)

½ unit

First subdivision into 8 cubes
(surface area 12 square units)

¼ unit

Second subdivision into 64 cubes
(surface area 24 square units)

B.

Attack on three sides

Attack on two sides

Attack on one side

Progressive reduction of cube to sphere

Figure 6.12 Weathering of Rock Cubes Subdivision and weathering of rock cubes. A. Each time a cube is subdivided by slicing it through the center of each of its edges, the aggregate surface area doubles. This greatly increases the speed of chemical reaction. B. Solutions moving along joints separating nearly cubic blocks of rock attack corners, edges, and sides at rates

that decline in that order, because the numbers of corresponding surfaces under attack are 3, 2, and 1. Corners become rounded, and eventually the blocks are reduced to spheres (Figure 6.11). Once a spherical form is achieved, the energy of attack becomes uniformly distributed over the whole surface, so that no further change in form occurs.

in determining the intensity of weathering, and that in this New England town marble weathers more rapidly than slate or granite. Were we then to travel west to an old desert mining town, we might find that the inscription on a mid-nineteenth-century marble tombstone in the local cemetery appears sharp and virtually unweathered. Here is a rock identical in composition to the marble in New England, and the tombstones are nearly the same age. Therefore, some other factor or factors must explain the contrast in weathering. In this case, the important factor is climate: the moist New England forest environment versus a dry desert.

Let's next examine how rock composition, climate, and other factors influence the intensity of weathering of surface rocks and sediments.

MINERALOGY

The resistance of a silicate mineral to weathering is a function of three principal things: (1) the chemical composition of the mineral; (2) the extent to which the silicate tetrahedra in the mineral are polymerized (see Chapter 3); and (3) the acidity of the waters with which the mineral reacts.

Ferromagnesian minerals with no polymerization, such as olivine, or with chain polymers such as pyroxene and amphibole, weather most easily at Earth's surface (Table 6.2). Calcium-rich plagioclase feldspar also weathers readily. Biotite, muscovite, and sodium-rich plagioclase are less easily weathered. Quartz, which has the highest degree of polymerization of all silicate minerals, is among the most resistant of rock-forming minerals and degrades chemically at an exceedingly slow rate. Nevertheless, given enough time, even quartz can be dissolved.

A glance at the mineral-stability sequence (Table 6.2) shows that the order in which the minerals are ranked is approximately the reverse of that in which minerals crystallize in Bowen's reaction series (Figure 4.20). This reversed order makes sense because the least polymerized minerals crystallized earliest in Bowen's series.

ROCK TYPE AND STRUCTURE

Because different minerals react differently to weathering processes, the mineralogical assemblage of a rock, and therefore the rock type, clearly must influence decomposition. Quartz is so resistant to chemical breakdown, for example, that rocks rich in quartz are also resistant. We can see the results in the Appalachian Mountains, where resistant quartzite beds form ridges that stand prominently above valleys underlain by more erodible rocks containing less quartz. Granite also is resistant to weathering, for it consists of minerals (such as quartz, muscovite, and potassium feldspar) that are more resistant to chemical breakdown than are most other silicate minerals (Figure 6.9; Table 6.2). Like quartzite, granite typically forms hilly or mountainous terrain.

The rate at which a rock weathers is also influenced by its texture and structure. Even a rock that consists entirely of quartz may break down rapidly if it contains closely spaced joints or other partings that make it susceptible to frost action.

Differences in the composition and structure of adjacent rock units can lead to contrasting rates of weathering and to landscapes that reflect such *differential weathering* (Figure 6.13). In a sequence of alternating shale and quartz-rich sandstone, for instance, the shale is likely to weather more easily, leaving the sandstone beds standing out in relief. If the beds are horizontal, the result will be a stepped topography, with the sandstone forming abrupt cliffs between more gentle slopes of shale. If the bedding is inclined, the sandstone will stand as ridges separated by linear depressions underlain by shale.

SLOPE ANGLE

A mineral grain loosened by weathering on a steep slope may be washed downhill by the next rain. On such a slope the solid products of weathering move quickly away, continually exposing fresh bedrock to renewed attack. As a result, weathered rock seldom extends far beneath the surface. On gentle slopes, however, weathering products are not easily washed away and in places may accumulate to depths of 50 m or more.

CLIMATE

Moisture and heat promote chemical reactions. Not surprisingly, therefore, weathering is more intense and generally extends to greater depths in a warm, moist climate than in a cold, dry one (Figure 6.14). In moist tropical

TABLE 6.2	Order of Stability of Common Minerals Under Attack by Chemical Weathering
MOST STABLE (Least susceptible to chemical weathering)	Ferric oxides and hydroxides
	Aluminum oxides and hydroxides
	Quartz
	Clay minerals
	Muscovite
	Potassium feldspar
	Biotite
	Sodium feldspar (albite-rich plagioclase)
	Amphibole
LEAST STABLE (Most susceptible to chemical weathering)	Pyroxene
	Calcium feldspar (anorthite-rich plagioclase)
	Olivine
	Calcite

Figure 6.13 Differential Weathering Differential weathering has differential erosion of Miocene marine strata at Nelson, New Zealand. Weathering has etched away erodible mudstone from between layers of harder siltstone, leaving the siltstone layers standing as ridges above a wave-eroded platform.

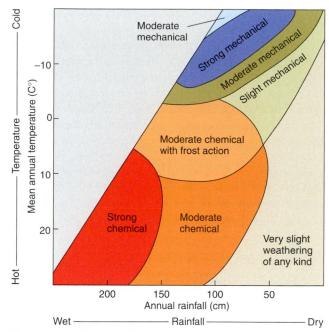

Figure 6.14 Climate and Weathering Climate (primarily temperature and rainfall) plays a major role in controlling the type and effectiveness of weathering processes. Mechanical weathering is dominant where rainfall and temperature are low. High temperature and precipitation favor chemical weathering.

lands, like Central America and Southeast Asia, obvious effects of chemical weathering can be seen at depths of 100 m or more. By contrast, in cold, dry regions like northern Greenland and Antarctica, chemical weathering proceeds very slowly. The effects of physical weathering are generally obvious, though, for bedrock surfaces typically are littered with rubble dislodged by frost action.

Dramatic contrasts in weathering can be seen in the case of carbonate rocks that crop out in different climatic regions. Rocks such as limestone and marble, which consist almost entirely of carbonate minerals, are highly susceptible to chemical weathering in a moist climate and commonly form low, gentle landscapes. In a dry climate, however, the same rocks form bold cliffs because, with scant rainfall and only patchy vegetation, little carbonic acid is present to dissolve carbonate minerals.

BURROWING ANIMALS

Large and small burrowing animals (for example, rodents and ants) bring partly decayed rock particles to the land surface, where they are exposed more fully to chemical

action. More than 100 years ago, Charles Darwin made careful observations in his English garden and calculated that every year earthworms bring particles to the surface at the rate of more than 2.5 kg/m^2 (25 tons per hectare). After a study in the Amazon River basin, geologist J. C. Branner wrote that the soil there "looks as if it had been literally turned inside out by the burrowing of ants and termites." Although burrowing animals do not break down rock directly, the amount of disaggregated rock they move over many millions of years must be enormous.

TIME

Studies of the decomposition of the stone in ancient buildings and monuments show that hundreds to thousands of years are required for a hard igneous or plutonic rock to decompose to depths of only a few millimeters. Granite and other hard bedrock surfaces in New England, Scandinavia, the Alps, and elsewhere still display polish and fine grooves that ice-age glaciers made before they disappeared about 10,000 years ago. In such regions, cool climate and successive glaciations have significantly reduced the rate and effectiveness of chemical weathering, so it takes many tens of thousands of years to create weathered regolith. However, in regions that escaped glaciation and have been continuously exposed to weathering, the zone of weathering often extends to great depths.

The rates at which rocks weather have been estimated in several ways. (1) Experiments have been designed in which the length of the experiment provides time control and the processes are speeded up by increasing temperature and available water, and by decreasing particle size. (2) Studies have been made of the degree of weathering of ancient architectural structures of known age. (3) The thickness of weathering rinds on stones exposed to weathering since prehistoric times can be used to estimate weathering rates over hundreds to thousands of years (see *The Science of Geology*, Box 6.2, *Using Weathering Rinds to Measure Relative Age*). Such investigations suggest that the rate of weathering tends to decrease with time as the weathering profile, or a weathering rind, thickens (Figure B6.3).

Before you go on:

1. How can differential weathering be recognized in landforms?

2. How does weathering in a warm, moist climate differ from weathering in a cold, dry one? Why?

3. How can rates of weathering be estimated?

SOIL: ORIGIN AND CLASSIFICATION

Soils are one of our most important natural resources. Soils support the plants that are the basic source of our nourishment and provide food for domesticated animals. Soils help maintain Earth's surface environment. They accomplish this by supporting vegetation, which, as it releases water to the atmosphere, is an integral part of the hydrologic cycle. In addition, soils store organic matter, thereby influencing how much carbon is cycled to the atmosphere as carbon dioxide, and they also trap pollutants.

ORIGIN OF SOIL

The physical and chemical breakdown of solid rock by weathering processes is the initial step in the formation of soil. However, as noted, soil also contains organic matter mixed in with the mineral component. This organic fraction is essential in the definition of **soil**: the part of the regolith that can support rooted plants.

The organic matter in soil is derived from the decay of dead plants and animals. Living plants are nourished by the nutrients released from decaying organisms, as well as by those released during weathering of mineral matter. Plants draw these nutrients upward, in water solution, through their roots. Therefore, through their life cycle, plants are directly involved in the manufacture of the fertilizer that will nourish future generations of plants. These activities are an integral part of a continuous cycling of nutrients between the regolith and the biosphere. With its

partly mineral, partly organic composition, soil forms an important bridge between Earth's lithosphere and its teeming biosphere.

SOIL PROFILES

As bedrock and regolith weather, the upper part of the regolith gradually evolves into soil. As a soil develops from the surface downward, an identifiable succession of approximately horizontal weathered zones, called **soil horizons**, forms. Each horizon has distinctive physical, chemical, and biological characteristics. Although soil horizons may resemble a sequence of deposits, or layers, they are not layers of different sediment. Instead, they represent physical and chemical changes in the upper part of the regolith that result from its closeness to the land surface. Taken together, the soil horizons constitute a **soil profile**, which consists of the succession of soil horizons that have developed within the regolith over time. The regolith within which the soil horizons have formed was the **parent material** for the soil. The parent material may be represented by still deeper regolith below the soil horizons, and below the regolith there may be fresh rock from which the regolith itself was formed.

Soil profiles generally display two or more horizons (Figure 6.15). The uppermost horizon may be a surface accumulation of organic matter (*O horizon*) that overlies mineral soil. The organic matter is in various stages of decomposition.

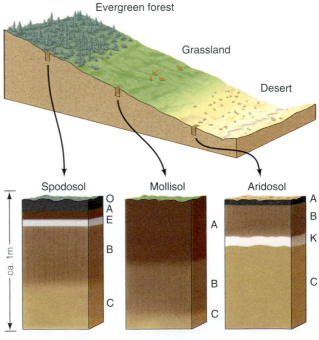

Figure 6.15 Climate and Soils Examples of soil profiles developed under different climatic and vegetation conditions (see Table 6.3).

BOX 6.2 THE SCIENCE OF GEOLOGY

USING WEATHERING RINDS TO MEASURE RELATIVE AGE

If you crack open a cobble of weathered basalt, you usually will see a light-colored rind surrounding a darker core of unaltered rock (Figure B6.2). The rind is composed of the solid products resulting from chemical weathering and is called a **weathering rind**. Weathering begins at the freshly exposed surface of a dark, unaltered cobble and proceeds slowly inward. Commonly, this involves oxidation of iron-rich minerals to produce goethite (Table 6.1, eq. 4), which imparts a light-brownish color to the developing rind. Such a rind forms on all but the most chemically stable rock types when chemical weathering attacks them. An exception is carbonate rocks, which dissolve under chemical attack (Table 6.1, eq. 6), thereby generally preventing the formation of an obvious rind.

In some rock types, the altered mineral matter in a developing rind crumbles away as the rock slowly weathers. In others, like fine-grained basalt, the rind tends to remain coherent and becomes thicker as weathering moves inward, attacking the solid, unaltered core.

Geologists have found that the thickness of weathering rinds on stones can be a useful measure of the relative ages of different bodies of sediment that contain the same rock types. However, care must be taken to compare measured rind thickness at sites that have similar climates. Remember that climate is one of the primary controls on weathering rate, and differences in rind thickness between different areas might reflect different climatic environments rather than different ages (Figure B6.3).

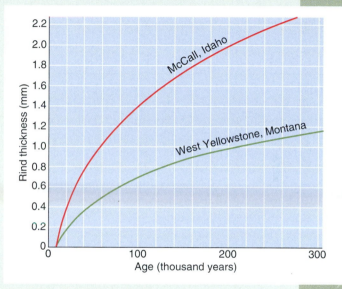

Figure B6.3 Weathering Rates Graph showing change in weathering rates through time for two localities in the northwestern United States. The thickness of weathering rinds on volcanic stones is plotted as a function of their estimated or known age. Weathering rates are rapid during the first few tens of thousands of years (steep parts of curves), but then decline steadily with increasing age; after several hundred thousand years, the rates become very slow. What factor or factors might explain why the curve for McCall, Idaho, is steeper than the one for West Yellowstone?

Figure B6.2 Weathering Rind A basaltic stone, measuring 35 x 45 mm, from the eastern Cascade Range, Washington, displays a well-developed weathering rind about 3 mm thick surrounding a black, unweathered core.

An **A horizon** may either underlie an O horizon or lie directly beneath the surface. Typically, the A horizon is dark grayish or blackish (at least near its top) because of the presence of *humus*, the decomposed residue of plant and animal tissues, which is mixed with mineral matter. The A horizon has lost some of its original substance through the downward transport of clay particles and, more important, through the chemical leaching of more soluble substances. An *E horizon*, sometimes present

beneath the A horizon, has a gray or whitish color. The light color is due mainly to the lack of a darker oxide coating on light-colored mineral grains. E horizons commonly are found in acidic soils that develop beneath evergreen forests.

The **B horizon** underlies the higher horizons and commonly is brownish or reddish. This horizon is enriched in clay and/or iron and aluminum hydroxides produced by the weathering of minerals within the horizon and also transported downward from overlying A and E horizons. Because the B horizon consists mostly of very fine particles, it tends to break into blocks or prisms when it dries. Where clay migration is an important process, each block or prism may be coated with clay. Although the B horizon generally is penetrated by plant roots, it contains less organic matter than the humus-rich A horizon.

A *K horizon*, present in some arid-zone soils beneath the B horizon, is densely impregnated with calcium carbonate, which coats other mineral grains and constitutes up to 50 percent of the volume of the horizon.

The **C horizon** is the deepest horizon and consists of rock in various stages of weathering. However, it lacks the distinctive properties of the A and B horizons. Oxidation of iron from the original rock material in the C horizon generally imparts a light-yellowish-brown color.

SOIL TYPES

An astute observer traveling across the landscape will note that soils are not everywhere the same. Different soils result from the influence of six soil-forming factors: climate, vegetation cover, soil organisms, composition of parent material, topography, and time. The soil forming under prairie grassland differs from soil in a boreal forest or that of a tropical rainforest. The character of a soil may change abruptly as we move from basalt to limestone or from a gentle slope to a steep slope, and it also will change with the passage of time.

Soil scientists classify soils according to their physical and chemical properties in much the same way that geologists classify rocks. Such classification makes it easier to map, study, and understand how soils are distributed on the landscape and how their properties reflect the six soil-forming factors (Figure 6.16). In the soil classification scheme now in standard use in the United States, soils are classified into 11 orders, distinguished on the basis of easily recognizable characteristics (Table 6.3). The eleventh order—the *Andisols*—is restricted to soils developed on tephra.

POLAR SOILS

In cold, high-latitude deserts, like those found in Greenland, northern Canada, and Antarctica, soils generally are dry and lack well-developed horizons (they are therefore classified as *Entisols*). Weakly oxidized parent material may underlie a layer of coarse frost-churned stones. In wetter high-latitude and high-altitude environments, mat-like tundra vegetation overlies perennially frozen ground that prevents water from percolating downward. This leads to water-logged soils that are rich in organic matter (*Histosols*). Only on well-drained sites do soils develop recognizable A and B horizons (*Inceptisols*). Because the cold climate retards chemical processes, well-drained soils generally do not develop the thick, clay-enriched B horizon typical of highly developed temperate-latitude soils.

TEMPERATE-LATITUDE SOILS

Soils of the temperate zone vary largely in response to differences in climate and in the resulting vegetation cover. *Alfisols*, characteristic of deciduous woodlands, typically have a clay-rich B horizon beneath a light-gray E horizon (Figure 6.17). Acidic *Spodosols* developed in cool, moist evergreen forests have an organic-rich A horizon, an ash-like E horizon, and an iron-rich B horizon. Mountainous

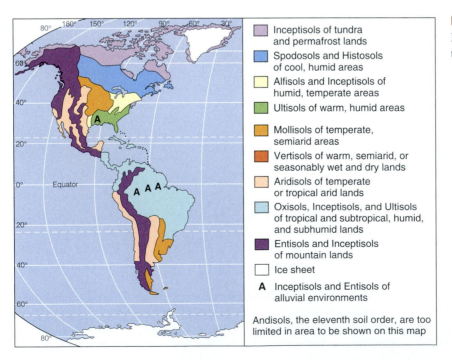

Inceptisols of tundra and permafrost lands

Spodosols and Histosols of cool, humid areas

Alfisols and Inceptisols of humid, temperate areas

Ultisols of warm, humid areas

Mollisols of temperate, semiarid areas

Vertisols of warm, semiarid, or seasonally wet and dry lands

Aridisols of temperate or tropical arid lands

Oxisols, Inceptisols, and Ultisols of tropical and subtropical, humid, and subhumid lands

Entisols and Inceptisols of mountain lands

Ice sheet

A Inceptisols and Entisols of alluvial environments

Andisols, the eleventh soil order, are too limited in area to be shown on this map

Figure 6.16 Distribution of Soil Types
Map showing distribution of the major soil types in North and South America.

TABLE 6.3 Orders of the Soil Classification System Used in the United States

Soil Order (Meaning of Name)	Main Characteristics
Alfisol (pedalfer[a] soil)	Thin A horizon over clay-rich B horizon, in places separated by light-gray E horizon. Typical of humid middle latitudes.
Aridosol (arid soil)	Thin A horizon above relatively thin B horizon and often with carbonate accumulation in K horizon. Typical of dry climates.
Entisol (recent soil)	Soil lacking well-developed horizons. Only a thin, incipient A horizon may be present.
Histosol (organic soil)	Peaty soil, rich in organic matter. Typical of cool, moist climates.
Inceptisol (young soil)	Weakly developed soil, but with recognizable A horizon and incipient B horizon lacking clay or iron enrichment. Generally occurs under moist conditions.
Mollisol (soft soil)	Grassland soil with thick dark A horizon, rich in organic matter. B horizon may be enriched in clay. E and K horizons may be present.
Oxisol (oxide soil)	Relatively infertile soil with A horizon over oxidized and often thick B horizon.
Spodosol (ashy soil)	Acidic soil marked by highly organic O and A horizons, an E horizon, and iron/aluminum-rich B horizon. Occurs in cool forest zones.
Ultisol (ultimate soil)	Strongly weathered soil characterized by A and E horizons over clay-rich B horizon. Characteristic of tropical and subtropical climates.
Vertisol (inverted soil)	Organic-rich soil having very high content of clays that shrink and expand as moisture varies seasonally.
Andisol (dark soil)	Soil developed on pyroclastic deposits and characterized by low bulk density and high content of amorphous minerals.

[a] Soils rich in iron and aluminum.

terrains, where cool climates cause low rates of soil formation and eroding slopes continually lose the uppermost soil, frequently have minimally developed profiles *(Entisols)* or

Figure 6.17 Soil Profile A soil profile developed in dune sand under a pine forest on Cape Cod, Massachusetts has a dark-gray A horizon near the surface and reddish-brown B horizon at depth. Separating these horizons is a thick light-gray E horizon, typical of acidic soils.

display weakly developed B horizons lacking clay enrichment *(Inceptisols)*. Grasslands and prairies typically develop *Mollisols* having thick, dark-colored, organic-rich A horizons. Soils formed in moist subtropical climates commonly display a strongly weathered B horizon *(Ultisols)*.

DESERT SOILS

In dry climates, where lack of moisture reduces leaching, carbonates accumulate in the profile during the development of *Aridosols*. These strongly alkaline soils contrast with the more acid soils of humid regions that lack a carbonate-rich K horizon. An important part of the carbonate accumulation results from evaporation of water that rises in the ground, bringing dissolved salts from below. Recently, soil scientists have found that windblown dust also contributes to the accumulation of salts in arid-land soils. Over extensive arid regions of the southwestern United States, carbonates have in this way built up in the soil profile a solid, almost impervious layer of whitish calcium carbonate known as *caliche* (Figure 6.18).

TROPICAL SOILS

Soils that form where rainfall is high and the climate is warm are characterized by extreme chemical alteration of the parent material. These soils *(Oxisols)* are intensely weathered and infertile because essential nutrients have been leached away. *Vertisols* of tropical regions, where the climate alternates between a wet and a dry season, have a high clay content that causes the soil alternately to swell and shrink in response to seasonal wetting and drying.

Figure 6.18 Caliche in Soil Profile A soil profile in semiarid central New Mexico includes whitish caliche forming a prominent K horizon at the top of a yellowish-brown C horizon, which underlies a reddish-brown B horizon above.

Figure 6.19 Temple Wall of Laterite Reddish-brown blocks of laterite were used by the ancient Khmer people to construct this temple wall at Angkor Wat in the Cambodian jungle. The color and texture of the laterite contrast with the smooth, grayish sandstone used to construct the adjacent temple building.

During the dry season, open cracks extend from the surface down into the soil profile.

Of the many minerals formed during chemical weathering, goethite and hematite are among the least soluble. In tropical regions where the climate is very wet and warm, rock-forming minerals are slowly leached away, leaving a soft, mottled, reddish-gray residue, rich in iron. Geologists generally refer to this product of deep weathering as **laterite**. One product of lateritic weathering is ferric hydroxide [$Fe(OH)_3$]. As a result of climatic change or deforestation, the upper part of a laterite may dry out and become hardened because the ferric hydroxide dehydrates to goethite [$Fe(OH)_3 \rightarrow FeO \cdot OH + H_2O$]. The resulting stone-like material, called *lateritic crust* or *ironstone*, is so hard that blocks of it can be used as construction material (Figure 6.19). The name laterite, which comes from the Latin word for brick (*latere*), recognizes this property.

RATE OF SOIL FORMATION

Although chemical weathering is part of the complex process of soil development, soil formation and weathering are not the same thing. Chemical weathering chiefly concerns the decomposition of bedrock, a process that takes a very long time. The time required to form a soil profile in regolith can be much shorter.

A soil profile can form rapidly in some environments. A study in the Glacier Bay area of southern Alaska shows us how soils develop when retreating glaciers leave unweath-

ered parent material consisting of finely ground rock flour exposed at the land surface. Despite the cold climate of Glacier Bay, within a few years after a glacier begins to recede, an A horizon develops on the newly exposed and revegetated landscape (Figure 6.20). In this area, moderate temperatures and high rainfall promote rapid leaching of the finely ground parent material, which is a carbonate-rich glacial sediment. As the plant cover becomes denser, carbonic and organic acids acidify the soil, and leaching becomes more effective. After about 50 years, a B horizon has appeared and the combined thickness of the A and B horizons reaches about 10 cm. The B horizon is not present in the youngest soils; it starts to form after about 30 years, and by 100 years the combined thickness of the A and B horizons is about 10 cm. Over the next 150 years, a mature forest develops on the landscape, the O horizon continues to thicken, and iron oxides accumulate in the developing B horizon (but the A and B horizons do not increase in thickness). This entire succession of vegetation and soils is clearly visible on the deglaciated landscape in front of the retreating glaciers.

In less humid climates, soil forms more slowly, and it may take thousands of years for a detectable B horizon to

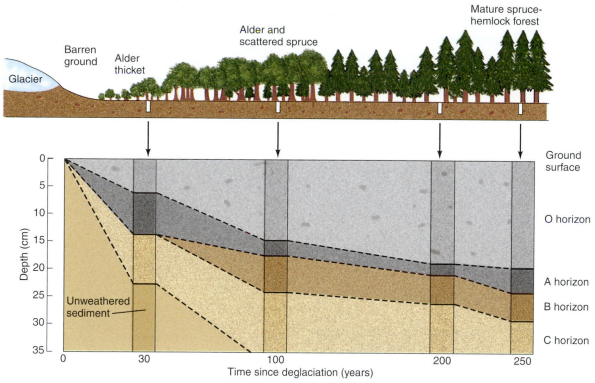

Figure 6.20   Evolution of a Soil Profile   Progressive soil development in Glacier Bay, Alaska, over the past 250 years can be seen in soil profiles from sites of different age. In the first 30 years, organic litter forming an O horizon increases in thickness and an A horizon develops. The B horizon then appears and after another 70 years has reached a thickness of 7 cm. As vegetation changes from alder thicket toward a mature spruce-hemlock forest, organic litter continues to thicken, reaching a thickness of 18 cm after 250 years. The age of each study site was determined by counting the annual growth rings of the oldest surrounding trees. (*Based on data from F. C. Ugolini.*)

appear. In the midcontinental United States, for example, B horizons that have developed during the last 10,000 years contain little clay and lack structure, indicating that they are not yet well-developed. By contrast, those dating back about 100,000 years generally are rich in clay and have a distinctive structure. As an extreme example of the effect of wetness on rate of soil formation, the glacier-free cold deserts of Antarctica are so dry and cold that sediments more than a million years old have only weakly developed soils.

PALEOSOLS

If a soil is buried, it becomes part of the geologic record. The soil is now a **paleosol**, defined as a soil that formed at the ground surface and subsequently was buried and preserved (Figure 6.21). Paleosols have been identified in rocks and sediments of many different ages; they are espe-

Figure 6.21   Paleosols   A thick reddish-brown paleosol (beside the figure) developed on a pumice layer near Guatemala City, Guatemala, is overlain by another layer of pumice and pyroclastic flows that are separated by thinner, brownish paleosols. A fault has displaced the layers about 2 m vertically.

cially common in unconsolidated deposits of the Quaternary Period (see Chapter 11).

Distinctive and widespread paleosols have been used to subdivide, correlate, and date sedimentary sequences. They also can provide important information regarding the nature of former landscapes, vegetation cover, and climate (see opening essay, *Chinese Soils and Monsoon Climate*).

> **Before you go on:**
>
> 1. What is a soil profile? What is a soil horizon? How are the two related?
>
> 2. Why do soils differ from place to place across the landscape?
>
> 3. What is a paleosol, and why do geologists find paleosols useful?

SOIL EROSION

As we have seen, it may take a very long time to produce a well-developed soil. Destruction of soil, however, may occur rapidly. Although soil erosion is a natural process, with rates of erosion determined by topography, climate, and vegetation cover, erosion related directly to human activity has far exceeded natural rates in many parts of the world.

SOIL EROSION DUE TO HUMAN ACTIVITY

Global agricultural production increased dramatically during the first half of the twentieth century, when world population was still less than 2.5 billion people. With world population now close to 6 billion, increasing deforestation and increasing agricultural use of land not well suited for agriculture are causing disastrous problems of soil erosion (Figure 6.22).

In many Third World nations, farmers have been forced beyond traditional farm and grazing lands onto steep, easily eroded slopes or into semiarid regions where periodic crop failure is a fact of life and plowed land is prone to severe wind erosion. At the same time, economic pressures in some of the more-developed countries have increasingly led farmers to shift from ecologically favorable land-use practices to the planting of profitable row crops that often leave the land vulnerable to increased rates of erosion.

Widespread felling of trees has led to accelerated rates of surface runoff and destabilization of soils due to loss of anchoring roots. Soils in the humid tropics, when stripped of their natural vegetation cover and cultivated, quickly lose their fertility (Figure 6.23). So widespread are soil erosion

Figure 6.22 Soil Loss Deforested hillslopes in Madagascar are deeply gullied by erosion. Streams draining this region turn reddish-brown as they carry away irreplaceable topsoil that is destabilized when anchoring tree roots disappear.

Figure 6.23 Amazon Deforestation Deforestation is destroying a luxurious rainforest in the Amazon basin near Maraba, Brazil. Stripped of their natural vegetation cover and planted with crops, soils on this landscape quickly lose their natural fertility.

UNDERSTANDING OUR ENVIRONMENT

THE SOIL EROSION CRISIS

The upper layers of a soil contain most of the organic matter and nutrients that support crops. When the O and A horizons are eroded away, not only the fertility but also the water-holding capacity of a soil diminishes. Because it generally takes between 80 and 400 years to form one centimeter of topsoil except under all but the most favorable circumstances, for all practical purposes, soil erosion is tantamount to mining the soil. Instead of remaining a continuously productive natural resource, it becomes an ever-decreasing and degraded resource. In the United States, the amount of farmland soil lost to erosion each year exceeds the amount of newly formed soil by more than 2 billion tons. It is estimated that farmers in the United States are now losing about 5 tons of soil for every ton of grain they produce. Soil loss in Russia is at least as rapid, whereas in India the soil erosion rate is estimated to be more than twice as high. Worldwide, the estimated loss exceeds 25 billion tons a year. Expressed another way, at the current rate of loss, the world's most productive soils are being depleted at the rate of 7 percent each decade. One recent estimate projected that, as a result of excessive soil erosion and increasing population, only two-thirds as much topsoil was available to support each person at the beginning of the present century as was available in 1984.

Soil erosion is a worldwide problem of massive proportions, and it affects each and every one of us, as the following numbers make clear. A person eats about 750 kg of food a year, and there now are more than 6 billion people on Earth. Food production leads to an annual loss of 25 billion tons of topsoil, or 4.2 tons (4200 kg) per person. Thus, for every kilogram of food we eat, the land loses 5.6 kg of soil.

---

and degradation that the problem has been described as "epidemic" (see *Understanding our Environment*, Box 6.3, *The Soil Erosion Crisis*).

Soil degradation and erosion obviously have direct economic effects, because they reduce or eliminate productive farmland. Because agriculture is the foundation of the world economy, progressive loss of soil signals a potential crisis that could undermine the economic stability of many societies.

In addition, there are indirect effects of soil erosion that often are not anticipated. Much of the topsoil eroded from agricultural lands is transported down rivers and deposited along valley floors, in marine deltas, or in reservoirs behind large dams. The resulting impact on society can be significant. For example, the designers of a major dam and reservoir in Pakistan projected a life expectancy for the reservoir of at least a century. However, increased popula-tion pressure on the region above the dam has resulted in greatly increased soil erosion, leading to such a high rate of sediment production that the reservoir is now expected to be filled with eroded soil within about 75 years, making it unusable.

CONTROL OF SOIL EROSION

Although soil erosion and degradation are severely impacting many countries, effective control measures can substantially reduce these adverse trends (Figure 6.24).

In places where crops are grown, the surface is ordinarily bare during part of each year. On nonvegetated, sloping fields, on pastures that are too closely grazed, and in areas planted with widely spaced crops such as corn, rates of erosion can be high. Wise farmers therefore reduce areas of bare soil to a minimum and prevent the grass cover on

Figure 6.24 Reducing Soil Erosion Contour plowing is one effective method of reducing soil erosion. On this farm, crops have been planted in belts that follow the natural contours of the land. This inhibits the development of rills and gullies that otherwise might form perpendicular to the contours.

Figure 6.25 Terraced Landscape A terraced hillside near Lanzhou, China, creates productive agricultural fields from steep hillslopes carved in deposits of erodible windblown dust. Where unterraced, hillslopes are rapidly eroding and deeply gullied.

pastures from being weakened by overgrazing. If crops such as corn, tobacco, and cotton must be planted on a slope, strips of such crops are often alternated with strips of grass or similar plants that help resist soil erosion.

Another method of reducing soil loss involves crop rotation. A study in Missouri showed that land that lost 49 tons of soil per hectare when planted continually in corn lost only 7 tons per hectare when corn, wheat, and clover crops were rotated. In this case, the bare land exposed between rows of corn is far more susceptible to erosion than land planted with a more continuous cover of wheat or clover.

The most serious soil erosion problems occur on steep hillslopes. In Nigeria, for example, land planted with cassava (a staple food source) and having a gentle 1 percent slope lost an average of 3 tons of soil per hectare each year. On a 5 percent slope, however, the annual rate of soil loss increased to 87 tons per hectare. At this rate, a 15 cm thickness of topsoil would disappear in a single generation (about 20 years). On a 15 percent slope, the annual erosion rate increased to 221 tons per hectare, a rate that would remove all topsoil within a decade.

Despite these grim statistics, steep slopes can be exploited through terracing. Terracing is a major factor in reducing soil loss in central China, where steep hillslopes are underlain by erodible deposits of windblown silt (Figure 6.25). In the arid regions of western China, where strong winds and restricted vegetation cover can cause rapid soil loss, rows of trees planted as wind screens can substantially reduce erosion on the downwind side.

Halting the worldwide loss of soils is a formidable task that will require effective programs at the national and international levels. Soils can be viewed as a renewable resource only over geologically long intervals of time. Over the lifetime of individuals, or even of nations, they must be considered nonrenewable resources that must be carefully utilized and preserved if we are to have a sustainable global economy.

Before you go on:

1. What natural factors control the rate of soil erosion?

2. How is soil erosion affected by growth of the human population?

3. What can be done to reduce soil erosion resulting from human activities?

REVISITING PLATE TECTONICS AND THE EARTH SYSTEM

GLOBAL WEATHERING RATES AND HIGH MOUNTAINS

Weathering is a key component of Earth's interconnected cycles (Figure 1.21). With so many interacting forces at work, determining the average global rate of chemical weathering is not easy. Perhaps the best approach is to measure the amount of dissolved substances delivered by rivers to the oceans, for the dissolved matter in a stream results from chemical weathering of rocks and sediments in the river's drainage basin. Although a multitude of rivers enter the sea, the three largest contributors of dissolved substances are the Yangtze River, which drains the high Tibetan Plateau of China; the Amazon River, which drains the northern Andes in South America; and the Ganges-Brahmaputra river system, which drains the Himalaya in eastern India and Nepal. Together, these three rivers deliver about 20 percent of the water and dissolved matter entering the oceans. If we add all the other rivers draining these three highland regions, we can conclude that chemical weathering and erosion in Tibet, the Andes, and the Himalaya must provide a substantial part of the total dissolved matter reaching the world's oceans. In other words, a direct relationship apparently exists between the occurrence of high-altitude landmasses and the global rate of chemical weathering. If true, then we might expect to see changes in weathering rates over geologic time as mountain systems are uplifted and then worn away.

High rates of chemical weathering and high mountains are related for several reasons. First, high mountains are areas of rapid uplift. They occur where plates of lithosphere converge. Rapid uplift goes hand in hand with rapid mechanical breakdown and erosion of rock. This disintegration exposes large quantities of rock and mineral debris to chemical weathering.

Second, the amount of dissolved matter in rivers is greatest in areas where easily eroded sedimentary rocks are exposed. If we examine the geology of high mountains like the Himalaya, the Andes, and the Alps, we find that these areas of rapid, active uplift are dominated by sedimentary rocks of the type that form along continental margins. Lowland regions in continental interiors tend to be dominated by ancient, less-erodible metamorphic and plutonic rocks.

Third, high mountains force moisture-bearing winds to rise and they generally receive large amounts of precipitation. This means high rates of river runoff and high erosion rates. This effect is especially pronounced in southern Asia, where intense monsoon rainfall on the southern slope of the Himalaya leads to unstable slopes, intense erosion, and a high discharge of dissolved matter and suspended sediment to the Indian Ocean (Figure 6.26). Studies in South America have shown that more than three-quarters of the dissolved substances carried by the Amazon River to the sea come from the Andean highlands.

Apparently, past weathering rates have not always been as high as they are today. Evidence from ocean sediments points to an increase in the amount of dissolved matter reaching the oceans in the past 5 million years. This implies an increase in global weathering rates. Furthermore, the major mountain systems of the world have not always been as high as we find them today. Evidence points to increased uplift rates in the Andes, Himalaya, Tibetan Plateau, and other tectonic regions like western North America and the Alps since about 5 million years ago. Sediments shed from the rising Himalaya coarsen upward from silts deposited about 5 million years ago to gravels about 1 million years old. This change

Figure 6.26 Rapid Erosion The Indus River, which drains the high western Himalaya and Karakorum ranges, carries a large load of dissolved weathering products to the Indian Ocean. This mountain region has one of the highest measured uplift rates in the world. High rates of erosion lead to high rates of chemical and physical weathering.

implies that rivers draining the mountains gained increasing energy with time as mountain slopes and stream channels became steeper.

The changes we can detect in the rates of mountain uplift parallel changes in weathering rates that we can infer from the record of sediments and sedimentary rocks. Mountain uplift rates are closely related to rates of seafloor spreading. Thus, global rates of chemical weathering must be inextricably linked to plate tectonics.

ANCIENT LATERITES AND THE NORTHWARD DRIFT OF AUSTRALIA

Most major shield areas of Earth are low in altitude and in relief. The Western Australia Shield is no exception, for its surface generally lies between 100 and 500 m altitude. Except for its extreme southwestern corner, the shield generally receives only 200–300 mm of rainfall a year. As a result, no large rivers flow off the shield and vegetation is sparse.

Weathering is very deep and in places is known to reach hundreds of meters. This suggests that the shield must have been exposed to erosion for a very long time, that it has experienced no significant uplift or downwarping, and that at some time in the past it must have had a wetter climate.

At many places on the shield it is possible to find remnants of an old reddish lateritic surface (Figure 6.27). As we have already seen, laterites are a product of deep weathering in warm, wet climates and they are forming today in places like southeastern Asia and the Amazon Basin of South America. The laterites of Western Australia are known to date from at least 30 to 40 million years ago. Because it is possible to show that they are not forming in today's arid climate, the landscape must have experienced a significant change of climate over time.

Today the shield lies between 20° and 35° S latitude. These are the latitudes where dry subtropical air causes major deserts to form (Chapter 17). However, Australia is slowly drifting directly north at a rate of about 5 cm a year. At this rate of movement, 40 million years ago the Western Australia Shield would have been about 2000 km south of its present position in a temperate zone of high rainfall like that of southern Chile today. Furthermore, the world climate about 30 to 40 million years ago was much warmer than it is today. A warm climate and high rainfall at that time probably explains the origin of the ancient Australian laterite. Like a ship slowly sailing north, the Western Australia Shield has been passing through different climatic zones. Today it lies in the dry subtropical belt, but if the present direction and rate of motion continue for another 20 million years, it will have entered a zone of tropical climate.

Figure 6.27 Australian Laterite This subdued region of the Western Australia Shield has a reddish color due to an ancient laterite at the land surface. The laterite is not forming today, and must have developed tens of millions of years ago when the climate was warmer and wetter and Australia lay far to the south of its present position.

What's Ahead?

Having explored the ways that physical and chemical weathering break down rocks at and near Earth's surface, we next discuss how these weathering products are mobilized, transported, and deposited as sediments that ultimately are transformed into sedimentary rocks. In this part of the rock cycle, we examine the settings in which sediments accumulate and generate a history of Earth's changing surface environments over hundreds of millions of years.

CHAPTER SUMMARY

1. Weathering extends to whatever depth air, water, and microscopic forms of life penetrate Earth's crust. Water solutions, which enter the bedrock along joints and other openings, attack the rock chemically and physically, causing breakdown and decay.

2. Physical and chemical weathering, although involving very different processes, generally work together.

3. Growth of crystals, especially ice and salt, along fractures and other openings in bedrock is a major process of physical weathering. Others include intense fires that cause spalling and fracturing, the wedging action of plant roots, and the churning of rock debris by burrowing animals. Each of these processes can have large cumulative effects over time.

4. Chemical weathering involves the transformation of minerals that are stable deep in the crust into forms that are stable at Earth's surface. The principal processes are hydrolysis, leaching, oxidation, hydration, and dissolution.

5. Subdivision of large blocks into smaller particles increases total surface area and thereby accelerates chemical weathering.

6. The effectiveness of weathering depends on rock type and structure, surface slope, local climate, and the time over which weathering processes operate.

7. Because heat and moisture speed chemical reactions, chemical weathering is far more active in moist, warm climates than in dry, cold climates.

8. Soils consist of regolith capable of supporting plants. Soils develop distinctive horizons, the characteristics of which are a function of climate, vegetation cover, soil organisms, parent material, topography, and time.

9. The classification system used in the United States places soils in 11 orders based on their physical characteristics.

10. Paleosols are ancient buried soils that can provide clues about former landscapes, plant cover, and climate, and are useful

for subdividing, correlating, and dating the strata in which they are found.

11. Soil erosion and degradation are global problems that have been increasing as world population rises. Effective control measures include crop rotation, plowing along contours rather than downslope, terracing, and tree planting, but halting widespread loss of soils is a formidable challenge.

12. Weathering is an integral part of the rock cycle. The silicate minerals least resistant to weathering are the ferromagnesian minerals with the least amount of polymerization of silicate tetrahedra.

THE LANGUAGE OF GEOLOGY

A horizon (p. 160)

B horizon (p. 160)

C horizon (p. 161)
carbonic acid (p. 152)
chemical weathering (p. 148)

dehydrate (p. 155)
dissolution (p. 153)

exfoliation (p. 156)

frost wedging (p. 151)

hydration (p. 155)
hydrolysis (p. 153)

joints (p. 149)

kaolinite (p. 153)

laterite (p. 164)
leaching (p. 154)

oxidation (p. 154)

paleosol (p. 165)
parent material (p. 160)
physical weathering (p. 149)

soil (p. 160)
soil horizon (p. 160)
soil profile (p. 160)
spheroidal weathering (p. 156)

weathering (p. 148)
weathering rind (p. 161)

QUESTIONS FOR REVIEW

1. How would you identify the effects of decomposition and disintegration on rocks in the field?

2. How does acid rain from industrialized regions contribute to weathering?

3. What causes flakes of rock to spall off exposed boulders during a forest fire?

4. Explain how weathering leads to concentration of minerals such as gold or platinum in sediments.

5. Why does the physical breakup of a rock increase the effectiveness of chemical weathering?

6. How does regolith formed on chemically weathered limestone differ from that formed on chemically weathered igneous rock, and why?

7. Describe the principal horizons in the profile of a well-developed soil in an evergreen forest; in an arid region.

8. If a buried soil contains a well-developed caliche at the top of the C horizon, what can you infer about the climate at the time the soil formed?

9. Under what conditions would high mountains experience high rates of chemical weathering?

10. How can movement of lithospheric plates lead to long-term changes in local climate and weathering?

Click on *Presentation* and *Interactivity* in the **Sedimentary Rocks** module of your CD-ROM to further explore resources and activities presenting concepts from this chapter. Select *Assessment* in the same module to test your understanding of this chapter.

Dipping sandstone strata, about 300 million years old, harbor a colony of nesting seabirds on Bonaventure Island, off the eastern Gaspé Peninsula, Canada.

Sediments and Sedimentary Rocks: Archives of Earth History

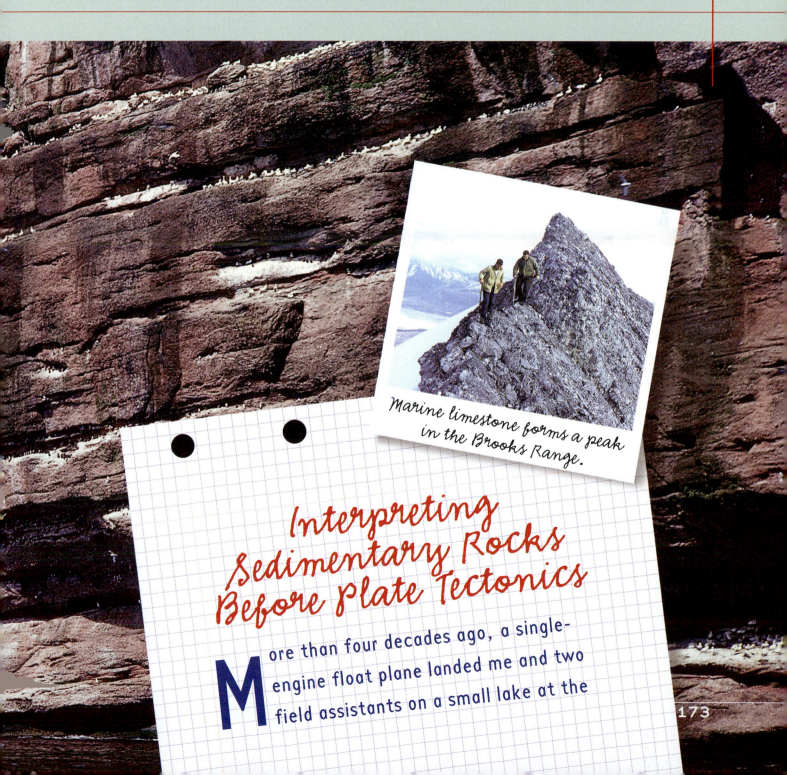

Marine limestone forms a peak in the Brooks Range.

Interpreting Sedimentary Rocks Before Plate Tectonics

More than four decades ago, a single-engine float plane landed me and two field assistants on a small lake at the drainage divide of the central Brooks Range of Arctic Alaska, a rugged region of high peaks and deep, glacially eroded valleys. A primary objective was to make a detailed geologic map of the rocks, determine their structure, and reconstruct their geologic history.

The thick succession of sedimentary rocks at the range crest includes marine deposits containing corals and other tropical sea creatures. The fossiliferous marine rocks apparently had been deposited in a warm, shallow sea about 350 million years ago. Our field area was close to latitude 68° North, far beyond the present range of tropical reef growth. Modern reefs are restricted to waters within 30° of the equator that have a mean annual temperature of 18°C or higher. In the 1950s, it was widely believed that the continents had occupied their present positions over long intervals of geologic time. I could not easily explain

how tropical conditions existed in the Brooks Range, where the ground now is permanently frozen to a depth of at least 600 m. To explain the warm-water fossils, I reasoned that Earth's climate must have become so warm that the tropical zone invaded the subpolar latitudes.

What I lacked, of course, was the concept of plate tectonics, which did not fully evolve until nearly a decade after our study was completed. Now we have evidence that the coral-bearing marine rocks in this part of the Brooks Range formed at much lower, presumably tropical, latitudes. Slow drift of Earth's continents over hundreds of millions of years not only can explain tropical corals in the Brooks Range, but subtropical coal deposits in Antarctica and evidence of ancient ice sheets in tropical India, South America, and Africa.

Stephen C. Porter

KEY QUESTIONS

1. **What is stratification?**

2. **What are the three broad classes of sediment and of sedimentary rock?**

3. **By what processes are sediments transported and deposited in various environments?**

4. **What four changes are involved in the transformation of deposited sediment to sedimentary rock?**

5. **What are the keys to classifying sedimentary rock?**

6. **How do sedimentary rocks give us information about the environmental history of Earth?**

7. **How are sediments related to plate tectonics?**

INTRODUCTION: SEDIMENTS IN THE ROCK CYCLE

In the previous chapter, we saw how weathering continually generates vast amounts of sediment. Nature ceaselessly sweeps this sediment from the solid rock beneath it, transports the sweepings by water, ice, wind, and gravity, and deposits them in many different environments. Sediments mantle nearly all land areas, as well as vast expanses of seafloor. Geologists use the physical and chemical properties of these sediments to learn about the geologic processes that deposited them, as well as the history of Earth's changing surface environments. Just as a good mystery writer provides clues to help the reader solve the story's puzzle, a single grain of sand or a fossil may offer a critical

clue for a geologist trying to reconstruct part of Earth's history. It is mainly from sediments and sedimentary rocks that geologists obtain data to construct maps showing the position of former ocean basins, continents, rivers, deserts, and mountain ranges. Studies of the biotic and isotopic variations in fossiliferous marine sediments provide crucial evidence about past ocean circulation, global changes in sea level, and the timing and causes of glacial ages.

The transport and deposition of sediments are initial phases of the rock cycle (Chapter 1), during which debris from elevated parts of Earth's crust moves inexorably to lower and lower altitudes under the pull of gravity. A particle of sediment may be dislodged and moved by more than a single transporting agent: initially by weathering and slope processes, then by ice, rivers, or wind, and finally by ocean currents. A particle may move progressively toward the ocean or, more likely, will be repeatedly deposited and stored for a time before resuming its journey to the sea. The ultimate fate of most sediment is burial and conversion to sedimentary rock. If the sedimentary rock remains relatively near the surface, the environmental clues it contains will be preserved and successive layers can be read like pages in a book. However, as sedimentary rocks are carried to greater depths in Earth's crust, heat and pressure transforms them into metamorphic rocks, and much of the environmental information is lost. Uplifted tectonically, these rocks will now have completed the rock cycle, as the processes of weathering and erosion begin anew.

SEDIMENTATION, STRATIFICATION, AND BEDDING

One of the most obvious features of sediments and the rocks that form from them is their layered appearance (Figure 7.1). Geologists refer to this arrangement of sedi-

Figure 7.1 Layering of Sedimentary Rocks The Colorado River, cutting downward through the Colorado Plateau over millions of years, has exposed many layers of sedimentary rock in the walls of the Grand Canyon.

mentary particles in layers as **stratification**, or **bedding**. Each **stratum** (plural = **strata**) or **bed** is a distinct layer of sediment, distinguishable from layers above and below it. The top or bottom surface of a bed is a **bedding plane**. Some strata are only a few millimeters thick, but others can reach a thickness of tens of meters.

A close look at sedimentary strata shows that they differ from one another because of differences in some characteristic of the particles, their thickness, or in the way they are arranged. For example, the average diameter of particles in one bed may differ from the average diameter in another, or their mineral composition or color may be different.

SEDIMENT TYPES AND CHARACTERISTICS

Geologists group sediments into three broad classes. *Clastic sediment* consists of loose fragments of rock debris produced by physical weathering; common examples are sand grains and clay particles. *Chemical sediment* precipitates from solution in water; common examples are calcium carbonate and salt. *Biogenic sediment* is composed of the fossilized remains of plants or animals, which, under suitable conditions, can form hydrocarbons like coal, oil, and natural gas. We will now examine each of these three types in more detail.

CHARACTERISTICS OF CLASTIC SEDIMENT

Along a stream, a close examination of the stream's sediment shows that the gravel, sand, and mud are fragments of rocks and minerals. The finest sedimentary particles, such

as those found in mud, are too small to see clearly. These also are derived from weathered rock, and they result from the chemical and mechanical processes we discussed in the last chapter. For example, chemical decomposition of feldspar results in the formation of clay minerals. Geologists refer to the loose, fragmental detritus produced by the mechanical breakdown of older rocks as **clastic sediment** (from the Greek word *klastos*, meaning "broken"). Any individual particle of clastic sediment is a *clast*.

PARTICLE SIZE: A BASIS FOR CLASSIFYING CLASTIC SEDIMENTS

Particles of clastic sediment can range in size from huge boulders down to submicroscopic clay particles. This range of particle size is the primary basis for classifying clastic sediments and clastic sedimentary rocks. Geologists classify sediments according to grain size in two ways. First, size range limits can be described in mm. In Table 7.1, clastic sediment is divided in this way into four main size classes. From coarsest to finest they are gravel, sand, silt, and clay. Gravel is further subdivided into boulder gravel, cobble gravel, and pebble gravel. If a gravel consists predominantly of clasts having diameters between 64 and 256 mm, we call it a cobble gravel. This classification scheme is called the Udden-Wentworth scale, after the geologists who first proposed it. Second, particle size can be stated in phi (ϕ) units, which are whole numbers rather than fractions (Table 7.1). In this scheme

$$\phi = -\log_2 d$$

where ϕ is the phi size and d is the particle diameter in millimeters. For example, a particle with a diameter of 8 mm

BOX 7.1 | THE SCIENCE OF GEOLOGY

GRAPHIC DISPLAY OF GEOLOGIC DATA

As in any other scientific field, advances in geology depend on the collection and analysis of data. Numerical data commonly are assembled in tabular form, which is a necessary but cumbersome manner of display. In most cases, however, these same data can be displayed graphically, making them far easier to comprehend and analyze. This is certainly true in the case of sediments and sedimentary rocks.

As an example, grain-size data are listed in tabular form in Table B7.1. The weight of each size-class of sediment is shown using φ-scale values. From the total weight of the sample (94 g), the weight percent of each size-class is calculated. Adding each successive weight percent value gives the cumulative weight percent. Among the common ways these data can be displayed are as a histogram, a frequency curve, or a curve of cumulative weight percent (Figures B7.1A, B, and C, respectively). From the latter curve, it is easy to determine the median grain size of the sample (i.e., the φ size equivalent to the 50 percent value, which in this case is about 1.75 φ, equivalent to a medium sand).

When a large set of data is assembled and compared with other similar data, if may be useful to know the average (or mean) value and the relative spread of values about the mean value. This involves calculating the *standard deviation* of the data. If a data set has a normal (i.e., bell curve) distribution of values about the mean (Figure B7.1D), 1 standard deviation (= 1 σ) encompasses the central 68 percent of the area beneath the curve. In other words, there is a 2 out of 3 chance that any single measured value lies within one standard deviation of the mean value. Mean values and standard deviations are commonly used to evaluate whether physical or chemical properties of several data sets are statistically similar or dissimilar.

When data sets involve three variable quantities, the information can be displayed as a ternary diagram (Figure

Table B7.1 Grain-Size Data for Sediment Sample

Φ Size	Weight (g)	Weight %[a]	Cumulative Weight %[b]
-1.0	0.7	0.7	0.7
-0.5	2.7	2.9	3.6
0.0	5.3	5.6	9.2
0.5	7.5	8.0	17.2
1.0	9.8	10.4	27.6
1.5	13.2	14.0	41.6
2.0	14.6	15.5	57.1
2.5	13.3	14.1	71.2
3.0	10.4	11.1	82.3
3.5	7.5	8.0	90.3
4.0	5.2	5.5	95.8
4.5	3.0	3.2	99
5.0	0.8	1.0	100
	94.0	100	

[a] Plotted in Figures B7.1A and B

[b] Plotted in Figure B7.1C

B7.1E). In the example illustrated, the relative percentages of sand, silt, and clay in multiple samples of three tills lie within three fields shown by different colors. Where fields overlap, the grain-size distributions are similar; if they do not, the grain-size distributions are dissimilar.

Time-series plots are a common type of graphic diagram designed to show variations in some physical or chemical variable as a function of time. In Figure B7.1F, variations in the amount of windblown dust accumulating in the deep Pacific Ocean are shown as a function of time. You will find a number of time-series plots in chapters appearing later in this book (for example, Figures 16.32 and 19.7).

Table 6.1 Definition of Clastic Particles Based on Grain Size

Clast Name	Udden-Wentworth Scale [a] Particle Size (mm)	Phi (φ) Scale
Boulder	≥ 256	≤-8
Cobble	64 to 256	-6 to -8
Pebble	2 to 64	-1 to -6
Sand	0.0625 to 2	4 to -1
Silt	0.0625 to 0.0039	8 to 4
Clay[b]	≤ 0.0039	≥8

[a] After C.K. Wentworth, 1922, "A Scale of Grade and Class Terms for Clastic Sediments," *Journal of Geology* 30, pp. 377-392.

[b] Clay, used in the context of this table, refers to particle size. The term should not be confused with clay minerals, which are definite mineral species.

has a phi value of –3, the power needed to raise the base of the logarithm ($\log_2$) to 8 (= 2^3). A particle 4 mm in diameter has a phi value of –2. This scheme is especially useful in statistical analyses and when plotting grain-size data on graphs (see *The Science of Geology*, Box 7.1, *Graphic Display of Geologic Data*, Figure B7.1).

SORTING BY SIZE, SPECIFIC GRAVITY, AND DURABILITY

Sorting describes sediment in terms of the variability in the size of its particles. Sediment having a wide range of particle size, as in a jumble of gravel, sand, and finer particles deposited by a flood, is *poorly sorted* (Figures 7.2 and 7.3); if the size range is small, as in a uniform gravel, the sediment is *well-sorted* (Figure 7.4).

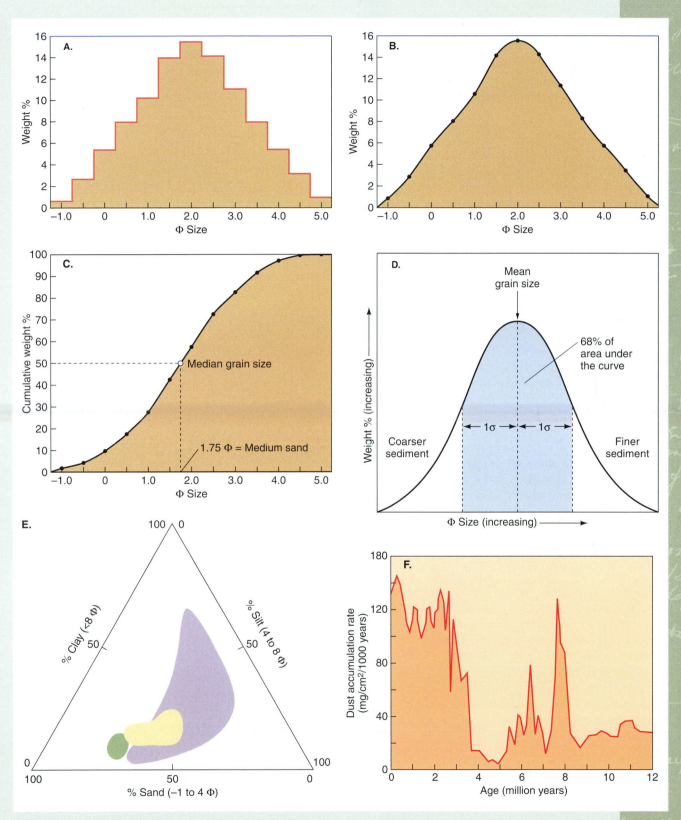

Figure B7.1 Graphic Display of Geologic Data A. Histogram. B. Frequency curve. C. Cumulative curve. D. Curve illustrating mean grain size and standard deviation. E. Ternary diagram. F. Time-series plot.

Figure 7.2 Sorting in Sediments Poorly sorted cobble-pebble gravel and interstratified well-sorted cross-bedded sand in glacial sediments exposed high on the slope of a Hawaiian volcano.

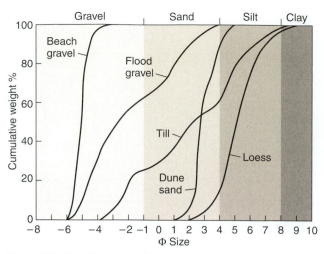

Figure 7.3 Size-Frequency Curve Curves of cumulative weight percent for some different sediment types. Which sediments are well sorted? poorly sorted?

Figure 7.4 Soring, Rounding, and Sphericity Well-sorted and well-rounded pebble gravel along the Columbia River, Oregon. Try to estimate the percentage of pebbles with high sphericity and of those with low sphericity.

For most clastic sediment, changes of grain size typically result from fluctuations in the velocity of the transporting agent, such as water or wind: the greater the speed and energy involved, the larger or heavier are the particles that can be transported. Consider a stream. Where the stream is flowing rapidly, it has enough energy to carry gravel, sand, silt, and clay particles. However, where it slows, heavier or larger particles fall to the bottom, followed by progressively smaller particles until the water may be transporting only very fine-grained silt and clay. Fluctuations in particle size in a stratum may therefore reflect fluctuations in the speed and energy of the transporting water or wind.

Sorting of particles also is related to differences in their specific gravity. Unusually dense minerals (e.g., gold, platinum, magnetite) are deposited first when stream velocity slows, whereas lighter particles are carried onward. During the California and Alaska gold rushes in the 1840s and 1890s, gold prospectors used this principle to separate nuggets and flakes of gold from river sediments (Figure 7.5). Nevertheless, most of the particles transported by water or wind are common rock-forming minerals, such as quartz and feldspar, that are sorted not by specific gravity but by *size*.

Long-continued movement of particles by turbulent water or air results in gradual destruction of the weaker particles, leaving behind more durable ones that can better survive in the turbulent environment. Commonly, the ultimate survivor is quartz because of characteristics you learned in Chapter 3: it is hard (7.0) and lacks cleavage. In this case, the relative increase, or enrichment, of a specific mineral in a sediment is based on *durability*.

NONSORTED SEDIMENT

In some sediments, the particles are a mixture of different sizes arranged chaotically, without obvious order. Such sediments are created, for example, by rockfalls, slow movement of debris down hillslopes, slumping of loose deposits on the seafloor, mudflows, and deposition of debris from glaciers or floating ice. Some nonsorted sedi-

Figure 7.5 Placer Nuggets One day's "take" of platinum and gold nuggets dredged from glacial gravels at a placer mine in Platinum, Alaska. Having a much higher specific gravity, these precious metals are easily concentrated from the enclosing stream sediments.

ments are given specific names; for example, *till* is a non-sorted sediment of glacial origin (Figure 7.6).

PARTICLE SHAPE

Mechanically weathered particles broken from bedrock tend to be angular because breakage typically occurs along grain boundaries, fractures, and bedding surfaces. Nearly all such weathered particles become smooth and rounded as they are transported by water or air and are abraded by other rock fragments. *Roundness*, as measured by the sharpness of a particle's edges, is not the same as *sphericity* which is a measure of how closely particle shape approaches that of a sphere (Figure 7.4). A flat particle bounded by cleavage or fracture surfaces may have well-rounded edges, but it also may have a low degree of sphericity. In general, the greater the distance of travel, the greater is the degree of rounding.

RHYTHMIC LAYERING

Some sediments display a distinctive alternation of parallel layers that have different properties. Such alternation suggests that some naturally occurring rhythm has influenced sedimentation. A pair of such sedimentary layers deposited over the cycle of a single year is termed a **varve** (Swedish: *cycle*). Varves are most commonly seen in deposits of high-latitude or high-altitude lakes, where there is a strong contrast in seasonal conditions. In spring, as a cover of winter ice melts away, the inflow of sediment-laden water increases and coarse sediment is then deposited throughout the summer. With the onset of colder conditions in the autumn, streamflow decreases and ice forms over the lake surface. During winter, very fine sediment that has remained suspended in the water column slowly settles to form a thinner, darker layer above the coarse, lighter-colored summer layer. Varved lake sediments are common in Scandinavia, southern Canada (Figure 7.7), and New England, where they formed beyond the retreating margins of ice-age glaciers. Varved marine sediments have been discovered in cores from a deep glacier-carved basin in southern British Columbia.

Geologists must be careful when interpreting laminated sediments, for they may not be varves. Rather than representing annual deposition, rhythmic laminations may result when the velocity of inflowing water varies in a cyclical manner, as during a succession of large storms.

Figure 7.6 Till Nonsorted till deposited by a lobe of the Cordilleran Ice Sheet in western Washington during the last glaciation. The pebbles, cobbles, and boulders represent various source regions and are encased in a mud-like mixture of light-gray sand and silt. Striations can be seen on the face of the largest clast.

Figure 7.7 Cyclical Sediment Varves deposited in a glacial-age lake in eastern Canada have a banded, cyclical appearance. Each pair of layers represents an annual deposit. Light-colored silty layers were deposited in summer, and the dark-colored clayey layers accumulated in winter. Each layer is well sorted.

CROSS BEDDING

Cross bedding refers to beds that are inclined with respect to a thicker stratum within which they occur (Figures 7.2 and 7.8). Cross bedding is the work of turbulent flow in streams, wind, or ocean waves and consists of particles coarser than silt. As the particles are transported, they tend to collect in ridges, mounds, or heaps in the form of ripples, waves, or dunes that migrate slowly forward in the direction of the moving current. Particles accumulate on the down-current slope of the pile to produce beds having inclinations as great as 30° to 35°. The direction in which cross bedding is inclined tells the direction in which the related current of water or air was flowing at the time of deposition.

GRADED BEDDING

The way particles are arranged within a layer provides important information about the conditions of sedimenta-tion. If a mixture of small solid particles having different diameters and about the same specific gravity is placed in a glass of water, shaken vigorously, and then allowed to stand, the particles will settle out and form a deposit on the bottom of the glass. The largest particles settle first, followed by successively smaller ones. The finest may stay in suspension for hours or days before they finally settle out at the top of the deposit. In the resulting **graded bed**, the particles are sorted more or less according to size and grade upward from coarser to finer (Figure 7.9). A graded bed also can form from a sediment-laden current. As the current slows, the heaviest and largest particles settle out first, followed by lighter and smaller ones.

MINERAL COMPOSITION

Most coarse clastic sediments consist of mineral grains and rock fragments, dominated by those least susceptible to chemical and physical breakdown. For example, sands typically have a high content of quartz and some have a good deal of potassium feldspar, because these are the common rock-forming minerals that are most resistant to weathering.

Figure 7.8 Ancient Sand Dunes Cross-bedded sand dunes converted to sandstone are exposed widely in Zion National Park, Utah. The inclined bedding dips to the left, in the direction toward which the prevailing wind blew when the dunes were active.

Figure 7.9 Graded Bed Mudflow sediment deposited during a prehistoric eruption of Mount St. Helens volcano displays a graded structure. Cobbles and small boulders in the lower part of the exposed section grade upward into pebbles and sand near the top.

Figure 7.10 Well-Rounded, Well-Sorted Sand Well-rounded grains of sand (~1 mm diameter) from the St. Peter Sandstone of Wisconsin have been sorted by size and polished by constant shifting and abrasion in surf along an ancient shoreline.

As detrital rock fragments and mineral grains are transported, they are subjected to continuous chemical and physical breakdown. After several cycles of erosion and deposition, the result can be a sediment that consists almost entirely of quartz, the most resistant of the rock-forming minerals (Figure 7.10).

CHARACTERISTICS OF CHEMICAL SEDIMENTS

Some sediment contains no clastic particles, yet the material composing the sediment has been transported from somewhere. How can this be? In such sediment, the components were fully dissolved, transported in solution, and then precipitated chemically. Sediment formed by precipitation of minerals from solution in water is **chemical sediment**, and it forms in two principal ways:

1. *Through biochemical reactions resulting from the activities of plants and animals in the water.* For example, tiny plants living in seawater can decrease the acidity of the surrounding water and thereby cause calcium carbonate to precipitate.

2. *Inorganic reactions in the water.* When water from a hot spring cools, it may precipitate opal (a hydrated silicate) or calcite (calcium carbonate) (Figure 7.11A). We can witness a similar effect when chemical sediment is deposited along the inside of a hot-water pipe, thereby constricting the flow of water and leading to expensive home repairs.

Another common example is evaporation of seawater or lake water. As the water evaporates, dissolved matter is concentrated and salts begin to precipitate as chemical sediment (Figure 7.11B). Examples of salts that precipitate from lake waters are sodium carbonate (Na_2CO_3), sodium sulfate (Na_2SO_4), borax ($Na_2B_4O_7 \cdot 10H_2O$), and *trona* ($Na_2CO_3 \cdot NaHCO_3 \cdot 2H_2O$). These and other salts have many uses, including the production of paper, soap, detergents, antiseptics, and chemicals for tanning and dyeing.

A.

B.

Figure 7.11 Hotspring and Evaporite Deposits A. Mammoth Hot Springs, at Yellowstone National Park, Wyoming, provides an example of the remarkable landforms created as hot spring water cools and precipitates chemical sediment. B. Evaporite salts encrust the surface of a desert playa on the floor of Death Valley, California. A shallow lake forms during rainy periods. As the water evaporates and the playa lake dries up, salts crystallize out of the brine. Polygonal fractures form as the drying sediment contracts.

The most important salts that precipitate from seawater are halite (rock salt; $NaCl$) and gypsum ($CaSO_4 \cdot 2H_2O$). Much of the common table salt we use daily and the gypsum used for plaster and construction materials are recovered from marine evaporite deposits.

CHARACTERISTICS OF BIOGENIC SEDIMENTS

Many sediments contain **fossils**, the remains of plants and animals that died and were incorporated and preserved as the sediment accumulated (Figure 7.12). A sediment composed mainly of fossil remains, and therefore produced directly by the physiological activities of organisms, is called **biogenic sediment**. Biogenic sediment is the direct result of the natural interactions of Earth systems. Living organisms, which are a part of the biosphere, depend on the hydrosphere and atmosphere for their existence. When they die, their remains may fossilize and become a part of the lithosphere.

BIOCLASTIC SEDIMENT

Sometimes the structure or skeleton of a plant or animal survives in the fossil record, but more commonly the remains are broken and scattered. The solid parts generally end up as fragments, or clasts. If the sediment is composed largely of such fragments, it is *bioclastic* sediment. The organic materials (soft tissues) of plants and animals commonly decompose, but in favorable environments they are transformed into new organic compounds that may evolve into fossil fuels (see Chapter 21).

DEEP-SEA OOZE: BIOGENIC SEDIMENT IN THE OCEAN DEPTHS

When floating microscopic marine organisms die, their remains settle and accumulate on the seafloor to form a muddy sediment called **deep-sea ooze**. If such a sediment consists mainly of carbonate remains, it is *calcareous ooze*. If the sediment is largely silica, it is *siliceous ooze*.

Calcareous ooze is formed of calcium carbonate and is widespread in the oceans. Although calcium carbonate can be precipitated chemically from seawater, most carbonate sediments are biogenic and form mainly in warm surface ocean waters. There, carbonate-secreting organisms precipitate calcite (or a polymorph, aragonite) in building their hard parts (Figure 7.13). In doing so, the calcium and bicarbonate ions in the water are combined to form solid calcium carbonate:

$$Ca^{2+} + (HCO_3)^{1-} = CaCO_3 + H^{1+}$$

$\quad\quad$ calcium $\quad\quad$ bicarbonate $\quad\quad$ calcium $\quad\quad$ hydrogen
$\quad\quad$ ion $\quad\quad\quad$ ion $\quad\quad\quad\quad$ carbonate $\quad\quad$ ion

Corals, algae, and other colonial organisms growing on the seafloor in shallow tropical waters precipitate calcium carbonate, thereby creating extensive carbonate reefs.

Siliceous ooze resembles calcareous ooze but is composed mainly of the siliceous remains of tiny floating protozoa (radiolarians) and algae (diatoms). The hard parts in this ooze are often delicate and lace-like (Figure 7.13). Some kinds of diatoms also live in lakes, where their remains accumulate as an important component of lake sediment.

Before you go on:

1. What are some ways in which sedimentary strata can differ?

2. What causes sorting, and what factors control the difference between well-sorted, poorly sorted, and nonsorted sediments?

3. How is chemical sediment formed?

SEDIMENTARY ENVIRONMENTS AND FACIES CHANGES

Examine a sequence of exposed sedimentary rocks and you will likely see that they change character in two dimensions, vertically and horizontally. Vertically, changes in strata reflect the passage of *time*. Successive layers of sandstone, siltstone, shale, and perhaps limestone were laid down during different periods of time. Horizontally, strata change over *distance*, but the changes occur within the same stratum. These differences reflect the change from one environment of deposition to another. Such a lateral change from one depositional environment to another is called a change of **facies** (pronounced *fay-seez*).

For example, suppose we examine a sequence of exposed sedimentary rocks in two locations several kilometers apart. If the strata are essentially flat-lying and not in an area where folding or faulting has occurred, the vertical sequences are likely to appear essentially the same. Horizontally, however, the story may be different. When the two outcrops are compared, we may find differences in thickness, particle size, sorting, fossil content, and other characteristics that reflect changes in the environment in which the sediments accumulated. Most sedimentary stra-

Figure 7.12 Fossil Fish Fossil fish exposed on a bedding plane of the Green River shale in Wyoming.

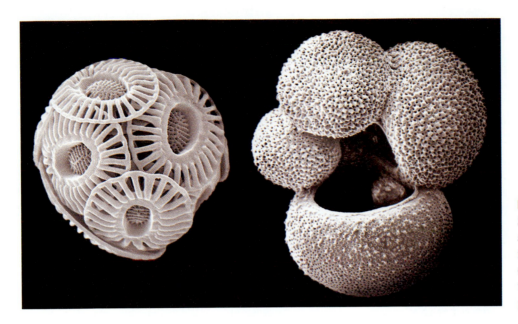

Figure 7.13   Deep-Sea Ooze
Skeletons of siliceous radiolarian (left) and a calcareous foraminifer from a deep-sea ooze, imaged by scanning electron microscope. Each could pass through the eye of a needle.

ta show such a change of facies through distances that range from meters to many kilometers.

Here is another example, on a much larger scale. If you were to travel across the edge of a continent and into the adjacent ocean basin, you would encounter a range of diverse environments (Figure 7.14). Each environment has its own distinctive sediments and associated organisms.

For each environment, you can identify distinctive physical, chemical, and biological characteristics that permit you to distinguish sediment accumulating there from sediment being deposited in another environment.

Therefore, a *sedimentary facies* can be thought of as sediment that accumulated at a certain time and in a specific depositional environment. This facies may be charac-

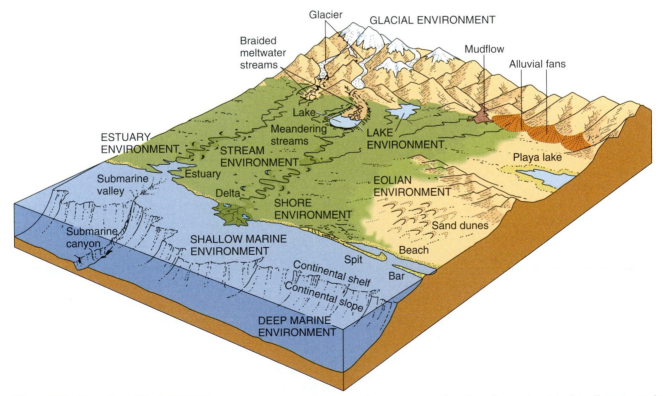

Figure 7.14   Depositional Environments   Various depositional environments are seen while traveling from the crest of a moun- tain range across the edge of a continent to the adjacent margin of a nearby ocean basin.

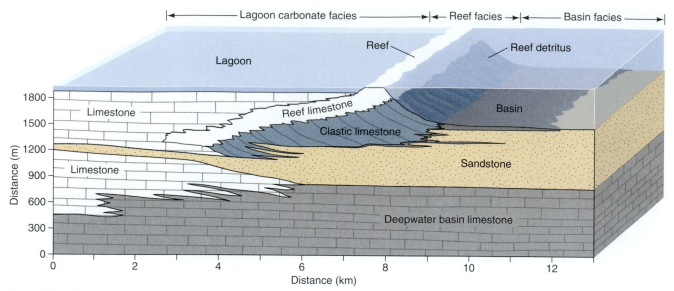

Figure 7.15 Sedimentary Facies A section through an accumulating ancient reef zone in Texas shows the relationship between different facies (lagoon, reef, and basin) that are accumulating across a shallow ocean margin. The vertical section shows the horizontal gradation between facies through time as the reef builds progressively farther toward the right of the diagram. In the center of the diagram, the stratigraphy indicates a change from deep-water limestone upward to sandstone, to clastic limestone, and then to reef limestone and lagoonal limestone. During one interval, basin sandstone was deposited across the area of the lagoon.

terized by distinctive grain size, grain shape, stratification, color, chemical composition, depositional structures, or fossils. Adjacent facies, deposited at the same time, can merge into one another either gradually or abruptly (Figure 7.15). For example, coarse gravel and sand of a beach may change very gradually offshore into finer sand, silt, and clay on the floor of the sea or lake. On the other hand, coarse, bouldery glacial sediment may end abruptly against stream sediments at the margin of a glacier.

By determining the distinctive characteristics of different sediments or sedimentary rocks, studying the relationships of different facies, and using these characteristics to identify original depositional settings, we can reconstruct a picture of varied environments at the time of deposition (Figure 7.15).

NONMARINE DEPOSITIONAL ENVIRONMENTS

Sediment derived from the mechanical and chemical breakdown of rocks is moved toward the sea or toward the lowest level of an inland basin. En route it is moved by water, ice, wind, or gravity. The sediment may be temporarily stored and then reworked repeatedly by one or several of these agencies before reaching its final resting place, where it is slowly converted to sedimentary rock.

STREAM SEDIMENTS

Streams constitute the principal agency for transporting sediment across the land (Figure 7.14 and Chapter 14).

Their deposits are widespread. Stream-deposited sediment differs from place to place depending on the type of stream, the energy available for doing work, and the nature of the sedimentary load. A typical large, smoothly flowing stream may deposit well-sorted layers of coarse and fine particles as it swings back and forth across its valley. Silt and clay are deposited on the valley floor adjacent to the stream during floods; clays and organic sediments may accumulate in abandoned sections of the stream channel that no longer carry flowing water. By contrast, a stream flowing from the front of a mountain glacier may divide into an intricate system of interconnected channels that change in size and direction as the volume of water fluctuates and as the stream copes with an abundance of detritus. The resulting sediments consist largely of coarse-grained stream-channel deposits. A large mountain stream flowing down a steep valley can transport an abundant load of sediment. When the stream reaches the mountain front and is no longer constrained by valley walls, it is free to shift laterally back and forth across the more gentle terrain as its load is dropped. The result is a fan-shaped deposit in which the sediments may range from coarse, poorly sorted gravels to well-sorted, cross-bedded sands.

LAKE SEDIMENTS

Sediments deposited in a lake accumulate both on the lakeshore and on the lake floor. Lakeshore deposits of generally well-sorted sand and gravel form beaches, bars, and spits across the mouths of bays. The sediment load of a stream entering a lake will be dropped as the velocity and transporting ability of the stream suddenly decrease. The resulting deposit, built outward into the lake, is a *delta* (Figure 7.14 and Chapter 14). Inclined, generally well-sorted layers on the front of a delta pass downward and outward into thinner, finer, and evenly laminated layers on the lake floor (Figure 7.7).

GLACIAL SEDIMENTS

Sedimentary debris eroded and transported by a glacier is either deposited along the glacier's base or released at the glacier margin as melting occurs. The sediment may then be subjected to reworking by running water. Debris deposited directly from ice commonly forms a random mixture of particles that range in size from clay to boulders and consist of all the rock types over which the ice has passed (Figure 7.6). Such sediment characteristically is neither sorted nor stratified, in contrast to most other non-marine sediments. Stones in glacially deposited sediment often are angular and some are striated (Figure 7.6). Sand grains typically show distinctive fractures due to crushing and abrasion.

EOLIAN (WINDBLOWN) SEDIMENTS

Both wind activity and the geologic results of it are referred to as **eolian**, after Aeolus, the Greek god of wind. Sediment carried by the wind tends to be finer than that moved by other erosional agents. Sand grains are easily moved where strong winds blow and where vegetation is too discontinuous to stabilize the land surface, as along seacoasts and in deserts (Figure 7.14). In such places, the sand may pile up to form dunes composed of well-sorted sand grains and with bedding inclined in the downwind direction (the direction toward which the air is flowing). Using these characteristics, geologists can identify ancient dune sands in the rock record (Figure 7.8).

Powdery dust picked up and moved by the wind is deposited as a blanket of sediment across the landscape (Chapter 17). Such sediment is thickest and coarsest near its source and becomes progressively thinner and finer with increasing distance downwind. Although common as a sediment in many parts of the world, windblown silt rarely forms sedimentary rock, probably because it is easily eroded from the landscape and therefore is unlikely to be widely preserved.

SHORELINE AND CONTINENTAL SHELF ENVIRONMENTS

The world's rivers continuously transport detritus to the edges of the sea where it can accumulate near the mouths of streams, be moved laterally along the coast by currents, or be carried seaward to accumulate on the continental shelves, sometimes to great thicknesses. In part spurred by the search for large undersea reservoirs of oil and gas,

geologists have learned a great deal about the sediments accumulating on the shelves.

ESTUARINE SEDIMENTS

Much of the load transported by a large river may be trapped in an **estuary**, a semienclosed body of coastal water within which seawater is diluted with fresh water (Figure 7.14). A wedge of marine water typically lies below the flowing fresh water, in which particles of clay are carried in suspension. When the stream water meets seawater, the clay particles tend to aggregate into clumps. Because of their greater mass, the aggregates settle more rapidly to the bottom. The resulting sediment is cohesive and therefore less susceptible to erosion and further transport. If the rate of sedimentation is high and the land is slowly subsiding, a thick body of estuarine sediment can form.

DELTAIC SEDIMENTS

Deltas are built not only into lakes, but also at places where streams reach the sea and deposit their sediment load (Figure 7.14). The largest marine deltas are complex deposits consisting of coarse stream-channel sediments, fine sediments deposited between channels, and still finer sediments deposited on the seafloor (Chapter 18).

BEACH SEDIMENTS

Quartz, the most durable of common minerals in continental rocks, is a typical component of beach sands. However, not all ocean beaches are sandy. Any beach consists of the coarsest rock particles contributed to it by the erosion of adjacent sea cliffs, or carried to it by rivers or by currents moving along the shore. Beach sediments tend to be better sorted than stream sediments of comparable coarseness and typically display cross stratification. Dragged back and forth by the surf and turned over and over, particles of beach sediment become rounded by abrasion. Although beach gravel and gravelly stream sediment may be similar in appearance, the pebbles and cobbles on many beaches acquire a distinctive flattened shape.

OFFSHORE SEDIMENTS

Fresh water flowing through an estuary or past a river mouth may continue seaward across the submerged continental shelf as a distinct layer overlying denser, salty marine water. Some fine-grained sediment, carried in suspension, thereby reaches the outer shelf. The sediment then either settles slowly to the seafloor or is ingested by floating organisms that excrete it as small pellets that fall to the bottom. On the continental shelf of eastern North America, up to 14 km of fine sediment has accumulated over the last 70 to 100 million years. To build the whole pile, an average of less than a millimeter of sediment would need to be deposited each year.

Most coarse marine sediment is deposited within 5–6 km of the land after being dispersed by currents that flow parallel to the shore. Coarse sediment is also found as far offshore as the seaward limits of the shelves. Its observed patchy distribution is mostly the result of changing sea level. At times when the sea fell below its present level, the shoreline migrated seaward across the shelves, exposing new land. Bodies of coarse sediment deposited near shore or on the land at such times were submerged as sea level again rose across the shelves. As much as 70 percent of the sediment cover on the continental shelves is probably a relict of such past conditions.

Only about 10 percent of the sediment reaching the continental shelves remains in suspension long enough to arrive in the deep sea, so it is clear that the great bulk of the Earth's sedimentary strata is shelf strata whose sediment originated on the continents. In effect, the shelves conserve continental crust, which is continually recycled within the continental realm.

CARBONATE SHELVES

Carbonate sediments of biogenic origin accumulate on the continental shelves wherever the influx of land-derived sediment is minimal and the climate and sea-surface temperature are warm enough to promote the abundant growth of carbonate-secreting organisms. Carbonate sediments accumulate mainly on broad, flat carbonate shelves that border a continent or rise as platforms off the seafloor (Figure 7.16). Most of the sediment consists of sand-sized skeletal debris, together with inorganic precipitates that form fine carbonate muds. Coarser debris is found mainly near coral and algal reefs or in areas of turbulence and strong currents.

Figure 7.16 Carbonate Shelf Carbonate sediments consisting of fine skeletal debris and inorganic precipitates accumulate in the warm, shallow marine waters of a broad, flat carbonate shelf surrounding the numerous islands of the Bahamas. The white carbonate sediments occur as shallow sand bars and along island beaches.

MARINE EVAPORITE BASINS

Ocean water occupying a basin with restricted circulation that lies in a region of very warm, dry climate will evaporate. This leads to the precipitation of soluble substances and the accumulation of a marine deposit called an **evaporite**. Such deposits are widespread in North America, where marine ancient evaporite strata underlie as much as 30 percent of the continent.

The Mediterranean Sea is an example of an evaporite basin (Figure 7.17). Were it not for continuous inflow of Atlantic water at its western end, the Mediterranean would gradually decrease in volume because of evaporation. It is estimated that if the Mediterranean were deprived of new water, evaporation would cause this landlocked sea to dry up completely in about 1000 years. In the process, a layer of salt about 70 m thick would be precipitated. Far thicker evaporites that underlie the Mediterranean basin are regarded as evidence of former periods when a high evaporation rate, together with a continuous inflow of Atlantic water to supply the necessary salt, allowed evaporites 2 to 3 km thick to accumulate.

CONTINENTAL SLOPE AND RISE ENVIRONMENTS

Along most of their length, the shallowly submerged continental shelves pass abruptly into continental slopes that descend to depths of several kilometers. Sediments that reach the shelf edge are poised for further transport down the slope and onto the adjacent continental rise.

TURBIDITY CURRENTS AND TURBIDITES

Thick bodies of coarse sediment of continental origin lie at the foot of the continental slope at depths as great as 5 km. The origin of these coarse accumulations at great depths in the oceans was difficult to explain until marine geologists demonstrated that the sediments could be deposited by **turbidity currents**. These are gravity-driven currents consisting of dilute mixtures of sediment and water having a density greater than the surrounding water. Such currents have been produced in the laboratory in water-filled tanks into which a dense mixture of water, silt, and clay has been introduced. They have also been documented moving across the floors of lakes and reservoirs (Figure 7.18). In the oceans, turbidity currents have been set off by earthquakes, landslides, and major coastal storms. Off the mouths of rivers, they can be set in motion by large floods.

The accumulated evidence leads us to believe that turbidity currents are effective geologic agents on continental slopes, where they can reach velocities greater than those of the swiftest streams on land. Some achieve a velocity of more than 90 km/h and transport up to 3 kg/m^3 of sediment, spreading it as far as 1000 km from the source.

A turbidity current typically deposits a graded layer of sediment called a **turbidite** (Figure 7.19). Such a graded

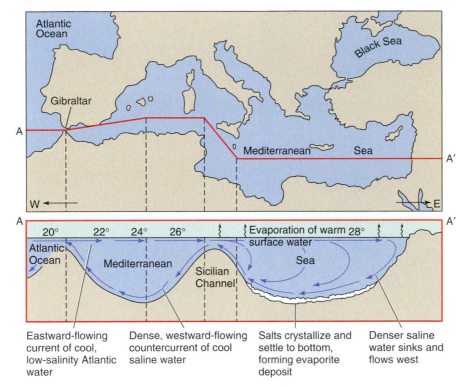

Figure 7.17 Currents in the Mediterranean Sea In the Mediterranean Sea, the inflow of fresh water from rivers in Europe and Africa is too small to offset the water evaporated from the sea surface in the hot eastern part of the basin. An inflowing current of Atlantic Ocean water balances the loss. As the relatively cool ocean water (summer temperature 20°C) enters the basin and moves eastward, it warms to (28°C) and evaporates. The salinity of the remaining water rises, increasing its density and setting up a westward-flowing current of cool saline water at depth. Salts crystallizing from the salty water settle to the bottom to form evaporite-rich sediments.

layer is the result of rapid, continuous loss of energy in the moving current. As a rapidly flowing turbidity current slows down, successively finer sediment is deposited. At any site on the continental rise or an adjacent abyssal plain, a turbidite is deposited very infrequently, perhaps only once every few thousand years. In these places, far distant from the sediment source, the deposits are mainly thin layers a few millimeters to 30 cm thick. Although deposition is infrequent, over millions of years turbidites can slowly accumulate to form vast deposits beyond the continental realm.

DEEP-SEA FANS

Some large submarine canyons on the continental slopes are aligned with the mouths of major rivers, like the Amazon, Congo, Ganges, and Indus. At the base of many such canyons is a huge **deep-sea fan**, a fan-shaped body of sediment that spreads downward and outward to the deep

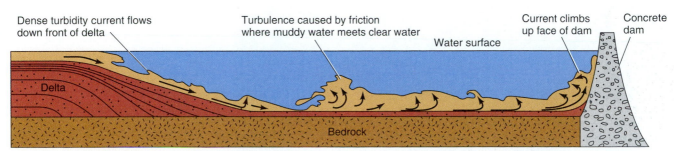

Figure 7.18 Turbidity Current A turbidity current generated by a surge of sediment-laden water enters the quiet water of a reservoir behind a dam. Moving rapidly down the face of a delta, the current passes along the lake floor and climbs up the face of the dam before subsiding. As the sediment settles to the bottom, it forms a graded turbidite layer.

Figure 7.19 Ancient Deep-Sea Turbidities Deep-sea turbidite beds that have been tilted and uplifted are exposed in a wave-eroded bench along the coast of the Olympic Peninsula, Washington.

seafloor (Figure 7.20). The sediments, which are derived mostly from the land, include fragments of land plants, as well as fossils of shallow- and deep-water marine organisms. Also present are many graded layers that are interpreted as turbidites.

Deep-sea fans are a major exception to the generalization that the final deposition of land-derived sediment in the ocean is largely confined to the continental shelves. When shelves are exposed at times of lowered sea level and rivers extend across them nearly to the continental slope, the stage is set for the rapid building of deep-sea fans.

SEDIMENT DRIFTS

Huge bodies of sediment up to hundreds of kilometers long, tens of kilometers wide, and two kilometers high have been discovered along the continental margins bordering the North Atlantic Ocean and Antarctica. Called *sediment drifts*, these massive deposits are associated with deep-ocean currents as much as 100 times the size of the Amazon River. Rippled sands along the axis of a sediment drift, where current velocities are relatively high, grade laterally into mud deposited where currents decrease in velocity. At most deep-ocean sites, rates of sediment accumulation reach only a few cm/1000 years, but on sediment drifts they are far higher; an exceptional rate of 200 cm/1000 years has been measured for a sediment drift near Bermuda. Such a rapid accumulation rate means that cores

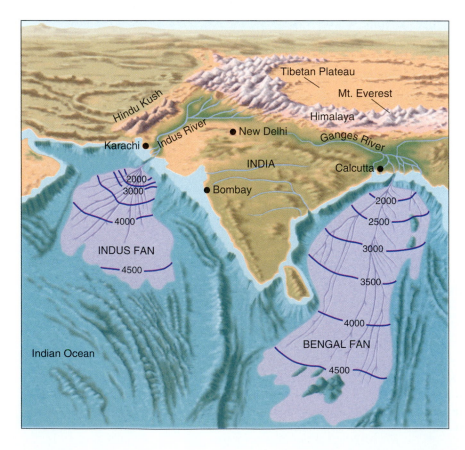

Figure 7.20 Deep-Sea Fans The Indus and Ganges–Bramaputra river systems have built the vast Indus and Bengal deep-sea fans on the seafloor adjoining the Indian subcontinent. Most of the sediment in the fans was transported from the high Himalaya to the north as the mountain system was uplifted over tens of millions of years. Contours are water depths in meters.

obtained from sediment drifts can provide a record of environmental change measurable at a decadal, rather than millennial, scale. Such sediments are of special importance because they link the marine, terrestrial, and atmospheric components of Earth's climate system (Chapter 19).

DEEP-SEA DEPOSITIONAL ENVIRONMENTS

Analyses of samples from sediment cores allow geologists to sort out the various sources of seafloor sediment (Figure 7.21). Study of a great number of samples indicates that all the sediments are mixtures. Although a large portion is produced by biologic activity in the surface waters, some sediment is transported over great distances from continental interiors and eventually reaches the deep sea.

DEEP-SEA OOZES

Calcareous ooze occurs over wide areas of ocean floor at low to middle latitudes where warm sea-surface temperatures favor the growth of carbonate-secreting organisms in the surface waters. However, calcareous ooze is rare in these same warm latitudes where the water is deeper than 3–4 km. The explanation is related to ocean chemistry. Cold deep-ocean waters are under high pressure and contain more dissolved carbon dioxide than shallower waters. When slowly sinking carbonate shells reach the depth where the water is undersaturated with respect to calcium carbonate, a level called the *carbonate compensation depth*, the shells are dissolved before they can reach the seafloor.

Other parts of the deep-ocean floor are mantled with siliceous ooze, most notably in the equatorial and northern Pacific and in a belt encircling the Antarctic region where siliceous organisms predominate. These are areas where surface waters have a high biological productivity, in part related to the upwelling of deep-ocean water rich in nutrients.

LAND-DERIVED SEDIMENT

In addition to biogenic oozes, deep-sea strata include sediments carried to the oceans by rivers, eroded from coasts by wave action, transported by wind (fine desert dust [Figure 17.20] and volcanic ash), and released from float-

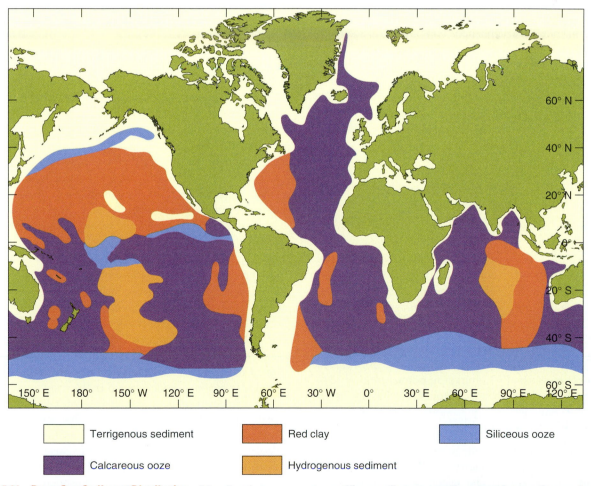

Terrigenous sediment	Red clay	Siliceous ooze
Calcareous ooze	Hydrogenous sediment	

Figure 7.21 Deep-Sea Sediment Distribution Map showing distribution of the major sediment types on the floor of the deep ocean. These sediments usually are a mixture of types, but on the map are named for their major component.

ing ice. Over much of the North Pacific and part of the central South Pacific the ocean floor lies at depths of more than 4 km and lacks calcareous ooze. In these regions, the seafloor is covered by oxidized reddish or brownish clay composed mainly of extremely fine clay minerals, quartz grains, and micas that originated on land.

Before you go on:

1. What is a sedimentary facies, and what can it tell us about the environment in which a sediment was deposited?

2. What are four general types of depositional environments? What differences serve to distinguish them?

3. How does changing sea level affect sediment deposition on continental shelves?

4. What are the chief kinds of sediments found on the seafloor?

Figure 7.22 Cementation of a Sandstone Sand grains (white to dark-gray) in a thin-section of a quartz sandstone from the Los Angeles Basin, California, are cemented by anhydrite (brightly colored matrix). The sandstone underlies beds of gypsum, which are the likely source of the cement.

DIAGENESIS: HOW SEDIMENT BECOMES ROCK

We can easily observe the transport of sediment to sites of deposition, but once deposited, how are these sediments turned into rock? **Lithification** (from the verb to *lithify*, meaning to turn to stone) is the overall process of creating sedimentary rock. **Diagenesis** is the collective term for all the chemical, physical, and biological changes that affect sediment as it goes from deposition through lithification. Processes involved include **compaction**, **cementation**, **recrystalization**, and chemical alteration.

Compaction. The first and simplest diagenetic change, compaction occurs as the weight of an accumulating sediment forces the grains together. As the *pore space* (space between grains) is reduced, water is forced from the sediment.

Cementation. When substances are dissolved in circulating pore water they precipitate to form a cement that binds the sediment grains together (Figure 7.22). Calcium carbonate is one of the most common cements. Silica, a particularly hard cement, may also bond grains together.

Recrystallization. As sediment accumulates, less stable minerals recrystallize into more stable forms. This process is especially common in porous tropical reefs. Over time, the mineral aragonite, which forms the skeletal structure of living corals, recrystallizes to its polymorph calcite.

Chemical alteration. In the presence of oxygen (*oxidizing environment*), organic remains are quickly converted to carbon dioxide and water. If oxygen is lacking (*reducing* or *anaerobic environment*), the organic matter does not completely decay but instead may be slowly transformed to solid carbon.

FORMATION OF PEAT: DIAGENESIS UNDER ANAEROBIC CONDITIONS

The formation of **peat** is a good example of diagenesis under reducing conditions. Geologists recognized long ago that the plants responsible for deposits of peat and coal must have lived in ancient swamps because (1) a complete physical and chemical gradation exists from peat to coal, and peat now accumulates mainly in swamps; and (2) only under swamp conditions is the conversion of plant matter to peat chemically probable.

On dry land and in running water, oxygen is abundant. Under these conditions, dead plant matter rots away very quickly, geologically speaking. However, in stagnant swamp water, oxygen is used up and not replenished, creating an anaerobic environment (lacking air or free oxygen). Thus, any plant matter lying in swamp water is attacked by anaerobic bacteria, which partly decompose it by splitting off oxygen and hydrogen from the plant's chemical compounds. The oxygen and hydrogen (combined in gases, such as methane) escape, and the carbon gradually becomes concentrated in the residue.

In this manner, plant matter is gradually converted to peat (Figure 7.23). This could not happen in a stream because the flowing water carries in fresh oxygen to decompose the plants. Although the anaerobic bacteria work to destroy the plant matter, before they can finish their work they are destroyed by poisonous acidic compounds liberated from the dead plants.

Figure 7.23 Cutting Peat A peat cutter harvests dark organic-rich peat from a bog in western Ireland. The peat has formed in a cool, moist climate that favors preservation of organic matter in a wet anaerobic environment. When dried, the peat provides fuel for heat and cooking.

SEDIMENTARY ROCKS: CLASTIC, CHEMICAL, AND BIOGENIC

Sedimentary rocks provide employment for most of the world's geologists, for in these rocks are vast resources of petroleum, coal, salt, gypsum, building stone, construction gravel, and many other useful materials. For as long as society requires these resources, we will need geologists who have learned to find them.

Sedimentary rocks, like the sediments from which they are derived, fall into three categories: clastic, chemical, and biogenic. Each has its own distinctive properties, environment and history of formation, and economic uses.

CLASTIC SEDIMENTARY ROCKS

Suppose you find a rock that you think is a sedimentary rock made up of igneous rock particles. How can you prove that the rock is actually sedimentary and not igneous? If the rock is in place in a large outcrop, you likely have the obvious clues of sedimentary layering (or lack

of). However, suppose you find the rock along a road or a stream bed. Here are some diagnostic characteristics you might look for:

Rock texture. Observe the mineral grains closely. In most igneous rocks, the mineral grains are irregular and interlocked. In sedimentary rocks the mineral grains and rock fragments commonly are rounded and abraded from transport.

Cement. Most igneous rocks consist of interlocking crystals. By contrast, clastic sedimentary rock particles are cemented together. The cement may or may not be easily visible, depending on the size and compactness of the particles.

Fossils. No organism can survive the high temperatures at which igneous rocks form, so the presence of ancient shells or similar evidence of past life is an important clue to sedimentary origin. (Of course, absence of fossils does not prove an igneous origin.)

Like the sediments from which they are derived, clastic sedimentary rocks are classified by predominant particle size. The four basic classes are conglomerate, sandstone, siltstone, and mudstone/shale, which are the rock equivalents of gravel, sand, silt, and clay (Figure 7.24).

Conglomerate is a lithified gravel (Figure 7.24A). Because the clasts in gravels can vary greatly in size, we often use a modifying term to indicate the predominant size of the particles, just as we do with the corresponding sediment (e.g., a "boulder conglomerate," a "cobble conglomerate," or a "pebble conglomerate"). The large clasts in a conglomerate generally are pieces of preexisting rock, whereas fine particles between the clasts (collectively known as the *matrix*) consist mainly of mineral fragments.

In a conglomerate, the clasts are rounded in shape; if the clasts are equally large but angular, the rock is called a *breccia* (Italian for *rubble*) (Figure 7.24B).

Sandstone consists mainly of sand grains, although coarser or finer particles also may be present (e.g., a "pebbly sandstone" or a "silty sandstone"). Different types of sandstone are recognized by their composition. Quartz, being very resistant to weathering, is a common mineral in sandstones (Figure 7.24C). A quartz sandstone consists predominantly or entirely of quartz grains. If feldspars are a major component, the sediment is called an *arkose* or *arkosic sandstone*. A dark-colored sandstone containing quartz, feldspar, and a large amount of tiny rock fragments (*lithic* particles) is a *lithic* sandstone (also called a *graywacke*).

Siltstone is composed mainly of silt-size mineral fragments, commonly quartz and feldspar (Figure 7.24D). It has the texture of shale, but lacks fissility, the tendency of a rock to split into thin layers or plates.

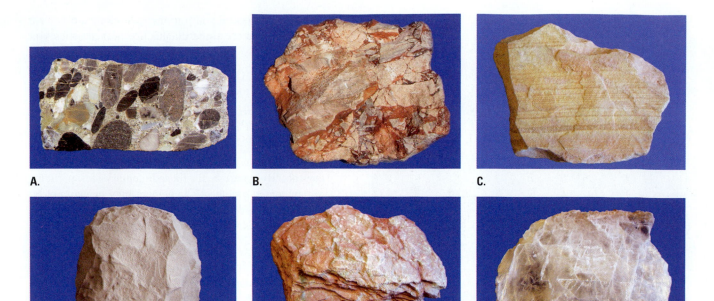

Figure 7.24 Some Common Sedimentary Rocks A. Conglomerate. B. Breccia. C. Sandstone. D. Siltstone. E. Shale. F. Rock salt (an evaporite deposit). The width of each sample is about 10 cm.

Mudstone and shale are clastic rocks of still-finer grain size (Figure 7.22E). Whereas mudstone breaks down into blocky fragments, shale is fissile, and cleaves into sheetlike fragments when it weathers. In mudstones and shales, the clay-size particles are so small that composition generally must be determined by X-ray methods. The most common components are clay minerals, quartz, feldspar, and calcite.

CHEMICAL SEDIMENTARY ROCKS

Chemical sedimentary rocks result from lithification of organic or inorganic chemical precipitates. Most of these rocks contain only one important mineral, which forms the basis for classification.

EVAPORITE DEPOSITS

Economically important concentrations of sedimentary minerals can result when lake or ocean water evaporates to form an evaporite deposit (Figure 7.24F). The Salina Group, a famous marine evaporite deposit in Michigan, contains multiple beds of salt having a cumulative thickness of nearly 500 m.

Salt domes, a characteristic feature of the Gulf Coastal Plain of the United States, are large plug-like bodies of salt that have risen through overlying sedimentary layers from salt beds 5–10 km beneath the top of the dome. Their rise reflects the low density of the salt compared with the overlying sedimentary layers. Because of their relatively low

permeability, salt domes have been proposed as potential sites for storage of high-level toxic and nuclear waste.

BANDED IRON DEPOSITS

Modern civilization depends on iron more than any other industrial metal. Some of Earth's most important iron concentrations are in sedimentary rocks that formed billions of years ago (e.g., in Brazil, Canada, Russia, South Africa, and Australia). These deposits are among the most unusual kinds of chemical sedimentary rocks known. They consist of sediment wholly chemical, or possibly biochemical, in origin and are free of detritus, although they are commonly interbedded with clastic sedimentary rocks (Figure 7.25)

Every aspect of the banded iron deposits indicates chemical precipitation, but the cause of precipitation remains a problem. The precipitation may be related to the chemistry of the ancient atmosphere. Very likely, the most important difference between modern seawater and the seawater in which the iron deposits accumulated was the oxygen content of the water. If the amount of oxygen in the surface waters was very low, large amounts of dissolved Fe^{2+} could have been transported and precipitated. This would further imply that the atmosphere then contained much less oxygen than it does today.

PHOSPHORUS DEPOSITS

Sedimentary deposits of phosphorus minerals, a primary source of fertilizer, form through the precipitation of apatite $[Ca_5(PO_4)_3(OH,F)]$ from seawater. The surface waters of the ocean are depleted in phosphorus because fish and other marine animals extract phosphorus to make bone, scales, and other body parts. When the animals die and sink to the seafloor, their bodies slowly decay and release phosphorus to the deep ocean water. If such phosphorus-rich waters are brought to the surface by upwelling currents, precipitation of apatite can occur. Phosphorus-rich sediments are forming today off the western coasts of Africa and South America, but the process was much more common in the past, especially when shallow seas inundated broad areas of the continents. Extensive marine phosphates are found in Florida, which has the world's highest annual production, western North America (especially Idaho), and North Africa (Morocco).

CHERT

Chert is a hard, very compact sedimentary rock composed almost entirely of very fine-grained, interlocking quartz crystals. It occurs as extensive continuous layers (bedded chert) or as nodules or layers in carbonate rocks (Figure 7.26A), and likely originated as siliceous ooze. Its typical fracture, which is splintery or conchoidal (like the curve of a conch shell), made it useful to early cultures for tool production (Figure 7.26B).

BIOGENIC SEDIMENTARY ROCKS

Biogenic rocks result from the lithification of biogenic sediment or of sediment having a high organic component.

LIMESTONE AND DOLOSTONE

Limestone, volumetrically the most important of the biogenic rocks, is formed chiefly of the mineral calcite (Figures 7.26A and 7.27A). Limestone is not easily classi-

Figure 7.25 Banded Iron Formation Iron-rich chemical sediment (dark) is interbedded with siliceous chert layers (white) in the Brockman Iron Formation, a 2-billion-year-old banded iron deposit in the Hamersley Range of Western Australia.

A.

Figure 7.26 Chert A. A light-gray shallow marine limestone, exposed on a mountain side in the central Brooks Range of Alaska, contains interstratified layers of black chert. B. Three

B.

spear points made of chert and found near St. Louis, Missouri, are about 4000 years old.

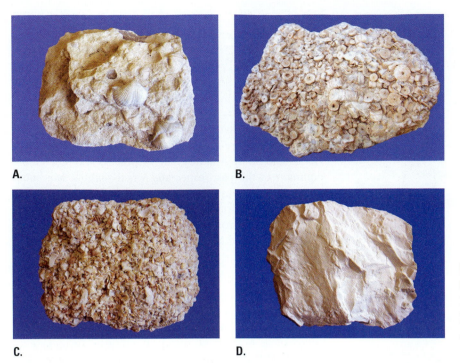

A.

B.

C.

D.

Figure 7.27 Some Common Carbonate Rocks A. Fossiliferous limestone. B. Bioclastic limestone. C. Coquina. D. Chalk. The width of each sample is about 10cm.

fied because it can be either clastic or chemical in origin. Bioclastic limestones consist of lithified shells or fragments of marine organisms that have carbonate hard parts (Figure 7.27B). One coarse-grained type composed entirely of shelly debris is called *coquina* (Figure 7.27C).

Other limestones consist of cemented reef organisms (*reef limestone*), the compacted carbonate shells of minute floating organisms (*chalk*: Figure 7.25D), and accumulations of tiny, round, calcareous accretionary bodies (ooliths) that are 0.5–1 mm in diameter (*oolitic limestone*). Some lime muds are transformed into *lithographic limestone*, a very compact, fine-grained rock used in an early color-printing technology (lithography, or "printing from stone").

Limestone accounts for a major proportion of the carbon dioxide stored in Earth's crust. If this CO_2 were somehow released, it would significantly change the composition of the atmosphere and cause the surface temperature of the planet to heat up dramatically (Chapter 19).

The rugged Dolomites of northern Italy, a frequent objective of experienced rock climbers, are a belt of steep rock pinnacles formed principally of **dolostone**. This rock consists of the mineral dolomite [$CaMg(CO_3)_2$], and is the product of diagenetic alteration of primary carbonate sediments and limestone. Dolostone has been observed to form in shallow tropical bays where marine water evaporates and the ions it contains are concentrated. The highly concentrated marine water, passing through porous carbonate sediments, reacts with the carbonate minerals aragonite and calcite ($CaCO_3$). Some of the calcium is replaced by magnesium to form dolomite. Buried limestone can also be transformed into dolostone through the action of circulat-

ing groundwater that is similarly enriched in magnesium ions.

COAL

Nearly everyone is familiar with *coal*, but few think of it as a sedimentary rock. This is because unlike other rocks, it is combustible and usually dark brown to jet black in color due to its high carbon content. More than 50 percent of coal is decomposed plant matter. Coal occurs in strata (miners call them *seams*) along with other sedimentary rocks, mainly shale and sandstone. Most coal seams are 0.5–3 m thick, although some reach more than 30 m (Figure 7.28), and they tend to occur in groups. A look through a magnifying glass at a piece of coal may reveal fossil wood, bark, leaves, roots, and other parts of land plants, chemically altered but still identifiable.

As we have seen, in water-saturated environments, such as bogs or swamps, plant remains accumulate to form peat, which has a carbon content of about 60 percent. Diagenetic alteration of peat brings about a series of progressive changes. The peat is compressed and water is squeezed out. Chemical changes cause water loss and generation of organic gases such as methane (CH_4), leaving an increased proportion of carbon. Under the right conditions, the peat is thereby converted into *lignite* and eventually into *bituminous coal*, both of which are sedimentary rocks.

Although peat can form even under subarctic conditions, the luxuriant plant growth needed to form thick, extensive coal seams requires tropical or subtropical climate. This implies either (1) that the global climate was warmer and wetter when the plant matter in coal accumu-

Figure 7.29   Oil Shale   This body of oil shale exposed near Debeque, Colorado contains dark layers that are enriched in hydocarbons. The formation records sediment accumulation in an ancient freshwater lake.

similar to those in petroleum. Extensive oil shales in Alberta, Colorado, Utah, and Wyoming can produce as much as 240 liters of oil per ton of rock (Figure 7.29).

Figure 7.28   Coal Seam   A thick seam of coal is mined from a pit near Douglas, Wyoming. Once the overlying rocks are stripped away, the exposed coal is easily and efficiently removed.

lated, or (2) that the wet, swampy environments in which most of the world's coal seams formed existed in the tropics, within about 20° of the equator. In fact, both conditions were likely involved. Coal deposits that we find today in the frigid lands of northern Alaska and Antarctica must have formed in warm, relatively low-latitude environments. They provide some of the most compelling evidence we have for the slow drift of continents over great distances.

OIL SHALE:
AN ORGANIC-RICH CLASTIC ROCK

Although shale is a clastic sedimentary rock, some shales have an unusually high content of organic matter. Dead organisms buried in marine or lake muds contain organic oils and fats. As the mud lithifies into shale, the organic matter remains and, under the right conditions, may be converted to hydrocarbon residues. These residues may ultimately form petroleum, but in some shales, the burial temperatures are never high enough to break down the organic molecules completely. Instead, an alteration process occurs in which wax-like substances are formed. If such a rock, called *oil shale*, is heated, the solid organic matter is converted to liquid and gaseous hydrocarbons

Before you go on:

1. What happens to organic remains in sediment when oxygen is present? when oxygen is lacking?

2. What is the basis for classifying the different kinds of sedimentary rocks?

3. What does the presence of fossils in a rock tell us about its origins?

4. How are the various kinds of limestone distinguished from one another?

5. Why is the formation of coal an example of diagensis?

ENVIRONMENTAL CLUES IN SEDIMENTARY ROCKS

Just as historians record the changing patterns and progress of civilization, sedimentary rocks record the history of our planet—the history of its lithosphere, hydrosphere, atmosphere, and biosphere. Layers of sediment, like pages in a book, can reveal the changing environmental patterns of Earth's surface and the progress of life over more than 3 billion years. If we can learn to read them, the sediments allow us to journey back through the ages and visualize how our planet has evolved during its long, dynamic history.

As already noted, the size, shape, and arrangement of particles in sediments, plus the geometry of sedimentary strata, provide us with evidence about the geologic environment in which sediments accumulate. These and other clues enable us to demonstrate the existence of ancient oceans, coasts, lakes, streams, glaciers, swamps, and other environments of transport and deposition of sediment that were important in the past.

CLUES ON BEDDING PLANES

Telltale patterns exist on the bedding planes of some sandstones and siltstones. These include wave-like irregularities formed by currents moving across the sediment, and cracks, grooves, and other minor depressions. Such features, collectively called *sole marks*, are useful in reconstructing past current directions and bottom conditions.

Bodies of sand that are being moved by wind, streams, or coastal waves are often rippled (Figure 7.30A), and such ripples may be preserved in sandstones and siltstones. These *ripple marks* (Figure 7.30B) tell us not only the environmental setting of the ancient surface, but also the direction of the current.

Some mudstones and siltstones contain layers that are cut by polygonal markings. By comparing them with similar features in modern sediments (Figure 7.31A), we infer that these are mud cracks, caused by shrinkage and cracking of wet mud as its surface dries (Figure 7.31B). Mud cracks imply former tidal flats, exposed stream beds, desert lake floors, and similar environments that dry out periodically.

Footprints and *trails* of animals are often found with ripple marks and mud cracks (Figure 7.32). Even *raindrop impressions* made during brief, intense showers may be preserved in strata. All provide evidence of environmental conditions at the time of formation.

CLUES FROM FOSSILS

Fossils provide important clues about former environments. Some animals and plants are restricted to warm, moist climates, whereas others are associated only with cold, dry climates. By using the climatic ranges of modern plants and animals as guides, we can infer the general character of the climate in which similar ancestral forms lived.

For example, plant fossils can provide estimates of past precipitation and temperature for sites on land, whereas fossils of tiny floating organisms can tell us about ancient surface temperatures and salinity conditions in the oceans. Fossils are also the chief basis for correlating strata in different parts of the world with respect to age; they have therefore been very important in reconstructing the past 600 million years of Earth history (for additional discussion, see Chapter 11).

CLUES FROM ROCK COLOR

The color of unweathered sedimentary rock is determined by the colors of its constituent minerals, rock fragments, and organic matter. Most dark colors are the result of iron sulfides and organic detritus buried with sediment, and

A.

B.

Figure 7.30 Modern and Ancient Ripples A. Ripples forming in shallow water near the shore of Ocracoke Island, North Carolina. B. Fossil ripple marks on the top surface of a sandstone bed exposed at Artist's Point in Colorado National Park, Colorado.

imply deposition in a reducing environment. Reddish and brownish colors result mainly from the presence of iron oxides, occurring either as powdery coatings on mineral grains or as very fine particles mixed with clay. These point to oxidizing conditions.

The weathered surface of a sedimentary rock may have a different color than a fresh, unweathered surface. For

A.

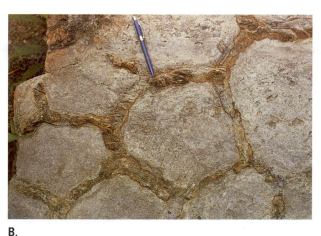

B.

Figure 7.31   Modern and Ancient Mud Cracks   A. Mud cracks formed on the surface of a dry lake floor. B. Ancient mud cracks preserved on the surface of a mudstone bed exposed at Ausable Chasm, New York.

Figure 7.32   Dinosaur Tracks   Footprints of a three-toed dinosaur are preserved in sandstone in the Painted Desert near Cameron, Arizona.

example, a sandstone that is pale gray on freshly broken surfaces may have a surface coating of iron oxide in weathered outcrop that gives it a yellowish-brown color. In this case, the surface color is derived from the chemical breakdown of iron-bearing minerals in the rock.

CLUES FROM ISOTOPES AND MAGNETIC PROPERTIES

Many environmental clues are available through the study of isotopes found in sediments and sedimentary rocks. For

example, the ratio of two isotopes of oxygen, oxygen-18 and oxygen-16, can provide important information about changing climate. As we will see in Chapter 16, this ratio measured in deep-sea sediments tells us about the volume of glacier ice on the planet during glacial and interglacial ages of the last several million years. When measured in cores from the Greenland and Antarctic ice sheets, the ratio allows us to determine the air temperature where falling snow reached the glacier surface. The ratio of two carbon isotopes, carbon-13 and carbon-12, from organic matter in sediments and rocks can tell us whether an ancient landscape was covered primarily by forest or by grassland.

Magnetic measurements in sediments and sedimentary rocks permit us to reconstruct changes in Earth's magnetic field through time. These changes, in turn, provide one of the most important tools we have for dating ancient sediments and rocks that contain clues to environmental change (Chapter 11).

Before you go on:

1. What are some environmental clues to be found on bedding planes?

2. How are isotopes in sediments and sedimentary rocks useful for gaining information about the past?

REVISITING PLATE TECTONICS AND THE EARTH SYSTEM

SEDIMENTATION AND PLATE TECTONICS

The energy that drives sedimentation ultimately comes from the two major energy sources of Earth: internal heat and the Sun. Internal heat energy drives the motions of the highly mobile lithosphere, including uplift of the land, and is the ultimate source of energy that drives plate tectonics. Sediment produced by weathering and erosion of rocks in uplifted regions is transferred downslope toward sea level under the pull of gravity, and most eventually reaches the ocean basins. Streams, glaciers, and ocean waves and currents, the major movers of sediment, are all part of the hydrologic cycle, which derives its energy from solar radiation.

Sedimentation rates are high adjacent to regions of active tectonic uplift, far lower on relatively stable continental interiors, and still lower in regions of the deep sea that are far from terrestrial sources of sediment. In the most tectonically active regions, uplift rates equal or even exceed erosion rates, and high mountain ranges persist as prominent features of the landscape. The magnitude of uplift can be impressive. Marine sedimentary rocks that crop out at the top of Mount Everest demonstrate at least 9 km of uplift since the sediments were deposited in a shallow sea about 100 million years ago. Here in the Himalaya, as in other major mountain ranges of the world, we see clear evidence that ancient sediments deposited on an ocean floor have been converted to rock, added to a continent, and uplifted to high altitudes by tectonic forces.

Exceptionally thick accumulations of strata are related to specific plate-tectonic settings. For example, wherever continents are split apart by a spreading center, thick sedimentary wedges may slowly accumulate along the new continental margins as streams transport sediment to the growing ocean basin; the Atlantic margin of North America is an example of this process (Figure 7.33A). A large portion of the thick pile of strata beneath the continental shelf consists of shallow marine sedimentary rocks, which means that this sedimentary wedge must have slowly subsided as accumulation took place.

In zones where continents collide, coarse stream sediments shed from a rising mountain system can accumulate to great thicknesses in adjacent sedimentary basins (Figure 7.33B). The Himalaya–Hindu Kush mountain arc of south-

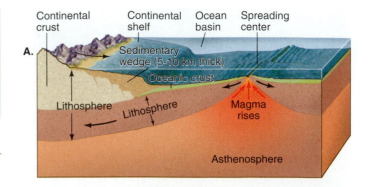

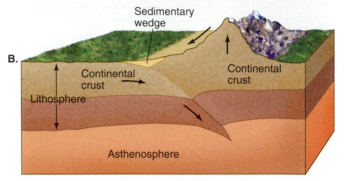

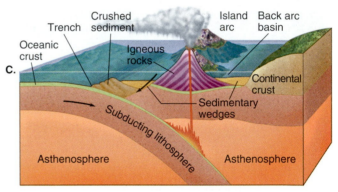

Figure 7.33 Sediments and Plate Tectonics Examples of thick sedimentary accumulations of sediment in different plate-tectonic settings. A. A thick wedge of sediment slowly accumulates along a new continental margin as it moves away from a spreading center. B. In a zone of continental collision, sediment is shed from a growing mountain system to form a thick sedimentary wedge. C. Sediment shed into a deep-sea trench from an adjacent continent, bordered by an arc of active volcanoes, forms a wedge that is compressed and crushed as the ocean plate is subducted.

central Asia is a dramatic example. Thick conglomerates and sandstones along the southern margin of the chain record the rise of these ranges. Finer-grained detritus has been transported by major streams to the Indian Ocean, where extensive submarine fans have accumulated since the uplift began (Figure 7.20).

At active subduction zones near continental margins, as along the western margin of South America, sediments are shed into deep-ocean trenches, where they may accumulate to a great thickness (Figure 7.33C). Because volcanoes are commonly associated with such tectonic belts, the accumulating clastic sediments contain a high proportion of volcanic detritus. As the moving plates slowly converge, the sediments are crushed against the continent and become part of it. In this way, sediments are cycled from continent to ocean and back to continent, where continuing uplift causes the process to begin anew.

LIMESTONE, CO_2, AND GLOBAL CLIMATE CHANGE

Along low-latitude passive plate margins, well away from growing mountain systems and their high flux of siliceous detrital sediment, warm shallow-water environments are sites of widespread carbonate accumulation. Past rates of carbonate sedimentation were especially great when the oceans stood high against the continents and flooded broad continental shelves. Tropical marine creatures require carbon dioxide to build their skeletons and shells (primarily $CaCO_3$), the ultimate source of which is in the atmosphere. As surface- and bottom-dwelling marine organisms die or shed their shells, these particles accumulate on the seafloor where they are stored as part of a growing pile of carbonate sediment.

Carbon dioxide is one of the most important greenhouse gases that control global temperature, and its removal from the atmosphere leads to cooling of world climate. We can reasonably infer that times when plate distribution and high sea level were favorable for widespread accumulation and burial of carbonate deposits were also times when world temperature should have been falling as the concentration of atmospheric CO_2 steadily decreased. The sedimentary record is consistent with this inference. Figure 7.34 displays a curve of modeled CO_2 variations over the past 600 million years. Times of high CO_2, well above today's value, characterized intervals from about 550–390, and 180–90 million years ago. These intervals, corresponding to much of the early Paleozoic and late Mesozoic eras (Figure 11.10), are well known for their extensive limestone formations. The former followed an important interval of cold climate about 600 million years ago that coincided with low atmospheric CO_2 values and important glaciation. The latter came after a major interval of widespread glaciation and low CO_2 values about 330–260 million years ago. The last 6–8 million years has also been a time of low atmospheric CO_2 and increasing glacier expansion.

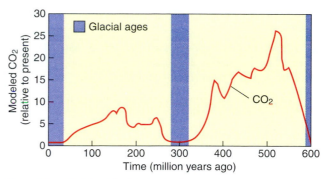

Figure 7.34 Carbon Dioxide in the Atmosphere Variations in atmospheric carbon dioxide over the past 600 million years. Times of high CO_2 were times when carbonate rocks formed in low latitudes. As atmospheric CO_2 levels fell due to storage of carbonate sediments in limestone and other carbonate rocks, the climate cooled, leading to glacial ages.

The important linkages between the atmosphere, the biosphere, the oceans, and plate tectonics are clearly evident in this example. Carbonate sedimentation, the extensive formation of limestone, and the attendant decrease in atmospheric CO_2 are all part of the climate puzzle, as we will see in greater detail in Chapter 19. Because climate helps control Earth's habitability, understanding the climate system is one of the major challenges in the Earth sciences.

What's Ahead?

In the next stage of the rock cycle, sedimentary rocks change form when they become deeply buried and are subjected to intense heat and pressure. In the process, many of the distinctive features of the original sediments and sedimentary rocks are lost. Fossils may be obliterated, the mineralalogy changes, and new structures are imposed. Although Earth history can be read from metamorphosed sedimentary rocks, it generally is a far more limited and fragmentary history than that contained in the original sediments.

CHAPTER SUMMARY

1. Sediment is transported by streams, glaciers, wind, slope processes, and ocean currents.

2. Stratification results from the arrangement of sedimentary particles in layers. Each bed or group of beds in a succession of strata is distinguished by its distinctive thickness or character.

3. Clastic sediment consists of fragmental rock debris resulting from weathering, together with organic remains. Chemical sediment forms where substances carried in solution are precipitated.

4. Particles of clastic sediment become rounded and sorted during transport by water and air; particles transported by glaciers are not sorted.

5. Various arrangements of the particles in strata are seen in

parallel strata and cross strata, graded bedding, and nonsorted layers.

6. An extensive body of strata may possess several facies, each representing a different depositional environment. The boundaries between facies may be abrupt or gradational.

7. Most sedimentary strata are built of continental detritus that is transported to the submerged continental margins. Some is trapped in basins on land where it is deposited by nonmarine processes. A small percentage reaches the deep sea.

8. Depositional environments of nonmarine and shallow-marine sediments can be inferred from such properties as texture, degree of sorting and rounding, character of stratification, and types of contained fossils.

9. Coarse land-derived sediment reaching the continental margins is deposited close to shore where it is reworked by longshore currents. Finer sediment is deposited on the continental shelves and slopes and in the deep sea. Extensive areas of the shelves are covered by relict sediments deposited at times of lowered sea level.

10. Carbonate shelves are found in low latitudes, where warm waters promote growth of carbonate-secreting organisms.

11. Marine evaporite deposits accumulate in restricted marine basins, where evaporation is high and continual inflow provides a supply of saline water.

12. By depositing turbidites, turbidity currents have built thick deposits of sediment at the base of the continental slope.

13. The chief kinds of sediment on the deep seafloor are brownish or reddish clay, calcareous ooze, and siliceous ooze. Their distribution is largely related to surface water temperature, water depth, and surface productivity.

14. During diagenesis, compaction, cementation, recrystallization, and chemical alteration act to transform sediment into sedimentary rock.

15. Clastic sedimentary rocks, like clastic sediments, are classified mainly on the basis of predominant particle size. Conglomerate, sandstone, siltstone, and mudstone or shale are the rock equivalents of gravel, sand, silt, and clay.

16. Limestone is an important and widespread biogenic rock that forms primarily in warm marine environments and stores carbon dioxide in Earth's crust.

17. Sediment is constantly being recycled, moving from continent to ocean and back to continent.

THE LANGUAGE OF GEOLOGY

bed (p. 175)
bedding (p. 175)
bedding plane (p. 175)
biogenic sediment (p. 182)

cementation (p. 190)
chemical sediment (p. 181)
chert (p. 193)
clastic sediment (p. 175)
compaction (p. 190)
cross bedding (p. 180)

deep-sea fan (p. 188)
deep-sea ooze (p. 182)

diagenesis (p. 190)
dolostone (p. 194)

eolian (p. 185)
estuary (p. 185)
evaporite (p. 186)

facies (p. 182)
fossil (p. 182)

graded bed (p. 180)

lithification (p. 190)

peat (p. 190)

recrystallization (p. 190)

sorting (p. 176)
stratification (p. 175)
stratum (pl. strata) (p. 175)

turbidite (p. 187)
turbidity current (p. 187)

varve (p. 179)

IMPORTANT ROCK NAMES TO REMEMBER

breccia (p. 191)

coal (p. 194)

conglomerate (p. 191)

limestone (p. 194)

sandstone (p. 192)
shale (p. 192)
siltstone (p. 192)

QUESTIONS FOR REVIEW

1. What characteristics are used in classifying sediments and sedimentary rocks?

2. What obvious clues could you use to tell a clastic sedimentary rock made up of igneous rock fragments from an igneous rock?

3. What chemical reactions can lead to precipitation of chemical sediments?

4. What features in a sediment or sedimentary rock are responsible for stratification?

5. Describe the processes involved in the conversion of sediment to sedimentary rock.

6. How can cross bedding and graded layers tell us about conditions of deposition?

7. Name some features on sedimentary bedding surfaces that provide clues about depositional environment.

8. How would you tell sediments deposited by running water from sediments deposited by wind or in a lake?

9. If estuaries are generally shallow bodies of water, how might you explain the occurrence of thick estuarine accumulations in the sedimentary record?

10. What explanation can you give for the occurrence of widespread *relict* sediments on the continental shelves?

11. If the Mediterranean Sea, which is very nearly an enclosed marine basin, were to evaporate completely, less than 100 m of evaporite sediments would be deposited. How, then, might *continuous* evaporite deposits under the Mediterranean that are more than 2 km thick have formed?

12. What distinctive features of the sediments in deep-sea fans provide clues about the way the sediments were transported?

13. Calcareous and siliceous oozes do not occur everywhere on the ocean floors. What factors explain their known distribution?

14. Where and why would you expect to find exceptionally thick accumulations of sedimentary rock?

15. Is volcanic ash a sediment? How is it similar to, yet different from, sediment derived by weathering?

16. On the island of Hawaii, some beaches consist of black sand, others of white sand, and still others of green sand. Can you suggest a geologic reason for such different-colored beach sands?

Click on *Presentation* and *Interactivity* in the **Sedimentary Rocks** module of your CD-ROM to explore resources and activities presenting concepts from this chapter. Select *Assessment* in the same module to test your understanding of this chapter.

Click on *Petroscope* for **Sedimentary Rocks** to learn more about the characteristics of sedimentary rocks from this chapter as well as to learn how to identify unknown sedimentary rock samples.

Chapter 8

Intense Metamorphism. Highly deformed layers of metamorphic rock exposed in the bed of the Sand River, South Africa. The area shown is about 15 m across.

Metamorphism and Metamorphic Rocks: New Rocks from Old

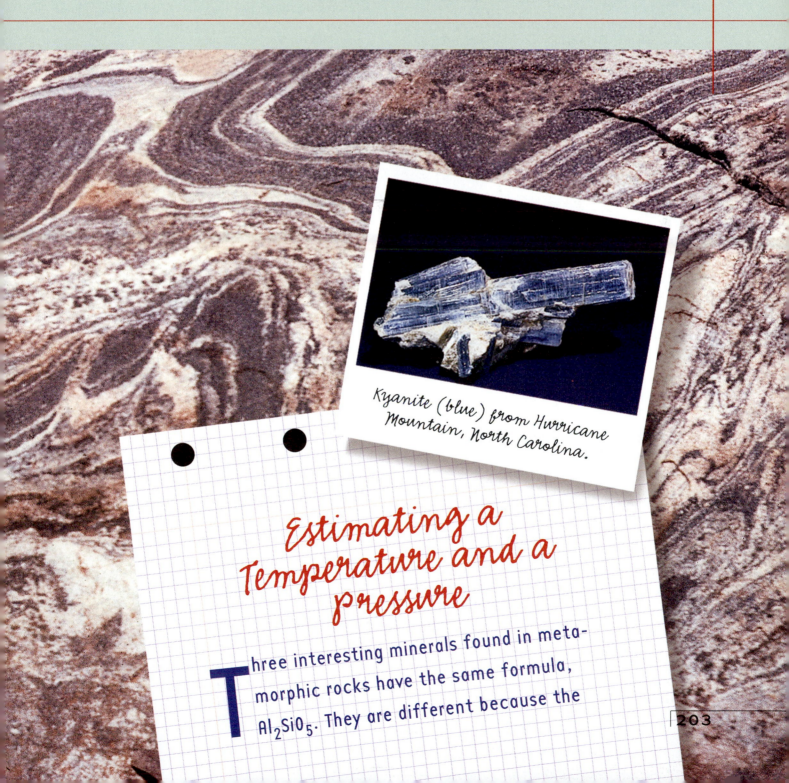

Kyanite (blue) from Hurricane Mountain, North Carolina.

Estimating a Temperature and a Pressure

Three interesting minerals found in metamorphic rocks have the same formula, Al_2SiO_5. They are different because the atoms are packed differently in each case. As we learned in Chapter 3, such minerals are called polymorphs. The three are andalusite, kyanite, and sillimanite.

Each mineral has its own specific range of temperature (*T*) and pressure (*P*). The laws of chemical thermodynamics forbid overlap in the ranges of *T* and *P* and require that adjacent ranges meet along a line. On such a line two polymorphs can coexist. Because there are three polymorphs, there are three lines of coexistence. Those three lines meet at a point, the triple point, where all three polymorphs can coexist. The triple point has a unique temperature and pressure.

I was mapping rocks in the Kanmantoo region of South Australia when I discovered rocks containing the three polymorphs of Al_2SiO_5. Such an occurrence is rare and important. If the triple point were known, the *T* and *P* of the conditions of metamorphism would also be known. Unfortunately, in the late 1950s, when I was doing the mapping, the triple point had not been determined. That set me to work measuring some of the properties of the polymorphs because it is possible, in principle, to calculate the triple point if all the needed properties are known. It took many years of work by many people, but the triple point of the Al_2SiO_5 polymorphs was finally determined to be at 500°C and a pressure of 4000 atmospheres. This means that the rocks I was studying had been buried and metamorphosed at a depth of about 14 kilometers. It also meant, of course, that those 14 kilometers of rock had been eroded away in order to expose the rocks I was mapping—compelling evidence of the rock cycle in action.

Brian J. Skinner

KEY QUESTIONS

1. **What is metamorphism?**

2. **What are the major factors in metamorphism?**

3. **What are the upper and lower limits of metamorphism?**

4. **How do rocks respond to temperature and pressure change in metamorphism?**

5. **What are the different kinds of metamorphic rock?**

6. **What are the different types of metamorphism?**

7. **What is metasomatism?**

INTRODUCTION: ALL THINGS METAMORPHOSE

Earth's processes constantly change from one form to another form. Caterpillars metamorphose into moths or butterflies. Eggs metamorphose into snakes or frogs or birds. Young things metamorphose into old. Living things metamorphose into dead. And rocks metamorphose into other rocks.

Earth's powerful tectonic forces compress and bend rocks, heat them, and force fluids through them, altering their appearance and character. This chapter is an adventure into the changing world of metamorphic rocks.

WHAT IS METAMORPHISM?

Cooking food is a simple metaphor for metamorphism. Through cooking, we metamorphose an "ingredient assemblage" (eggs, milk, flour, etc.) into something very different: a cake. In cooking, what you get depends on what you start with and on the cooking conditions. So too with rocks: the end product is controlled by the initial composition of the rock (ingredients) and by the metamorphic (cooking) conditions. The cooking conditions are determined by temperature (*T*) and pressure (*P*). Another critical factor is time.

Metamorphism (*meta* = change; *morph* = form) is the change in form that happens in Earth's crustal rocks in response to changes in temperature and pressure. Any type of rock can become metamorphosed: igneous rocks, sedimentary rocks, or existing metamorphic rocks.

Metamorphism is of particular interest because it does not happen in the liquid state, as with igneous rocks, but in the solid state. How can rocks change in the solid state? As tectonic plates move and portions of the crust are crushed together, the rocks are squeezed, stretched, bent, and heated. This causes changes in complex ways: both their mineral assemblages and textures are altered and the rocks are said to have been metamorphosed. Further, rocks can be metamorphosed repeatedly through successive tectonic events.

The igneous rocks you studied in Chapter 4 began life from a melt, so they cannot "remember" their prior form, just as molten steel cannot "remember" its pre-melting shape. However, if you heat and hammer a piece of steel and thereby change its shape, the steel does carry a memory of its pre-hammering state. Similarly, metamorphic rocks, formed in the solid state, usually preserve some tell-tale evidence of their former conditions and of the events that changed them. Thus, metamorphic rocks are the most complex and can be the most interesting among the three rock families.

In metamorphic rocks is preserved the saga of many tectonic events. Deciphering this record is an exceptional challenge for geologists. For example, when tectonic plates collide, distinctive metamorphic rocks form along the plate edges. Consequently, geologists attempt to discern the boundaries of ancient continents by studying metamorphic rocks.

Geologists are also trying to use evidence derived from metamorphic rocks to determine how long plate tectonics has been active on the Earth. So far, the evidence suggests that plate tectonics has been operating for at least 2 billion years, or nearly half of our planet's 4.6-billion-year lifetime.

MAJOR FACTORS IN METAMORPHISM

There are six major factors in metamorphism. In a rock that is being metamorphosed, (1) chemical composition is a major control over the new mineral assemblage. So is (2) the change in temperature, and (3) the change in pressure, although they are less straightforward—because they are strongly influenced by (4) the presence or absence of fluids, (5) how long a rock is subjected to high pressure or high temperature, and (6) whether the rock is simply compressed or is twisted and broken during metamorphism.

CHEMICAL COMPOSITION OF ORIGINAL ROCK

The chemical composition of the original (premetamorphic) rock is the greatest factor in determining the mineral assemblage of a metamorphic rock. For example, if the original rock is a pure quartz sandstone, the chemical elements you have to work with are Si, O, and whatever elements are in the cementing material between the sand grains.

Therefore, whatever metamorphic rock derives from this sandstone can contain only those same elements. In this example, the resulting metamorphic rock is quartzite and consists of quartz (SiO_2), with some minor impurities from elements in the cement.

If the original rock were a shale, it would contain clay minerals, probably bits of mica, chlorite, quartz, and other minerals. A great many elements would be present and metamorphism could produce a rock containing minerals such as biotite, garnet, kyanite, quartz, and feldspar.

Thus, the chemical composition of the original rock controls the mineralogy of the metamorphosed rock.

TEMPERATURE AND PRESSURE

When a mixture of flour, eggs, salt, sugar, yeast, and water is baked in an oven, the high temperature causes a series of chemical reactions. New compounds form and the result is a loaf of bread. Similarly, when rocks are heated, some minerals recrystallize and others react to form new minerals, and the result is a metamorphic rock.

For rocks, the heat source is Earth's internal heat. Rock can be heated by burial or by a nearby igneous intrusion. But burial, being underground, is inevitably accompanied by an increase in pressure due to the weight of the overlying rocks. An intrusion may be shallow—low pressure—or deep, which means a high pressure. Therefore, whatever the cause of the heating, metamorphism is only rarely due entirely to a temperature rise. The effects of changing temperature and pressure must be considered together, as discussed in *The Science of Geology*, Box 8.1, *Pressure–Temperature–Time Paths: The Tectonic History of Metamorphic Rock.*

The combined influence of temperature and pressure on the melting properties of rocks and minerals is discussed in Chapter 4, Box 4.1, *How Rock Melts*. The dual effects of temperature and pressure also control metamorphic transformations of mineral assemblages. By lab measurements of temperature and pressure effects on mineral assemblages, we can delineate the ranges of pressure and temperature conditions under which metamorphism has occurred in the crust. (We will return to the important issue of the controls exerted by temperature and pressure on mineral assemblages later in the chapter.)

Geologists find the concept of "metamorphic grade" very useful. **Low-grade metamorphism** refers to metamorphic processes that occur at temperatures from about 100°C to 500°C and at relatively low pressures (Figure 8.1). **High-grade metamorphism** refers to metamorphic processes at high temperature (above about 500°C) and high pressure.

When talking about deformation in a solid and development of new textures, we use the term *stress* instead of pressure. We do so because stress has the connotation of *direction*. As we discussed in Box 2.1, stress is measured as force/unit area. Pressure is the special case in which stress is equal in all directions; that is, it is a **uniform stress**, as in a liquid. Uniform stress, by either squeezing or stretching equally in all directions, changes the size of a solid, but not its shape. But rocks are solids, and solids can be squeezed more strongly in one direction than another— that is, stress in a solid, unlike a liquid, can be different in different directions. Stress that is different in different directions induces a change in the shape of a solid, and we refer to such stress as shear or **differential stress**. The textures in many metamorphic rocks record differential stresses during metamorphism. Most igneous rocks, by contrast, have textures formed under uniform stress because igneous rocks crystallize from liquids.

Box 8.1

THE SCIENCE OF GEOLOGY

PRESSURE–TEMPERATURE–TIME PATHS: THE TECTONIC HISTORY OF METAMORPHIC ROCK

Rock subjected to metamorphism develops a distinctive mineral assemblage characteristic of the grade of metamorphism attained. To reach the temperature and stress at which the highest grade of metamorphism occurs in a given rock, that rock must earlier have been subjected to lower temperatures and stresses, and thus it previously must have contained different mineral assemblages. *Prograde metamorphism is a dynamic process, and mineral assemblages replace one another in succession.*

A triumph of modern scientific instrumentation has been the development of techniques for examining tiny mineral fragments. In many metamorphic rocks, microscopic relicts of earlier mineral assemblages remain—see Figure B8.1. By analyzing these relicts, geologists can decipher how stress—here called pressure (*P*) for simplicity—and temperature (*T*) changed over time (*t*) during metamorphism.

Surprisingly, metamorphic rocks are rarely subjected *simultaneously* to the highest pressures and highest temperatures. This means that, over time, geothermal gradients change in regions where metamorphism is occurring.

For example, the *P–T* path shown in Figure B8.2 might apply to rock that is buried rapidly as a result of subduction, if the downward movement ceases and the rock at depth continues to warm over a long period of time. When cool rock moves downward relatively rapidly because of subduction, pressure increases in proportion to depth, but temperature will not increase in accord with a normal geothermal gradient because warming of the subducted slab is a very slow process.

If the downward movement slows and then ceases—for example, because of a tectonic change from subduction to continental collision—the rock will experience a maximum pressure at point B in the figure. Because the rock will still be cooler than the ambient for its depth, temperature will continue to increase, even as uplift starts and for some time afterward, until point C is reached.

Thus, the highest temperature and the highest-grade mineral assemblages will be produced at point C. If uplift continues until the rock is eventually exposed at the surface, the remainder of the *P–T* path will be a curve like C–A.

Through another advance in instrumentation, it is sometimes possible to obtain radiometric ages for the metamorphic mineral assemblages (Chapter 11). Knowing the *P–T* path of metamorphism and two or more time points

Figure B8.1 Record of Changing Conditions A grain of garnet (2 mm across) in a garnet-biotite gneiss. The small inclusions are relicts from an earlier mineral assemblage formed under *P–T* conditions different from those that formed the gneiss.

on the path, we can calculate rates of burial and uplift and thereby compare the subduction rate of former tectonic plates with present subduction rates. They turn out to be very similar. Bodies of metamorphic rock are sensitive monitors of large-scale tectonic processes.

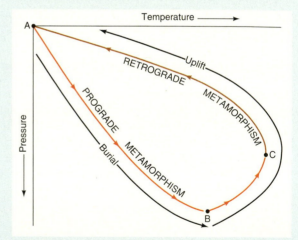

Figure B8.2 A Record of Changing *P* and *T* *P–T* path for a body of rock undergoing metamorphism along a subduction zone.

The most visible effect of differential stress during metamorphism is the orientation of silicate minerals like micas and chlorites, because they contain polymerized sheets of silica tetrahedral $(Si_4O_{10})^{4-}$. Figures 8.2A and B show the striking difference. Figure 8.2A is granite with a typical texture of randomly oriented mineral grains that grew under uniform stress. Figure 8.2B is a high-grade metamorphic rock containing the same minerals, except

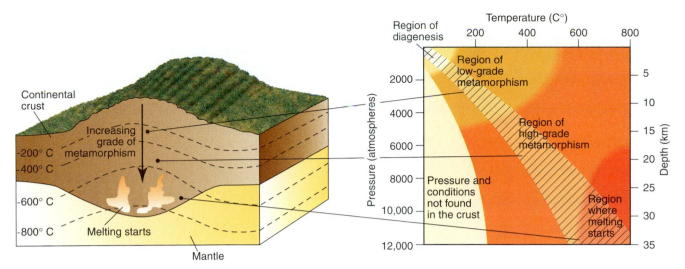

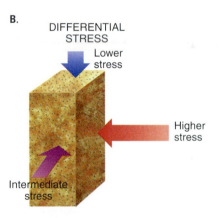

Figure 8.1 **Temperature and Pressure Conditions of Metamorphism** This diagram shows the conditions of pressure and temperature in which diagenesis, metamorphism, and melting occur. *Left:* where in the crust metamorphism and melting occur. *Right:* plot of pressure versus temperature, indicating the pressure-temperature regimes of metamorphism. The broad, curved band shows how pressure and temperature increase with depth through much of the continental crust. Pressure, as measured in atmospheres, is shown along the lefthand side of the diagram; the equivalent depth in the crust, measured in kilometers, is shown on the righthand side.

Figure 8.2 **Uniform and Differential Stress** Contrast the textures of these rocks, which have the same composition, but which formed under uniform stress (left) and differential stress (right). A. This granite (quartz, feldspar, and biotite) crystallized in a uniform stress field. Note the biotite grains are randomly oriented. B. This high-grade metamorphic rock (also quartz, feldspar, and biotite) crystallized in a differential stress field. Here the biotite grains are roughly parallel, giving the rock a distinct foliation. This rock, *gneiss*, is discussed later in the chapter. Both specimens are 10cm across.

Figure 8.3 Flattened Conglomerate Deformation of a conglomerate during metamorphism. Sandstone pebbles, originally round, have been flattened.

they formed under differential stress. Note in Figure 8.2B that the mica grains are roughly parallel, giving the rock a distinctive texture.

Figure 8.3 shows another example of differential stress during metamorphism: uniformly rounded pebbles in a conglomerate have been squeezed to flattened discs.

Note that it is the *texture*, not the mineral assemblage, that is controlled by differential versus uniform stress. For this reason, geologists often use the terms *stress* and *pressure* interchangeably when describing mineral assemblages and metamorphic grades.

FLUIDS FACILITATE METAMORPHISM

Sedimentary rocks have innumerable open spaces between their grains, and many igneous rocks have tiny open fractures. These tiny open spaces are called *pores*, and most pores are filled by a watery **intergranular fluid**. This fluid is never pure water, for it always contains small amounts of dissolved gases and salts, plus traces of all the mineral constituents in the enclosing rock. The amount of fluid present in pores is never very large but the fluid plays an important role in metamorphism. The fluid comes from many sources, but two are especially important; some of the fluid is retained surface water buried with the rocks, and some is released when hydrous minerals (containing water in the formula), such as clays, micas, and amphiboles, decompose and lose water as the temperature increases on burial.

At the temperatures of high-grade metamorphism, the water is technically a vapor, but high pressure forces the vapor to be virtually as dense as liquid water, so it is just as effective a solvent as liquid water. Thus, even at high temperatures, intergranular fluid plays a vital role in metamorphism.

When the temperature and pressure change in a rock that is undergoing metamorphism, so does the composition of the intergranular fluid. Some of the dissolved con-

stituents leave the fluid to become part of the new minerals growing in the metamorphic rock. Other constituents do the opposite, leaving the minerals to join the fluid. Thus, the intergranular fluid is an important *transporting medium* that facilitates metamorphism by speeding chemical reactions, in much the same way that water in a stew pot hastens the cooking of a tough piece of meat.

When intergranular fluids are absent (or present only in trace amounts), metamorphic reactions are very slow. Thus, when a dry rock is heated, few changes occur, because the growth of new minerals requires atoms to move by diffusing through the solid minerals, and diffusion through solids is exceedingly slow. For example, at 500°C it takes more than 100 million years for a K^+ ion to diffuse 1 cm through a dry feldspar crystal. But if an intergranular fluid is introduced—perhaps by the rock being crushed, creating pores—diffusion of atoms can occur through the intergranular fluid. This is vastly faster—at least a million times in the K^+ ion example—allowing new minerals to grow rapidly, enhancing the metamorphic effects.

When pressure increases due to burial of a rock, and as metamorphism proceeds, the amount of pore space decreases, and the intergranular fluid is slowly squeezed from the rock. As a rock's temperature increases, hydrous minerals recrystallize to anhydrous minerals, releasing water. This released water joins the intergranular fluid and is also slowly driven from the metamorphic rock. For this reason, rock subjected to high-grade metamorphism contains fewer hydrous minerals—and less intergranular fluid—than does low-grade metamorphic rock.

Because intergranular fluid is a solution, any fluid that escapes during metamorphism will carry with it small amounts of dissolved mineral matter. If such a fluid flows through fractures formed as a result of tectonic forces, some of the dissolved minerals may be precipitated in the fracture. The result is a **vein**—a mineral-filled fracture (Figure 8.4).

Figure 8.4 Vein in a Metamorphic Rock Quartz vein in a metamorphic rock. Veins are thought to be deposited from escaping intergranular fluid during metamorphism.

The metamorphic changes that occur while temperatures and pressures are rising (and usually while abundant intergranular fluid is present) are termed **prograde metamorphic effects**. Those that occur as temperature and pressure are declining (and usually after much of the intergranular fluid has been expelled) are called **retrograde metamorphic effects**.

Not surprisingly, because of the lack of fluid, retrograde metamorphic effects are slower and less pronounced than prograde effects. Indeed, it is only because retrograde reactions happen so slowly that we see high-grade metamorphic rocks at Earth's surface. If retrograde reactions were rapid, the minerals that are characteristic of high-grade metamorphism could not persist and we would lose much of the history about what has happened to rocks under intense metamorphism. In many places around the world it is possible to see high-grade metamorphic rocks that were metamorphosed 2 billion or more years ago and little or no retrograde changes have happened in all that time.

ROLE OF TIME IN METAMORPHISM

Chemical reactions involve energy, and two compounds will react only if a lower state of energy is reached in the process. When the lowest energy state possible under the prevailing temperature and pressure has been reached, change due to chemical reaction ceases, and we say a *state of equilibrium* has been attained. A certain amount of time is needed for any chemical reaction to reach equilibrium. Some reactions, such as burning methane gas (CH_4) to yield carbon dioxide and water, can happen so rapidly that they create explosions. At the other end of the scale are reactions that require tens or hundreds of millions of years to proceed to completion.

Many of the chemical reactions that occur in rocks undergoing metamorphism are of the latter kind—ponderously slow. No reliable ways have yet been developed to determine exactly how long a given metamorphic rock has remained at a given temperature and pressure. However, laboratory demonstrations show that high temperature, high pressure, and long reaction times produce large mineral grains.

Thus, it is possible to draw the interesting general conclusion that coarse-grained rocks are the products of long-sustained metamorphic conditions (possibly over millions of years) at high temperatures and pressures. Fine-grained rocks are products of lower temperatures, lower pressures, or, in some cases, short reaction times.

Before you go on:

1. What is metamorphism?

2. How do metamorphic rocks differ from igneous rocks?

3. What factors control the mineral assemblage of a metamorphic rock?

4. What is the role of fluids in metamorphism?

5. How do veins form in metamorphic rocks?

THE UPPER AND LOWER LIMITS OF METAMORPHISM

It is important to know the lower and upper limits of metamorphism.

At the lower end, metamorphism occurs in sedimentary and igneous rocks that are subjected to temperatures greater than about 100°C, usually under pressures of hundreds of atmospheres, caused by the weight of a few thousand meters of overlying rock. Metamorphism does not refer to changes caused by weathering or by diagenesis (processes that affect sediment as it goes from deposition through lithification—Chapter 7), because both weathering and diagenesis take place at temperatures below 100°C and low pressures.

At the upper end, metamorphism ceases to occur at temperatures that melt rock. Again, metamorphism refers only to changes in solid rock, not to changes caused by melting. Melting changes are igneous phenomena (Chapters 4 and 5).

ROLE OF WATER IN DETERMINING LIMITS

At least a small amount of water is present in most rocks, and this water has a very important role in metamorphism. The upper limit of metamorphism in the crust is determined by the onset of *wet partial melting* (Figure 8.1 and Box 4.1). The water present controls the temperature at which wet partial melting commences and the amount of magma that can form from a metamorphic rock. The upper limit of metamorphism is therefore a temperature range that depends on the amount of water present.

As shown in Figure 8.1, the upper limits of metamorphism overlap with the region of temperature and pressure where magma formation commences. When a tiny amount of water is present, only a small amount of melting occurs and the melt stays trapped as small pockets in the metamorphic host. In many rocks we see evidence that melting started in water-rich layers of metamorphic rock, even though adjacent, drier layers do not show any sign of melting. **Migmatites** are composite rocks that contain an igneous component formed by a small amount of melting plus a metamorphic portion.

When large volumes of magma develop by partial melting, their lesser density causes them to rise and intrude the metamorphic rock above. Eventually, a large volume of rising magma formed by wet partial melting will solidify as an intrusive igneous rock. As a result, we observe that batholiths of granitic rock and large volumes of metamorphic rock tend to be closely associated. The geologic setting of this igneous-metamorphic rock association is along subduction and collision margins of tectonic plates. The relationship is well displayed in the Coast Range Batholith of Alaska and British Columbia, and the Sierra Nevada Batholith (Figure 4.14).

> **Before you go on:**
> 1. At approximately what depth does metamorphism start in a pile of sedimentary rocks?
> 2. What determines the upper limit of metamorphism?

HOW ROCKS RESPOND TO TEMPERATURE AND PRESSURE CHANGE IN METAMORPHISM

During metamorphism, rocks respond by changing their texture, changing their mineral assemblage, or both.

TEXTURE CHANGE

Most metamorphic rocks form under differential stress. They respond by developing conspicuous directional textures. As metamorphism proceeds and sheet-structure minerals like mica and chlorite start to grow, the minerals are literally squeezed and forced to grow with their sheets perpendicular to maximum stress, as shown in Figure 8.2B.

These new, parallel flakes of mica form a planar texture called **foliation** (Figure 8.5). The term comes from the Latin *folium*, meaning *leaf*, because many foliated rocks tend to split into thin, leaf-like flakes. Much foliation is not leaf-like at all and is better described as a "wavy pattern." Foliation may be pronounced or subtle, but when present, it provides strong evidence of metamorphism.

LOWER-GRADE METAMORPHISM: SLATY CLEAVAGE

During the earliest stages of low-grade metamorphism, the weight of the overlying rock is the only cause of stress. Predictably, the newly forming sheet-structure minerals create foliation that tends to be parallel to the bedding planes of the sedimentary rock being metamorphosed. But deeper burial along a continental margin typically involves crustal thickening and compression from a plate collision. This deforms the flat sedimentary layers into folds and the sheet-structure minerals and the foliation are no longer parallel to the bedding planes (Figure 8.6).

Low-grade metamorphic rocks tend to be so fine-grained that the new mineral grains can be seen only with a microscope. The foliation is then called **slaty cleavage**,

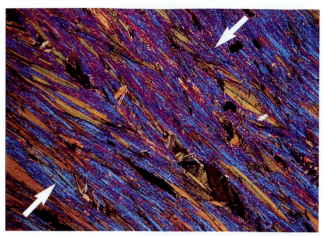

Figure 8.5   Foliation Seen Through a Microscope   Microscopic thin section of a metamorphic rock showing pronounced foliation due to the roughly parallel arrangement of mica grains. The dark-colored grains are quartz. Direction of maximum stress is indicated by arrows. The sample is 1 cm wide.

the property by which a low-grade metamorphic rock breaks into plate-like fragments along planes (Figure 8.7).

HIGHER-GRADE METAMORPHISM: SCHISTOSITY

At intermediate and high grades of metamorphism, grain size increases and individual mineral grains can be seen with the unaided eye. Foliation in coarse-grained metamorphic rocks is called **schistosity** (from Greek *schistos*, "cleaves easily"). Schistosity refers to the parallel arrangement of coarse grains of the sheet-structure minerals. It differs from slaty cleavage mainly in grain size.

Intermediate- and high-grade metamorphic rocks tend to break along wavy or slightly distorted surfaces, reflecting the presence and orientation of grains of quartz, feldspar, and other minerals.

MINERAL ASSEMBLAGE CHANGE

In addition to new textures, metamorphism also produces new mineral assemblages. As temperature and pressure rise, one mineral assemblage "morphs" into another. Each assemblage is characteristic of a given range of temperature and pressure for any given rock composition. Some minerals in these assemblages occur only rarely (or not at all) in igneous and sedimentary rocks. Thus, their presence is usually sufficient evidence that the rock has been metamorphosed.

Examples of these almost-uniquely metamorphic minerals are chlorite, serpentine, epidote, talc, and the three polymorphs of Al_2SiO_5 (kyanite, sillimanite, and andalusite). Figure 8.8 illustrates how mineral assemblages change with grade of metamorphism as a shale is metamorphosed.

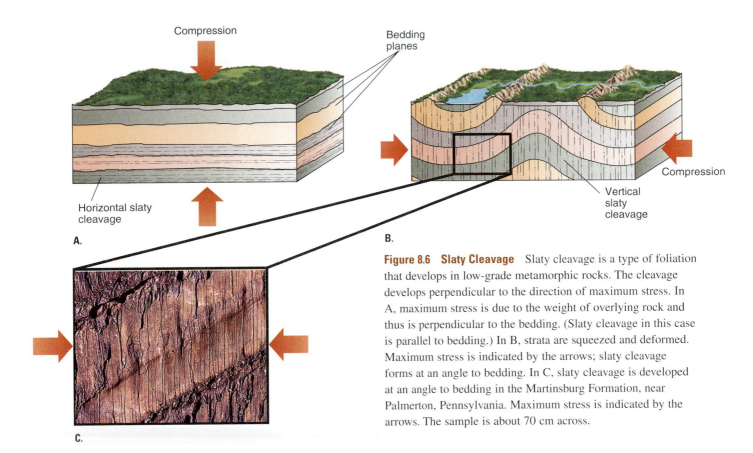

Figure 8.6 **Slaty Cleavage** Slaty cleavage is a type of foliation that develops in low-grade metamorphic rocks. The cleavage develops perpendicular to the direction of maximum stress. In A, maximum stress is due to the weight of overlying rock and thus is perpendicular to the bedding. (Slaty cleavage in this case is parallel to bedding.) In B, strata are squeezed and deformed. Maximum stress is indicated by the arrows; slaty cleavage forms at an angle to bedding. In C, slaty cleavage is developed at an angle to bedding in the Martinsburg Formation, near Palmerton, Pennsylvania. Maximum stress is indicated by the arrows. The sample is about 70 cm across.

METAMORPHISM OF SHALE AND MUDSTONE

SLATE (LOW GRADE)

The low-grade metamorphic product of shale (obvious bedding planes) or mudstone (bedding planes poorly developed) is **slate**. As you can see in Figure 8.9A, the minerals present in shale and mudstone usually include quartz, various clays, and calcite (and sometimes feldspar). Under low-grade metamorphic conditions, muscovite and/or chlorite crystallize. Although the rock may still look a lot like shale or mudstone, the tiny new mineral grains produce slaty cleavage (in the figure, see "Foliation" at "Low Grade"). The presence of slaty cleavage is clear proof that a rock has gone from being sedimentary to metamorphic.

PHYLLITE (INTERMEDIATE GRADE)

Continued metamorphism of a slate to intermediate grade produces both larger mica grains and a changing mineral assemblage. The rock develops a pronounced foliation (see Figure 8.8 photo) and is called **phyllite** (from the Greek, *phyllon*, meaning a leaf). In a slate, you cannot see the new mica grains with your unaided eye. But in a phyllite, the new mica grains are just visible.

Figure 8.7 **Splitting Slate.** Slate being split along the cleavage direction by using a broad chisel and a mallet.

TYPES OF METAMORPHIC ROCK

Please refer to Figure 8.9 as you read the following. Part A shows the progressive change of shale into schist or gneiss. Part B shows the progressive change of basalt into granulite.

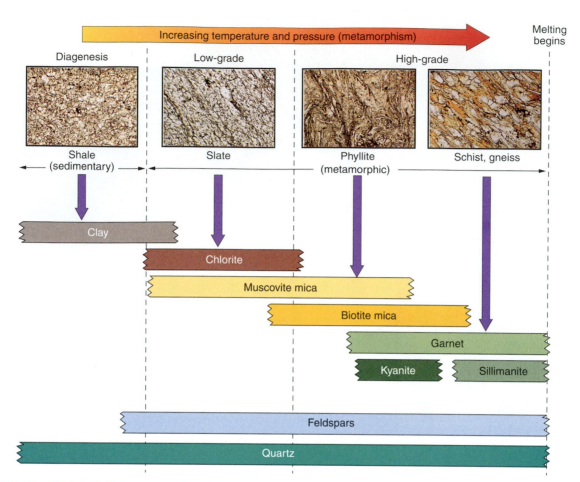

Figure 8.8   From Shale to Schist   The bars show mineral assemblage changes as shale progressively metamorphoses from low to high grade. Before metamorphism, the shale is a sedimentary rock made of clay particles and quartz grains. The first metamorphic rock to develop is a low-grade slate, then a phyl-lite, and finally a high-grade schist or gneiss. Each photomicrograph shows about 3 mm of rock. The minerals kyanite and sillimanite have the same composition (Al_2SiO_5) but different crystal structures —they occur only in metamorphic rocks.

SCHIST AND GNEISS (HIGH GRADE)

Still further metamorphism creates a coarse-grained rock with pronounced schistosity, called **schist**. The most obvious difference among slate, phyllite, and schist is grain size, but as you can see in Figure 8.9A, there are several changes. At the high grades of metamorphism characteristic of schists, minerals may start to segregate into distinct bands.

Gneiss is a high-grade rock with coarse grains and pronounced foliation, but with layers of micaceous minerals segregated from layers of minerals such as quartz and feldspar (Figure 8.2B). (Gneiss, pronounced *nice*, is from the German, *gneisto*, meaning to sparkle.)

A WORD ON NAMING METAMORPHIC ROCKS

Metamorphic rocks are named partly for *mineral assemblage* and partly for *texture*. The basic names of the coarse-grained rocks—schist and gneiss—are derived from their textures, but mineral names are commonly added as adjec-

tives because the minerals are large enough to see and readily identify. For example, we might refer to a *quartz-plagioclase-biotite-garnet gneiss*.

The names *slate* and *phyllite* also describe textures, but the mineral grains are so small they are difficult to identify, so the terms commonly are used without adding mineral-name adjectives.

METAMORPHISM OF BASALT

GREENSCHIST

In Figure 8.9B, you can see that the main minerals in basalt are olivine, pyroxene, and plagioclase. Each mineral is anhydrous. Thus, when a basalt is metamorphosed under conditions where water can enter the rock and form hydrous minerals, distinctive mineral assemblages develop.

At low grades of metamorphism, mineral assemblages such as chlorite + epidote + plagioclase + calcite form. The

A.

	Not metamorphosed	Low grade	Intermediate grade	High grade
Rock name	Shale →	Slate →	Phyllite →	→ Schist / → Gneiss
Foliation	None	Subtle: slaty cleavage	Distinct; schistosity apparent	Conspicuous; schistosity and compositional layering
Size of mica grains	Microscopic	Microscopic	Just visible with hand-held magnifier	Large and obvious
Typical mineral assemblage	Quartz, clays, calcite	Quartz, chlorite, muscovite, plagioclase	Quartz, muscovite, biotite, garnet, kyanite, plagioclase	Quartz, biotite, garnet, sillimanite, plagioclase

Intensity of metamorphism →

B.

	Not metamorphosed	Low grade	Intermediate grade	High grade
Rock name	Basalt — +H$_2$O →	Greenschist →	Amphibolite →	Granulite
Foliation	None	Distinct schistosity	Indistinct; when present due to parallel grains of amphibole	Indistinct because of absence of micas
Size of grains	Visible with hand-held magnifier	Visible with hand-held magnifier	Obvious by eye	Large and obvious
Typical mineral assemblage	Olivine, pyroxene, plagioclase	Chlorite, epidote, plagioclase, calcite	Amphibole, palgioclase, epidote, quartz	Pyroxene, plagioclase, garnet

Intensity of metamorphism →

Figure 8.9 Progressive Metamorphism Progressive metamorphism of shale and basalt. Mineral assemblage and foliation change as a result of increasing temperature and differential stress. A. Shale. B. Basalt.

resulting rock is equivalent in metamorphic grade to a slate, but it has a very different appearance. It has pronounced foliation like phyllite, but also a very distinctive green color because of its chlorite content, giving it the name *greenschist*.

AMPHIBOLITE AND GRANULITE

When a greenschist is subjected to intermediate-grade metamorphism, amphibole replaces the chlorite. The resulting rock is generally coarse-grained and is called an **amphibolite** (Figure 8.10). Foliation is present in amphibolites but is not pronounced because micas and chlorites are usually absent.

At the highest grade of metamorphism, amphibole is replaced by pyroxene, and an indistinctly foliated rock called a granulite develops.

METAMORPHISM OF LIMESTONE AND SANDSTONE

Marble is the metamorphic derivative of limestone, and *quartzite* is the metamorphic derivative of quartz sandstone. Neither limestone nor quartz sandstone contains the necessary ingredients (when pure) to form sheet- or chain-

Figure 8.10 Metamorphosed Basalt: Amphibolite
Amphibolite resulting from metamorphism of a pillow basalt. Compare this with Figure 5.22. The pillow structure was compressed during metamorphism but can be discerned by the borders of pale-yellow epidote formed from the original glassy rims of the pillows. The outcrop is in Namibia.

A.

A.

B.

B.

Figure 8.11   Marble, a Nonfoliated Metamorphic Rock
Texture of marble seen in a 10 cm-wide hand-sized specimen (A) and in a 1 cm wide microscopic thin section viewed in polarized light (B). Notice the interlocking grain structure produced by recrystallization during metamorphism. Marble is composed entirely of calcite. All vestiges of sedimentary structure have disappeared.

Figure 8.12   Quartzite, Another Nonfoliated Metamorphic Rock
Texture of quartzite, like that of marble (Figure 8.11), lacks foliation. A. Hand-sized specimen of quartzite 9 cm wide. B. Microscopic thin section of quartzite. Arrows point to faint traces of the original rounded quartz grains in some of the grains. Field of view is 1.5 cm wide.

structure minerals (mica, chlorite). As a result, marble and quartzite commonly lack foliation.

MARBLE

Marble has a coarsely crystalline, interlocking network of calcite grains. During recrystallization of a limestone, its bedding planes, fossils, and other features of sedimentary rocks are largely obliterated. The end result (Figure 8.11) is an even-grained rock with a distinctive, somewhat sugary texture.

Pure marble is snow white and consists entirely of pure grains of calcite. Such marbles are favored for statuary, marble gravestones, and cemetery statuary, perhaps because white symbolizes purity to many. Many marbles contain impurities—organic matter, pyrite, limonite, and small quantities of silicate minerals, that impart various colors.

QUARTZITE

Quartzite is derived from quartz sandstone by filling of the spaces between the original grains with silica and by recrystallization of the entire mass (Figure 8.12). Sometimes the

ghost-like outlines of the original sedimentary grains can still be seen, even though recrystallization may have completely rearranged the original grain structure.

Before you go on:

1. What is foliation and how does it form?
2. How does slaty cleavage differ from schistosity?
3. What changes in texture accompany progressive metamorphism of a shale from a slate to a gneiss?
4. What rock-type is formed by low-grade metamorphism of a basalt?

TYPES OF METAMORPHISM

Processes that cause metamorphic changes in rocks are *mechanical deformation* and *chemical recrystallization*. The processes usually proceed together, but sometimes the change is due more to one than the other.

Understanding Our Environment

CONCRETE: THE ARTIFICIAL METAMORPHIC ROCK THAT CHANGED OUR ENVIRONMENT

The word *cement* comes from the Latin *caementum*, meaning a substance that binds rock particles together. When cement is used to bind an aggregate of crushed rock, gravel, or oyster shells, the result is concrete—derived from another Latin word, *concrescere*, "grow together."

Roman engineers discovered the forerunner of modern cement more than 2000 years ago. They found that water added to a mixture of quicklime (CaO, made by heating limestone) and glass-rich volcanic ash from Pozzuoli, near Naples, produced chemical reactions that caused new compounds to form, resulting in a hardening of the mixture. The hardened mass was stable in air and water, and the Roman engineers soon discovered that, by mixing their Pozzuolan cement with crushed rock, they could make a concrete that was physically strong and chemically stable.

The Romans eventually discovered that cement could be made from materials other than glassy volcanic ash. They found that ordinary shale, when heated to a sufficiently high temperature, would partially melt, forming a product similar to volcanic glass. When mixed with quicklime and water, the heated shale produced a tough, stable cement.

The Roman formula for cement disappeared during Europe's Dark Ages. Then, in 1756, John Smeaton, a British engineer, was told to build a lighthouse on the treacherous Eddystone Rocks off England's Plymouth Harbor. As he searched ancient Latin documents for ideas on building underwater foundations for the lighthouse, Smeaton rediscovered the formula for Roman cement.

Concrete quickly became a popular building material. In 1824 another English engineer, Joseph Aspidin, patented a formula for *portland cement*, so-called because its color resembles Portland stone, a limestone widely used in England for building.

Today, concrete made from portland cement and rock aggregate is the most widely used building material around the world. Concrete has changed the environment of our cities, motorways, harbors, and airports in ways that are unmatched by any other material (Figure B8.3). In fact, it is difficult to imagine a concrete-free environment.

Limestone and shale, the starting materials for cement, are common and widely available, so concrete is made in nearly every country worldwide. When an appropriate mixture of limestone and shale is heated to about 1480°C, carbon dioxide and water vapor are expelled, the limestone becomes quicklime, and the shale partially melts to glass. When water is added to the mixture, the quicklime reacts with the glass to form an interlocking mesh of calcium-aluminum silicate compounds—in short, an instant metamorphic rock.

Figure B8.3 A Concrete Landscape La Grande Arche in the Place de La Defense, Paris. The building was erected between 1982 and 1990. It is made of reinforced concrete and faced with marble.

Mechanical deformation includes grinding, crushing, and the development of foliation. The deformed conglomerate shown in Figure 8.3 is an example in which mechanical deformation probably played the dominant role.

Chemical recrystallization includes all the changes that occur as rock is heated: mineral composition, new crystal growth, and additions/losses of H_2O and CO_2. (An example of change induced solely by chemical recrystallization is presented in *Understanding Our Environment*, Box 8.2, *Concrete: The Artificial Metamorphic Rock That Changed Our Environment*.)

Different types of metamorphism reflect the different levels of involvement of the two processes. The different types are:

Cataclastic metamorphism (dominated by mechanical deformation).

Contact metamorphism (dominated by recrystallization due to contact with magma).

Burial metamorphism (dominated by recrystallization aided by plentiful water).

Regional metamorphism (both mechanical deformation and chemical recrystallization).

CATACLASTIC METAMORPHISM

Sometimes, mechanical deformation of a rock can occur with only minor chemical recrystallization. Such deformation is usually localized and seen in igneous rocks. For example, when a coarse-grained granite undergoes intense differential stress, individual mineral grains may shatter and pulverize. This deformation is called **cataclastic metamorphism** (*cataclastic*, Greek: to break down).

A.

B.

Figure 8.13 **Cataclastic Metamorphism in Granite** A. Undeformed granite consisting of quartz, feldspar, and biotite. The dark oval is a xenolith of amphibolite. Note the absence of foliation. B. The original granitic texture has been completely changed, and the granite has been cataclastically transformed to a gneiss with a distinct foliation. Amphibolite xenoliths have been flattened and elongated. (From Groothoek, South Africa.)

As cataclastic metamorphism proceeds, grain and rock fragments become elongated and a foliation develops. Figure 8.13 illustrates a granite changed by cataclastic metamorphism.

CONTACT METAMORPHISM

When bodies of hot magma intrude into cool rocks of the crust, **contact metamorphism** occurs in the cool rocks adjacent to the magma. Such metamorphism involves vapors given off by the intruding magma, and by a pronounced increase in temperature and therefore chemical recrystallization. Mechanical deformation is minor or absent, because the stress around a mass of magma tends to be uniform. Rock adjacent to the intrusion becomes heated and metamorphosed, developing a well-defined shell—a **metamorphic aureole**—of altered rock (Figure 8.14).

How thick is a metamorphic aureole? It depends on the size of the intrusive body and the amount of H_2O in the rock being metamorphosed, as well as on vapors released by the solidifying magma. With a small intrusion—a dike or sill a few meters thick—and very little fluid, the aureole may be only a few centimeters thick. The rock in the aureole is usually hard and fine-grained, composed of interlocking, uniformly sized mineral grains, called a **hornfels**.

A large intrusion contains much more heat energy than a small one, and may also give off a lot of water vapor. Around some large intrusions more than a kilometer in diameter, aureoles reach more than 100 m thickness and the metamorphic rocks are typically coarse-grained.

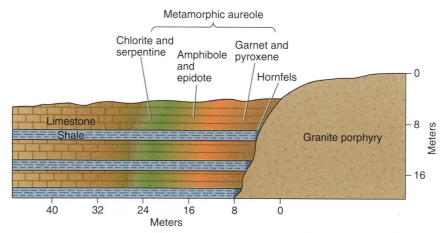

Figure 8.14 **Conact Metamorphism Around a Granite Porphyry Intrusion** Fluids released from the cooling granite created a metamorphic aureole up to 20 m wide and changed the limestone composition. New minerals form a series of concentric shells in the aureole, each with a distinct mineral assemblage. Shale between limestone layers was impervious to fluids and was unaffected, except for a narrow band of hornfels immediately adjacent to the granite. (Site near Breckenridge, Colorado.)

Within a large metamorphic aureole through which a lot of fluid has passed, several zones of different mineral assemblages—roughly concentric—can usually be identified (Figure 8.14). Each zone is characteristic of a certain temperature range.

Immediately adjacent to the intrusion, where temperatures are very high, we find anhydrous minerals such as garnet and pyroxene. Beyond them we find hydrous minerals such as epidote and amphibole, and beyond them micas and chlorites. The exact assemblage of minerals in each zone depends, of course, on the chemical composition of the intruded rock and on that of the invading fluid, as well as on temperature and pressure. As we discuss later in this chapter, metamorphism that involves a lot of fluid and a large change in rock's composition is called **metasomatism**.

BURIAL METAMORPHISM

When buried deeply in a sedimentary basin, sediments (together with interlayered pyroclastics) may attain temperatures of a few hundred degrees Celsius, causing **burial metamorphism**. Buried sediment contains abundant pore water, which accelerates chemical recrystallization and helps new minerals to grow. But water-filled sediment is weak and behaves more like a liquid than a solid. Thus, the stress during burial metamorphism tends to be uniform.

As a result, little mechanical deformation is involved in burial metamorphism. Hence the texture of the resulting metamorphic rock looks the same as essentially unaltered sedimentary rock, despite substantial change in the mineral assemblage.

Zeolites particularly characterize the conditions of burial metamorphism. Zeolites are a group of minerals with fully polymerized silicate structures containing the same chemical elements as feldspars, plus water.

Burial metamorphism is the first stage of metamorphism following diagenesis. It occurs in deep sedimentary basins, such as trenches on the margins of tectonic plates. As temperatures and pressures increase, burial metamorphism grades into regional metamorphism.

REGIONAL METAMORPHISM—A CONSEQUENCE OF PLATE TECTONICS

Regional metamorphism results from tectonic forces that build mountains. Therefore, regional metamorphism occurs over large areas of present and former mountains—tenthousands of square kilometers. Thus it creates the most common metamorphic rocks in the continental crust—slate, greenstone, phyllite, amphibolite, schist, and gneiss.

Unlike contact or burial metamorphism, regional metamorphism results from pronounced differential stresses and extensive mechanical deformation in addition to chemical recrystallization. During a continental collision, sedimentary rock along the continental margin experiences intense differential stress. The foliation so characteristic of slates, schists, and gneisses is a consequence of this stress.

Regional metamorphism also produces greenschists and amphibolites. These tend to occur where segments of ancient basaltic oceanic crust have become incorporated into the continental crust and metamorphosed.

To understand what happens during regional metamorphism, consider a segment of crust that is subjected to horizontal compression from mountain-building tectonic forces:

1. First, rock in the crust folds and buckles.

2. The folding and buckling cause the crust to thicken locally (Figure 8.1).

3. This pushes the bottom of the thickened mass deeper, where temperatures are higher. This increases the stress on, and temperature of, the rocks near the bottom of the thickened pile. New mineral assemblages grow in response.

4. However, rocks are poor conductors of heat. Therefore, the heating process can be very slow. The temperatures reached depend both on depth and on how long a rock is buried in the thickened pile.

5. If the folding and thickening happen very slowly, heating of the pile keeps pace with the temperature of adjacent parts of the crust and mantle (i.e., a normal continental geothermal gradient is maintained). The result is a normal progression of metamorphic rocks—slate, phyllite, schist, gneiss.

6. However, if burial is very fast, as with sediment dragged down in a subduction zone, the pile has insufficient time to heat and conditions of high pressure but rather low temperature prevail. The result is a progression of unusual metamorphic rocks containing high pressure minerals such as jadeite, the main component of precious jade.

7. So, depending on the rate of burial, the same starting rock can yield two quite different metamorphic rocks because different pressures and temperatures are reached.

REGIONAL METAMORPHISM, METAMORPHIC ZONES, AND INDEX MINERALS

The Scottish Highlands saw the first systematic study of a regionally metamorphosed terrain in the 1890s (Figure 8.15B). Geologists observed that rocks having the same overall chemical composition (which happened to be that of shale) could be divided into a sequence of zones—each zone had a distinctive mineral assemblage. Each assemblage was characterized by the appearance of new minerals.

They selected five characteristic *index minerals* that marked the appearance of each new mineral assemblage. Proceeding from low-grade rocks to high-grade rocks, their index minerals were chlorite, biotite, garnet, kyanite, and sillimanite.

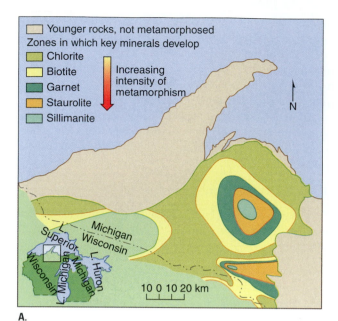

A.

B.

Figure 8.15 Regional Metamorphism and Metamorphic Zones
Metamorphic zones resulting from regional metamorphism. A. Upper Peninsula of Michigan. B. Scottish Highlands.

By plotting on maps where each index mineral first appeared in rocks having the chemical composition of shale, the Scottish Highlands geologists defined a series of isograds. An **isograd** is a line on a map connecting points of first occurrence of a mineral in metamorphic rocks. The concept of isograds is now widely used to study metamor-

phic rocks of all kinds. It is just as applicable to burial and contact metamorphism as to regional metamorphism.

The regions on a map between isograds are known as **metamorphic zones**. We speak of the chlorite zone, the biotite zone, and so forth, and these zones are commonly depicted on maps that show relationships among metamorphic rocks (Figure 8.15A, B).

> **Before you go on:**
> 1. What are the two kinds of processes that produce changes in metamorphic rocks?
> 2. There are four kinds of metamorphism. What are they?
> 3. What are index minerals and how are they related to metamorphic zones?

METAMORPHIC FACIES

Worldwide study demonstrates that the chemical composition of most metamorphic rocks changes little with metamorphism. The main changes that occur are the addition or loss of volatiles such as H_2O and CO_2, and the loss of some SiO_2 removed by the intergranular fluid. Most of the constituents of rocks, such as Al_2O_3, and CaO, remain fixed, or approximately so. The main change brought about during metamorphism, then, is *change in the mineral assemblage*, not large changes in the overall chemical composition of the rocks.

The conclusion to be drawn from this observation is that the temperatures and stresses to which the rocks are subjected during metamorphism determine the mineral assemblages in the metamorphic derivatives of common sedimentary and igneous rocks.

Based on this conclusion, Finnish geologist Pennti Eskola proposed, in 1915, the concept of **metamorphic facies**. *Facies* (French: face) refers to the overall appearance and characteristics of a rock unit.

According to this concept, for given conditions of temperature and stress, the mineral assemblages that form in rocks of different composition belong to the same metamorphic facies. Eskola drew his conclusions from studies of metamorphosed basalts that were interlayered with rocks of entirely different composition.

Metamorphic facies were originally described by recurring mineral assemblages, assuming that a specific set of temperature/pressure conditions caused each assemblage. But the realization that temperature, stress, and rock composition *each* play a role in determining the mineral assemblage provided the link geologists needed. They could now determine the conditions under which rocks were metamorphosed by comparing their mineral assemblages to assemblages produced in laboratory experiments.

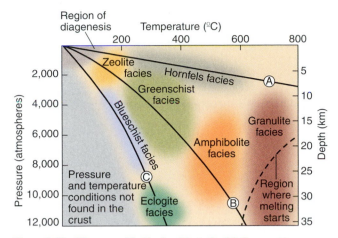

Figure 8.16 Metamorphic Facies Plotted with Respect to Temperature and Depth Curve A is a typical thermal gradient around an intrusive igneous rock that is causing contact metamorphism. Curve B is a normal continental geothermal gradient. Curve C is the geothermal gradient developed in a subduction zone. Note that the zeolite facies overlaps in temperature and pressure with the conditions of diagenesis.

It is now possible to prepare a detailed *petrogenetic grid* in which the mineral assemblage of almost any rock composition can be plotted for a wide range of temperatures and pressures. The principal metamorphic facies, together with geothermal gradients to be expected in the continental crust under three different geologic conditions, are shown in Figure 8.16.

Table 8.1 shows facies names and their characteristic minerals. Because Eskola was studying metamorphosed basalts when he proposed the facies concept, most of his facies names reflect basaltic mineral assemblages.

However, the mineral assemblages in the table result just as much from original rock composition as from temperature and pressure. When comparing mineral assemblages of rocks subjected to different grades of metamorphism, therefore, be certain that the rocks have the same overall chemical composition.

METASOMATISM

The metamorphic processes we have discussed involve essentially fixed compositions and relatively small amounts of fluid. There is little fluid because the pore volume in rocks undergoing metamorphism tends to be small and because the release of H_2O and CO_2 from minerals in metamorphic reactions happens slowly, not all at once.

Geologists describe the metamorphic environment as having "a small water–rock ratio," meaning that the weight ratio of fluid (mainly water) to rock is about 1:10 or less. This is enough fluid to serve as a metamorphic juice, but not enough to dissolve a lot of the rock and noticeably change the rock composition.

Under a few circumstances, however, "large water–rock ratios" occur. For example, when a lot of fluid flows through an open rock fracture, a water–rock ratio can be 10:1 or even 100:1, and rocks adjoining the fracture can be drastically altered by the addition of new ions, removal of material in solution, or both.

Metasomatism (*meta* = change; *soma* = juice), as we learned earlier, is the process in which rock compositions are distinctively altered through exchange with ions in solution. Mineral assemblages formed during metasomatism do not match those of metamorphic facies.

Metasomatism is common in contact metamorphism, especially where premetamorphic rocks are limestones. In

TABLE 8.1   Characteristic Minerals of Differing Metamorphic Facies for Selected Rocks

Facies Name[a]	Precursor Rock Type	
	Basalt	Shale
Granulite	Pyroxene, plagioclase, garnet	Biotite, K-feldspar, quartz, sillimanite
Amphibolite	Amphibole, plagioclase, garnet, quartz	Garnet, biotite, muscovite, kyanite or sillimanite, quartz
Epidote—Amphibolite	Amphibole, epidote, plagioclase, garnet, quartz	Garnet, chlorite, muscovite, biotite, quartz
Greenschist	Chlorite, amphibole, plagioclase, epidote	Chlorite, muscovite, plagioclase, quartz
Blueschist	Glaucophane,[b] chlorite, Ca-rich silicates	Glaucophane,[b] chlorite quartz, muscovite, lawsonite[c]
Eclogite	Pyroxene (variety jadeite), garnet, kyanite	Not yet observed
Hornfels	Pyroxene, plagioclase	Andalusite,[d] biotite, K-feldspar, quartz
Zeolite	Calcite, chlorite, zeolite (variety laumontite)	Zeolite, pyrophyllite,[e] Paragonite[f]

[a] For temperature and pressureconditions of each facies, refer to Figure 8.16.

[b] Glaucophane is a bluish-colored amphibole, $Na_2(Mg,Fe)_3Al_2Si_8O_{22}(OH)_2$.

[c] Lawsonite is a high-pressure mineral similar in composition to anorthite; its formula is $CaAl_2Si_2O_7(OH)_2H_2O$.

[d] Andalusite is a polymorph of kyanite and sillimanite.

[e] Pyrophyllite is $AlSi_2O_5(OH)$.

[f] Paragonite is a sodium mica, $NaAl_2(AlSi_3)O_{10}(OH)_2$.

Figure 8.17 Metasomatism Metasomatically altered marble. The white mineral is calcite, brown is garnet, green is a pyroxene, and purple is fluorite. Elements needed to form the garnet, pyroxene, and fluorite were added to the marble by metasomatic fluids. The sample is 8 cm wide and comes from King Island, Australia.

Figure 8.14, metasomatic fluids released by a cooling magma pass outward through the limestone, causing contact metamorphism. Because the fluids may carry constituents such as silica, iron, and magnesium in solution, the composition of the limestone close to the cooling magma can be drastically changed, even though the limestone distant from the magma, beyond reach of invading fluids, remains unchanged.

Figure 8.17 shows a contact metamorphic rock that was originally limestone. Without addition of new elements, the limestone would simply have become marble. But through metasomatism, the limestone was changed to an

assemblage of garnet, diopside (a green pyroxene), fluorite, and calcite. Metasomatic fluids may also carry valuable metals and form mineral deposits.

> **Before you go on:**
> 1. What is the metamorphic facies concept?
> 2. What is meant by a water–rock ratio?
> 3. How is a water–rock ratio involved in metasomatism?

REVISITING PLATE TECTONICS AND THE EARTH SYSTEM

PLATE TECTONICS AND METAMORPHISM

Plate tectonics has given us our first comprehensive explanation for the distribution of metamorphic zones in regionally metamorphosed rocks. Figure 8.18 shows a convergent plate boundary and the different kinds of metamorphism around it. Look first at the contours of temperature and see how the cold slab of sinking oceanic lithosphere drags them down. The five settings where metamorphism occurs are numbered on the figure as follows:

1. Burial metamorphism happens in the thick pile of sediment that accumulates on the continental shelf and continental slope. Such metamorphism is happening today in the sediment accumulated in the trench off the coasts of Peru and Chile.

2. When oceanic crust with a covering of sedimentary rocks is dragged down by a rapidly subducting plate, pressure increases faster than temperature, subjecting the rock

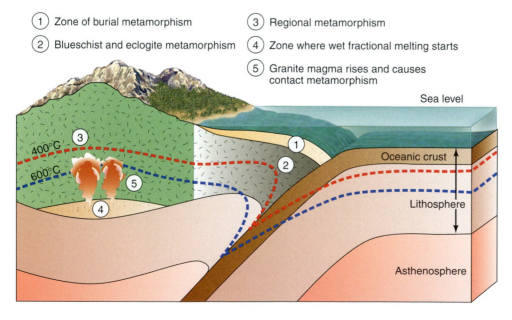

① Zone of burial metamorphism
② Blueschist and eclogite metamorphism
③ Regional metamorphism
④ Zone where wet fractional melting starts
⑤ Granite magma rises and causes contact metamorphism

Figure 8.18 Where Metamorphism Occurs Diagram of a convergent plate boundary showing the different regions of metamorphism. Dashed lines indicate temperature contours.

to high pressure but relatively low temperature. Thus, temperatures and pressures characteristic of *blueschist* and *eclogite facies* metamorphism are reached.

Rock that has been subjected to blueschist and eclogite facies metamorphism is widespread in the Coast Ranges of California, related to ancient subduction. Blueschist metamorphism is probably happening today along the subducting margin of the Pacific Plate where it plunges under the coast of Alaska and the Aleutian Islands.

3. Regional metamorphism occurs over large areas in the continental crust above the subduction boundary of the plate. Where continental crust is thickened by plate convergence and heated by rising magma (5), *greenschist* and *amphibolite facies* metamorphic conditions occur.

Continental collision is the most common setting for regional metamorphism, and broad areas of regionally metamorphosed rocks can be observed throughout the Appalachians and the Alps, both of which are ancient collision zones. Such metamorphism is no doubt occurring today beneath the Himalayas, where the continental crust has been thickened by collision, and beneath the Andes, where it has been both thickened and heated by rising magma.

4. If the crust is sufficiently thick, rocks subjected to amphibolite facies or higher-grade metamorphism can reach temperatures at which wet partial melting commences. Initially, localized melting leads to the formation of migmatites, but when 10 percent or more of the crust has melted, the magma so formed will rise buoyantly, heating and intruding the overlying metamorphic rocks. The magma has a granitic composition and the intruded body is a stock or batholith.

5. As the granitic magma formed by wet partial melting rises, it heats and metamorphoses the rocks with which it comes in contact.

TECTONICS, METALS, AND METASOMATISM

The fluids that cause metasomatism are rich in H_2O, and they are hot—250°C or higher. Such fluids are called **hydrothermal solutions**. Metasomatism and the generation of hydrothermal solutions can be linked to plate tectonics because, as we have just seen, metasomatism is closely related to metamorphism and magmatic activity. A striking example can be seen in the distribution of chalcopyrite-rich copper deposits in North and South America. Figure 8.19 shows a pronounced belt of deposits formed in, or associated with, old stratovolcanoes that lie along the western margin of the Americas.

The magmas that produced the stratovolcanoes were formed by wet partial melting of mantle rocks that lie above subducted oceanic crust. The magmas served as the heat sources for the hydrothermal solutions that formed the mineral deposits. The magmas may also have carried up from the mantle the metals now found in the mineral deposits.

Figure 8.19   Subduction and Rich Mineral Deposits
Chalcopyrite-rich copper deposits formed by hydrothermal solutions create a well-defined belt parallel to the subduction edges of the South and North American plates. Subduction is still active in South America and northern North America but has ceased in the southern part of North America.

The actual source of the metals, whether from the mantle-generated magmas or from the continental crust and gathered by the magmas as they rose upward, is a matter of ongoing research. Regardless of the exact source, the abundant mineral resources we enjoy are due to the combined effects of magmatic, metamorphic, and metasomatic processes, all of which occur because of plate tectonics.

METAMORPHISM AND THE ROCK CYCLE

Please refer back to the rock cycle diagram in Figure 1.19 and see that metamorphism is an essential part of the cycle.

Now consider what happens as sediment progresses to metamorphic rock. Sediment is always porous; because most sediment is water-laid, the pores are filled with water—sediment may contain 20 percent water or more. As sediment piles up, the bottom of the pile compacts and some of the water is squeezed out. Conversion of the sediment to sedimentary rock reduces pore space still further, and by the time a sedimentary rock is deep enough—and therefore hot enough—for metamorphism to start, the water content of the rock will be less than 10 percent.

Some of the water will be pore water, and some will be present in the crystal structures of hydrous minerals. In fact, by the time metamorphism starts it is incorrect to say water is present, because some of the H_2O is present as vapor, some as a constituent of minerals, and none may actually be liquid water.

As metamorphism proceeds to successively higher grades, pore space continually decreases, and the metamorphic mineral assemblages that form contain fewer and fewer hydrous minerals. A high-grade metamorphic rock like a granulite may contain less than 1 percent of H_2O.

What happens to the H_2O driven off during the long process of metamorphism? The answer is that it slowly makes its way back toward Earth's surface where it once again becomes part of the hydrosphere. The residence time of H_2O in a body of rock undergoing regional metamorphism may be hundreds of millions of years, but eventually the rock cycle will bring it back to the surface.

Fast or slow, the cycles of the Earth system continually move materials between Earth's spheres. The next glass of water you drink will almost certainly contain some water molecules that spent a hundred million years or so buried in a mass of metamorphic rock.

What's Ahead?

Metamorphism and metamorphic rocks are the result of interactions between the rock cycle and the tectonic cycle. But there is another kind of metamorphism—impact metamorphism—that has nothing to do with either cycle. Impact metamorphism is all the changes produced by meteorites when they crash into Earth's surface. Impact-metamorphosed rocks are known on Earth, but they are not common because erosion tends to remove evidence of impacts. On the Moon, Mars, and other bodies of the solar system there are surfaces covered by impact craters. Such surfaces are as old as 4 billion years and have not been eroded. On such surfaces evidence of impact metamorphism must be widespread.

But how is it possible to know that a planetary surface or a geologic feature is 4 billion years old? In the next chapter we discuss how rocks are deformed by tectonic events, then two chapters ahead we address the issue of geologic time, and discuss how it is possible to determine the ages of rocks formed millions or billions of years ago.

CHAPTER SUMMARY

1. Metamorphism involves changes in mineral assemblage and rock texture and occurs in the solid state as a result of changes in temperature and pressure.

2. Mechanical deformation and chemical recrystallization are the two processes that affect rock during metamorphism.

3. The presence of intergranular fluid greatly speeds up metamorphic reactions.

4. Foliation, as expressed by directional textures such as slaty cleavage and schistosity, arises from parallel growth of minerals formed during metamorphism and from the mechanical deformation of materials under differential stress.

5. Cataclastic metamorphism involves mechanical deformation together with chemical recrystallization, but mechanical deformation is the predominant effect.

6. Heat given off by bodies of intrusive igneous rock causes contact metamorphism and creates contact metamorphic aureoles. Contact metamorphism involves chemical recrystallization but little mechanical deformation.

7. Regional metamorphism, which involves both mechanical deformation and chemical recrystallization, is a result of plate tectonics. Regionally metamorphosed rocks are produced along subduction and collision edges of plates.

8. Rocks of the same chemical composition that are subjected to identical metamorphic environments react to form the same mineral assemblages. For given conditions of metamorphism, the equilibrium assemblages of minerals that form during the metamorphism of rocks of different composition belong to the same metamorphic facies.

9. Metasomatism involves the changes in rock composition that occur when material in solution is added to the rock, or material is taken away, as the result of fluids flowing through a rock.

10. Hydrothermal solutions are naturally formed hot-water solutions that are capable of dissolving and transporting substances and precipitating them to form new minerals.

11. Metamorphism can be explained by plate tectonics. Burial metamorphism occurs within the thick piles of sediment at the base of the continental slope; regional metamorphism occurs in regions of subduction and continental collision. High-pressure and low-temperature metamorphism happens within a subducted plate of oceanic lithosphere, and in any sediment carried down by a subducting plate.

12. Metasomatism due to hydrothermal solutions is linked to plate tectonics because the solutions tend to form in, or be associated with, stratovolcanoes. Stratovolcanoes are formed above subduction zones.

THE LANGUAGE OF GEOLOGY

amphibolite (p. 213)

burial metamorphism (p. 217)

cataclastic metamorphism (p. 216)
contact metamorphism (p. 216)

differential stress (p. 205)

foliation (p. 210)

gneiss (p. 212)

high-grade metamorphism (p. 205)
hornfels (p. 216)
hydrothermal solution (p. 221)

intergranular fluid (p. 208)
isograd (p. 218)

low-grade metamorphism (p. 205)

marble (p. 214)
metamorphic aureole (p. 216)
metamorphic facies (p. 218)
metamorphic zones (p. 218)
metamorphism (p. 204)
metasomatism (p. 219)
migmatite (p. 209)

phyllite (p. 211)
prograde metamorphic effects (p. 209)

quartzite (p. 214)

regional metamorphism (p. 217)
retrograde metamorphic effects (p. 209)

schist (p. 212)
schistosity (p. 210)
slate (p. 211)
slaty cleavage (p. 210)

uniform stress (p. 205)

vein (p. 208)

QUESTIONS FOR REVIEW

1. Briefly discuss the factors that control metamorphism.

2. How and why does slaty cleavage form?

3. What is schistosity? How does it differ from slaty cleavage?

4. What is the difference between a schist and a gneiss?

5. How does a quartzite differ from a sandstone?

6. What is a metamorphic aureole?

7. What is regional metamorphism?

8. What is the geologic setting of regional metamorphism? Can you name two places in the world where regional metamorphism is probably happening today?

9. What is burial metamorphism? Suggest some place on the Earth where it is probably happening today.

10. Geologists have found the concept of metamorphic zones to be very helpful in studying regional metamorphism. Suggest a reason why.

11. What is the metamorphic facies concept, and how does it help in the study of metamorphic rocks?

12. Under what conditions of pressure and temperature does blueschist facies metamorphism occur? What is the geologic environment where such temperatures and pressures are found? Suggest some place on the Earth where blueschist metamorphism is probably happening today.

13. Name three minerals that are found only in metamorphic rocks.

14. What are the two main volatiles that are added to, or lost from, rocks undergoing metamorphism? What roles do volatiles play in metamorphism?

15. What is cataclastic metamorphism? How would you distinguish it from contact metamorphism of an impure limestone?

16. Discuss the importance of metasomatism and describe how metasomatic processes might form valuable mineral deposits.

Click on *Presentation* and *Interactivity* in the **Metamorphic Rocks** module of your CD-ROM to further explore resources and activities presenting concepts from this chapter. Select *Assessment* in the same module to test your understanding of this chapter.

Click on *Petroscope* for **Metamorphic Rocks** to learn more about the characteristics of metamorphic rocks from this chapter as well as to learn how to identify unknown metamorphic rock samples.

Chapter 9

Intense Deformation of Strata. Folded strata at Mount Cascade, Banff National Park, Canada. Originally horizontal, these strata of sedimentary rocks have been folded and contorted by intense compressive stresses.

How Rock Bends, Buckles, and Breaks

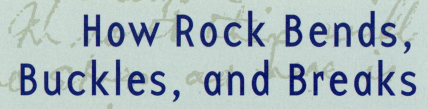

Sterling Hill Mine, New Jersey, 1964; geologist is Brian Skinner.

The Ore Body That Sank

Forty miles northwest of New York City, in the hills of New Jersey, are the mined-out remnants of two extraordinary ore bodies, Franklin Furnace and Sterling Hill. The ore bodies are remarkable for their unusual minerals—willemite (zinc silicate), franklinite (zinc, manganese, iron oxide), and zincite (zinc oxide)—and for their richness, in places up to 40 percent zinc by weight.

In the 1960s, when I worked for the U.S. Geological Survey (USGS), a colleague and I were given permission to map part of the Sterling Hill ore body (the Franklin ore body had already been mined out and was inaccesible). We sought to learn how the unusual ore had formed, and why the ore body had a curious shape, as if a flat layer, like a sheet of newspaper, had been folded in two, crumpled, and squeezed. When we started work it was known from studies by others that the ore body had been metamorphosed at high temperature (at least 700°C) and high pressure (depths of at least 20 km). It was also known that the deformation of the ore body happened at the same time as the metamorphism.

Our mapping established that a stratigraphy of thin layers that looked like bedding existed in the ore body. This suggested that the ore body may have formed by some process of sedimentation. But the ore body was encased in marble, and lay at the base of the marble layer where it was in contact with gneiss. Furthermore, the folded shape of the ore body was not reflected by similar folding in the marble or the gneiss. Both the location and the shape of the ore body were puzzles.

A sedimentary ore body would be more likely to form on top of a limestone than at the bottom, be a chemical sediment, and have a sheet-like structure. From our mapping we felt confident that the ore body must once have been sheet-like. But how could it have been so intensely folded, and the adjacent gneiss have escaped the folding? One night when we were working underground, my USGS colleague, Paul Barton, and I developed a hypothesis to explain the details. Calcite, the mineral present in the marble, has little strength, and at high temperature marble is very ductile. The density of calcite is 2.7 g/cm^3 but the density of the ore body is much higher—the densities of the two main ore minerals, willemite and franklinite, are 4.0 and 5.1 g/cm^3 respectively, and we estimated that the entire ore body would have a density of at least 3.5g/cm^3. A dense ore body sitting at the top of a weak marble layer would be unstable. Considering the density difference and the weakness of the marble, we hypothesized that, during metamorphism, the ore body sank through the marble until it came into contact with a gneiss layer strong enough to prevent further movement. As the originally flat layer sank, it became folded and crumpled. We cannot prove our hypothesis beyond doubt, but it does explain the shape of both the Franklin and Sterling Hill ore bodies. It explains many small features in the ore, such as fractured layers of willemite, and trails of ore minerals left behind in the marble, like particles in a slowly flowing stream.

Brian J. Skinner

KEY QUESTIONS

1. **How is it possible for tough rocks in the lithosphere to break?**
2. **Why does rock sometimes weaken and bend rather than break?**
3. **What's the connection between rock deformation and plate tectonics?**

INTRODUCTION

At many places in this book we have referred to rocks that have been folded, stretched, fractured, and deformed. In Chapter 2, for example, we explained that it is the rigidity or toughness of the lithosphere, as opposed to the weakness or ductility of the asthenosphere, that allows plate tectonics to happen. In Chapter 8, where we discussed metamorphism, we pointed out how uniform stress and differential stress can deform rocks and cause distinctive textures to develop. We have now reached a point where we need to look more closely at deformation and ask the question, "How do rocks deform?"

HOW IS ROCK DEFORMED?

Greece is well known for its earthquakes. One destroyed Sparta, the famous city of antiquity, in the fifth century B.C., and there have been many more since. Recently, a team of Greek and British scientists discovered that the tectonic forces that cause the earthquakes are also stretching Greece and slowly making it grow larger.

A century ago, the distances between a series of Greek survey monuments were measured very accurately. In 1988, a scientific team repeated the measurements and found that Greece is now a meter longer. They also discovered that Greece is being twisted so that the southern end, the Peloponnesus, is moving to the southwest relative to the rest of Greece. The reason for the stretching and twisting is plate tectonics. Africa is moving north and slowly forcing a slice of Mediterranean seafloor under Greece. The resulting forces are stretching and twisting rock in the overlying continental crust.

Evidence that rocks can be deformed is easy to find. If you look at a photograph of the Alps, the Rockies, the

Appalachians, or any other mountain range, you will see strata, once horizontal, that are now tilted and bent. Enormous forces are needed to deform such huge masses of rock. Tectonic forces continuously squeeze, stretch, bend, and break rock in the lithosphere. The source of energy is Earth's heat energy—convection converts the Earth's heat energy to mechanical energy. The huge, slow convective flows of hot rock in the mesosphere and asthenosphere continuously buckle and warp the lithosphere. It is those convective forces that are ultimately the cause of the rock deformation we observe in mountain ranges and that are stretching Greece.

In order to discuss deformation in rocks, such as bending, twisting, and fracture, it is helpful to review some of the elementary properties of solids. We have discussed some of the properties in earlier chapters, but they are so important we will revisit those earlier discussions here.

Knowledge of the factors controlling rock deformation comes largely from laboratory experiments in which cylinders or cubes of rock are squeezed and twisted under controlled conditions.

STRESS AND STRAIN

We learned in Chapters 2 and 8 that when rock deformation is being discussed the term *stress* is used rather than pressure. We also learned that *uniform stress* describes the situation where the stress is equal in all directions, such as the stress on a small body immersed in a liquid or gas. Uniform stress in rocks is also called **confining stress** because any body of rock in the lithosphere is confined by the rocks around it and is uniformly stressed by the weight of the overlying rocks. Confining stress causes a body to change size but not shape. *Differential stress*, by contrast, is stress that is not equal in all directions. The stress that causes rocks to change shape is differential stress. The three kinds of differential stress are shown in Figure 9.1. **Tensional stress** stretches rocks, **compressional stress** squeezes them, while **shear stress** causes slippage and translation. Differential stresses are produced by tectonic forces. An example of the different effects caused by uniform and differential stress is illustrated in Figure 8.2.

The term used to describe the deformation of a rock is **strain**, which is defined as the change in size or shape, or both, in a solid as a result of stress. Uniform stress causes a solid to change size but not shape. Differential stress causes a solid to change shape, but it may or may not also cause a change in size.

STAGES OF DEFORMATION

When a rock is subjected to increasing stress, it passes through three stages of deformation in succession:

1. Elastic deformation is a reversible, or nonpermanent, change in the volume or shape of a stressed rock. Elastic

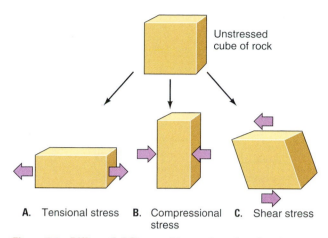

A. Tensional stress   **B.** Compressional stress   **C.** Shear stress

Figure 9.1   Differential Stress   Shape of a cube of rock deformed by differential stress. Arrows indicate directions of maximum stress. A. Tensional stress. B. Compressional stress. C. Shear stress.

deformation is a familiar process. When you sit on a bed you compress the bedsprings elastically. Stand up and the springs return to their original position. So too with a rock or any other solid—squeeze or stretch it, and it will deform. When the stress causing the squeezing or stretching is removed, the rock returns to its original size and shape. There is a limiting stress, called the **elastic limit**, beyond which a solid suffers permanent deformation and does not return to its original size or shape once the stress is removed. Permanent deformation can be a fracture, or a permanent change of shape.

A famous British scientist, Sir Robert Hooke (1635–1703), was the first to demonstrate that, for all materials, provided the elastic limit is not exceeded, a plot of stress versus strain is always a straight line. Hooke proved his point by using a spring, as in Figure 9.2A. However, Hooke's law is equally true for rocks, as illustrated in Figure 9.2B.

2. Ductile deformation is an irreversible change in shape and/or volume of a rock that has been stressed beyond the elastic limit. If a cylinder of rock is stressed by a compressional stress applied parallel to the long axis of the cylinder, an interesting result is obtained. As shown in Figure 9.3, the stress–strain curve for the cylinder rises first through the elastic region, and then, at the elastic limit (point Z), the curve flattens and additional stress causes ductile deformation. If the stress is removed at point X′, the cylinder partially returns to its original shape—the strain decreases along the curve X′Y. A permanent strain, equal to XY, has been induced in the rock. The permanent strain XY is due to ductile deformation.

3. Fracture occurs in a solid when the limits of both elastic and ductile deformation are exceeded. Consider again the stress–strain curve in Figure 9.3. If, instead of releasing the stress at point X′, we continued to increase the stress,

A.

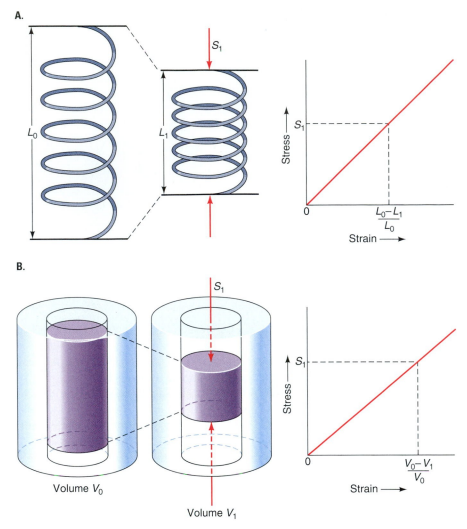

B.

Figure 9.2 Hooke's Law Hooke's law states that for elastic solids, strain is proportional to stress. A. A spring, shortened by a compressional stress S_1, has its length reduced from L_0 to L_1. A plot of the strain $(L_0 - L_1)/L_0$ as a function of stress produces a straight line. B. A cylinder of rock subjected to a confining stress by a tight metal jacket, and shortened by a compressional stress S_1, has its volume reduced from V_0 to V_1. A plot of strain $(V_0 - V_1)/V_0$ as a function of stress produces a straight line.

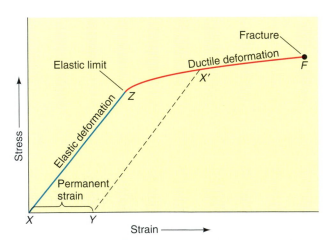

Figure 9.3 Stress and Strain Typical stress–strain curve for a cylinder of rock tested in the laboratory. Following elastic deformation (X to Z), the elastic limit Z marks the onset of ductile deformation. If, at point X', the stress is removed, the solid will return to an unstressed state at point Y by following the path $X'Y$. The distance XY is a measure of the permanent, irreversible strain produced by ductile deformation. If the stress is not released at X', but is increased and maintained, the strength of the solid is exceeded when rupture occurs at the point of fracture, F.

the stress–strain curve would continue to point F, where the cylinder breaks by fracturing. Obviously, fracture, like ductile deformation, is an irreversible kind of deformation.

DUCTILE DEFORMATION VERSUS FRACTURE

A brittle substance tends to deform by fracture, while a ductile substance deforms by a change of shape. Drop a piece of chalk on the floor and it will fracture, but if you drop a piece of butter it will bend but not break. Chalk is brittle; butter is ductile.

A typical stress–strain curve for a brittle substance is shown in Figure 9.4A. Note that Z, the elastic limit, is very close to F, the point of fracture. Thus, little ductile deformation occurs in a brittle substance. In contrast, in a ductile substance the elastic limit and fracture point are far apart, as shown in Figure 9.4B.

Examples of deformation of brittle and ductile rocks are shown in Figures 9.5A and B. The strata in Figure 9.5A have fractured cleanly with little or no evidence of bending due to ductile deformation. The strata in Figure 9.5B, by contrast, have been intensely twisted and bent by ductile deformation.

A.

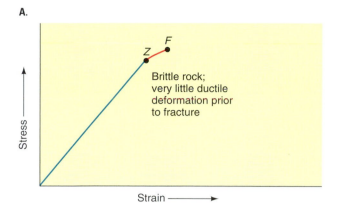

B.

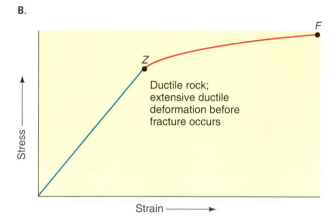

Figure 9.4 Stress–Strain Curves Comparison of stress–strain curves. Z is the elastic limit, F the point where fracture occurs. A. Curve for a brittle substance. B. Curve for a ductile substance.

To evaluate deformation in rocks, it is necessary to estimate the relative importance of brittle properties versus ductile properties. The essential conditions controlling the relative importance of the two kinds of properties are (1) temperature, (2) confining stress, (3) time and strain rate, and (4) composition.

TEMPERATURE

The higher the temperature, the more ductile and less brittle a solid becomes. A rod of glass is difficult to bend at room temperature; if we try too hard, it will break because it is brittle. However, a glass rod becomes ductile and can be readily bent if it is heated to redness over a flame. Rocks are like glass rods. They are brittle at the Earth's surface, but at depth, where temperatures are high because of the geothermal gradient, rocks become ductile.

CONFINING STRESS

The effect of confining stress on deformation is not familiar in common experience. Confining stress is a uniform squeezing of rock owing to the weight of all the overlying strata. High confining stress hinders the formation of fractures and so reduces brittle properties. At high confining stress, it is easier for a solid to bend and flow than to break and fracture. Reduction of brittleness by high confining stress is a second reason why solid rock can be bent and folded by ductile deformation.

TIME AND STRAIN RATE

The effect of time on rock deformation is vitally important, but as with confining stress, it is not obvious from common experience. All the constituent atoms of the solid transmit stress applied to a solid. If the stress exceeds the strength of the bonds between atoms, either the atoms must move to another place in the crystal lattice in order to relieve the stress or the bonds must break, which means fracture occurs. Atoms in solids cannot move rapidly. Nevertheless,

A.

Figure 9.5 Brittle versus Ductile Examples of rock deformation. A. Fracture of strata by brittle deformation; Triassic-aged

B.

Bunter Sandstone, Merseyside, U.K. B. Bending of strata by ductile deformation; limestones in Crete.

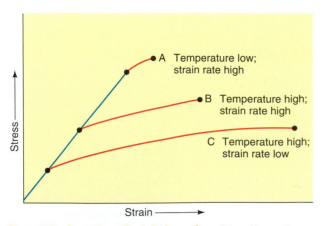

Figure 9.6 Controls on Rock Deformation The effects of temperature and strain rate on rock deformation. Curve A: low temperature, high strain rate. Curve B: high temperature, high strain rate. Curve C: high temperature, low strain rate.

if the stress builds up slowly and gradually and is maintained for a long period, the atoms have time to move, and the solid can slowly readjust and change shape by ductile deformation. The important point to appreciate is that the *rate* at which a solid is strained is just as significant as how long a stress is active.

The term used for time-dependent deformation of a rock is **strain rate**, which is the rate at which a rock is forced to change its shape or volume. Strain rates are measured in terms of change of volume per unit volume per second. For example, a strain rate that is sometimes used in laboratory experiments is 10^{-6}/s, by which is meant a change in volume of one-millionth of a unit volume per unit volume per second. Strain rates in the Earth are much lower than this—about 10^{-14} to 10^{-15}/s. The lower the strain rate, the greater the tendency for ductile deformation to occur.

A comparison of the influences of temperature, confining pressure, and strain rate on rock properties can be seen in Figure 9.6. Low temperature, low confining stress, and high strain rate enhance brittle properties. These conditions are characteristic of the crust (especially the upper crust), and as a result failure by fracture is common in rocks of the upper crust. High temperature, high confining stress, and low strain rates, which are characteristic of the deeper crust and mantle, reduce brittle properties and enhance the ductile properties of rock.

COMPOSITION

The composition of a rock has a pronounced effect on its properties. Composition has two aspects. First, the kinds of minerals in a rock exert a strong influence on properties because some minerals (such as quartz, garnet, and olivine) are very brittle, while others (such as mica, clay, calcite, and gypsum) are ductile. Second, the presence of water in a rock reduces brittleness and enhances ductile

properties. Water affects properties by weakening the chemical bonds in minerals and by forming films around mineral grains, thereby reducing the friction between grains. Thus, wet rocks have a greater tendency to be deformed in a ductile fashion than do dry rocks.

Rocks that readily deform by ductile deformation are limestone, marble, shale, slate, phyllite, and schist. Rocks that tend to be brittle rather than ductile are sandstone and quartzite, granite, granodiorite, and gneiss.

BRITTLE–DUCTILE PROPERTIES OF THE LITHOSPHERE

Rock strength in Earth does not change smoothly with depth. As is evident in Figure 9.7, there are two peaks in the plot of rock strength with depth. The reason is that strength is determined by composition, temperature, and pressure.

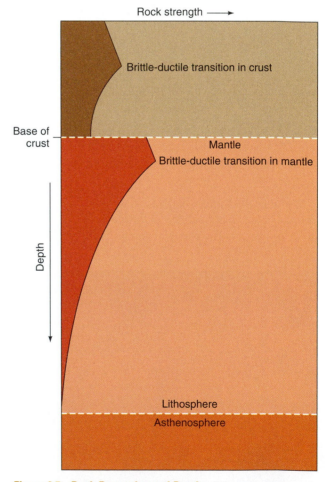

Figure 9.7 Rock Properties and Depth Schematic diagram showing the way rock strength changes with depth. Two brittle–ductile transitions are present. In the crust, the transition is controlled by the properties of quartz-rich rocks; in the mantle, olivine-rich rocks determine the properties. The depth at which failure by brittle deformation disappears is the lithosphere–asthenosphere boundary.

Rocks in the crust are quartz-rich, so the strength properties of quartz play an important role in the strength properties of the crust. Rock strength increases down to a depth of about 15 km. Above 15 km rocks are strong; they fracture and fail by brittle deformation. Below 15 km fractures become less common as quartz weakens and rocks become increasingly ductile. The depth in the crust where ductile properties start to predominate over brittle properties is known as the *brittle–ductile transition*.

Quartz is absent from the mantle. Rocks in the mantle are olivine-rich. Olivine is stronger than quartz, and the brittle–ductile transition of olivine-rich rock is reached only at a depth of about 40 km, just below the crust-mantle boundary. Thus, a second maximum in rock strength is reached at this depth (Figure 9.7). Below the brittle–ductile transition in the mantle, rock strength again declines. As discussed in Chapter 2 in the section on geothermal gradients, by about 1300°C, rock strength is very low; brittle deformation is no longer possible. The disappearance of all brittle deformation properties marks the lithosphere–asthenosphere boundary.

Figure 9.8 Active Fault An orange grove in southern California planted across the San Andreas Fault. Movement on the fault displaced the originally straight rows of trees. The direction of motion is such that trees in the background moved from left to right relative to the trees in the foreground.

Before you go on:

1. Why do we use the term *stress* instead of pressure when we discuss rock deformation?

2. Look at the rocks in the area in which you live. Have they been deformed? If they have, was it by brittle deformation, ductile deformation, or both?

DEFORMATION IN PROGRESS

Most deformation in the crust is too slow or too deeply buried to be observed. Large movements happen so slowly, which means at such low strain rates, that they can be measured only over a hundred or more years. The deformation occurring in Greece, discussed at the beginning of this chapter, is an example. Nevertheless, deformation does sometimes happen fast enough to be detected and measured. For convenience we divide large-scale, observable deformation of the crust into two groups: abrupt movement that involves fracture, in which blocks of the crust suddenly move a few centimeters or a few meters in a matter of minutes or hours, and gradual movement involving ductile deformation, in which slow, steady motions occur without any abrupt jarring.

ABRUPT MOVEMENT

Abrupt movement involves the fracture of brittle rocks and movement along the fractures. A fracture in a rock along which movement occurs is a **fault**. Once fracturing has started, friction inhibits continuous slippage. Instead, stress again builds up slowly until friction between the two sides of the fault is overcome. Then abrupt slippage occurs again. If the stresses persist, the whole cycle of slow buildup fol-

lowed by an abrupt movement repeats itself many times. Although movement on a large fault may eventually total many kilometers, this distance is the sum of numerous small, sudden slips. Each sudden movement may cause an earthquake and, if the movement occurs near Earth's surface, may disrupt and displace surface features.

Figure 9.8 is an example of horizontal abrupt movement. Abrupt vertical movements are also well documented. The largest abrupt vertical displacement ever observed occurred in 1899 at Yakutat Bay, Alaska, during an earthquake. A stretch of the Alaskan shore (including the beach, barnacle-covered rocks, and other telltale features) was suddenly lifted as much as 15 m above sea level. This visible vertical displacement may be less than the total amount, because the fault is hidden offshore and the block of crust on the other side of it, entirely beneath the sea, may have moved downward, thus adding to the total displacement of the stretch of beach.

Abrupt movements in the lithosphere are commonly accompanied by earthquakes and therefore can be hazardous to people living nearby. Earthquakes and earthquake hazards are discussed in Chapter 10.

GRADUAL MOVEMENT

Movement along faults is usually, but not always, abrupt. Measurements along the San Andreas Fault in California reveal places where gradual slipping occurs, sometimes reaching a rate as high as 5 cm a year. Because the San Andreas Fault coincides with a plate boundary that cuts through the entire lithosphere—that is, well below the

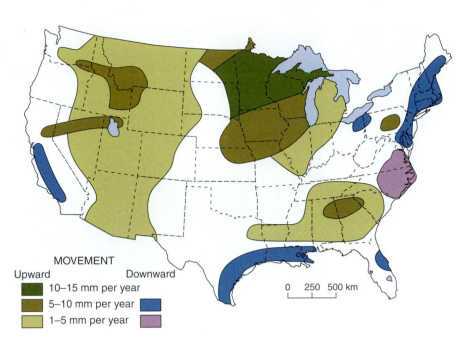

Figure 9.9 The Moving Land Surface
Accurate measurements over a 100-year period show that in large areas of the United States, the surface is slowly moving either upward or downward. Subsidence along the coasts of California and the Gulf of Mexico is believed to have been caused in part by withdrawal of gas, oil, and water, which allows subsurface reservoirs to collapse. Uplift near the Great Lakes is a rebound effect following the melting of the last ice sheet. The causes of movements in other areas are not known with certainty. Those areas in which no movement is shown are not necessarily stationary. They are simply areas in which measurements are very few or movement is so slow it has not yet been detected.

MOVEMENT
Upward   Downward
■ 10–15 mm per year
■ 5–10 mm per year ■
■ 1–5 mm per year   ■

0   250   500 km

brittle–ductile transition—it is probable that deformation of rocks is ductile at depth but brittle near the surface.

Possibly no spot on Earth is completely stationary. Measurements by U.S. government surveyors over 100 years, for example, reveal great areas of the United States where the land is slowly sinking and other places where it is slowly rising (Figure 9.9). The causes of these vast, slow, vertical movements are not well understood, but the movements do prove that the solid Earth is not as rigid as it seems at first sight and that great internal forces are continuously deforming its crust.

EVIDENCE OF FORMER DEFORMATION

With such convincing evidence of present-day deformation of Earth's crust, we might reasonably expect to find a great deal of evidence of former deformation. Studies of land and sea-bottom topography provide abundant evidence of vertical movements. In some areas the distribution of various kinds of rock provides clear evidence that horizontal movements have occurred through distances as great as several hundred kilometers.

Not all evidence of movement and deformation observed in bedrock is as obvious as the examples cited. But once we learn to recognize it, evidence of deformation is seen to be very widespread—so much so that a special branch of geology, *structural geology*, has the study of rock deformation as its primary focus.

STRIKE AND DIP

The law of original horizontality (Chapter 1) tells us that sedimentary strata and lava flows were initially horizontal.

Where we observe such rocks to be tilted, we can conclude that deformation has occurred. In order to decipher and explain this deformation, a geologist starts by measuring the angle and direction of tilting.

In order to measure the orientation of a tilted plane, we need to remember the two principles of geometry shown in Figure 9.10A: (1) the intersection of two planes defines a line and (2) in an inclined plane all horizontal lines are parallel. The line formed by the intersection of an inclined plane with a horizontal plane is always horizontal. Such a line can be visualized as the waterline on an inclined stratum along the shore of a lake, as shown in Figure 9.10B. The lake surface is a convenient horizontal plane. The waterline marks the **strike**, which is the compass direction of the horizontal line formed by the intersection of a horizontal plane and an inclined plane.

Once we know the strike, we need only one more measurement to fix the orientation of an inclined plane. That is the **dip**, the angle in degrees between a horizontal plane and the inclined plane, measured down from horizontal. The direction and angle of dip are indicated in Figure 9.10.

GEOLOGIC MAPS

Rarely is it possible for a geologist to see all the structural details of deformed rocks in a given area. Soil and vegetation usually cover much of the evidence. A geologist must therefore decipher the structures by making observations and measurements at a number of individual *outcrops* (places where bedrock is exposed at the surface). When the observations made at each outcrop are plotted on a map, and inferences are made as to what has happened beneath the cover of soil and vegetation, the result is a geologic map. An example of a simple geologic map is shown in Figure 9.11. Note the way the geologist recorded the strike

A.

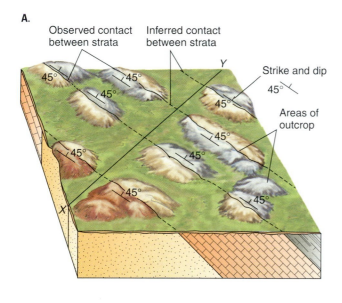

Figure 9.10 Dip and Strike A. The geometric principles used to measure the direction and angle of tilt of an inclined plane. B. Strike and direction and angle of dip. Note the symbol used by geologists to indicate strike and dip.

B.

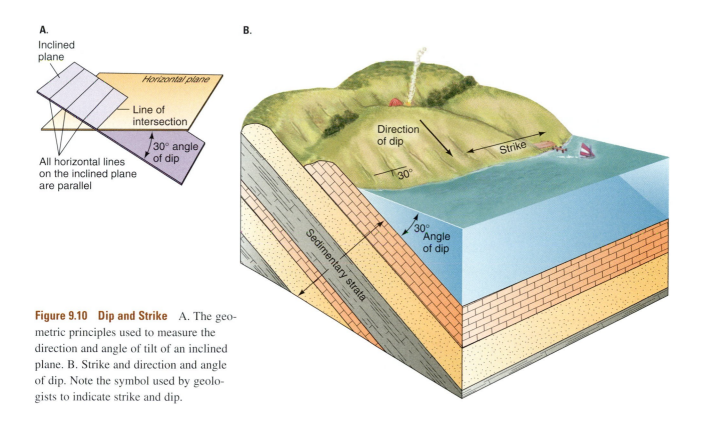

A.

B.

Figure 9.11 Geologic Map A. Solid lines denote areas of outcrop where the geologist identified rock types and strata. Dashed lines indicate inferred contacts. B. Section along the line *XY* in part A.

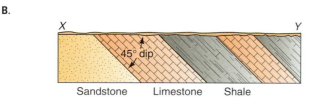

and the angle and direction of dip of the contacts between strata at each outcrop.

DEFORMATION BY FRACTURE

Rock in the crust, especially rock close to the surface, tends to be brittle. As a result, rock at or near Earth's surface tends to be cut by innumerable fractures called either *joints* or *faults*. As discussed in Chapter 6, a joint is a special kind of fracture because there has not been any movement along it. A fault, as we learned earlier, is a fracture along which visible displacement has occurred.

RELATIVE DISPLACEMENT

Generally, it is not possible to tell how much movement has occurred along a fault, nor which side of the fault has moved. In an ideal case—for example, if a single mineral grain or a single pebble in a conglomerate has been cut through by the fault and the halves have been carried apart a measurable distance—the amount of movement can be determined. Yet even then it is not possible to say whether one block stood still while the other moved past it, or whether both sides moved. In classifying fault movements, therefore, geologists can determine only relative displacements; that is, one side of a fault has moved in a given direction relative to the other side. For example, in Figure 9.5A it is apparent that there has been displacement on a fault, but all we can say about the movement is that the lefthand side moved down relative to the righthand side.

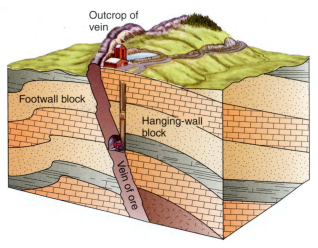

Figure 9.12   Old Mining Terms   Hanging-wall and footwall.

Similarly, in Figure 9.8 all we can say is that the orange trees in the rear moved to the right relative to those in the front.

HANGING-WALL AND FOOTWALL

Most faults are inclined like the faults in Figure 9.5A. (That is, to use a geologic term, they dip.) To describe the inclination, geologists have adopted two old mining terms. From a miner's viewpoint, the rocks above an inclined vein overhang him, while the rocks below the vein are beneath his feet (Figure 9.12). Because veins occupy openings created by faults, we use the old miner's terms in the following way. The **hanging-wall block** is the block of rock above an inclined fault; the block of rock below an inclined fault is the **footwall block**. These terms, of course, do not apply to vertical faults.

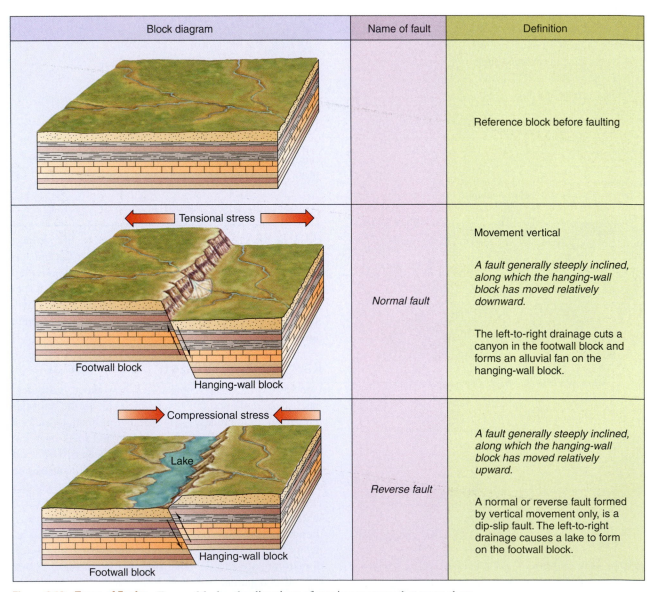

Block diagram	Name of fault	Definition
		Reference block before faulting
Tensional stress	*Normal fault*	Movement vertical *A fault generally steeply inclined, along which the hanging-wall block has moved relatively downward.* The left-to-right drainage cuts a canyon in the footwall block and forms an alluvial fan on the hanging-wall block.
Compressional stress	*Reverse fault*	*A fault generally steeply inclined, along which the hanging-wall block has moved relatively upward.* A normal or reverse fault formed by vertical movement only, is a dip-slip fault. The left-to-right drainage causes a lake to form on the footwall block.

Figure 9.13   Types of Faults   Types of faults, the directions of maximum stress that cause them, and some of the topographic changes they produce.

CLASSIFICATION OF FAULTS

Faults are classified according to (1) the dip of the fault and (2) the direction of relative movement. The common classes of faults, together with the changes in local topography they sometimes create, are shown in Figure 9.13. The standard planes of reference in classifying faults are the vertical and the horizontal. Along many faults, movement is entirely vertical or entirely horizontal, but along some faults combined vertical and horizontal movements occur.

NORMAL FAULTS

Normal faults are caused by tensional stresses that tend to pull the crust apart, as well as by stresses created by a push from below that tend to stretch the crust. Movement on a normal fault is such that the hanging-wall block moves down relative to the footwall block. The faults shown in Figure 9.5A are normal faults.

Commonly, two or more normal faults with parallel strikes but opposite dips enclose an upthrust or down-dropped segment of the crust. As shown in Figure 9.14, a down-dropped block is a **graben**, or **rift**, if it is bounded by two normal faults and a **half-graben** if subsidence occurs along a single fault. An upthrust block is a **horst**. The central, steep-walled valley that runs down the center of the Mid-Atlantic Ridge and cuts through Iceland is a graben. Perhaps the world's most famous system of grabens and half-grabens is the African Rift Valley seen in Figure 9.15, which runs north–south through the countries of East Africa for more than 6000 km. Within parts of the

Block diagram	Name of fault	Definition
	Strike-slip fault	Movement horizontal *A fault on which displacement has been horizontal.* Movement of a strike-slip fault is described by looking directly across the fault and by noting which way the block on the opposite side has moved. The example shown is a *left-lateral fault* because the opposite block has moved to the left. If the opposite block had moved to the right, it would be a *right-lateral fault.* Notice that horizontal strata show no vertical displacement
	Oblique-slip fault	Combined horizontal and vertical movement *A fault on which movement is both horizontal and vertical.* Forces are a combination of forces causing strike-slip and normal faulting.
	Hinge fault	Combined horizontal and vertical movement *A fault on which displacement dies out (perceptibly) along strike and ends at a definite point.* Forces are the same as those causing normal faulting.

Figure 9.13 *continued.*

A.

Footwall block —— Hanging-wall block —— Footwall block

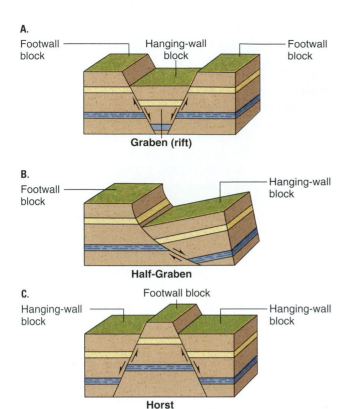

Graben (rift)

B.

Footwall block —— —— Hanging-wall block

Half-Graben

C.

Footwall block

Hanging-wall block —— —— Hanging-wall block

Horst

Figure 9.14 Horsts and Grabens Horsts and grabens form when tensional stresses produce normal faults.

Rift Valley magma has followed channels that lead upward along the fault surfaces, creating volcanoes.

Normal faults are innumerable. Therefore, horsts and grabens are also very common, although none is more spectacular than the African Rift Valley. The north–south valley of the Rio Grande in New Mexico is a graben. The valley in which the Rhine River flows through western Europe follows a series of grabens. A spectacular example of normal faulting is found in the Basin and Range Province in Utah, Nevada, and Idaho. There, movement on a series of parallel and nearly parallel north–south striking, normal faults has formed horsts and half-grabens. The horsts are now mountain ranges, and the grabens and half-grabens are sedimentary basins. As seen in Figure 9.16, the province, which is bounded in the east by the western edge of the Wasatch Range and continues westward to the eastern edge of the Sierra Nevada, contains some spectacularly beautiful scenery.

REVERSE FAULTS AND THRUST FAULTS

Reverse faults arise from compressional stresses. Movement on a reverse fault is such that a hanging-wall block moves up relative to a footwall block. Reverse fault movement shortens and thickens the crust.

A special class of faults, called **thrust faults**, are low-angle reverse faults with dips less than 15° (Figure 9.17). Such faults, common in great mountain chains, are note-

Figure 9.15 Spreading Center Splitting a Continent The African Rift Valley, which extends from the Red Sea in the north to Malawi in the south, is a gigantic rent in Earth's surface marking the place where a spreading center is splitting Africa into two pieces. The photo is of a portion of the Rift Valley in central Kenya. To the east (righthand side) is a plateau bounded by a jagged fracture. The two hills (rear and lefthand edge) are volcanoes.

Figure 9.16 Basin and Range Province Scenery in the northern part of the Basin and Range Province. Looking east toward the Lost River Range, Idaho, from Pioneer Mountain. Valley in the foreground is a graben; the Lost River Range is a horst.

worthy because along some of them the hanging-wall block has moved many kilometers over the footwall block. In most cases the hanging-wall block, thousands of meters thick, consists of rocks much older than those adjacent to the thrust on the footwall block (Figure 9.18).

STRIKE-SLIP FAULTS

Strike-slip faults are those in which the principal movement is horizontal and therefore parallel to the strike of the fault (Figure 9.13). Strike-slip faults arise from shear stresses. One strike-slip fault is so famous almost everyone has heard of it—the San Andreas Fault. In Figure 9.8 it is strike-slip movement on the San Andreas that is offsetting the rows of orange trees.

Movement on a strike-slip fault is designated as follows: to an observer standing on either fault block, the movement of the other block is *left lateral* if it is to the left

and *right lateral* if it is to the right. The sense of relative motion is the same regardless of which block the observer is standing on. The San Andreas is a right-lateral strike-slip fault. Apparently, movement has been occurring along it for at least 65 million years. The total movement is not known, but some evidence suggests that it now amounts to more than 600 km.

Many of the largest and most active faults are strike-slip faults. This is so because strike-slip faults, spreading centers, and subduction zones are the three kinds of margins that bound tectonic plates (Chapter 2). The three plate-margin components link together to form continuous networks encircling Earth. Where one plate margin terminates, another commences; their junction point is called a *transform*. J. T. Wilson, the Canadian scientist who first recognized the network relation, proposed that the special class of strike-slip faults that form plate boundaries be called **trans-**

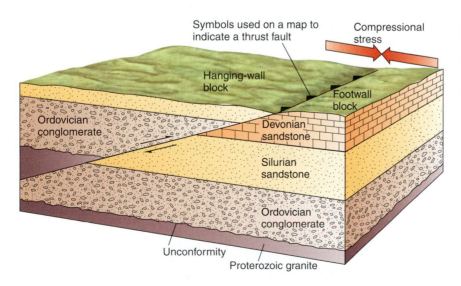

Figure 9.17 Thrust Fault—Older Over Younger Example of a thrust fault. Note that thrusting causes older strata in the hanging-wall block to lie locally above younger units in the footwall block.

Figure 9.18 Keystone Thrust Keystone Thrust, west of Las Vegas, Nevada. A. Air view northward defines this thrust fault by a color contrast in the strata. Light-colored Jurassic sandstone, forming a cliff nearly 600 m high, lies below the fault and forms the footwall block; dark-colored Paleozoic limestones and dolostones lie above the fault and form the hanging-wall block. B. Section drawn across area shown in the photograph.

A.

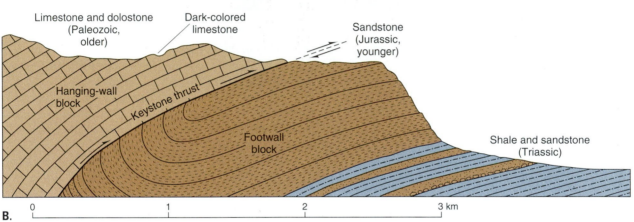

B.

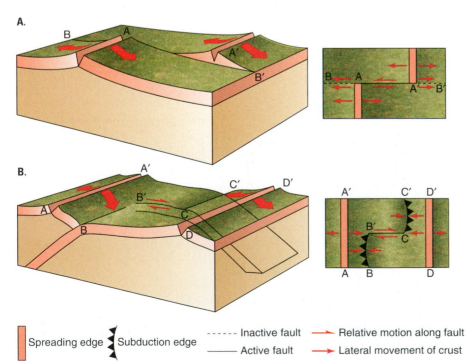

Figure 9.19 Transform Faults

Transform faults as they appear on a map. A. A spreading center offset by a transform fault. Crust on both sides of the two spreading center segments moves laterally away from the spreading center. Between segments of the spreading center, along *A–A′*, movement on the two sides of the fault is in opposite directions. Beyond the spreading center, however, along segments *A–B* and *A′–B′*, movement on both sides of the projected fault is in the same direction. Therefore, a transform fault does not cause the spreading center segments to move continuously apart. B. Transform fault joining two subduction edges.

form faults. Two possible configurations for transform faults are shown diagrammatically in Figure 9.19.

EVIDENCE OF MOVEMENT ALONG FAULTS

Often we find fractures in rock but cannot tell at first glance whether movement has occurred along them; in other words, we don't know whether we are looking at a fault or a joint. For example, in uniform, even-grained rock such as granite, or in a pile of thin-bedded strata no one of which is unique or distinctive, it may not be possible to see the dis-

placement of any obvious features. However, examination of either the fault surface or rock immediately adjacent to it sometimes reveals signs of local deformation, indicating that movement has occurred. Under special circumstances, even the relative direction of movement can be deciphered.

Movement of one mass of rock past another can cause the fault surfaces to be smoothed, striated, and grooved. Striated or highly polished surfaces on hard rocks, abraded by movement along a fault, are called *slickensides*. Parallel grooves and striations on such surfaces record the direction of the most recent movement (Figure 9.20A).

Figure 9.20 Slickensides and Breccias The effects of faulting. A. Slickensides on a fault surface, Borrego, California. The hanging-wall rocks have been removed by erosion, thus exposing the fault on the footwall. B. Fault breccia, Titus Canyon, Death Valley. Angular gneiss fragments (dark) broken by faulting are set in a matrix of rock flour and calcite.

A.

B.

Not all fault surfaces have slickensides. In many instances, fault movement crushes rock adjacent to the fault into a mass of irregular pieces, forming *fault breccia* (Figure 9.20B). Intense grinding breaks the fragments into such tiny pieces that they may not be individually visible even under a microscope. Some contain such tiny fragments they resemble chert.

DEFORMATION BY BENDING

Bending may consist of broad, gentle warping that extends over hundreds of kilometers, or it might be close, tight flexing of microscopic size, or anything in between. Regardless of the volume of rock involved or the degree of warping, the bending of rocks is referred to as *folding*. Folding is most easily recognized in layered rocks. An individual bend or warp in layered rock is called a **fold**.

Regardless of their size, folds are formed by ductile deformation as a result of compressional and shear stresses. At very low strain rates, even shallow rocks that are well above the brittle–ductile transition can be folded. That is apparently what is happening in Greece today; apparently too, it is what happened when the great oil fields of the Middle East were formed, as discussed in *Understanding Our Environment*, Box 9.1, *Rock Deformation and Oil Pools*. However, the very intense folding that we can observe so widely in mountain ranges probably occurred below the brittle–ductile transition when the rocks were deeply buried and thus were subjected to high temperature and high confining stress.

TYPES OF FOLDS

The simplest fold is a **monocline**, a local steepening in an otherwise uniformly dipping pile of strata. An easy way to visualize a monocline is to lay a book on a table and then drape a handkerchief over one side of the book and out onto the table. So draped, the handkerchief forms a mono-

cline. Most folds are more complicated than monoclines (Figure 9.21). An upfold in the form of an arch is an **anticline**; a downfold with a trough-like form is a **syncline**. Anticlines and synclines are usually paired.

GEOMETRY OF FOLDS

As shown in Figure 9.21A, the sides of a fold are the **limbs**, and the median line between the limbs, along the crest of an anticline or the trough of a syncline, is the **axis** of the fold. A fold with an inclined axis is said to be a *plunging fold*, and the angle between a fold axis and the horizontal is the **plunge** of a fold (Figure 9.21B). An imaginary plane that divides a fold as symmetrically as possible, and that passes through the axis, is the *axial plane*.

Many folds, such as those shown in Figure 9.21, are nearly symmetrical. Others, however, are not symmetrical; intense stress can create complex shapes. The common forms of folds are shown in Figure 9.22.

An *open fold*, such as that depicted in Figure 9.22A, is one in which the two limbs dip gently and equally away from the axis. The more intense the compressional stress, the less open a fold will be. When stress is very intense, the fold closes up and the limbs become parallel to each other; such a fold is said to be isoclinal (Figure 9.22B). Intense stress is also the reason why a fold becomes either asymmetric (Figure 9.22C), in which limbs have unequal dips, or overturned, in which limbs dip in the same direction (Figure 9.22D). Eventually, an overturned fold may become recumbent, meaning the two limbs are horizontal or nearly so (Figure 9.22E). Recumbent folds are common in mountainous regions, such as the Alps and the Himalaya, that were produced by continental collisions.

If only fragmentary exposures of bedrock are available, it is apparent that difficulties might arise in deciding whether a given fold is overturned. As is apparent in Figures 9.22D and E, it is necessary to know whether a

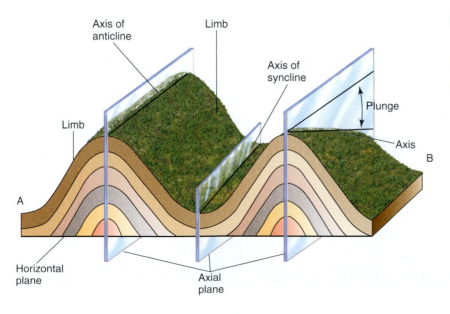

Figure 9.21 Geometry of a Fold Features of simple folds. Note that the strata dip away from the axis of an anticline but toward the axis of a syncline. A. Fold axis horizontal. B. Fold axis plunging.

BOX 9.1   # UNDERSTANDING OUR ENVIRONMENT

ROCK DEFORMATION AND OIL POOLS

Most of the world's largest oil pools are where they are because of the way rocks deform. "Oil pool" is a somewhat confusing term because it is not a pool in the usual sense of a "lake"; the term actually refers to a body of rock in which oil occupies the pore spaces.

Oil and natural gas occur together, and for a pool to form, five essential requirements must be met:

1. A *source rock* rich in organic matter, such as shale, must provide the oil. Most shales contain some organic matter from dead plants and animals and can serve as source rocks. Conversion of organic matter to oil and gas happens spontaneously when shale is buried, and, in response to the geothermal gradient, the temperature of the source rock rises.

2. A permeable and porous *reservoir rock* must be present so that the oil can percolate in from the source rock.

3. Oil floats on water, and so oil in a reservoir rock will float upward to the water table unless the reservoir rock is covered by an impermeable *roof rock*.

4. Like groundwater in an aquifer, oil will move laterally in a reservoir rock and eventually escape. An important requirement for oil-pool formation is that the reservoir and roof rocks must form a trap that can hold the oil and prevent it from being flushed out by groundwater. As shown in Figure B9.1, traps are structural—either anticlines or faults—or stratigraphic—either unconformities or changes in sedimentary facies. Structural traps are far more common than stratigraphic traps.

5. The final, and in some ways most important, requirement for pool formation is that the deformation that forms a trap must occur before all the oil and gas has escaped from the reservoir rock. This is important because conversion of organic matter to oil occurs soon after a shale is buried. If the trap were formed long after oil formation, no oil pool would form.

The great oil pools of Saudi Arabia, Kuwait, Iraq, and other parts of the Middle East are in structural traps. The deformation that produced the traps was caused by plate-tectonic movements. At the right moment for oil trapping, the sedimentary rocks in the Middle East were deformed by compressional stresses due to Africa's colllision with Asia.

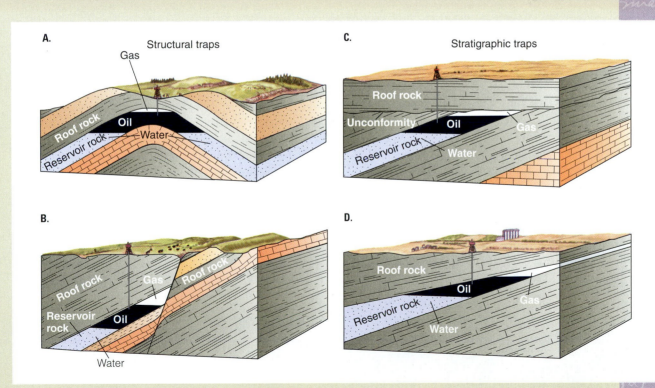

Figure B9.1   Oil Traps   Four types of oil traps. A, B, structural traps; C, D, stratigraphic traps. Folds (part A) are traps formed by ductile deformation. They are the most important of all oil traps. Faults (part B) are traps formed by fracturing of brittle rocks. In C an unconformity marks the top of the reservoir; in D a porous stratum (reservoir) thins out and is overlain by an impermeable roof rock. Gas overlies oil, which floats on groundwater, saturates the reservoir rock, and is held down by a roof of shale. Oil fills only the pore spaces in the rock.

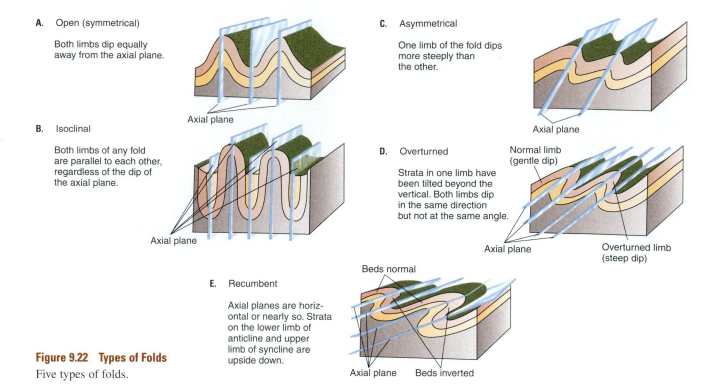

A. Open (symmetrical)

Both limbs dip equally away from the axial plane.

Axial plane

B. Isoclinal

Both limbs of any fold are parallel to each other, regardless of the dip of the axial plane.

Axial plane

C. Asymmetrical

One limb of the fold dips more steeply than the other.

Axial plane

D. Overturned

Strata in one limb have been tilted beyond the vertical. Both limbs dip in the same direction but not at the same angle.

Normal limb (gentle dip)

Axial plane

Overturned limb (steep dip)

E. Recumbent

Axial planes are horizontal or nearly so. Strata on the lower limb of anticline and upper limb of syncline are upside down.

Beds normal

Axial plane   Beds inverted

Figure 9.22   Types of Folds
Five types of folds.

stratum is right-side up or upside down in order to decide which limb of a fold it is in. Figuring this out is not always possible, but in some cases sedimentary structures, such as mud cracks and graded layers, do record whether strata are in their original orientation or inverted (Figure 11.2). In other examples, only careful, thorough mapping of all bedrock exposures can provide the answer.

RELATIONSHIP BETWEEN FOLDS AND FAULTS

Folds and faults do not continue forever. Faults tend to die out as folds. Note in Figure 9.13 that the hinge fault dies out as a small fold. Folds die out by becoming smaller and smaller wrinkles, in much the same way as wrinkles in a bedsheet die out.

When two kinds of rock are subjected to the same stresses, one kind may be brittle and deform by fracture, and the other by ductile deformation. Certain monoclines form as a result of such differences. Monoclines commonly result from movement on a fault that causes flat-lying, ductile strata to bend as shown in Figure 9.23.

Some of the great thrust faults in the Alps probably started as recumbent folds. As depicted in Figure 9.24, when stress continues to build up and forms a recumbent fold, the overturned limb may become so stretched and strained that it eventually breaks and becomes a thrust fault. Movement on some of the great recumbent-fold and thrust-fault structures in the Alps is in excess of 50 km.

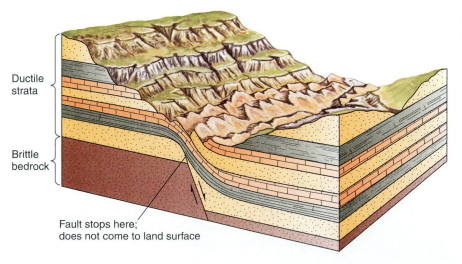

Ductile strata

Brittle bedrock

Fault stops here; does not come to land surface

Figure 9.23   Faulting Causes a Monocline to Form   Monocline developed in soft, ductile strata due to movement on a fault in the underlying bedrock.

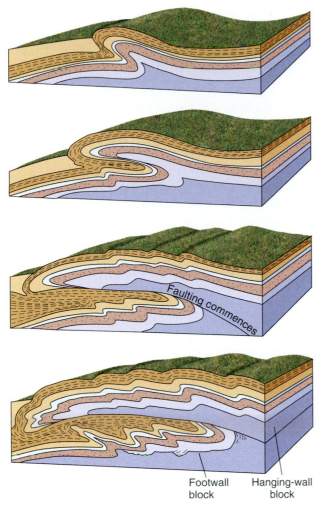

Figure 9.24 **Recumbent Fold** The evolution of a recumbent fold into a thrust fault. This kind of structure is important in the Alps and many other mountain ranges.

FOLDS AND TOPOGRAPHY

Differences in resistance to erosion by adjacent strata can lead to distinctive topographic forms, and commonly these topographic forms reveal the presence of folds. It is important to remember, however, that anticlines do not necessarily make ridges or synclines valleys, as Figure 9.25 shows. A particularly striking example of the relationship between plunging folds, erosion, and topography can be seen in the Valley and Ridge province of Pennsylvania (Figure 9.26). There, a series of plunging anticlines and synclines were created during the Paleozoic Era by a continental collision between North America, Africa, and Europe. The Appalachian mountain chain was the result. Now deeply eroded, the folded rocks determine the pattern of the topography because soft, easily eroded strata (shales) underlie the valleys, while resistant strata (sandstones) form the ridges.

Before you go on:

1. Suppose you were working in the African Rift Valley in Kenya. How would you try to determine the direction of motion on the faults that bound the valley?

2. What happens to faults as they proceed downward through the lithosphere?

3. What are your four choices of places to go to see giant recumbent folds and thrust faults of the kind seen in the Alps?

Figure 9.25 **Folding Revealed by Topography** Distinctive topographic forms and patterns resulting from differing resistance to erosion of different kinds of rock reveal the presence of plunging folds. A. Block diagram showing topographic effects. Note that resistant strata (layers 3 and 5) make topographic highs in both anticlines and synclines, while easily eroded strata (layers 2 and 4) make topographic lows in both anticlines and synclines. B. Geologic map of area shown in Part A. Compare with Figure 9.26.

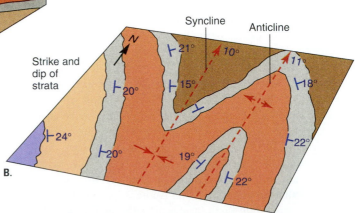

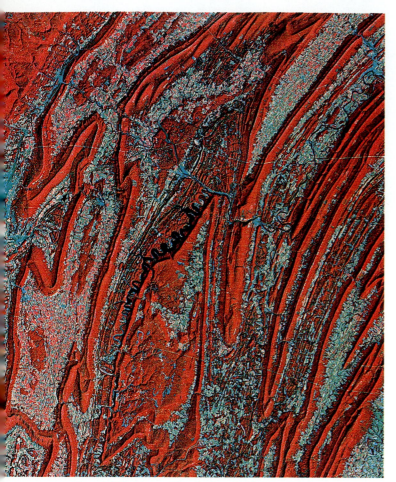

Figure 9.26 Scars of an Ancient Collision   Strata, once horizontal, are folded and contorted into plunging anticlines and synclines as a result of collisions during the assembly of Pangaea. These eroded roots of a once much grander mountain range are today's central Appalachians in Pennsylvania.

REVISITING PLATE TECTONICS AND THE EARTH SYSTEM

STRIKE-SLIP FAULTS

Say the words *strike-slip fault* and the San Andreas Fault in California comes immediately to mind. The San Andreas Fault is so well known because two large cities, Los Angeles and San Francisco, are so close to the fault they are frequently rattled by earthquakes when movement occurs along the fault. Even though they are not so well known as the San Andreas, there are many other strike-slip faults, some active, others dormant. One active fault, called the Alpine Fault, is part of the boundary between the Pacific plate and the Australian–Indian plate, and slices through the south island of New Zealand. Motion on the Alpine Fault, like that on the San Andreas, is right lateral. The North Anatolian Fault, also with right-lateral motion,

slices through Turkey in an east–west direction, and is the cause of many dangerous earthquakes (Figure 2.18).

Movement can shatter rock adjacent to a fault. Because shattered rock erodes more rapidly than solid rock, strike-slip faults are often marked on Earth's surface by long, thin valleys. One such valley marks the trace of the Great Glen Fault of Scotland. The Great Glen Fault was active during the Paleozoic Era, and is now thought to be dormant, though of course it is a zone of weakness and could again become active if disturbed by tectonic stresses. The Great Glen Fault slices through Scotland in a northeast–southwest direction, and in the valley that marks its trace lies Loch Ness, home to the presumably mythical Loch Ness monster.

TECTONISM AND ITS EFFECT ON CLIMATE

The fact that a landscape influences the local climate is almost too obvious to need mention. Anyone who has climbed or driven across a mountain range can hardly fail to be aware of a decrease in temperature with altitude, of a change in plant life, and, if high enough, a decrease in air pressure.

Mountains are the result of tectonic processes. Great fold mountain ranges such as the Alps in Europe, the Himalaya in Asia, and the Sierra Nevada in North America are the result of compressive stresses and associated magmatic activity generated by plate motions. The Sierra Nevada provides a classic example of the influence of a range on local climate. Along the west coast of North America moisture-laden winds flow inland from the sea. The Sierra run north–south, and impose a topographic barrier to flow that forces the winds upward, causing them to precipitate rain and snow on the western slopes. When the winds flow down the eastern side of the mountains, most of the moisture has been stripped from them. The lands east of the Sierra are in a rain shadow, and the result is the dry lands, in places so dry they are deserts, of Nevada and Utah.

What's Ahead?

Earth is a very dynamic planet. Things may change slowly but they change continually. Along plate boundaries, as you read these lines, rocks in the lithosphere are being folded and fractured, while deeper down, in the asthenosphere, rock is slowly moving by ductile flow. Much of the deformation, especially the ductile deformation, is difficult to detect from Earth's surface, but fracturing and movement along faults releases elastically stored energy, and if the movement is sudden, an earthquake is the result. In the next chapter, therefore, we explore the topic of earthquakes and what we can learn from them about the Earth's interior.

CHAPTER SUMMARY

1. Rocks can be deformed in three ways: by elastic deformation (no permanent change), by ductile deformation (folds), and by fracture (faults, joints).

2. High confining stress and high temperatures enhance ductile properties. Low temperatures and low confining stress enhance elastic properties and failure by fracture when the elastic limit is exceeded.

3. The rate at which a solid is deformed (strained) also controls the style of deformation. High strain rates lead to fractures; low strain rates lead to folding.

4. Weak rocks (limestone, marble, slate, phyllite, and schist) enhance ductile properties. Strong rocks (sandstone, quartzite, and granite) enhance brittle properties. Dry rocks are stronger than wet rocks.

5. The orientations of contacts between strata, faults, joints, or any other inclined plane surfaces are determined by the strike (direction of intersection of the inclined plane and a horizontal plane) and the dip (angle between the inclined plane and the horizontal).

6. Fractures in rocks along which slippage occurs are called faults.

Normal faults are caused by tensional stresses that tend to pull the crust apart, whereas thrust and other types of reverse faults arise from compressional stresses that squeeze, shorten, and thicken the crust. Strike-slip faults are caused by shear stresses; they are vertical fractures that have lateral motion.

7. It is usually possible to determine only relative motions of rocks on either side of a fault.

8. Ductile deformation of strata causes bends or warps, which are called folds. Folding is due to compressional stress. An upward, arched fold is an anticline; a downward, trough-like fold is a syncline. The sides of a fold are called limbs.

9. Overturned folds (meaning both limbs dip in the same direction) are common in mountain ranges formed by continental collision. In some ranges, such as the Alps, overturned folds have nearly horizontal limbs, in which case the folds are said to be recumbent.

10. Faults die out by becoming folds, and folds die out by becoming smaller and smaller wrinkles.

THE LANGUAGE OF GEOLOGY

anticline (p. 240)
axis (of a fold) (p. 240)
compressional stress (p. 227)
confining stress (p. 227)
dip (p. 232)
ductile deformation (p. 227)
elastic deformation (p. 227)
elastic limit (p. 227)

fault (p. 231)
fold (p. 240)
footwall block (of a fault) (p. 234)
fracture (p. 227)
graben (p. 235)
half-graben (p. 235)
hanging-wall block (of a fault)
 (p. 234)

horst (p. 235)
limbs (of a fold) (p. 240)
monocline (p. 240)
normal faults (p. 235)
plunge (of a fold) (p. 240)
reverse fault (p. 236)
rift (p. 235)
shear stress (p. 227)

strain (p. 227)
strain rate (p. 230)
strike (p. 232)
strike-slip fault (p. 237)
syncline (p. 240)
tensional stress (p. 227)
thrust fault (p. 236)
transform faults (p. 237)

QUESTIONS FOR REVIEW

1. What are stress and strain, and what is the relationship between them?

2. Discuss the ways a seemingly rigid solid, such as a rock, can be deformed. In what sequence do deformation properties come into play?

3. What properties determine whether a rock is brittle and fails by fracture or is deformed by ductile deformation?

4. Name three rock types that readily deform by ductile deformation and three that tend to deform by fracture.

5. Identify three prominent topographic features on Earth made up of a system of grabens and horsts. What kind of stresses cause horsts and grabens to form?

6. Describe the way a transform fault works. Why are they called *transform* faults? Cite an example of a transform fault that is still active.

7. What kind of stresses produce folds? Draw a sketch of an anticline and mark the axis, axial plane, and limbs.

8. How do folds and faults die out?

9. What is a recumbent fold and where would you expect to find large recumbent folds? Describe the way a recumbent fold can become a thrust fault.

10. Draw a geologic map of a plunging anticline. Mark strike and direction of dip at several places around the fold. Draw the fold axis and indicate, if you can, the direction of plunge.

Click on *Presentation* and *Interactivity* in the **Plate Tectonics** and **Metamorphic Rocks** modules of your CD-ROM to further explore resources and activities presenting concepts from this chapter. Select *Assessment* in the same module to text your understanding of this chapter.

Chapter 10

A portion of the Nimitz Freeway in Oakland, California, which collapsed in the 1989 Loma Prieta earthquake. Seismologists discovered that a match between the natural resonance periods of the double-level freeway and the underlying sediments amplified the shaking experienced by the freeway structure, increasing its vulnerability.

Earthquakes and Earth's Interior

House shaken off its foundation, Marina District, San Francisco.

A Temblor Strikes Home

I first heard the news in a supermarket. A cashier had the baseball World Series on TV, the first ever cross-bay match between the San Francisco Giants and the

247

Oakland A's. The players were not playing, just looking around puzzled and frightened. An earthquake had struck nearby, deep beneath the Santa Cruz Mountains. People worried for their lives, not for strikeouts and stolen bases.

My sister Jennifer was unlucky. Her house lay directly over the point where, 20 kilometers down, a sudden rock fracture began the earthquake in the late afternoon of October 18, 1989. Fault rupture raced upward toward the surface at nearly 3 kilometers per second, but halted short of the surface. No geologist would later find the fault at the surface, just cracks, broken roads, and landslides in the mountains. Deep in the crust, the faces of the San Andreas Fault scraped a meter or more past each other over an area of 200 square kilometers. Violent shaking froze my sister's kitchen clock at 5:04, never to run again. Older buildings in downtown Santa Cruz collapsed, and dozens of workers suffered disabling injuries.

Sometimes people are unaccountably cheerful after surviving a natural disaster. My sister and her husband pitched a tent on their front lawn to sleep in. A month later I toured the local devastation with them. Scores of wooden houses had been sheared off their concrete foundations, but could be easily slid back in place. Downtown Santa Cruz would not be rebuilt for several years. Central cities are often built on coastal plains, sediment-filled river valleys, and landfill, all of which amplify seismic vibrations. San Francisco and Oakland are 90 km distant from Santa Cruz, but suffered greater damage and more casualties. Sediment-amplified shaking took down the Marina residential district and collapsed the two-level Nimitz Freeway. In earthquake country, geology and natural hazard risk are closely linked.

Jeffrey Park

KEY QUESTIONS

1. **How do we estimate earthquake size, location, and character?**

2. **Where are earthquake hazards a significant concern for public policy?**

3. **How can we use seismic waves to probe the internal properties of Earth?**

4. **What can we predict about future earthquakes?**

5. **What does the presence or absence of earthquake activity tell us about rock deformation in the crust and mantle?**

6. **What role do earthquakes play in the Earth system?**

INTRODUCTION

When Earth quakes, the energy stored in elastically strained rocks is suddenly released. The more energy released, the stronger the quake. Do an experiment yourself. Have a friend hit one end of a wooden plank or the top of a wooden table with a hammer while you press your hand on the other end. You will feel vibrations set up in the plank or tabletop by the energy of the blow. The harder the blow, the stronger the vibrations. You can feel those vibrations because some of the energy imparted by the hammer is transferred to your hand by elastic waves through the whole length of the solid wood.

Although gigantic hammers don't pound our planet from above, a bomb blast or a violent volcanic explosion will serve just as well. More commonly, massive bodies of rock slip along fault surfaces deep underground. Most of the time, rocks break apart and slip on faults in a brittle response to the forces of plate tectonics. Earthquakes are key indicators of plate motion, as well as a key mechanism for causing uplift of Earth's surface. Earthquakes are also a major natural hazard to lives and property, responsible for $3–4 billion in damage on average per year in the United States alone. Because large earthquakes occur infrequently, these yearly average costs underestimate the economic and psychological devastation that can afflict regions where earthquake risk is high.

HOW EARTHQUAKES ARE STUDIED

The name given to the study of earthquakes is **seismology**, a word that comes directly from the ancient Greek term for earthquakes, *seismos*.

SEISMOMETERS

The device used to record the shocks and vibrations caused by earthquakes is a **seismometer**. An ideal way to record the vibrations and motions of Earth would be to put the seismometer on a stable platform that is not affected by the vibrations. But a seismometer must stand on Earth's vibrating surface, and it will therefore vibrate with the surface. This makes the act of measurement difficult because there is no fixed frame of reference against which to make measurements. The problem is the same one that a sailor in a small boat faces when attempting to measure waves at sea.

To overcome the frame-of-reference problem, all seismometers make use of **inertia**, which is the resistance of a stationary mass to sudden movement. If you suspend a heavy mass, such as a block of iron, from a light spring and then suddenly lift the spring, you will notice that the block remains almost stationary while the spring stretches (Figure 10.1). The block's inertia hampers its movement. This is the principle used in *inertial seismometers*. Vertical motion is measured by a device similar to the one in Figure 10.2A. When the ground vibrates, the spring expands and contracts but the mass remains almost stationary. The distance between the ground and the mass can be used to sense vertical displacement of the ground surface.

Horizontal displacement can be similarly measured by suspending a heavy mass from a wire to make a pendulum (Figure 10.2B). Because of its inertia, the mass does not keep up with the horizontal ground motion, and the difference between the pendulum and ground movement records the horizontal ground motion. Three inertial seismometers are commonly used in one instrument housing to measure up–down, east–west, and north–south motions simultaneously.

MODERN SEISMOMETER DESIGN

Modern seismometers have improved on the mechanical designs shown in Figure 10.2 by the use of sophisticated electronics. Seismic motion caused by earthquakes varies greatly, from the smallest tremors to the largest earthquakes of the twentieth century (see *The Science of Geology*, Box 10.1, *Calculating a Richter Magnitude*). In order to record the full range of seismic motion without distorting their springs, the most sensitive seismometers use electronic feedback to move the inertial mass with the rest of the seismometer. Rather than measure the motion of the mass relative to the vibrating ground, the seismometer measures the electric current needed to make the mass and ground move together.

Because modern seismic measurement is electronic rather than mechanical, motion can be measured with remarkable accuracy. Digital audio on music CDs plays sound vibrations with 16-bit accuracy. This allows roughly

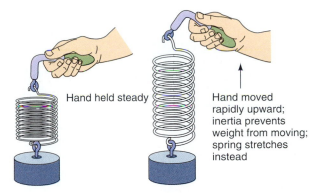

Figure 10.1 How a Seismometer Works The mechanical principle of the inertial seismometer.

a factor of 32,000 between the loudest and softest sounds. Modern seismometers can record 24-bit Earth vibrations, which allows roughly a factor of 8,300,000 (8.3×10^6) between the largest and smallest recorded signals. Vibrational movements as tiny as one-hundred-millionth (10^{-8}) of a centimeter can be detected. Indeed, many seismometers are so sensitive they can detect the effects of ocean waves and tides many kilometers from the seashore.

EARTHQUAKE FOCUS AND EPICENTER

The **earthquake focus** is the point where an earthquake starts to release the elastic strain of surrounding rock. An earthquake focus lies at some depth below Earth's surface. The **epicenter** is the point on Earth's surface that lies vertically above the focus of an earthquake (Figure 10.3). A convenient way to describe the location of an earthquake focus is to state the location of its epicenter and its depth.

Fault slippage begins at the focus and spreads across a fault surface in a *rupture front*. The rupture front travels at roughly 3 kilometers per second for earthquakes in the

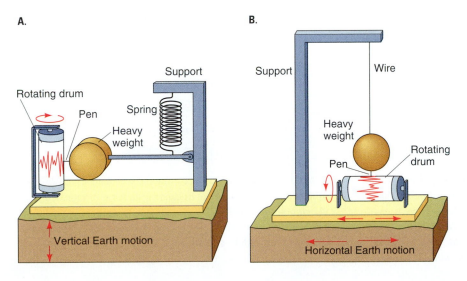

Figure 10.2 Recording Earthquake Activity Seismometers measure vibrations sent out by earthquakes. A. An inertial seismometer for measuring vertical motions. B. An inertial seismometer for measuring horizontal motions.

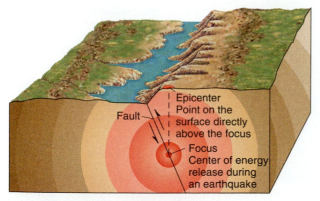

Figure 10.3 Location of an Earthquake The focus of an earth-quake is the site of first movement on a fault and the center of energy release. The epicenter of an earthquake is the point on Earth's surface that lies vertically above the focus.

crust. Because large earthquakes can rupture faults for 100 kilometers or more, it can take 30 seconds or more for all slippage to occur.

SEISMIC WAVES

When an earthquake occurs, energy is transmitted to other parts of Earth. Vibrational waves spread outward initially from the focus, and continue to radiate from elsewhere on the fault as rupture proceeds. The waves, called **seismic waves**, spread out in all directions, just as sound waves spread in all directions when a gun is fired.

There are several kinds of seismic waves, and they belong to two families. **Body waves** can transmit either compressional motion (P waves) or shear motion (S waves), corresponding to the two ways a solid can be elastically deformed. **Surface waves**, on the other hand, are vibrations that are trapped near Earth's surface, similar to the way that light waves can be trapped within a fiber-optic cable. There are two types of surface waves: *Love waves* and *Rayleigh waves*. Body waves are analogous to light or sound waves in clear air, both of which travel outward in all directions from their point of origin. Surface waves are analogous to ocean waves and are restricted to the vicinity of Earth's surface, including the seafloor in oceanic regions.

BODY WAVES

Rocks can be elastically deformed by either a change of volume or a change of shape. The first kind of body wave, a *compressional wave*, deforms rocks largely by change of volume and consists of alternating pulses of contraction and expansion acting in the direction of wave travel (Figure 10.4A). Compressional waves are the first waves to be recorded by a seismometer after an earthquake, so

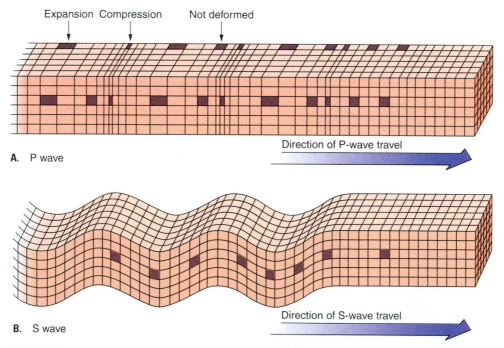

Figure 10.4 Seismic Body Waves Seismic body waves of the P (compressional) and S (shear) types. A. P waves cause alternate compressions and expansions in the rock the wave is passing through. An individual point in a rock will move back and forth parallel to the direction of P-wave propagation. As wave after wave passes through, a square will repeatedly expand to a rectangle, return to a square, contract to a rectangle, return to a square, and so on. B. S waves cause a shearing motion. An individual point in a rock will move up and down, perpendicular to the direction of S-wave propagation. A square will repeatedly change to a parallelogram, then back to a square again.

Box 10.1

THE SCIENCE OF GEOLOGY

CALCULATING A RICHTER MAGNITUDE

In 1935 Charles Richter at the California Institute of Technology devised a convenient calibration scheme, now called the Richter scale, for describing the size of an earthquake. Designed to be easily communicated to the news media, Richter used logarithmic scaling to keep the magnitudes of significant earthquakes between the numbers zero and 10. As earthquakes increase in size, they occur over larger faults, require more time to rupture, and therefore generate seismic waves of longer period. Thus, the energy released by an earthquake affects both the amplitude X and the oscillation period T of the P wave. Richter's original magnitude formula depends on the logarithm of the ratio X/T. One unit of Richter magnitude corresponds to a tenfold increase in X/T as measured by a seismometer. Richter developed correction factors to adjust for the distance between the earthquake and the seismometer, and for the geologic properties of California. Later seismologists devised more general magnitude estimates for worldwide use, based either on P waves (T near 1 s), surface waves trapped in the crust ($T = 20$ s), or surface waves trapped in the uppermost mantle ($T = 200$ s).

The Richter scale is still used to describe earthquake size, but the method for estimating earthquake magnitude has lately been connected more directly to motions on a fault, rather than to motions on a seismometer. In 1977 Hiroo Kanamori, also at the California Institute of Technology, proposed a relation between Richter magnitude M and **seismic moment** M_0 that is used today by all seismologists. The moment M_0 is directly related to the mechanical energy released by fault motion. The seismic moment can be computed from the formula $M_0 = \mu A D$, where μ is the shear stiffness of rock surrounding the fault, A is the area of the fault, and D is the average slip during the earthquake. D and A are expressed in meters and square-meters, respectively. μ is expressed in newtons per square-meter, that is, force per unit area. Seismic moment has the units of newton-meters, equivalent to joules, the metric unit of energy. In these units, Kanamori's relation between moment and magnitude is

$$M = (2/3)\log_{10}M_0 - 6.0$$

Most crustal rocks have shear stiffness μ near 3×10^{10} newtons per square meter. An earthquake that slips $D = 3$ meters on a vertical fault 50 km long that extends from the surface to 15 km depth will have a seismic moment (Figure B10.1)

$$M_0 = (3 \times 10^{10} \text{ nt/m}^2)(50{,}000 \text{ m})(15{,}000 \text{ m})(3 \text{ m})$$

$$= 6.75 \times 10^{19} \text{ joules}$$

Using the moment-magnitude formula, this earthquake would have Richter magnitude $M = 7.2$.

Seismologists cannot measure the fault area and earthquake slip directly. However, sophisticated computer programs can generate *synthetic seismograms* to match observed data. The amplitude of a computer-generated seismogram is proportional to seismic moment M_0, so one can estimate M_0 by matching the observed seismic data to computer-generated vibrations.

According to Kanamori's formula, the increase in seismic moment corresponding to one Richter scale unit is a factor of $10^{1.5} = 31.6$, roughly a thirtyfold increase. This increase could occur if the fault area A were 10 times larger and the slip distance D were 3.16 times larger. Taking this comparison a step further, the difference in energy released between an earthquake of magnitude $M = 4$ and one of magnitude $M = 7$ is $(10^{1.5})^3 = 10^{4.5} \approx 31.6 \times 31.6 \times 31.6 \approx 31{,}600$ times! How big can earthquakes get? The largest earthquake in the twentieth century had a Richter magnitude of about 9.5, breaking along an 800-km-long fault in 1960 in the subduction zone beneath Chile in South America. The average fault slip was estimated to be 25 meters. It is theoretically possible for larger earthquakes to occur, but there are few potential locations for such enormous events. Earth's plate boundaries do not have many segments that extend 1000 km or more without interruption of some kind.

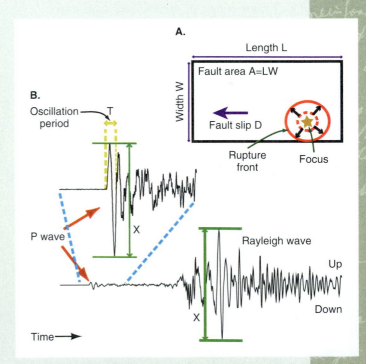

Figure B10.1 Seismic Moment and Richter Magnitude
A. Schematic representation of parameters used to interpret seismic moment. B. Measurements used for determining Richter magnitude (*M*) from seismic waves.

they are called **P (for primary) waves**. The second kind of body wave is a *shear wave*. Shear waves deform materials by change of shape. Because shear waves are slower than P waves and reach a seismometer some time after a P wave arrives, they are called **S (for secondary) waves**.

Compressional (P) waves can pass through solids, liquids, or gases. Sound waves are P waves. Movement in a compressional wave is back and forth in the direction that the wave travels. P waves have the greatest speed of all seismic waves: 6 km/s is typical for the crust, 8 km/s is typical of the uppermost mantle.

Shear (S) waves consist of an alternating series of sidewise movements, each particle in the deformed solid being displaced perpendicular to the direction of wave travel (Figure 10.4B). Gases and liquids will not rebound elastically to their original shape after such a distortion. Therefore, shear waves can travel only within solid matter. A typical speed for a shear wave in the crust is 3.5 km/s, 5 km/s in the uppermost mantle.

Seismic body waves, like light waves and sound waves, can be *reflected* and *refracted* by changes in material properties (Figure 10.5). Reflection is the familiar phenomenon of light bouncing off the surface of a mirror or a glass of water. Seismic body waves are reflected by numerous surfaces within Earth where material properties change abruptly. Refraction is less familiar in everyday life, but for seismic waves it occurs in tandem with reflection. At any change in material properties that involves a change in wavespeed, refraction bends the direction of wave travel. When a ray of light in air strikes a water surface, some of the light is reflected, but some also crosses the surface and travels through the water. The speed of light is less in water than in air, and the wave path is sharply bent, or refracted, at the surface.

For seismic waves within Earth, the changes in wavespeed and wave direction can be either gradual or abrupt, depending on changes in chemical composition, pressure, and mineralogy. If Earth had a homogeneous composition and mineralogy, rock density and wavespeed would increase steadily with depth as a result of increasing pressure. For this case, refraction would curve seismic wave paths into smooth arcs. Measurements show that wave paths are indeed curved, owing to gradual refraction. In addition, measurements reveal that the seismic waves are also refracted and reflected by several abrupt changes in wavespeed. The largest such change occurs at the transition from silicate rock to liquid iron at the core–mantle boundary (Figure 10.6).

SURFACE WAVES

Surface waves travel more slowly than P and S waves, but are often the largest vibrational signals in a seismogram. Surface waves lag behind because they are trapped near Earth's surface, where wavespeeds are lower, and because they travel around Earth's surface rather than through

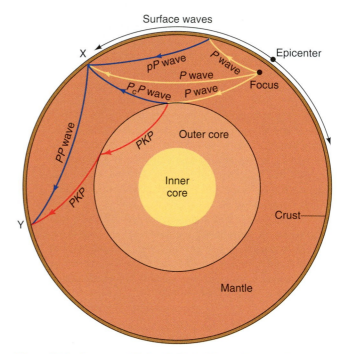

Figure 10.6 Passage of Seismic Body Waves Through Earth
Various paths of P waves in Earth, which is a compositionally layered planet. Within the mantle and core, seismic wavespeed increases gradually, so that wave paths curve as they refract. At the core–mantle boundary, seismic waves reflect and refract at sharp angles. Seismometers at some places (locations X and Y, for example) receive both direct P waves (yellow) as well as reflected (blue) and refracted (red) P waves. The P wave reflected off Earth's surface and observed at location Y is called a PP wave; the "pegleg" reflected wave observed at location X is called a pP wave.

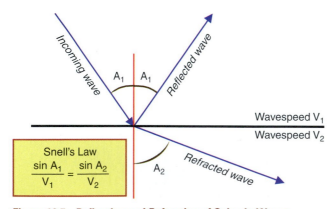

Figure 10.5 Reflection and Refraction of Seismic Waves
Seismic waves reflect at flat interfaces between different materials. Reflection preserves the angle between vertical and horizontal, as shown in the diagram. Seismic waves refract toward the horizontal as the seismic wavespeed increases. The angle of the refracted wave can be computed from Snell's Law, as shown in the diagram.

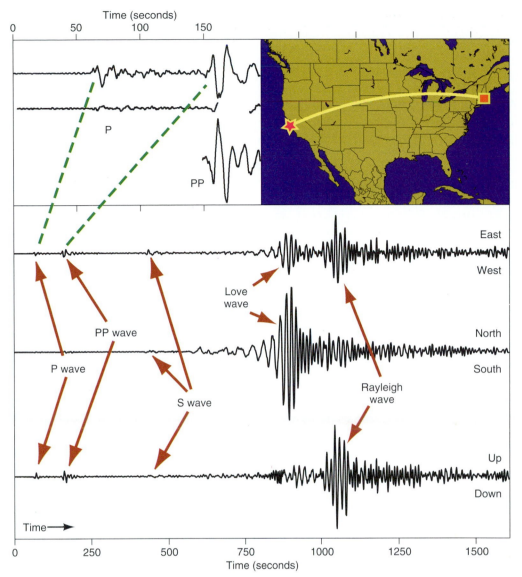

Figure 10.7   Seismic Waves in Three Dimensions   Different directions of motion on a seismometer record different waves. An earthquake (star) in California approaches a seismic observatory (square) in the northeast United States from the west. The up–down and east–west sensors record P wave and Rayleigh surface wave motion. The north–south sensors record S and Love-surface wave motion. Note the dispersion of vibrational periods in the Love and Rayleigh waves. Long-period surface waves arrive first; short-period surface waves arrive later.

Earth's interior. At the same time, surface waves spread out in two dimensions (Earth's surface) rather than three dimensions (Earth's interior), which helps explain why surface wave signals are larger than body wave signals in most cases. Love waves consist entirely of shear vibrations in the horizontal plane, analogous to an S wave that vibrates horizontally. Rayleigh waves combine shear and compressional vibration types, and involve motion in both the vertical and horizontal directions (Figure 10.7).

One important difference between body waves and surface waves concerns a property called dispersion. Body waves travel through Earth at the same velocity regardless of wavelength, but this is not true for surface waves. The longer the wavelength of a surface wave, the deeper the wave motion penetrates Earth. Because Earth is compositionally layered and pressure increases with depth, long-wavelength surface waves penetrate to deeper, denser, higher-velocity zones, while short-wavelength surface waves do not. In this way, surface waves of different wavelengths develop different velocities. This behavior is called *dispersion*. Because some wavelengths reach a depth of several hundred kilometers, dispersion of surface waves is a very important way to determine Earth's properties in the lithosphere and asthenosphere.

DETERMINING THE EPICENTER

An earthquake's epicenter can be located from the arrival times of the P and S waves at a seismometer. The farther a seismometer is away from an epicenter, the greater the time difference between the arrival of the P and S waves (Figure 10.8). After using a graph like that shown in Figure 10.8B to determine how far an epicenter lies from a seismometer, one draws a circle on a map around the station with a radius equal to the calculated distance to the epicenter. The epicenter can be determined when data from three or more seismometers are available—it lies where the circles intersect (Figure 10.9).

The depth of an earthquake focus below an epicenter can also be determined. If a local earthquake is recorded by several nearby seismometers, the focal depth can be determined in the same way that the epicenter is determined, by using P–S time intervals. For distant earthquakes a different method is employed. Note in Figure 10.6 that a seismometer stationed at position X would record both a direct P wave and a P wave reflected from Earth's surface (labeled pP). The direct P wave, having a shorter path length, arrives before the pP wave. The travel-time difference between P and pP is a measure of the focal depth of the earthquake.

EARTHQUAKE MAGNITUDE AND FREQUENCY

In principle, measurements of elastically deformed rocks before an earthquake, and of undeformed rocks after an earthquake, could provide an estimate of the amount of energy released. Such measurements are impractical in most cases, however. Therefore, seismologists have developed a way to estimate the energy of an earthquake by measuring the amplitudes of seismic waves. The **Richter magnitude scale** is divided into steps called magnitudes with numerical values M. Each step in the Richter scale, for instance, from magnitude $M = 2$ to magnitude $M = 3$, represents approximately a thirtyfold increase in earthquake energy. To see how a Richter magnitude M is calculated, see Box 10.1.

In earthquake-prone regions like Southern California, earthquakes of magnitude $M = 3$ are almost a daily occurrence, but their modest vibrations cause little or no damage. An earthquake with $5 < M < 6$ can cause significant

A.

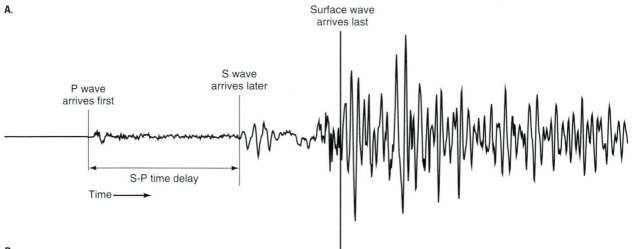

Surface wave arrives last

S wave arrives later

P wave arrives first

S-P time delay

Time ⟶

B.

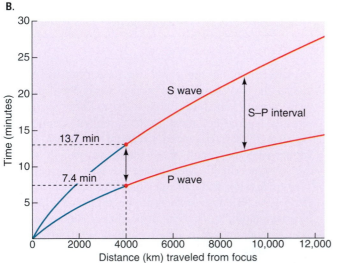

Figure 10.8 Travel Times of Seismic Waves Different travel times of P, S, and surface waves. A. Typical record made by a seismometer. The P and S waves leave the earthquake focus at the same instant. The fast-moving P waves reach the seismometer first, and some time later the slower-moving S waves arrive. The delay in arrival times increases with the distance traveled by the waves. The surface waves travel more slowly than either P or S waves. B. Average travel-time curves for P and S waves, used to locate an epicenter. For example, when seismologists at a station measure the S–P time interval to be 13.7 min – 7.4 min = 6.3 min, they know the epicenter is 4000 km away from their station.

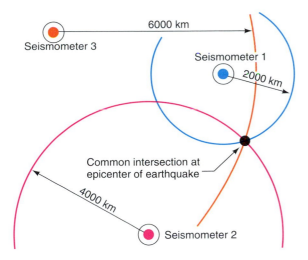

Figure 10.9 Locating Site of an Earthquake Locating an epicenter. The effects of an earthquake are felt at three different seismometer stations. The time differences between the first arrival of the P and S waves depend on the distance of a station from the epicenter. The following distances are calculated by using the curves in Figure 10.8B.

	Time Difference	Calculated Distance
Seismometer 1	8.8 min – 4.7 min = 4.1 min	2000 km
Seismometer 2	13.7 min – 7.4 min = 6.3 min	4000 km
Seismometer 3	17.5 min – 9.8 min = 7.7 min	6000 km

On a map, a circle of appropriate radius is drawn around each station. The epicenter is where the three circles intersect.

property damage if it occurs close to a population center. Each year there are roughly 200 earthquakes worldwide with magnitude $M = 6.0$ or higher. Larger earthquakes are more rare. Each year on average there are 20 earthquakes with $M = 7.0$ or larger, and one "great" earthquake with $M = 8.0$ or larger. Seismologists at the U.S. Geological Survey estimate that four earthquakes in the twentieth century met or exceeded magnitude 9.0: 1952 in Kamchatka ($M = 9.0$), 1957 in the Aleutian Islands ($M = 9.1$), 1964 in Alaska ($M = 9.2$), and 1960 in Chile ($M = 9.5$).

To appreciate the energy associated with earthquakes, we can compare them with underground nuclear explosions, which are the largest man-made seismic events. A fission-bomb explosion comparable in size to the nuclear bomb dropped in 1945 on the Japanese city of Hiroshima generates seismic waves about as large as an earthquake of magnitude $M = 5.3$. One of the last nuclear tests of the twentieth century was a thermonuclear, or fusion-bomb, test in China in May 1992, with $M = 6.6$. Thus, even the most destructive man-made devices are small in comparison with the largest earthquakes.

EARTHQUAKE HAZARD

Seismic events are most common along plate boundaries. In the United States there are plate boundaries at or near the Pacific coastlines of California, Oregon, Washington, and Alaska. In addition, the U.S. island territories of Guam, Puerto Rico, and the Virgin Islands lie along plate boundaries. The last earthquake with $M = 8.0$ to strike the United States occurred near Guam on August 8, 1993. Earthquakes associated with hot spot volcanism pose a hazard to Hawaii. Earthquakes are common in much of the intermontane western United States, primarily Nevada, Utah, and Idaho, because east–west stretching is occurring as a byproduct of tectonic uplift (see Chapter 12).

Not all large earthquakes involve plate boundaries, however. Several large earthquakes jolted central and eastern North America in the nineteenth century. Near New Madrid, Missouri, three earthquakes of great size occurred on December 16, 1811, and January 23 and February 7, 1812. The exact magnitude of these earthquakes is unknown because seismometers did not exist at the time. However, judging from the local damage caused, as well as the fact that tremors were felt and minor damage occurred as far away as New York and Charleston, South Carolina, the most recent estimates place their magnitudes M between 7.0 and 7.7. On September 1, 1886 an earthquake with $M \approx 7$ devastated Charleston, South Carolina. Several events of comparable size have shaken sparsely populated portions of Quebec Province in Canada.

Based on known geologic structures (mainly faults) and on the location and intensity of past earthquakes, the U.S. Geological Survey prepares seismic-risk maps. To design roadways and public buildings, builders need to know just how strongly the ground is likely to shake and what the frequency of large earthquakes might be. To answer this need, the U.S. Geological Survey publishes a seismic-risk map of the kind shown in Figure 10.10. The strength of a possible earthquake is compared to the acceleration due to gravity, which is 980 cm/s^2 (that is to say, 1.0 G). The maximum acceleration is likely to occur with California earthquakes, reaching 80 percent of gravity (which would be 0.8 G). Damage begins at about 0.1 G.

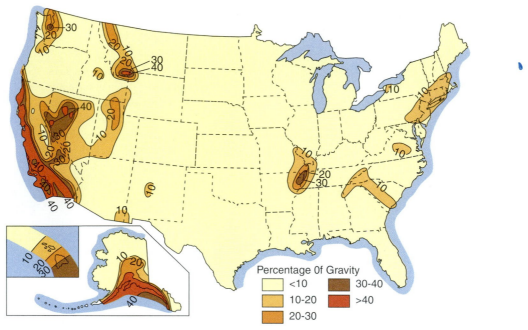

Percentage Of Gravity

☐ <10	☐ 30-40
☐ 10-20	☐ >40
☐ 20-30	

Figure 10.10 Seismic Risk and City Planning A seismic-risk map based on horizontal acceleration. Because acceleration during ground movement is the most important factor to be considered in the design of an earthquake-stable building, this type of risk map is preferred by builders and city planners. Numbers on contours refer to the maximum horizontal acceleration during an earthquake, expressed as a percentage of the acceleration due to gravity (980 cm/s^2). The probability that an acceleration of the amount indicated will occur in any given 50-year period is 1 chance in 10.

EARTHQUAKE DISASTERS

Every year, Earth experiences many hundreds of thousands of earthquakes. Fortunately, only a handful are large enough, or close enough to major centers of population, to cause catastrophic loss of life. In the industrialized Western nations, urban areas that are known to be earthquake-prone have special building codes that require structures to resist earthquake damage. However, building codes are absent or ignored in many developing nations. All too often an unexpected earthquake will devastate an urban area where buildings are not adequately constructed, as in the T'ang Shan earthquake in 1976 in China, where 240,000 people lost their lives. Other earthquakes caused 9500 deaths in 1985 in Mexico City (Figure 10.11), 25,000 deaths in 1988 in Armenia (Figure 10.12), 17,000 deaths in 1999 in Turkey, and 20,000 deaths in 2001 in Rajastan, India.

Eighteen earthquakes are known to have caused 50,000 or more deaths apiece (Table 10.1). The most disastrous one on record occurred in 1556, in Shaanxi Province, China, where an estimated 830,000 people died. Many of those people lived in cave dwellings excavated in loess (see Chapter 17), which collapsed as a result of the quake. On average since 1950, one earthquake per year worldwide has caused 1000 or more deaths. By contrast, no earthquake in the United States has been associated with as many as 200 deaths since the great 1906 San Francisco earthquake (estimated 3000 deaths).

Figure 10.11 Not Designed to Withstand an Earthquake A building that was not constructed to withstand expected earthquakes. A high-rise apartment house was one of the buildings that collapsed during the earthquake that struck Mexico City in 1985. Proper building design can minimize damage. Nearby buildings of sturdier construction withstood the shaking.

Figure 10.12   Collapse of City Buildings   When a magnitude $M = 6.8$ earthquake struck Armenia on December 7, 1988, poorly constructed buildings with inadequate foundations collapsed like houses of cards. The principal cause of collapse was ground motion amplified by soft sediments.

EARTHQUAKE DAMAGE

The dangers of earthquakes are profound, and the havoc they can cause is often catastrophic. Their effects are of six principal kinds. The first two, ground motion and faulting, are primary effects, which cause damage directly. The other four effects are secondary and cause damage indirectly as a result of processes set in motion by the earthquake.

1. Ground motion results from the movement of seismic waves, especially surface waves, through surface-rock layers, regolith, and thicker sediment cover. The motions can damage and sometimes completely destroy buildings, especially buildings on unlithified sediment, because this soft material tends to amplify seismic wave vibrations. Proper design of buildings (including such features as steel framework and foundations tied to bedrock) can do much to prevent damage from seismic shaking, but in a very strong earthquake even the best-engineered buildings may suffer some damage. Much damage from the 1985 Mexican, 1988 Armenian, and 1989 Loma Prieta earthquakes was due to ground motion amplified by surface layers of sediment.

2. Where a fault breaks the ground surface itself, buildings can be split, roads disrupted, and any feature that crosses or sits on the fault broken apart.

3. A secondary effect, but one that is sometimes a greater hazard than moving ground, is fire (Figure 10.13). Ground movement displaces stoves, breaks gas lines, and loosens electrical wires, thereby starting fires. Ground motion also breaks water mains, so that no water is available to put out fires. In the earthquakes that struck San Francisco in 1906, and Tokyo and Yokohama in 1923, more than 90 percent of the damage to buildings was caused by fire.

4. In regions of steep slopes, earthquake vibrations may cause regolith to slip, cliffs to collapse, and other rapid

TABLE 10.1   Earthquakes During the Past 800 Years That Have Caused 50,000 or More Deaths

Place	Year	Estimated Number of Deaths
Silicia, Turkey	1268	60,000
Chihli, China	1290	100,000
Naples, Italy	1456	60,000
Shaanxi, China	1556	830,000
Shemaka, USSR	1667	80,000
Catania, Italy	1693	60,000
Beijing, China	1731	100,000
Calcutta, India	1737	300,000
Lisbon, Portugal	1755	60,000
Calabria, Italy	1783	50,000
Messina, Italy	1908	160,000
Gansu, China	1920	180,000
Tokyo and Yokohama, Japan	1923	143,000
Gansu, China	1932	70,000
Quetta, Pakistan	1935	60,000
T'ang Shan, China	1976	240,000
Iran	1990	52,000

Figure 10.13 Broken Gas Lines Fire caused by gas lines broken as a result of the Loma Prieta earthquake in 1989. This photo shows a fire in the Marina district of San Francisco.

mass-wasting movements to start. This is particularly true in Alaska, parts of southern California, China, and hilly places such as Iran and Turkey. Houses, roads, and other structures are destroyed by rapidly moving regolith.

5. The sudden shaking and disturbance of water-saturated sediment and regolith can turn seemingly solid ground to a liquid mass similar to quicksand. This process is called liquefaction, and it was one of the major causes of damage during the earthquake that destroyed much of Anchorage, Alaska, on March 27, 1964 (Figure 10.14), and that caused apartment houses to sink and collapse in Niigata, Japan, in an earthquake that same year.

6. Earthquakes generate **seismic sea waves**, called *tsunami*, which have been particularly destructive in the Pacific Ocean. About 4.5 hours after a severe submarine earthquake near Unimak Island, Alaska, in 1946, a tsunami struck Hawaii. The wave traveled at a velocity of 800 km/h. Although the amplitude of the tsunami in the open ocean was less than 1 m, the amplitude increased dramatically as the wave approached land. When it hit Hawaii, the Unimak tsunami had a crest 18 m higher than normal high tide. This destructive wave demolished nearly 500 houses, damaged a thousand more, and killed 159 people.

MODIFIED MERCALLI SCALE

Because damage to the land surface and to human property is important, seismologists developed the **Modified Mercalli Scale** to calibrate it. This scale is based on the amount of vibration people feel during low-magnitude quakes and the extent of building damage during high-magnitude quakes. The correspondence between Mercalli intensity and Richter magnitude is listed in Table 10.2.

Figure 10.14 Shaking Causes Liquefaction Destruction of part of Anchorage, Alaska, caused liquefaction as a result of the earthquake of 1964. The building in the foreground was the home of the publisher of the *Anchorage Times*.

TABLE 10.2 **Earthquake Magnitudes, Frequencies for the Entire Earth, and Damaging Effects**

Richter Magnitude	Number per Year	Modified Mercalli Intensity Scale[a]	Characteristic Effects of Shocks in Populated Areas
<3.4	800,000	I	Recorded only by seismometers
3.5–4.2	30,000	II and III	Felt by some people who are indoors
4.3–4.8	4,800	IV	Felt by many people; windows rattle
4.9–5.4	1,400	V	Felt by everyone; dishes break, doors swing
5.5–6.1	500	VI and VII	Slight building damage; plaster cracks, bricks fall
6.2–6.9	100	VII and IX	Much building damage; chimneys fall; houses move on foundations
7.0–7.3	15	X	Serious damage, bridges twisted, walls fracture; many masonry buildings collapse
7.4–7.9	4	XI	Great damage; most buildings collapse
>8.0	1	XII	Total damage, waves seen on ground surface, objects thrown in the air

[a] Mercalli numbers are determined by the amount of damage to structures and the degree to which ground motions are felt. These depend on the magnitude of the earthquake, the distance of the observer from the epicenter, and whether an observer is in or out of doors.

Note that an earthquake has only one Richter magnitude, but its damage (and hence the Mercalli Scale estimate) varies from place to place, largely depending on distance from the epicenter.

WORLD DISTRIBUTION OF EARTHQUAKES

Although no part of Earth's surface is exempt from earthquakes, plate boundaries are most subject to frequent earthquakes. (Figure 10.15), and subduction zones tend to have the largest quakes. The *circum-Pacific belt*, where about 80 percent of all recorded earthquakes originate, follows the subduction zones on the Pacific coast of the Americas from Cape Horn to Alaska, and crosses to Asia where it extends southward down the coast, through Japan, the Philippines, New Guinea, and Fiji, where it loops around Tonga far southward to New Zealand. Next in prominence, giving rise to 15 percent of all earthquakes, is the Mediterranean-Himalayan belt, extending from Gibraltar to Southeast Asia along convergent plate boundaries, mostly continental in character, between Eurasia and the plates that border it.

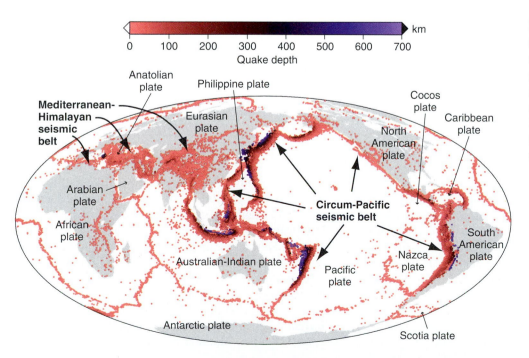

Figure 10.15 Earthquakes and Plate Margins Earth's seismic activity outlines plate margins. This map shows earthquakes of magnitude 5.5 or greater from 1960 to 1989. Earthquake foci are color-coded for depth.

The depths of earthquake foci around plate edges are also informative. Most foci are no deeper than 100 km, because brittle behavior occurs only in the upper part of the lithosphere. However, where cool, brittle lithosphere subducts into the deeper mantle, earthquakes define the **Benioff zone** that extends from the surface to as deep as 700 km (Figure 10.16A). The reason why earthquakes have not been detected at depths below 700 km remains unexplained. Two hypotheses for this are (1) sinking lithosphere warms sufficiently to become entirely ductile at 700 km depth, and (2) the slab undergoes a mineral phase change near 670-km depth (discussed later in this chapter) and loses its tendency to fracture.

Earthquakes worldwide mark the locations of plate boundaries, as we saw in Chapter 2, and most fault slip can be related to the relative motion between plates. However, earthquakes do not always occur from the top to the bottom of the plate boundary. Plate boundaries in continental regions often exhibit earthquakes only in the upper half of the crust. For transform plate boundaries this behavior is at first quite puzzling. The San Andreas Fault in California, which divides the Pacific and North American plates, experiences very few earthquakes deeper than 20 km (Figure 10.16B). In fact, most earthquakes occur in the upper 12 km of the fault. In order for plate motion to occur, we must conclude that the lower crust deforms in a ductile manner beneath California. Similar behavior at continental plate boundaries worldwide suggests that continental crust exhibits a transition between brittle and ductile deformation at depths of 15–20 km.

How can continental crust become ductile without losing the stiffness that we associate with tectonic plates? As discussed in Chapter 9, laboratory experiments demonstrate that quartz loses strength as temperature and pressure increase, especially in the presence of water. Quartz-rich rocks of the continental crust, brittle at Earth's surface, should become ductile at the depths where earthquakes become scarce. Laboratory experiments also show that olivine-rich rocks, like the ultramafic peridotites of the upper mantle, remain stiff at the temperatures and pressures that cause quartz-rich rocks to weaken.

The continental lithosphere has been compared to a jelly sandwich: ductile lower crust between brittle upper crust and stiff upper mantle. Ductile flow in the lower crust helps to explain why plate boundaries in continental regions tend to be broad, with many parallel faults. Much of the seismic hazard in California, for example, is associated with faults that lie 10–200 km away from the plate boundary defined by the San Andreas Fault. Los Angeles residents can blame the ductile lower crust for the wide variety of active earthquake faults under their feet.

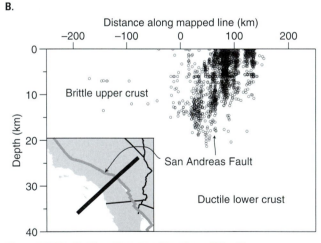

A.

B.

Figure 10.16 Earthquakes as a Function of Depth A. Earthquake foci in cross-section beneath the Kurile Trench, Pacific Ocean, during 1964–1998. Each dot represents a single earthquake. The earthquakes define the Benioff zone and are generated by downward movement of a comparatively cold slab of lithosphere. B. Earthquake foci in cross section across the San Andreas Fault System in California. Foci are scarce in the lower crust along this plate boundary, because quartz-rich crustal rocks at this depth deform in a ductile, rather than brittle, manner.

FIRST-MOTION STUDIES OF THE EARTHQUAKE SOURCE

By careful study of earthquake waves recorded by seismometers, it is possible to tell the fault orientation and the direction of fault slip during an earthquake. The required information is contained in the seismic body waves. As discussed in Chapter 2, this information is a key confirmation of the relation between earthquakes and plate motion.

Consider the P-wave. If the first motion of the arriving P wave pushes the seismometer upward, then fault motion at the earthquake focus is *toward* the seismometer. Incredible as it might seem, the seismometer senses the "push" of fault motion toward it (Figure 10.17A).

A.

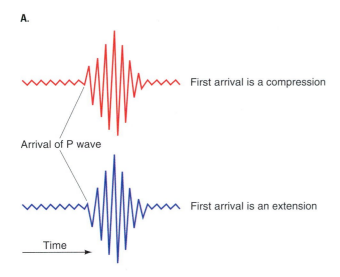

First arrival is a compression

Arrival of P wave

First arrival is an extension

Time

B.

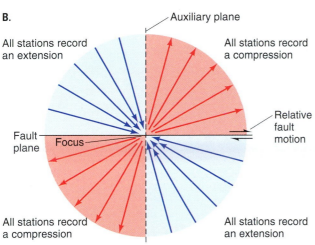

Auxiliary plane

All stations record an extension

All stations record a compression

Fault plane

Focus

Relative fault motion

All stations record a compression

All stations record an extension

Figure 10.17   Motion on a Fault   Initial motion of seismic body waves used to determine the direction of movement on a fault. A. Initial motion of a P wave detected by a seismometer is either a push away from the focus (i.e., arrival of a compression) or a pull toward the focus (arrival of an extension). B. Plotting the first motions detected at a number of seismometer stations allows the direction of movement on a fault to be determined. The example shown is for right-lateral motion on a strike-slip fault.

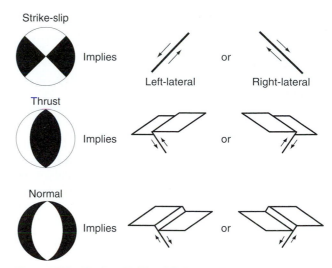

Strike-slip

Implies

Left-lateral   or   Right-lateral

Thrust

Implies   or

Normal

Implies   or

Figure 10.18   Earthquake Focal Spheres   First-motion information is plotted graphically as a focal sphere, which distinguishes at a glance between strike-slip, thrust, and normal earthquakes.

Conversely, if the first motion of the P wave is downward, the fault motion must be *away* from the seismometer. Figure 10.17B shows the effect of an earthquake caused by movement on a strike-slip fault. It is apparent that the first P-wave motion depends on the location of the seismometer relative to the moving blocks of rock at the fault. The fault movement can be determined by plotting first motions from several seismometers that can sense the earthquake focus from a variety of directions.

The radiation pattern of body waves has an important ambiguity, which can be seen in Figure 10.17. Although slip on the **fault plane** generates a distinctive pattern of P-wave

first motions, an identical pattern of first motions can be generated by slip along an **auxiliary plane** that lies at a right angle to the fault plane. Although nearly all earthquakes actually slip on a single fault plane, the first motions of P waves cannot distinguish between the two possible slip planes. Because two possibilities for fault orientation exist, additional information is needed. In practice most large earthquakes occur on faults whose orientation is already known from field mapping or from a chain of past earthquake foci.

S-waves and surface waves also carry the signature of earthquake slip and fault orientation and can provide independent estimates of motion at the earthquake focus. However, these data also suffer from the ambiguity between slip on the fault and auxiliary planes. To adapt to this uncertainty, seismologists represent earthquake sources with a symbol called a **focal sphere** (Figure 10.18). Nicknamed a "beachball," the focal sphere represents graphically the pattern of P-wave first motion that radiates from the earthquake focus. Two planes divide the focal sphere into quarters like the sections of an orange, but neither plane is identified as the fault plane in the symbol itself. The quarters of the focal sphere identify what we can observe about the earthquake focus from far-traveling seismic waves: two positive (up) first-motion quadrants, and two negative (down) first-motion quadrants.

EARTHQUAKE FORECASTING AND PREDICTION

Some of the worst natural disasters have been caused by earthquakes. Policymakers could protect lives and property more effectively if seismologists could predict where and when large earthquakes are likely to occur. We distin-

UNDERSTANDING OUR ENVIRONMENT

A GREAT EARTHQUAKE IN THE PACIFIC NORTHWEST?

Subduction zones are places of intense seismic activity. Most of the largest earthquakes ever recorded occurred in subduction zones: the Chilean earthquake of 1960, $M = 9.5$, for instance, and the Alaskan earthquake of 1964, $M = 9.2$. Such high-magnitude earthquakes in convergent boundaries are called *megathrusts*. These spread colossal destruction to communities hundreds of kilometers apart.

From northern California to central British Columbia, the Juan de Fuca Plate (named for an early Spanish explorer) has been slipping beneath the North American Plate at a rate of 3 to 4 cm/yr for the past million years (Figure B10.2). Despite this convergence, for the past 300 years no great earthquake has occurred on the Cascadia subduction zone. Furthermore, seismometers tell us that seismic activity along the interface between the two plates, where megathrusts occur, is currently weak and infrequent. A large damaging earthquake ($M = 6.8$) rocked the Seattle–Tacoma metropolitan area on February 28, 2001. However, this earthquake was caused by bending in the subducting Juan de Fuca plate, not by thrusting on the plate interface. Is the Pacific Northwest a likely spot for a gigantic quake sometime in the future? Is the Juan de Fuca Plate locked to the overriding North American Plate, or is it sliding smoothly downward?

Geologists of the U.S. Geological Survey and other institutions are attempting to confront these questions by looking at the stratigraphic record. Great earthquakes in subduction zones tend to cause abrupt elevation changes along nearby coastlines. Coastal subsidence up to 2 m or more can change well-vegetated coastal lowlands into intertidal flats. Careful mapping along the coastline of Washington reveals a number of places (labeled in Figure B10.2) where marshes and swamps suddenly became barren tidal flats in the geologically recent past (Figure B10.3).

When lush coastal lowlands flood abruptly, organic matter is buried beneath sediment. Radiocarbon dating of

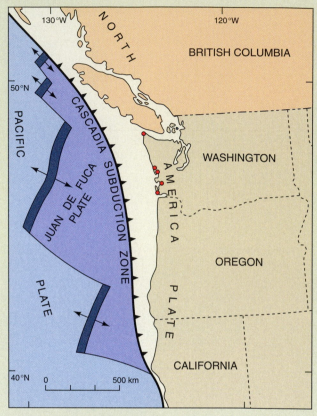

Figure B10.2 Interacting Plates The Cascadia subduction zone separates the downgoing Juan de Fuca Plate from the overriding North American Plate. Evidence of sudden, catastrophic sinking of near-coastal lowlands has been discovered along the Washington coast at the five sites marked.

such organic matter shows that a catastrophic change in the coastline occurred about 300 years ago in southwestern Washington. Evidence also suggests that catastrophic

---

guish between two scientific activities with similar names: **earthquake forecasting** and **earthquake prediction**. "Forecasting" identifies both earthquake-prone areas and man-made structures that are especially vulnerable to damage from shaking. Policymakers can use earthquake forecasts to develop cost-effective building codes and emergency-response plans. Earthquake "prediction" refers to attempts to estimate precisely when the next earthquake on a particular fault is likely to occur.

Earthquake forecasting is based largely on plate tectonics and elastic rebound theory. The **elastic rebound theory** suggests that if fault surfaces do not slip easily past each other,

energy will be stored in elastically deformed rock, just as in a steel spring that is compressed. When the fault finally does slip, the strained rock rebounds to its original size and shape.

The first evidence supporting the elastic rebound theory came from studies of the San Andreas Fault. During long-term field observations in central California, beginning in 1874, scientists from the U.S. Coast and Geodetic Survey measured the positions of many points both adjacent to and distant from the fault (Figure 10.19). As time passed, relative movement of the points revealed that the crust was slowly being bent. For reasons unknown at the time, the fault was locked near San Francisco and did not

coastline changes occurred about 1700, 2700, 3100, and 3400 years ago in that area.

Corroborative evidence has come from Japan. In a nation acutely vulnerable to tsunami hazard, the Imperial government has kept records of natural disasters for many centuries. In their archives for January 1700, researchers have found reports of damage from a tsunami 2 to 3 meters high, measured at several locations on the Pacific coast of Japan, but no connection to any nearby earthquake. Computer simulations of earthquake-generated tsunami waves established that an earthquake with magnitude $M \approx 9$ in the Pacific Northwest could explain the signal. The timing of the pre-

sumed earthquake, at 9 P.M. local time, agrees with stories of a great shaking event on a winter night, told by coastal Indian tribes to the early European explorers of the region.

Historical evidence and geologic research suggest that the Pacific Northwest, which now lies above a strangely quiet subduction zone, suffered a $M = 9+$ subduction-zone earthquake in the past, and probably could do so in the future. Since 1700, plate motion has accumulated 9–12 meters of potential slip beneath the region. If all this slip were released at once along the entire subduction zone, much of coastal Washington and Oregon would experience shaking comparable to 1700.

Figure B10.3 Evidence of a Prehistoric Earthquake   Geologist Brian Atwater stands at the high-tide level on the surface of a brackish tidal marsh in Willapa Bay on the coast of Washington. The marsh plants overlie roots of a former spruce forest that was killed by sudden subsidence of the land during a major earthquake about 300 years ago.

slip. On April 18, 1906, the two sides of this locked fault shifted abruptly. The fault released elastic energy as the bent crust snapped back, thereby creating the great 1906 San Francisco earthquake. Repetition of the survey later revealed that the bending across the fault had disappeared.

Nowadays seismologists use plate tectonic motions and Global Positioning System (GPS) measurements to monitor the accumulation of strain in rocks near active faults. Reasoning from elastic rebound theory, one can estimate that larger earthquakes can be expected on faults that have been accumulating strain for a long time. *Seismic gaps* are places where earthquakes have not occurred for a long time and

where elastic strain is steadily increasing. Seismic gaps receive a lot of research attention from seismologists because they are considered to be the places most likely to experience large earthquakes (Figure 10.20). Evidence from one prominent seismic gap is discussed in *Understanding Our Environment*, Box 10.2, *A Great Earthquake in the Pacific Northwest?* For policymakers, population centers in a seismic gap are treated with a higher priority when allocating governmental resources for the reinforcement of structures and emergency-response planning.

Earthquake prediction has had few successes so far. Because large earthquakes on any one fault are infrequent,

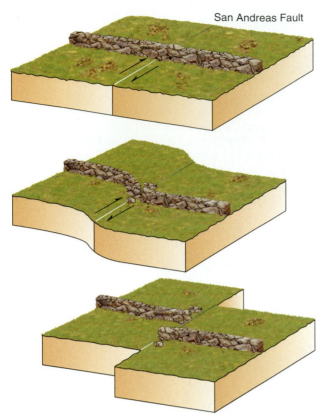

San Andreas Fault

Figure 10.19 Elastic Rebound Theory An earthquake, caused by sudden release of energy. Sketch based on detailed surveys near the San Andreas Fault, California, before and after the abrupt movement that caused the earthquake of 1906. A stone wall crosses the fault and is slowly bent as rock strains elastically. After the earthquake, two segments of the wall are offset 7 m.

scientists usually are forced to hypothesize **earthquake precursors** after an earthquake has already happened. For instance, sometimes people report that animals behave in an unusual manner before an earthquake. In China on July 18, 1969, zookeepers reported suspicious animal behavior to authorities before a 7.4 earthquake struck nearby. However, there are few other documented cases where people associated animal behavior with an impending large earthquake *before* its occurrence. Before the 1989 Loma Prieta earthquake, a scientist by chance recorded unusual electrical signals near the San Andreas Fault in an experiment unrelated to seismology. Despite strenuous efforts to measure electric fields near active faults, no similar electric signal has yet been observed as a precursor to any later earthquake. Many other hypothesized earthquake precursors have failed to be reliable predictors of imminent fault slip when seismologists monitor fault zones continuously.

There is a serious barrier to successful earthquake prediction. Seismologists lack a successful, agreed-upon theory for how earthquakes start. Without such a theory, it is difficult to hypothesize what would be a reliable earthquake precursor. (For instance, why should animals be sensitive to imminent earthquakes?) Seismologists have formulated many hypotheses for how earthquakes start, but none of them has yet been validated convincingly by observations.

Many large earthquakes are preceded by smaller earthquakes called **foreshocks**. Chinese authorities used an ominous series of foreshocks to anticipate a $M = 7.3$ earthquake near the city of Haicheng in 1975. Local residents were induced to sleep outside their houses on the night of the

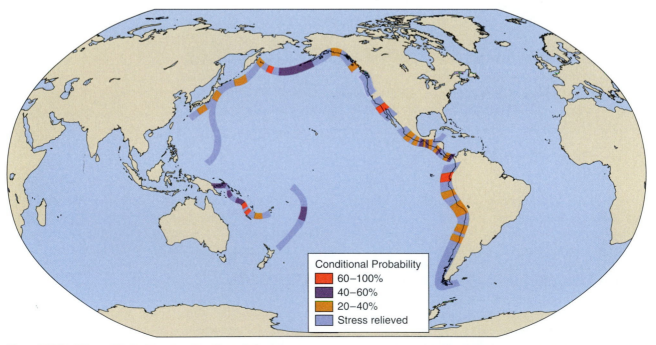

Conditional Probability
- 60–100%
- 40–60%
- 20–40%
- Stress relieved

Figure 10.20 Where Big Earthquakes Are Expected Seismic gaps in the circum-Pacific belt. In the areas indicated, earthquakes of magnitude $M = 7.0$ or greater are known but have not occurred in recent times. Strain is now building up in each seismic gap, raising the probability that a large quake will occur soon.

earthquake, which kept the death toll under 2000. Unfortunately, not all large earthquakes are preceded by strong foreshocks. In 1976, only one year after the Haicheng quake, a stronger earthquake struck the Chinese city of Tangshan without warning and killed 240,000 people.

One hypothesis for earthquake prediction was based on faults where earthquakes occur at regular intervals. Elastic rebound theory explains the long-term risk of future earthquakes well, and could explain short-term risk if the following conditions were satisfied:

- The rocks along a fault have a well-defined critical threshold for strain, above which the fault will slip.

- Strain accumulates in the fault zone in a steady manner.

If the strain accumulation rate is known from the motions associated with plate tectonics, one can project when the next earthquake will occur on the fault.

A small segment of the *San Andreas Fault* north of Los Angeles, California, seemed to satisfy these criteria. Near the town of Parkfield, the San Andreas Fault has had a remarkably regular pattern of fault slip, with earthquakes in the years 1857, 1881, 1901, 1922, 1934, and 1966, each with $M = 6$ or larger. Fault slip in 1857 continued southward to cause the last magnitude-8 earthquake to strike Los Angeles, so Parkfield seemed to hold the key to predicting the next truly devastating earthquake in southern California. U.S. seismologists made a tentative prediction that the next Parkfield earthquake would occur sometime in a 10-year interval centered on 1988. Numerous seismometers and other monitoring instruments were concentrated near Parkfield to detect earthquake precursors that might precede fault slip by days or weeks. The San Andreas Fault, however, did not cooperate. No earthquake larger than $M = 4$ occurred at Parkfield during the decade of intensive monitoring. As of 2002, seismologists are still waiting for the expected $M = 6$ quake.

Seismologists learned from the Parkfield experiment that the simplest hypotheses about earthquake fault behavior were inadequate to predict earthquakes on the time scale of a few years or less. The economic and social benefits of predicting even some damaging earthquakes, however, have kept scientists interested in the problem. Although we still lack an accepted theory for how earthquakes start, there are several promising hypotheses:

- *Fault interaction.* After a fault slips during an earthquake, the stresses on all neighboring faults are affected. Every large earthquake is followed by numerous **aftershocks**, which are smaller earthquakes that occur in response to the sudden release of strain in surrounding rock. Seismologists have documented cases where aftershocks concentrate in volumes of rock where computer models predict that stresses have increased, and are less numerous where stresses have decreased (Figure 10.21). Moreover, there are several cases where the next large earthquake occurs, sometimes decades

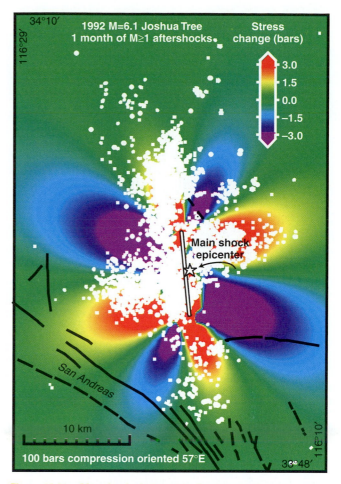

Figure 10.21 Aftershocks Induced by Earthquake Stress The locations of small (down to magnitude 1) aftershocks of a 1992 earthquake in Joshua Tree, California. Red and yellow indicate areas where a computer model has predicted that stress increased slightly after the earthquake. Purple and blue indicate areas where stress should have decreased slightly. Changes in stress are small, up to three bars or so, comparable to a pressure change of three atmospheres. The white bar indicates the ruptured fault.

later, in the region where the last earthquake has increased the local stress. For instance, the focus of the 1994 Northridge earthquake in Los Angeles ($M = 6.8$) was located where stress had been increased by a 1971 earthquake in nearby San Fernando.

- *Weak fault behavior.* Fault zones are weak surfaces within subsurface rock. Friction on the fault prevents slip as strain and stress accumulate in surrounding rock. In the laboratory, geologists have determined that friction on many rock surfaces decreases greatly once the surfaces start to slip. This effect, called **velocity-weakening behavior**, allows slip to accelerate and to release all the pent-up strain of the rock. Geologists have developed special equations to describe laboratory rock behavior and have incorporated them into computer simulations of fault slip. In these simulations small patches of the fault surface can be stressed by slip on neighboring patches, sometimes causing large portions of the fault to slip simultaneously. These computer simulations can match the behavior of real faults remarkably well, displaying earthquakes of all sizes at irregular intervals.

- *Fluids in faults.* Subsurface faults form a network of pathways for water, carbon dioxide, and other volatiles in the brittle upper crust. There are three main sources for these volatiles: (1) *rainwater*, which percolates downward through surface fractures and porous rock, (2) *mantle outgassing*, principally a byproduct of magma migration, eruption, and emplacement, and (3) *metamorphic dehydration* reactions. Fluids in faults will decrease the friction that holds their rock faces in place, much like wet leaves can make an autumn sidewalk slippery. Many observations suggest that fluids participate in the faulting process. Water well levels have risen or fallen just before earthquakes, some open faults have gushed water after an earthquake, and small earthquakes tend to occur near newly filled reservoirs. Several seismologists have hypothesized that many earthquakes deeper than 100 km in subducting slabs are induced by the release of water from hydrated minerals.

Some recent research has suggested that, if the detailed interaction of small fault patches is critical in determining when a large earthquake starts, we are unlikely ever to have sufficient information to make reliable earthquake predictions. Nevertheless, a better understanding of how earthquakes start is probably necessary if we are ever to predict one.

Before you go on:

1. What local factors can intensify damage from earthquakes?

2. How do we determine the fault orientation and slip of an earthquake from body-wave first motions?

3. Can you explain the difference between earthquake forecasting and earthquake prediction?

4. What do we know about how earthquakes start?

USING SEISMIC WAVES AS EARTH PROBES

We have seen that P and S waves travel through rock at different speeds. These two types of waves also respond differently to changing rock properties. The arrival times of P and S waves at seismometers around the world provide records of waves that have traveled along many different paths. From such records, it is possible to calculate how rock properties change and where distinct boundaries occur between layers having sharply different properties.

Seismic waves are the most sensitive probes we have to measure the properties of the unseen parts of the crust, mantle, and core.

LAYERS OF DIFFERENT COMPOSITION

If Earth's composition were uniform, and if no polymorphic changes occurred in Earth's minerals, the velocities of P and S waves would increase smoothly with depth. This is so because higher pressure leads to an increase in the density and rigidity of a solid, and it is these two properties that control the wave velocities. However, distinct boundaries (or discontinuities, as they are more commonly called) can be readily detected by refraction and reflection of body waves deep within Earth. Two major compositional boundaries have been detected. The first is the boundary between the crust and the mantle, and the second is that between the mantle and the core.

THE CRUST

Early in the twentieth century, the boundary between Earth's crust and mantle was demonstrated by a Croatian scientist named Mohorovičić. Mohorovičić noticed that, for earthquakes whose focus lay within 40 km of the surface, seismometers about 800 km from the epicenter recorded two distinct sets of P and S waves. He concluded that one pair of waves must have traveled from the focus to the station by a direct path through the crust, whereas the other pair represented waves that had arrived slightly earlier because they had been refracted by a boundary at some depth. Evidently, the refracted waves had penetrated a zone of higher velocity below the crust, had traveled within that zone, and then had been again refracted upward to the surface (Figure 10.22). Mohorovičić hypothesized that a distinct compositional boundary separates the crust from this underlying zone of different composition. Scientists now refer to this boundary as the **Mohorovičić disconti-**

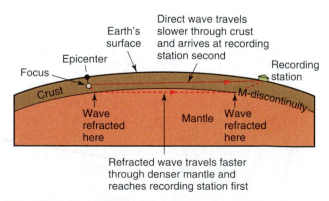

Figure 10.22   Travel Paths   Travel paths of direct and refracted seismic waves from shallow-focus earthquake to nearby seismometer station.

nuity and recognize it as the seismic discontinuity that marks the base of the crust. The feature is commonly called the **M-discontinuity**, and in conversation it is shortened still further to **Moho**.

THICKNESS AND COMPOSITION OF THE CRUST

Seismologists can determine the thickness of crust by seismic methods, supplemented with measurements of gravity (Figure 10.23). Seismic wave speeds can be measured for different rock types in both the laboratory and the field. When the arrival times of waves received at a number of seismometers are combined with laboratory measurements of common rocks, seismologists estimate the depth of the Moho and the probable composition of the crust. For instance, beneath ocean basins the crust is 8 km thick on average. Elastic properties of the oceanic crust are those characteristic of basalt and gabbro.

The thickness and composition of continental crust vary greatly from place to place. Thickness ranges from 20 to nearly 70 km and tends to be thickest beneath major continental collision zones, such as Tibet. Velocities in the continental crust differ from those in the oceanic crust. They indicate elastic properties like those of rock such as granite and diorite, although at some places just above the Moho, velocities close to those of oceanic crust are often observed. These conclusions agree well with what is known about the composition of the crust from other lines of evidence, such as geologic mapping and deep drilling.

THE MANTLE

The mantle is huge and controls much of what happens in the crust, but it cannot be seen. P-wave speeds in the crust range between 6 and 7 km/s. Beneath the Moho, speeds are greater than 8 km/s. Laboratory tests show that rocks common in the crust, such as granite, gabbro, and basalt, all have P-wave speeds of 6 to 7 km/s. But rocks that are rich in dense minerals, such as olivine, pyroxene, and garnet, have speeds greater than 8 km/s. We therefore infer that the most common such rock, called peridotite, must be among the principal materials of the mantle. This inference is consistent with what little direct evidence is available concerning the composition of the upper part of the mantle. For example, some evidence can be obtained from rare samples of mantle rocks found in *kimberlite pipes*—narrow pipe-like masses of intrusive igneous rock, sometimes containing diamonds, that intrude the crust but originate deep in the mantle (Figure 10.24). Smaller pieces of peridotite are sometimes found as xenoliths in mantle-derived magmas, and offer a broader sampling of the mantle than kimberlites.

THE CORE

Both P and S waves are strongly influenced by a pronounced boundary at a depth of 2900 km. When P waves reach that boundary, they are reflected and refracted so strongly that the boundary casts a P-wave shadow, which is an area of Earth's surface, opposite the epicenter, where no P waves are observed (Figure 10.25). Because this 2900-km boundary is so pronounced, geologists infer that it is the boundary between the mantle and the core. The same boundary casts an S-wave shadow. Here, however, the reason is not merely refraction, but the fact that shear waves cannot traverse liquids. No seismic waves have been found that can be explained with an S wave traveling in the outer part of the core, so we conclude that the *outer core* is liquid.

Seismic waves cannot tell us the composition of the core directly, but they help us to deduce what it might be. Seismic-wave speeds calculated from travel times indicate that rock density increases from about 3.3 g/cm^3 at the top of the mantle to about 5.5 g/cm^3 at the base of the mantle. The mean density of Earth is 5.5 g/cm^3. Therefore, to balance the less-dense crust and the mantle, the core must be composed of material with a density of at least 10 to 13 g/cm^3. The only common substance that comes close to fitting this requirement is iron. Suggestive evidence also comes from meteorites. *Iron meteorites* are samples of material believed to have come from the core of ancient, tiny planets, now disintegrated. All iron meteorites contain a little nickel, and Earth's core presumably does too.

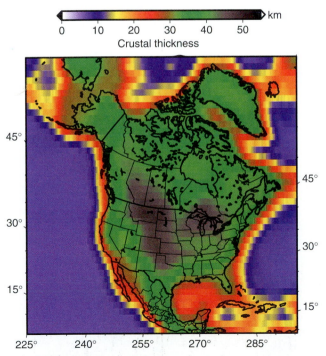

Figure 10.23 Thickness of the Crust Crustal thickness beneath North America and surrounding oceanic lithosphere. This map is compiled from measurements of seismic waves and measurements of gravity. Crustal rocks have lower density than mantle rocks. Therefore the force of gravity tends to be slightly weaker over regions of thickened crust.

A.

Figure 10.24 Samples from Deep in the Mantle The intrusion of kimberlite brings up fragments of rock from deep in the mantle. A. The shape and size of a kimberlite pipe can be judged from the hole left after the mining at Kimberley, South Africa. B. Kimberlite containing a crystal of diamond plus rounded fragments of mantle peridotite. The grayish-colored background material formed by crystallization of kimberlite magma.

B.

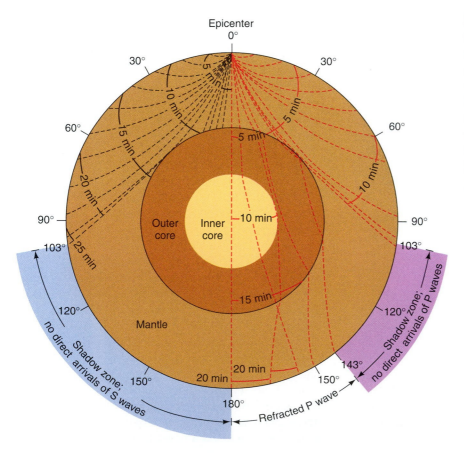

Figure 10.25 Seismic Shadow Zones Paths of P waves from an earthquake focus with epicenter at 0° shown in red in right half of section only. Paths of S waves shown in black in the left half. Reflection and refraction of P waves at the mantle–core boundary create a P-wave shadow zone from 103° to 143°. Because S waves cannot pass through a liquid, an S-wave shadow exists between 103° and 180°.

THE INNER CORE

P-wave reflections indicate the presence of a *solid inner core* enclosed within the molten outer core. The two cores appear to be essentially identical in composition. The reason for the change from liquid to solid probably relates to the effect of pressure on the melting temperature of iron. As Earth's center is approached, pressure rises, to a value millions of times greater than atmospheric pressure. Temperature also rises but not steeply enough to offset the effect of pressure. From the base of the mantle (at a depth of 2900 km) to a depth of 5350 km, geologists estimate that temperature and pressure are so balanced that iron is molten. Below this depth, at which P waves sense a reflecting and refracting boundary, rising pressure must overcome rising temperature, so that iron in the deepest layer of Earth is more stable as a solid.

LAYERS OF DIFFERENT PHYSICAL PROPERTIES IN THE MANTLE

As far as we can determine, there are no pronounced compositional boundaries within the mantle. Nevertheless, as shown in Figure 10.26, seismic-wave velocities do not increase regularly from the base of the crust to the core–mantle boundary. There are sudden changes in wave speed that are apparently due to changes in the physical properties of the mantle.

THE LOW-VELOCITY ZONE

The P-wave velocity at the top of the mantle is about 8 km/s, and it increases to 14 km/s at the core–mantle boundary. This increase is neither smooth nor constant. From the base of the crust to a depth of about 100 km, the P-wave velocity rises slowly to about 8.3 km/s. The velocity then starts to drop slowly to a value just below 8 km/s, and it remains low to a depth of about 350 km. This zone of reduced velocity between 100 and 350 km does not have a sharp top or bottom and is best developed beneath the oceans. Beneath continents, the zone of low seismic velocity is often deeper and weaker. The low-velocity zone can be seen as a small blip in both the P- and S-wave velocity curves in Figure 10.26.

An integral part of the theory of plate tectonics is the idea that stiff plates of lithosphere slide over a weaker zone in the mantle called the *asthenosphere*. The *low-velocity zone* has many characteristics that one might expect of the mantle asthenosphere. No evidence exists to suggest significant changes in rock composition in the low-velocity zone. To account for the velocity changes, therefore, we infer that the zone is closer to its melting point than the rock above or below it. Closer to its melting point, rock in the low-velocity zone is less rigid, less elastic, and more ductile than the adjacent regions. The mineral assemblage of mantle rock is subject to partial melting over a range of temperatures (see Chapters 4 and 5). It is therefore possible a small amount of liquid develops in the low-velocity zone and forms very thin films around the mineral grains, thus serving as a lubricant. This would increase the tendency of rock to flow in mantle convection, linking low seismic velocity with low viscosity.

Evidence for the low-velocity zone comes from studies of the dispersion of surface waves. Long-wavelength surface waves reach down more than 100 km and are affected by a slow region in the mantle. The amplitude of seismic waves decays more rapidly in the low-velocity zone than in the layers above or below, which agrees with the notion that the rock is close to its melting temperature. The amount of melting, if it occurs at all, must be very small, because the low-velocity zone does transmit S waves and we know that S waves cannot pass through liquids.

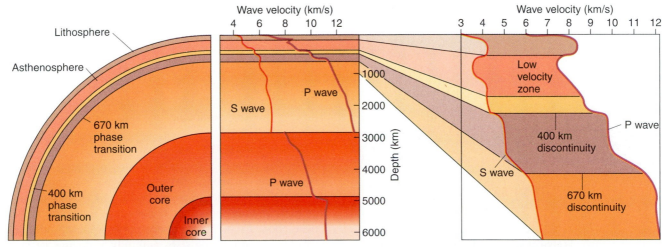

Figure 10.26 Velocity Changes with Depth Variation in seismic-wave velocity within Earth. Changes occur at the boundaries between crust and mantle and between mantle and core, owing to change in composition. Another change occurs at a depth of 100 km and corresponds to the lithosphere–asthenosphere boundary. Changes also occur at 400 km and 670 km.

THE 400-KM SEISMIC DISCONTINUITY

From the P- and S-wave curves in Figure 10.26 it is apparent that the velocities of both P and S waves increase in a small jump at about 400 km. A probable explanation is suggested by laboratory experiments. When olivine is squeezed at a pressure equal to that at a depth of 400 km, the atoms rearrange themselves into a denser polymorph. This process of atomic repacking caused by changes in pressure and temperatures is called a *polymorphic transition* (Chapter 3). In the case of olivine, the repacking involves a change to a structure resembling that found in a family of minerals called the *spinels*, of which magnetite is a well-known example. The structural repacking involves a 10 percent increase in density. The increase in seismic-wave velocities at 400 km is likely caused by the olivine–spinel polymorphic transition.

THE 670-KM SEISMIC DISCONTINUITY

An increase in seismic-wave velocities also occurs at a depth of 670 km. This velocity jump is difficult to explain in detail. It is not yet possible to determine from either seismic evidence or laboratory experiment whether the boundary is due to a polymorphic transition, to a small compositional change, or to both. Some geologists originally suggested that the increase at 670 km results solely from a polymorphic transition involving the repacking of atoms in the pyroxene minerals present in mantle rocks. The high-pressure form of pyroxene has the *perovskite* crystal structure, which involves the rearrangement of silicon and oxygen to create denser anions in which each silicon atom is surrounded by six oxygen atoms rather than four. Experiments suggest that high-pressure spinel minerals also rearrange themselves into the perovskite crystal structure, but must also form high-pressure oxide minerals with their extra Fe, Mg, and O atoms. So the 670-km discontinuity may correspond to a polymorphic change affecting all silicate minerals present.

Before you go on:

1. What is the Moho? How does the thickness of the crust vary from place to place on Earth?

2. How do we deduce that the outer core is liquid iron?

3. Why is the low-velocity zone associated with the asthenosphere?

REVISITING PLATE TECTONICS AND THE EARTH SYSTEM

Plate tectonics is Earth's method for cooling off. Hot asthenosphere rises close to Earth's surface at midocean ridges and cool lithosphere sinks into the mesosphere at subduction zones. Seismic wavespeed is affected by temperature. Seismic waves travel faster in cooler rock and slower in warmer rock, with wavespeed changes ranging from a few percent to nearly 10 percent. Seismic body waves and surface waves speed up (or slow down) in cold (or hot) regions of the mantle, and arrive at the seismometer a little ahead of (or a little behind) the time that they are expected. The discrepancy in arrival time ranges from a fraction of a second to several seconds for body waves. Surface waves can suffer larger travel-time discrepancies, which indicates that the largest temperature variations occur in the top layers of the mantle, corresponding to the lithosphere and asthenosphere.

Seismologists translate such travel-time discrepancies into maps of "fast" and "slow" regions of Earth's interior using **seismic tomography**. The method is similar to that used in medicine, where a three-dimensional picture of the interior of the human body is developed from slight differences in the intensities of X-rays passing through in different directions. (CAT scan is the common name for X-ray tomography.) In a similar manner, variations within the mantle can be revealed by measuring slight differences in the arrival times of seismic waves.

IMAGES OF EARTH'S CONVECTING INTERIOR

Earth's temperature variations (and, to a smaller extent, compositional variations) are revealed in color-coded maps of seismic wavespeed variations. The strongest contrasts in wavespeed are found at shallow depths, where young, warm lithosphere near midocean ridges is slow relative to the older, cooler lithosphere of the abyssal ocean and the continents (Figure 10.27A). Faster wavespeeds beneath most continents extend to depths of 200–400 km, which indicates that continental lithosphere is thicker than oceanic lithosphere (Figure 10.27B).

In the mantle transition zone between the 420- and 670-km mantle interfaces, regions of fast wavespeed are found where cool subducting lithosphere sinks into the deeper mantle (Figure 10.27C). Evidence of cool slabs persists even at the base of the mantle (Figure 10.27D), where a broad circular region of faster wavespeeds around the Pacific has been called the "slab graveyard" by some researchers. More remarkable are two broad patches of slow wavespeed, one under central Africa and one under the Central Pacific Ocean. Researchers hypothesize that these "slow" regions are the hot source rocks of most mantle plumes.

Where earthquakes and seismometers are sufficiently numerous, tomography can be applied with greater detail. Near active volcanoes, seismologists have interpreted travel-time discrepancies to reconstruct the location of hot and partially molten rock that supplies lava for eruptions. In plate-boundary zones, fast seismic wavespeed variations follow cool subducting slabs faithfully. In Figure 10.28, a curved line of relatively fast wavespeed follows the Kurile and Aleutian trenches at 150-km depth, showing where slabs penetrate into the asthenosphere. Wavespeeds

A. 60 km

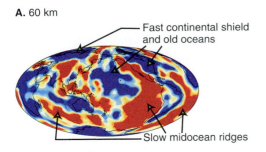

Fast continental shield and old oceans

Slow midocean ridges

B. 290 km

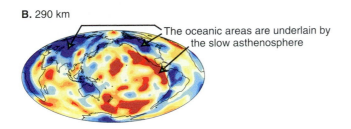

The oceanic areas are underlain by the slow asthenosphere

C. 700 km

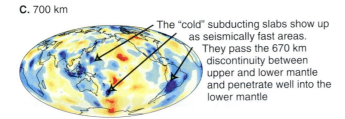

The "cold" subducting slabs show up as seismically fast areas. They pass the 670 km discontinuity between upper and lower mantle and penetrate well into the lower mantle

D. 2770 km

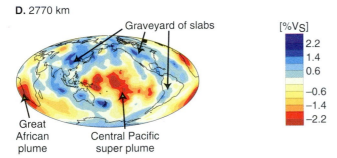

Graveyard of slabs

[%V$_S$]

2.2
1.4
0.6
−0.6
−1.4
−2.2

Great African plume

Central Pacific super plume

Figure 10.27   Seismic Tomography of the Deep Mantle Lateral contrasts in the shear wavespeed V_s in the mantle at four depths, as revealed through seismic tomography. Seismic waves travel faster through cooler, more rigid material (shown in blue), and more slowly in hotter, less rigid material (red).

beneath the western Aleutian Islands are less fast than elsewhere along the trench. This can be explained by the curve of the Aleutians: subduction has ceased here because the Pacific plate slides past the North American Plate, rather than converging against it.

EARTHQUAKES INFLUENCE GEOCHEMICAL CYCLES

Earthquakes play an important role in the transport of volatiles through Earth's solid interior. Fluids percolate through the subsurface fault systems that earthquakes cre-

Variation in shear wavespeed at 150 km depth reference wavespeed is 4.45 km/sec

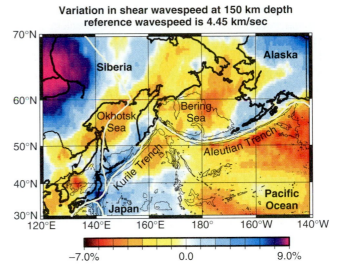

−7.0%   0.0   9.0%

Figure 10.28   Seismic Tomography of a Plate Boundary Zone Lateral variations in shear wavespeed V_s in the upper mantle of the North Pacific region at a depth of 150 km revealed through seismic tomography. Seismic waves travel faster through cooler, more rigid material (shown in blue), and more slowly in hotter, less rigid material (red).

ate and maintain. As a result, earthquakes facilitate the concentration of many important metals into ore deposits. Many deep-seated fault zones have been exhumed by later erosion. Field studies have established clear links between earthquake faults, fluid transport, and rock metamorphism. Fluids from dehydration reactions in the middle crust return to the surface via fault zones, and cause further metamorphism in the rocks that surround the faults.

In the mantle, the carbon and hydrologic cycles are fed when the subducting slab releases water, CO_2, and other volatiles at roughly 100-km depth beneath the overriding plate. The upward path for fluids from the slab to the island-arc volcanoes in the overriding plate traverses a volume of rock with no significant earthquakes, so the existence of faults in the "mantle wedge" is highly unlikely. However, some seismologists speculate that water released from the slab helps to cause brittle fracture in the slab itself, and that water may be necessary for deep earthquakes to occur in the **Benioff zone**. A critical question is how rapidly metamorphic dehydration occurs. If dehydration is slow and gradual, water may diffuse away before it can participate in rock fracture and fault slip. If dehydration occurs quickly, fluids may accumulate rather than escape, and help fracture rock.

Evidence for rapid slab dehydration was reported by Japanese seismologists in Spring 2002. Japan is bordered on the east and south by active subduction zones. The Japanese government has installed numerous seismometers to monitor tectonic activity. Researchers combined data from seismometers throughout Japan to dampen surface noise and signals from known earthquakes and search for small signals from the slab. They discovered that steady vibrational tremors were emanating from a part of the subduction zone where the downgoing slab is expected to release volatiles. If these tremors actually are caused by metamorphic fluid release, seismology has now allowed us to eavesdrop on one of the most important processes in the Earth system.

What's Ahead?

Earthquakes can cause devastation in seconds, but in thousands or millions of years they help to form mountains. A Changing Landscapes diorama follows this chapter, showing recent earthquakes in the Western US with reference to its uplifted topographic

relief. Motion on faults is one process that leads to tectonic uplift. Uplift combined with erosion causes topographic relief and the development of distinctive landscapes. You will read about this topic in Chapter 12, *The Changing Face of the Land*.

Changes in the landscape occur gradually, so human beings rarely observe the process from start to finish. Geologic observations and hypotheses are necessary to estimate the rate of landscape development. A critical tool is the measurement of time from geologic materials. You will learn about determining the age of rocks in the next chapter, *Geologic Time and the Rock Record*.

CHAPTER SUMMARY

1. Abrupt movements of faults that release elastically stored energy cause earthquakes.

2. Earthquake vibrations are measured with seismometers.

3. Energy released at an earthquake's focus radiates outward as body waves, which are of two kinds—P waves (primary waves, which are compressional) and S waves (secondary waves, which are shear waves). Earthquake energy also causes Earth's surface to vibrate. These vibrations travel laterally as surface waves.

4. The focus and epicenter of an earthquake can be located by measuring the differences in travel times between P and S waves.

5. The amount of energy released during an earthquake can vary by very large amounts. The standard measure of earthquake size is the Richter magnitude scale. The Richter scale is logarithmic in order to span seismic activity from small local tremors (magnitude $M \approx 0$) to great earthquakes (magnitude M > 8). Seismologists estimate Richter magnitude from seismometer records of either seismic body waves or surface waves. Seismologists also quantify earthquake size with the seismic moment, whose definition relates directly to fault size and the distance a fault slips during an earthquake.

6. Ninety-five percent of all earthquakes originate in the circum-Pacific belt (80%) and the Mediterranean-Himalayan belt (15%). The remaining 5 percent are widely distributed along the midocean ridges and elsewhere. Seismic activity outlines the tectonic plates. Seismic-wave first-motion studies are used to determine the fault orientation and the direction of fault slip.

7. Earthquake forecasting is a long-term estimate of future earthquake activity. Based largely on the records of past earthquakes and present plate motions, earthquake forecasting helps policymakers develop land-use policies that decrease overall seismic hazard to lives and property.

8. Earthquake prediction is a short-term estimate of future earthquake activity. Earthquake prediction relies on earthquake precursors, which must be observed in time for public officials to take appropriate actions. Despite some documented successes, no earthquake precursor, or set of precursors, has been found that can predict future earthquakes reliably.

9. Earthquake prediction is hampered by our ignorance of what

causes a particular earthquake to happen *when* it does. Recent research has established several useful facts that may help overcome this ignorance: (a) the stresses of a large quake may increase stress on neighboring faults, and cause them to slip either soon (in an aftershock) or after years or decades; (b) fault surfaces lose their frictional strength when earthquake slip starts, and (c) the presence of fluid in fault zones tends to lubricate their motion and encourage earthquake activity.

10. Seismic body waves can be refracted and reflected just as sound and light waves are. From the study of seismic-wave refraction and reflection, scientists infer Earth's internal structure by locating boundaries, or discontinuities, in its composition and physical properties. Pronounced compositional boundaries occur between the crust and mantle and between the mantle and outer core.

11. The base of the crust is a pronounced seismic discontinuity called the Mohorovičić discontinuity. Thickness of the crust ranges from 20 to 70 km in continental regions but is less than 8 km on average beneath oceans.

12. Within the mantle there are two zones, at depths of 400 and 670 km, where small jumps in seismic wavespeed produce seismic-wave reflections. The change at 400 km is probably produced by a polymorphic transition of olivine. The 670-km change might be due to either a polymorphic transition, a compositional change, or a combination of both.

13. The core has a high density and is inferred to consist of iron plus small amounts of nickel and other elements. The outer core must be molten because it does not transmit S waves. The inner core is solid.

14. From a depth of 100 km to 350 km there is a zone of low seismic-wave velocity that also causes pronounced dispersion of surface waves. This low-velocity zone coincides with the asthenosphere. The lithosphere, which is rigid and on average 100 km thick, overlies the asthenosphere.

15. Seismic tomography reveals geographic variations in seismic wavespeed. Lateral variations in mantle wavespeed are caused mostly by variations in temperature. Hotter regions have slower wavespeed; colder regions have higher wavespeed. Seismic wavespeed variations agree with the temperature variations expected from plate tectonics and mantle convection.

THE LANGUAGE OF GEOLOGY

aftershock (p. 265)
auxiliary plane (p. 261)

benioff zone (p. 260)
body waves (p. 250)

earthquake focus (p. 249)
earthquake forecasting (p. 262)
earthquake precursor (p. 264)
earthquake prediction (p. 262)
elastic rebound theory (p. 262)
epicenter (p. 249)

fault plane (p. 261)
focal sphere (p. 261)
foreshock (p. 264)

inertia (p. 248)

M-discontinuity (p. 267)
Modified Mercalli Scale (p. 258)
Moho (p. 267)
Mohorovičić discontinuity (p. 266)

P (for primary) waves (p. 252)

Richter magnitude scale (p. 254)

S (for secondary) waves (p. 252)
seismic sea waves (p. 258)
seismic moment (p. 251)
seismic tomography (p. 270)
seismic waves (p. 250)
seismometer (p. 248)
seismology (p. 248)
surface waves (p. 250)

velocity-weakening behavior (p. 265)

QUESTIONS FOR REVIEW

1. Explain how most earthquakes are thought to occur and why there seems to be a limit on earthquake magnitudes.

2. What is the relationship between an earthquake focus and the corresponding epicenter?

3. How are seismic waves recorded and measured? How would you locate an epicenter from seismic records? Explain how a focus is determined.

4. What are the differences between seismic body waves and surface waves? Identify two kinds of body waves and explain the differences.

5. Explain how seismologists use the seismic moment to estimate earthquake size. Suppose a crustal fault 3 km long and 2 km wide slips 10 cm? If you assume that the shear stiffness μ of the surrounding rock is 3×10^{10} newtons per square meter, what is the seismic moment? What is the Richter magnitude? How much will seismic moment and magnitude increase if the length, width and slip of the fault each grows by a factor of 3.16?

6. Earthquakes can cause damage in many ways; name four. Where was the most disastrous earthquake on record and how did the people die? Where was the biggest known earthquake in the United States?

7. What are reflection and refraction, and how do they affect the passage of seismic waves? How can refraction and reflection be used to define the base of the crust? the core–mantle boundary?

8. Briefly describe how seismic waves can be used to infer that the outer core is molten while the inner core is solid. Why is the composition of the core thought to be largely metallic iron?

9. Under what circumstance is it possible to obtain samples of rocks from the mantle? How can such samples be used to check information about the mantle inferred from seismic-wave velocities?

10. Why is earthquake prediction more difficult than earthquake forecasting?

11. How do subsurface fluids influence earthquakes? What evidence do we have?

12. How do earthquakes influence subsurface fluids? What evidence do we have?

Click on *Presentation* and *Interactivty* in the **GeoHazards** module of your CD-ROM to further explore resources and activities presenting concepts from this chapter. Select *Assessment* in the same module to test your understanding of this chapter.

Changing Landscape

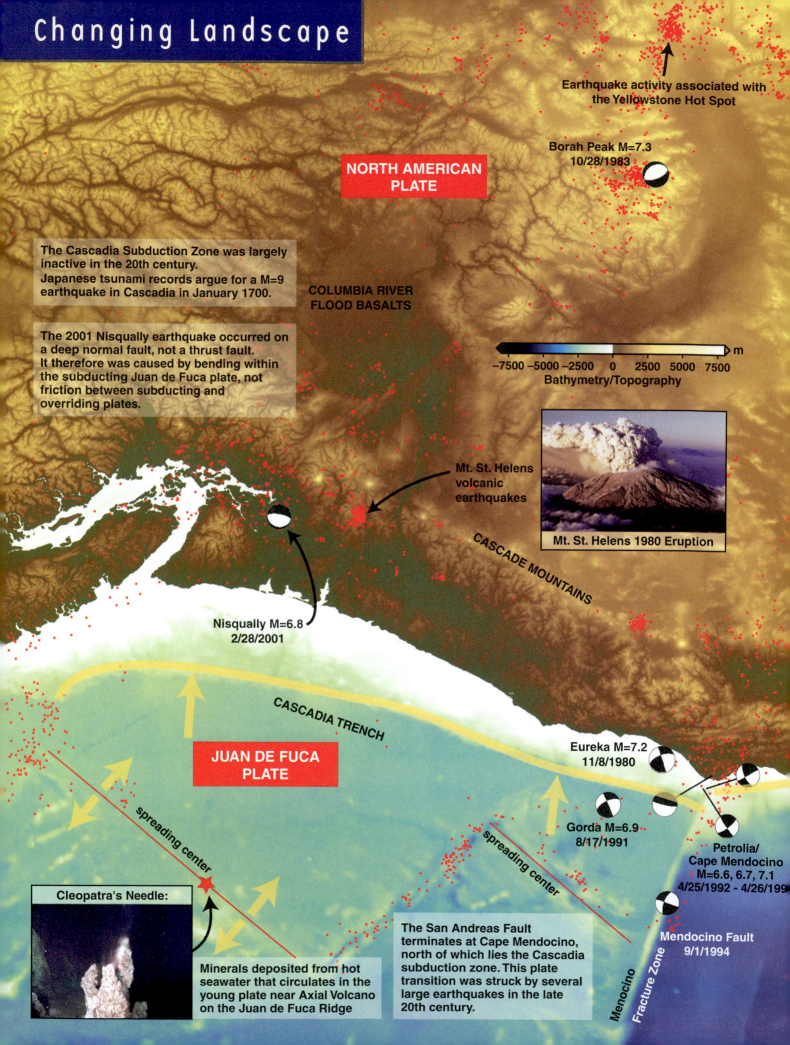

Earthquake activity associated with the Yellowstone Hot Spot

Borah Peak M=7.3
10/28/1983

NORTH AMERICAN PLATE

The Cascadia Subduction Zone was largely inactive in the 20th century. Japanese tsunami records argue for a M=9 earthquake in Cascadia in January 1700.

The 2001 Nisqually earthquake occurred on a deep normal fault, not a thrust fault. It therefore was caused by bending within the subducting Juan de Fuca plate, not friction between subducting and overriding plates.

COLUMBIA RIVER FLOOD BASALTS

Bathymetry/Topography
−7500 −5000 −2500   0   2500   5000   7500   m

Mt. St. Helens volcanic earthquakes

Mt. St. Helens 1980 Eruption

CASCADE MOUNTAINS

Nisqually M=6.8
2/28/2001

CASCADIA TRENCH

JUAN DE FUCA PLATE

spreading center

Eureka M=7.2
11/8/1980

spreading center

Gorda M=6.9
8/17/1991

Petrolia/ Cape Mendocino M=6.6, 6.7, 7.1 4/25/1992 - 4/26/199

Mendocino Fault 9/1/1994

Mendocino Fracture Zone

Cleopatra's Needle:

Minerals deposited from hot seawater that circulates in the young plate near Axial Volcano on the Juan de Fuca Ridge

The San Andreas Fault terminates at Cape Mendocino, north of which lies the Cascadia subduction zone. This plate transition was struck by several large earthquakes in the late 20th century.

The location of earthquakes, and the sense of motion on earthquake faults, tell us how surface rocks are responding to the motions of tectonic plates and to mantle convection. Three plates have their boundaries along the Pacific coast of the United States, and warm buoyant mantle beneath the Yellowstone hotspot and the Basin and Range Province has lifted up the Western US to form broad plateaus. The red dots mark earthquakes that have been catalogued by seismologists from 1964 to 1998. We plot large earthquakes since 1950 with their focal spheres to indicate the sense of fault motion. Continental crust deforms over a broader area than does oceanic crust, so the transform-fault motion of the San Andreas Fault is spread among dozens of secondary faults. A major bend in the San Andreas leads to a complicated pattern of thrust faults and recent mountain-building in Southern California. Broad deformation associated with uplift and extension in the Western US causes earthquakes inland. Rivers incise deep canyons in uplifted plateaus throughout the region. The Grand Canyon, in Arizona, is one example of this.

Wasatch Fault system: A series of normal faults that form the western boundary of the Basin and Range province.

BASIN AND RANGE EXTENSIONAL PROVINCE

Grand Canyon

Several segments of the San Andreas Fault were largely inactive in the 20th century. South of Parkfield, only secondary faults have had large strike-slip earthquakes since 1857. Several large earthquakes occurred in the late 20th century on neighboring blind-thrust faults. Blind-thrust faults are common near the "big bend" of the San Andreas, where the Pacific and North American plates converge as well as slide past each other.

Gulf of California

Superstition Hills M=6.6 11/24/1987

Hector Mine M=7.1 10/16/1999

Imperial Valley M=7.0 10/15/1979

Mammoth Lakes M=6.1 5/25/1980

Landers M=7.5 6/28/1992

Fault scarp formed by 1992 Landers quake

Kern County M=7.8 7/21/1952

Joshua Tree M=6.3 4/23/1992

San Andreas "Big Bend"

Coalinga M=6.5 5/2/1983

Sylmar M=6.5 2/9/1971

Northridge M=6.8 1/17/1994

Loma Prieta M=7.1 10/18/1989

The San Andreas Fault, and several active faults that parallel it, connect with a set of spreading ridges (not shown) in the Gulf of California.

Parkfield M=6.0 6/28/1966

SAN ANDREAS FAULT

PACIFIC PLATE

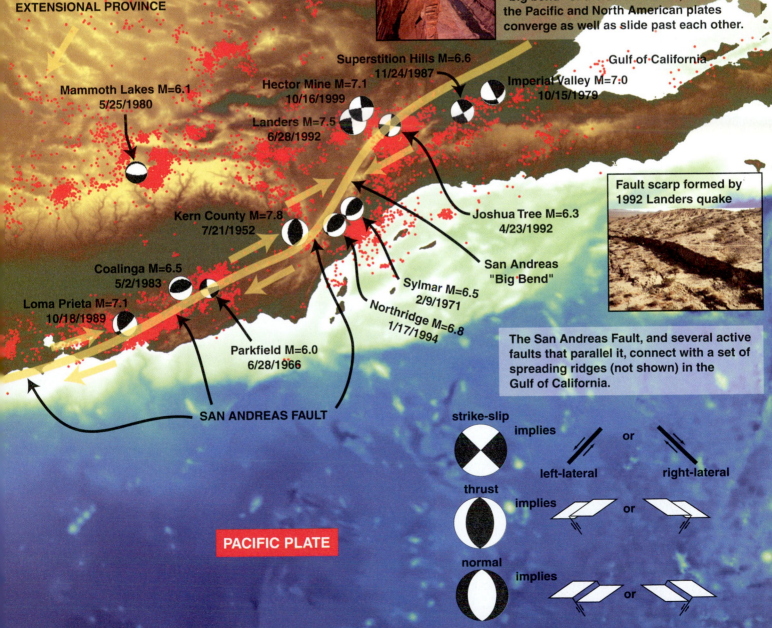

strike-slip implies left-lateral or right-lateral

thrust implies

normal implies

Chapter 11

Geologic Wonderland. Strata of sedimentary rocks exposed in the Grand
Staircase, Escalante National Monument, Kanab, Utah.

Geologic Time and the Rock Record

Midsummer view of the Coorong, South Australia. White sediment is dolomite.

Determining How Rapidly Carbonate Sediments Accumulate

In the southeast of South Australia, one of the six states of Australia, there is a long, thin arm of the sea called the

Coorong. This body of water, which is separated from the sea by a high sand dune, is about 100 miles in length and a mile or two wide. Water enters the Coorong at the northern end and leaves both the Coorong and associated lakes by evaporation. As a result, the salinity of water in the southern end of the Coorong is very high during summer months.

In the special environment of the Coorong a fine, white precipitate of carbonate minerals separates from the saline water. In some places the precipitate is calcite; in others there exists a magnesian calcite in which some magnesium replaces calcium in the calcite structure; and in a few places, dolomite precipitates. Dr. H. Catherine Skinner, my wife, was studying the chemistry of the precipitates when the question of the rate of accumulation arose. We carefully sampled the accumulated carbonate sediments in one of the saline lakes, and subjected the samples to carbon-14 analysis in order to determine their ages. The results were published in the journal *Science*, v. 139, pages 335–336, in 1963. Our analysis showed that carbonates are accumulating at a rate of 0.2 mm/year—a rate that we suspect may be rapid for such a geologic process. A subsequent analysis from sediment further up the Coorong, published a year later in the *American Journal of Science*, confirmed the accumulation rate.

Brian J. Skinner

KEY QUESTIONS

1. **What can be learned from layered rocks?**
2. **What is the stratigraphic record?**
3. **What are the geologic column and geologic time scales?**
4. **What are radiometric ages?**
5. **How can magnetism be used to date rocks?**

INTRODUCTION

The concept that most geologic processes happen very slowly—so slowly that we can barely detect changes over the span of a human life—and that Earth's history must involve immensely long spans of time, was discussed in Chapter 1. The person who first articulated this point of view and backed up his argument with evidence, was James Hutton. The evidence at Siccar Point, in Scotland (Figure 1.4) particularly impressed Hutton, and to make the point about what we today call the rock cycle, he would take friends and colleagues to see the evidence themselves. Siccar Point and evidence elsewhere in Scotland moved Hutton to write his famous comment, that the rock record contains "no vestige of a beginning, no prospect of an end."

Hutton's observations did more than open people's eyes to the immensity of geologic time. They also pointed the way to unraveling Earth's history from a study of the rock record. In this chapter we will discuss how geologists read the rock record, and how they are able to use their observations to sort Earth's history into a sequence of events. The events are placed in an order relative to each other— we refer to the ages of events as *relative ages*. Naturally, anyone with an enquiring mind would ask how ages in terms of years—*numerical ages*—might be determined. Then, and only then, can questions like "How long ago did the dinosaurs die out?" or "How long ago did the Atlantic Ocean first open?" be answered. The path to determining numerical ages came from the discovery of radioactivity by physicists. The physics of natural radioactive decay allows us to determine *radiometric ages* in terms of years, and an age in years is, of course, a numerical age.

READING THE RECORD OF LAYERED ROCKS

The historical information that geologists have to work with is largely in the form of layered sedimentary or volcanic rocks that crop out at Earth's surface or that can be penetrated by drilling. If we examine the rocks exposed in the upper walls of the Grand Canyon (Figure 11.1A), where the Colorado River has cut nearly 2 km into Earth's crust, we can see many nearly horizontal layers. These strata formed one atop the other as sediment accumulated on the floor of a shallow sea and, at certain times when the sea withdrew, on the land. Such rocks contain important clues about past environments at and near Earth's surface. Their sequence and relative ages provide the basis for reconstructing much of Earth history.

The study of strata is called **stratigraphy**. Knowledge of stratigraphic principles and the relative ages of rock sequences makes it possible to work out many of the fundamental principles of physical geology. Two straightforward and simple, but nevertheless very powerful, laws govern stratigraphy: the law of original horizontality, and the principle of stratigraphic superposition.

THE LAWS OF STRATIGRAPHY

ORIGINAL HORIZONTALITY

Most sediment is laid down in the sea, generally in relatively shallow waters, or by streams on the land. Under such conditions, each new layer is laid down horizontally over older ones. These observations form the basis of the **law of original horizontality**, which states that water-laid sediments are deposited in strata that are horizontal or nearly horizontal. From this generalization we can infer

A.

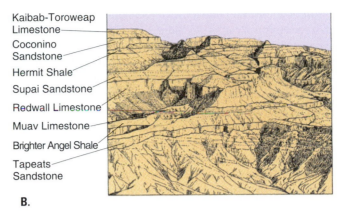

Kaibab-Toroweap Limestone
Coconino Sandstone
Hermit Shale
Supai Sandstone
Redwall Limestone
Muav Limestone
Brighter Angel Shale
Tapeats Sandstone

B.

Figure 11.1 The Principle of Stratigraphic Superposition The Grand Canyon of the Colorado River. A. Flat-lying strata, nearly 2000 m thick and accumulated over 300 million years, were laid down on older strata, now tilted and tectonically deformed. B. Sequence of horizontal sedimentary strata laid down on the tilted and deformed basement rocks of the Grand Canyon, from the Tapeats Sandstone (oldest) to the Kaibab-Toroweap Limestone (youngest), illustrates both the principle of stratigraphic superposition and the law of original horizontality.

that layers of sedimentary rock, now inclined, or even buckled and folded, must have been disturbed since the time when they were deposited as sediments.

STRATIGRAPHIC SUPERPOSITION AND THE RELATIVE AGES OF STRATA

Another principle follows from the law of original horizontality. Consider the following example. Toward the end of a cold winter, it is often possible to see layers of old snow that are compact and perhaps also dirty, overlain by fresh, looser, clean snow deposited during the latest snowstorm. Here are layers, or strata, that were deposited in sequence, one above the other. The simple principle involved here also applies to layers of sediment and sedimentary rock. The **principle of stratigraphic superposi-**

tion states that, in any sequence of sedimentary strata, the order in which the strata were deposited is from bottom to top. The relative ages of any two strata can therefore be determined according to whether one of the layers lies above or below the other. Figures 11.1A and 11.1B offer examples of the principle of stratigraphic superposition. The horizontal strata at the top of the Grand Canyon are younger than the horizontal strata below.

The principle of stratigraphic superposition must be employed with a certain amount of care. Observe in Figure 11.2 that tilting and buckling of strata, such as happens

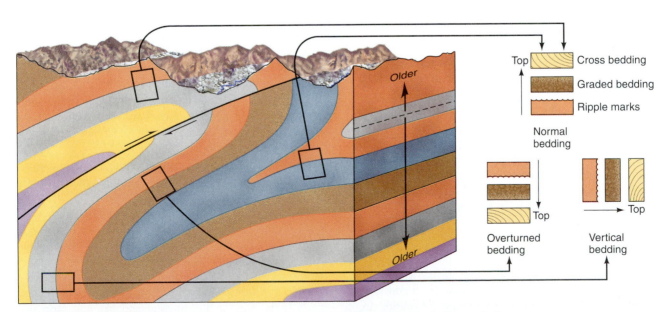

Top — Cross bedding
Graded bedding
Ripple marks

Normal bedding

Overturned bedding   Vertical bedding

Figure 11.2 Facing of Strata Sketch illustrating how some sedimentary features can indicate whether strata are normal (right side up), vertical, or overturned as a consequence of tectonic deformation.

during continental collisions, can sometimes be so severe that overturning can bring older strata to overlie younger strata. Observations made on overturned strata would obviously lead to incorrect conclusions concerning the relative timing of deposition unless the overturning were recognized. Evidence such as ripple marks, graded beds, and cross-stratified beds, as discussed in Chapter 7, can be used to determine whether strata are correctly oriented or overturned.

GAPS IN THE STRATIGRAPHIC RECORD

Lyell and other geologists of the nineteenth century speculated that it might be possible to determine numerical ages by using the stratigraphic record. If one measures the rate of sedimentation in the sea, they argued, and if one determines the thickness of all strata in a given body of rock, it should be possible to calculate how long it took for all the sediments represented by the strata in question to accumulate. Two assumptions must be correct for the method to work.

1. It must be assumed that the rate of sedimentation was constant throughout the time of sediment accumulation.

2. It must be assumed that all strata exhibit **conformity**, meaning they have been deposited layer after layer without interruption. If there are any gaps in the stratigraphic record due to erosion or nondeposition, the elapsed time must somewhere be represented by strata that can be placed in correct sequence by correlation (demonstration of time-equivalance).

Estimates of hundreds of millions of years for all strata that have been deposited throughout geologic time were made—but they are wrong because both assumptions are false. The first assumption is false because it can be observed today that sedimentation rates vary widely from place to place and time to time. Furthermore, there is a lot of evidence to suggest that rates have varied even more widely throughout geologic history.

The second and even more important assumption is false because sedimentation can be disrupted periodically by major environmental changes, such as sea level changes and tectonic activity that lead to intervals of erosion or nondeposition. In many cases there may not be strata elsewhere that can be used to fill the missing time gap—the process of correlating strata in different places is discussed later in the chapter. Many breaks due to erosion or nondeposition are found in the sedimentation record, and we have no way of knowing how much time the breaks represent.

KINDS OF UNCONFORMITIES

An **unconformity** is a substantial break or gap in a stratigraphic sequence. It records a change in environmental conditions that caused deposition to cease for a considerable time. Three important kinds of unconformities are found in sedimentary rocks (Figure 11.3).

The most obvious kind of unconformity is the **angular unconformity**, which is marked by angular discontinuity between older and younger strata. It is labeled (2) in Figure 11.3. An angular unconformity implies that the older strata were deformed and then cut off by erosion before the younger layers were deposited across them. The outcrop at Siccar Point (Figure 1.4) observed by James Hutton is obviously an angular unconformity.

The second kind of unconformity is called a **disconformity**; it is an irregular surface of erosion between parallel strata. The surface numbered (3) in Figure 11.3 is a disconformity. A disconformity implies a cessation of sedimentation, plus erosion, but no tilting. Disconformities can be hard to recognize because the strata above and below are parallel. The way disconformities are usually discovered is through recognition that fossils of very different ages are present in adjacent strata.

The third kind of unconformity, labeled (1) in Figure 11.3, is a **nonconformity**, in which strata overlie igneous or metamorphic rocks. Note that in the panel second from the bottom of the page, the nonconformity has been folded and distorted by tectonic forces.

The three types of unconformity can be seen in the Grand Canyon, as shown in Figure 11.4. At the base of the sedimentary sections is a nonconformity, and some distance above it is an obvious angular unconformity; still higher are three disconformities. Some of the same unconformities can be seen in Figures 11.1A and 11.1B; see if you can pick them out.

THE SIGNIFICANCE OF UNCONFORMITIES

A study of unconformities brings out the close relationship between tectonics, erosion, and sedimentation. All of Earth's land surface is a potential surface of unconformity. Some of today's surface will be destroyed by erosion, but other parts will be covered by sediment and preserved as a record of the present landscape.

For example, the Swiss Alps, which were elevated by plate-tectonic movements, are being rapidly eroded away. Meanwhile, the eroded material is being carried away by streams and deposited in the Mediterranean Sea. The Mediterranean seafloor was once dry land, but tectonic forces depressed it, just as tectonic forces elevated the Alps. A surface of unconformity separates the young, river-transported sediments and the older rocks of the seafloor on which the sediments are being piled. In a sense, accumulation in one place compensates for destruction in another.

The many surfaces of unconformity exposed in rocks of Earth's crust are evidence that former seafloors were uplifted by tectonic forces and exposed to erosion. Preservation of a surface of erosion occurs when later tectonic forces depress the surface, so it, in turn, becomes a site of deposition of sediment. Unconformities testify that interactions between the internal and external processes have been going on throughout Earth's long history.

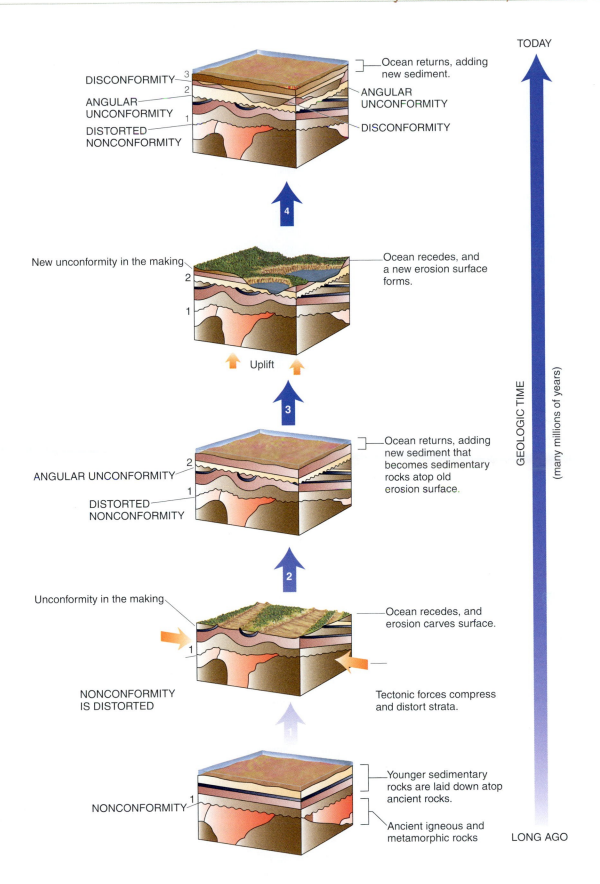

TODAY

Ocean returns, adding new sediment.

DISCONFORMITY
ANGULAR UNCONFORMITY
DISTORTED NONCONFORMITY

ANGULAR UNCONFORMITY
DISCONFORMITY

4

New unconformity in the making

Ocean recedes, and a new erosion surface forms.

Uplift

3

ANGULAR UNCONFORMITY
DISTORTED NONCONFORMITY

Ocean returns, adding new sediment that becomes sedimentary rocks atop old erosion surface.

2

Unconformity in the making

Ocean recedes, and erosion carves surface.

NONCONFORMITY IS DISTORTED

Tectonic forces compress and distort strata.

1

NONCONFORMITY

Younger sedimentary rocks are laid down atop ancient rocks.

Ancient igneous and metamorphic rocks

LONG AGO

GEOLOGIC TIME
(many millions of years)

Figure 11.3 Gaps in the Rock Record Sequence of geologic events leading to the three kinds of unconformity; (1) nonconformity, (2) angular unconformity, and (3) disconformity. Note that in the second panel from the bottom the nonconformity is distorted as a result of tectonic deformation.

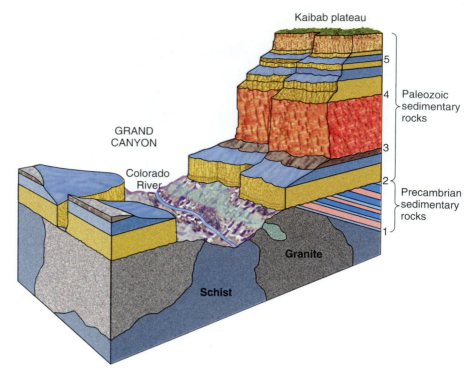

Figure 11.4 Unconformities in the Grand Canyon Geologic section through rocks exposed in the Grand Canyon. The lowest unconformity (1), separating tilted sedimentary strata from older crystalline rocks, is a nonconformity. An angular unconformity (2) separates the tilted strata from horizontally layered strata above, while three disconformities (3, 4, and 5) are seen still higher in the section. Unconformities 2, 3, 4, and 5 are visible in Figure 11.1A.

Before you go on:

1. What are the two laws of stratigraphy?

2. What is meant by the term conformity?

3. What is an angular unconformity? a disconformity?

STRATIGRAPHIC CLASSIFICATION

Every stratum can tell us something about the physical and biological character of a part of the Earth system at some time in the geologic past. Anyone counting strata would quickly realize that the rock record is like a vast library consisting of thousands upon thousands of volumes. Like a library too, the rock record is a complex catalogue that is employed by geologists who work on the record of past events. Only the most important cataloging terms used in stratigraphic classification are introduced here.

Three related concepts are employed in stratigraphic classification. The first two are based on the rock units, while the third is somewhat abstract in that it concerns intervals of geologic time.

THE ROCK-STRATIGRAPHIC RECORD

The first kind of unit employed in stratigraphic classification uses any distinctive stratum that differs from the strata above and below. An example of a *rock-stratigraphic unit* is the Navajo Sandstone in Zion National Park, Utah, seen in Figure 11.5. This striking sandstone is very differ-

ent from the strata immediately above and below and can be easily mapped in the field.

The basis of rock stratigraphy is the **formation**, which is a collection of similar strata that are sufficiently different from adjacent collection of strata so that on the basis of physical properties they constitute a distinctive, recognizable unit that can be used for geologic mapping over a wide area. The Navajo Sandstone is a formation. (Each formation is given a name. In North America it typically is the name of a geographic locality near which the unit is best exposed.)

THE TIME-STRATIGRAPHIC RECORD

The second kind of unit used in stratigraphic classification is one representing all the rocks that formed during a specific interval of geologic time. Each of the boundaries of a *time-stratigraphic unit*, upper and lower, is everywhere the same age.

Recall that a formation is defined only on the basis of its material characteristics, and its upper and lower boundaries lie where a recognizable change in physical properties occurs. As illustrated in Figure 11.6, the ages of the boundaries of a formation can differ from place to place. In contrast, a time-stratigraphic unit may include more than one rock type, and its upper and lower boundaries may not necessarily coincide with a formational boundary, but as shown in Figure 11.6, each boundary is everywhere the same age.

The primary time-stratigraphic unit is a **system**, which is chosen to represent a time interval sufficiently great so that such units can be used all over the world. Names for systems arose from nineteenth-century studies of strata, mainly in Europe, and often derive from geographic localities. Examples are the Devonian System, named for Devon in England, and the Jurassic System, named for the Jura Mountains of Switzerland and France. Most systems are now known to encompass numerical time intervals of tens of millions of years.

GEOLOGIC TIME INTERVALS

The final units used in stratigraphic classification are the intervals of geologic time during which strata in time-stratigraphic units accumulated. Units of geologic time are nonmaterial, while time-stratigraphic units are material. The primary unit of geologic time is a geologic *period*; it is the time during which a geologic system accumulated. For example, the Cambrian System accumulated during the Cambrian Period.

As we have seen, the geologic record is incomplete and punctuated by numerous unconformities; a given area may not contain a complete depositional record for a particular geologic time interval (Figure 11.3). However, we can bridge many of the gaps by piecing together sequences of strata from different geographic areas, thereby providing a more complete picture of Earth history. The piecing together involves correlation, the subject to which we turn next.

BRIDGING THE GAPS: CORRELATION OF ROCK UNITS

WHAT IS CORRELATION?

William Smith was an English land surveyor who was active around the beginning of the nineteenth century. His profession gave him an ideal opportunity to observe not only the landscape but also the rocks that underlie it. While surveying for the construction of new canals in western England, he

Figure 11.5   Rock-Stratigraphic Unit   Spectacular view of the Navajo Sandstone, Zion National Park, Utah. The Navajo Sandstone is an example of a rock-stratigraphic unit.

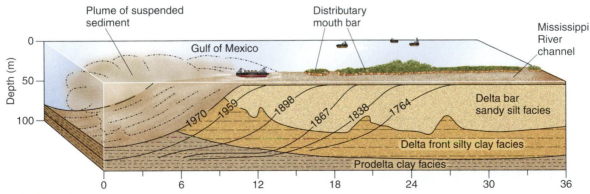

Figure 11.6   Time-Stratigraphic Unit   Section through the Southwest Pass lobe of the Mississippi River Delta, which is expanding outward into the Gulf of Mexico. Dated time lines show successive positions of the delta front between 1764 and 1979. The boundaries separating three sedimentary facies inter-

sect each time line. Should this deposit be preserved and eventually converted to sedimentary rock, the resulting formations, corresponding to the three facies, will each be younger in the direction of delta growth.

observed the sedimentary strata and realized that they lay, as he put it, "like slices of bread and butter" in a definite, unvarying sequence. He became familiar with the physical characteristics of each layer, with the fossils each contained, and with the sequence of the layers. By looking at a specimen of sedimentary rock collected from anywhere in southern England, he could name the layer from which it had come and, of course, the position of the layer in the sequence.

Smith did not believe that his discovery reflected any particular scientific principle; it was purely practical. Nevertheless, it opened the door to the correlation of sedimentary strata over increasingly wide areas. **Correlation** means the determination of equivalence in time-stratigraphic or rock-stratigraphic units of the succession of strata found in two or more different places. Smith correlated strata on the dual basis of physical similarity and fossil content initially over distances of several kilometers, and later over tens of kilometers. By means of fossils alone, it ultimately became possible to correlate through hundreds and then thousands of kilometers.

HOW CORRELATION IS ACCOMPLISHED

Correlation involves two main tasks. The first task is to determine the relative ages, one to another, of units exposed within a local area being studied. To accomplish the goal, a geologist employs various physical and biological criteria; one is not necessarily more dependable or precise than the others.

When a geologist starts mapping an area, he or she is almost always faced with the task of correlating from one outcrop to the next. Continuous outcrops of bedrock are not common—regolith, vegetation, lakes, even buildings may cover rocks—and so we are often faced with correlating between widely spaced outcrops. The task is to try and identify the same formation wherever it crops out. The physical matching of pieces of a formation generally involves the use of rock characteristics such as grain size, color, and sedimentary structures that permit the unit to be distinguished from others. The task must be carried out with care because, in some instances, quite different formations look almost identical. The Navajo Sandstone in Zion National Park shown in Figure 11.5, for example, is very similar to the Coconino Sandstone in the Grand Canyon (Figure 11.1B), 120 km to the south. However, correlation of the two formations based on physical criteria would be incorrect because careful mapping reveals that the Navajo Sandstone sits much higher in the stratigraphic section than the Coconino Sandstone.

When geologists get lucky, they discover thin and generally widespread sedimentary beds with characteristics so

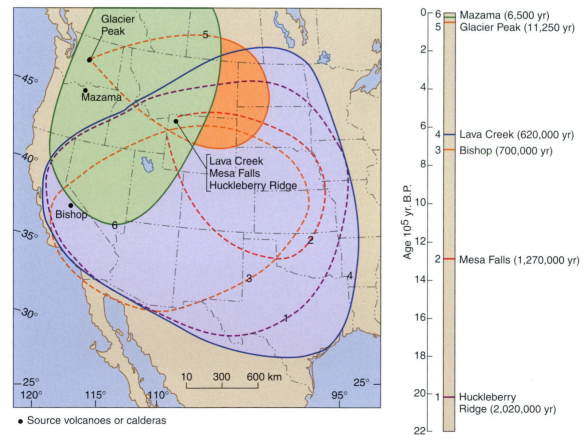

Figure 11.7 Key Beds Map showing distribution of widespread tephra layers in western North America which erupted during the last 2 million years. Each tephra layer is a key bed. The time bar at the side gives the age of each eruption.

distinctive that they can be easily recognized but not confused with any other bed. Such a distinctive unit is called a *key bed*, and it is exceedingly useful for correlation. In areas of volcanic activity, ash layers can serve as distinctive key beds for purposes of regional correlation as shown in Figure 11.7.

The second task of correlation is to establish the ages of the local rock units relative to a standard scale of geologic time. Distinctive fossils are especially useful for this purpose. We call a fossil that can be used to identify and date the stratum in which it is found an *index fossil*. To be most useful, an index fossil should have common occurrence, a wide geographic distribution, and a very restricted age range. The best examples are swimming or floating organisms that underwent rapid evolution and quickly became widely distributed (Figure 11.8). If a distinctive index fossil is recognizable at an outcrop, a rapid and reliable means of correlation is available (Figure 11.9). Although some genera and species permit long-range correlation of rocks in different geographic areas or even on different continents, more often close dating and correlation involve using assemblages of fossils of as many different types as possible.

Figure 11.8 Index Fossil *Homotelus bromidensis*, a trilobite that lived about 500 million years ago, during the Ordovician Period, is a distinctive index fossil. These specimens came from the Bromide Formation in Oklahoma.

> ### Before you go on:
>
> 1. What is the difference between a rock-stratigraphic record and a time-stratigraphic record?
>
> 2. What is correlation? What are the two main tasks accomplished by correlation?

THE GEOLOGIC COLUMN AND THE GEOLOGIC TIME SCALE

Earlier in the chapter we noted that nineteenth-century geologists began to study strata in terms of time-stratigraphic units. These geologists demonstrated, through stratigraphic correlation, that time-stratigraphic sequences are the same on all continents. Through worldwide correlation they assembled a **geologic column**, which is a composite column containing, in chronological order, the succession of known strata, fitted together on the basis of their fossils or other evidence of relative age. This worldwide standard is still being added to and refined as more rock units are described and mapped.

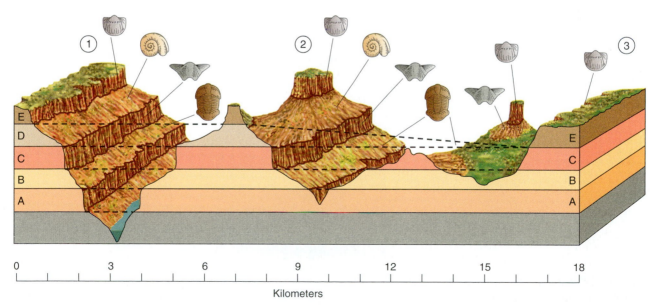

Figure 11.9 Correlating with Fossils Correlation of strata exposed at three localities, many kilometers apart, on the basis of similarity of the fossils they contain. The fossils show that at Locality 3 stratum D is missing because E directly overlies C. Either D was never deposited there, or it was deposited and later removed by erosion before the deposition of layer E.

You can think of the geologic column as a column of rock. The corresponding column of time is the **geologic time scale**, which chronologically orders geologic time units. Standard names have evolved for the subdivisions of the geologic time scale. These units include eons, eras, periods, and epochs as shown in Figure 11.10.

The names of the geologic time scale constitute the standard time language of geologists the world over. By learning about them you can begin to understand many details of Earth history that led to the discovery of the important principles of physical geology discussed in this book.

EONS

An **eon** is the largest interval into which geologic time is divided, and there are four of them. The term **Hadean** (Greek for "beneath the Earth") is given to the oldest eon. This is the earliest time unit in Earth's history, a time for which no rock record has been discovered because the rock cycle has apparently reworked whatever rocks did exist at that time. However, rocks of the Hadean Eon are present on other planets and moons of the solar system because the rock cycles on those planets and moons are much less effective than the rock cycle on Earth. Some of the samples

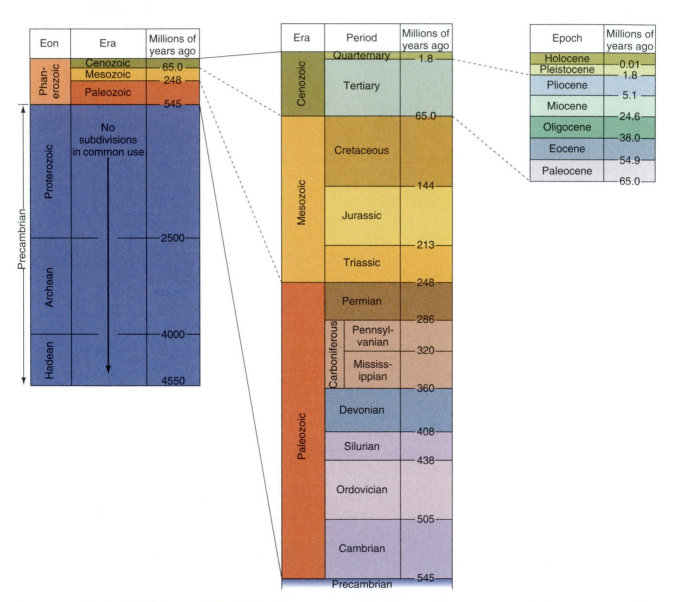

Figure 11.10 The Geologic Column and Time Scale The geologic time scale is the list of numerical ages in each of the right-hand columns. Numerical ages were obtained from radiometric dating. Note that the Pennsylvanian and Mississippian periods are equivalent to the Carboniferous period of Europe. The time boundary between the Archean and Hadean is a matter of definition because no rocks of the Hadean Eon are known on Earth. Hadean rocks are known to exist on other planets in the solar system.

brought back from the Moon were formed during the Hadean Eon.

The **Archean** (Greek for "ancient") **Eon** follows the Hadean. Archean rocks are the oldest rocks we know of on Earth, and they contain microscopic life forms of primitive character. Archean rocks are widespread on Earth and are found on all of the continents.

The **Proterozoic** ("earlier life") **Eon** follows the Archean. Proterozoic rocks include evidence of multicelled organisms that lacked preservable hard parts. Understandably, the record of the ancient Archean and Proterozoic is not as well known as the record of younger rocks because many of these ancient rocks have been intensely deformed, metamorphosed, and eroded.

The **Phanerozoic** ("visible life") is the most recent of the four eons. Phanerozoic rocks often contain plentiful evidence of past life in the form of well-preserved hard parts. Most examples of fossils that we see displayed in museums or illustrated in books are from the Phanerozoic Eon.

ERAS

Each of the eons is subdivided into shorter time units called **eras**. Though shorter than eons, eras nevertheless encompass major spans of time. No formal eras are yet widely recognized for subdivision of the Archean and Proterozoic Eons. In the Phanerozoic Eon, the eras are defined on the basis of the life-forms found in the corresponding rocks. As result, the Phanerozoic Eon is divided into the **Paleozoic** ("old life"), **Mesozoic** ("middle life"),

and **Cenozoic** ("recent life") eras, each name reflecting the relative stage of development of the life of these intervals (Figure 11.10).

Paleozoic forms of life progress from marine invertebrates to fishes, amphibians, and reptiles. Early land plants also appeared, expanded, and evolved. The Mesozoic Era saw the rise of the dinosaurs, which became the dominant vertebrates on land. Mammals first appeared during the Mesozoic Era, and later dominated the Cenozoic Era. The Mesozoic Era also witnessed the evolution of flowering plants, while during the Cenozoic Era grasses appeared and became an important food for grazing mammals.

PERIODS

The eras of the Phanerozoic Eon are divided into **periods**. The periods are defined on the basis of the fossils contained in the equivalent rocks, and because widespread fossils are confined to rocks of the Phanerozoic Eon, eras of the Proterozoic and Archean eons are not subdivided into periods.

The geologic periods have a haphazard nomenclature. They were defined over an interval of nearly 100 years on the basis of strata that crop out in Britain, Germany, Switzerland, Russia, and the United States. The names are partly geographic in origin, but in some cases they are based on characteristics of the strata in the place of original study (Table 11.1).

The lowermost period of the Paleozoic Era, the Cambrian Period, is the time when animals with hard

TABLE 11.1 Origin of Names for Periods of the Paleozoic, Mesozoic, and Cenozoic Eras, and the Epochs of the Quaternary and Tertiary Periods

Era	Period	Epoch	Origin of Name
Cenozoic	Quaternary[a]	Holocene	Greek for wholly recent
		Pleistocene	Greek for most recent
	Tertiary[a]	Pliocene	Greek for more recent
		Miocene	Greek for less recent
		Oligocene	Greek for slightly recent
		Eocene	Greek for dawn of the recent
		Paleocene	Greek for early dawn of the recent
Mesozoic	Cretaceous	↑	Latin for chalk, after chalk cliffs of southern England and France
	Jurassic	Epoch	Jura Mountains, Switzerland and France
	Triassic	Names	Threefold division of rocks in Germany
Paleozoic	Permian	Used	Province of Perm, Russia
	Pennsylvanian	Only	State of Pennsylvania
	Mississippian	By	Mississippi River
	Devonian	Specialists	Devonshire, county of southwest England
	Silurian		Silures, ancient Celtic tribe of Wales
	Ordovician	↓	Ordovices, ancient Celtic tribe of Wales
	Cambrian		Cambria, Roman name for Wales

[a] Derived from eighteenth- and nineteenth-century geologic time scale that separated crustal rocks into a fourfold division of Primary, Secondary, Tertiary, and Quaternary, based largely on relative degree of lithification and deformation.

shells first appear in the geologic record. Prior to the Cambrian Period, all animals were soft bodied and the fossil evidence they left is very sparse. Rocks formed during the Archean and Proterozoic eons cannot be readily differentiated on the basis of the fossils they contain, so geologists often refer to the entire time preceding the Cambrian as simply Precambrian.

EPOCHS

Periods are further subdivided into **epochs**, and the basis of subdivision is the fossil record. Although all of the periods of the Paleozoic and Mesozoic eras are subdivided into epochs, only the epochs of the Tertiary and Quaternary periods are in common enough use to warrant inclusion here.

The epochs of the Tertiary Period were defined in an interesting manner. Tertiary marine strata in sedimentary basins of France and Italy contain fossils that are identical, in some cases, to marine creatures that live today. The youngest Tertiary strata contain fossils of many living creatures. The oldest Tertiary strata have fossils of few living creatures. Charles Lyell subdivided the Tertiary strata into groupings based on the percentage of their fossils that are represented by still-living species, and named the epochs accordingly (Figure 11.10 and Table 11.1). The Quaternary Period is divided into the Holocene Epoch, which is the epoch in which we are now living, and which stretches back in time to the withdrawal of the last great ice sheet from Europe and North America, and the Pleistocene Epoch, which is the time before the Holocene when the great ice sheets waxed and waned.

Each of the various periods of the Paleozoic and Mesozoic eras is also subdivided into epochs, the names of which are primarily geographic in origin. They are used mainly by specialists concerned with detailed studies of these strata and their contained fossils.

Before you go on:

1. What is the geologic time scale? How is it related to the geologic column?

2. What are the names of the four eons? What is their order of ages, starting with the oldest?

3. What are the three eras of the Phanerozoic Eon? the periods of the Phanerozoic Eon?

MEASURING GEOLOGIC TIME NUMERICALLY

The scientists who worked out the geologic column and time scale were challenged by the question of numerical time. They knew the relative ages of the different systems in considerable detail, but they also wished to know whether the strata in each system had accumulated over the same length of time. They sought answers to questions such as these: "How much time elapsed between the end of the Cambrian Period and the beginning of the Permian Period?" "How long was the Tertiary Period?" "How long did the dinosaurs dominate the lands?" and "When did human ancestors first appear in the fossil record?" The question of numerical time is as important as the geologic column and the geologic time scale. Numerical ages must be determined in order to answer such challenging and important questions as the age of Earth, the age of the ocean, and how fast mountain ranges rise.

EARLY ATTEMPTS TO MEASURE GEOLOGIC TIME NUMERICALLY

During the nineteenth century, many attempts were made to develop a numerical scale of years for the geologic time scale. One widely used method was discussed earlier in this chapter. It consisted of estimates of the time during which the rock cycle has been at work based on rates of sedimentation and the thickness of sedimentary strata. Estimates for the age of Earth, which was presumed to be the same as the duration of the rock cycle, ranged from 3 million to 1.5 billion years!

A clever suggestion for estimating the age of the ocean concerned the saltiness of seawater. Because sea salts come from the erosion of common rocks and reach the sea dissolved in river water, it was argued, why not measure the salts in modern river water and calculate the time needed to transport all the salts now in the sea? Edmund Halley (for whom the comet is named) suggested, in 1715, that sea salt might be used to date the ocean. John Joly finally made the necessary measurements and calculations in 1889. His answer for the ocean's age, 90 million years, does not represent the age of the ocean at all, however. The composition of the ocean is essentially constant because all the salts in the sea, like all other chemical constituents, are cyclic. Salts are added both by erosion and by submarine volcanism, but salts are also removed from solution. Some are removed as evaporite minerals (Chapter 7), whereas others are removed by reaction with hot volcanic rocks on the seafloor. Both Halley and Joly were misled because they did not understand the workings of the Earth system.

Perhaps the most interesting estimates were those made by Lord Kelvin, a physicist, who attempted to calculate the time Earth has been a solid body. Earth started as a very hot object, he argued. Once it had cooled sufficiently to form a solid outer crust, it could continue to cool only by the conduction of heat through solid rock. By measuring the thermal properties of rock and estimating the present temperature of Earth's interior, he calculated the time for Earth to cool to its present state.

Kelvin's logic was faultless, and his mathematical calculations were correct. However, his estimate of 100 million years for the maximum age of Earth is incorrect because he made an incorrect assumption. Kelvin assumed that no additional heat had been added since Earth was formed. Because radioactivity was not known at the time, Kelvin could not have realized that radioactivity continuously supplies heat to Earth's interior. Instead of cooling rapidly, Earth's interior is cooling so slowly it has a nearly constant temperature over periods as long as hundreds of millions of years.

RADIOACTIVITY AND THE MEASUREMENT OF NUMERICAL TIME

Resolving the dilemma of numerical time required finding a way to measure geologic time by some process that runs continuously, is not reversible, is not influenced by other processes and other cycles, and leaves a continuous record without gaps. In 1896, the discovery of radioactivity provided the needed method.

NATURAL RADIOACTIVITY

We learned in Chapter 3 that the atomic number of a given element—that is, the number of protons in the atomic nucleus of the element—is constant and characteristic of that element. However, an atomic nucleus also contains neutrons, and the number of neutrons can vary without changing the number of protons. For example, all carbon atoms contain six protons, but the protons can be joined by six, seven, or eight neutrons.

Different kinds of atoms of an element that contain different numbers of neutrons are called **isotopes**. An isotope is identified by its **mass number**, which is the sum of the neutrons plus the protons. Carbon, for example, has three isotopes (Figure 11.11) with mass numbers 12, 13, and 14, which are written ^{12}C, ^{13}C, and ^{14}C, respectively. Most of the common chemical elements have several isotopes.

Isotopes of the chemical elements found in Earth are generally stable and not subject to change. However, a few, such as ^{14}C, are radioactive. Radioactivity arises because of instability within an atomic nucleus. If the ratio of the number of neutrons (n) to the number of protons (p) is too high or too low, the atomic nucleus of a radioactive isotope will transform spontaneously to a nucleus of a more stable isotope of a different chemical element. Even though the process is one of transformation—from an unstable nucleus to the nucleus of an atom of a different kind—it has become common practice to call the process **radioactive decay**. An atomic nucleus undergoing radioactive decay is

said to be a **parent**; the product arising from radioactive decay is called a **daughter**. For example, ^{14}C decays to ^{14}N and ^{238}U decays to ^{206}Pb; ^{14}C and ^{238}U are parents, ^{14}N and ^{206}Pb daughters.

KINDS OF RADIOACTIVE DECAY

Radioactive decay can happen in five ways:

1. *Beta decay.* By emission of an electron from the nucleus; such an electron is referred to as a β^- (beta) particle, and emission occurs because a neutron becomes a proton:

$$n \rightarrow p + \beta^-$$

By a beta emission, the number of neutrons in a nucleus is decreased by one and the number of protons is increased by one, thus reducing the ratio n/p.

2. *Positron emission.* By emission of a particle with the same mass as an electron but with a positive charge; such a particle is called a *positron* and is written β^+. Positron emission decreases the number of protons in a nucleus by one and increases the number of neutrons by one, according to the reaction:

$$p \rightarrow n + \beta^+$$

Thus, by positron emission the ratio of n/p is increased.

3. *Electron capture.* By capture into the nucleus of one of the orbital electrons, a process that decreases the number of protons in the nucleus by one, according to the reaction:

$$p + \text{electron} \rightarrow n$$

Note that the term *beta particle* is used only for electrons emitted from the nucleus. Electron capture, like positron emission, increases the ratio of n/p.

4. *Alpha decay.* By emission from the nucleus of a heavy atomic particle consisting of two neutrons and two protons $(2p + 2n)$, called an α (alpha) particle. Loss of an alpha particle reduces the mass number by four and the atomic number by two. Alpha decay is a process that happens with isotopes that have large atomic numbers. Because isotopes with large atomic numbers tend to have more neutrons than protons in the nucleus, emission of an alpha particle causes an increase in n/p.

5. *Gamma ray emission.* By emission of γ rays (gamma rays), which are very short-wavelength, high-energy electromagnetic rays. Gamma rays have no mass, so gamma-

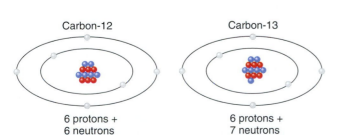

Carbon-12

6 protons +
6 neutrons

Carbon-13

6 protons +
7 neutrons

Carbon-14

6 protons +
8 neutrons

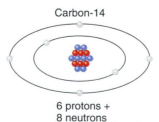

Figure 11.11 Naturally Occurring Isotopes of Carbon The three isotopes of carbon. Note that in each case there are six protons in the nucleus and six electrons in the energy-level shells. The differences lie in the numbers of neutrons in the nucleus.

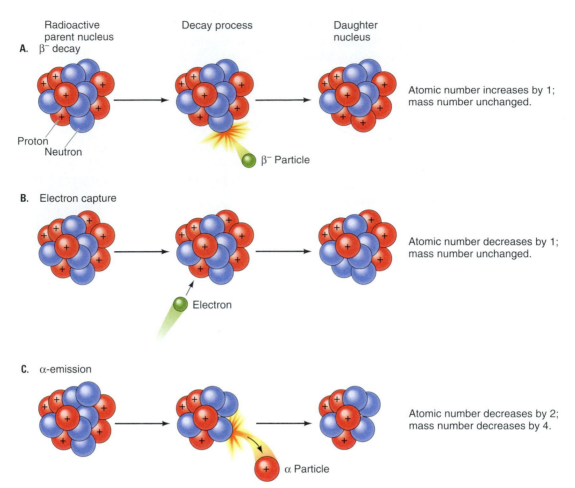

Radioactive parent nucleus   Decay process   Daughter nucleus

A. β⁻ decay

Proton
Neutron

β⁻ Particle

Atomic number increases by 1; mass number unchanged.

B. Electron capture

Electron

Atomic number decreases by 1; mass number unchanged.

C. α-emission

α Particle

Atomic number decreases by 2; mass number decreases by 4.

Figure 11.12 Radioactive Decay Three of the ways radioactive isotopes decay. Note that in each case the atomic number (number of protons) of the daughter isotope differs from the atomic number of the parent isotope.

ray emission does not affect either the atomic number or the mass number of an isotope. Emission of a gamma ray happens when decay by one of the four previous methods produces an isotope with a high-energy state. By emitting a gamma ray, the newly formed isotope moves to a more stable, lower-energy state.

All of these radioactive decay schemes occur in isotopes found in Earth, but three of the schemes—beta decay, electron capture, and alpha decay, illustrated in Figure 11.12—are especially important for determining numerical ages of geologic samples.

RATES OF DECAY AND THE HALF-LIVES OF ISOTOPES

The rate at which radioactive decay occurs varies among isotopes. Many of the radioactive isotopes that were once in Earth have decayed away and are no longer present because their rates of spontaneous decay are fast. A few radioactive isotopes that transform very slowly are still present, however.

Careful study of radioactive isotopes in the laboratory has shown that decay rates are unaffected by changes in the chemical and physical environment. Thus, the decay rate of a given isotope is the same in the mantle, or in a magma, as it is in a sedimentary rock. This is a particularly important point because it leads to the conclusion that rates of radioactive decay are not influenced by geologic processes, like erosion, metamorphism, or the melting of rock to form magma.

In the least complicated process of radioactive decay, the number of radioactive parent atoms continuously decreases, while the number of nonradioactive daughter atoms continuously increases. More complicated processes of radioactive decay involve daughter atoms that are themselves radioactive and so undergo further decay until a stable daughter is finally produced. Regardless of the complications involved in the decay scheme, all decay timetables follow the same basic law: the *proportion*— fraction or percentage—of parent atoms that decay during each unit of time is always the same (see Figure 11.13).*

The rate of radioactive decay is measured in terms of the **half-life**, which is the amount of time needed for the

---

* The application of this law is valid only when the number of atoms is large enough (at least tens of thousands) so that statistical variations in the counting procedure are negligible.

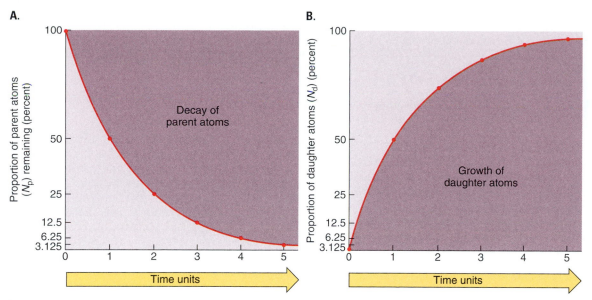

Figure 11.13 Radioactivity and Time Curves illustrating the basic law of radioactivity. The curves are drawn for the circumstance where a radioactive parent isotope decays to a stable daughter isotope. A. At time zero, a sample consists of 100 percent radioactive parent atoms. During each time unit, half the atoms remaining decay to daughter atoms. B. At time zero, no daughter atoms are present. After one time unit corresponding to a half-life of the parent atoms, 50 percent of the sample has been converted to daughter atoms. After two time units, 75 percent of the sample is daughter atoms, 25 percent parent atoms. After three time units, the percentages are 87.5 and 12.5, respectively. Note that at any given instant N_p, the number of parent atoms remaining, plus N_d, the number of daughter atoms, equals N_0, the number of parent atoms at time zero.

number of parent atoms to be reduced by one-half. For example, let's say that the half-life of a radioactive isotope is 1 hour. If we started an experiment with 1,000,000 parent atoms and the daughter atoms are not radioactive, only 500,000 parent atoms would remain at the end of an hour, and 500,000 daughter atoms would have formed. At the end of the second hour, another half of the parent atoms would be gone, so there would be 250,000 parent atoms and 750,000 daughter atoms. After the third hour, another half of the parent atoms would have decayed, leaving 125,000 parent atoms and 875,000 daughter atoms. The proportion of parent atoms that decay during each time interval is 50 percent. (The same kind of law governs compound interest paid on a bank account.)

In the graphic illustration of radioactive decay in Figure 11.13, the time units marked are half-lives. At the end of each unit the number of parent atoms, and therefore the radioactivity of the sample, has decreased by exactly one-half. Figure 11.13, which is drawn for the case where a radioactive isotope decays directly to a stable daughter isotope, also shows that the growth of daughter atoms just matches the decline of parent atoms. When the number of remaining parent atoms (N_p) is added to the number of daughter atoms (N_d), the result is N_0, the number of parent atoms that we started with. That fact is the key to the use of radioactivity as a means of measuring geologic time and determining ages.

We can think of the radioactivity in a mineral as a built-in clock. The length of time this clock has been ticking—the age, expressed in years, since radioactive decay in the mineral began—is the mineral's **radiometric age**. Knowing when radioactive decay began can tell us when the mineral formed, as we will soon demonstrate.

What is actually determined when a radiometric age is calculated? If some of the daughter atoms are lost, an incorrect radiometric age is the result. At high temperature, for example, daughter atoms might diffuse out of a mineral. The most effective mineral samples for determining radiometric ages are those from extrusive igneous rocks. Mineral formation as a result of crystallization of lava or volcanic ash, and cooling to surface temperature, are rapid processes—a matter of days or weeks, which is instantaneous in terms of geologic time spans. With extrusive igneous rocks, loss due to diffusion is not a problem. A radiometric mineral age in such a case, therefore, is the age of the volcanic rock. Intrusive igneous rocks and metamorphic rocks, by contrast, cool over much longer time spans than extrusive igneous rocks. Radiometric ages for such slow-cooling bodies are less certain because daughter atoms may well diffuse out of the mineral at the highest temperatures and lead to false ages. Fortunately there is one mineral, zircon ($ZrSiO_4$), from which daughter atoms do not diffuse out, even at magmatic or high-grade metamorphic temperatures, and so can be used to determine radiometric ages of intrusive igneous and metamorphic rocks.

Many natural radioactive isotopes can be used for **radiometric dating**—the determination of radiometric age—but six predominate in geologic studies: two radioactive isotopes of uranium plus radioactive isotopes of thorium,

THE SCIENCE OF GEOLOGY

POTASSIUM–ARGON (^{40}K/^{40}Ar) DATING

To illustrate how minerals can be dated radiometrically to determine when they formed, we have selected one of the naturally radioactive isotopes, potassium-40 (^{40}K). Potassium has three natural isotopes: ^{39}K, ^{40}K, and ^{41}K. Only one, ^{40}K, is radioactive and its half-life is 1.3 billion years. The decay of ^{40}K is interesting because two different decay schemes occur. Twelve percent of the ^{40}K atoms decay by electron capture to ^{40}Ar, an isotope of the gas argon. The remaining 88 percent of the ^{40}K atoms change by beta decay to ^{40}Ca. The appropriate equations are:

$$^{40}\text{K} + \text{electron} \rightarrow {}^{40}\text{Ar}$$

and

$$^{40}\text{K} \rightarrow {}^{40}\text{Ca} + \beta^-$$

It is important to know that the fraction of ^{40}K atoms decaying to ^{40}Ar is always 12 percent; the percentage is not affected by changes in physical or chemical conditions.

When a potassium-bearing mineral crystallizes from a magma or grows within a metamorphic rock, it includes some ^{40}K in its crystal structure. As soon as the mineral is formed, ^{40}Ar and ^{40}Ca daughter atoms start accumulating in the mineral because, like the parent ^{40}K atoms, they are trapped in the crystal structure.

Because the ratio of ^{40}Ar to ^{40}Ca daughter atoms is always the same, it is only necessary to measure either ^{40}Ar or ^{40}Ca daughter atoms in order to know how many ^{40}K atoms have decayed. Argon is the daughter isotope selected for analysis. There are several reasons for doing so; one is that ^{40}Ca is the main calcium isotope in nature so samples are essentially always contaminated. The main reason to select ^{40}Ar for analysis, however, concerns the unusual atomic properties of argon. The electron energy-level shells of argon are filled, so atoms of argon do not readily form chemical bonds. Thus, ^{40}Ar atoms are not chemically bound in minerals; they are trapped by the crystal lattice but do not form bonds. The trapping is effective only at low temperatures. Thus at high temperatures argon rapidly diffuses out of a mineral but at lower temperatures it cannot diffuse out and stays trapped. When a potassium-bearing mineral crystallizes it will not retain any initial argon because magmatic temperatures are far above trapping temperatures. As a result, all the ^{40}Ar atoms in a potassium-bearing mineral in an extrusive igneous rock such as a rhyolite or andesite must have come from decay of ^{40}K and must have accumulated since the temperature fell below the trapping temperature. Because extrusive igneous rocks cool very quickly, the trapping time and eruption time are essentially the same.

All that has to be done to determine the numerical age of eruption of an extrusive igneous rock, then, is to select a potassium-bearing mineral and measure the amount of parent ^{40}K that remains and the amount of trapped ^{40}Ar.

Since the half-life of ^{40}K is known, it is a straightforward matter to calculate the radiometric age.

Flexibility of ^{40}K/^{40}Ar Dating

Dating by ^{40}K is not limited to minerals that contain potassium as a major element. Even minerals that contain small amounts of potassium will serve the purpose. Thus, hornblende, a calcium–iron–magnesium silicate, can be used for ^{40}K/^{40}Ar dating because it generally contains a small quantity of potassium present by ionic substitution.

^{40}K/^{40}Ar dating is most successfully applied to volcanic rocks because their solidification and cooling are rapid. As a result, they have solidification ages that are essentially the same as their trapping ages. Because argon analyses can be performed with great sensitivity and because contamination by initial argon at the time of crystallization is generally not a problem, the method can be used for volcanic rocks as young as 20,000 years. For this reason, ^{40}K/^{40}Ar dating has proved very useful in studies of archaeology as well as geology (see Box 11.2).

Calculation of a ^{40}K/^{40}Ar Age

Chemical analysis of a potassium feldspar sample from a pyroclastic rock shows that for every 20,000 parent atoms of ^{40}K present there are 1200 atoms of ^{40}Ar. We know that the ratio ^{40}Ar/^{40}Ca is constant, so 1200 atoms of ^{40}Ar means that 8800 atoms of ^{40}Ca are also present. This in turn means that N_d, the number of daughter atoms, is 1200 + 8800, or 10,000. The equation for radioactive decay (i.e., the equation for the curve in Figure 11.13) is:

$$\frac{N_p}{N_0} = (1 - \lambda)^y$$

where N_p is the number of parent atoms now, in our example 20,000. N_0 is the number of parent atoms when the mineral formed. Because each parent atom that decays produces only one daughter atom,

$$N_0 = N_p + N_d = 20,000 + 10,000 = 30,000$$

λ is the decay constant, which is the fraction of parent atoms that decays per unit time. y is the number of time units.

We can simplify the calculation in two ways. First, we select the unit of time to be equal to the half-life, which means $\lambda = 0.5$. This in turn makes y equal to the number of half-lives since the mineral formed, so that:

$$\frac{N_p}{N_0} = (1 - 0.5)^y = 0.5^y$$

The second simplification is to put the equation in a logarithmic form:

$$\log N_0 - \log N_p = 0.3y$$

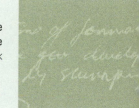

Thus

$$\log 30{,}000 - \log 20{,}000 = 0.3y$$

or

$$y = \frac{4.477 - 4.301}{0.3} = 0.587$$

The feldspar therefore formed 0.587 half-lives ago. The half-life of ^{40}K has been measured in the laboratory to be 1300 million years. Thus, the age of the mineral is $1300 \times 0.587 = 760$ million years.

potassium, and rubidium, each of which has a long half-life and can be used to determine igneous and metamorphic radiometric ages, and carbon, which is used to determine radiometric ages of organic matter. These isotopes, the materials they can be used with, and the radiometric age spans they are useful for, are further described in Table 11.2. Because these isotopes occur widely in different mineral and rock types, and because they have a wide range of half-lives, many geologic samples can be dated radiometrically through their use. In practice, an isotope can be used for dating samples that are no older than about six half-lives of the isotope—in samples older that six half-lives there is too little of the parent isotope remaining for analysis.

Here, we focus on two of the isotopes in Table 11.2. *The Science of Geology*, Box 11.1, *Potassium–Argon ($^{40}K/^{40}Ar$) Dating*, illustrates how an isotope is typically used in radiometric dating of a mineral. The following section discusses carbon-14 (^{14}C), which is unique among the isotopes in Table 11.2 in that it is used principally for dating organic remains.

RADIOCARBON DATING

Among the radiometric dating methods listed in Table 11.2, the one based on ^{14}C (also known as radiocarbon) is especially useful for dating geologically young samples, by which we mean samples from the Holocene and the latest part of the Pleistocene. The dating of such young samples is possible because the half-life of radiocarbon is short—5730 years—by comparison with the half-lives of most isotopes used for radiometric dating of geologic samples (Table 11.2).

Radiocarbon is continuously created in the atmosphere through bombardment of ^{14}N by neutrons created by cosmic radiation (Figure 11.14). The ^{14}C mixes with ^{12}C and ^{13}C and diffuses rapidly through the atmosphere, hydrosphere, and biosphere. Because the rates of mixing and exchange are rapid compared with the half-life, the proportion of ^{14}C is nearly constant throughout the atmosphere and biosphere. As long as the production rate remains constant, the radioactivity of natural carbon in the

TABLE 11.2 Some of the Principal Isotopes Used in Radiometric Dating

Isotopes			Half-Life of Parent (years)	Effective Dating Range (years)	Minerals and Other Materials That Can Be Dated
Parent	Decay System	Daughter			
^{238}U	α and β^- decay	^{206}Pb	4.5 billion	10 million–4.6 billion	Zircon and uraninite
^{235}U	α and β^- decay	^{207}Pb	710 million	10 million–4.6 billion	Zircon and uraninite
^{232}Th	α and β^- decay	^{208}Pb	14 billion	10 million–4.6 billion	Zircon and uraninite
^{40}K	Electron capture β^- decay	^{40}Ar ^{40}Ca	1.3 billion	50,000–4.6 billion	Muscovite Biotite Hornblende Whole volcanic rock
^{87}Rb	β^- decay	^{87}Sr	47 billion	10 million–4.6 billion	Muscovite Biotite Potassium feldspar Whole metamorphic or igneous rock
^{14}C	β^- decay	^{14}N	5730 ± 30	100–70,000	Wood, charcoal, peat, grain, and other plant material Bone, tissue, and other animal material Cloth Shell Stalactites Groundwater Oceanwater Glacier ice

A. ^{14}C created by neutron capture

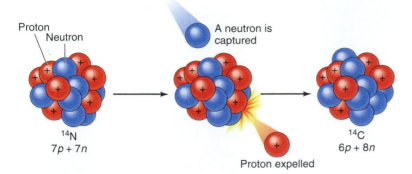

B. ^{14}C decays to ^{14}N by β$^-$ decay

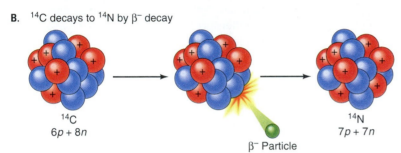

Figure 11.14 Radiocarbon Production and decay of radiocarbon. A. The nucleus of a ^{14}N atom captures a neutron that was created by cosmic ray bombardment of the atmosphere and expels a proton, thereby changing ^{14}N to ^{14}C. B. The nucleus of a ^{14}C atom is radioactive. By beta decay, ^{14}C reverts back to ^{14}N.

atmosphere remains constant because the rate of production balances the rate of decay.

Radiocarbon dating is particularly effective for determining the time of death of organic remains. While an organism is alive, it will continuously take in carbon from the atmosphere through activities such as feeding, breathing, and photosynthesis, and so will contain the balanced proportion of ^{14}C. However, at death the balance is upset, because replenishment by life processes ceases. The ^{14}C in dead tissues continuously decreases by radioactive decay.

Analysis of a sample for its radiometric age by the radiocarbon method poses a special problem. As we noted, ^{14}C is formed from ^{14}N, but it also decays back to ^{14}N by beta decay—that is, the daughter product of ^{14}C decay is ^{14}N. Because nitrogen is the most abundant gas in the atmosphere, and ^{14}N is the most abundant isotope of nitrogen, it is impossible to get a sample for dating that is not contaminated with ^{14}N. This means that radiocarbon dating cannot be done by measuring the daughter atoms. Instead, analysis for the radiocarbon date of a sample involves a determination of the radioactivity level of the carbon it contains. The method works because the fraction of carbon that was radiocarbon at the time of death was the same as the fraction in the atmosphere. No radiocarbon is added after death, so by measuring the radioactivity remaining in an organic sample, we can calculate how many half-lives ago the organism died.

Because of its application to organisms (by dating fossil wood, charcoal, peat, bone, and shell material), radiocarbon has proved to be enormously valuable in establishing dates for prehistoric human remains and for recently extinct animals. In this way it is of extreme importance in archaeology.

It is also of great value in dating the most recent part of geologic history, particularly the latest glacial age. For example, many samples of wood have been taken from trees overrun by the advance of the latest great ice sheet and buried in rock debris. Dating these samples has revealed that the ice reached its greatest extent in the Ohio–Indiana–Illinois region about 18,000 to 21,000 years ago. It is even possible to date young ice, such as that in the Greenland Ice Sheet, directly. As the ice forms, bubbles of air are trapped in it. The carbon dioxide in the air bubbles can be liberated in the laboratory and dated, providing a date for ice formation.

Similarly, radiocarbon dates afford the means for determining rates of geologic processes, such as the rate of advance of the last ice sheet across Ohio; the rate of rise of the sea against the land while glaciers melted throughout the world; average rates of circulation of water in the deep ocean; the rates of local uplift of the crust that raised beaches above sea level; and even the frequency of volcanism. For an example of the way radiocarbon dating can be combined with other techniques to solve a geologic problem, see *Science of Geology*, Box 11.2, *K/Ar Dating of Hawaiian Glaciations and African Hominids*.

RADIOMETRIC DATING AND THE GEOLOGIC COLUMN

Radiometric dating has been particularly effective in solving the problem of adding numerical ages to the geologic column. The standard units of the geologic column consist of sedimentary strata containing characteristic fossils, but the typical rocks from which radiometric ages (other than radiocarbon dates) are determined are igneous rocks.

Box 11.2 THE SCIENCE OF GEOLOGY

K/Ar Dating of Hawaiian Glaciations and African Hominids

The ability to attach dates to events in the geologic record has been an enormous boon for geologists. It is now possible to obtain quantitative answers to many questions that a few years ago could be approached only in a descriptive way. Two examples demonstrate how radiometric dating can be used.

During the recent ice age, ice caps and glaciers existed in many places that are ice-free today. The problem is to determine whether the age of the ice was everywhere the same. The age problem is especially severe on oceanic islands. On Mauna Kea, a 4200 m high shield volcano on Hawaii, an ice cap formed, then expanded and contracted several times. Mauna Kea is now dormant, but it was active when the ice cap existed. On the slopes of the volcano, flows of potassium-rich lavas are interlayered with sediment deposited by the ice as shown in Figure B11.1. The lavas have been dated by the ^{40}K/^{40}Ar method; organic matter in sediment from a postglacial lake has been dated by the ^{14}C method.

The deepest exposed glacial sediments, which are part of a buried moraine (Chapter 16), are close to 150,000 years old based on K/Ar ages of underlying and overlying lavas. A younger moraine lying partly exposed at the surface is overlain by lavas with ages close to 65,000 years, while the youngest moraines are associated with lavas and lake sediments that are between about 40,000 and 13,000 years old. These ages indicate that the glaciers on Mauna Kea advanced at essentially the same time as glaciers in the mountains of western North America.

A second dating example concerns the Haddar region of northern Ethiopia, one of the most productive places in the world for finding fossils of ancestral human beings. The sediments give good magnetic signals (Figure B11.2). The problem is to know where in the magnetic polarity time scale the Haddar reversals fall. ^{40}K/^{40}Ar dates were obtained on a tuff and a basalt flow. The two radiometric dates determine the magnetic reversal ages unambiguously and indicate that early hominids lived in the region between 3 and 4 million years ago; that is, they lived there during the Pliocene Epoch.

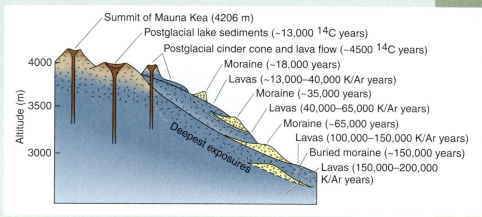

Figure B11.1 Radiometric Dating An example of radiometric dating. Glacial deposits on Mauna Kea Volcano, Hawaii, are interlayered with basaltic lava flows that have been dated by the ^{40}K/^{40}Ar method. See text for discussion.

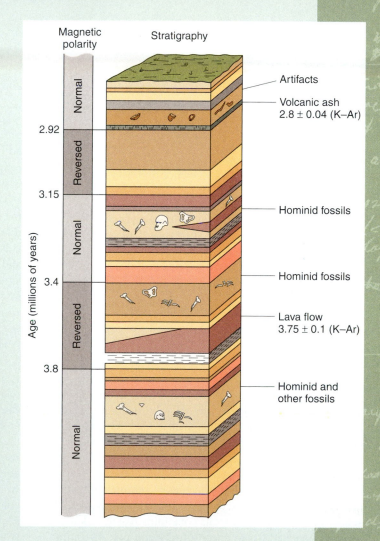

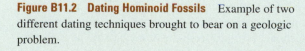

Figure B11.2 Dating Hominoid Fossils Example of two different dating techniques brought to bear on a geologic problem.

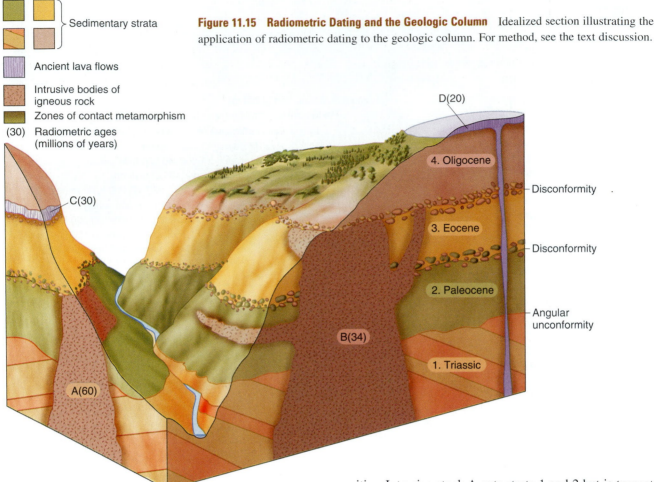

Sedimentary strata

Ancient lava flows

Intrusive bodies of igneous rock

Zones of contact metamorphism

(30) Radiometric ages (millions of years)

Figure 11.15 Radiometric Dating and the Geologic Column Idealized section illustrating the application of radiometric dating to the geologic column. For method, see the text discussion.

D(20)

4. Oligocene

Disconformity

3. Eocene

C(30)

Disconformity

2. Paleocene

Angular unconformity

B(34)

1. Triassic

A(60)

Stratum	Age (millions of years)	Interpretation
4	<34 (younger than B) <30 (younger than C) >20 (older than D)	Age lies between 20 and 30 million years
3	<60 (younger than A) >34 (older than B) >30 (older than C)	Age lies between 34 and 60 million years
2	>60 (older than A)	Age of both is greater
1	>60 (older than A)	than 60 million years

Through various methods of radiometric dating, geologists have determined the dates of solidification of many bodies of igneous rock. When such bodies have identifiable positions in the geologic column, it is possible to date, approximately, a number of the sedimentary layers in the column.

Figure 11.15 shows how the ages of sedimentary strata are approximated from the ages of igneous bodies. A sequence of sedimentary strata containing fossils of known ages is separated by an unconformity and two disconfor-

mities. Intrusive stock A cuts strata 1 and 2 but is truncated by the disconformity at the top of stratum 2. Thus, A must be younger than strata 1 and 2 but older than stratum 3, which was laid down on the erosion surface at the top of stratum 2 and contains weathered fragments of A among the sedimentary particles. Similarly, the combination of dikes and sills that make up the intrusive igneous complex B is truncated by the disconformity at the top of stratum 3, and they must be younger than stratum 3 but older than stratum 4. Lava flow C above the disconformity at the top of stratum 3 must also be younger than stratum 3 and younger than the dike–sill complex B. Lava flow C must be older than stratum 4, however, because it is covered by stratum 4, and lava flow D must be even younger because it overlies stratum 4.

From the radiometric dates of the igneous bodies and the relative ages of the rock units shown in Figure 11.15, we can draw inferences concerning the ages of the sedimentary strata, as shown in the table that accompanies the figure.

Through a combination of geologic relations and radiometric dating methods, twentieth-century scientists were able to fit a scale of numerical time to the geologic column worked out in the nineteenth century. The scale is being continuously refined, and so the numbers given in Figure 11.10 should be considered the best available now. Further work will make them more accurate.

It is a great tribute to the work of geologists during the nineteenth century that radiometric dating has fully confirmed the geologic column they established by the ordering of strata into relative ages. It is interesting, too, to see just how wrong Lord Kelvin was in his estimate of 100 million years for Earth's age.

How Old Is Earth?

The earliest record of rocks on Earth, as indicated in Figure 11.10, comes from the great assemblage of metamorphic and igneous rocks formed during the Precambrian. Of the many radiometric dates obtained from them, the youngest are around 600 million years and the oldest about 4.0 billion years. The Precambrian portion of the geologic column, then, existed during a *minimum* time equal to 4.0 billion minus 600 million years, or 3.4 billion years—a span about five times as long as the entire Phanerozoic Eon.

Given that some Precambrian rocks are about 4 billion years old, the beginning of Earth's history must be still further back in time. The oldest radiometric dates, 4.1 billion years, have been obtained on individual mineral grains in clastic sedimentary rocks from Australia. Dates that are almost as old—4.0 billion years—have been obtained from granitic rocks from Canada. The existence of ancient granite proves that continental crust was present 4.0 billion years ago, whereas the 4.1-billion-year-old sedimentary grains prove that rock was being weathered, sediments were forming, and the rock cycle was operating then.

Further confirmation of the ancient age of continental crust comes from another body of very ancient Precambrian rock, a 3.6-billion-year-old granite in South Africa. Although itself an igneous rock, this ancient granite contains xenoliths of quartzite. At an earlier time, before it became enveloped by the granite magma, the quartzite must have been part of a layer of sandstone. Before that, it must have been part of a layer of loose sand. Earlier still, an igneous rock must have been subjected to weathering and erosion to produce the grains of quartz sand. Clearly, therefore, the rock cycle must have been operating in its present manner well before the granite magma solidified. Hence, as far back as we can see through the geologic column, we find evidence of the rock cycle. Furthermore, because we see ancient sediment that must have been transported by water, we know that when that sediment was deposited there must have been a hydrosphere.

The ancient Precambrian rocks we have been discussing are all from the Archean Eon. So far, at least, no rocks from the still older Hadean Eon have been discovered on Earth. How long did the Hadean Eon last, and therefore how much older might our planet be? Strong evidence suggests that Earth formed at the same time as the Moon, the other planets, and meteorites (small independent bodies that have "fallen" onto Earth). Through various methods of radiometric dating—in particular, the Rb/Sr and U/Pb systems—it has been possible to determine the oldest ages of meteorites and of rock fragments in "Moon dust" (brought back by astronauts) as 4.55 billion years. By inference, the time of formation of Earth, and indeed of all the other planets and meteorites in the solar system, is believed to be approximately 4.55 billion years ago. The Hadean Eon, therefore, lasted for 550 million years, from 4.55 to 4.0 billion years ago.

Before you go on:

1. What is the half-life of an isotope?

2. What is a radiometric age?

3. How can organic remains be dated?

THE MAGNETIC POLARITY TIME SCALE

Rocks have magnetic properties that provide yet another dating method. Earth is like a gigantic magnet. It has an invisible magnetic field that permeates everything. If a small magnet is allowed to swing freely in Earth's magnetic field, the magnet will become oriented so that its axis is parallel to Earth's magnetic lines of force. Therefore the axis will point to the direction of Earth's magnetic poles (Figure 11.16). This is true for all places on Earth. All free-swinging magnets will point to the magnetic poles.

Certain rocks become permanent magnets as a result of the way they form, and like free-swinging magnets, the direction of their magnetic axis points to the magnetic poles. That property leads them to be useful as dating tools in the following manner.

Magnetism in Rocks

Magnetite and certain other iron-bearing minerals can become permanently magnetized. This property arises because orbital electrons spinning around a nucleus create a tiny atomic magnetic field. In minerals that can become permanent magnets, the atomic magnets line up in parallel arrays and reinforce each other. In nonmagnetic iron minerals, the atomic magnets are oriented in random directions.

Let's look more closely at how magnetism occurs. Temperature plays a key role. Above a certain temperature (called the Curie point), the thermal agitation of atoms is such that permanent magnetism is impossible; the magnetic fields of all the iron atoms are randomly oriented and cancel each other out. Below that temperature, however, the magnetic fields of adjacent iron atoms reinforce each other (Figure 11.17). When an external magnetic field is present, all magnetic domains (regions) parallel to the external magnetic field become larger and expand at the expense of adjacent, nonparallel domains. Quickly, the parallel domains become predominant, and a permanent magnet is the result.

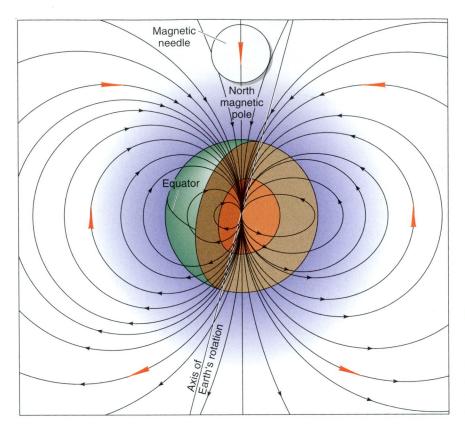

Figure 11.16 Earth's Magnetic Field
Earth is surrounded by a magnetic field generated by the magnetism of the core. A small magnet, if allowed to swing freely, will always line up parallel to the magnetic field and point to the magnetic north pole.

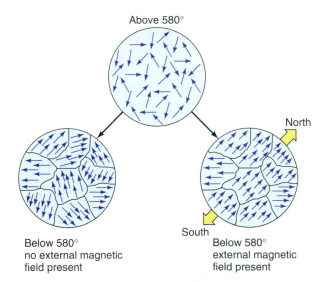

Figure 11.17 Magnetization of Magnetite Magnetization of magnetite. Above 580°C (the Curie point), the thermal motion of atoms is so great that the magnetic poles of individual atoms, shown as arrows, point in random directions. Below 580°C, atoms in small domains influence one another and form tiny magnets. In the absence of an external field, the domains are randomly oriented. In the presence of a magnetic field (lower right), most pole directions tend to be parallel to that of the external field and magnetite becomes permanently magnetized. The example in the lower right is that for a magnetite grain in a crystallizing lava.

Consider what happens when lava cools and solidifies. All the minerals crystallize at temperatures above 700°C—well above the Curie points of any magnetic minerals present. As the solidified lava continues to cool, the temperature will drop below 580°C, the Curie point for magnetite. When the temperature drops below the Curie point, all the magnetite grains in the rock become tiny permanent magnets with the same polarity as Earth's field. Grains of magnetite locked in lava cannot move and reorient themselves the way a freely swinging bar magnet can. Therefore, as long as that solidified lava lasts (until it is destroyed by weathering or metamorphism), it will carry a record of Earth's magnetic field at the moment it passed through the Curie point.

Sedimentary rocks can also acquire weak but permanent magnetism through the orientation of magnetic grains during sedimentation. As clastic sedimentary grains settle through ocean or lake water, or even as dust particles settle through the air, any magnetite particles present will act as freely swinging magnets and orient themselves parallel to the magnetic lines of force caused by Earth's magnetic field. Once locked into sediment, the grains make the rock a weak permanent magnet.

THE POLARITY-REVERSAL TIME SCALE

How is magnetism useful in dating? From a study of magnetism in lavas, geologists discovered that some rocks contain a record of reversed polarity. That is, when their magnetism was measured, some lava indicated a south magnetic pole where the north magnetic pole is today, and vice versa (Figure 11.18). Just why Earth's magnetic field reverses polarity is not yet understood, but evidence has now been discovered proving that many polarity reversals have happened through geologic time.

All lavas formed at the same time record the same magnetic polarity information, regardless of where on Earth they solidified. As we have seen, the ages of lavas can be accurately determined using radiometric dating techniques. Through combined radiometric dating and magnetic polarity measurements in thick piles of lava extruded over several million years, it has been possible to determine precisely when magnetic polarity reversals occurred. A detailed record of all changes back to the Jurassic Period has now been assembled, and still earlier reversals are the topic of ongoing research.

The polarity record for the past 20 million years is shown in Figure 11.19. Time is divided into periods of predominantly normal polarity (as at present), or predominantly reversed polarity. These periods are called **magnetic chrons**. The four most recent chrons have been named for scientists who made great contributions to studies of magnetism: Brunhes, Matuyama, Gauss, and Gilbert.

Use of polarity reversals for geologic dating is particularly effective for sediments drilled from the seafloor, but the way the dating is carried out differs from other dating methods. One magnetic reversal looks like any other in the rock record. When evidence of a magnetic reversal is found in a sequence of strata, the problem is to know which of the many reversals it actually is. Additional information, such as the presence of fossils, is needed to indi-

cate approximately where in the geologic column the sequence may lie. When one reversal has been discovered and dated in a section of strata undisturbed by disconformities, it is simply a matter of counting backward or forward to other reversals. This is the technique used in the dating of oceanic crust discussed in Chapter 20.

Magnetism in sedimentary rocks has proved to be a very sensitive and important correlation and dating technique. Sediment cores recovered from the seafloor can be correlated very accurately using a combination of fossils and magnetic reversals. The measurements are so good that magnetic reversals can even provide an accurate way to measure rates of sedimentation in the world ocean.

In the later chapters of this book, many examples of actual rates of geologic processes are mentioned. Few, if any, examples would be possible without the numerical dates obtained through radiometric dating. The ability to determine numerical dates has, more than any other contribution by geologists, changed the way we humans think about the world.

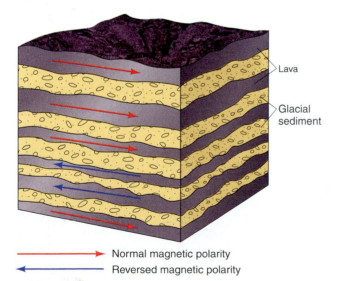

Normal magnetic polarity
Reversed magnetic polarity

Figure 11.18 Lavas and the Record of Earth's Magnetic Field
Lavas retain a record of the polarity of Earth's magnetic field at the instant they cool through the Curie point. At Tjornes, in Iceland, a series of lava flows, separated by glacial sediment, demonstrate the reversibility of the magnetic field.

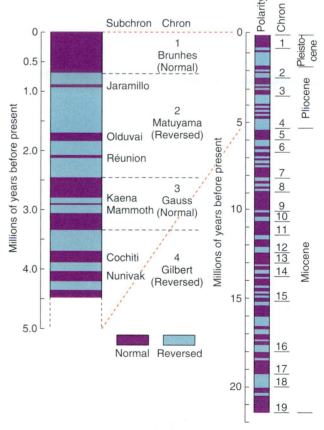

Figure 11.19 Magnetic Reversal Record Polarity reversals dating back 20 million years. Periods of normal polarity, as today, and periods of primarily reversed polarity are called magnetic chrons. Nineteen magnetic chrons have been identified to the beginning of the Miocene. Within each chron, one or more subchrons may occur. During a normal chron, subchrons are times of reversed polarity. Polarity reversals have been dated back to the mid-Jurassic, approximately 162 million years ago.

REVISITING PLATE TECTONICS AND THE EARTH SYSTEM

WHEN DID THE ROCK CYCLE BEGIN?

Evidence is clear that Earth was a planet by 4.55 billion years ago. We don't really know what it looked like, but we can form hypotheses and direct research to test the hypotheses. We know that one reason Earth looks the way it does today is because it has a rock cycle. We also know that the rock cycle operates the way it does because Earth is a tectonically active planet that has an atmosphere and a hydrosphere. Most of the water in the hydrosphere is in the ocean, so another way to ask when the rock cycle began would be to ask "When did the ocean form?"

Today, we know that water enters the mantle through subduction, and returns to the surface by volcanism. The two processes are in balance, so the ocean is neither shrinking nor growing. Water recycling through the ocean negates a hypothesis that all water comes from volcanoes. The ocean may have formed by slow outgassing as Earth heated up, but recycling prevents us from proving the point. Studies of volcanic gases provide other clues, however. Three gases, ^{40}Ar (daughter of ^{40}K), ^{3}He, and ^{36}Ar (both primordial gases trapped in Earth from the solar nebula), are being outgassed, but they are not being recycled—they accumulate in the atmosphere. From studies of the rates of outgassing, and the amounts in the atmosphere, long-continued outgassing is insufficient, and it seems, instead, that a gas such as ^{36}Ar must have arrived at Earth's surface in the very earliest days. If outgassing was not the origin, some form of initial capture must have happened. Considering the volume of water in the hydrosphere, geologists argue that ^{36}Ar and water must share a common history, and that water must also have been on Earth from the beginning.

Where could the water have come from? There are two possibilities. The first is by way of the solar nebula. We know that some planets and moons far from the Sun, such as the moons of Jupiter, have huge quantities of ice around them. The ice must have condensed from the solar nebula as part of the planetary growth system. As the nebula cooled down, Earth probably captured some of the water vapor too, and that water vapor condensed and fell as rain. A second possible source of water is comets. Modern studies have shown that comets are essentially dirty snowballs. There are vast numbers of comets in the outer solar system and the impact of comets on the early Earth suggests them as a viable source. How might we select between the two hypotheses? Geologists who are researching the topic are looking at isotopes of elements like neon, which are not recycled by the Earth system but have different isotopic signatures from the nebula and comets. The question of the water source cannot yet be answered in a definitive manner.

Whatever the origin of the first water of Earth, the evidence seems very strong that is has been here since the beginning. That means that the ocean must be nearly as old as Earth, and that the rock cycle must also be very ancient. Indeed, the antiquity of the rock cycle is possibly the very reason that we have not, so far, found rocks of the Hadean Eon on Earth.

What's Ahead?

We have now completed that portion of the book devoted to the solid materials of Earth—rocks and minerals—and the processes driven primarily by Earth's internal heat energy. We turn next to a discussion of the processes whereby Earth's surface is continually changed and modified. Surficial processes are driven mainly by energy from the Sun.

CHAPTER SUMMARY

1. Strata provide a basis for reconstructing Earth history and past surface environments. Most strata were horizontal when deposited (law of original horizontality), and all strata accumulated in sequence from bottom to top (principle of stratigraphic superposition).

2. Stratigraphic superposition concerns relative time. The relative ages of two strata can be fixed according to whether one of the layers lies above or below the other.

3. Unconformities are physical breaks in a stratigraphic sequence marking a period of time when sedimentation ceased and erosion removed some of the previously laid strata. Angular unconformity results when rocks are disturbed by tectonic activity prior to deposition of overlying strata.

4. A formation is a fundamental rock unit for field mapping distinguished on the basis of its distinctive physical characteristics and usually named for a geographic locality.

5. Systems are rock sequences and are the primary time-stratigraphic units used to construct the geologic column.

6. Geologic time units are based on time-stratigraphic units and represent the time intervals during which the corresponding systems accumulated.

7. Correlation of strata from place to place is based on physical and biological criteria that permit demonstration of time equivalence. Reliability of correlation is greatest if several criteria are used.

8. The geologic column is a composite section of all known strata, arranged on the basis of their contained fossils or other age criteria.

9. The geologic time scale is a hierarchy of time units established on the basis of corresponding time-stratigraphic units. Systems (time-stratigraphic units) and periods (geologic-time units) are based on type sections or type areas in Europe and North America. The geologic time scale constitutes the global standard to which geologists correlate local sequences of strata.

10. Decay of radioactive isotopes of various chemical elements is the basis for radiometric dating. The main radioactive isotopes and their daughters are $^{40}K/^{40}Ar$, $^{238}U/^{206}Pb$, $^{235}U/^{207}Pb$, $^{232}Th/^{208}Pb$, $^{87}Rb/^{87}Sr$, and $^{14}C/^{14}N$.

11. A sedimentary rock layer can be dated when it is bracketed between two bodies of igneous rock to which a radiometric dating method can be applied.

12. Radiocarbon dating is effective only for relatively young carbon-bearing materials (Holocene and the latest Pleistocene).

13. The age of Earth is 4.55 billion years.

14. Magnetism in rocks and the polarity-reversal time scale are useful for dating oceanic crust, lavas, and young sedimentary rocks.

THE LANGUAGE OF GEOLOGY

angular unconformity (p. 280)
Archean Eon (p. 287)
Cenozoic Era (p. 287)
conformity (p. 280)
correlation (of strata) (p. 284)
daughter (from radioactive decay) (p. 289)
disconformity (p. 280)
eon (p. 286)

epoch (p. 287)
era (p. 287)
formation (p. 282)
geologic column (p. 285)
geologic time scale (p. 286)
Hadean Eon (p. 286)
half-life (p. 290)
isotope (p. 289)
magnetic chron (p. 299)

mass number (p. 289)
Mesozoic Era (p. 287)
nonconformity (p. 280)
original horizontality (law of) (p. 278)
Paleozoic Era (p. 287)
parent (radioactive) (p. 289)
period (p. 287)
Phanerozoic Eon (p. 287)

Proterozoic Eon (p. 287)
radioactive decay (p. 289)
radiometric age (p. 291)
radiometric dating (p. 291)
stratigraphic superposition (principle of) (p. 279)
stratigraphy (p. 278)
system (p. 282)
unconformity (p. 280)

QUESTIONS FOR REVIEW

1. How do the law of original horizontality and the principle of stratigraphic superposition help geologists unravel the history of deformed belts of sedimentary rock? Is the history so determined known in numerical or relative time?

2. What geologic events are implied by an angular unconformity? by a disconformity?

3. How does a rock-stratigraphic unit differ from a time-stratigraphic unit?

4. How can strata be correlated from place to place? Can you identify an important geological advance that came about through correlation?

5. What is the geologic column? Is the column in North America the same as the column in Australia? How could you prove that your answer is correct?

6. How is a newly discovered formation placed in its correct position in the geologic column?

7. The geologic time scale is divided into four eons. Name them in the correct order, starting with the most ancient.

8. What are the three eras of the Phanerozoic Eon, and what do their names mean?

9. What features make radioactivity an ideal way to measure geologic time? Would the radiometric ages of a rock on the Moon and one on Earth, formed at the same instant, be the same? Why?

10. What radiometric dating method would be suitable for obtaining the age of (a) a rhyolite thought to be about 100 million years old; (b) an

Archean granite containing uraninite; and (c) charcoal from an archaeological site thought to be about 10,000 years old?

11. How has ^{14}C been used to date the advance of the last great ice sheet in central North America?

12. What is the Curie point, and why is it important for magnetic dating?

13. Polarity reversals of Earth's magnetic field can be recorded in rocks in two different ways. What are they?

14. How can magnetic polarity reversals be used to determine the timing of past geologic events?

15. Can you explain why the vast time span known as the Precambrian does not have a detailed geologic time scale of periods and epochs? Which dating schemes would you consider using in Precambrian igneous rocks?

16. The radiometric age for a sheet-like mass of igneous rock is 20 million years. The sheet of igneous rock is parallel to the layering of a sandstone below, a shale above. How could you tell if the sheet is a lava flow or a sill? What could you say about the age of the sandstone if the sheet is a sill? a lava flow?

17. Uranium has an atomic number of 92. When ^{238}U decays to lead, it does so in a series of steps, emitting in the process 8 α-particles and 6 β-particles. What are the atomic and mass numbers of the daughter isotope of lead?

18. The half-life of ^{235}U decaying to ^{207}Pb is 710 million years. What is the age of a grain of zircon that contains ^{235}U/^{207}Pb atoms in the ratio of 3:1? What would the age be if the ratio were 1:3?

Click on *Presentation* and *Interactivity* in **The Rock Record and Geologic Time** module of your CD-ROM to further explore resources and activities presenting concepts from this chapter. Select *Assessment* in the same module to text your understanding of this chapter.

Yellowstone National Park in Wyoming is centered on a volcanic plateau, which has been raised isostatically by the buoyancy of hot mantle beneath. Rainfall runoff from the plateau has cut steep valleys in its flanks, most notably the Grand Canyon of the Yellowstone River, shown here.

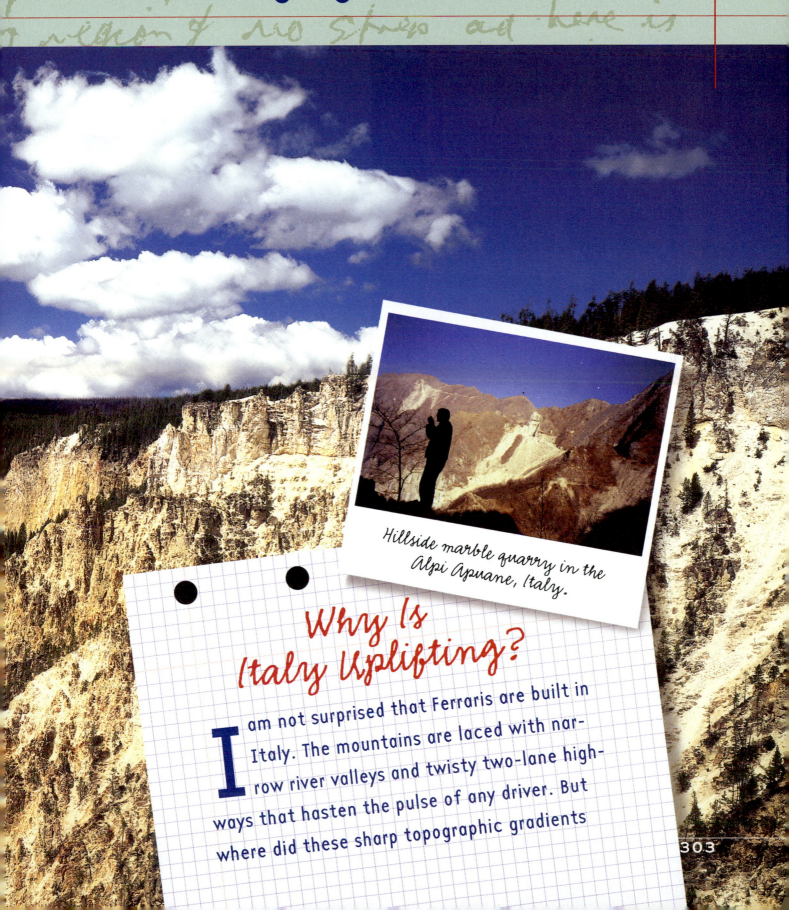

The Changing Face of the Land

Hillside marble quarry in the Alpi Apuane, Italy.

Why Is Italy Uplifting?

I am not surprised that Ferraris are built in Italy. The mountains are laced with narrow river valleys and twisty two-lane highways that hasten the pulse of any driver. But where did these sharp topographic gradients come from? They imply an uplift of the landscape so rapid that erosional processes are straining to catch up.

On a recent field trip to the northern Apennine Mountains, local geologists showed me landforms indicating that a flat depositional surface had risen from just below sea level to 1 km elevation and higher. Here and there along the mountaintops lay this flat surface that could not have formed from erosion at high altitude. At many places, the surface was topped by a few meters of limestone containing Pliocene marine fossils only a few million years old. This is strong evidence that, far from being ancient in a geologic sense, Italy has only recently risen from the sea.

Rivers tumble through the uplifted rocks of the northern Apennines, pausing to deposit sediments in the flatter valley segments. Some older sediment layers are stranded high above the present streams, forming terraces. From the ruins of an Etruscan cemetery, we gained a panoramic view of these terraces above the Reno River, terraces split by the river as it chewed into the uplifting crust—all evidence of recent uplift.

Landscape evolution and classical architecture came together on the next day of our field trip, in the mountainous Alpi Apuane. Here erosion has stripped away the surface rocks—yet further evidence of dramatic uplift—exhuming large pockets of fine white marble from depths approaching 30 km. Master builders from the days of classical Rome through the Renaissance have plucked these metamorphic rocks from the mountains, leaving the yawning cavities of old quarries.

But why was Italy uplifted? The lithosphere beneath the Adriatic Sea, between Italy and the Balkan coast, subducts beneath the Apennines, but so slowly that no island arc volcanoes have formed beneath the northern half of the range. Is hot, buoyant asthenosphere invading the region and uplifting it? Or is the crust acting independently, behaving like a large sand pile pushed by a plow? We have not yet figured it out.

Jeffrey Park

KEY QUESTIONS

1. **Why is Earth's landscape so varied?**

2. **What are the competing geologic forces behind landscape evolution?**

3. **What are the principal models of landscape evolution?**

4. **How can we calculate rates of landscape evolution?**

5. **Does landscape evolution ever end?**

6. **How do uplift and weathering relate to plate tectonic motions and the carbon cycle?**

INTRODUCTION: WHENCE EARTH'S VARIED LANDSCAPES?

When hikers in the mountains reach the end of a wilderness trail, they often acquire a strong sense of relief. Not relief from steep climbs and buzzing mosquitoes, but rather the relief of the landscape. In breathtaking views from the crest of a mountain or a topographic ridge, the contrast of high peaks and plunging valleys fills most human observers with awe.

If the human observer is also a geologist, the relief of a dramatic landscape raises some obvious questions. What geologic processes have produced such steep slopes, craggy peaks, and deeply exhumed river valleys? What makes such places different from surrounding landscapes of low relief?

How can we explain the variety of Earth's landscapes? If they have changed through time, what processes caused and controlled the changes? What clues do landscapes hold to the history of Earth's mobile lithosphere and past climates?

In this chapter, we describe *uplift* processes that raise topography and *exhumation* and *denudation* processes that erode it, all acting together to form the landscape you see out your window.

Uplift usually occurs as a byproduct of plate tectonics and thermal convection, so it can be influenced by both the mantle and the crust beneath our feet. Exhumation and denudation are closely related terms that imply different perspectives on erosional processes. **Exhumation** refers to the gradual exposure of subsurface rocks by stripping off (eroding) the surface layers. **Denudation** refers to the transport of eroded material to another location. Erosional processes are influenced strongly by climate and by the physical properties of surface rocks and sediments.

Geomorphology is the study of Earth's landforms. Directly translated from Greek, geomorphology is the study of Earth's shapes. Processes that change the land surface operate on widely varied time and spatial scales. Some act rapidly, even abruptly, and may cause only local changes. Such landform changes can directly and adversely impact people, for example, earthquakes, volcanic eruptions, storms, floods, and landslides.

Other processes operate on long geologic time scales that are difficult or impossible to observe directly. They may also occur over extensive areas. Nevertheless, we can see and measure their long-term effects, both in ancient landscapes that survive and in the stratigraphic record. Clever interpretations of radioactive decay enable geologists to estimate rates of exhumation directly from present-day surface rocks.

We also can use computer models to help understand how slow-acting processes influence landscape evolution.

Because we live on Earth's surface, its processes affect us and our property. In turn, we are everywhere changing the face of the land with our own processes of agriculture, mining, and development. By constructing dams across rivers, excavating for buildings or waste-disposal sites, and build-

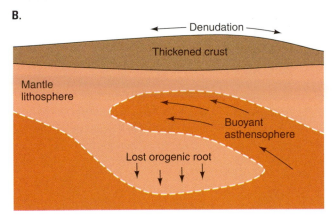

A.

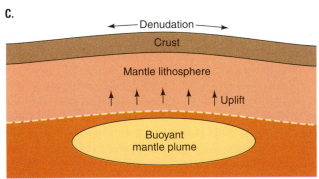

B.

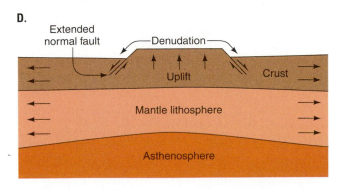

C.

D.

ing highways, we modify the landscape. Even the choice of farming methods we use to grow our food influences the landscape. Individual actions might seem insignificant, but they interact and accumulate on a global scale until, over several human generations, their effect on the landscape is substantial. For one perspective on the decline of agriculture in the Roman Empire, see *Understanding Our Environment,* Box 12.1, *Human Activity as an Agent of Denudation.*

UPLIFT AND DENUDATION: COMPETING GEOLOGIC FORCES

The continuing changes in Earth's landscapes reflect an ongoing contest between forces that raise the lithosphere (tectonic and isostatic uplift, volcanism) and the force that attempts to level it—gravity, aided by physical and chemical weathering that break apart rock, causing various erosional agents to transfer rock debris from high places to low places.

We perceive landscape evolution as only a snapshot in time. Because Earth's surface appears stable to us, it is natural to hypothesize that uplift and denudation processes are equally balanced. But careful observation tells geologists that such a balance is not as common as we might expect. Tectonic uplift varies over space and in time. Climatic changes have altered denudation rates.

- When uplift dominates denudation, the unbalanced processes may lead to rough, steeply sloped topography, typically with high **relief**. Relief is quantified as the difference in altitude between the highest and lowest points on a landscape. Uplift-dominated landscapes are often characterized as "young" landscapes.

- When denudation dominates uplift, topographic slopes become gentler and relief may decrease as topographic lows become filled with sediment. Eroded landscapes of subdued relief are often characterized as "old" landscapes.

FACTORS CONTROLLING UPLIFT

Uplift occurs as a byproduct of mantle convection and plate tectonics. Without the churning heat transfer from our planet's interior, surface tectonics would be a tame affair. Both uplift and subsidence of surface rocks can occur within a region, depending on the circumstances illustrated in Figure 12.1. We can identify several types of uplift:

Figure 12.1   Uplift Processes   A. Collisional uplift in a subduction zone. B. Isostatic uplift that would follow the loss of a cool dense lithospheric "root" of a collisional mountain range. C. Isostatic uplift over a rising plume of hot buoyant mantle. D. Extensional uplift can arise from the combination of factors. Local relief arises from relative motion along normal faults in the upper crust. Broader uplift occurs if the cool dense mantle lithosphere thins as extension progresses, so that the lithosphere has less mass per unit area.

UNDERSTANDING OUR ENVIRONMENT

HUMAN ACTIVITY AS AN AGENT OF DENUDATION

A traveller who visits the depths of an Alpine ravine…and calculates the mass of rock required to fill the vacancy, can hardly believe that the humble brooklet which purls at his feet has been the principal agent in accomplishing this tremendous erosion. Closer observation will often teach him, that the seemingly unbroken rock which overhangs the valley is full of cracks and fissures, and is really in such a state of disintegration that every frost must bring down tons of it.

So wrote Charles Perkins Marsh in his 1864 book, *Man and Nature*. Born in 1801 and educated as a lawyer, Marsh ran for Congress and became the U.S. Minister to Turkey and, later, Italy. While traveling in the Mediterranean, Marsh observed and noted the geologic processes that shape landscapes. In the pages of *Man and Nature*, he estimates how the volume of material eroded from surrounding mountains would spread to form a layer of sediment in the broad valley of the Po River in northern Italy. To persuade his readers that such large-scale denudation of a mountain range was possible, Marsh described an example from his own experience. He had witnessed in his native Vermont the transport of soil, pebbles, and large stones along a stream bed after a local pond burst out of its dam. Using the principle of uniformitarianism, Marsh extended his American observation to the entire Italian Alps.

Marsh's focus in *Man and Nature* was not on natural geologic processes, but rather on the effect of human civilization on the landscape. Everywhere he traveled, Marsh saw how human activities, such as clearing forests, farming, and draining water from bogs, altered the landscape and usually accelerated its erosion. Marsh identified human activity as a powerful geologic process. His warnings about the destruction of natural forests and wetlands inspired the conservation movement in the United States. The establishment of our national parks and forests owes a direct debt to Marsh's writings.

Figure B12.1   Exhausted Topsoil from Roman Empire Agricultural Practices   Many landscapes near Rome have the grassy steep hillsides characteristic of "badlands" topography. Intensively farmed since the dawn of human history, topsoil is now scarce in these landscapes.

- *Collisional uplift*
- *Isostatic uplift*
- *Extensional uplift*

Isostatic uplift occurs through several mechanisms, as we will illustrate.

After reading Chapter 2 (*Global Tectonics: Our Dynamic Planet*), you might expect the most natural uplifted landscape would be a long, narrow **orogen** that forms along a convergent plate boundary. Such **collisional uplift** may result from compression of relatively light crustal rock atop a continental plate, which thickens in a plate collision zone due to underthrusting just above the mantle portion of the lithosphere. Crustal rocks, especially sedimentary rocks, are weaker than mantle rocks, and thus can detach from the underlying plate and accumulate in the orogen. When an oceanic plate subducts beneath a continental plate, an orogen can form from a rising wedge of marine sediments scraped from the downgoing plate. The island of Taiwan and the Olympic mountains of Washington State are examples of such orogens (Figure 12.2).

Broad, uplifted plateaus sometimes accompany an orogen, especially where two continental plates collide. Geologists see in such plateaus evidence for **isostatic uplift**. Consider the Indian subcontinent, which is colliding with Eurasia, raising the Himalayan Mountains. The Himalayan orogen forms the southern border of the Tibetan Plateau, which approaches 1000 km width (Figure 12.3). The width of Tibet and similar regions, like the Altiplano in South America, suggests that the underlying mantle plays a key role in plateau uplift.

But how? What is the mechanism for uplift over such a broad area? When continents collide, their plate boundaries tend to crumple somewhat, so the lithosphere locally thickens. Many geologists argue that the dense mantle lithosphere of a thickened plate boundary can peel off and be replaced by hot buoyant asthenosphere. Isostatic uplift would follow. The shallow mantle beneath Tibet appears to

Marsh supported his call for better management of natural resources by claiming that wasteful agricultural practices hastened the decline and fall of the Roman Empire. He cited in particular the consumption of forests for timber and fuel, which left bare hillsides naked to erosional processes. Modern Italy has many examples of treeless **badlands** landscapes that seem to confirm Marsh's argument (Figure B12.1). In a badlands landscape, grass-covered hills are eroded into steep slopes (>35°) and gullies by rainwater runoff. Without the roots of trees to hold it in place, any soil that develops washes away before it can accumulate. Central Italy was reported by ancient historians to be rich farmland when it was annexed by the armies of Rome in the centuries before Christ. However, the productivity of the region declined markedly by the time Rome fell to barbarian rule in the fifth century A.D.

In some ways the Romans were unlucky in their homeland provinces. Most surface rocks in central Italy are easy to erode, because they are composed mainly of young sedimentary rocks and pyroclastic deposits from nearby volcanoes. Agriculture remained productive in the eastern parts of the Roman Empire, in modern Greece, Turkey, and Syria, where the landscape was less fragile. The capital of the empire shifted from Rome to Constantinople (modern Istanbul in Turkey) and prospered for several more centuries.

In our own times, farming in the Italian badlands can be profitable with modern practices, that is, with fertilizers and tractors. Even so, studies have shown that careless bulldozing of badlands to form gently sloping wheat fields can actually increase soil erosion. Aggressive land management, designed to increase short-term productivity, may be weakening the long-term sustainability of agriculture. The message of Charles Perkins Marsh still applies in our own time. A society that cares about its future should manage its landscapes carefully. (See Figure B12.2.)

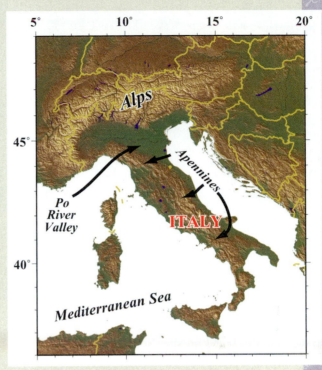

Figure B12.2   Uplift and Relief in Italy   Relief map of Italy and surrounding regions. Note the juxtaposition of the Alps with the Po River valley. Denudation of the Alps contributes to thick sedimentary deposits in the Po valley.

Figure 12.2   Orogens Built from Subduction Zone Sediments
Relief images of (A) the Olympic Mountains of Washington State, formed where the Juan de Fuca Plate subducts beneath the North American Plate; and (B) the mountains of the island of Taiwan, where the Eurasian Plate subducts beneath the Philippine Plate.

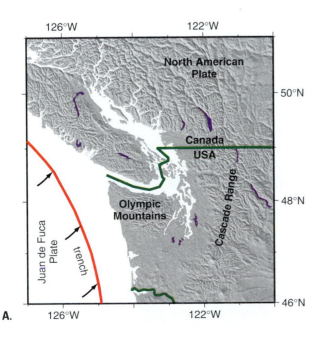

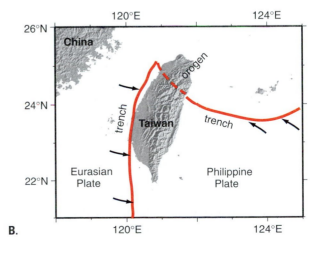

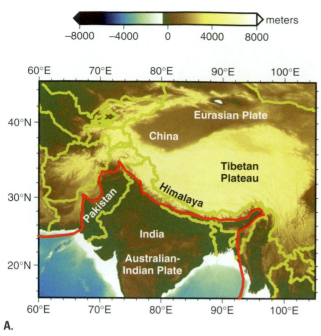

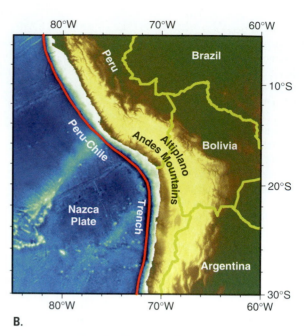

Figure 12.3 **Continental Plateaus Along Convergent Plate Boundaries** Color-coded elevation maps of (A) the Tibetan Plateau, formed along the collision between the Australian–Indian Plate and the Eurasian Plate; and (B) the Altiplano ("high flat surface" in Spanish), formed along the subduction boundary between the Nazca Plate and the South American Plate.

be hot and buoyant, based on slow seismic wave speeds and patches of geologically recent volcanism. Thus, mantle-driven uplift seems plausible.

How else can isostatic uplift occur? Recall the plume swell in Pacific Ocean bathymetry above the Hawaii hot spot in Figure 2.21. Isostatic uplift occurs wherever a hot, buoyant mantle plume rises to the base of the lithosphere, as beneath the Hawaiian Islands or Yellowstone National Park. On a smaller scale, topography elevates locally wherever magmatic activity creates individual volcanoes.

Erosion itself can drive a more focused isostatic uplift. If a mountain range is able to adjust to mass imbalances through vertical movement, uplift will occur to compensate for the mass of any eroded material.

Extensional uplift can occur in regions where the ascent of hot buoyant mantle at the base of continental lithosphere induces broad uplift and stretching of crust. Broad isostatic uplift can induce short-scale topographic relief by extension and normal faulting. As an example, bathymetric scarps parallel the midocean ridges, where buoyancy and plate divergence occur together. The scarps are associated with normal faults and extension of young oceanic crust. On land, extensional faulting caused by crustal stretching near the Yellowstone hot spot is responsible for the spectacular relief of the Grand Teton Range of northwest Wyoming (Figure 12.4). The Basin and Range Province, a region of uplifted topography and crustal extension that encompasses much of Nevada, Utah, Arizona, and portions of surrounding states, gains relief from numerous north–south-trending horst-and-graben

structures, built by the normal faults that accommodate its extension (Figure 12.5).

Figure 12.4 **Grand Tetons** Rapidly uplifting mountain range in Wyoming. Caused by a region of crustal extension that surrounds the Yellowstone hot spot, the Grand Teton range rises on the footwall of a major normal fault. The rolling meadows of Jackson, Wyoming, home to thick herds of bison, elk, and moose, lie on the hanging wall of the fault. Erosional debris from the Grand Teton range accumulates on the Jackson side, forming its gentle topography.

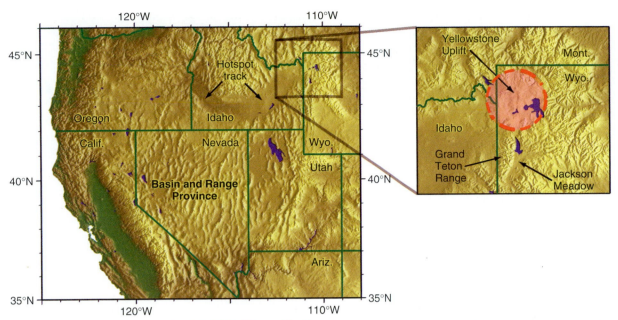

Figure 12.5 Isostatic and Extensional Uplift in Western North America Relief image of the western United States. The curved region of flat relief in southern Idaho is the track of the Yellowstone hot spot over the past 12 million years. Although most of the intermontane West lies at elevations greater than 2000 m, the elevation in Yellowstone National Park is several hundred meters greater than surrounding areas, owing to the buoyancy of hot mantle beneath the crust. South of the hot spot track, north–south-trending mountain ranges cover much of Nevada and Utah in the Basin and Range Province. These ranges have formed from normal faulting associated with crustal extension.

FACTORS CONTROLLING DENUDATION

Denudation is a slow dance of unlikely partners. Physical and chemical weathering, fostered by the volatile compounds of the hydrosphere and atmosphere, break down the solid rocks at and just beneath Earth's surface, as was discussed in Chapter 6. Erosion of surface sediment is facilitated by winds, running water, and glaciers, which will be examined in detail in Chapters 14, 16, and 17.

CLIMATE

Climate helps determine which surface processes are active in any area. In humid climates, streams may be the primary agent that moves and deposits sediment, whereas in an arid region wind may locally assume the dominant role. Glaciers are largely restricted to high latitudes and high altitudes where frigid climates prevail. Because climate also controls vegetation cover, it further controls the effectiveness of some important erosive processes. For example, a hillslope stabilized by plant roots that anchor the soil may become prone to slumping if the plant cover is destroyed. (See Box 12.1, *Human Activity as an Agent of Denudation*.)

Just as we identify distinctive climatic regions of Earth, we also can identify regions containing distinctive landforms generated by one or more surface processes. However, because climates have changed through time, the active surface processes in some regions also may have changed. Thus, many modern landscapes reflect former conditions—like glaciation—rather than current processes.

LITHOLOGY

Within any climatic zone, a surface process may interact differently with various exposed rocks, depending on their **lithology**. Some rock types are less erodible and thus will produce greater relief and steeper landforms than rocks more susceptible to erosion—granite or hard sandstone can maintain steep slopes for a long time, whereas a soft shale easily erodes into low-relief debris. However, a specific rock type may behave differently under different climatic conditions. For example, limestone in a moist climate may dissolve and erode into a terrain of sinkholes, but in a desert, the same limestone may form bold cliffs.

The ease with which streams erode a rock formation depends chiefly on its composition and structure. Because of differential erosion, folded or faulted beds may stand in relief or control the drainage to impart a distinctive pattern to the landscape that discloses the underlying structure (Figure 12.6). An experienced geologist can often use drainage patterns to infer several things: rock type, the orientation of a dipping rock unit, how the rocks are folded or offset, and the pattern and spacing of joints.

RELIEF

Relief of the land exerts a strong influence on denudation rates. Tectonically active regions with high rates of uplift

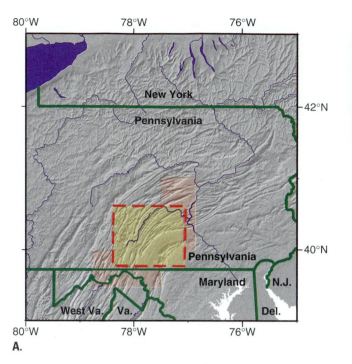

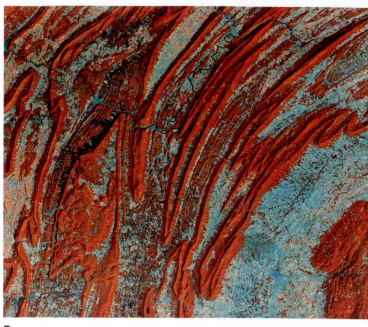

A.

B.

Figure 12.6 Differential Erosion A. The folded structure of rocks in central Pennsylvania leads to topographic relief between less erodible and more erodible strata. B. A false-color satellite image of the boxed region.

tend to be extremely dynamic and generate high erosion rates (Figure 12.7). In areas far from active tectonism, relief typically is low, erosion rates are much slower, and landscape changes are more gradual. Even in tectonically inactive areas, however, a rapid change of sea level or small regional isostatic movements related to changing ice or water loads may significantly change streams and the landscapes they influence.

TECTONIC AND CLIMATIC CONTROL OF CONTINENTAL DIVIDES

Except for ice-covered Antarctica, we can divide all the continents into large regions from which major rivers flow to the ocean. The line separating any two such regions is a **continental divide**, a major landscape element of our planet.

In North America, continental divides lie at the head of major streams that drain into the Pacific, Atlantic, and Arctic oceans (Figure 12.8). In South America, a single continental divide extends along the crest of the Andes and splits the continent into two drainage regions of very unequal size. Streams draining the western (Pacific) slope of the Andes are steep and short, whereas to the east the streams take much longer routes along gentler gradients to the Atlantic. An example of the latter is the Amazon River, second longest on Earth after the Nile, about 6500 km in length.

Because continental divides often coincide with the crests of mountain ranges and because mountain ranges are the

Figure 12.7 Erosion in High-Relief Terrain A stream cuts a deep gorge in the Hindu Kush of northern Pakistan. It transports a coarse-gravel bed load, contributed largely by rockfalls and landslides from the steep adjacent cliffs.

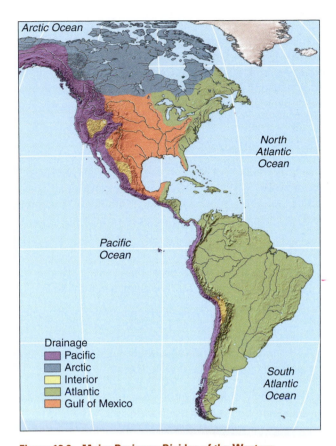

Figure 12.8   **Major Drainage Divides of the Western Hemisphere**   A great continental divide follows the crest of the high cordillera near the Pacific Coast. In North America, it separates streams that drain to the Pacific, Arctic, and Gulf of Mexico. In South America, it divides streams that drain to the Pacific and Atlantic. In eastern North America, another divide follows the Appalachian Mountains and separates Atlantic and Gulf of Mexico drainage.

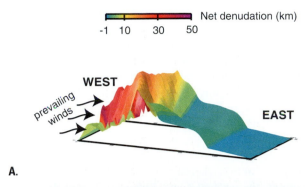

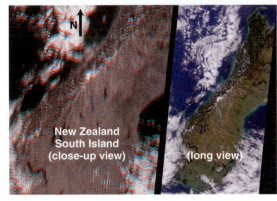

Figure 12.9   **Asymmetric Topography Maintained by Rainfall Patterns**   Weather systems at the latitude of New Zealand's southern island travel from west to east, so that rainfall is greatest on the western coastline of the island. As a result, erosion and denudation on the western side is relatively large. A. Computer simulations for the isostatic uplift induced by erosion predict that rapidly uplifting mountains and the rainfall divide should be found on the western (wet) side. B. The mountains of New Zealand's southern island do lie on the western side, consistent with the effects of coupled denudation and uplift.

result of uplift associated with the interaction of tectonic plates, a close relationship exists between plate tectonics and the location of primary stream divides and drainage basins. Climate can also influence the location of the divide. Prevailing winds may cause one side of a divide to be wetter than the other, and therefore more subject to denudation. Erosion of the divide's wet side may shift it into the drier region where the topography is more stable. Erosion-induced isostatic uplift, however, may cause high relief to persist on the wet side, stabilizing the divide there (Figure 12.9).

Before you go on:

1. What are uplift, exhumation, and denudation?

2. What are the three types of uplift? Can you give an example of each?

3. How do climate and lithology influence denudation rates?

HYPOTHETICAL MODELS FOR LANDSCAPE EVOLUTION

Hypotheses to explain uplift–denudation relations in a landscape often focus on a single geologic process or an idealized concept. Although most landscapes are influenced by many factors, hypothetical reasoning helps us to frame questions about what we observe in the field. Here we will first describe an influential early theory of geographic cycles, then the recent concept of steady-state landscapes, and finally catastrophic landscape evolution, in the form of threshold effects.

THE GEOGRAPHIC CYCLE OF W. M. DAVIS

Mountain ranges and high plateaus are massive, rocky landforms. How long can these features persist? As mountains are built, erosion simultaneously attacks the rising land, so the average land altitude at any time should reflect this ongoing contest between uplift and erosion.

American geographer William Morris Davis (1850–1934) proposed an influential theory of landscape evolution in the late nineteenth century. Davis called his model the **geographic cycle**, implying that it had a beginning and an end. From the landforms he observed, Davis deduced that a landscape passes through stages (Figure 12.10). In the initial stage, "youthful" streams downcut vigorously into the uplifted land surface to produce sharp, V-shaped valleys, thereby increasing the local relief. Gradually, the original gentle upland surface is eroded as the drainage system expands and valleys grow deeper and wider.

During the next stage, the land achieves its greatest local relief. As the cycle progresses, streams begin to "mature" and meander in their gentler valleys, and valley slopes are gradually worn down by mass wasting (Chap. 13) and erosion. In the final stage, the "old age" landscape consists of broad valleys containing wide floodplains. Stream divides are low and rounded, and the landscape is slowly worn down ever closer to sea level.

Davis envisioned interruptions of erosion cycles related to climatic fluctuations or to renewed uplift. We can guess that the crust's isostatic response to erosional removal of sediment would prolong the cycle that Davis proposed (Figure 12.10).

For many decades, Davis's concept was the basis for most interpretations of landscape evolution. But with renewed interest in surface-process studies and the widespread acceptance of plate tectonics theory, it has become harder to reconcile his concept of a simple cycle with the large variety of potential uplift processes.

Current *geomorphologists* (geologists who specialize in the shape of the land) have discarded Davis's landscape cycles because of questions about landscape equilibrium, the relative importance of uplift and erosion rates, and the complex feedbacks among the atmosphere, hydrosphere, biosphere, and lithosphere. Relationships between uplift and denudation processes dictate the character and evolution of the major landscape elements of Earth.

STEADY-STATE LANDSCAPES

Change is implicit in landscape evolution, so landscapes should evolve whenever a change takes place in any controlling variable (process, climate, lithology, structure, relief). Change may be triggered by several things:

- A tectonic event that lifts up a landmass.

- A substantial sea level drop that causes streams to steepen and incise the newly emerged coastal zone.

- A climate shift that changes the proportional influence of the different surface processes.

- A stream that erodes downward through weak rock and encounters massive, hard rock beneath.

If rates of denudation and uplift are stable for long periods, the landscape can adjust itself to a steady state (Figure 12.11A). A *steady-state landscape* maintains a constant altitude and topographic relief while still undergoing change. If uplift ceases altogether, the average rate of change gradually decreases as the land surface is progressively eroded toward sea level (Figure 12.11B).

Is it common for a landscape to achieve a state in which no further change takes place? The answer apparently is no, for we have abundant evidence that Earth's surface is now—and likely always has been—quite dynamic, experiencing continual changes in response to the natural motions of the lithosphere, hydrosphere, and atmosphere. Nevertheless, it is apparent that near-equilibrium conditions can be reached—for a while.

RAPID LANDSCAPE CHANGES: THRESHOLD EFFECTS

In many natural systems, sudden changes can occur once a critical threshold condition is reached. For example, sand grains in a stream channel may remain at rest until a critical current velocity is reached, at which point they begin to move. Certain glaciers apparently start to surge when the buildup of water pressure at their base attains a critical threshold value that forces the ice to float off its bed (Chapter 16).

A **threshold effect** implies that landscape development, rather than being progressive and steady, can be punctuated by occasional abrupt changes. A landscape in near-equilibrium may undergo a sudden change if a process operating on it reaches a threshold level. The landscape is thrown

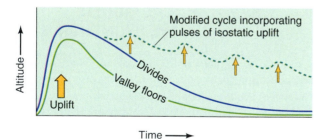

Figure 12.10 W. M. Davis's Geographic Cycle Davis's geographic cycle begins with a pulse of uplift, during which the uplift rate increases, then diminishes. Initially, the landscape's altitude increases rapidly. As the uplift rate decreases, the land is lowered by erosion. Valleys deepen and local relief (the distance between tops of divides and valley floors) reaches maximum. During later stages of the cycle, the land is progressively lowered toward sea level and relief decreases. If isostatic adjustments are also considered, a long time is required to erode the landscape to lower altitude (dashed line).

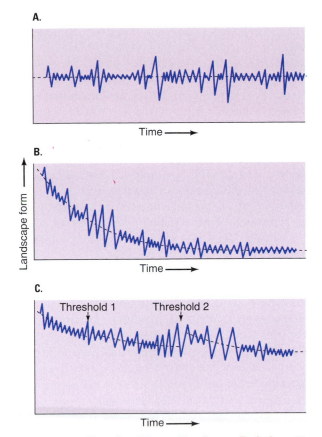

A.

Time ⟶

B.

Landscape form ⟶

Time ⟶

C.

Threshold 1   Threshold 2

Time ⟶

Figure 12.11 Three Conditions of Landscape Evolution Three possible equilibrium conditions in landscape evolution. A. Steady-state equilibrium, with the landscape fluctuating about some average condition. B. Dynamic equilibrium, with the landscape oscillating about an average value that is declining with time. C. Dynamic equilibrium, with sudden changes to the landscape occurring whenever a critical threshold is reached.

out of equilibrium and it changes as it moves toward a new equilibrium condition (Figure 12.11C).

For example, studies in the western United States have shown that gully development on some alluvial valley floors is related to valley slope. Where slopes reach a certain critical value—depending on the basin area—they become unstable and gullying begins. Where the slope is gentler, valley floors remain ungullied. The critical slope value therefore constitutes a threshold; when it is exceeded, valley floor erosion begins.

Before you go on:

1. What is the geographic cycle of landscape evolution?

2. What is the steady-state model of landscape evolution?

3. Can you describe a threshold effect?

HOW CAN WE CALCULATE RATES OF UPLIFT AND DENUDATION?

It is difficult to obtain reliable geologic information about long-term rates of landscape change:

• First, we must determine how mobile a landmass has been through time. Is it rising? How rapidly? Has the rate of uplift changed through time?

• Second, we must be able to calculate changing rates of denudation—the combined destructive effects of weathering, mass wasting, and erosion.

Knowing uplift and denudation rates for a region, we can then infer how the landscape might have evolved and predict how it might evolve in the future. However, uplift and denudation are not easy to estimate independently.

HOW CAN WE CALCULATE UPLIFT RATES?

Three methods are used to calculate uplift rates: extrapolation from earthquake displacements, measurement of warped strata, and measurement of river terraces.

EXTRAPOLATION FROM EARTHQUAKE DISPLACEMENTS

One way to calculate uplift rates is to measure how much local uplift occurred during historical large earthquakes, then estimate the recurrence interval of such earthquakes, and finally extrapolate the recent rates of uplift back in time. Of course, this assumes that the brief historical record is representative of longer intervals of geologic time—a risky assumption. For example, seismologists estimate that many thrust faults in continental regions have recurrence intervals of several thousand years, longer than most historical records.

Geologists have developed methods for estimating past earthquake occurrence that are useful in special circumstances. In New Zealand, for instance, lichens grow on newly exposed rock surfaces at a predictable rate for several centuries. By examining lichen that grows on rock that was fragmented by past earthquakes, geologists can estimate when the earthquakes occurred.

CALCULATION FROM WARPED STRATA

This technique measures the warping (vertical dislocation) of originally horizontal strata of known age. Examples are flood basalts that have been deformed since extrusion, and uplifted coral reefs along a tectonically active coast. For this method, we need to know (1) the altitude difference between where the stratum is today and where it was at its formation and (2) a radiometric age for the stratum.

CALCULATION FROM RIVER TERRACES

As discussed in Chapter 14, rivers tend to carry material eroded from steep slopes and deposit it along flatter portions of river valleys. In an uplifting region, a river may form a sequence of abandoned depositional surfaces called *river terraces*, as described in the introductory essay to this chapter. As uplift raises a river valley, the river erodes into its former depositional surface and its terraces are abandoned. If we can determine the age of the last sediment deposited, abandonment times and uplift rates can be calculated. However, climate changes can alter streambed erosion rates and cause river terraces to form as well, so interpretations require some caution.

Measured uplift rates in tectonic belts are quite variable—around 1 to 10 mm/yr, averaged over intervals of several thousand to several million years. For any region, we can assume that average values have changed through time as rates of seafloor spreading and plate convergence have varied. Each such change likely leads to compensating adjustment in landscapes as they evolve toward a new steady-state condition.

How Can We Calculate Exhumation and Denudation Rates?

The exhumation rate is equal to the erosion rate. Sometimes the local erosion rate is equated to the local uplift rate as a starting assumption, although uplift equals erosion only for a steady-state landscape. The denudation rate we can estimate from the volume of rock removed and deposited elsewhere.

Exhumation Rate

What is the local exhumation rate? Because erosion is often too slow to measure directly, we must ask a slightly different question: When were the surface rocks last at a particular depth? If we can make a measurement of such a depth at a specific time in the past, the simple ratio "depth over time" estimates the rate at which erosion has stripped off the rock that lay atop the present surface.

Radioactive-Decay Techniques

Several techniques use radioactive decay to measure how long a rock has been buried no deeper than a particular distance beneath the surface. These techniques exploit a flaw in some decay systems. Specifically, we must hypothesize, when using radioactive decay to determine the age of a rock, that both the original radioactive isotopes and their decay products are retained within the rock throughout its lifetime. Any loss of daughter isotopes or other decay indicator can throw off the age calculation. The loss of a daughter isotope, for instance, the escape of argon (^{40}Ar) produced from the decay of ^{40}K, will make a rock appear to be younger than it really is. Some radioactive products, such as **fission tracks** and **radiogenic helium**, will not accumulate if the rock is too hot. They are retained only if the rock is cooler than a certain critical temperature, called

the **closure temperature** (see *The Science of Geology*, Box 12.2, *Estimating Exhumation Rates with Radioactive Decay*).

Radioactive decay products use thermal energy to escape their host rock. Each decay product has an independent closure temperature. Below the closure temperature, the available thermal energy is too scarce for significant escape to occur. Escape is easier if the decay product is a noble gas, like helium (He) or argon (Ar). Escape is much more difficult for daughter elements that form chemical bonds, like lead (Pb). Fission tracks damage the crystalline structure of minerals. This damage will repair itself by annealing if the temperature is sufficiently high. Different minerals have different propensities to repair their crystalline structures, and so have different closure temperatures for fission-track preservation.

Because temperature increases steadily from the surface inward toward the core, deeply buried rocks usually exceed the closure temperature for one or more radioactive-decay systems. While deep and hot, therefore, these radiometric clocks fail to measure the passing of time. When exhumation brings these deep rocks close to Earth's surface, heat conduction cools them, and radioactive products start to accumulate. Fission tracks and radiogenic helium do not give reliable ages for the original formation of surface rocks, but rather tell us how many years the rocks have been at or below their closure temperatures. If the closure temperature (°C) and the geothermal gradient (°C/km) are known, we can compute the exhumation rate (see Box 12.2).

We can make a similar estimate of exhumation rate from the abundance of **cosmogenic nuclides** in surface rocks. Cosmogenic nuclides are radioactive isotopes that are not present in a rock at its formation, but rather are generated by high-energy subatomic particles emitted by the Sun. Their abundance in a surface rock can tell us how long a rock has been exposed to such radiation.

Denudation Rates

To calculate long-term denudation rates, we must know how much rock debris has been removed from an area during a specified length of time. The total sediment removed from a mountain range, for example, will include sediment currently in transit, sediment temporarily stored on land on its way to the sea, and sediment deposited in the adjacent ocean basin, mostly on or near the continental shelf. If we can measure the volume of all this sediment, we can calculate its solid-rock equivalent and the average thickness of rock eroded from the source region. Finally, if we can determine the duration of the erosional interval, we can calculate the average denudation rate.

For areas drained by major streams, the volume of sediment reaching the ocean each year is a measure of the modern erosion rate. Most of this sediment ultimately is deposited in deltas, across the continental shelves, and in vast submarine sediment fans. We can estimate the volume of sediment deposited during a specific time interval by using drill-core and seismic records of the seafloor. Calculating the equivalent rock volume of this sediment and averaging it over the area of the drainage basin(s) from which it was derived then gives an average denudation rate for the source region.

WORLD SEDIMENT YIELDS

The highest measured sediment yields are from the humid regions of southeastern Asia and Oceania and from basins that drain steep, high-relief mountains of young orogenic belts, such as the Himalaya, the Andes, and the Alps (Figure 12.12). Low sediment yields characterize deserts and the polar and subpolar sectors of the northern continents. As one might expect, rates are high on steep slopes, and in areas underlain by erodible clastic sediments or sed-

imentary rocks, or by low-grade metamorphic rocks. Denudation rates are low in areas where crystalline or highly permeable carbonate rocks crop out.

Structural factors also play a role, because rocks that are more highly jointed or fractured are more susceptible to erosion than massive rock formations. Denudation is surprisingly high in some dry-climate regions, often because the surface lacks a protective cover of vegetation. Soil protection by vegetation is likely a threshold phenomenon; below a certain vegetation density, erosion rates may increase. This threshold effect explains why soil erosion is a serious problem after forested land is cleared for farming or livestock grazing.

The dominant erosional process can also strongly influence sediment yield. Values are unusually high in south-coastal Alaska, the most extensively glaciated temperate mountain region in the world, where denudation rates exceed those of comparable-sized nonglaciated basins in other regions by factor of 10 or more. Under climates favorable for the expansion of temperate valley glaciers, both chemical and physical weathering rates tend to be substantially above the global average.

Asia discharges the most sediment to the ocean of any continent. It is also the source of the greatest measured average stream sediment loads. Asian rivers entering the sea between Korea and Pakistan contribute nearly half the total world sediment input to the oceans (Figure 12.13). Second to Asia is the combined area of the large western Pacific islands of Indonesia, Japan, New Guinea, New Zealand, the Philippines, and Taiwan. Taiwan is especially remarkable because it produces only slightly less sediment than the entire contiguous United States! Each of these land masses lies on or near a convergent plate boundary, and so experiences rapid tectonic uplift.

HUMAN IMPACT ON DENUDATION

One important factor that influences denudation rates is human activity—especially clearing forests, developing cultivated land, damming streams, and building cities. All have affected erosion rates and sediment yields in drainage basins where they have occurred.

Sometimes the results are dramatic: in parts of the eastern United States, areas cleared for construction produce 10 to 100 times more sediment than comparable rural areas or natural areas that are vegetated. On the other hand, in

Figure 12.12 The Ganges River, which drains the western Himalaya and Tibetan Plateau, is responsible for one of the highest sediment yields in the world. This space photo shows the main river and tributary streams entering it from the eastern Nepalese Himalaya. Sediment transported by the river forms a muddy discharge into the Bay of Bengal.

BOX 12.2 | **THE SCIENCE OF GEOLOGY**

ESTIMATING EXHUMATION RATES WITH RADIOACTIVE DECAY

Several products of radioactive decay measure geologic time only when rocks are near the surface and relatively cool. These examples of radioactive decay function as "cool-rock clocks."

Fission-Tracks

A very small fraction of uranium atoms decay into highly charged nuclear fragments by spontaneous fission, rather than by alpha-particle decay. When these fragments career through a mineral, they damage the internal arrangement of atoms, leaving tiny pathways (called fission tracks) that can be detected under a microscope (Figure B12.3). At high temperatures, fission tracks quickly anneal and disappear, leaving no permanent record. Only when the cooling mineral falls below its *closure temperature* does annealing cease, and fission tracks persist as scars in a mineral crystal. The number of tracks in a mineral increases with time, so track density can be used to date the time elapsed since the mineral cooled below the closure temperature. The minerals zircon ($ZrSiO_4$) and apatite accumulate fission tracks efficiently, but with different closure temperatures (240°C for zircon, 110°C for apatite). Both zircon and apatite can be found in many plutonic rocks.

Radiogenic Helium

The alpha particle happens to be the nucleus of a helium atom, specifically the ^4He isotope. Helium is a noble gas and does not form chemical bonds, but its atoms can be trapped within the crystal lattices of minerals. ^4He accumulates in a mineral as a radioactive decay product if the temperature is sufficiently low (closure temperature for apatite is 70°C). Trapped helium atoms can be extracted in the laboratory and counted with a mass spectrometer.

Cosmogenic Nuclides

Near-surface rocks are exposed to cosmic rays from the Sun and other galactic sources. Cosmic rays are high-energy protons and other subatomic particles that can break apart the earthbound atoms they collide with. Cosmic rays convert some atoms to rare isotopes of chlorine, aluminum, and beryllium, such as ^{36}Cl, ^{26}Al, and ^{10}Be. Only 10–100 atoms of these isotopes per gram of rock are produced each year, but even this tiny amount can be measured to estimate the exposure time of the rock. The accumulation of cosmogenic nuclides in a rock is not limited by temperature, but rather by the depth of burial. Surface rocks absorb incoming cosmic rays efficiently, so that only half of them penetrate to 45 cm depth. Half of the surviving cosmic rays are absorbed by the next 45 cm of rock, and half again in the next layer, and so on. As a result, cosmogenic nuclide production is too small to measure at depths beyond a few meters.

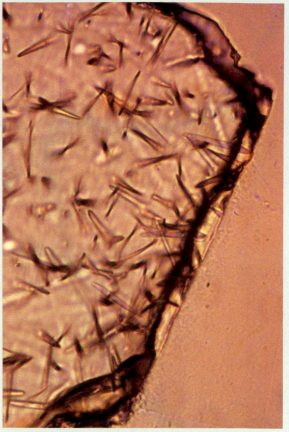

Figure B12.3 Fission Tracks Picture of fission tracks in a zircon crystal, as seen through a microscope. The damage to a mineral crystal associated with a fission track occurs on an atomic scale. Therefore mineral samples are treated with substances that react chemically with the atoms that border the fission tracks and enlarge the tracks to be visible.

The closure temperatures of fission tracks and radiogenic helium can be related to specific depths in the crust. In the example shown in Figure B12.4, we assume a geothermal gradient of 25°C/km. If the exhumation rate R is steady, the different radioactive-decay systems in the surface rocks will give ages that are proportional to the depths at which closure temperatures occur. For instance, suppose a geologist collects a surface rock, formerly from the deep crust, that contains both zircon and apatite. If fission tracks in zircon indicate an age of 5 million years, one can assume that the zircon lay at a depth of 9.6 km (or 10 km, more roughly) at 5 million years ago (Myr). This implies an exhumation rate

$$R = 10 \text{ km}/(5 \text{ Myr}) = 2 \text{ mm/yr}$$

If this rate has held steady during the past 5 Myr, then fission tracks in apatite minerals should indicate an age slightly less than half the zircon estimate, roughly 2.3 million years.

Fission-track dating studies tell us that erosion exhumes the crests of mountain ranges more rapidly than the surrounding foothills. The current local uplift rate for one high region of the western Himalaya, based on fission-track measurements, is as high as 5 mm/year (0.2 in/yr). If sustained for only 2 million years, the total uplift would be 10 km (33,000 ft). Such a high rate of uplift is generally associated with steep slopes and high relief. This is certainly true in the Himalaya, where glacially eroded mountain slopes near the crest of the range typically exceed angles of 30° and local relief can exceed 5 km (16,500 ft). At lower altitudes, where streams are the dominant erosional process, slopes generally decline to between 15 and 20°. Mean erosion rates in these high mountains are estimated to be about 3 to 4 mm/year (0.1 to 0.2 in/yr), which is close to the average uplift rate of the Himalaya.

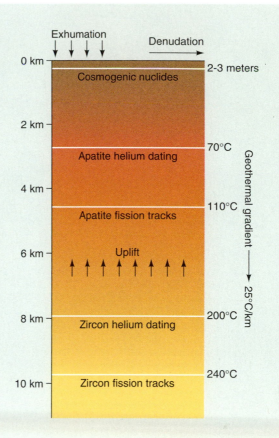

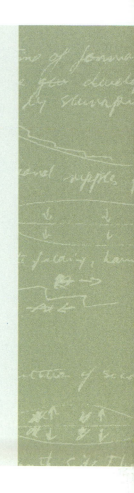

Figure B12.4 Exhumation Estimation Methods
Schematic figure of the depths that correspond to the closure temperatures of several common radioactive-decay systems. If exhumation is steady, the measured ages of the different systems should be proportional to their closure temperatures.

urbanized areas, sediment yield tends to be low because the land is almost completely covered by buildings, sidewalks, and roads that protect the underlying rocks and sediments from erosion.

In many drainage basins, the measured and estimated sediment yields reflect conditions that are probably quite different from those of only a few decades ago because much of the sediment that formerly reached the sea is now being trapped in reservoirs behind large dams. For example, the high Aswan Dam on the Nile River, in Egypt, now intercepts most of the sediment that the Nile used to carry to the Mediterranean Sea.

Before you go on:

1. How can we use earthquakes and river terraces to estimate the uplift rate?

2. What is a closure temperature?

3. How can the number of fission tracks in a mineral be used to estimate the exhumation rate?

4. What regions have the largest denudation rates?

5. How does human activity influence landscape evolution?

ANCIENT LANDSCAPES OF LOW RELIEF

The ultimate reduction of a landmass to low elevation, completing the tectonic cycle envisioned by Davis, is likely to occur only if changes in plate motion lead to cessation of tectonic uplift. Denudation can then "catch up" and gradually lower the relief. Examples of such landscapes are in the world's oldest continental areas, where the roots of ancient mountain systems have been exposed as the crust was thinned by long erosion and compensating isostatic adjustment.

Australia is widely believed to have some of the oldest landscapes of any of the continents. Measurements of cosmogenic nuclides just beneath the surface of the granitic inselbergs have shown that the tops of these domes are eroding at a rate of less than a meter per million years. A combination of relatively arid climate, long-term tectonic stability, and low erodibility of the exposed granite contribute to the exceptional stability of these landforms. Based on these measurements, it seems likely that this landscape predates the Quaternary Period and may have originated well back in the Tertiary Period, or even earlier.

Widespread erosional landscapes of low relief and altitude are uncommon. This must mean either that Earth's

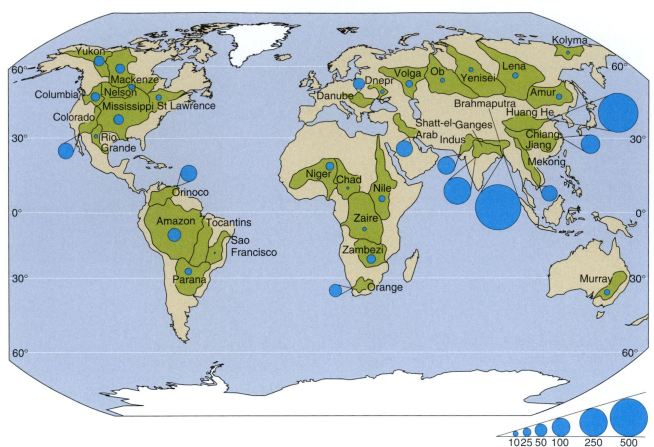

Figure 12.13 Present Denudation Rates for Major Drainage Basins Compare with Figure 12.12.

Denudation rate (mm/1000 yr)
10 25 50 100 250 500

crust has been very active in the recent geologic past or that such landscapes take an extremely long time to develop. By extrapolating current denudation rates into the past, estimates have been made of the time required to erode a landmass to or near sea level.

Such estimates must consider two important factors. Studies show that rates of denudation are strongly related to altitude. A river in a broad lowland valley will erode a landscape less than a river in a steeper upland valley. This implies that, as the land is lowered, the rate of denudation will decline. Further, eroding land rises isostatically as the lithosphere adjusts to the transfer of sediment from land to sea. Both factors tend to increase the time needed for final reduction of a landmass to low altitude. Estimates of the time required to reduce a landmass about 1500 m high to near sea level (assuming no tectonic uplift) range from about 15 to 110 million years, depending on climate and rock type.

You might argue that much of Earth's crust has been unusually active during the last 15 million years or more. But could there have been earlier intervals of relative crustal quiet when low-relief surfaces did develop? Many ancient land surfaces are preserved in the geologic record as unconformities (Chapter 11). We can trace some over thousands of

square kilometers and show them to possess only slight relief (Figure 12.14). Associated with such surfaces are weathering profiles that imply that the land surface was continuously exposed for long intervals at relatively low altitude. Such buried ancient landscapes are evidence of times when broad areas were eroded to low relief.

REVISITING PLATE TECTONICS AND THE EARTH SYSTEM— UPLIFT, WEATHERING, AND THE CARBON CYCLE

The long-term climate trend during the past 100 million years generally resembles the reconstructed trend in atmospheric CO_2 concentration (as discussed in Chapter 19). Explaining this long-term trend is important to understanding how Earth's climate system has evolved. The carbon cycle on Earth is driven by plate tectonic motions. This involves many different processes and feedbacks, some of which we examine here. Many think that uplift and exhumation are among the most important factors in determining the level of CO_2 in the atmosphere.

Figure 12.14 Angular Unconformity in the Grand Canyon
Nearly flat-lying sedimentary strata in the upper walls rest on tilted older strata. The angular unconformity separating the two rock groups is a low-relief surface traceable for a considerable distance. It represents a subdued ancient land surface.

Controls Over the Carbon Cycle

Global volcanism adds CO_2 to the atmosphere from Earth's interior. How can we estimate the history of volcanic CO_2 emissions? According to one popular hypothesis, the global rate of seafloor spreading controls the global rates of subduction, continental collision, and by logical extension, the transfer rate of CO_2 to the atmosphere (Figure 12.15).

Another source of atmospheric CO_2, possibly larger than volcanic outgassing, is metamorphism of carbon-rich ocean sediments carried downward in subduction zones. This process strips CO_2 from the subducting plate before it can reach the mantle source of island-arc magmas. Even

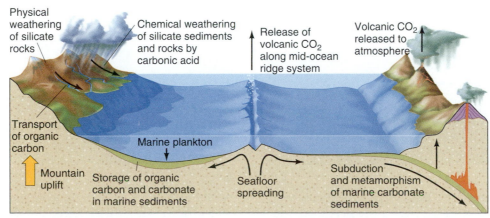

Figure 12.15 The Carbon Cycle The primary natural sources of atmospheric CO_2 are volcanism and metamorphism. CO_2 is removed by surface weathering, transport, and long-term storage of organic carbon in marine sediments.

though this CO_2 source is not volcanic, it should increase and decrease with subduction rates. Subduction and seafloor spreading are fundamental aspects of plate tectonics. It follows that variations in plate motion play a large role in controlling the rate of CO_2 buildup in the atmosphere.

However, at the same time, CO_2 is removed from the atmosphere by the weathering of surface silicate rocks. As you saw in Chapter 7, rainwater combines with CO_2 to form carbonic acid (H_2CO_3), and this acid weathers silicate rocks:

$$CaSiO_3 + H_2CO_3 \rightarrow CaCO_3 + SiO_2 + H_2O$$
silicate rock carbonic acid   weathering products   water

Land plants synthesize other carbon-based acids in their roots to leach chemical nutrients from rocks. This biologic activity is thought to accelerate the consumption of CO_2 by rock weathering.

Streams transport the weathering products to the ocean. There the carbonate and silica are used by marine organisms, and their remains accumulate on the seafloor and are stored as sediment (Figure 12.15). Weathering of silicate rocks, therefore, is a negative feedback in the climate system it moderates.

If seafloor spreading accelerates, more CO_2 enters the atmosphere, Earth's surface warms, atmospheric water vapor increases, and vegetation density increases. This speeds the chemical weathering of silicate rocks, which removes CO_2 from the atmosphere, thereby keeping the system in a more balanced state. If spreading slows, less CO_2 enters the atmosphere, the climate cools, weathering rates decrease, and the consumption of atmospheric CO_2 decreases.

A further important negative feedback is burial of organic carbon. High erosion rates can lead to rapid deposition in sedimentary basins and in submarine fans. Isolated from the weathering environment at the land surface, organic carbon is quickly buried rather than returned to the atmosphere by surface weathering. Recent studies suggest that carbon burial may be even more important than silicate weathering in long-term carbon storage.

ROLE OF UPLIFT IN CO_2 LEVELS

An alternative hypothesis has been proposed in which tectonic uplift is a major driving force behind changes in atmospheric CO_2 levels. Uplift could increase the removal rate for atmospheric CO_2 because it generates faulting, which exposes fresh, fractured rock to chemical weathering. Runoff from high summer rainfall transfers physically weathered products from high and middle altitudes, where slopes are steepest, to lower altitudes, where they are chemically weathered in floodplains and deltas in warm, moist climates.

A.

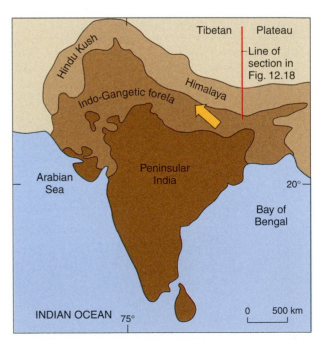

B.

Figure 12.16 Himalaya and Tibetan Plateau   A. Oblique view of snow-covered Himalaya and adjacent Tibetan Plateau rising above the Gangetic lowland (dark green) of India, seen from an orbiting spacecraft. B. Relationship of the Himalaya and Hindu Kush mountain systems to the Tibetan Plateau and the Indian subcontinent. Arrow shows direction of view in A.

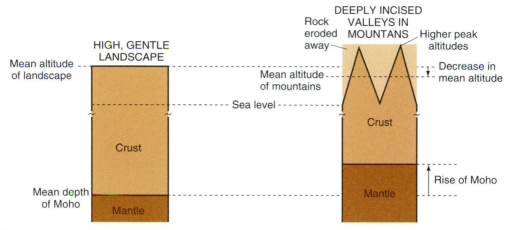

Figure 12.17   Effect of Erosion on Mean Altitude   Simplified sections through the continental crust show the effects of erosion on mean altitude, the depth of the Moho (crust/mantle boundary—see Chapter 10), and uplift of rock. If a climate change causes a high, gentle landscape to become deeply incised by streams that erode valleys toward sea level, the remaining crustal rocks and the Moho will rise (due to isostasy), the mean altitude of the mountainous area will decrease slightly, and the highest peaks will be much higher than before.

Thus, instead of operating somewhat passively as a negative feedback, chemical weathering could be a major factor in controlling the long-term concentration of CO_2 in the atmosphere. The uplift hypothesis focuses attention on tectonically active areas as primary influences on Earth's carbon budget.

Reconstructions of plate tectonics, atmospheric CO_2, and climate over the past 200 million years of Earth history support—but do not prove—the claim that uplift and chemical erosion are important agents in the carbon cycle. Reconstructions suggest that CO_2 levels and global average temperatures increased to peak levels near 100 million years ago, during a long interval when continents split apart to form the present Atlantic and Indian Oceans. Global sea levels were generally high during this time, submerging many continental landscapes and thereby limiting global erosion rates.

A trend over the past 100 million years toward cooler global temperatures and reduced atmospheric CO_2 has occurred at the same time as the uplift and erosion of several major continental plateaus, including Tibet in central Asia, the Altiplano in South America, and the Laramide uplift (Rocky Mountains) in North America. (See Figure 12.16.) Many other factors also influenced the overall rise and fall of atmospheric CO_2 and global temperatures over this time, of course, which makes the relative importance of chemical weathering harder to assess.

RECENT MOUNTAIN UPLIFT AND ISOSTATIC FEEDBACKS

The great altitude and relief of major mountain ranges like the Himalaya and Andes have led most geologists to assume that these impressive topographic features are the result of ongoing plate convergences that have led to accelerated uplift during the late Cenozoic. But is plate tectonics the most likely cause or the only cause?

Isostatic adjustment to erosion can generate topographic relief through important feedbacks in the Earth system. If tectonic forces uplift a mountain system (by plate subduction or collision), erosion and mass wasting will tear away at the rising landmass, transferring sediment to adjacent basins and to the ocean. Removal of this rock mass creates an isostatic imbalance, the adjustment to which causes the mountains to rise further because of their reduced load on the underlying mantle (Figure 12.17). Erosion tears away at the gaps between mountains in the range, however, so that individual mountains increase in altitude while the mean altitude of the mountain range actually decreases.

This somewhat surprising result is confirmed by recent studies of the topography of the Tibetan Plateau and the Himalaya. Although individual mountains rise to altitudes of more than 8000 m, the mean Himalayan altitude of approximately 5000–5500 m is nearly identical to that of the adjacent relatively undissected Tibetan Plateau (Figure 12.18). This argues that isostatic adjustment is an important factor.

The contrasting hypotheses illustrate the complexity of the problem. The entire Earth system is involved—not only the basic mechanisms, but also the complex positive and negative feedbacks. These leave us with difficult questions and challenging research problems. Our single certainty at this writing is that Earth's surface is not static, and that it will change in response to recognizable geologic processes.

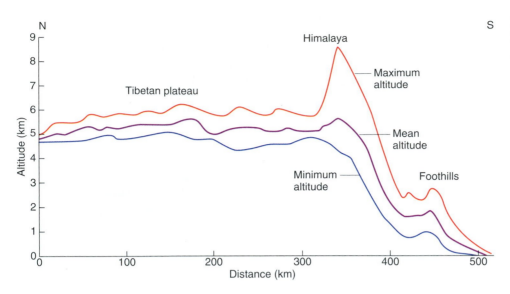

Figure 12.18 Topographic Profile Across Tibetan Plateau and the Himalaya See Figures 12.3A and 12.16 for orientation. Despite much higher altitudes, the average altitude of the Himalaya is nearly equal to that of the plateau.

What's Ahead?

The internal processes behind uplift, such as plate tectonics, volcanism, rock deformation, and earthquakes, have been discussed in previous chapters. Rock weathering, an important surface process in exhumation, has been discussed in Chapter 6. Other surface processes, such as mass wasting, stream and river systems, glaciers, and climate, will be discussed in the chapters that follow.

CHAPTER SUMMARY

1. Changes in the landscape are generated by the competing processes of uplift, exhumation, and denudation. Each of these processes can change the relief of the landscape. Geomorphology is the study of landforms and their evolution.

2. Uplift is tied to plate tectonics and mantle convection, because it involves the deformation of lithosphere.

3. Exhumation refers to processes that expose rock that lies beneath the surface. Denudation refers to processes that remove debris from the surface. Although exhumation and denudation represent different perspectives on the erosion of a landscape, the terms are roughly equivalent in usage.

4. Uplift can occur from collisional, isostatic, and extensional mechanisms. Collisional uplift generally takes the form of an orogen (mountain chain) along a convergent plate boundary, constructed largely by motion on thrust faults. Extensional uplift occurs when the lithosphere is stretched and its brittle surface rocks break apart in normal faults.

5. Isostatic uplift takes several forms. The buoyancy of a mantle hot spot may cause the overlying lithosphere to rise. A portion of cool mantle lithosphere, thickened like toothpaste in a plate collision, may drop off into the mantle, leaving the residual lithosphere intact but more buoyant. Finally, erosion removes material from a landscape, leaving the residual lithosphere more buoyant.

6. Earth's surface topography reflects sculpture by different erosional processes and the deposition of resulting sediment.

Water, ice, and wind are the principal agents that erode the land and transfer sediment toward the ocean basins. The principal factors influencing denudation rate are climate, lithology, relief, and time.

7. Exhumation rates can be estimated from radioactive decay systems in which the daughter products accumulate only when the rock is cool, and therefore close to the surface. These systems include the accumulation of the noble gas helium from alpha-decay of uranium and thorium, and damage to mineral crystals from alpha particles, called "fission tracks."

8. Another estimate of exhumation rate can be obtained from cosmogenic nuclides, rare isotopes that are generated by the absorption of the Sun's cosmic rays in the top meter or so of exposed rock.

9. Natural denudation processes are constantly changing the character of the landscape, but so, too, are human activities. Although generally operating at slow rates, natural processes can produce dramatic changes on geologic time scales.

10. Landscapes are constantly adjusting to changes in the factors that control their development and likely are never in a complete state of equilibrium.

11. In the "geographic cycle" of W. M. Davis, uplift dominates in young landscapes, and denudation becomes more dominant as the landscape matures. The steady-state hypothesis proposes that uplift and denudation remain in balance for long periods of time.

12. Ancient landscapes with low relief imply long intervals of erosion that slowly lowered the land surface to low altitude. Although not common in modern landscapes, extensive surfaces of low relief are preserved in some ancient rocks as widespread unconformities.

13. Silicate weathering and burial of organic carbon remove CO_2 from the atmosphere and place it in long-term storage. Removal of CO_2 from the atmosphere reduces the greenhouse effect and influences world climate. Mountain uplift may have exerted a strong influence on atmospheric CO_2 levels in Earth history, because uplift exposes fresh rock to chemical weathering.

THE LANGUAGE OF GEOLOGY

badlands (p. 307)

closure temperature (p. 314)
collisional uplift (p. 306)
continental divide (p. 310)
cosmogenic nuclide (p. 314)

denudation (p. 304)

exhumation (p. 304)

extensional uplift (p. 308)

fission tracks (p. 314)

geographic cycle (p. 312)
geomorphology (p. 304)

isostatic uplift (p. 306)

orogen (p. 306)

radiogenic helium (p. 314)
relief (p. 305)

steady-state landscape (p. 312)

threshold effects (p. 312)

uplift (p. 304)

QUESTIONS FOR REVIEW

1. How might one calculate the volume of sediment removed from the land surface during the last 2 million years by a major stream that has built a large delta where it enters the ocean?

2. Contrast the denudation potential of a landscape in a rainy climate, as compared to a landscape in a dry climate.

3. The closure temperature for helium dating in apatite (70°C) and zircon (200°C) differ because the noble gas helium is trapped more effectively in the zircon crystal lattice. In an eroding landscape, suppose the helium age for zircon is 4.0 Myr. If the exhumation rate has been constant over time, what should the helium age for apatite be at the same location?

4. Suppose the geothermal gradient in the location of question 3 is 20°C/km. What is the exhumation rate consistent with the helium ages given?

5. How do bedrock geology and structural history control the position of the continental divide in western North America?

6. What geologic evidence points to periods of relative landscape equilibrium at times in the past?

7. How does the concept of thresholds apply to landscape evolution?

8. Describe two ways that you might be able to determine the uplift rate in a coastal mountain system.

9. How can human activity influence sediment yield from a drainage basin?

10. How might biologic factors help control the denudation rate of a drainage basin?

11. How could mountain uplift be related to geologic factors other than plate tectonics?

Click on *Presentation* and *Interactivity* in the **Plate Tectonics** and **GeoHazards** modules of your CD-ROM to further explore resources and activities presenting concepts from this chapter. Select *Assessment* in the same modules to test your understanding of this chapter.

In September 2002, a massive rockfall avalanche blocked the main road in the Karmadon Gorge, near Vladikavkas in southern Russia, killing more than 150 persons.

Mass Wasting

Front of 1717 rock-avalanche deposit below Mt. Blanc massif.

The Night the Mountain Fell

Large rockfalls that give rise to huge avalanches of rocky debris have been an ever-present hazard in populated mountains like the Alps and the Andes. Although infrequent, they can be devastating and lead to great loss of life and property. In many mountain regions, however, inhabitants are unaware of the potential danger.

In 1977, an Italian colleague, Giuseppe Orombelli, and I were studying the history of recent glacier variations in the Alps on the southern flank of the Mont Blanc range. We were dating glacial deposits using lichenometry, a method based on the fact that lichens (plants) growing on glacial boulders increase in diameter with time. By establishing the growth rate for lichens on granite boulders and finding the largest lichens on them, we could determine the time since a deposit formed. Most of the deposits date from the late eighteenth and early nineteenth centuries, but one was unusually old, apparently having formed early in the eighteenth century. Furthermore, it had some unusual characteristics: its lateral margin was very flat, unlike that of local mountain glaciers, and it consisted only of granite, whereas the local glacial deposits contained a mixture of rock types. Other geologists regarded the deposit as glacial in origin, but we concluded that it likely was the result of a massive rockfall. Two early eighteenth-century texts told us that a catastrophic fall of ice-covered rocks from the mountain occurred on the night of September 12, 1717. Pulverizing on impact, the debris moved rapidly downvalley for 7 km to the point where our anomalous glacial deposit lay. As the debris overwhelmed two small mountain villages, killing all the inhabitants and their livestock, its estimated speed was at least 125 km/h. Such a rapid travel time means that escape was impossible.

Further study disclosed evidence of many other large rockfall events in this region, enabling us to make a map of potential rockfall hazards (Figure 13.28). In several cases, repetition of such rockfall events would spread debris across communities with dozens to several thousand inhabitants.

Stephen C. Porter

KEY QUESTIONS

1. **What is mass wasting and what is its primary cause?**

2. **What are the two broad categories of mass-wasting movements and how do they differ?**

3. **What factors influence the type and velocity of mass-wasting movements?**

4. **What mass-wasting processes are especially important in cold climates? in underwater environments?**

5. **What sorts of triggering occurrences are often related to large mass-wasting events?**

6. **What hazards do mass-wasting events pose for human populations, and how can these hazards be mitigated?**

wooded hillslope, the movement would be imperceptibly slow, but on a steep rock wall in the mountains, movement can be extremely rapid. A giant rockfall or avalanche of debris can move downward and onto a valley floor in only a few minutes. Because we are unlikely to witness such rapid, rare events in the course of our lives, and are mostly unaware of slow, everyday movements, it is difficult to perceive how important mass wasting is in the evolution of landscapes. Nevertheless, far more rock and detritus is directly removed from hillslopes by mass wasting than by streams or by glaciers.

Mass wasting is a basic part of the rock cycle: weathering, mass-wasting, and other aspects of erosion constitute a continuum of interacting processes, the end result of which is the gradual breakdown of solid bedrock and the redistribution of its weathered components.

In this chapter, we will explore the many types of mass-wasting processes, how they operate, how they alter the landscape, how they are triggered, and how some pose a major hazard for humans.

INTRODUCTION: WHAT IS MASS WASTING?

The landscapes we see about us may outwardly appear fixed and unchanging, but if we were to make a time-lapse motion picture of almost any hillslope, the slope would seem almost alive and constantly changing. Much of the recorded motion would be the result of **mass wasting**, the downslope movement of regolith and masses of rock under the pull of gravity. In mass wasting, other erosional processes such as water, wind, and ice are not directly involved as sediment-transporting agents. On a gentle

DOWNSLOPE MOVEMENT OF ROCK DEBRIS

A smooth, vegetated slope may appear outwardly stable and show little obvious evidence of geologic activity. Yet if we examine the regolith beneath the surface, it is likely we will find some rock particles derived from bedrock farther upslope. We can deduce, therefore, that the particles have moved down the slope.

A particle's journey downslope can be imperceptibly slow, very fast, or at some rate in between, but in any case, the movement is caused primarily by gravity. Under most

Box 13.1 THE SCIENCE OF GEOLOGY

DOWNSLOPE MOVEMENT AND THE SAFETY FACTOR

Whether a body of rock or sediment on the land surface remains stable or begins to move downslope depends on several factors. On a horizontal surface, gravitational force acts to hold objects in place by pulling on them in a direction perpendicular to the surface (Figure B13.1). On any slope, however, the total gravitational force can be resolved into two component forces. The *perpendicular component of gravity* (*gp* in Figure B13.1A) acts at right angles to the slope and tends to hold objects in place. The *tangential component of gravity* (*gt* in Figure B13.1A) acts down a slope, and it is this force that causes objects to move downhill. As a slope becomes steeper, the tangential component increases relative to the perpendicular component, making it easier for forces restraining the object to be overcome. In considering a rock mass resting on a slope, *shear stress* is the downslope component of the total stress involved and the perpendicular component is the *normal stress*.

Properties in rock and sediment that resist stresses related to gravity are collectively called *shear strength*. Shear strength is governed by factors inherent in aggregates of rock or regolith that include frictional resistance and cohesion between particles, and the binding action of plant roots.

Another way to express the relationship between shear strength and shear stress is as a ratio, known as the *safety factor (Fs)*:

$$Fs = \text{shear strength/shear stress}$$

When this factor is less than 1, mass movement will occur (Figure B13.1B). This will happen if shear stress is increased or shear strength is reduced. Steepening a slope by erosion, jolting it by an earthquake, or shaking it by blasting can each cause an increase in shear stress.

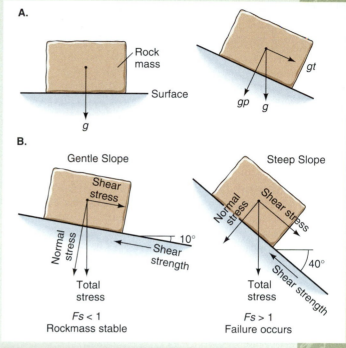

Figure B13.1 Gravity A. Effects of gravity on a rock lying on a flat surface and on a hillslope. Gravity acts vertically, and on a slope can be resolved into two components. One is perpendicular (*gp*) and one tangential (*gt*). B. A rock mass on a sloping surface is stable if shear strength exceeds shear stress (*Fs* <1). On a steeper slope, shear stress exceeds shear strength (*Fs*>1) and the rock mass will move downslope.

Shear strength, on the other hand, can be reduced by weathering, by the decay of plant roots, or by saturation during a heavy rain.

---

conditions, a slope evolves toward an angle that allows the quantity of regolith reaching any point from upslope to be balanced by the quantity that is moving downslope. Such a slope is said to be in a balanced, or *steady-state*, condition.

ROLE OF GRAVITY AND SLOPE ANGLE

Two opposing forces determine whether any mass of rock or rock debris located on a slope will remain stationary or will move. Gravity tends to pull any mass on a sloping surface downhill, whereas friction tends to retard such movement. On a flat surface, the pull of gravity is directly downward, but as the slope angle increases, gravity exerts a pull

in the downslope direction, which means that conditions favoring mass movement increase as slope angle increases. Steep slopes, of course, are most common in mountainous areas, so it is not surprising that mass wasting is especially important in high mountains. As we have seen, the distribution and altitude of mountains is related to the interaction of lithospheric plates, and therefore mass wasting on mountain slopes is a natural consequence of plate tectonics.

The forces involved in slope processes are most clearly shown if shear stress and shear strength are considered (see *The Science of Geology*, Box 13.1, *Downslope Movement and the Safety Factor*).

BOX 13.2   THE SCIENCE OF GEOLOGY

THE BEVERAGE-CAN EXPERIMENT

An innovative experiment to show, by analogy, how large rock masses can move easily due to high fluid pressure was described by geologists M. King Hubbert and William Rubey in 1959. An empty beverage can is placed in an upright position on the wetted surface of a sheet of glass (Figure B13.2A). If the glass is then slowly tilted, the can will not begin to slide until a critical angle (approximately 17°) is reached. This angle is characteristic for the substances used in this demonstration (metal and wet glass). Next, the can is chilled in a freezer, and the experiment is repeated. With the can placed on the wetted glass with its open end upward, the angle at which sliding begins is seen to be the same. Finally, the cold can is placed with its open end downward on the wetted glass, which is tilted so it has a 1° slope (Figure 13.2B). The can will slide the length of the glass and then stop abruptly at the edge.

The reason it now moves on such a gentle slope is that the conditions at the base of the can have changed. As the can warms, the air inside expands, increasing the pressure. The increased air pressure at the base buoys up the can, reducing the friction between metal and glass, and allows the can to glide down the gentle slope. The can stops at the edge of the glass as the pressure is suddenly released. This simple experiment constitutes an analogy for natural geologic conditions in which high water pressure at the base of a large rock mass can promote downslope movement of the rock. Although increased air pressure caused the can to travel easily down a gentle slope, if we substitute water for air and apply the same principle to a rock or debris mass on a slope, the results will be similar. High fluid pressure reduces friction (shear strength) at the base of the mass until the safety factor is exceeded. The rock mass then moves readily downslope.

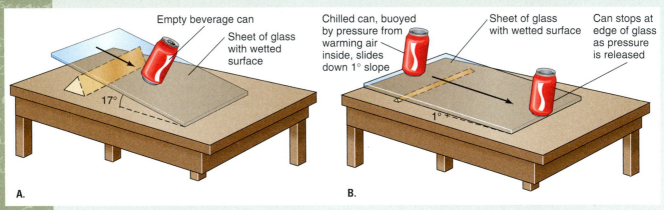

A.

B.

Figure B13.2 Interesting Experiment An experiment illustrating how compressed air reduces friction at the base of a mass resting on a slope. A. An empty beverage can placed open-end up on the wetted glass will begin to slide down the surface when the angle reaches about 17°. B. A chilled empty can, placed upside down on wet glass, will slide when the glass has a slope of only 1° because friction at the base is reduced as warming air in the can expands. Sliding ceases when the can reaches the edge of the glass and pressure at the bed is suddenly released. The experiment shows conditions analogous to a rock mass beneath which trapped water generates sufficient fluid pressure to overcome friction and promote downslope movement.

THE ROLE OF WATER

Water is almost always present within rocks and regolith near Earth's surface. As mentioned earlier, water is not a transporting agent in mass wasting and in some types of mass wasting it plays no significant role. However, in others its role can be significant.

Unconsolidated sediments behave in different ways depending on whether they are dry or wet, as anyone knows who has constructed a sand castle at the beach. Dry sand is unstable and difficult or impossible to mold, but when some water is added, the sand gains strength and can be shaped into vertical castle walls. The water and sand grains are drawn together by a force called *capillary attraction*. The attraction results from surface tension, a property of liquids that causes the exposed surface to contract to the smallest possible area. This force tends to hold the wet sand together as a cohesive mass. However, the addition of too much water saturates the sand and turns it into a slurry that easily flows away, as the sand-castle builder sees with dismay when the rising tide on the beach destroys the elaborate work of an afternoon.

Moist or weakly cemented fine-grained sediments, such as fine silt and clay, may be so cohesive that they can stand in near-vertical cliffs, like the walls of a sand castle at the beach. However, if the silt or clay becomes saturated with water and the fluid pressure of this water rises above a critical limit, the fine-grained sediment will lose strength and begin to flow.

The movement of some large rock masses has been attributed to the effects of increased water pressure in voids in the rock. If the voids along a contact between two rock masses of low permeability are filled with water, and the water is under pressure, a supporting effect may result. In other words, the water pressure bears part of the weight of the overlying rock mass, thereby reducing friction along the contact. The result can be sudden *failure*, the collapse of a rock mass due to reduced friction (i.e., shear strength).

An analogous situation, in which water pressure buoys up a heavy object, can make driving in a heavy rainstorm extremely dangerous. When water is compressed beneath the wheels of a moving vehicle, the increasing fluid pressure can cause the tires to "float" off the roadway. The driver quickly loses control, a condition known as hydroplaning. An experiment illustrating this same principle is described in *The Science of Geology*, Box 13.2, *The Beverage-Can Experiment*.

As seen from these examples, water can be instrumental in reducing friction and shear strength, thereby promoting movement of rock and sediment downslope under the pull of gravity. It does so in two important ways: (1) by reducing the natural cohesiveness between grains and (2) by reducing friction at the base of a rock mass through increased fluid pressure.

Before you go on:

1. How is shear stress related to slope angle?

2. What two components of gravity act on a rock lying on a slope? How does the slope's steepness affect these two components?

3. What is shear strength? What properties govern the shear strength of a rock mass on a slope?

4. In what ways can water promote mass wasting?

MASS-WASTING PROCESSES

We now turn to the variety of processes involved in mass wasting. These processes all share one characteristic: they take place on slopes. Any perceptible downslope movement of a mass of bedrock, regolith, or both is commonly referred to as a **landslide**. However, we can recognize many different kinds of slope movements, and because they often grade into one another, no simple classification exists. The composition and texture of the material involved, the amount of water and air mixed with the solid material, and the steepness of slope all influence the type and velocity of movement.

The approach we will take here is to describe mass-wasting processes in terms of two broad categories: (1) the sudden failure of a slope that results in the downslope transfer of relatively coherent masses of rock or rock debris by slumping, falling, or sliding, and (2) the downslope flow of mixtures of solid material, water, and air, which are distinguished on the basis of velocity and the concentration of particles in the flowing mixture.

SLOPE FAILURES

Failure, as noted earlier, refers to the collapse of rock or sediment mass. The constant pull of gravity makes all hillslopes and mountain cliffs susceptible to failure. When failure occurs, rock debris is transferred downslope and a more stable slope condition is reestablished. We look first at several types of slope failure: slumps, falls, and slides.

SLUMPS

A **slump** is a type of slope failure in which a downward and outward rotational movement of rock or regolith occurs along a curved concave-up surface (Figure 13.1A). The top of the displaced block usually is tilted backward, producing a reversed slope. Slumps may occur singly or in groups, and they can range in size from small displacements only a meter or two in dimension to large complexes that cover hundreds or even thousands of square meters.

Slumps are one of the types of mass movement that we are most likely to see, for many result from artificial modification of the landscape. They are numerous along roads and highways where bordering slopes have been oversteepened by construction activity. We can also see them along river banks or seacoasts where currents or waves have undercut the base of a slope (Figure 13.2).

Slumps frequently are associated with heavy rains or sudden shocks, such as earthquakes. Distinct episodes of slumping may be related to changing climatic conditions. Slumping may recur seasonally and be associated with seepage of water into the ground during the rainy season. In parts of the western United States, increased slumping during recent decades appears to be correlated with an overall rise in average rainfall. A similar increase about 5000 years ago apparently was related to a shift from a warm, dry climate in the middle Holocene to cooler and wetter conditions since.

ROCKFALLS AND DEBRIS FALLS

Ask a mountain climber about the greatest dangers associated with his sport and he likely will place falling rock near the top of the list. **Rockfall**, the free falling of detached bodies of

Slope Failures

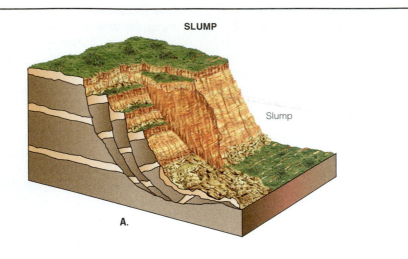

SLUMP

Slump

A.

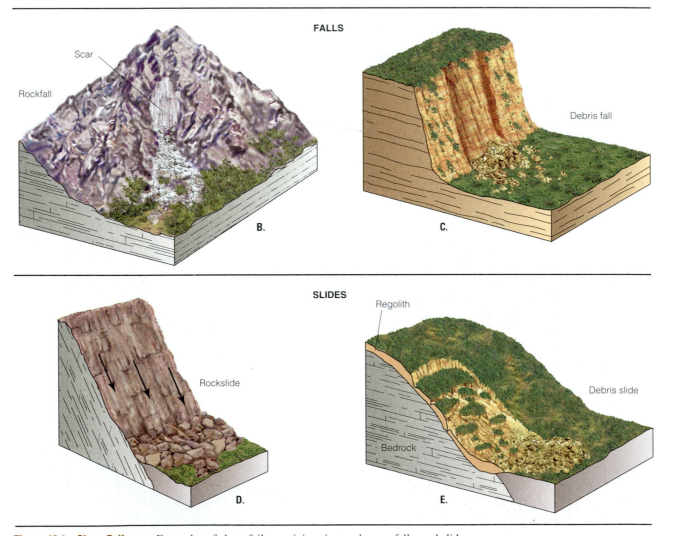

FALLS

Scar

Rockfall

B.

Debris fall

C.

SLIDES

Rockslide

D.

Regolith

Debris slide

Bedrock

E.

Figure 13.1 Slope Failures Examples of slope failures giving rise to slumps, falls, and slides.

bedrock from a cliff or steep slope, is common in precipitous mountainous terrain where rockfall debris forms conspicuous deposits at the base of steep slopes (Figure 13.1B).

As a rock falls, its speed increases. Knowing the distance of fall (*h*), we can calculate the velocity (*v*) on impact as

$$v = 2\sqrt{gh}$$

where g is the acceleration due to gravity. This formula tells us that a rock of a given size will be traveling at a much higher velocity if it falls free from a high cliff than from a low cliff.

A rockfall may involve the dislodgement and fall of a single rock fragment, or it may involve the sudden collapse of a huge mass of rock that plunges hundreds of meters, gathering speed until it breaks on impact into a vast number of smaller pieces. These pieces continue to bounce, roll, and slide downslope before friction and decreasing slope angle bring them to a halt.

When a mountain slope collapses, not only rock but overlying regolith and plants are generally involved. The resulting **debris fall** is similar to a rockfall, but it consists of a mixture of rock and weathered regolith, as well as vegetation (Figure 13.1C).

ROCKSLIDES AND DEBRIS SLIDES

Rockslides, like rockfalls, often involve the rapid displacement of masses of rock. A **rockslide** is the sudden downslope movement of a detached mass of bedrock (or of debris, in the case of a **debris slide**) along an inclined surface, such as a bedding plane (Figure 13.1D, E). Like falling rock and debris, rock slides and debris slides are common in high mountains where steep slopes abound (Figures 13.3 and 13.4). When large rockslides occur, the resulting deposit generally is a chaotic, jumbled mass of

Figure 13.2 Slump A large slump in a high gravel terrace beside the Yakima River in central Washington has broken up a major highway and displaced it more than 100 m laterally into the river channel.

Figure 13.3 Rockslide A jumble of boulders, many the size of houses, forms a massive rockslide deposit that descended a steeply sloping bedrock surface on a high Andean mountain in Argentina. Other large masses of jointed rock are poised at the top of this slope, ready to produce future rockslides.

Figure 13.4 Debris Slide A debris slide from a wooded hillside above a neighborhood of San Salvador buried hundreds of homes and killed nearly 400 people shortly after a magnitude 7.6 earthquake in January 2001.

Figure 13.5 Talus A talus at the base of a steep mountain slope in the Brooks Range, northern Alaska. Rockfall debris moving down a steep gully spreads at its base to form a conical talus.

Figure 13.6 Angle of Repose Coarse, angular limestone blocks stand at the angle of repose (about 30°) in a talus below steep cliffs in the central Brooks Range, Alaska.

rock, with individual boulders that may measure tens of meters across.

Accumulations of angular rock fragments are a common sight at the bases of steep cliffs. The rock debris typically ranges in size from sand grains to large boulders. Such a body of debris sloping outward from the cliff that supplies it is a **talus** (Figure 13.5). The movement, from cliff to talus, is chiefly by falling, sliding, bounding, and rolling. The rock fragments come to rest at the steepest angle at which the debris remains stable (Figure 13.6). This angle, referred to as the **angle of repose**, typically lies between 30° and 37°. Fine particles falling onto a talus tend to settle into open voids between coarser fragments. Large falling rocks have more momentum than small particles and tend to move farther down the slope. Some may bound beyond the toe of the talus and form a scattered array of isolated boulders.

SEDIMENT FLOWS

There are many mass-wasting processes in which solid particles move in such a way that the overall motion of the

entire mass can be described as a flowing motion. The material that flows is usually a mixture of solid particles, water, and sometimes air, but the presence of water is not required for flow to occur (consider, for example, the flow of dry sand in an hourglass). Mass-wasting processes that involve the flow of such mixtures are called **sediment flows**.

FACTORS CONTROLLING FLOW

At above-freezing temperatures, the way a sediment flows depends on the relative proportion of solids, water, and air, as well as on the physical and chemical properties of the sediment. We will focus here on the proportion of particles and water.

First, let's consider what distinguishes a sediment flow from a flow of sediment-laden water. All streams carry at least some sediment, but if the sediment becomes so concentrated that the water no longer can transport it, a sediment-laden stream will change into a very fluid sediment flow. Because the water in a sediment flow is not a transporting medium, a sediment flow, by definition, is a mass-wasting process. The water helps promote flow, but the pull of gravity on the solid particles remains the primary reason for their movement. In contrast, in a stream particles are moved largely by the force of flowing water, which flows because of the pull of gravity on the water.

Figure 13.7 divides sediment flows into two classes based on sediment concentration. A **slurry flow** is a moving mass of water-saturated sediment. A **granular flow** is a mixture of sediment, air, and water. Unlike a slurry flow, it is not saturated with water; instead, the full weight of the flowing sediment is supported by grain-to-grain contact or by collision between grains. Each of these two classes is further subdivided into several processes on the basis of flow velocity. For example, *creep* is a very slow type of granular flow, measured in millimeters or centimeters per

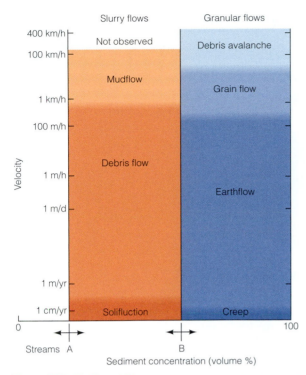

Figure 13.7 **Sediment Flows** Classification of sediment flows on the basis of their velocity and sediment concentration. The transition from a sediment-laden stream to a slurry flow occurs when the sediment concentration becomes so high that the stream no longer acts as a transporting agent; instead, the direct action of gravity becomes the primary force causing the saturated sediment to flow. As the percentage of water decreases further, a transition from slurry flow to granular flow takes place. Now the sediment may contain water and/or air. The boundaries between muddy streams and slurry flows (A) and between slurry and granular flows (B) are not assigned sediment-concentration percentages because the position of the boundaries can shift to the left or right (arrows) depending on the physical and compositional characteristics of the sediment+water+air mixture.

year, whereas a *debris avalanche* is measured in kilometers per hour. In this classification of sediment flows, the boundaries between processes are only approximate and depend on the grain-size distribution, sediment concentration, and other factors.

SLURRY FLOWS

In slurry flows, the sediment mixture is often so dense that large boulders can be suspended in it. Boulders too large to remain in suspension tend to roll along with the flow. When flow ceases, fine and coarse particles remain mixed, resulting in a nonsorted or very poorly sorted sediment. Next we'll consider several key types of slurry flows.

Solifluction. The very slow downslope movement of saturated soil and regolith is known as **solifluction**. As you

can see in Figure 13.7, this process lies at the lower end of the velocity scale for flowing sediment–water mixtures. Rates of movement are less than about 30 cm/yr, generally so slow as to be detectable only by measurements made over several seasons. The slow movement results in distinctive surface features, including lobes and sheets of debris that sometimes override one another (Figures 13.8A and 13.9). Solifluction occurs on hillslopes in temperate and tropical latitudes, where regolith remains saturated with water for long intervals.

Debris Flows. A **debris flow** involves the downslope movement of unconsolidated regolith, the greater part being coarser than sand, at rates ranging from only about 1 m/yr to as much as 100 km/h (Figures 13.7 and 13.8B). In some cases a debris flow begins with a slump or debris slide, the lower part of which then continues to flow downslope. A typical debris flow, once mobilized, may flow far enough to enter a stream channel and then spread across the surface of an alluvial fan, where it forms a poorly sorted deposit.

Debris flow deposits commonly have a tongue-like front. They also have a very irregular surface, often with concentric ridges and depressions that resemble the surfaces of deposits left by mountain glaciers. Debris flows are frequently associated with intervals of extremely heavy rainfall that lead to saturation of the ground.

Mudflows. A **mudflow** is a rapidly moving debris flow with a water content sufficient to make it highly fluid. The predominant particles are no coarser than sand. In Figure 13.7, the velocity range of mudflows lies above the upper range of debris flows (more than about 1 km/h). Most mudflows are highly mobile, tend to travel rapidly along valley floors, and spread laterally when no longer confined by valley walls (Figure 13.8C).

If you were to scoop up a handful of moving mudflow sediment, its consistency could range from that of freshly poured concrete to a soup-like mixture only slightly denser than very muddy water. After heavy rain in a mountain canyon, a mudflow can start as a muddy stream that continues to pick up loose sediment. It becomes a moving dam of mud and rubble that extends to each wall of the canyon and is urged along by the force of the muddy sediment behind it (Figure 13.10). On reaching open country at the mountain front, the moving dam collapses, and mud mixed with boulders is spread as a wide, thin sheet. Sediment fans at the base of mountain slopes in such arid regions as the central Andes and California's eastern slope of the Sierra Nevada may consist largely of superposed sheets of mudflow sediments interstratified with stream sediments (Figure 13.11).

On active volcanoes in wet climates, layers of tephra and other volcanic debris commonly cover the surface and are easily mobilized as mudflows. Such highly fluid mudflows can travel great distances and at such high velocities

Sediment Flows

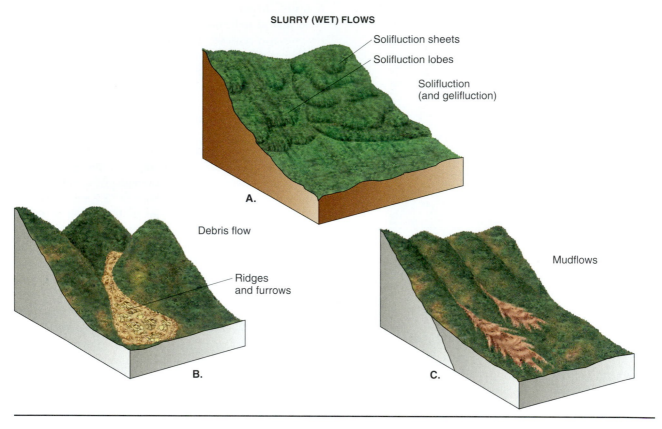

SLURRY (WET) FLOWS

Solifluction sheets
Solifluction lobes
Soliflucion (and gelifluction)

A.

Debris flow

Ridges and furrows

B.

Mudflows

C.

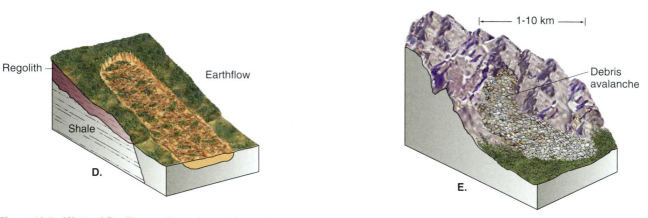

GRANULAR (DRY) FLOWS

Regolith

Earthflow

Shale

D.

1-10 km

Debris avalanche

E.

Figure 13.8 Wet and Dry Flows   Examples of slurry flows and granular flows.

that they constitute one of the major hazards associated with volcanic eruptions. Mudflow sediments in valleys surrounding many of the stratovolcanoes in the Cascade Range contain much of the total volume of tephra and volcanic rocks that had earlier accumulated closer to the vents. A particularly large mudflow that originated on the slopes of Mount Rainier about 5700 years ago traveled at least 72 km. The sediment spread out beyond the mountain front as a broad lobe as much as 25 m thick. Its volume is estimated to be well over a billion cubic meters. Thousands of people now live on top of this mudflow deposit, which lies in the path of comparably large future mudflows. Mount St. Helens, an unusually active volcano, has produced mudflows throughout much of its history. The most

Figure 13.9 Solifluction Two solifluction lobes, 1–2 m thick, have crept slowly downslope and covered glacial deposits on the floor of the Orgière Valley in the Italian Alps.

A.

B.

C.

Figure 13.10 Muddy Flow Passage of a muddy debris flow along a canyon bottom near Farmington, Utah, in June 1983. A. The bouldery front of a muddy debris flow advances from left to right along a stream channel in the wake of an earlier surge of muddy debris. B. The steep bouldery front, about 2 m high and advancing at 1.3 m/s, acts as a moving dam, holding back the flow of muddy sediment upstream. C. The main slurry, having a sediment concentration of about 80 percent and now moving at about 3 m/s, is viscous enough to carry cobbles and boulders in suspension.

recent occurred during the huge eruption of May 1980 (Figure 13.12).

GRANULAR FLOWS

The sediment of granular flows may be largely dry, with air filling the pores, or it may be initially saturated with water but have a range of grain sizes and shapes that allows water to escape easily. Next we describe the important kinds of granular flows.

Creep and Colluvium. Most of us have seen old fences, telephone poles, or gravestones leaning at an angle on hillslopes, or have seen evidence of the downslope displacement of fractured road surfaces (Figure 13.13). All are common evidence of **creep**, the imperceptibly slow downslope movement of regolith. Steeply inclined rock strata are sometimes bent over in the downslope direction just below the ground surface, showing further evidence of creep (Figure 13.14).

A number of factors contributing to creep are listed in Table 13.1. Although creep occurs at a rate too slow to be seen, careful measurements of the downslope displacement of objects at the surface record the rates involved (Figure 13.15). As might be expected, rates tend to be higher on steep slopes than on gentle slopes. Measurements in Colorado, for example, document a creep rate of 9.5 mm/yr on a slope of 39° but only 1.5 mm/yr on a 19° slope. Rates also tend to increase as soil moisture increases. However, in wet climates vegetation density also increases and the roots of plants, which bind the soil, tend to inhibit creep. The creep rate measured on one grassy hillside in England having a slope of 33° is only

Figure 13.11 Mudflow Deposit A roadcut in the Chilean Andes near Santiago exposes a poorly sorted prehistoric volcanic mudflow in which large boulders cluster near the base of the deposit and smaller stones are scattered near the top.

Figure 13.12 Mount St. Helens Mudflows During the 1980 eruption of Mount St. Helens in Washington, volcanic mudflows were channeled down valleys west and east of the mountain. Some mudflows reached the Columbia River, having traveled more than 90 km. Flow velocities were as high as 40 m/s and averaged 7 m/s.

0.02 mm/yr. Despite a slow rate of movement, creep affects all hillslopes covered with regolith, and its cumulative effect is therefore very great.

Loose, incoherent deposits on slopes that are moving mainly by creep are termed **colluvium**. The particles in colluvium tend to be angular and lack obvious sorting. These characteristics generally make it possible to distinguish colluvium from sediments deposited by flowing water (alluvium), which tend to consist of rounded particles, sorted and deposited in layers.

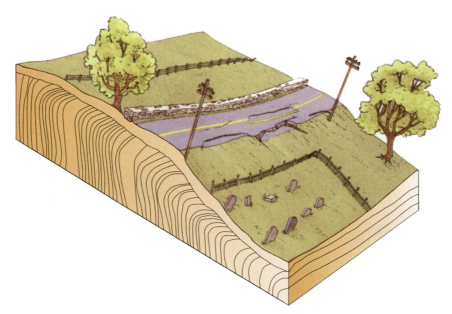

Figure 13.13 Creep Effects of creep on surface features and on bedrock. Steeply inclined strata have been deformed near the surface by differential creep, so they appear folded. Telephone poles and fence posts affected by creep are tilted, stone fences are deformed, roadbeds are locally displaced, tree trunks may be bent, and gravestones are tilted or fallen.

Figure 13.14 Shale Overturned by Creep Beds of shale have been overturned by slow downslope creep on a hillslope in the Laramie Basin, Wyoming.

TABLE 13.1 Factors Contributing to Creep of Regolith

Frost action	Freezing and thawing, without necessarily saturating the regolith, cause lifting and subsidence of particles.
Wetting and drying	Cause expansion and contraction of clay minerals.
Heating and cooling without freezing	Cause volume changes in mineral particles.
Growth and decay of plants	Cause wedging, moving particles downslope. Cavities formed when roots decay are filled from upslope.
Activities of animals	Worms, insects, and other burrowing animals displace particles, as do animals trampling the surface.
Dissolution	Mineral matter taken into solution creates voids in bedrock that tend to be filled from upslope.
Effects of snow cover	Seasonal snow cover tends to creep downslope and drag with it particles from the underlying ground surface.

Earthflows. Among the more common mass-wasting features on the landscape are **earthflows**, granular flows having a velocity in the range of about 1 cm/day to several hundred m/h (Figures 13.7 and 13.8D). Earthflows may remain active for several days, months, or even years. Even after initial motion ceases, they may be highly susceptible to renewed movement. Like debris flows, earthflows are often made up of weak regolith, predominantly of silt and

A.

B.

Figure 13.15 Creep on a Greenland Hillslope Colored targets placed in a straight line across a hillslope in Greenland (A) had moved differentially downslope by creep when photographed a year later (B). The maximum recorded movement along the slope averaged 12 cm/yr.

clay-sized particles, and occur on gentle to moderately steep slopes (2° to 35°). They occur where the ground is saturated, at least intermittently, and they frequently are associated with intervals of excessive rainfall.

At the top of a typical earthflow is a steep *scarp*, which is a cliff formed where slide material has moved away from undisturbed ground upslope (Figure 13.16). An earthflow generally has a narrow, tongue-like shape and a rounded, bulging front. Earthflows range in size from several meters long and wide and less than a meter deep, to as much as several hundred meters wide, more than 1 km long, and more than 10 m deep. Many occur as parts of large earthflow complexes. In a longitudinal profile from head (top) to toe (leading edge), an earthflow is concave upward near the head and convex upward near the toe. This profile implies thinning of the sediment in the upper part of the displaced mass and thickening in the lower part during movement.

Field studies show that earthflows become mobilized when water pressure along shear surfaces rises. Pressures can reach such high values that movement will occur on slopes of only a few degrees.

A special type of earthflow occurs in wet, highly porous sediment consisting of clay- to sand-size particles. Such units may weaken if shaken suddenly, as by an earthquake. An abrupt shock increases shear stress and may cause a momentary buildup of water pressure in pore spaces, which decreases the shear strength. The result is rapid fluidization of the sediment and abrupt failure, a process known as **liquefaction**. Any structure built on such sediments, or in their path, may be quickly demolished (Figure 13.17).

Grain Flows. If you have ever walked along the crest of a sand dune and stepped too close to the steep slope that faces away from the wind, your footstep likely started a cascade of sand flowing down the dune face. This example illustrates a type of mass-wasting called *grain flow*, which involves movement of a dry or nearly dry granular sediment with air filling the pore spaces. Such grain flow occurs naturally when accumulating sand grains produce a slope that exceeds the angle of repose, leading to failure. During flow, moving grains collide frequently. Velocities of the moving sediments typically range between 0.1 and 35 m/s.

Debris Avalanches. A large, rapidly moving **debris avalanche** commonly constitutes a huge mass of falling rock and debris that breaks up, pulverizes on impact, and then

Figure 13.17 Liquefaction Chaotically tilted trees and houses in suburban Anchorage, Alaska, show how violent shaking of the ground during the great 1964 earthquake caused sudden liquefaction of underlying clays and widespread slumping.

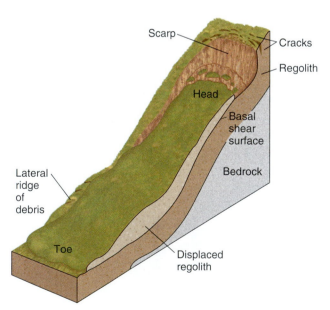

Figure 13.16 Earthflow In this section through an idealized earthflow, regolith moves downslope across a basal shear surface, leaving a scarp at the head where the regolith separated from the slope above. A bulging toe protrudes beyond ridges of sediment that have piled up along the lower margins of the earthflow.

Figure 13.18 Prehistoric Debris Avalanche The prehistoric Blackhawk Landslide northeast of Los Angeles, California, involved 300 million m^3 of fractured rock that collapsed from the summit of Blackhawk Mountain about 18,000 years ago. The rock mass traveled about 9 km and formed a lobate tongue of debris 2 km wide, 7 km long, and 10–30 m thick.

Figure 13.19 Debris Avalanche, Mount Shasta A massive prehistoric debris avalanche from the northwestern flank of Mount Shasta volcano in northern California left a chaotic deposit (hills in middle distance) that extends 34 km from the volcano and covers at least 450 km^2.

continues to travel downslope, often for great distances (Figure 13.18). Fortunately, these spectacular events are rare, for they travel at high velocity (tens to hundreds of km/h) and can be extremely destructive (Figure 13.8E and Table 13.2).

Because large debris avalanches are infrequent and extremely difficult to study while they are moving, few observational data about the process are available. Their extreme mobility has been attributed to the debris riding on a layer of compressed air, a process described in Box 13.2. If true, debris avalanches behave somewhat like a commercial hovercraft that travels across land or water on air compressed by a large propeller. Alternatively, air trapped and compressed within the moving debris may reduce friction between particles and cause the mass to behave in a highly fluid manner.

The flanks of steep stratovolcanoes are especially susceptible to collapse that can lead to the production of debris avalanches. The deposits of these events can be difficult to recognize because of their huge dimensions. A broad valley that extends some 40 km north of towering Mount Shasta in northern California contains a complex of hills and mounds of volcanic rock that many geologists had interpreted as relics of numerous minor eruptions from isolated vents. Recognition of the close similarity between these deposits and those of a huge debris avalanche associated with the 1980 Mount St. Helens eruption prompted a reassessment. A new study led to the remarkable conclusion that the entire array of features resulted from a similar, but far larger, collapse of a flank of the Shasta volcano about 300,000 years ago. In this case, the volume of rock involved was at least 26 km^3—almost 10 times the volume of the landslide on Mount St. Helens. Whereas the St. Helens deposits cover an area of 60 km^2, the Shasta debris avalanche overwhelmed at least 450 km^2 (Figure 13.19).

TABLE 13.2 Characteristics of Some Large Debris Avalanches

Locality	Date	Volume (million m^3)	Vertical Movement (m)	Horizontal Movement (km)
Huascaran, Peru	1971	10	4000	14.5
Sherman Glacier, Alaska	1964	30	600	5.0
Mount Rainier, Washington	1963	11	1890	6.9
Madison, Wyoming	1959	30	400	1.6
Elm, Switzerland	1881	10	560	2.0
Triolet Glacier, Italy	1717	20	1860	7.2
Blackhawk, California	Prehistoric	280	1220	8.0
Saidmarreh, Iran	Prehistoric	2000	1650	14.5

Before you go on:

1. What is the angle of repose?

2. What controls the movement in a sediment flow?

3. Describe common types of granular flows and slurry flows and tell how they differ.

MASS WASTING IN COLD CLIMATES

We have reviewed the basic processes involved in mass wasting, and turn next to environments where mass wasting is a dominant feature of the landscape. Mass wasting is especially active at high latitudes and high altitudes, where average temperatures are very low. These are regions where much of the landscape is underlain by perennially frozen ground and where frost action is an important geologic process.

FROST HEAVING AND CREEP

When water freezes, it increases in volume. Ice forming in saturated regolith therefore pushes the ground surface up. This lifting of regolith by the freezing of contained water is called **frost heaving**.

Frost heaving strongly influences downslope creep of regolith in cold climates. When water freezes within the pores of regolith, its expansion as it changes to ice increases the sizes of the pores and the ground surface is forced upward. As the ground thaws, the regolith returns to a more natural state as particles settle into the oversized pores that had been created by the ice. During this settling, some horizontal movement takes place (Figure 13.20). A particle's net motion during each freeze–thaw cycle is a very short distance downslope. Thus, the net result of repeated episodes of freezing and thawing, during which a particle experiences a succession of up and down movements, is slow but progressive downslope creep.

GELIFLUCTION

In cold regions underlain year-round by frozen ground, a thin surface layer thaws in summer and then refreezes in winter. During the summer, this thawed layer becomes saturated with meltwater and is very unstable, especially on hillsides. As gravity pulls the thawed sediment slowly downslope, distinctive lobes and sheets of debris are produced. This process, known as **gelifluction**, is similar to solifluction in temperate and tropical climates.

Although measured rates of movement are low, generally less than 10 cm/yr, gelifluction is so widespread on high-latitude landscapes that it constitutes a highly important agent of mass transport. Hillslopes in arctic Alaska and Canada, for example, commonly are mantled by superimposed sheets or lobes of geliflucted regolith.

ROCK GLACIERS

A **rock glacier**, another characteristic feature of many cold, relatively dry mountain regions, is a tongue or lobe of ice-cemented rock debris that moves slowly downslope in a manner similar to glaciers (Figure 13.21). Rock glaciers generally originate below steep cliffs, which provide a source of rock debris. Active rock glaciers may reach a thickness of 50 m or more and advance at rates of up to about 5 m/yr. They are especially common in high interior mountain ranges like the Swiss Alps, the Argentine Andes, and the Rocky Mountains.

MASS WASTING UNDER WATER

As geologists have extended their search for petroleum to the continental shelves and slopes, their explorations have shown that mass wasting is an extremely common and widespread means of sediment transport on the seafloor. Mass wasting also has been documented in lakes. As on land, so also in subaqueous (underwater) environments, the potential for gravity-induced movement of rock and sediment exists.

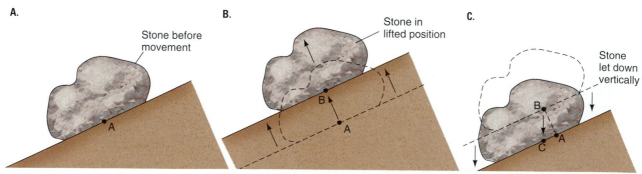

A. Stone before movement

B. Stone in lifted position

C. Stone let down vertically

Figure 13.20 Freeze–Thaw Cycle Stone moved downslope by alternate freezing and thawing of the ground. As freezing occurs (B), the stone is raised perpendicular to the ground surface, which also rises. When the ground thaws and settles (C), gravity pulls the stone down approximately vertically, giving it a small but significant component of movement downslope (A–C).

Figure 13.21 Rock Glacier A jumbled mass of angular rock debris, supplied from a steep cliff on Sourdough Mountain in the Wrangell Mountains of southern Alaska, moves slowly downslope as a rock glacier.

MASS WASTING ON MARINE DELTAS

Major marine deltas, built beyond the mouths of large rivers, commonly display surface features and sediments attributable to slope failures. In such environments, failure can occur even on slopes as low as 1°. The slope failures generally display three distinct zones: a source region where sinking or downward settling and slumping take place, a central channel where sediment is transported, and a zone where sediment is deposited, often in the form of overlapping lobes of sediment. These and other features are common on the submarine slopes of the Mississippi Delta, which is among the best-studied deltas in the world (Figure 13.22). Slides and sediment flows are extremely

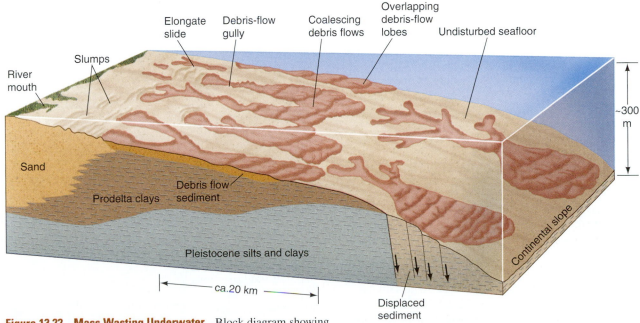

Figure 13.22 Mass Wasting Underwater Block diagram showing mass-wasting features on the submarine surface of the Mississippi Delta.

active on the delta front. In places, sediment flows have deposited more than 30 m of sediment in the last 100 years.

MASS WASTING IN THE WESTERN NORTH ATLANTIC

Extensive studies of the continental slope of eastern North America using a variety of modern techniques (deep-ocean drilling, piston coring, side-scan sonar, echograms, submersible vessels) have shown that vast areas of the seafloor are disrupted by submarine slumps, slides, and flows (Figure 13.23). Some large slide complexes cover areas of more than 40,000 km^2 and reach depths as great as 5400 m. Series of slumps heading in abrupt scarps up to 10 m high and 50 km long lie upslope from slide complexes. The slides generally have affected the uppermost 50 m of seafloor sediment. Cores taken within the slide masses show disturbed and contorted layering, overlain by silty turbidite beds. Buried deposits have been identified within slide complexes, pointing to multiple episodes of movement.

Submarine slope failures often occur when shear strength of the sediments is reduced by an abrupt increase in the pore pressure. One way this can happen is by a sudden earthquake shock. Several major turbidity currents and slumps off eastern North America occurred during strong earthquakes, for example, Grand Banks, Newfoundland, in 1929; Charleston, South Carolina, in 1886; and Cape Ann, Massachusetts, in 1775.

The dating of numerous sediment cores from submarine slump masses and debris flows has shown that most of the mass-wasting events occurred during the last glacial age when world sea level was at least 100 m lower than now, and also during the subsequent rise of world sea level that accompanied the melting of continental ice sheets. At such times, large quantities of sediment were dumped in the ocean by rivers crossing the emergent continental shelf and were then transported by mass wasting into the deep sea. Because there have been numerous periods of low sea level during the ice ages, such processes have slowly built a thick wedge of sediment along the base of the Atlantic continental slope.

HAWAIIAN SUBMARINE LANDSLIDES

The tallest Hawaiian volcanoes, the largest on the Earth, rise 7000 m or more from the floor of the Pacific Ocean. Like their smaller counterparts on the continents, they are composed of huge piles of well-jointed lava flows and vol-

Figure 13.23 Submarine Landslides Map of a region off the eastern coast of the United States showing the distribution of large blocky landslide and debris-flow deposits on the continental slope and rise.

canic rubble. Chaotic topography along the submerged lower margins of the Hawaiian volcanoes has been interpreted as evidence of repeated massive landslides on the volcano flanks (Figure 13.24). Irregular terrain on the seafloor adjacent to Molokai suggests that much of the northern half of this volcano has slid downslope from a

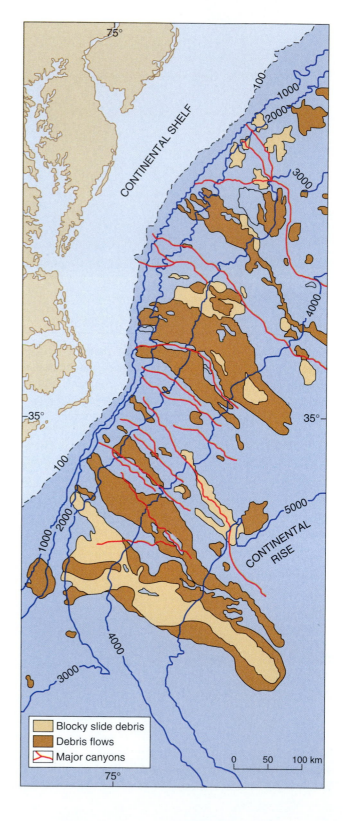

■ Blocky slide debris
■ Debris flows
◢ Major canyons

0   50   100 km

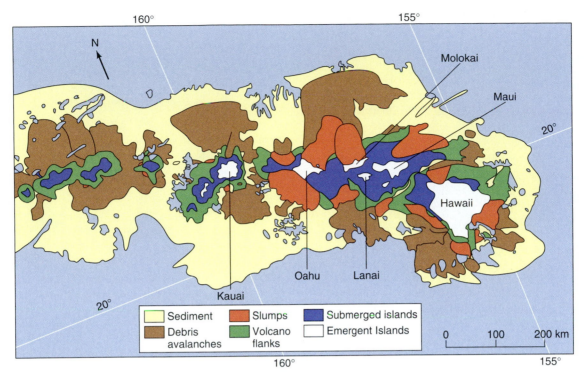

Figure 13.24 Massive Hawaiian Landslides A broad belt of chaotic terrain extending hundreds of kilometers across the seafloor adjacent to the Hawaiian Islands is interpreted as deposits of massive landslides that originated on the steep, unstable submarine flanks of the volcanoes.

steep landslide scarp visible on submarine surveys. Anomalous coral-bearing gravels found up to altitudes of 326 m on Lanai and nearby islands have been attributed to a giant wave that deposited the coral fragments high above sea level. The wave is believed to have resulted from a huge submarine landslide off the western coast of the island of Hawaii that moved downward from its source near sea level to depths of 4800 m and traveled nearly 100 km along the deep ocean floor. Based on dating of the corals on Lanai, the landslide occurred about 105,000 years ago.

Before you go on:

1. Why is mass wasting especially significant at high latitudes and high altitudes?

2. How does frost heave result in downslope movement of rock particles?

3. How do earthquakes contribute to mass wasting underwater?

WHAT TRIGGERS MASS-WASTING EVENTS?

Mass-wasting events sometimes seem to occur at random, with no apparent reason. However, most events, particularly the largest and most disastrous, are related to some extraordinary activity or occurrence. These occurrences include shocks and modification of slopes.

SHOCKS

A sudden shock, such as an earthquake, may release so much energy that slope failures of many types and sizes are triggered simultaneously. For example, in 1929 a major earthquake in northwestern South Island, New Zealand, triggered at least 1850 landslides larger than 2500 m^2 within an area of 1200 km^2 near the quake's center. An estimated 210,000 m^3 of debris was displaced, on average, in each 1 km^2 of land. Landslides were reported to be most numerous on well-bedded and well-jointed mudstones and fine sandstones. The Alaska earthquake of 1964 triggered many rockfalls. One became a huge rock avalanche that swept across the surface of Sherman Glacier, burying it with up to several meters of coarse, angular debris.

SLOPE MODIFICATION

Landslides often result when natural slopes are modified by human activities. Slides may occur, for example, where roads have been cut into the ground, creating artificially steep slopes that are much less stable than the more gentle original slopes (Figure 13.25). Such landslides are especially common along the coastal cliffs of California, where roads have been carved into deformed sedimentary rocks.

A.

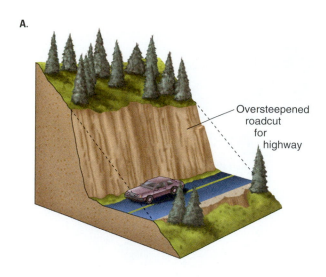

Oversteepened roadcut for highway

B.

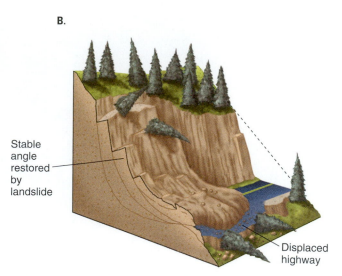

Stable angle restored by landslide

Displaced highway

Figure 13.25 Roadcut Failure When a natural slope is oversteepened in building a road, failure can result. A. A highway cut exceeds the natural angle of the slope, producing an unstable situation. B. The oversteepened slope fails, and a landslide buries the road. The slope angle of the resulting deposit now is similar to the slope prior to modification.

Retaining walls may reduce the likelihood of landslides in thick colluvium, but unless such barriers are very strong, persistent downslope creep of the colluvial debris may ultimately cause them to fail.

UNDERCUTTING

Slumps and other types of landslides can be triggered by the undercutting action of a stream along its bank or by surf action along a coast. Coastal landslides are often associated with major storms that direct their energy against rocky headlands or the bases of cliffs of unconsolidated sediment. Windward coasts of the Hawaiian Islands retreat as pounding surf removes jointed lava from the base of steep seacliffs, causing them to collapse (Figure 13.26).

EXCEPTIONAL PRECIPITATION

Landslides are often associated with heavy or persistent rains that saturate the ground and make it unstable. Such was the case in 1925, when prolonged rains, coupled with melting snow, started a large debris flow in the Gros Ventre River basin of western Wyoming. The water saturated a porous sandstone that overlies impermeable shale and dips toward the valley floor. This saturated condition was an ideal trigger for slope failure. An estimated 37 million m^3 of rock, regolith, and organic debris moved rapidly downslope and created a natural dam that ponded the river. Two years later the dam failed, causing a flood that led to several deaths. Today, more than 75 years after the debris flow began, the scar at the head of the slide is still quite obvious, as is the distinctive chaotic topography downslope.

VOLCANIC ERUPTIONS

Volcanic eruptions are another means of initiating mass-wasting events. Large stratovolcanoes consist of interstratified lava flows, rubble, and pyroclastic layers that form steep slopes. On high, ice-clad volcanoes, slopes may be further steepened by glacial erosion. During eruptions,

Figure 13.26 Coastal Landslide Steep seacliffs of jointed basalt along the windward coast of Hawaii are undercut by pounding surf. When a cliff collapses, the resulting landslide debris is rapidly reworked by surf and currents, and the process begins anew.

slope failure is often widespread. Large volumes of water, released when summit glaciers and snowfields melt during eruption of hot lavas or pyroclastic debris, can combine with unconsolidated deposits to form mudflows or debris flows that move rapidly downslope and often continue for many kilometers downvalley.

SUBMARINE SLOPE FAILURES

As we have seen, unstable conditions on continental slopes and delta fronts can promote the formation of large submarine landslides. Factors involved in such slope failures include:

1. High internal water pressures resulting from rapid deposition of sediment and the inability of water trapped in accumulating sediment to escape.

2. Generation of methane gas from organic matter deposited with sediments, which increases pressure in pores between grains.

3. Local oversteepening of slopes, a result of high rates of sedimentation.

4. Displacement along vertical fractures in seafloor sediments or rocks, which leads to oversteepening of slopes.

5. Shocks induced by earthquakes.

HAZARDS TO LIFE AND PROPERTY

As the human population increases and cities and roads expand across the landscape, the likelihood that mass-wasting processes will affect people increases. Landslides occur worldwide, and their impact, in terms of loss of lives and property, can be devastating (Table 13.3). In the United States alone, landslides in a typical year cause more

than $1 billion in economic losses and 25 to 50 deaths, and the figures are rising. Although it may not always be possible to predict accurately the occurrence of significant mass-wasting events, knowledge of the processes and their relationship to local geology can lead to intelligent planning that will help reduce the loss of lives and property.

ASSESSMENTS OF HAZARDS

Planning begins with assessments of potential hazards resulting from major mass-wasting events. Such assessments are based mainly on reconstructions of similar past events aimed at evaluating their magnitude and frequency. From such information, it is possible to calculate how often an event of a certain magnitude is likely to recur.

Maps showing potential areas of impact of mass-wasting events are important tools for land-use planners. For example, large debris avalanches and small rockfalls are ever-present hazards in the Italian Alps (Figure 13.27). Field studies have shown that large rock avalanches similar to that of 1717, described in the opening essay, have repeatedly blanketed valley floors with rocky debris during the last 3000 years. From this evidence, a map has been constructed showing areas that could be affected by future rockfalls having various trajectories and distribution patterns (Figure 13.28). A number of small communities are at hazard from small- to intermediate-size rock avalanches

TABLE 13.3	Fatalities Resulting from Some Major Landslides During the Twentieth Century	
Year	Location	Fatalities
1916	Italy, Austria	10,000
1920	China[e]	200,000
1945	Japan[f]	1200
1949	Former USSR[e]	12,000–20,000
1954	Austria	200
1962	Peru	4000–5000
1963	Italy	2000
1970	Peru[e]	70,000
1985	Colombia[v]	23,000
1987	Ecuador[e]	1000

Landslides related to earthquakes (e), floods (f), and volcanic eruptions (v).
Source: National Research Council (1987).

Figure 13.27 Lucky Escape A new apartment building at the base of a steep mountain slope in the Italian Alps was struck by a large boulder falling from the cliffs above. This relatively small rockfall, which occurred just one day before the owners were scheduled to move in, demolished the bedroom and most of the living room.

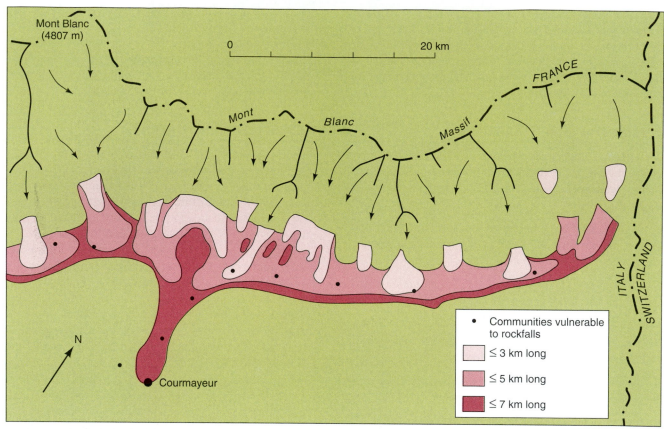

Figure 13.28 Rockfall Hazards Map   This map, based on a study of historic and prehistoric rockfall avalanches in the Alps, outlines zones of potential hazard from future rockfalls traveling up to 3, 5, and 7 km from likely places of origin. Several com- munities lie within 5 km of such sources, and 10 lie within the zone that includes a 2700-year-old rockfall deposit on which the town of Courmayeur has been built.

(traveling 3 to 5 km), while several large communities, including one village with a population of several thousand, could be affected by large rock avalanches (traveling up to 7 km) like some recorded in deposits on the valley floors.

Valleys in the Cascade Range of Washington and Oregon contain deposits of large mudflows that repeatedly spread from high volcanoes during the last 10,000 years. Based on the number and extent of such deposits, hazards maps have been prepared, like that shown for Mount Rainier and vicin- ity in Figure 13.29. This map shows that risk from mudflows is high within about 25 km of the volcano's slopes, and that risk exists even at distances of 100 km or more along dense- ly populated valley floors. A similar map prepared prior to the 1980 eruptions of Mount St. Helens proved prophetic, for mudflows generated during that series of eruptions had distributions very similar to those predicted on the basis of geologic studies of past events.

MITIGATION OF HAZARDS

Once potential hazards have been identified, careful plan- ning can often reduce or even eliminate the impact of mass- wasting processes on human environments. Slopes subject to creep can be stabilized by draining or pumping water from saturated sediment, while oversteepened hillslopes can be prevented from slumping if they are regraded to angles equal to or less than the natural angle of repose. In some mountain valleys subject to mudflows from active volcanoes, water-filled reservoirs can be quickly emptied so that the dams will pond potentially destructive mudflows before they reach population centers (Figure 13.29). We generally have no way of anticipating or preventing large rockfalls and debris avalanches. Eliminating or restricting human activities in possible impact zones offers the best means of mitigating these hazards.

Before you go on:

1. How can road building lead to slope instabilities?

2. What causes undercutting?

3. How can potential mass-wasting hazards be assessed?

4. What are some ways that potential mass-wasting hazards can be mitigated?

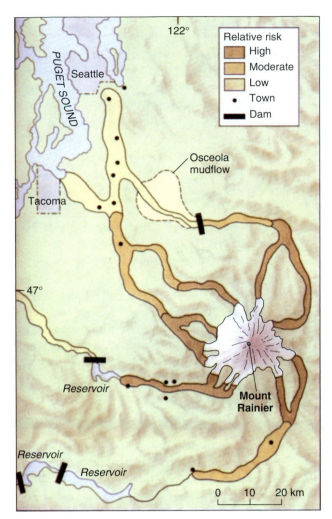

Figure 13.29 Mudflows from Mount Rainier Map of the southeastern Puget Lowland and adjacent Cascade Range of Washington, showing areas of low, moderate, and high risk from mudflows and floods originating at Mount Rainier volcano. Also shown is the extent of the huge prehistoric Osceola mudflow that was associated with a summit eruption about 5700 years ago.

REVISITING PLATE TECTONICS AND THE EARTH SYSTEM

FROM ALPINE PEAKS TO THE DEEP OCEAN

Large rivers and mighty glaciers obviously erode and transport large volumes of sediment toward the sea. We can see the sediment in transit and we can observe its deposition. We conclude that these transporting agencies form a key element of the rock cycle, in which rocks are broken down, their debris is transported to the oceans, and there it is buried to form new rocks that are uplifted and recycled in turn. It is easy to forget the significance of mass-wasting processes in this cycle, but quantitatively they are extremely important. Along the path from mountain peak

to the deep ocean, a particle of sediment will almost certainly be moved by several of these processes, sometimes very slowly by creep, but often rapidly, as in a debris fall or a mudflow (Figure 13.30). A particle may initially be dislodged by frost action on a mountain peak, fall onto a glacier, be transported beyond the glacier terminus by a meltwater stream, be temporarily stored in a terrace deposit, then slump into an active channel and move along its way. Deposited in a delta at the ocean shore, the particle may then travel deeper underwater in slumps or currents to reach the edge of the continental slope. There it may travel down a submarine canyon in a turbidity current or be transferred downslope in a submarine landslide eventually to end up on the deep-ocean floor, many thousands of meters below its alpine starting point. Other particles may take quite different paths, be affected by other mass-wasting processes, and move at very different rates. However, in the end, the result is the same—the inexorable pull of gravity transfers the particles ever lower on the landscape.

LANDSLIDES, FLOODS, AND PLATE TECTONICS

If we were to plot on a map the location of the world's major historic and prehistoric landslides, we would find that most tend to cluster along belts that lie close to the boundaries between converging lithospheric plates. They do so for two main reasons.

First, the world's highest mountain chains lie at or near plate boundaries, and on steep mountain slopes the safety factor often lies close to 1. The rocks of many mountain ranges consist of well-jointed strata that have been strongly fractured and deformed as they were uplifted. Both the joint planes and the bedding surfaces are potential zones of failure. Furthermore, along some of these belts lie the world's highest stratovolcanoes, the slopes of which also tend to lie at steep angles.

Second, it is along the boundaries between plates, where plate margins slide past or over one another, that most large earthquakes occur. Earthquakes also are associated with upward-moving magma that feeds volcanic eruptions at Earth's surface. The major landslides listed in Tables 13.2 and 13.3 occurred in active tectonic zones near plate margins, and several are known to have been directly related to major earthquakes.

Stream valleys carved deeply into the lofty Himalaya of Pakistan, India, and Nepal are the site of frequent landslides (Figure 13.31). The collision that resulted when the northward-moving Australian–Indian Plate encountered the vast Eurasian Plate has led to uplift of the mountain range. Measured uplift rates here are the highest in the world. Near lofty Nanga Parbat (8125 m) in the western Himalaya, which lies at the boundary between the converging plates, uplift rates are as high as 5 mm/yr. At this rate, the mountain should increase in altitude by 5000 m

Figure 13.30   From Mountain Peaks to Ocean Floor
Mass wasting is a primary mechanism for transferring rock debris from high mountains toward the oceans and then down continental slopes to the deep seafloor. The vertical distance involved is as great as 12 km, measured from Himalayan summits (7500–8000 m high) to the Bengal and Indus deep-sea fans (>4500 m deep) (Figure 7.28). Mass wasting therefore is a key component of the rock cycle.

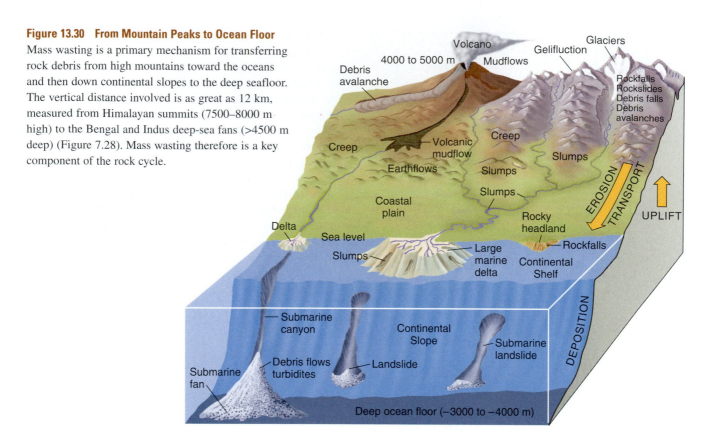

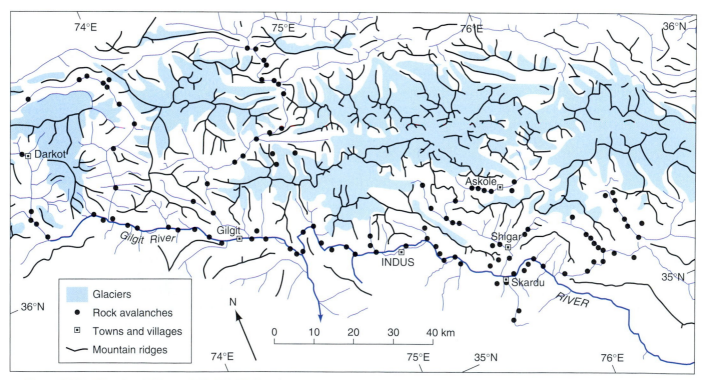

Figure 13.31   Giant Landslides and High Uplift Rates   A map showing the location of major rockfall avalanche deposits in the southern sector of the lofty Karakorum Himalaya of Pakistan, where measured uplift rates are among the highest on Earth. Numerous large rockfall events have occurred along steep cliffs bordering the Indus River, which has cut a deep gorge through the rising mountain range.

between highest and lowest points on a landscape) measures nearly 7000 m. High relief, steep slopes, fractured rocks, persistent downcutting by active streams and glaciers, and frequent large earthquakes all make Nanga Parbat an obvious place for landslides. In fact, landslides are probably the most effective of the agents involved in tearing down the mountain. One such landslide and its resulting human impact have been vividly recorded.

In early June 1841, a Sikh army was camped upstream from the town of Attock, which lies beyond the mouth of the steep gorge where the Indus flows out of the Himalaya. Suddenly, in midafternoon, a huge wall of muddy debris came rushing out of the gorge and overwhelmed the army. An eyewitness described the terrible scene: "It was a horrible mess of foul water, carcasses of soldiers, peasants, war-steeds, camels, prostitutes, tents, mules, asses, trees, and household furniture, in short every item of existence jumbled together in one flood of ruin.... As a woman with a wet towel sweeps away a legion of ants, so the river blotted out the army of the Raja."

The cause of the disaster lay 400 km up the river canyon (Figure 13.32) where, in January of that same year, an earthquake caused a spur of Nanga Parbat to collapse, triggering a massive landslide that rushed downslope toward the river. The slide debris quickly dammed the Indus, forming a lake that steadily grew until it was 150 m deep and more than 30 km long. As the rising water reached the top of the landslide dam, it overflowed and rapidly cut through the unconsolidated debris. The gigantic flood of water that was released rushed swiftly downstream, sweeping everything before it, toward the unsuspecting army on the plains below.

Figure 13.32 Indus River Canyon The Indus River near the place where a great landslide from Nanga Parbat dammed the river in January 1841, leading to a catastrophic flood.

every million years. However, high mountains also mean high erosion rates, and erosion is tearing down the mountain about as rapidly as it is rising. Much of this destruction is the result of mass wasting.

At the base of Nanga Parbat, only 21 km from the summit, lies the deep gorge of the Indus River, one of the two largest streams draining the Himalaya. From the river to the top of the mountain, the *relief* (the difference in altitude

What's Ahead?

Although mass wasting in its many forms is very important in the transfer of detritus downslope, streams account for most observable sediment movement and landscape evolution at Earth's surface. Following decades of intensive studies, geologists know a great deal about how rivers transport their sediment and how they shape landscape. In the next chapter we will look at stream channels, their patterns, how rivers erode the land, and the variety of deposits they produce. We also will explore the evolution of drainage systems and the factors that control them.

CHAPTER SUMMARY

1. Mass wasting is the downslope movement of rock debris under the pull of gravity without a transporting medium. Mass wasting occurs both on land and beneath the sea.

2. The composition and texture of debris, the amount of air and water mixed with it, and the steepness of slope influence the type and velocity of downslope movements.

3. Mass-wasting processes include sudden slope failures (slumps, falls, and slides) and downslope flow of mixtures of regolith, water, and air.

4. Failures occur when shear stress reaches or exceeds the shear strength of slope materials. High water pressure in rock voids or sediment reduces shear strength and increases the likelihood of failure.

5. Slumps involve a rotational movement along a concave-up surface that results in backward-tilted blocks of rock or regolith.

6. Falling and sliding masses of rock and debris are common in mountains where steep slopes abound.

7. Rockfall debris accumulates at the base of a cliff to produce a talus with slopes that stand at the angle of repose.

8. Slurry flows involve dense moving masses of water-saturated sediment that form nonsorted deposits when flow ceases. Flow velocities range from very slow (solifluction) to rapid (debris flows).

9. In granular flows, sediment is in grain-to-grain contact or grains constantly collide. The sediment may be largely dry, or it may be saturated with water that can escape easily.

10. Although creep is imperceptibly slow, it is widespread and therefore quantitatively important in the downslope transfer of debris.

11. Large, rapidly moving debris avalanches are relatively infrequent but potentially hazardous to humans.

12. In regions of perennially frozen ground, frost heaving, creep, and gelifluction are important mass-wasting processes.

13. Large areas of seafloor on the continental slopes show evidence of widespread slumps, slides, and flows. Mass wasting on submarine slopes was especially active during glacial ages, when sea level was lower and large quantities of stream sediment were transported to the edge of the continental shelves.

14. Slope failures can be triggered by earthquakes, undercutting by streams, heavy or prolonged rains, or volcanic eruptions. Subaqueous slope failures are frequently related to earthquake shocks and to oversteepening of slopes caused by rapid deposition of sediments.

15. Loss of life and property from mass-wasting events can be prevented or mitigated by adequate assessment and planning based on geologic studies of previous occurrences.

THE LANGUAGE OF GEOLOGY

angle of repose (p. 332)

colluvium (p. 336)
creep (p. 335)

debris avalanche (p. 338)
debris fall (p. 331)
debris flow (p. 333)
debris slide (p. 331)

earthflow (p. 337)

frost heaving (p. 340)

gelifluction (p. 340)
granular flow (p. 332)

landslide (p. 329)
liquefaction (p. 338)

mass wasting (p. 326)
mudflow (p. 333)

rockfall (p. 329)
rock glacier (p. 340)
rockslide (p. 331)

sediment flows (p. 332)
slump (p. 329)
slurry flow (p. 332)
solifluction (p. 333)

talus (p. 332)

QUESTIONS FOR REVIEW

1. How does mass wasting differ from weathering? from stream erosion?

2. What primary factors influence shear stress and shear strength on hillslopes? How can the shear strength of sediment or a rock mass on a hillslope suddenly change?

3. Why can the presence of water in rock or regolith promote downslope movement?

4. What distinctive landscape features might enable you to identify an area where numerous slumps have occurred?

5. What conspicuous type of deposit generally is found at the base of a cliff subject to frequent rockfalls? How would you expect the particles in the deposit to be sorted?

6. Why would regolith that has been artificially excavated so that the surface slope exceeds the natural slope angle be likely to fail?

7. How might you prove that creep is occurring on a slope, and how can its rate be measured?

8. Why are lava flows that are erupted from stratovolcanoes largely restricted to the volcanic cones, while mudflows originating on the same mountains are often distributed for many tens of kilometers beyond the volcanoes along adjacent valleys?

9. Explain why large-scale mass wasting on the continental slopes apparently was far more active during the last glacial age than it is today.

10. Why are large debris avalanches and volcanic mudflows generally far more dangerous to people than are lava flows?

11. How might prolonged and heavy rainfall affect the shear strength of a body of regolith and make it susceptible to failure?

12. What geologic conditions make high mountains especially prone to landslide activity?

Click on *Presentation* and *Interactivity* in the **Natural Hazards** module of your CD-ROM to further explore resources and activities presenting concepts from this chapter. Select *Assessment* in the same module to test your understanding of this chapter.

Chapter 14

The Columbia River flows through a scenic gorge it carved through the techtonically uplifted Cascade Range.

Streams and Drainage Systems

Qin Dynasty terra-cotta statues buried in Holocene alluvial fan.

A Chinese Emperor's Buried Army

Among the astonishing historical sites in China, one on the outskirts of Xi'an is among the most famous. There a museum displays a partly excavated assemblage of greater-than-life-size terracotta statues of warriors and horses arrayed like a defending army beyond a high artificial hill where China's first emperor, Qin Shi Huang, is believed to lie entombed. The buried army came to light in 1974 when farmers drilling a shallow well unearthed remnants of these statues in the middle of an agricultural field. Subsequent excavations uncovered one the world's great archeological treasures. It is clear that the army was designed to impress. It's ranks of soldiers and officers likely number close to 6000. Was it the emperor's intention to bury this make-believe army beneath the land surface? Or were the emperor's geologists too inexperienced to recognize the sedimentary environment in which the army was placed?

In 1988 I was studying well-developed Holocene stream terraces in the Ba He valley, a tributary of the Wei River, with two young Chinese field assistants. We wished to understand the relationship of stream activity to changing climate during the last 5000 years. In mapping the distribution and stratigraphy of the stream sediments, we found that the Qin army lay beneath the surface of several sediment fans that had been accumulating a mix of stream deposits, mudflow sediments, and windblown dust for several thousand years. The climate during this interval was marked by strong monsoon summer rainfall, favorable conditions for the deposition of stream sediments beyond the mouths of nearby mountain valleys. As the sediment began to accumulate on the lower parts of the fans, it encroached upon and then completely buried the terracotta army. Centuries passed and the site was forgotten until a chance discovery brought it to light. If the emperor decreed that his army was to be underground, he selected one of the best possible sites. If not, at least it was preserved largely intact by the accumulating stream sediment for us to marvel at today.

Stephen C. Porter

KEY QUESTIONS

1. **In what ways are streams important as geologic agents?**

2. **What is a channel? What accounts for differences in the patterns of channels?**

3. **What five factors control the way a stream behaves?**

4. **What changes characterize a stream as it flows from source to mouth?**

5. **What governs the ability of streams to erode?**

6. **What is a stream's load, and what are the load's components?**

7. **Where do stream deposits form, and why?**

8. **How are stream systems organized?**

INTRODUCTION: STREAMS IN THE LANDSCAPE

Almost anywhere we travel over the land surface, we can see evidence of the work of running water. Even in places where no rivers flow today, we are likely to find deposits and landforms that tell us water has been instrumental in shaping the landscape. Most of these features can be related to the activity of streams that are part of complex drainage systems.

A **stream** is a body of water that flows downslope along a clearly defined natural passageway, in the process transporting detrital particles and dissolved substances. The passageway is called the stream's **channel**, and for most streams the detritus constitutes the bulk of its **load**, which is the sediment and dissolved matter the stream transports. The quantity of water passing by a point on the streambank in a given interval of time is the stream's **discharge**. As a stream moves sediment from place to place, its channel is continually being altered. A stream and its channel are closely related and form an ever-changing, interrelated system.

Streams play important roles in our lives. Large streams, like the Amazon, the Rhine, and the Mississippi, are important avenues of transportation. Many of the world's great cities are built in stream valleys; New Orleans, St. Louis, Cairo, London, Paris, Rome, and Moscow are examples (Figure 14.1). People choose to live near streams because valley floors are flat and easy to build on, soils tend to be deep and fertile, and water is available. But stream valleys have drawbacks as well. They can be threatened by floods, and as cities grow, human and industrial wastes begin to pollute the water. How to achieve an acceptable balance between human needs and the capacities of streams to maintain a safe, clean water supply is one of the major issues facing society.

In addition to their immediate practical and aesthetic importance, streams are vital geologic agents:

- Streams carry most of the water that goes from land to sea and so are an essential part of the hydrologic cycle.

- Streams transport billions of tons of sediment to the oceans each year; there, the sediment is deposited and eventually becomes part of the rock record.

- Streams carry billions of tons of soluble salts, released by weathering, to the oceans each year. These salts play an essential role in maintaining the saltiness of seawater.

- Streams shape the surface of Earth. Most landscapes consist of stream valleys separated by higher ground and are the result of weathering, mass wasting, and stream erosion working in combination.

In this chapter we will examine streams to understand the factors that control their shapes and flow characteristics, their patterns on the landscape, their sediment load, and their deposits; we will consider floods and the hazards resulting from them; and we will look at the place of streams in the rock cycle, how drainage systems evolve, and how streams are influenced by the rock across which they flow.

STREAM CHANNELS

A stream's channel is an efficient conduit for running water. The discharge varies both along the channel and through time, mainly because of changes in precipitation and melting of winter snow cover. In response to varying discharge and load, the channel continuously adjusts its shape and orientation. Therefore, a stream and its channel are dynamic elements of the landscape. Two ways to characterize a channel are by measuring its cross-sectional shape and its long profile.

CROSS-SECTIONAL SHAPE

The size and shape of any particular channel cross section reflect the typical stream conditions at that place. Very small streams may flow in channels that are as deep as they are wide, whereas very large stream channels usually have widths many times greater than their depths (Figure 14.2). Because the volume of water moving through a channel generally increases downstream, it follows that the ratio of

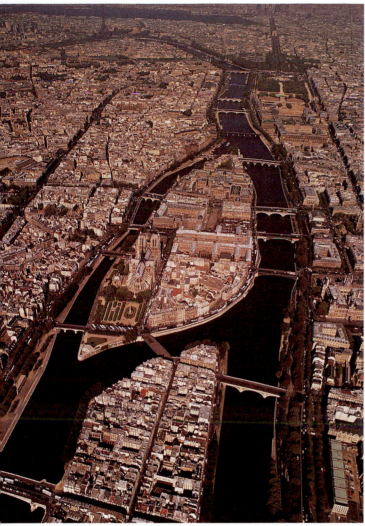

Figure 14.1 An Urban Stream Paris, like many of the world's great cities, was founded along the banks of a major river. The Seine provides water for human and industrial use, is an avenue of transportation, and has great aesthetic and recreational value. However, under the stress of a growing population, the Seine, like other urban rivers throughout the world, is increasingly susceptible to pollution.

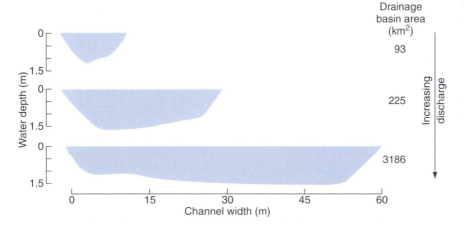

Figure 14.2 Stream Channels Cross sections of three natural streams in the drainage basin of the upper Green River, Wyoming, demonstrating that the ratio of width to depth increases with increasing drainage basin area and discharge. In each cross section, the vertical scale is five times the horizontal scale.

channel width to channel depth is likely to change downstream as the volume of water increases.

LONG PROFILE

If we measure the vertical distance that a stream channel falls between two points a known distance apart along its course, we will have obtained a measure of the stream's **gradient**. The gradients of steep mountain streams, such as the Sacramento River in the Trinity Mountains of California, are typically greater than 30 m/km, whereas near the mouth of a large stream like the Missouri River, the gradient is typically about 0.1 m/km (Figure 14.3).

Usually, the gradient of a river decreases downstream, and so the stream's **long profile** (a line drawn along the surface of a stream from its source to its mouth) is a curve that decreases in gradient downstream (Figure 14.3). However, a long profile is not a perfectly smooth curve because irregularities in the gradient occur along the channel. For example, a local change in gradient may occur where a channel passes from a bed of resistant rock into one that is more erodible, or where a landslide or lava flow forms a temporary dam across the channel. Abrupt increases in the gradient cause water to flow rapidly and turbulently through a stretch of rapids or to plunge over a steep drop as a waterfall. A hydroelectric dam also introduces an irregularity in the long profile of a stream channel and may create an extensive reservoir upstream.

DYNAMICS OF STREAMFLOW

The average annual rainfall on the area of the United States is equivalent to a layer of water 76 cm thick covering this same land surface. An amount equivalent to 45 cm returns to the atmosphere by evaporation and transpiration (Figure 2.18), and 1 cm infiltrates the ground; the remaining 30 cm forms **runoff**, the portion of precipitation that flows over the land surface. By standing outside during a heavy rain, you can see that water initially tends to move down slopes in broad, thin sheets, a process called **overland flow**. You will also notice, however, that after traveling a short distance overland flow begins to concentrate into well-defined channels, thereby becoming **streamflow**. Runoff is a combination of overland flow and streamflow.

FACTORS IN STREAMFLOW

Several basic factors control the way a stream behaves: (1) *gradient*, expressed in meters per kilometer; (2) *stream cross-sectional area* (width × average depth), expressed in square meters; (3) *average velocity* of waterflow, expressed in meters per second; (4) *discharge*, expressed in cubic meters per second; and (5) *load*, expressed in kilograms per cubic meter. Unlike the sediment of a stream's load, dissolved matter generally does not affect stream behavior.

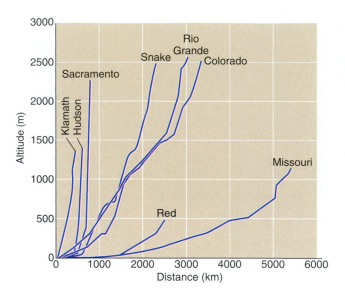

Figure 14.3 Stream Profiles Long profiles of some streams in the United States. The Klamath, Hudson, and Sacramento are relatively short, steep-gradient streams, whereas the Missouri is a long stream with a low average gradient.

The relationship among discharge, velocity, and channel shape for a stream can be expressed by the equation

$$Q = A \times V$$

Q	=	A	×	V
Discharge (m^3/s)		Cross-sectional area of stream (width × average depth) (m^2)		Average velocity (m/s)

This equation tells us that when discharge changes, one or both of the factors on the right side of the equation also must change. For example, if discharge increases, both velocity and cross-sectional area are likely to increase in order to accommodate the added flow. Conversely, a decrease in discharge can lead to a corresponding decrease in stream-channel dimensions and velocity.

Because gravity pulls on water, just as it pulls on rock and regolith, a stream's gradient—steep or gentle—is a factor in stream behavior. A stream plunging over a waterfall or cascading down a series of steep rapids obviously is behaving differently from the same stream where it reaches gentle terrain.

CHANGES DOWNSTREAM

Traveling down a typical stream from its head to its mouth, we can see that orderly changes occur along the channel: (1) discharge increases; (2) stream cross-sectional area increases; (3) velocity increases slightly; and (4) gradient decreases (Figure 14.4).

The fact that velocity increases downstream seems to contradict the common observation that water rushes turbu-

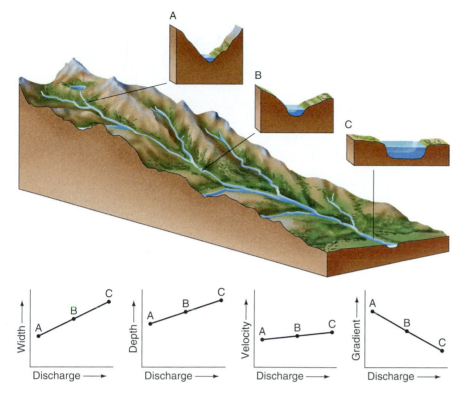

Figure 14.4 Downstream Changes
Changes in the downstream direction along a stream system. Discharge increases as new tributaries join the main stream and as groundwater seeps in. Width and depth of the stream are shown by cross sections A, B, and C. Graphs show the relationship of discharge to stream width and depth, to velocity, and to stream gradient at the same three cross sections.

lently down steep mountain slopes and flows smoothly over nearly flat lowlands. However, the physical appearance of a stream is not a true indication of its velocity. Most of the time, discharge is low in the headward reaches of a stream, so the flowing water is shallow. The stream bed causes much more resistance to the flow of shallow water than it would for deep water, so much so that the velocity of shallow mountain streams is low even though their gradients are steep. Discharge increases downstream as each **tributary** (a stream joining a larger stream) and inflow of groundwater introduce more water. To accommodate the greater volume of water, velocity increases accordingly, together with the cross-sectional area of the stream.

FLOODS

The uneven distribution of rainfall through the year causes many streams to rise seasonally in flood. A *flood* occurs when a stream's discharge becomes so great that it exceeds the capacity of the channel, thereby causing the stream to overflow its banks (Figure 14.5). People affected by floods are frequently surprised and even outraged at what a ram-

Figure 14.5 Flood A devastating flood on the Riberia River in 1997 inundated the town of Eldorado, southwest of São Paolo, Brazil.

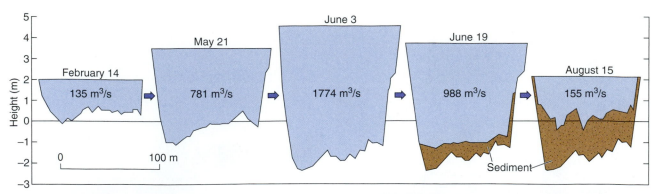

Figure 14.6 Changes in the Colorado River Changes in the cross-sectional area of the Colorado River at Lees Ferry, Arizona. As discharge increased from February to June, the channel floor was scoured and deepened and the water level rose higher against the banks. During the falling-water phase, the river level fell and sediment was deposited in the channel, decreasing its depth.

paging stream has done to them. Geologists, however, tend to view floods as normal and expected events, and they believe that floods have been occurring as long as rain has been falling on Earth's surface.

A dramatic example of changes in stream factors can be seen when floods occur. During a flood on the Colorado River at Lees Ferry, Arizona, a major change in stream dimensions occurred as discharge increased and then fell (Figure 14.6). Prior to the flood, the river averaged about 2 m deep and 100 m wide. As discharge increased in late spring, the water rose in the channel and erosion scoured the bed until at peak flow the river was about 7 m deep and 125 m wide. Together with an increase in velocity, the enlarged channel was now able to accommodate the increased flood discharge and carry a greater load. As discharge fell, the stream was unable to transport as much sediment, and the excess load was dropped in the channel, causing its floor to rise. At the same time, decreasing discharge caused the water level to fall, thereby returning the cross-sectional area close to its preflood dimensions.

As we can see from this example, a stream and its channel are intimately related. The channel is always responsive to changes in discharge, so that the system, at any point along the stream, moves to a more balanced condition.

Unusually large discharges associated with floods appear as major peaks on a *hydrograph*, a graph that plots stream discharge against time. In the example shown in Figure 14.7, a passing storm generated a brief interval of intense rainfall. As the runoff moved into the stream channel, the discharge rose quickly. The crest of the resulting flood, when peak flow was reached, passed the point where the discharge was being measured about 3 hours after the peak rainfall. It took an additional 7.5 hours before all of the storm runoff passed through the channel at that point and discharge decreased to the normal nonflood level.

The example just cited is for a small stream basin, so the times are short. For large basins, times are much longer. Regardless of the size of the stream basin, however, as discharge increases during a flood, so does velocity.

This velocity increase has the double effect of enabling a stream to carry not only a greater load, but also larger particles. The collapse of the large St. Francis Dam in southern California in 1928 provides an extreme example of the exceptional force of floodwaters. When the dam gave way, the water behind it rushed down the valley as a spectacular flood, moving blocks of concrete weighing as much as 9000 metric tons through distances of more than 750 m. Because natural floods are also capable of moving very large objects as well as great volumes of sediment, they are able to accomplish considerable geologic work.

FLOOD PREDICTION

Major floods can be disastrous events, causing both loss of life and extensive property damage (Table 14.1). Therefore, it is highly desirable to be able to predict their occurrence. By plotting the occurrence of past floods of

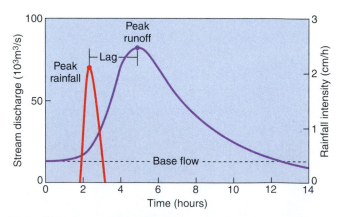

Figure 14.7 Storm-Related Discharge Hydrograph of a stream following a brief storm. An interval of intense rainfall causes a rise in stream discharge as the runoff passes through the stream. Peak discharge lags about 3 hours behind the peak rainfall. It takes an additional 7.5 hours for all of the storm runoff to pass and the discharge to fall to its pre-storm value (the stream's base flow).

TABLE 14.1   Fatalities from Some Disastrous Floods

River	Date	Fatalities	Remarks
Huang He, China	1887	ca. 900,000	Flood inundated 130,000 km^2 and swept away many villages.
Johnstown, Pennsylvania	1889	2200	Dam failed. Wave 10–12 m high rushed downvalley.
Yangtze, China	1911	ca. 100,000	Formed lakes 130 km long and 50 km wide in flooded terrain .
Yangtze, China	1931	ca. 200,000	Flood extended from Hankow to Shanghai (>800 km), leaving tens of millions homeless.
Huange He, China	1938	ca. 900,000	Chinese troops destroyed dikes to divert river as a means of blocking advance of Japanese army.
Vaiont, Italy	1963	2000	Landslide into lake caused wave that overtopped dam and inundated villages below.

Source: Encyclopedia Americana (1983).

different sizes on a probability graph, a *flood-frequency curve* can be produced (Figure 14.8). The measure of how often a flood of a given magnitude is likely to occur is called the *recurrence interval*.

We can use a flood-frequency curve to estimate the probability that a flood of a certain magnitude will occur in any one year. In the case of the Skykomish River at Gold Bar, Washington (Figure 14.8), a flood having a discharge of 1750 m^3/s has a 1 in 10 (10%) probability of occurring in any given year, whereas a larger flood of 2500 m^3/s has a 1 in 50 (2%) probability. We refer to a flood having a recurrence interval of 10 years as a 10-year flood, whereas if the interval is 50 years, it is a 50-year flood, and so forth.

One potential problem with using flood-frequency curves to estimate the probability of future floods is that the curves are based on floods of the recent past. If Earth's climate changes during the next several decades (Chapter 19), present flood-frequency curves may be of little value in predicting future floods.

CATASTROPHIC FLOODS

Exceptional floods—those well outside a stream's normal range—occur very infrequently, perhaps only once in several centuries. Even greater floods, evidence for which we can find in the geologic record, can be viewed as catastrophic events that occur very rarely even on geologic time scales.

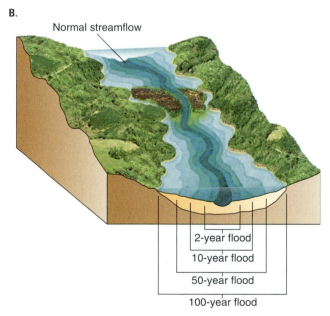

Figure 14.8   Flood Prediction   A. A graph shows how often floods of different sizes occur on the Skykomish River at Gold Bar, Washington, plotted on a probability graph. A flood with a discharge of 1750 m^3/s has a recurrence interval of 10 years and, thus, a 1-in-10 chance of occurring in any given year (a 10- year flood). B. The landscape diagram shows the normal, non- flood discharge for a hypothetical stream in the darkest blue. In lighter blues, the diagram shows what larger floods might be like: the 2-year, 10-year, 50-year, and 100-year floods for this stream.

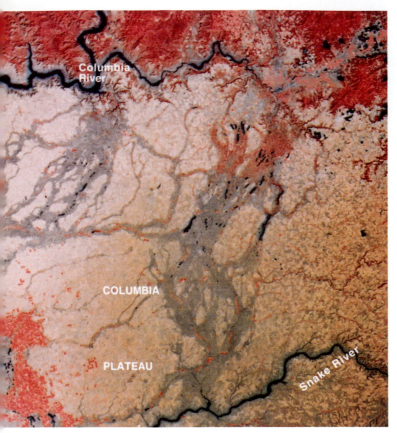

A.

B.

Figure 14.9 Channeled Scabland A. Satellite view of the Columbia Plateau showing dark channels of the Channeled Scabland where floodwaters stripped away a deposit of light-colored windblown dust from underlying dark-gray basaltic lava flows. The floodwaters traveled from the upper right to lower left. The area shown measures approximately 100×100 km. B. Giant ripples on a point bar formed by raging floodwaters as they swept around a bend of the Columbia River. Composed of coarse gravel, the ripples are up to several meters high, and their crests are as much as 100 m apart.

In the 1920s, geologist J. Harlen Bretz began a study of a curious landscape in eastern Washington locally called the Channeled Scabland. An array of dark, channel-like features mark places where bare lava flows lie exposed at the surface, stripped of the fertile topsoil that makes this a prime wheat-producing region (Figure 14.9A). Bretz carefully documented the character and distribution of a variety of landforms that provide evidence of the Scabland's origin: dry coulees (canyons) with abrupt cliffs marking sites of former huge waterfalls, deep rock basins carved in the basalt, massive piles of gravel containing enormous boulders, linear deposits of gravel in the form of huge current ripples (Figure 14.9B), and upper limits of water-eroded land that lie hundreds of meters above valley floors.

Bretz considered different hypotheses to explain this array of features, but he was led inescapably to conclude that they could be accounted for only by a catastrophic event—a truly gigantic flood, far larger than any historic flood. The source of the enormous volume of floodwater was resolved with the discovery that the continental ice sheet covering western Canada during the last glaciation had advanced across the Clark Fork River, damming it to create a huge lake in the vicinity of Missoula, Montana. The glacier-blocked lake contained between 2000 and 2500 km^3 of water when it was filled and remained in existence only as long as the ice dam was stable. When the glacier retreated or its front began to float in the rising lake water, the dam failed, and water was released rapidly from the basin, as though a plug had been pulled from a giant bathtub. The main exit route lay across the Channeled Scabland region and down the Columbia River to the sea. Recent geologic studies have shown that the lake formed and drained repeatedly, creating numerous floods. The array of features scattered throughout the Scabland region thus provides us with dramatic evidence that the geologic work accomplished by catastrophic floods can be prodigious.

BASE LEVEL

As a stream flows downslope, its potential energy decreases and finally falls to zero as it reaches the sea. At this level the river no longer has the ability to deepen its channel.

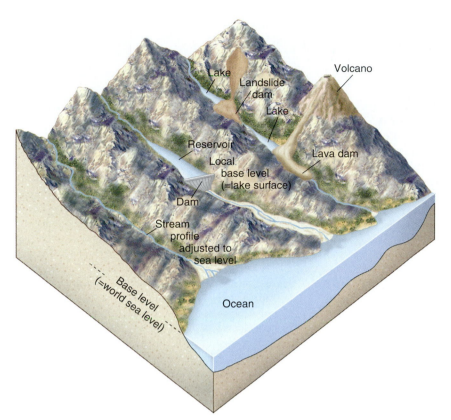

Figure 14.10 Base Levels Relationship of streams to base level (the world ocean) and to local base levels. The stream on the left flows directly to the ocean, which is its base level. The dammed middle stream flows into a reservoir, which acts as a local base level; the base level of the stream segment below the dam is the ocean. The stream to the right is dammed in two places, by landslide sediment and by a lava flow, each of which forms a lake upstream that acts as a local base level; the stream below the lava flow has the ocean as its base level.

The limiting level below which a stream cannot erode the land is called the **base level** of the stream. The base level for most streams is global sea level (Figure 14.10). Exceptions are streams that drain into closed interior basins having no outlet to the sea. Where the floor of a tectonically formed basin lies below sea level (for example, Death Valley, California), the base level coincides with the basin floor. A stream flowing into such a basin can erode its channel below the level of the world ocean.

In relating base level of most streams to sea level, we must recognize that world sea level fluctuates over geologically long periods of time because of changes in the shape and capacity of the ocean basins and the growth and shrinkage of glaciers on the continents. Thus, base level is always slowly changing.

When a stream flows into a lake, the surface of the lake acts as a local base level (Figure 14.10). However, the lake outlet may be lowered by erosion, causing the water to drain away. Once the local base level is destroyed, the stream will adjust its long profile to the changed conditions.

NATURAL AND ARTIFICIAL DAMS

Not all streams flow steadily from their headwaters to the sea. The courses of many are interrupted by lakes that have formed behind natural dams consisting, for example, of landslide sediments, glacial deposits, lava flows, or even glacier ice (Figure 14.10). Such a dam acts as a local base level and creates an irregularity in a stream's long profile. However, any natural dam is a temporary feature, for eventually it will be breached and eroded away.

Large artificial dams also disrupt the normal flow of water in a stream. They are being constructed in ever-increasing numbers to provide water storage, flood control, and hydroelectric power.

An artificial dam built across a stream creates a reservoir that traps nearly all the sediment that the stream formerly carried to the ocean. Therein lies one of the major long-term problems in the generation of hydroelectric power: accumulating sediment will eventually fill a reservoir, often within 50 to 200 years, making it useless. Lake Nasser, ponded behind the Aswan Dam on the Nile River, will be half filled with silt in the next century. Thus, although water power is continuous, the reservoirs needed to convert this power to electricity have limited lifetimes. (See Revisiting Earth Systems and Plate Tectonics *Tampering with the Nile*.)

HYDROELECTRIC POWER

Hydroelectric power is recovered from the potential energy of water in streams as they flow downslope to the sea. Water ("hydro") power is just another expression of solar power because it is the Sun's heat energy that drives the water cycle. That cycle is continuous, and so energy obtained from flowing water is also continuously available.

UNDERSTANDING OUR ENVIRONMENT

TAMING THE YANGTZE RIVER

When completed, it will be the largest and costliest dam ever built. It will benefit millions of consumers by generating at least 18,000 megawatts of electricity, the equivalent of 10 major coal-fired power plants. It will help control one of the world's most flood-prone rivers, which has killed more than 300,000 people during the twentieth century alone.

The Three Gorges Dam on China's 6300-km-long Yangtze River is a monumental undertaking that will cost at least $25 billion and take more than a decade to build. When completed in 2009, it will change the map of central China. The 185-m-high dam will back up a reservoir 600 km long and as much as 175 m deep. It will be more visible from an orbiting spacecraft than China's famous Great Wall.

However, the water that the dam impounds will submerge more than 450 villages and towns, and displace as many as 1.5 million people, thereby creating a massive resettlement problem. Half these people will be urban dwellers and the other half rural residents, all of whom will need new jobs or new farmland to begin their lives anew. Resettlement is not the only problem caused by the dam. Despite its strong promotion by the government, the project has generated a storm of adverse criticism, in large part focused on environmental issues.

According to some critics, the Three Gorges project may become one of the worlds' largest environmental disasters. Not only will the vast reservoir submerge one of China's most scenic and historically important areas, it will threaten rare plant and animal species along the river, including the giant sturgeon, giant salamander, freshwater jelly fish, alligators, monkeys, and various birds. In addition, the reservoir will receive large amounts of industrial wastes and as much as 1 billion tons of sewage each year, creating a toxic soup of human waste, industrial chemicals, and heavy metals. The dam will slow the flow of water, reducing the natural self-cleaning process that now flushes waste products downstream toward the East China Sea. Silting of the reservoir, some critics claim, could block drainage outlets in Chongqing, home to 30 million people, causing sewage to back up and slosh through the streets. Some geologists fear that the weight of the huge lake on the Earth's crust could trigger an earthquake. In a worst-case scenario, this could cause the dam to fail, inundating millions of people downstream in a flood far larger and more disastrous that the natural floods of historical times.

Whether or not the benefits and environmental problems are as great as the opposing sides profess, it is clear that China has embarked on a great environmental and social experiment, the consequences of which will not be known until well into the present century.

Water power has been used in small ways for thousands of years, but only in the twentieth century has it been widely used for generating electricity. All the water flowing in the streams of the world has a total recoverable energy estimated as 9.2×10^{19} J/yr. This is an amount of energy equivalent to burning 15 billion barrels of oil per year. Unlike coal and oil, however, hydropower cannot be used up; it is a renewable resource. Nevertheless, hydropower cannot solve all the world's energy problems, for streams powerful enough to provide large amounts of power are few and do not always flow near places where power is needed. Furthermore, reservoir siltation limits the useful life of power dams and will eventually give rise to the problem of dealing with huge obsolete concrete structures and the vast accumulation of sediment behind them. Sometimes the sites of power dams are in places with high population densities. As Box 14.1, *Taming the Yangtze River*, makes clear, the consequences are not easy to predict.

Before you go on:

1. What is a stream's long profile, and what is its usual shape?

2. Why does velocity of a stream tend to decrease downstream?

3. What causes floods, and what methods are available to predict floods?

4. What is meant by global base level and local base level?

CHANNEL PATTERNS

From an airplane, it is easy to see that no two streams are alike: they vary in size and shape. The variety of channel patterns on the landscape can be explained if we understand the relationships among stream gradient, discharge, and sediment load.

STRAIGHT CHANNELS

Straight channel segments are rare. Generally, they occur for only brief stretches before the channel begins to curve. If a stream channel has many curves, we refer to the channel pattern as a *sinuous* one. Close examination of a straight segment of natural channel shows that it has some of the features of sinuous channels. A line connecting the deepest parts of the channel typically does not follow a

straight path equidistant from the banks but wanders back and forth across the channel (Figure 14.11A). This pattern may result from random variations in channel depth. At places where the deepest water lies at one side of a channel, a deposit of sediment (a *bar*) tends to accumulate on the opposite side, where velocity is lower. The sinuous flow of water within a channel causes a succession of bars to form on alternate sides of the channel.

MEANDERING CHANNELS

In many streams, the channel forms a series of smooth bends that are similar in size and resemble in shape the switchbacks of a mountain road (Figure 14.12). Such a bend in a stream channel is called a **meander**, after the Menderes River (in Latin, *Meander*) in southwestern Turkey, which is noted for its winding course. Meanders are not accidental. They occur most commonly in channels that lie in fine-grained stream sediments and have gentle gradients. The meandering pattern reflects the way in which a river responds to resistance to flow and the way energy is dissipated as uniformly as possible along its course.

Try to wade or swim across a meandering stream and it quickly becomes apparent that the velocity of the flowing water is not uniform. Velocity is lowest along the bed and walls of the channel because here the water encounters maximum frictional resistance to flow. The highest velocity along a straight channel segment usually is found near the surface in midchannel (Figure 14.13). However, wherever the water rounds a bend, the zone of highest velocity swings toward the outside of the bend.

Over time, meanders migrate slowly down a valley. As water sweeps around a meander bend, the zone of highest velocity swings toward the outer streambank. Strong turbulence causes undercutting and slumping of sediment where the fast-moving water meets the steep bank.

A. Straight channel

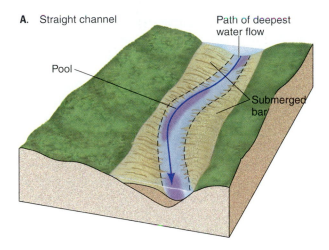

B. Meandering channel

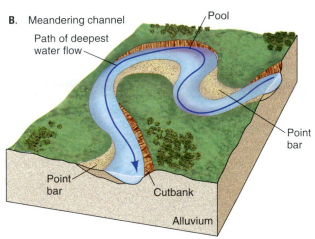

C. Braided channel

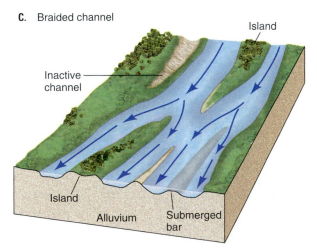

Figure 14.11 Channel Patterns Features associated with (A) straight, (B) meandering, and (C) braided channels. Pools are places along a channel where the water is deepest. Arrows indicate the direction of streamflow and trace the path of the deepest water.

Figure 14.12 Meandering Stream A meandering stream near Phnom Penh, Cambodia, flows past agricultural fields that cover the river's floodplain. Light-colored point bars, composed of gravelly alluvium, lie opposite cutbanks on the outsides of meander bends. Two oxbow lakes, the product of past meander cutoffs, lie adjacent to the present channel.

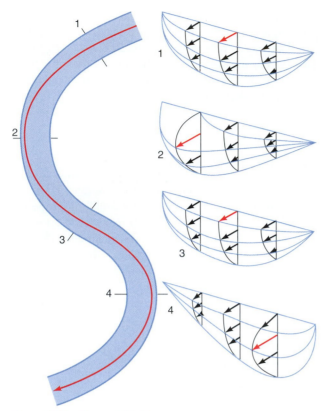

Figure 14.13 Stream Velocity Velocity distribution in cross sections through a sinuous channel (lengths of arrows indicate relative flow velocities). The zone of highest velocity (red arrow) lies near the surface and toward the middle of the stream where the channel is relatively straight (sections 1 and 3). At bends, the maximum velocity swings toward the outer bank and lies below the surface (sections 2 and 4).

Meanwhile, along the inner side of each meander loop, where water is shallow and velocity is low, coarse sediment accumulates to form a *point bar* (Figures 14.11B and 14.12). As a result, meanders slowly change shape and shift position along a valley as sediment is removed from and added to their banks.

The behavior of artificial streams and their channels has been studied in laboratory experiments using large sedimentation tanks. The experiments show that if the streambank sediment has a uniform particle size, meanders are symmetrical and all tend to migrate downstream at the same rate. In nature, however, bank sediment generally is not uniform, as shown in Figure 14.14, which is an example from the lower Mississippi River. Wherever the downstream part of a meander that is cutting into sandy sediment encounters less-erodible sediment, such as clay, its migration can be slowed. Meanwhile, the next meander upstream, migrating more rapidly, may intersect the slower moving meander. The water in the channel now can take a shorter route downstream over a steeper gradient, so the stream bypasses the meander, which is cut off. As sediment

is deposited along the margin of the new channel route, the cutoff meander is blocked and converted into a curved *oxbow lake* (Figures 14.12 and 14.14).

Nearly 600 km of the Mississippi River channel has been abandoned through such cutoffs since 1776. In 1883, former steamboat pilot Mark Twain, then a best-selling humorist and an observant amateur geologist, speculated about the future history of the Mississippi River in his book *Life on the Mississippi* (1883):

The Mississippi between Cairo and New Orleans was 1215 miles long 176 years ago. It was 1180 after the cutoff of 1722. It was 1040 after the American Bend cutoff. It has lost 67 miles since. Consequently, its length is only 973 miles at present.

Now, if I wanted to be one of those ponderous scientific people, and "let on" to prove what had occurred in the remote past by what had occurred in a given time in the recent past, or what will occur in the far future by what has occurred in late years, what an opportunity here! Geology never had such a chance, nor such exact data to argue from! Please observe:

In the space of 176 years the lower Mississippi has shortened itself 242 miles. That is an average of a trifle over one mile and a third per year. Therefore, any calm person, who is not blind or idiotic, can see that in the Old Oolitic Silurian Period, just over a million years ago next November, the Lower Mississippi River was upwards of 1,300,000 miles long, and stuck out over the Gulf of Mexico like a fishing rod. And by the same token any person can see that 742 years from now the Lower Mississippi will be only a mile and three-quarters long, and Cairo and New Orleans will have joined their streets together, and be plodding comfortably along under a single mayor and a mutual board of aldermen. There is something fascinating about science. One gets such wholesale returns of conjecture out of such a trifling investment of fact.

Despite Mark Twain's perceptive analysis, the river's length has not changed appreciably over the last two centuries because the loss of channel due to cutoffs has been balanced by lengthening of the channel as other meanders have enlarged.

BRAIDED CHANNELS

The intricate geometry of a **braided stream** resembles the pattern of braided hair, for the water repeatedly divides and reunites as it flows through two or more adjacent but interconnected channels separated by bars or islands (Figure 14.11C). The cause of braiding is related to the stream's ability to transport sediment. If a stream is unable to move all the available load, it tends to deposit the coarsest sediment as a bar that locally divides the flow and concentrates it in the deeper segments of channel to either side. As the bar builds up, it may emerge above the surface as an island

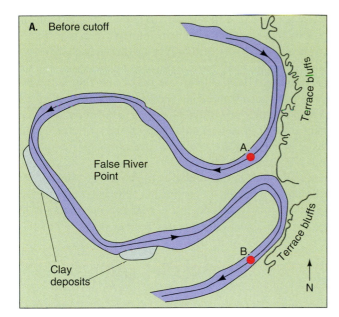

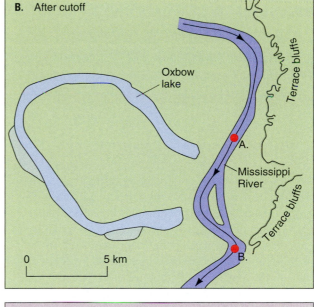

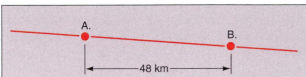

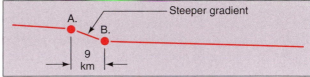

Figure 14.14 Meander Cutoffs Cutoff of a meander loop of the Mississippi River in Louisiana. A. The downvalley migration of a meander loop was halted when the channel encountered a mass of clay in the floodplain sediments. This allowed the next meander loop to advance and finally cut off the river segment surrounding False River Point. B. The new, shorter channel had a steeper gradient than the abandoned course, and a braided pattern developed.

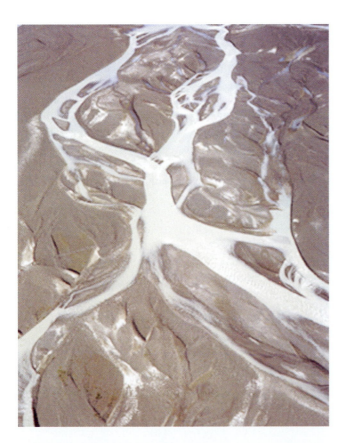

and become stabilized by vegetation that anchors the sediment and inhibits erosion.

A braided pattern tends to form in streams having highly variable discharge and easily erodible banks that can supply abundant sediment load to the channel system. Streams of meltwater issuing from glaciers generally have a braided pattern because the discharge varies both daily and seasonally, and the glacier supplies the stream with large quantities of sediment. The braided pattern, therefore, seems to represent an adjustment by which a stream increases its efficiency in transporting sediment.

Large braided rivers typically have numerous constantly shifting shallow channels (Figure 14.15). Although at any moment the active channels may cover no more than 10 percent of the width of the entire channel system, within a single season all or most of the surface sediment may be reworked by the laterally shifting channels.

Figure 14.15 Braided River Intricate braided pattern of the Tasman River below Tasman Glacier in New Zealand's Southern Alps. The constant lateral shifting of channels is related to variations in discharge and load in this steep-gradient stream.

EROSION BY RUNNING WATER

Erosion by water begins even before a distinct stream has formed. It occurs in two ways: by impact as raindrops hit the ground and by overland flow during heavy rains, a process known as **sheet erosion**. As raindrops strike bare ground, they dislodge small particles of loose soil, spattering them in all directions. On a slope the result is net displacement downhill. One raindrop has little effect, but the number of raindrops is so great that together they can accomplish a large amount of erosion.

The effectiveness of raindrops and overland flow in eroding the land is greatly diminished by a protective cover of vegetation. The leaves and branches of trees break the force of falling raindrops and cushion their impact on the ground. More important, the intricate network of roots forms a tight mesh that holds soil in place, greatly reducing erosion. The root network also holds water, letting it percolate slowly down through the soil. As a consequence, in vegetated areas there is less overland flow than in areas of bare ground.

The ability of streams to erode is related to the way water moves through a stream channel. If the velocity is very slow, the water particles travel in parallel layers, a motion called **laminar flow**. With increasing velocity, the movement becomes more erratic and complex, giving rise to the swirls and eddies that characterize **turbulent flow**. The velocity in stream channels is sufficiently high for turbulent flow to dominate. Only in a very thin zone along the bed and channel walls, where frictional drag is high, is velocity low enough for laminar flow to occur.

The ability of a stream to pick up particles of sediment from its channel and move them along depends largely on the turbulence and velocity of the water. Figure 14.16, based on experimental data, shows the velocities required to erode particles of different size from a stream bed, the range of velocity in which particles can be transported, and the velocities at which particles can no longer be moved and will settle to the bottom. In general, as velocity increases, so does the ability of the turbulent water to lift ever-larger particles of sediment. Silt and clay are an exception, for they tend to be cohesive and difficult to erode except under conditions of high velocity.

THE STREAM'S LOAD

The solid portion of a stream's load consists of two parts. The first part is the coarse particles that move along the stream bed (the **bed load**), while the second is the fine particles that are suspended in the water (the **suspended load**). Wherever they are dropped, these solid particles constitute **alluvium**, which is any detrital sediment deposited by a stream.

Streams also carry dissolved substances (the **dissolved load**) that are chiefly a product of chemical weathering.

BED LOAD

The bed load generally amounts to between 5 and 50 percent of the total load of a stream. Bed-load particles move at a slower velocity than the stream water, for the particles are not in constant motion. Instead, they move discontinuously by rolling or sliding. Where forces are sufficient to lift a particle, it may move short distances by **saltation**, a motion that is intermediate between suspension and rolling or sliding. Saltation involves the progressive forward movement of a particle in a series of short intermittent jumps along arcuate paths (Figure 14.17). Saltation continues as long as currents are turbulent enough to lift particles from the bed.

The distribution of bed-load sediment in a stream channel is related to the velocity distribution (Figure 14.18). Coarse-grained sediment is concentrated where the velocity is high, whereas finer-grained sediment is found in zones of progressively lower velocity.

PLACER DEPOSITS

The famous California gold rush of 1849 followed the discovery that the sand and gravel in the bed of a small stream contained bits of gold. Similar gold-bearing gravels are found in many other parts of the world. The gravels themselves are sometimes rich enough to be mined, but even when they are too lean, the gold is a clue that a source must

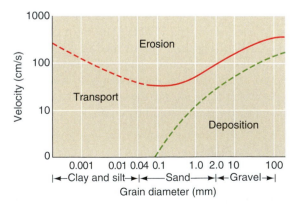

Figure 14.16 Velocity Controls Graph showing how stream velocity controls erosion, transport, and deposition of sediment particles of different sizes.

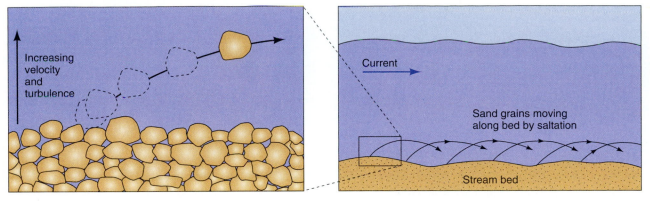

Figure 14.17 Saltation A sandy bed load moves by saltation when sand grains are carried up into a stream at places where turbulence locally reaches the bottom or where suspended grains impact other grains on the bed. Once raised into the flowing water, the grains are transported along arc-shaped trajectories as gravity pulls them toward the stream bed where they impact other particles, which in turn are set in motion.

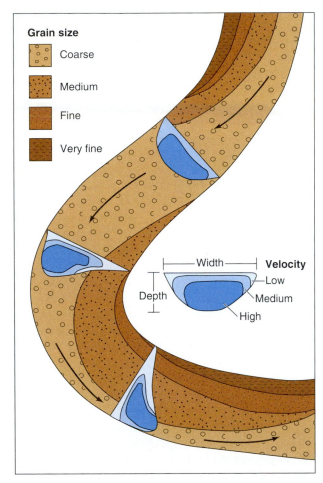

Figure 14.18 Bed Load and Velocity Relationship of bed-load grain size to velocity in a section of meandering channel. The coarsest sediment is associated with the zone of highest velocity; both lie on the outside of a bend adjacent to the cutbank, but in the center of the channel between bends. The finest sediment is associated with the zone of lowest velocity and lies on the inside of a meander bend, opposite the cutbank.

lie upstream. In fact, many mining districts have been discovered by following trails of gold and other minerals upstream to their sources in veins in bedrock.

Because pure gold is dense (specific gravity = 19), it is deposited from the bed load of a stream very quickly, while quartz, with a specific gravity of only 2.65, is washed away. As most silicate minerals have specific gravities that are low by comparison with gold, grains of gold become mechanically concentrated in places where the velocity of streamflow is high enough to remove the low-density particles but not high enough to remove the higher-density ones. Such concentration occurs, for example, behind rock bars or in bedrock holes along the channel, below waterfalls, on the inside of a meander bend, and downstream from the point where a tributary enters a main stream. A deposit of heavy minerals concentrated mechanically is a **placer**.

Many heavy, durable minerals other than gold also form placers. These include minerals that occur as pure metals, such as platinum and copper, as well as tinstone (cassiterite, SnO_2) and nonmetallic minerals such as diamond, ruby, and sapphire. Even if a vein contains a low percentage of a mineral, the placer it yields may be quite rich. In order to become concentrated in placers, the minerals must be not only dense but also resistant to chemical weathering and not readily susceptible to cleaving as the mineral grains are tumbled in the stream.

Every phase of the conversion of gold in a vein to placer gold has been traced. Chemical weathering of the exposed vein releases the gold, which then moves slowly downslope by mass wasting. In some places mass wasting alone concentrates the gold sufficiently to justify mining. More commonly, however, the mineral particles enter a stream, which concentrates them more effectively than mass wasting.

Most placer gold occurs as grains the size of silt particles, the "gold dust" of miners. Some of it is coarser. A

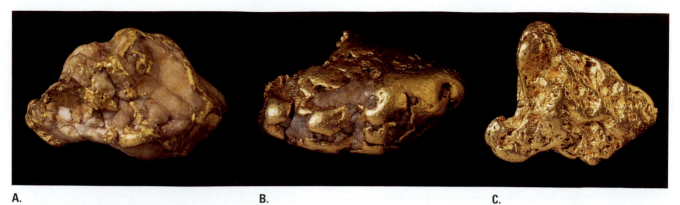

A.   B.   C.

Figure 14.19 Nuggets Formation of a nugget. A. A vein of metallic gold cutting through a pebble of vein quartz. Stream abrasion causes the brittle quartz to chip and be reduced in size, while the malleable gold deforms but is not reduced in size. B.

The ratio of gold to quartz increases as the quartz is abraded away. Eventually a nugget of almost solid gold forms. C. A nugget of metallic gold. No quartz remains. Each of the specimens has a diameter of about 4 cm.

lump of pebble or cobble size is a *nugget* (Figure 14.19). The largest nugget ever recorded weighed 80.9 kg.

In following placers along stream channels, prospectors have learned that rounding and flattening (by pounding) of nuggets increase downstream. Therefore, when they find rough, angular nuggets, prospectors know the primary source is near.

SUSPENDED LOAD

The muddy character of many streams is due to the presence of fine particles of silt and clay moving in suspension (Figure 14.20). Most of the suspended load is derived from

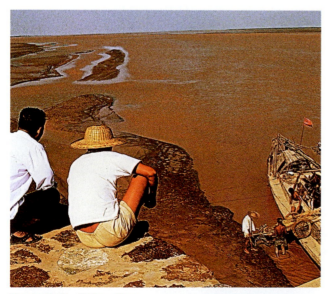

Figure 14.20 Yellow River A large suspended load, eroded from extensive deposits of windblown silt, gives China's Huang He a very muddy appearance and its English name (Yellow River).

fine-grained regolith washed from areas unprotected by vegetation and from sediment eroded and reworked by the stream from its own banks. China's Yellow River received its name for the great load of yellowish silt it erodes and transports seaward from widespread deposits of windblown dust that underlie much of its basin. Prior to construction of several major dams along its course, the Colorado River was also extremely muddy; those who lived along it sometimes remarked that the river was "too thin to plow, but too thick to drink."

Because upward-moving currents within a turbulent stream exceed the velocity at which particles of silt and clay can settle toward the bed under the pull of gravity, such particles tend to remain in suspension longer than they would in nonturbulent waters. They settle and are deposited only where velocity decreases and turbulence ceases, as in a lake or in the sea.

DISSOLVED LOAD

All stream water contains dissolved chemical substances that constitute part of its load. The bulk of the dissolved content of most rivers consists of seven ionic species: bicarbonate (HCO_3^{1-}), calcium (Ca^{2+}), sulfate (SO_4^{2-}), chloride (Cl^{1-}), sodium (Na^{1+}), magnesium (Mg^{2+}), potassium (K^{1+}), plus dissolved silica as $Si(OH)_4$.

Although in some streams the dissolved load may represent only a small percentage of the total load, in others it amounts to more than half. Streams that receive large contributions of underground water (Chapter 15) generally have higher dissolved loads than those whose water comes mainly from surface runoff.

DOWNSTREAM CHANGES IN PARTICLE SIZE

The size of the particles a stream can transport is related mainly to the flow velocity. Therefore, we might expect the

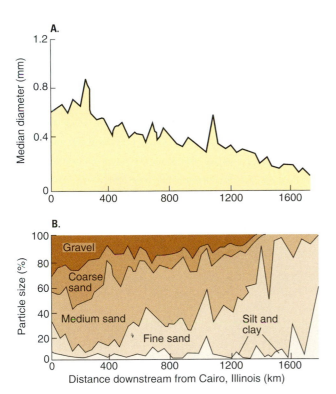

Figure 14.21   Sediment Sizes in the Mississippi River   Change in sediment size along the Mississippi River downstream from Cairo, Illinois. A. Through the lower 1600 km of the channel, median diameter decreases from 0.7 to 0.2 mm. B. At Cairo, about 30 percent of the stream's channel sediment is gravel, 60 percent is sand, and 10 percent is silt and clay. At a point 1600 km downstream, the sediment is almost entirely finer than medium sand.

maximum size of sediment to increase in the downstream direction as velocity increases. In fact, the opposite is true; sediment normally decreases in coarseness downstream. The explanation for this unexpected result involves both sorting and abrasion. In mountainous headwaters of large rivers, tributary streams flow mostly through channels floored with coarse gravel that may include large boulders. Because fine sediment is easily moved, even by streams having low discharge, it is readily carried away by small mountain streams, leaving the coarser sediment behind. Through time, the coarse bed load is gradually reduced in size by abrasion and impact as it moves slowly along. When the stream eventually reaches the sea, its bed load may consist mainly of sediment no coarser than sand.

We can see such a progressive downstream change in sediment size along the channel of the Mississippi River below Cairo, Illinois (Figure 14.21). Sampling along the channel has shown that the median diameter of sediment decreases from about 0.7 mm near Cairo to only 0.2 mm at the river mouth, more than 1600 km downstream. In the upper 300 km of this section of channel, at least 50 percent of the channel sediment consists of gravel and coarse sand,

but in the lower several hundred kilometers, more than 90 percent of the channel sediment is fine sand, silt, and clay.

DOWNSTREAM CHANGES IN COMPOSITION

Large streams generally cross a variety of exposed rocks. Therefore, the stream load changes composition along the channel system as sediments of different composition are introduced. The Nile River provides a good example. Flowing through lower Egypt toward its delta, the Main Nile includes water contributed by three major tributaries: the White Nile, the Blue Nile, and the Atbara (Figure 14.22). The White Nile, which contributes nearly a third of the total discharge, is responsible for only 3 percent of the bed load in the Main Nile. In the mineral component of this load, the ratio of amphibole (eroded from metamorphic bedrock of the central African plateau) to pyroxene is 97:3. The Blue Nile, which drains the highlands of Ethiopia, contributes more than half the discharge and nearly three-quarters of the bed load. Its amphibole:pyroxene ratio is 79:21, reflecting the volcanic character of the source region. The more northerly Atbara contributes 14 percent of the discharge and a quarter of the bed load. In this stream pyroxene is quite abundant, and the amphibole:pyroxene ratio is 9:91. These differing mineral components mix together as they enter the Main Nile, resulting in an amphibole:pyroxene ratio of 59:41. This ratio, as we might expect, largely reflects the major contribution of amphibole-rich sediment from the Blue Nile, which accounts for the largest sediment input to the Main Nile.

SEDIMENT YIELD

Some streams run clear nearly all the time; others are constantly muddy and obviously are transporting a considerable load. Such contrasts imply that land areas are eroded at different rates. These differences are related to geologic, climatic, and topographic factors that include rock type and structure, local climate, and relief and slope. Together, these factors control the *sediment yield*, which is the amount of sediment/unit area eroded from the land and transported by streams (Figure 14.23).

Climate influences erosion in different ways. We might guess that the greater the precipitation, the greater the erosion. In fact, the highest sediment yields are seen in some equatorial and tropical lands where high precipitation and temperature combine to promote rapid chemical weathering (Figure 14.23). However, in moist regions, plant roots tend to anchor the soil, thereby curtailing erosion. In temperate eastern North America and Western Europe, for example, vegetation cover is more or less continuous and erosion rates are low. In drier regions, reduced precipitation limits vegetation, making the land vulnerable to erosion. Areas receiving abundant precipitation may actually experience less erosion than some relatively dry regions.

A.

B.

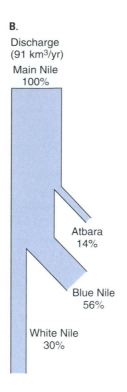

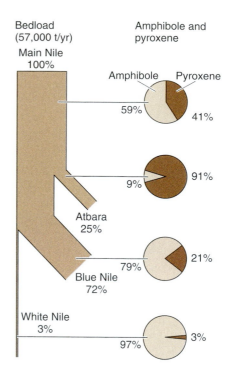

Figure 14.22 Sediment in the Nile River   Change in channel sediment composition along the Nile River. A. Map of the Nile River and its principal tributaries. B. Discharge, bed load, and amphibole:pyroxene ratio of the Nile and its main tributaries. The White Nile, which flows from Lake Victoria, contributes less than a third of the Nile's discharge and only 3 percent of its bed load. The greatest percentages of discharge and bed load are supplied by the Blue Nile, which originates in the highlands of Ethiopia. Different percentages of minerals are contributed by each tributary, causing a change in sediment composition of the Main Nile as each tributary joins it.

Field measurements suggest that some of the greatest local sediment yields are from landscapes that are traditional between desert and grassland.

Some of the highest measured sediment yields are from basins that drain steep mountains along plate boundaries (Figure 14.23). In these young orogenic belts, continental crust is generally less than 250 million years old. Mountains in the monsoon region of southeastern Asia, and many maritime ranges along the western coast of the Americas, receive abundant precipitation that generates high runoff (Figure 12.12). In southern Alaska and the southern Andes, large active glaciers contribute to high sediment yields. An additional reason why sediment yields are high in these regions is because they contain erodible clastic sediments or sedimentary rocks. Structural factors also play a role, for rocks that are highly jointed or fractured are more susceptible to erosion than massive rocks.

The clearing of forests, cultivation of land, damming of streams, construction of cities, and numerous other human activities also affect erosion rates and sediment yields. Sometimes the results are dramatic. In parts of the eastern United States, areas that have been cleared for construction produce between 10 and 100 times more sediment than comparable rural or natural areas that remain vegetated.

However, once cleared land is fully developed, sediment yield is greatly reduced because a nearly complete cover of buildings, sidewalks, and roads protects the underlying rocks and sediments from erosion.

> *Before you go on:*
> 1. How do bed-load particles move in a stream?
> 2. What factors account for the formation of placers?
> 3. What are the primary sources of the dissolved load of a stream?
> 4. How and why do particle size and sediment composition change downstream?

STREAM DEPOSITS

When a stream's turbulence decreases, usually because of a decrease in velocity caused by a reduction in gradient or in discharge, its transporting power drops and it deposits part of its load. Distinctive stream deposits form along channel margins, valley floors, mountain fronts, and the

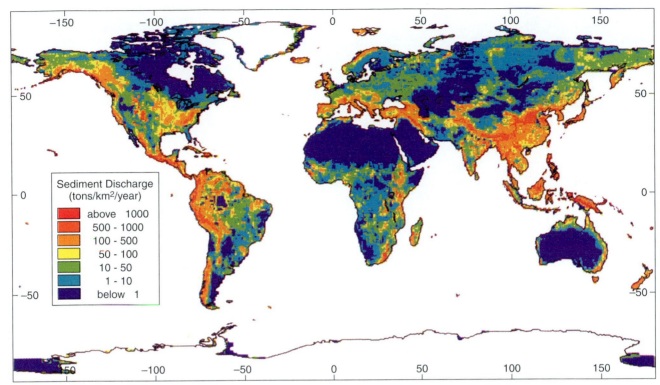

Figure 14.23 **Sediment Yields on the Continents** Average sediment yields from drainage basins of the world. Areas of highest sediment yields lie along mountain belts: the western cordilleras of the Americas and the Alps-Caucasus-Himalaya–Tien Shan belt of Eurasia. Low yields are concentrated in desert regions (north Africa, Arabia, central Asia, Australia) and in low-lying high-latitude regions (northern Canada, Siberia, Antarctica).

margins of lakes and the ocean, for these all are places where changes in stream energy take place.

FLOODPLAINS AND LEVEES

When a stream rises during a major flood, the water overflows the banks and inundates the adjacent **floodplain**, which is the part of a stream valley that is inundated by floodwater (Figure 14.24). The boundary between channel and floodplain may be the site of a **natural levee**—a broad, low ridge of alluvium built along the side of a channel by debris-laden floodwater. As sediment-laden water flows out of the completely submerged channel during a flood, the depth, velocity, and turbulence of the water decrease abruptly at the channel margins. The abrupt decrease in these factors results in sudden, rapid deposition of the coarser part of the suspended load (usually fine sand and coarse silt) along the margins of the channel, building up the natural levee. Farther away, finer silt and clay settle out in the quiet water covering the floodplain.

TERRACES

Many stream valleys contain one or more relatively flat alluvial terraces that lie above the floodplain (Figures

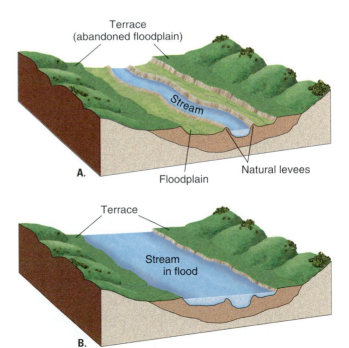

Figure 14.24 **Alluvial Landforms** Main features of an alluvial valley. A. Landforms visible during average or low-water conditions. B. Same valley during flood conditions. Terraces, by definition, lie above the highest modern flood levels.

Figure 14.25 Stream Terraces Terraces of the Strathmore River in northern Scotland record a succession of former flood-plains, each abandoned as the stream cut downward into alluvium underlying the floor of its valley.

14.24 and 14.25). A **terrace** is a remnant of an abandoned floodplain. It may be underlain by either sediments or bedrock, or by both. In other words, a terrace is a landform rather than a deposit, and it is distinct from the materials that underlie it.

A terrace forms when a stream erodes downward to a level such that floods no longer reach the former floodplain. Typically, such downcutting occurs in response to a change in discharge, load, gradient, or base level. Many stream valleys contain a series of terraces that record a complex history of alternating depositional and erosional events.

ALLUVIAL FANS

When a stream flowing through a steep upland valley emerges suddenly onto a nearly level valley floor or an alluvial plain, it experiences a decrease in slope, a corresponding drop in velocity, and a decrease in its ability to carry sediment. It therefore deposits the part of its load that cannot be transported on the gentler slope. As deposition takes place in and beside the channel, the stream is locally diverted to one side or the other by the accumulating sediment. In the process, the abundant load may lead the stream to assume a braided pattern. Shifting to new courses across slightly lower ground, the stream continues to deposit its load. Over time, the stream channels shift later-

ally, back and forth, in response to continuing deposition. As a result, the deposit assumes the shape of an **alluvial fan**, which is a fan-shaped body of alluvium typically built where a stream leaves a steep mountain valley (Figure 14.26).

Deposition on a fan surface may also occur as water percolates down into the underlying porous sediments, thereby reducing surface discharge. In some cases, such percolation can cause a stream to disappear near the top of a large fan, only to reappear again near the fan's base.

The profile of most fans, from top to base in any direction, has the same curved form characteristic of the long profiles of streams. The exact form of the profile depends chiefly on discharge and on the size of particles in the bed load. Hence, no two fans are exactly alike. A small stream carrying a load of coarse particles builds a shorter, steeper fan than a larger stream carrying a load of finer particles. The area of a fan generally is closely related to the size of the upstream area from which its sediments are derived.

DELTAS

When a stream enters the standing water of the sea or a lake, its speed drops rapidly, its ability to transport sediment decreases markedly, and it deposits its solid load. As we learned in Chapter 7, a sedimentary deposit that forms where a stream flows into standing water is a **delta**, named

Figure 14.26 Alluvial Fan A symmetrical alluvial fan has formed at the margin of Death Valley, California, at the mouth of a steep mountain canyon.

for the Nile delta, which has a crudely triangular shape that resembles the Greek letter delta (Δ) (Figure 14.27). All other deltas derive their name from the Nile delta.

Deltas built by streams transporting coarse sediments are of two types. A gravel-rich delta that is formed where an alluvial fan is building outward into a standing body of

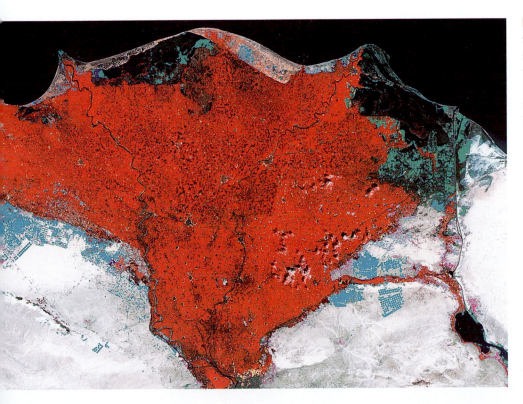

Figure 14.27 Nile Delta The delta of the Nile River, as seen from an orbiting spacecraft. Crops covering the well-watered agricultural land of the delta surface appear red in this false-color image and contrast sharply with the barren desert landscape beyond.

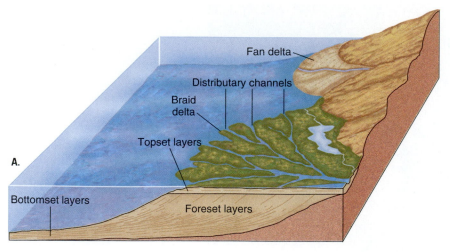

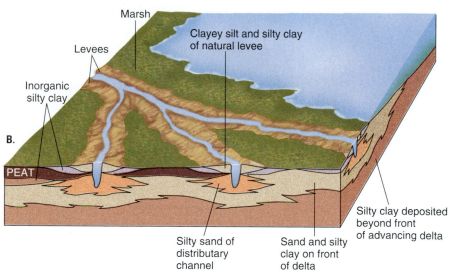

Figure 14.28 Features of Deltas

Main features of deltas. A. A braid delta built into a lake displays topset, foreset, and bottomset layers. A nearby fan delta is an alluvial fan that is building out into the body of water. B. Part of a large delta built into the sea shows the intertonguing relationship of coarse channel deposits and finer sediments deposited on the delta front and elsewhere.

water (a *fan delta*, Figure 14.28A) typically is built adjacent to a mountain front. Stream-channel, sheet-flood, and debris-flow sediments characteristic of alluvial fans form the upper parts of such deltas. The sediments in a fan delta show evidence of highly variable currents and abrupt changes of facies.

A *braid delta* is a coarse-grained delta constructed by a braided stream that builds outward into a standing body of water (Figure 14.28A). Its upper part displays features characteristic of braided streams. Braid deltas are especially conspicuous where braided glacial meltwater streams flow into lakes or the sea.

As a stream enters standing water, particles of the bed load are deposited first, in order of decreasing weight or density. Then the suspended sediments settle out. A layer representing one depositional event (such as a single flood) therefore grades from coarse sediment at the stream mouth to finer sediment offshore. The accumulation of many successive layers creates an embankment that grows progressively outward (Figure 14.28A). The coarse, thick, steeply sloping part of a depositional layer in a delta is a *foreset*

layer. Traced away from the shore, the same layer changes facies, becoming thinner and finer, and covering the bottom over a wide area. This part of a depositional layer in a delta is called a *bottomset layer*.

As deposition proceeds and the delta builds outward, the coarse foreset layers progressively overlap the bottomset layers. Thus, the stream gradually extends its channel outward over the growing delta. Coarse channel deposits and finer sediment deposited between channels, together called *topset layers*, overlie the foreset layers in a delta (Figure 14.28A).

Many of the world's largest streams, among them the Ganges-Brahmaputra, the Huang He (Yellow River), the Amazon, and the Mississippi, have built massive deltas at their mouths. Each delta has its own peculiarities, determined by such factors as the stream's discharge, the character and volume of its load, the shape of the bedrock coastline near the delta, the offshore submarine topography, and the intensities and directions of currents and waves.

Most major streams transport large quantities of fine suspended sediment, the bulk of which is carried seaward as the fresh stream water overrides denser saltwater at the coast. The fine sediment then settles out to form a gently sloping delta front.

Where strong currents and wave action redistribute sediment as quickly as it reaches the coast, delta formation may be inhibited. However, if the rate of sediment supply exceeds the rate of coastal erosion, then a delta will be built seaward. The Mississippi River delivers a huge sediment load to the margin of the Gulf of Mexico each year. Much of the load is deposited along and around numerous *distributaries*, which are long finger-like channels that branch from the main channel. The coarsest sediment is found along the channels. Finer sediment reaches the front of the delta and also accumulates between distributary channels during floods. The result is a complex intertonguing of facies and an intricate delta margin (Figure 14.28B).

Before you go on:

1. What are the primary depositional features associated with river floodplains?

2. What factors control the formation of alluvial fans? of deltas?

DRAINAGE SYSTEMS

To city dwellers used to a structured, artificial environment, nature sometimes seems to lack any obvious organization or pattern. But organization does exist, even though we may have to look for it. Streams are not distributed randomly across the landscape but are organized into intricate drainage systems, the geometry of which can provide clues about the underlying geology and the evolution of continents.

DRAINAGE BASINS AND DIVIDES

Every stream is surrounded by its **drainage basin**, the total area that contributes water to the stream. The line that separates adjacent drainage basins is a **divide**. Drainage basins range in size from less than a square kilometer to vast areas of near-continental dimensions (Figure 12.13). The huge drainage basin of the Mississippi River encompasses an area that exceeds 40 percent of the area of the conterminous United States (Figure 14.29). Not surprisingly, the area of any drainage basin is related to both the length and the mean annual discharge of the stream that drains the basin.

STREAM ORDER

The arrangement and dimensions of streams in a drainage basin tend to be orderly. This can be verified by examining a stream system on a map and numbering the observed stream segments according to their position, or order, in the system. The smallest segments lack tributaries and are classified as first-order streams. Where two first-order streams join they form a second-order stream, which has only first-order tributaries. Third-order streams are formed by the joining of two second-order streams and can have first-order and second-order tributaries, and so forth for successively higher stream orders (Figure 14.30).

If for any drainage basin, the number of streams assigned to each order is counted, we quickly see that the sums increase with decreasing stream order. In its pattern of a main stream joined by increasingly greater numbers of successively smaller tributaries, a stream system resembles a tree with its trunk and increasing numbers of successively smaller branches. This orderliness is like that inherent in a stream's long profile, in which gradient decreases systematically from head to mouth, while discharge, velocity,

Figure 14.29 Drainage Basin of the Mississippi River The drainage basin of the Mississippi River encompasses a major portion of the central United States. In this diagram, the width of the river and its major tributaries reflect discharge values.

550 m³/s
1440 m³/s
2800 m³/s
4250 m³/s
8500 m³/s

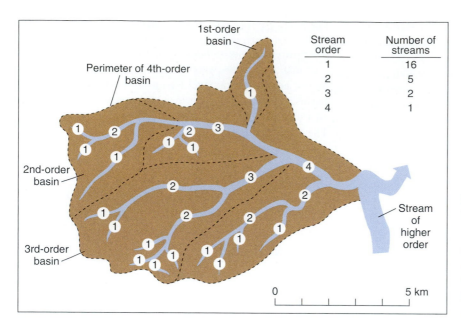

1st-order basin

Perimeter of 4th-order basin

2nd-order basin

3rd-order basin

Stream order	Number of streams
1	16
2	5
3	2
4	1

Stream of higher order

0   5 km

Figure 14.30 Stream Orders Drainage basin of a fourth-order stream in the Appalachian region showing tributary channels numbered according to stream order. Both the number of tributaries and their length are related to stream order. Basins are classified according to the order of the largest stream they contain.

and channel dimensions increase. All these relationships imply that in response to a given quantity of runoff, stream systems develop with just the size and spacing required to move the water off each part of the land with greatest efficiency.

EVOLUTION OF DRAINAGE

A stream system can develop quickly, as indicated by the following example. In August 1959, an earthquake raised and tilted the bed of Hebgen Lake, near West Yellowstone, Montana, exposing a large area of silt and sand. With the first rain, small stream systems began to develop on the newly exposed lakebed. Sample areas were surveyed and mapped one and two years after the earthquake. The results showed the same basic geometry that characterizes much larger and older stream systems. The small, newly formed valleys, together with the areas between them, were disposing of the available runoff in a highly systematic way, and all within a period of two years after the surface had emerged from beneath the lake.

A similar result has been obtained experimentally using sprinkler systems that subject large sediment-filled containers to artificial rainfall (Figure 14.31). As erosion proceeds, the stream network spreads upslope, eventually encompassing the entire basin. In the course of drainage

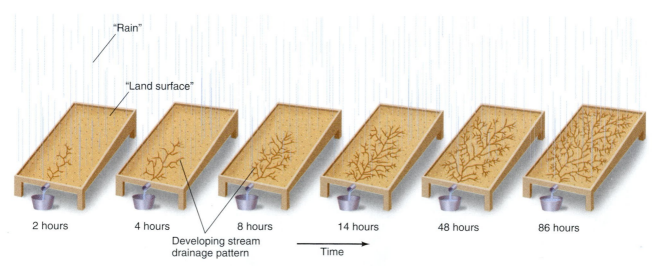

"Rain"

"Land surface"

2 hours   4 hours   8 hours   14 hours   48 hours   86 hours

Developing stream drainage pattern

Time

Figure 14.31 Evolution of a Stream System Evolution of a drainage network in an experimental rainfall-erosion container. The initial channel, which directed runoff toward the lower end of the container, grew headward and developed new tributaries as it spread to encompass the drainage basin.

evolution, both the length of each channel and the number of tributaries increase.

As a system of drainage develops, details of its pattern change. Streams acquire new tributaries, and some old tributaries are lost as a result of *stream capture*, which is the interception and diversion of one stream by another stream that is expanding its basin by erosion in the head-

ward direction. When stream capture occurs, some stream segments are lengthened and others are shortened. Just as the hydraulic factors within a stream are constantly adjusting to changes, so too is the drainage system constantly changing and adjusting as it grows. Like a stream channel, a drainage system is a dynamic system tending toward a condition of equilibrium.

DRAINAGE PATTERNS, ROCK STRUCTURE, AND STREAM HISTORY

One of the best ways to view stream systems on the landscape is from the window of an orbiting spacecraft; next best is from an airplane. From an altitude of 8 or 9 km, stream patterns can tell us a great deal about geologic structure and landscape history.

The ease with which a formation is eroded by streams depends chiefly on its composition and structure. The course a stream takes across the land bears a close relationship to these factors. Thus, drainage patterns we can see from an airplane or those that can be traced on a topographic map give us information about underlying rock type and structure. Figure 14.32 shows some of the most common drainage patterns and the geologic factors that control them. An experienced geologist can use these drainage patterns to infer such things as rock type, the direction and degree of slope of an inclined rock unit, the manner in which the rocks are folded or offset, and the orientation and spacing of joints.

A close relationship between streams and the rock units across which they flow can provide important insights regarding the structure and geologic history of an area. Geologists classify such relationships into several categories that reflect distinctive stream histories (Figure 14.33).

Dendritic — *Branching of channels ("tree-like") in many directions.* Common in massive rock and in flat-lying strata. In such situations, differences in rock resistance are so slight that their control of the directions in which valleys grow headward is negligible.

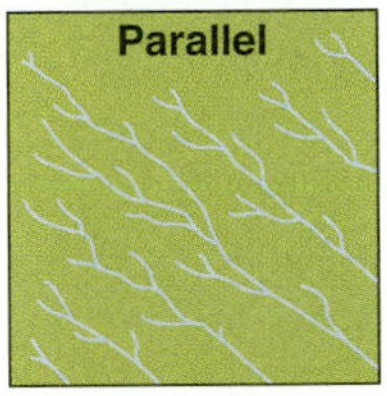

Parallel — *Parallel or subparallel channels that have formed on sloping surfaces underlain by homogeneous rocks.* Parallel rills, gullies, or channels are often seen on freshly exposed highway cuts or excavations having gentle slopes.

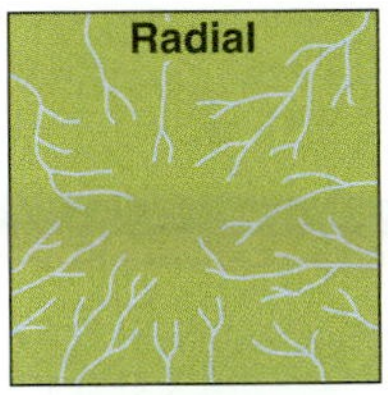

Radial — *Channels radiate out, like the spokes of a wheel, from a topographically high area,* such as a dome or a volcanic cone.

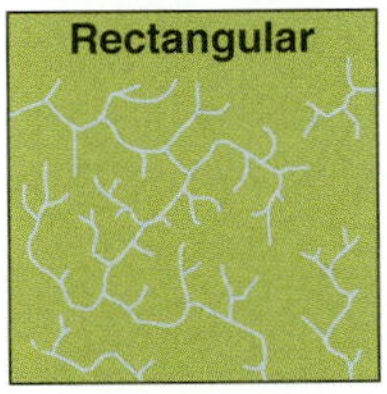

Rectangular — *Channel systems marked by right-angle bends.* Generally results from the presence of joints and fractures in massive rocks, or foliation and fractures in metamorphic rocks. Such structures, with their cross-cutting patterns, have guided the directions of valleys.

Trellised — *Rectangular arrangement of channels in which principal tributary streams are parallel and very long,* like vines trained on a trellis. This pattern is common in areas where the outcropping edges of folded sedimentary rocks, both weak and resistant, form long, nearly parallel belts.

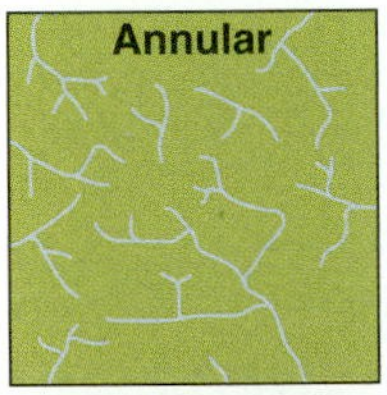

Annular — *Streams follow paths that are segments of circles* that ring a dissected dome or basin where erosion has exposed successive belts of rock of varying degrees of erodibility.

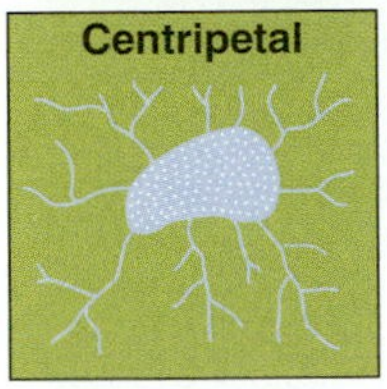

Centripetal — *Streams converge toward a central depression,* such as a volcanic crater or caldera, a structural basin, a breached dome, or a basin created by dissolution of carbonate rock.

Figure 14.32 Stream Patterns Some common stream patterns and their relationship to rock type and structure.

Before you go on:

1. What is a drainage basin, and what determines its area?
2. How are streams organized in a drainage basin?
3. What can drainage patterns tell us about a stream's history?

REVISITING PLATE TECTONICS AND THE EARTH SYSTEM

DRAINAGE AND LANDSCAPE EVOLUTION NEAR RIFTED PLATE MARGINS

The fragmentation of landmasses that accompanies the rifting of lithospheric plates can leave a strong imprint on drainage and subsequent landscape development. A good

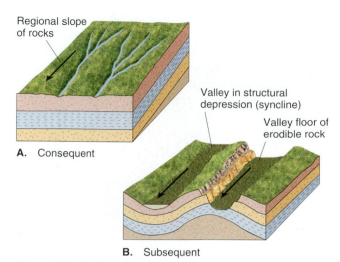

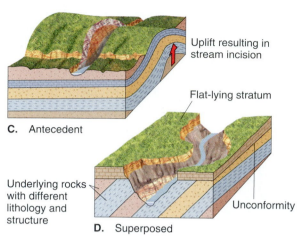

Figure 14.33 Relationship Between Streams and Geology
The relationship of streams to geology provides information about the structural history of an area. A. The course of a *consequent stream* is determined by the slope of land surface. B. A *subsequent stream* occupies belts of weak rock having taken a course dictated by geologic structure. C. An *antecedent stream* maintains its course across rocks that have been raised across its path; typically, it occupies a gorge that crosses a structural ridge rather than taking a course around the obstruction. D. A *superposed stream* has cut downward through strata until its channel lies in underlying rock of different lithology or structure; the initial course of the stream was not determined by the rocks across which it is now flowing.

example is the rift margin in southwestern Africa, where the early Mesozoic breakup of ancient Gondwana led to a stream pattern typical of rift margins.

Inland from the Atlantic coast of Namibia and South Africa lies the Great Escarpment, a major topographic upwarp that approximately parallels the coastline. Its crest generally coincides with the drainage divide from which relatively short subparallel streams flow into the Atlantic. Inland from the escarpment, streams drain into the broad, relatively flat continental interior, except where they have breached the upwarp and have been deflected toward the ocean.

A general hypothesis for landscape evolution along continental margins having features like the Great Escarpment involves the reorientation of pre-rift drainage systems. Extension during rifting causes significant uplift adjacent to the rift margin, disrupts the preexisting drainage, and leads to two new stream systems (Figure 14.34A). An exterior system of short, steep streams oriented perpendicular to the new coastline drains the rift flank. Vigorous erosion by these streams produces high local relief (Figure 14.34B, C). An interior system flows away from the uplifted rift margin toward sedimentary basins in the continental interior and likely resembles the pre-rift drainage pattern. Relief in this region remains low because the interior streams lie a great distance from their base level.

As steep streams of the exterior system erode the rift flank, the divide at their head retreats toward the continental interior (Figure 14.34B, C). Some vigorous headward-cutting streams may intersect streams of the interior region

and deflect them. The interior drainage is thereby "captured" and connected to the exterior drainage and its much lower base level (Figure 14.34C).

If rocks of differing resistance to erosion are exposed during retreat of the rift flank, a steplike topography may result. The most-resistant rock units will form bold escarpments, like the Great Escarpment, that retreat progressively inland toward the drainage divide.

If this hypothesis of landscape evolution is correct, then topography and drainage resulting from rifting may be constantly changing. The landscape at any time will reflect

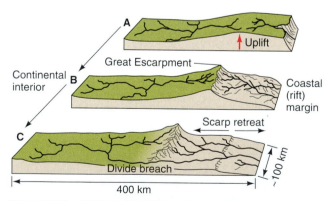

Figure 14.34 Drainage Evolution Uplift adjacent to a rifted margin results in development of rift-margin and interior drainage systems. A rift-facing scarp retreats inland and a stepped topography evolves as resistant rock units are exposed by denudation. Breaching of the scarp at the drainage divide results in capture of interior drainage by a steep coastal stream.

Figure 14.35 Aswan Dam View of Aswan High Dam, which impounds the Nile River (right) to form Lake Nasser (to south behind dam). Sediment formerly carried northward to the Mediterranean Sea and now settling out in the lake will eventually fill the reservoir and make it unusable.

the contest between differing erosion rates in the exterior and interior stream basins, a static or landward-shifting drainage divide, and the variable erosion rates of different lithologies that are exposed at the land surface.

TAMPERING WITH THE NILE

For more than seven millennia people have lived along the banks of the lower Nile River, their fields forming a ribbon of green that crosses the vast desert of Egypt. Long before the pyramids were built, people learned to live with the Nile's annual cycle. During part of each year the river flowed peacefully within its banks, but by late summer it began to rise during the annual period of flooding. Although floods could devastate those living along the stream's margins, at the same time the floods were beneficial, for they carried an abundant load of silt that added annual nourishment to agricultural fields.

To provide an adequate water supply and stabilize flow through the lower, densely populated Nile basin, a high dam was constructed at Aswan in the 1960s (Figures 14.35 and 14.22). The Aswan Dam was designed to reduce downstream the immense difference in discharge between the flood and low-flow seasons, and to permit full utilization of the Nile water. It was intended to provide water not only for human consumption, but also for irrigation, hydroelectric power, and inland navigation. These primary goals have been realized, for the dam led to a marked change in the river's annual discharge pattern. No longer is there a large seasonal flood below the dam; instead, the annual hydrograph reflects the more uniform, controlled discharge of water from the reservoir. Despite the success of the project, some important geologic consequences resulted from tampering with the Nile's hydrology.

The Nile, like all other large streams, is a complex natural system, and its behavior reflects a balance between discharge and sediment load. Ninety-eight percent of the Nile's load is suspended sediment. Prior to construction of the Aswan Dam, an average of 125 million metric tons of sediment passed downstream each year, but the dam reduced this value to only 2.5 million metric tons. Nearly 98 percent of the suspended sediment is now deposited in the reservoir behind the dam; within a few hundred years, the reservoir will be filled with silt and no longer be usable. Under natural conditions, this sediment was carried downstream by floodwaters, where much of it was deposited over the floodplain and delta, thus adding to the rich agricultural soils at a rate of 6 to 15 cm/century. With this natural source of nourishment eliminated, farmers must now resort to artificial fertilizers and soil additives to keep the land productive.

Before the Aswan Dam was constructed, the annual Nile flood transported at least 90 million metric tons of sediment to the Mediterranean Sea, adding it to the front of the delta. The shoreline at that time reflected a balance between sediment supply and the attack of waves and currents that redistributed the sediment along the coast. Because the annual discharge of sediment has now been cut off, the coast has become increasingly vulnerable to erosion. Over the long term, we can expect to see continuing changes to the delta as the shoreline adjusts to the reduction in Nile sediment and a new balance is reached.

What's Ahead?

We tend to be quite conscious of water that falls as rain onto the land surface, and the portion of this rain that collects and moves downhill in streams. We are less aware of the vast amounts of water beneath our feet, a huge reservoir of groundwater. Groundwater also moves, albeit at a much slower pace than streams, and it also can modify the rocks through which it flows, producing distinctive landscapes. However, its importance to humans cannot be over-looked, for in many regions it provides a large percentage of the usable water for society. Of particular concern, therefore, is the contamination of groundwater by human and industrial wastes.

In the next chapter we will examine all these aspects of groundwater and also look at its role in the rock cycle. We will learn that groundwater, like soils, is an invaluable resource that in many places is being severely depleted.

CHAPTER SUMMARY

1. Streams are part of the hydrologic cycle and the chief means by which water returns from the land to the sea. They help shape Earth's surface and transport sediment to the oceans.

2. A stream's long profile decreases in gradient downstream.

3. The discharge of a stream at any place along its course is equal to the product of its cross-sectional area and its average velocity. When a stream deepens because of an increase in discharge, its velocity also increases.

4. As discharge increases downstream, stream width and depth increase, and velocity increases slightly.

5. Streams experiencing major floods are capable of transporting large loads and moving large boulders. Exceptional floods can do a great deal of geologic work, but they have a low recurrence interval.

6. World sea level constitutes the base level for most stream systems. A local base level, such as a lake, may temporarily halt downward erosion upstream.

7. Straight channels are rare. Meandering channels form where gradients are low and the load is fine-grained. Braided patterns develop in streams with highly variable discharge and a large load to transport.

8. Stream load is the sum of bed load, suspended load, and dissolved load. Bed load is usually a small fraction of the total load of a stream. Most suspended load is derived from erosion of fine-grained regolith or from streambanks. Streams that receive large contributions of underground water commonly have higher dissolved loads than those deriving their discharge principally from surface runoff.

9. Sediment size decreases downstream because of sorting and abrasion of particles. The composition of a stream's load changes downstream as sediments of different compositions are introduced.

10. Sediment yield is influenced by rock type and structure, climate, and topography. The greatest sediment yields are recorded in mountainous terrain with steep slopes and abundant runoff, and in small basins that are transitional from grassland to desert conditions. In moist climates, vegetation anchors the surface, thereby inhibiting erosion.

11. During floods, streams overflow their banks and construct natural levees, which grade laterally into silt and clay deposited on the floodplain. Terraces result from the abandonment of a floodplain as a stream erodes downward.

12. Alluvial fans are constructed where a stream experiences a sudden decrease in gradient. The area of a fan is closely related to the size of the area upstream that supplies sediment to the fan.

13. A delta forms where a stream enters a body of standing water and loses its ability to transport sediment. The shape of a delta reflects the balance between sedimentation and erosion along the shore.

14. A drainage basin encompasses the area supplying water to the stream system that drains the basin. Its area is related to the stream's length and annual discharge.

15. Stream systems possess an inherent orderliness, with the number of stream segments increasing with decreasing stream order.

16. Drainage patterns are related to underlying rock type and structure, and often can reveal information about a stream's history.

THE LANGUAGE OF GEOLOGY

alluvial fan (p. 372)
alluvium (p. 366)

base level (p. 361)
bed load (p. 366)
braided stream (p. 364)

channel (p. 354)

delta (p. 372)
discharge (p. 354)
dissolved load (p. 366)
divide (p. 375)
drainage basin (p. 375)
floodplain (p. 371)

gradient (p. 356)

laminar flow (p. 366)
load (p. 354)
long profile (p. 356)

meander (p. 363)

natural levee (p. 371)

overland flow (p. 356)

placer (p. 367)

runoff (p. 356)

saltation (p. 366)
sheet erosion (p. 366)
stream (p. 354)
streamflow (p. 356)
suspended load (p. 366)

terrace (p. 372)
tributary (p. 357)
turbulent flow (p. 366)

QUESTIONS FOR REVIEW

1. What evidence leads us to think that streams must be a major force in shaping Earth's landscapes?

2. Describe how overland flow differs from streamflow.

3. Why does vegetation decrease the effectiveness of sheet erosion?

4. How do a stream's dimensions (depth, width) and velocity adjust in response to changes in discharge?

5. What controlling factors would have to change in order for a braided channel system to change to a meandering system?

6. Why is it that stream velocity generally increases downstream, despite a decrease in stream gradient?

7. When a large dam is built across a stream to create a reservoir upstream, the channel immediately below the dam commonly experiences erosion. Why should this happen?

8. What is meant by a "200-year flood"? What is the likely effect of such a flood on landscape evolution compared to the cumulative effect of annual floods?

9. What factors in a stream would have to change to cause fine gravel being moved as bed load to be transported as suspended load? In what way would the rate of forward movement of such particles change?

10. Why, if velocity increases downstream, does the maximum size of bed load particles typically decrease in size in that direction?

11. How does increasing urbanization affect the amount of sediment eroded from a drainage basin, and why?

12. How might internal stratification and sedimentary characteristics permit you to distinguish between a delta and an alluvial fan that are preserved in the stratigraphic record?

Click on *Presentation* and *Interactivity* in **The Water Cycle** module of your CD-ROM to further explore resources and activities presenting concepts from this chapter. Select *Assessment* in the same module to test your understanding of this chapter.

Chapter 15

Karst towers carved in thick limestone strata near Guilin, China create a distinctive jagged landscape.

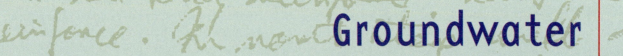

Excavation of paleolithic remains in a limestone cave in the Spanish Pyrenees.

Ice-Age Cave Dwellers in the Pyrenees

It was very cold at the culmination of the last glacial age in northern Spain, where frigid winter winds arrived after passing over the ice-choked North Atlantic Ocean.

There was little incentive to "camp" out of doors, and so our hardy ancestors sought refuge in natural caves. Along the northern foothills of the Spanish Pyrenees, as in the famous Dordogne region of southwestern France, ice-age people left behind a record of their culture in refuse piles on cave floors and in impressive wall paintings colored using natural pigments. In both regions, the caves are excavated in carbonate rocks, and display features that reflect the role of groundwater in their formation.

The study of paleolithic cave sites in western Europe is generally an interdisciplinary effort. In other words, archeologists team up with botanists, vertebrate paleontologists, geologists, geochronologists, and other specialists who contribute their expertise in reconstructing ice-age environments. In the late 1970s, I joined such a team that was exploring the archeological record in El Juyo, a 17,000-year-old cave site not far from the coastal city of Santander.

While the archeologists excavated the remains of ice-age banquets that included red deer and local shellfish, I focused on understanding the geologic factors that contributed to the cave's formation. After studying the bedrock stratigraphy and surrounding landforms, it became clear that the long, sinuous cave had been excavated by groundwater flowing along a bedding surface of a thick limestone formation. The cave lay adjacent to several sinkholes, roughly circular depressions in the landscape, that indicated local collapse of the underlying limestone. In the inner recesses of the cave, actively forming stalactites and stalagmites demonstrated the continuing role of groundwater in the cave's evolution. The cultural debris lay beneath a capping of precipitated carbonate that sealed in the archeological evidence once the cave was abandoned at the end of the ice age.

Stephen C. Porter

KEY QUESTIONS

1. **What is groundwater, and where does it come from?**

2. **What is the water table, and why is it important?**

3. **How fast does groundwater flow? What factors control this movement?**

4. **What causes springs?**

5. **What are aquifers? What are the two types of aquifers?**

6. **What factors affect the quality of groundwater used for human water supplies?**

7. **How can water supplies be safeguarded from contamination?**

8. **By what three processes does groundwater produce its geologic effects? How do these processes work?**

INTRODUCTION: THE IMPORTANCE OF WATER

Access to water, whether from streams, lakes, springs, and direct rainfall or from underground, is a vital human need. Most early cities and towns were founded close to streams that would provide a reliable source of water. As the population of these towns and cities grew, the streams often became insufficient. People then either resorted to bringing water from a more distant source through canals, or they dug wells to obtain water from underground.

As society has become increasingly more populous and industrialized, communities have generated ever-larger amounts of human and industrial wastes, a good deal of which has inevitably found its way into the very water that people must rely on for their existence. In many places water is dwindling in both quantity and quality, creating important questions for the communities involved: Will there be enough clean water to sustain future needs? Is the quality adequate for the uses to which we put this water? Is the water being used with a minimum of waste?

In this chapter, we address some of these questions. We begin by discussing general characteristics of groundwater and the water table. Next, we describe how groundwater moves and look in more detail at major sources of groundwater for human use: springs, wells, and aquifers. We then discuss the effects of human activities that deplete or contaminate groundwater. Finally, we turn to the geologic activity of groundwater, which can produce distinctive and interesting landscapes.

WATER IN THE GROUND

Groundwater is defined as all the water in the ground occupying the pore spaces within bedrock and regolith. Although the volume of groundwater sounds small, it is 40 times larger than the volume of all the water in fresh-water lakes or flowing in streams and nearly a third as large as the water contained in all the world's glaciers and polar ice.

ORIGIN OF GROUNDWATER

Less than 1 percent of the water on Earth is groundwater. Most groundwater originates as rainfall. Rainwater that soaks into the ground and moves down to the saturated zone becomes part of the groundwater system and moves slowly toward the ocean, either directly through the ground or by flowing out onto the surface and joining streams (Figure 1.17B).

That groundwater comes from rain was established on a quantitative basis in the seventeenth century, when Pierre Perrault, a French physicist, measured the mean annual rainfall for part of the drainage basin of the Seine River in eastern France and the mean annual stream runoff in the same basin area. After estimating the loss by evaporation, Perrault concluded that the difference between the amounts of rainfall and runoff was ample enough, over a period of years, to account for the amount of water in the ground.

DEPTH OF GROUNDWATER

Water is present everywhere beneath the land surface, but more than half of all groundwater, including most of what is usable, occurs above a depth of 750 m. The volume of water in this zone is estimated to be equivalent to a layer of water approximately 55 m deep spread over the world's land areas. Below a depth of about 750 m, the amount of groundwater gradually, though irregularly, diminishes. Holes drilled for oil have found water as deep as 9.4 km, and one deep experimental hole drilled on the Kola Peninsula by Russian scientists encountered water at more than 11 km. However, even though water may be present in crustal rocks at such depths, the pressure exerted by overlying rocks is so high and openings in rocks are so small that it is unlikely that much water is present.

THE WATER TABLE

Much of what we know about the occurrence of groundwater has been learned from the accumulated experience of generations of people who have dug or drilled wells. This experience tells us that a hole penetrating the ground ordinarily passes first through a layer of moist soil and then into a zone in which open spaces in regolith or bedrock are filled mainly with air (Figure 15.1). This is the **zone of aeration** (also called the *unsaturated zone*, for although water may be present, it does not saturate the ground).

The hole then enters the **saturated zone**, a zone in which all openings are filled with water. We call the upper surface of the saturated zone the **water table**. Normally, the water table slopes toward the nearest stream or lake, but in deserts, for example, it may lie far underground. Just as water is present at some depth everywhere beneath the land surface, so too is the water table always present.

In fine-grained sediment, a narrow fringe as much as 60 cm thick immediately above the water table is kept wet by capillary attraction. *Capillary attraction* is the adhesive force between a liquid and a solid that causes water to be drawn into small openings. This is the same force that

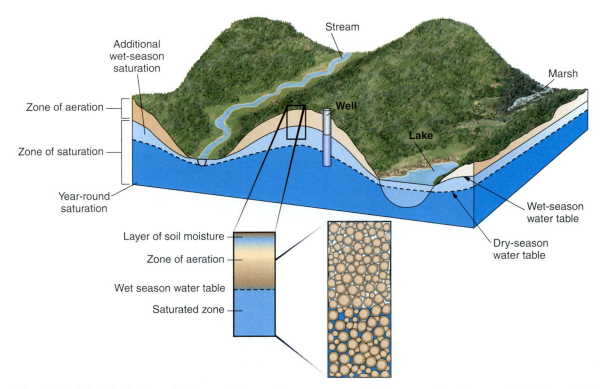

Figure 15.1 Water in the Ground In a typical groundwater system, the water table separates the zone of aeration from the saturated zone and fluctuates in level with seasonal changes in precipitation. Corresponding fluctuations are seen in the water level in wells that penetrate the water table. Lakes, marshes, and streams occur where the water table intersects the land surface. A layer of moisture coincides with the surface soil, and a thin capillary fringe lies immediately above the water table. In shape, the water table is a subdued imitation of the land surface. (The vertical scale is greatly exaggerated).

draws ink through blotting paper and kerosene through the wick of a lamp.

In humid regions, the water table is a subdued imitation of the land surface above it (Figure 15.1). It is high beneath hills and low beneath valleys because water tends to move toward low points in the topography under the influence of gravity. If all rainfall were to cease, the water table would slowly flatten and gradually approach the levels of the valleys; water seepage into the ground would diminish and then cease, and streams would dry up as the water table fell beneath valleys. In times of drought, when rain may not fall for several weeks or even months, we can sense the flattening of the water table in the drying up of wells. When a well becomes dry, we know that the water table has dropped to a level below the bottom of the well. It is repeated rainfall, dousing the ground with fresh supplies of water, that maintains the water table at a normal level.

Whatever its depth, the water table is a significant surface, for it represents the upper limit of all readily usable groundwater. For this reason, a major aim of groundwater geologists and well drillers alike is to determine the depth and shape of the water table.

Before you go on:

1. What is the depth range of most groundwater?

2. What is the zone of aeration? the saturated zone?

HOW GROUNDWATER MOVES

Groundwater operates continuously as a small but integral part of the hydrologic cycle (Figure 1.17). Water, evaporated mainly from the oceans and falling on the land as rain, seeps into the ground and enters the groundwater reservoir. Some of this slowly moving underground water reaches stream channels and contributes to the water they carry to the ocean, and the cycle continues.

Most of the groundwater within a few hundred meters of the surface is in motion. Unlike the swift flow of rivers, however, which is measurable in kilometers per hour, groundwater moves so slowly that velocities are expressed in centimeters per day or meters per year. The reason for this contrast is simple. Whereas the water of a stream flows unimpeded through an open channel, groundwater must move through small, constricted passages, often along a tortuous route. Therefore, the flow of groundwater to a large degree depends on the nature of the rock or sediment through which the water moves. More specifically, it depends on the material's porosity and permeability.

POROSITY AND PERMEABILITY

Porosity is the percentage of the total volume of a body of regolith or bedrock that consists of open spaces, called *pores*. It is porosity that determines the amount of water that a given volume of regolith or bedrock can contain.

The porosity of sediments is affected by the sizes and shapes of the rock particles, as well as by the compactness of their arrangement and by the weight of any overlying material (Figure 15.2A, B). In some well-sorted sands and

A.

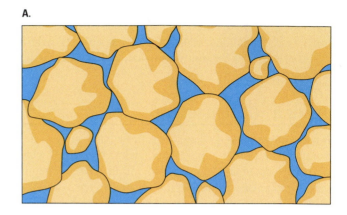

B.

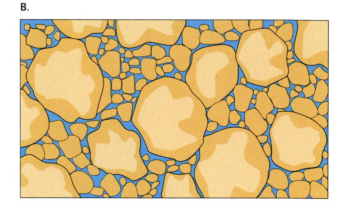

C.

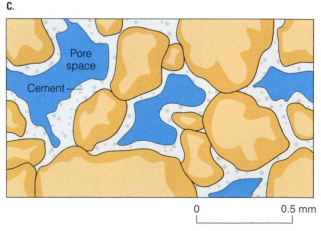

Figure 15.2 **Porosity in Sediments** Porosity in different sediments. A. A porosity of 30 percent in a reasonably well-sorted sediment. B. A porosity of 15 percent in a poorly sorted sediment in which fine particles fill spaces between larger grains. C. Reduction in porosity in an otherwise very porous sediment due to cement that binds particles together.

gravels that are not deeply buried, the porosity may reach 20 percent, while some very porous clays have porosities as high as 60 percent.

The porosity of a sedimentary rock is affected by several factors: the sizes and shapes of the rock particles, the compactness of their arrangement, the weight of any overlying rock or sediment, and the extent to which the pores become filled with the cement that holds the particles together (Figure 15.2C). The porosity of igneous and metamorphic rocks generally is low. However, if such rocks have many joints and fractures, or if a lava is vesicular, the porosity will be higher.

Permeability is a measure of how easily a solid allows fluids to pass through it. A rock of very low porosity is also likely to have low permeability. However, a high porosity does not necessarily mean a correspondingly high permeability. The sizes of pores, how well they are connected, how crooked or straight a path water must follow as it travels through porous material, all determine the permeability of a rock or sediment.

An example of a sediment with high porosity and low permeability is clay. Clay particles have diameters of less than 0.004 mm (Table 7.1), yet clay may have a very high porosity because the percentage of pore space is high. However, because the pores are very small, the permeability is low.

By contrast, in a sediment with grains at least as large as sand (grain diameters of 0.06 to 2 mm), the pores commonly are wide, and the water in the pores is free to move. Such sediment is permeable. As a general rule, then, as the diameters of the pores increase, permeability increases. Gravel, with very large pores, is more permeable than sand and can yield large volumes of water to wells.

RECHARGE AND DISCHARGE OF GROUNDWATER

We have seen that porosity and permeability exert strong control on the flow of groundwater. Next we examine features of the flow itself. In doing so, we look at the process by which groundwater is replenished, called *recharge*, and the process by which groundwater reaches and flows from the surface, called *discharge*.

RECHARGE AND DISCHARGE AREAS

An area of the landscape where precipitation seeps downward beneath the surface and reaches the saturated zone is called a **recharge area**. The water continues to move slowly toward **discharge areas**, which are areas where subsurface water is discharged to streams or to lakes, ponds, or swamps. The surface extent of recharge areas is invariably larger than that of discharge areas.

In humid regions, recharge areas encompass nearly all the landscape beyond streams and their adjacent floodplains (Figure 15.3). In more arid regions, recharge occurs mainly in mountains and in the alluvial fans that border them. In such regions, recharge also occurs along the channels of major streams that are underlain by permeable alluvium; the water leaks downward through the alluvium and recharges the groundwater (Figure 15.4).

The time water takes to move through the ground from a recharge area to the nearest discharge area depends on rates of movement and on the travel distance. It may take only a few days, or possibly thousands of years in cases where water moves through the deeper parts of a groundwater body (Figure 15.5). We look next at how that flow occurs.

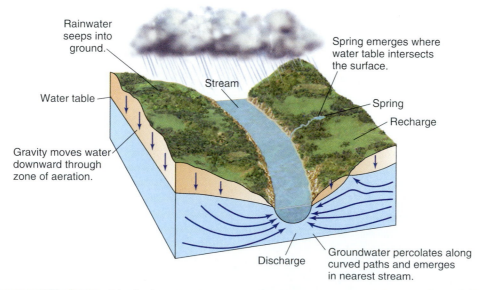

Figure 15.3 Recharge and Discharge Distribution of recharge and discharge areas in a humid landscape. The time required for groundwater to reach the discharge area from the recharge area depends on the path and distance of travel. Downward and upward percolation is faster and more direct in the most permeable pathways.

Figure 15.4 Water Loss from Streams   In arid regions, direct recharge is minimal and the water table lies below the bed of a river. During times of low flow, large throughflowing streams and intermittent streams lose water, which seeps downward to resupply groundwater in the saturated zone.

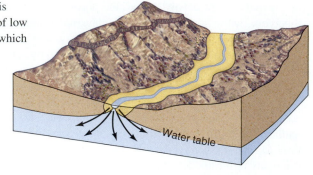

MOVEMENT IN THE ZONE OF AERATION

Recharge begins when rainfall or snowmelt enters the ground. Water initially soaks into the soil, which usually contains clay resulting from the chemical weathering of bedrock. Because of these fine clay particles, the soil is generally less permeable than underlying coarser regolith or rock. The low permeability and the fine clay particles cause part of the water to be retained in the soil by forces of molecular attraction. This is the layer of soil moisture shown in Figure 15.1. Some of this moisture evaporates directly into the air, but much of it is absorbed by the roots of plants, which later return it to the atmosphere through transpiration (Figure 1.18).

Because of the pull of gravity, water that cannot be held in the soil by molecular attraction seeps downward until it reaches the saturated zone.

MOVEMENT BY PERCOLATION IN THE SATURATED ZONE

Once in the saturated zone, groundwater moves by **percolation**, which is similar to the flow of water when a saturated sponge is squeezed gently. Percolating water moves slowly through very small pores along parallel, thread-like paths. Movement is easiest through the central parts of the spaces but diminishes to zero immediately adjacent to the sides of each space because there the water is retarded by its tendency to wet the surfaces of mineral grains.

Responding to gravity, water percolates from areas where the water table is high toward areas where it is lowest. In other words, it generally percolates toward surface streams or lakes (Figure 15.5). Much of it flows along innumerable long, curving paths that go deep through the ground. Some of the deeper paths turn upward and enter the stream or lake from beneath. This upward flow is possible because water tends to flow toward points where pressure is least. However, most of the groundwater entering a stream travels along shallow paths not far beneath the water table.

Groundwater does not move everywhere at a constant rate. We can demonstrate this by injecting colored dye at some point in a recharge area and measuring the time it takes for the dye to appear in nearby springs or wells. Experiments conducted with materials of uniform permeability have shown that the velocity of groundwater flow increases as the slope of the water table increases. In other words, the steeper the slope, the faster the water moves.

The reason that flow rates of groundwater tend to be very slow is because percolating groundwater encounters a large amount of frictional resistance. Normally, velocities range between half a meter a day and several meters a year. The highest rate yet measured in the United States, in exceptionally permeable material, was only about 250 m/yr. For a discussion of the factors that control the rate of groundwater flow, see *The Science of Geology*, Box 15.1, *How Fast Does Groundwater Flow?*

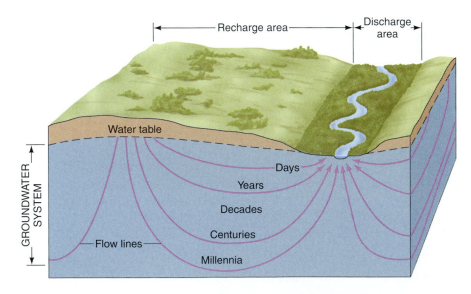

Figure 15.5 Groundwater Flow   Paths of groundwater flow in a humid region in uniformly permeable rock or sediment. Long, curved arrows represent only a few of many possible paths. Springs are located where the water table intersects the land surface.

Box 15.1 THE SCIENCE OF GEOLOGY

HOW FAST DOES GROUNDWATER FLOW?

As discussed in the text, the velocity of groundwater flow increases with the slope of the water table. This slope can be determined by measuring the difference in altitude of two points (h_1 and h_2) on the water table and dividing this figure by the horizontal distance (l) between the points (Figure B15.1). The resulting slope value is generally referred to as the **hydraulic gradient**, and the velocity of groundwater (V) is proportional to the hydraulic gradient:

$$V \propto \frac{h_1 - h_2}{l}$$

In 1856, Henri Darcy, a French engineer, concluded that the velocity of groundwater must be related not only to the slope of the water table (the hydraulic gradient), but also to the permeability of the rock or sediment through which the water is flowing. He proposed an equation in which permeability, together with the acceleration due to gravity and the density and viscosity of water, is expressed as a coefficient (K). This coefficient, referred to as the *coefficient of permeability* or the *hydraulic conductivity*, is simply a measure of the ease with which water moves through a rock or sediment. The equation Darcy proposed can be expressed as:

$$V = \frac{K(h_1 - h_2)}{l}$$

In Chapter 14 we learned that discharge (Q) in streams varies as a function of both stream velocity (V) and cross-sectional area (A). The discharge of groundwater through a rock or sediment also depends on the velocity of flow and cross-sectional area of flow. In this case, however, the cross-sectional area is not that of an open channel but

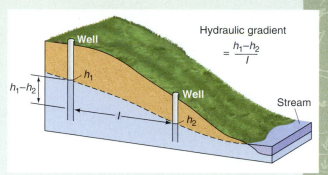

Hydraulic gradient
$$= \frac{h_1 - h_2}{l}$$

Figure B15.1 Hydraulic Gradient The hydraulic gradient is the slope of the water table. In the example illustrated, the hydraulic gradient is found by subtracting the altitude of h_2, the level of water in the downslope well, from h_1, the level of the water in the upslope well, and dividing the difference by the distance (l) between the two wells.

rather that of an interconnected system of pores. We can express discharge by using an equation we learned in Chapter 14: $Q = AV$. If we substitute the value for V given in the first equation into this equation from Chapter 14, we arrive at a new equation,

$$Q = \frac{AK(h_1 - h_2)}{l}$$

This relationship is known as **Darcy's Law**. If we take the cross-sectional area (A) as constant for any given situation, by measuring any two of the remaining three variables [discharge (Q), coefficient of permeability (K), and hydraulic gradient ($h_1 - h_2$)/l], we can calculate the value of the third.

SPRINGS AND WELLS

People generally obtain supplies of groundwater either from springs or by excavating wells that reach the saturated zone underground. In this section, we examine characteristics of each.

SPRINGS

A **spring** is a flow of groundwater emerging naturally at the ground surface. The simplest kind of spring is one that issues from a place where the land surface intersects the water table (Figure 15.3). Small springs are found in all kinds of rocks, but almost all large springs issue from lava flows, limestone, or gravel.

A vertical or horizontal change in permeability is a common explanation for the location of springs. Often this

change involves the presence of an **aquiclude**, a body of impermeable or distinctly less permeable rock adjacent to a permeable one (Figure 15.6A). If sand overlies a relatively impermeable clay aquiclude, water percolating down through the sand will be forced to flow laterally when it encounters the clay. It then will emerge as a spring where the stratigraphic boundary between the sand and the aquiclude intersects the land surface, as along the side of a valley or a coastal cliff (Figure 15.6B). Springs may also issue from lava flows, especially where a jointed lava flow overlies an aquiclude (Figure 15.6C), or along the trace of a fault (Figure 15.6D).

WELLS

A well will supply water if it intersects the water table (Figure 15.1). Figure 15.7 shows that a shallow well can

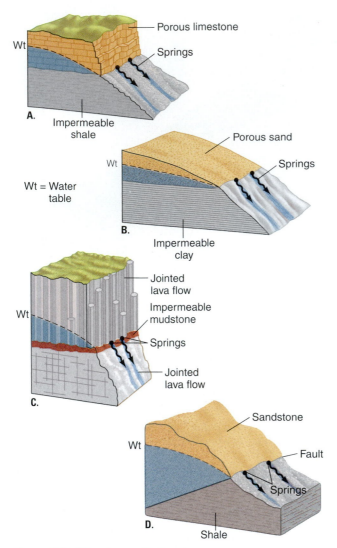

Figure 15.6 What Causes Springs? Examples of springs formed in different geologic conditions. A. Two springs discharge water at the contact between a porous limestone and an underlying impermeable shale. B. Springs lie at the contact between a porous sandy unit and an underlying impermeable clay. C. Springs issue along the contact between a highly jointed lava flow and an underlying impermeable mudstone. D. Springs issue along the trace of a fault where it intersects the land surface.

become dry at times when the water table is low, whereas a nearby deeper well may yield water throughout the year.

When water is pumped from a new well, the rate of withdrawal initially exceeds the rate of local groundwater flow. This imbalance in flow rates creates a conical depression in the water table immediately surrounding the well called a **cone of depression** (Figure 15.7). The locally steepened slope of the water table increases the flow of water to the well, consistent with Darcy's Law (see Box 15.1). Once the rate of inflow balances the rate of withdrawal, the hydraulic gradient stabilizes, but it will change

if either the rate of pumping or the rate of recharge changes. In most small domestic wells the cone of depression is hardly discernible. Wells pumped for irrigation and industrial uses, however, withdraw so much water that the cone can become very wide and deep and can lower the water table in all wells of a district.

If the source of a groundwater supply is rock or sediment that varies in its porosity or permeability, water yields from wells may vary considerably within short distances. For example, igneous and metamorphic rocks generally are not very permeable because the spaces between their mineral grains are extremely small and constricted. On the other hand, many massive igneous and metamorphic rock bodies contain numerous fissures, joints, and other openings that permit free circulation of groundwater. A well reaching such a conduit may produce water, whereas one nearby that does not intersect any openings may be dry (Figure 15.8A).

Isolated or discontinuous bodies of permeable and impermeable rock or sediment can result in very different water yields from wells. In the example shown in Figure 15.8B, an impermeable layer of clayey sediment in the zone of aeration produces a *perched water body* (a water body perched atop an aquiclude that lies above the main water table). The impermeable layer catches and holds the water reaching it from above.

AQUIFERS

When people wish to find a reliable supply of groundwater, they search for an **aquifer** (Latin for "water carrier"), which is a body of highly permeable rock or regolith that can store water and yield sufficient quantities to supply wells. Bodies of gravel and sand generally are good aquifers, for they tend to be highly permeable and often have large dimensions. Many sandstones are also good aquifers. Aquifers are of two types: confined and unconfined. Confined aquifers may give rise to artesian systems.

UNCONFINED AND CONFINED AQUIFERS

An aquifer may be bounded by aquicludes, in which case it is a **confined aquifer**. An aquifer that is not overlain and confined by an aquiclude is called an **unconfined aquifer**.

An example of an unconfined aquifer is the High Plains aquifer, which lies at shallow depths beneath the High Plains of the United States (Figure 15.9). The aquifer, which averages about 65 m thick and is tapped by about 170,000 wells, consists of a number of sandy and gravelly rock units of late Tertiary and Quaternary age. The water table at the top of this aquifer slopes gently from west to east, and water flows through the aquifer at an average rate of about 30 cm/day. Recharge comes directly from precipitation and seepage from streams.

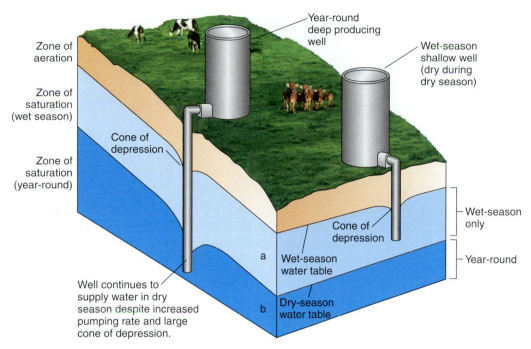

Zone of aeration

Zone of saturation (wet season)

Zone of saturation (year-round)

Year-round deep producing well

Wet-season shallow well (dry during dry season)

Cone of depression

Cone of depression

Wet-season only

Year-round

a

Wet-season water table

b

Dry-season water table

Well continues to supply water in dry season despite increased pumping rate and large cone of depression.

Figure 15.7 Changes in the Water Table Effect of seasonal changes in precipitation on the position of the water table. During the wet season, recharge is high and the water table is high, so water is present both in a shallow well and in a deeper well upslope. During the dry season, the water table falls, the hydraulic gradient decreases, and the shallow well is dry. The deeper well continues to supply water, but increased pumping during the dry season enlarges the cone of depression.

About 30 percent of the groundwater used for irrigation in the United States is obtained from the High Plains aquifer. The use of groundwater for irrigation in the region was spurred by severe droughts in the 1930s and the 1950s. Today, annual recharge of the High Plains aquifer from precipitation is much less than the amount of water being withdrawn; the inevitable result is a long-term fall in the level of the water table. In parts of Kansas, New Mexico, and Texas, the water table has dropped so much over the past half century that the thickness of the saturated zone has declined by more than 50 percent. The resulting decreased water yield and increased pumping costs have led to major concern about the future of irrigated farming on the High Plains.

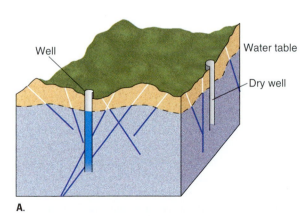

Well

Water table

Dry well

A.

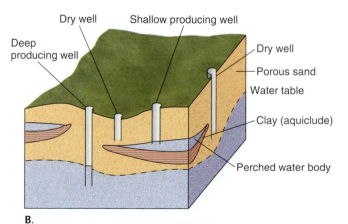

Dry well

Shallow producing well

Deep producing well

Dry well

Porous sand

Water table

Clay (aquiclude)

Perched water body

B.

Figure 15.8 Successful and Unsuccessful Wells Yield to wells from nonhomogeneous rocks can be highly variable. A. Wells that penetrate fractures in metamorphic and igneous rocks produce water. Dry wells result if no water-bearing fractures are encountered. B. Perched water bodies above the main water table are held up by aquicludes and provide shallow sources of groundwater. Wells that miss the perched water body and do not reach the deeper water table are dry.

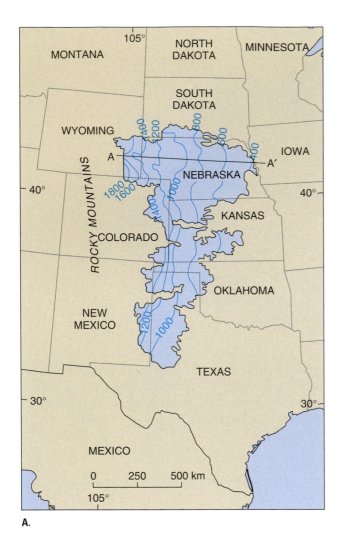

A.

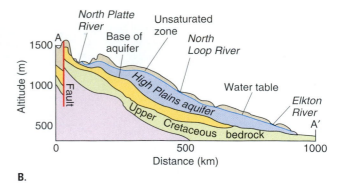

B.

Figure 15.9 High Plains Aquifer The High Plains aquifer, an example of an unconfined aquifer. A. Regional extent of the aquifer and contours (in meters) on the water table. Water flow is generally east, perpendicular to the contour lines. B. Cross section along profile A–A′ showing the slope of the water table and the relation of the High Plains aquifer to underlying bedrock units.

rocks around which permeable strata and bounding aquicludes form the land surface. Rain falling on the permeable units where they reach the surface recharges the groundwater, which flows down the inclined rock layers toward the east.

ARTESIAN SYSTEMS

Water that percolates into a confined aquifer flows downward under the pull of gravity. As it flows to greater depths, the water is subjected to increasing hydrostatic pressure. If a well is drilled to the aquifer, the difference in pressure between the water table in the recharge area and the level of the well intake will cause water to rise in the well. Potentially, the water could rise to the same height as

The Dakota aquifer system in South Dakota provides a good example of a confined aquifer (Figure 15.10). It lies to the east of the Black Hills, an elongate dome of uplifted

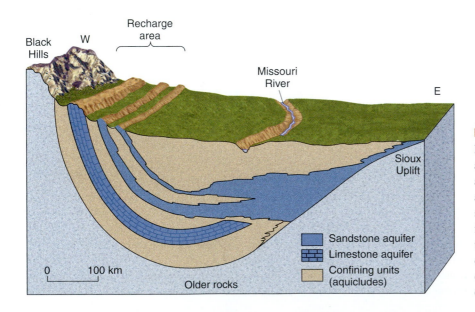

Figure 15.10 Dakota Aquifer The Dakota aquifer system in South Dakota, an example of a confined aquifer. The aquifer units consist of porous sandstones and a permeable limestone that reach the surface in the recharge area where they form linear ridges lying beyond the central core of the Black Hills. Intervening confining layers are aquicludes. In this diagram, the vertical scale is greatly exaggerated.

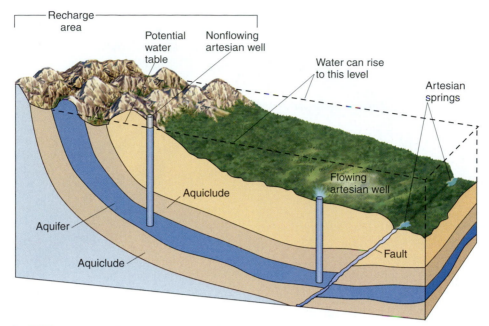

Figure 15.11 Artesian Wells Two conditions are necessary for an artesian system: a confined aquifer and sufficient water pressure in the aquifer to make the water in a well rise above the aquifer. The water in a well drilled into the aquifer rises to the height of the water table in the recharge zone, indicated by the dashed line. If this height is above the ground surface, the water will flow out of the well without being pumped.

the water table in the recharge area. If the top of the well is lower in altitude than the recharge area, the water will flow out of the well without pumping.

Such an aquifer is called an **artesian aquifer**, and the well is called an *artesian well* (Figure 15.11). Similarly, a freely flowing spring supplied by an artesian aquifer is an *artesian spring*. The term *artesian* comes from a French town, Artois (called Artesium by the Romans), where artesian flow was first studied.

Under unusually favorable conditions, artesian water pressure can be great enough to create fountains that rise as much as 60 m above ground level. Such wells and springs are naturally attractive, for the costs of pumping are avoided as long as the amount of recharge to the system is sufficient to maintain the necessary water pressure.

THE FLORIDAN AQUIFER

A vertical sequence of highly permeable carbonate rocks along the peninsula of Florida provides an example of a complex regional aquifer system in which both confined and unconfined units are present and in which water locally reaches the surface by artesian flow (Figure 15.12). The aquifer system is restricted mainly to middle and late Tertiary limestones. In the upper part of these are numerous caves and smaller openings that have been dissolved in the rock. The permeable beds are interconnected to varying degrees, and their permeability is at least 10 times greater than that of strata above and below.

The Floridan aquifer gives rise to numerous springs. Their concentration and discharge probably exceed those of any country in the world. Most of the major springs are artesian and occur where the overlying impermeable beds have been breached and the water pressure is high enough to allow water to rise above the land surface.

The age of groundwater in the Floridan aquifer system has been determined by radiocarbon dating of carbonate molecules dissolved in the water. Most of the radiocarbon enters the ground with rain falling on the recharge area and moves through the aquifer with the groundwater. The age of the water was found to increase with increasing distance from the recharge area. Water in the well farthest from the recharge area is calculated to have been in the ground for at least 19,000 years.

Before you go on:

1. What is porosity? What factors affect the porosity of a mass of rock or regolith?

2. What is the relation between porosity and permeability?

3. How does recharge occur? Where are recharge areas commonly found in humid areas? in arid regions?

4. How is groundwater movement through the zone of aeration different from movement through the saturated zone?

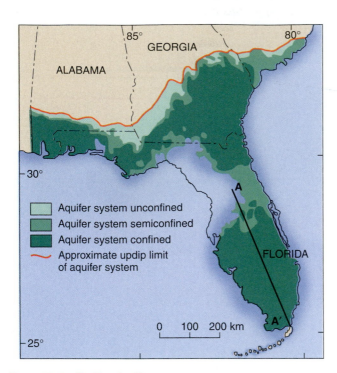

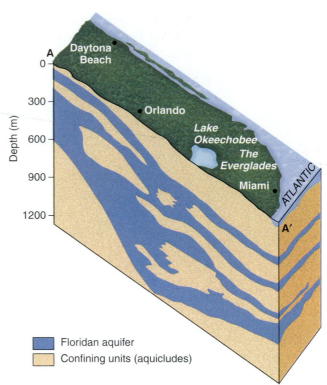

Figure 15.12 Floridan Aquifer The Floridan aquifer system. The map shows the distribution of the system and areas where it is unconfined, semiconfined (overlying confining unit is less than 30 m thick), and confined (overlying confining unit is more than 30 m thick). The cross section along line A–A′ shows the relation of aquifer to overlying, underlying, and intervening aquicludes.

MINING GROUNDWATER, AND ITS CONSEQUENCES

In the dry regions of western North America, where streams are few and average discharge is low, groundwater is a major source of water for human consumption. In many of these dry regions, withdrawal exceeds natural recharge. Thus, the volume of stored water is steadily diminishing. In the same way that petroleum is being steadily withdrawn from the most accessible oil pools and minerals are being mined from the most accessible rocks of the upper crust, groundwater also is being mined from the best aquifers. We regard fossil fuels and minerals as nonrenewable resources, for they form only over geologically long intervals of time. We don't often stop to think that groundwater can also be a nonrenewable resource. In some regions, natural recharge would take so long to replenish a depleted aquifer that formerly vast underground water supplies have essentially been lost to future generations. Even where the problem has been recognized and measures have been taken to stem the loss, centuries or millennia of natural recharge will be required to return aquifers to their original state.

Next we will examine some of the potentially adverse effects of this depletion of groundwater, and measures that can reduce them.

LOWERING OF THE WATER TABLE

As we have seen in the case of the High Plains aquifer, when groundwater withdrawal exceeds recharge, the water table falls. This fall in level can lead to the drying up of springs and streams if the water table no longer intersects the land surface. It can also cause shallow wells to run dry and necessitate the drilling of still deeper wells. Under these conditions, the groundwater reservoir is steadily depleted while the cost of pumping water from an ever-deepening water table continues to increase.

To halt the fall of the water table, groundwater sometimes can be artificially recharged. Artificial recharge might involve, for example, spraying biodegradable liquid wastes from a food-processing or sewage-treatment plant over the land surface. The pollutants are removed by biologic processes as the liquid percolates downward through the soil, and the purified water then recharges the groundwater system. Runoff from rainstorms in urban areas can be channeled and collected in basins where it will seep into permeable strata below, raising the water table. In addition, groundwater withdrawn for nonpolluting industrial use may be pumped back into the ground through injection wells, thereby recharging the saturated zone.

SUBSIDENCE OF THE LAND SURFACE

The water pressure in the pores of an aquifer helps support the weight of the overlying rocks or sediments. When groundwater is withdrawn, the pressure is reduced, and the particles of the aquifer shift and settle slightly. As a result, the land surface subsides. The amount of subsidence depends on how much the water pressure is reduced and on the thickness and compressibility of the aquifer. Such land subsidence is widespread in the southwestern United States, where withdrawal of groundwater has caused disruptions of the land surface (Figure 15.13); structural damage to buildings, roads, and bridges; damage to buried cables, pipes, and drains; and an increase in areas subject to flooding.

Land subsidence can be especially damaging where water is pumped from beneath cities. A well-known example is Mexico City, built on the site of the ancient Aztec capital of Tenochtitlán, which lay in the middle of a shallow lake. As groundwater was exploited, the porous lake sediments slowly compressed, and many buildings began to shift and tilt as the land subsided. Another example is Pisa's famous Leaning Tower. The tower was built on unstable fine-grained floodplain sediments and began to tilt when construction began in 1174 (Figure 15.14). The tilting increased rapidly during the present century as groundwater was withdrawn from deep aquifers. Recent strengthening of the foundation is designed to keep the tower stable in the future, but it will do so only if groundwater withdrawal is strictly controlled.

Before you go on:

1. What is an aquiclude? How is it related to the location of springs?

2. Why is it possible for water yields from wells to vary a good deal within short distances?

3. How can confined aquifers give rise to artesian systems?

Figure 15.13 Subsidence Fissure located near Chandler Heights, Arizona, caused by subsidence of the ground due to removal of large quantities of underground water.

Figure 15.14 Leaning Tower Tilting of the Leaning Tower of Pisa, Italy accelerated as groundwater was withdrawn from aquifers to supply the growing city.

WATER QUALITY AND GROUNDWATER CONTAMINATION

Citizens of modern industrialized nations take it for granted that when they turn on a faucet, water that is safe and drinkable will flow from the tap. Throughout much of the world, however, drinking water is barely adequate for human consumption. Not only do natural dissolved substances make some water unpalatable, but many water supplies have become severely contaminated by human and industrial waste products. We'll look first at substances occurring naturally in groundwater, and then at various sources of contamination.

CHEMISTRY OF GROUNDWATER

Analyses of many wells and springs show that the compounds dissolved in groundwater are mainly chlorides, sulfates, and bicarbonates of calcium, magnesium, sodium, potassium, and iron. We can trace these substances to the common minerals in the rocks from which they were weathered.

As might be expected, the composition of groundwater varies from place to place according to the kind of rock in which the water occurs. In much of the central United States, for instance, the water is rich in calcium and magnesium bicarbonates dissolved from local carbonate bedrock. Taking a bath in such water, termed *hard water*, can be frustrating because soap does not lather easily and a crust-like ring forms in the tub. Hard water also leads to deposition of scaly crusts in water pipes, eventually restricting water flow. By contrast, water that contains little dissolved matter and no appreciable calcium is called *soft water*. Such water is found, for example, in parts of the northwestern United States where volcanic rocks and graywacke sandstones are common. With soft water, we can easily get a nice soapy lather in the shower.

Groundwater can dissolve noxious elements from rocks it flows through, making the water unsuitable for consumption. Water circulating through sulfur-rich rocks may contain dissolved hydrogen sulfide (H_2S) that, though harmless to drink, has the disagreeable odor of rotten eggs. In some arid regions, the concentration of dissolved sulfates and chlorides is so great that the groundwater is unusually noxious. In very dry regions, groundwater moving through porous sedimentary rocks will dissolve salts that then are deposited as the water evaporates in the zone of aeration. The resulting saline soils are thereby rendered unsuitable for agriculture (see *Understanding Our Environment*, Box 15.2, *Toxic Groundwater in the San Joaquin Valley*).

POLLUTION BY SEWAGE

The most common source of water pollution in wells and springs is sewage. Drainage from septic tanks, broken sewers, privies, and barnyards contaminates groundwater. If water contaminated with sewage bacteria passes through sediment or rock with large pores, such as coarse gravel or cavernous limestone, it can travel long distances and remain polluted (Figure 15.15).

If the contaminated water percolates through sand or permeable sandstone, however, it can become purified within short distances, in some cases less than about 30 m from where the pollution occurred (Figure 15.15). Sand is an especially suitable cleansing agent because it promotes purification by (1) mechanically filtering out bacteria (water gets through, but most of the bacteria do not); (2) oxidizing bacteria so they are rendered harmless; and (3) placing bacteria in contact with other organisms that consume them. For this reason, purification plants that treat municipal water supplies and sewage percolate these fluids through sand.

CONTAMINATION BY SEAWATER

Along coasts, fresh groundwater is separated from seawater by a thin transition zone of brackish water (Figure 15.16A). Any pumping from an aquifer near the coast will reduce the flow of fresh groundwater toward the sea. This reduced fresh-water flow may then allow saltwater to move landward

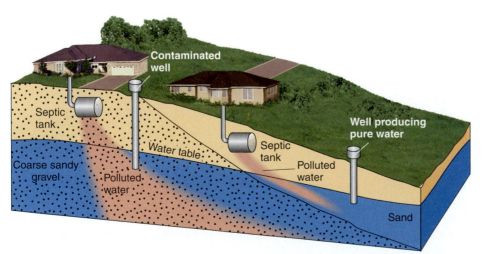

Figure 15.15 Contaminated Groundwater Purification of groundwater contaminated by sewage. Pollutants percolating through a highly permeable sandy gravel contaminate the groundwater and enter a well downslope from the source of contamination. Similar pollutants moving through permeable fine sand higher in the stratigraphic section are removed after traveling a relatively short distance and do not reach a well downslope.

A.

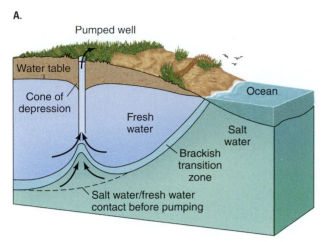

Figure 15.16 Contamination by Seawater Seawater contaminates a pumping well in a coastal area. A. Near the coast, a body of fresh groundwater overlies salty marine water. If pumping is not excessive, the well will draw only fresh water from

the aquifer. B. Heavy pumping of groundwater forms a pronounced cone of depression both at the top and at the base of the groundwater body, eventually permitting salty water to enter and contaminate the well.

through permeable strata. Excessive pumping that exceeds the natural flow of fresh groundwater toward the sea may eventually permit saline water to encroach far inland and reach major pumping centers. Such seawater intrusion can then contaminate water supplies (Figure 15.16B). Once intrusion occurs, it is very difficult to reverse.

TOXIC WASTES AND AGRICULTURAL POISONS

Vast quantities of human garbage and industrial wastes are deposited each year in open basins or excavations at the

land surface. When such a landfill site reaches its capacity, it generally is covered with dirt and then revegetated.

Many of the waste products, now underground, are mobilized when rainwater seeps downward through the site and carries away soluble substances. In this way, harmful chemicals slowly leach into groundwater reservoirs and contaminate them, making them unfit for human use. The pollutants travel from landfill sites as plumes of contaminated water in directions that depend on the regional groundwater flow pattern and often are dispersed at the same rates as the percolating water (Figure 15.17). The pollutants often are toxic to humans as well as to plants and animals.

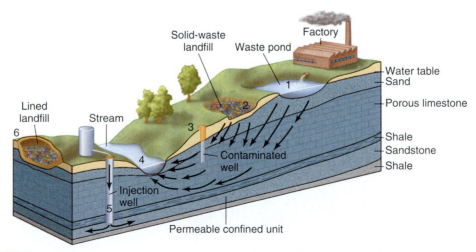

Figure 15.17 Toxic Wastes A groundwater system contaminated by toxic wastes. Toxic chemicals in an open waste pond (1) and an unlined landfill (2) percolate downward and contaminate an underlying aquifer. Also contaminated are a well downslope (3) and a stream (4) at the base of the hill. Safer, alternative approaches to waste management include injection into a

deep confined rock unit (5) that lies well below aquifers used for water supplies, and a carefully engineered surface landfill (6) that is fully lined to prevent downward seepage of wastes. Because neither of the latter approaches is completely foolproof, constant monitoring at both sites would be required.

UNDERSTANDING OUR ENVIRONMENT

TOXIC GROUNDWATER IN THE SAN JOAQUIN VALLEY

The only time most people ever hear of selenium is in their high school chemistry course when they learn that it is thirty-fourth in the periodic list of the elements. Selenium is a naturally occurring element and a necessary trace nutrient in our diet, as well as in the diet of livestock and many wild animals. However, in excessive amounts, selenium can be toxic.

In 1983, selenium attracted national attention when the U.S. Fish and Wildlife Service reported fish kills and high incidences of mortality, birth defects, and decreased hatching rates in nesting waterfowl at the Kesterson National Wildlife Refuge in California's San Joaquin Valley (Figure B15.2A, B). Laboratory studies showed high concentrations of selenium in fish from Kesterson Reservoir in the wildlife refuge. Birds using the reservoir were found to have high concentrations of selenium and obvious symptoms of selenium poisoning. Federal agencies quickly began investigations to learn how these toxic levels of selenium were entering the ecological system.

Geologists showed that the poisoning of wildlife at Kesterson resulted from a combination of geologic, hydrologic, and agricultural factors. The western San Joaquin

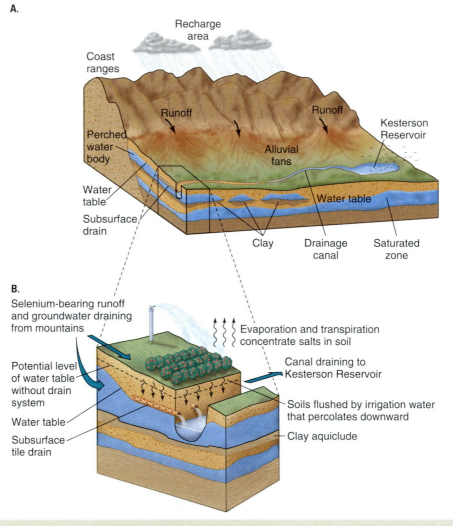

Figure B15.2 Contamination of a Reservoir Geologic setting of Kesterson Reservoir in western San Joaquin Valley, California. Runoff carries dissolved selenium to alluvial fans where irrigation water flushes it into a drainage system that concentrates it in the Kesterson Reservoir. Subsurface drains lower the water table, which otherwise would lie at shallow depths because the groundwater is perched above a clay aquiclude.

Valley is a prime agricultural area, but because the climate is arid, the land is irrigated. The irrigation artificially raised the water table to such a degree that a system of subsurface drains was established to remove excess water. The drainage system carries subsurface water to a surface canal that funnels the water northward along the valley to Kesterson Reservoir.

A natural source of selenium lies in the Coast Ranges immediately to the west of the San Joaquin Valley. Rainwater falling in these mountains dissolves selenium-bearing salts from marine sedimentary strata. Surface runoff then carries the dissolved salts to broad alluvial fans in the valley where the water seeps into the ground and recharges a shallow regional aquifer.

Evaporation in this arid region, where an annual rainfall of less than 250 mm is greatly exceeded by an annual evaporation rate of about 2300 mm, concentrates the salts in the soil. An irrigation system on the fan surfaces is designed to supply water to crops and also to flush salts out of the soil and into the drainage canal that leads toward Kesterson Reservoir. Because the reservoir has no outlet, the selenium is concentrated there and now has reached toxic levels.

Groundwater conditions in the western San Joaquin Valley contribute to the Kesterson Reservoir problem. Most of the shallow groundwater in this area is alkaline and slightly to highly saline. Under these conditions, selenium is very soluble and is carried with the flowing groundwater. This mobility greatly increases the ease with which selenium moves from the soils into the artificial drainage system.

A further factor is the presence of a clay layer 3–23 m beneath the land surface. Being impermeable, the clay layer restricts the downward percolation of selenium-bearing groundwater and produces a perched water body. It is this perched water body above the clay aquiclude that necessitates the drain system, which in turn creates the environmental hazard at Kesterson Reservoir.

Although we are constantly alerted to the environmental impact of pesticides and other manufactured poisons that are introduced into natural ecosystems, the Kesterson saga illustrates how natural substances that pose no special hazard under normal conditions can reach toxic levels through human intervention. The solution to such problems, which are increasing in number as an expanding human population places greater demands on limited natural resources, rests on an understanding of the complex interrelationship of the factors involved. In seeking these solutions, geologists have an increasingly important role to play.

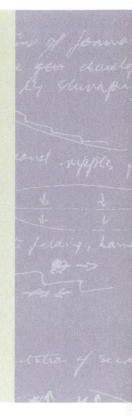

In the United States, the pollution problems associated with landfill wastes have become so severe that the government has begun a major long-term program to clean up such sites and render them environmentally safe. However, the identified sites number in the tens of thousands, and it is difficult to judge how much time and money will be required to accomplish this formidable task.

In addition, each year pesticides and herbicides are sprayed over agricultural fields and suburban gardens to help improve quality and productivity. Some of these chemicals have been linked with cancers and birth defects in humans, and some have led to disastrous population declines of wild animals. For example, a dramatic drop in the bald eagle population in the United States has been linked to the introduction of pesticides (primarily DDT) into the natural food chain. Because of the manner in which they are spread, such toxic chemicals invade the groundwater over wide areas as precipitation flushes them into the soil.

UNDERGROUND STORAGE OF HAZARDOUS WASTES

One of the leading environmental concerns of industrialized countries is the necessity of dealing with highly toxic industrial wastes. Experience has demonstrated that surface dumping quickly leads to contamination of surface and subsurface water supplies and thereby to the possibility of serious and potentially fatal health problems (Figure 15.18). In addition, countries with nuclear capacity have the special problem of disposing of high-level radioactive waste products. Some of the isotopes involved (e.g., ^{90}Sr and ^{137}Ce) are so highly radioactive that even minute quantities can prove fatal to people if released to the surface environment.

Most studies concerning disposal of hazardous wastes—both toxic and radioactive—have concluded that underground storage is appropriate, provided safe sites can be found. In the case of high-level nuclear wastes, which can remain dangerous for tens or hundreds of thousands of years because of the long half-lives of some of the radioactive isotopes, a primary requirement is that a site will be stable over a very long time interval. Therefore, the only completely safe sites for disposing of radioactive wastes and their containers are those that will not be affected chemically by groundwater, physically by earthquakes or other disruptive events, or accidentally by people.

The placement of hazardous wastes underground, even far underground, immediately raises concerns about groundwater. Water is a nearly universal solvent, and the weakly acidic character of most groundwater means that any container of toxic or radioactive substances eventually is likely to corrode, so that the contents will then dissolve and be transported away from the storage site. Water is

Figure 15.18 Wastes from a Landfill Pollutants leak from rusting containers and soak into the groundwater system beneath this site near the town of Kotzebue, Alaska.

present in crustal rocks to depths of many kilometers, and in many of these rocks it is circulating at rates of 1 to 50 m/yr. Over tens or hundreds of thousands of years, even such slow rates can move dissolved substances over great distances and introduce them to more rapidly flowing parts of the hydrologic system.

Geologists generally agree that the ideal underground storage site for radioactive wastes should possess the following characteristics:

- The enclosing rock should have few fractures and low permeability.

- The enclosing rock should have no present or future economic mineral potential.

- Local groundwater flow should be away from plant and animal life.

- Only very long paths of groundwater flow should be directed toward places accessible to humans.

- The area should have low rainfall.

- The zone of aeration should be thick.

- The rate of erosion should be very low.

- The probability of earthquakes or volcanic activity should be very low.

- Future change of climate in the region should be unlikely to affect groundwater conditions substantially.

The safe long-term storage of toxic and nuclear wastes at underground sites provides a major challenge for geolo-

gists. Historically, geologists have studied past events, but now they are being asked to predict possible future events. To do so with any confidence requires considerable knowledge of local and regional groundwater conditions. It also demands a much better understanding of how complex groundwater systems might respond to crustal movements, local and global climatic change, and other natural factors that can affect the stability of a storage site.

Before you go on:

1. What naturally occurring dissolved compounds are most common in groundwater? Where do they originate?

2. What is the most common source of water pollution in wells and springs?

3. How can pumping from an aquifer near a coastline lead to contamination of groundwater by seawater?

4. What characteristics are important in selecting underground storage sites for hazardous wastes?

GEOLOGIC ACTIVITY OF GROUNDWATER

In regions underlain by rocks that are highly susceptible to chemical weathering, groundwater creates distinctive landscapes that are among the most interesting and picturesque on our planet. It does this through the processes of dissolution, cementation, and replacement. Among the results are caves, with their characteristic deposits; sinkholes; and karst topography.

DISSOLUTION

As soon as rainwater reaches Earth's surface, it begins to react with minerals in regolith and bedrock and weathers them chemically. An important part of chemical weathering occurs when minerals and rock materials pass directly into solution through dissolution and hydrolysis.

Of all the rocks in the Earth's crust, the carbonate rocks are among the most readily attacked by this process (Figure 15.19). Limestone, dolostone, and marble are the most common carbonate rocks and underlie millions of square kilometers of the Earth's surface. Although carbonate minerals are nearly insoluble in pure water, they are readily dissolved by weak carbonic acid (HCO_3^{1-}),

formed by the interaction of carbon dioxide and water vapor in the atmosphere (Table 6.1, reaction 1):

$$H_2O + CO_2 \rightleftarrows H_2CO_3 \rightleftarrows H^{1+} + HCO_3^{1-}$$

As a result of downward-percolating rainwater, the groundwater becomes charged with calcium cations and bicarbonate anions (Table 6.1, reaction 6).

$$CaCO_3 + H_2CO_3 \rightarrow Ca^{2+} + 2(HCO_3)^{1-}$$

The weathering attack occurs mainly along joints and other partings in the carbonate bedrock. The result is impressive. When limestone weathers, nearly all its volume can be dissolved away in slowly moving groundwater.

By measuring over a period of time the amount of dissolution observed on small, precisely weighed limestone tablets placed at open sites in various areas, geologists have obtained estimates of the average rate at which limestone landscapes are being lowered by dissolution. In tem-

Figure 15.19 Dissolving Marble A marble balustrade of the Forbidden City in Beijing, China, shows the effects of more than 300 years of dissolution. The original sharply carved design has become smooth and indistinct as chemical weathering, enhanced by acid rainfall, has dissolved the stone.

perate regions with high rainfall, a high water table, and a nearly continuous cover of vegetation, carbonate landscapes are being lowered at average rates of up to 10 cm/1000 years. In dry regions with scanty rainfall, low water tables, and discontinuous vegetation, rates are far lower. Measured rates of dissolution by groundwater in carbonate terrains of the United States show that the dissolution rate can exceed the average erosional reduction of the surface by mass wasting, sheet erosion, and streams.

CHEMICAL CEMENTATION AND REPLACEMENT

The conversion of sediment into sedimentary rock is primarily the work of groundwater. A body of sediment lying beneath the sea is generally saturated with water, as is sediment lying in the saturated zone beneath the land. Substances in solution in the water are precipitated as cement in the spaces between rock and mineral particles of the sediment. This process transforms the loose sediment into firm rock. Calcite, quartz, and iron compounds (mainly hydroxides such as limonite) are, in that order, the chief cementing substances.

Less common than the deposition of cement between the grains of a sediment is **replacement**, the process by which a fluid dissolves matter already present and at the same time deposits from solution an equal volume of a different substance. Evidently, replacement takes place on an approximately volume-for-volume basis, because the new material preserves the most minute textures of the material replaced. Both mineral and organic substances can be replaced. In *petrified wood*, a common example of replacement, the basic structure of the wood is retained even though the original organic matter is gone (Figure 15.20).

CARBONATE CAVES AND CAVERNS

People have long been interested in caves. The earliest evidence we have of human dwellings comes from limestone caves in Europe and Asia that provided shelter for paleolithic peoples during the Pleistocene glacial ages. The walls of these caves were the rocky canvases of prehistoric artists, whose polychrome paintings provide us with superb renditions of the prey of ice-age big-game hunters.

Caves come in many sizes and shapes. Although most caves are small, some are of exceptional size. A very large cave or system of interconnected cave chambers is often called a *cavern*. The Carlsbad Caverns in southeastern New Mexico include one chamber 1200 m long, 190 m wide, and 100 m high. Mammoth Cave, in Kentucky, consists of interconnected caverns with an aggregate length of at least 48 km. The recently discovered Good Luck Cave on the tropical island of Borneo has one chamber so large that it could accommodate not only the world's largest previously known chamber (in Carlsbad Caverns), but also the largest chamber in Europe (in Gouffre St. Pierre Martin, France) and the largest chamber in Britain (Gaping Ghyll).

Figure 15.20 Petrified Wood Logs of petrified wood weather out of mudstone layers in Petrified Forest National Park, Arizona.

Cave formation is mainly a chemical process involving the dissolution of carbonate rock by circulating groundwater. The usual sequence of development is thought to involve these steps:

1. Initial dissolution along a system of interconnected open joints and bedding planes by percolating groundwater.

2. Enlargement of a cave passage along the most favorable flow route by water that fully occupies the opening.

3. Deposition of carbonate formations on the cave walls while a stream occupies the cave floor.

4. Continued deposition of carbonate on the walls and floor of the cave after the stream has stopped flowing.

Although geologists have argued for years as to whether caves form in the zone of aeration or in the saturated zone, available evidence favors the idea that most caves are excavated in the shallowest part of the saturated zone, along a seasonally fluctuating water table.

The rate of cave formation is related to the rate of dissolution. Where the water is acidic, the rate of dissolution increases with increasing flow velocity. Therefore, as a passage increases in size and the flow changes from very slow laminar flow to more rapid turbulent flow, the rate of dissolution will rise. The development of a continuous passage by slowly percolating waters has been estimated to take up to 10,000 years, while the further enlargement of the passage by more rapidly flowing water into a fully developed cave system may take an additional 10,000 to 1 million years.

Although limestone caves are generally believed to involve dissolution by carbonic acid, chemical evidence suggests that at least some caves, including Carlsbad Caverns, may involve dissolution by sulfuric acid. The proposed agent is hydrogen-sulfide-bearing solutions derived from petroleum-rich sediments. The solutions rise along joints, where they meet and interact with oxygenated water to form sulfuric acid, which then dissolves the limestone.

CAVE DEPOSITS

Some caves have been partly filled with insoluble clay and silt, originally present as impurities in limestone and gradually concentrated as the limestone was dissolved. Others contain partial fillings of **dripstone** and **flowstone**, deposits chemically precipitated from dripping and flowing water, respectively, in an air-filled cavity. Both kinds of deposits are commonly composed of calcium carbonate. The carbonate precipitates take on many curious forms, which are among the chief attractions for cave visitors. Among the most common shapes are *stalactites* (icicle-like forms of dripstone hanging from ceilings), *stalagmites* (blunt mounds or pinnacles of dripstone projecting upward from cave floors), *columns* (stalactites joined with stalagmites, forming connections between the floor and roof of a cave), and crenulated or curtain-like formations of flowstone (Figure 15.21).

As its name implies, dripstone is deposited by successive drops of water. As each drop forms on the ceiling of a cave, it loses a tiny amount of carbon dioxide gas and precipitates a particle of calcium carbonate (Figure 15.22). In flowstone, a similar loss of carbon dioxide from water flowing across carbonate surfaces also leads to precipitation. This chemical reaction is the reverse of the one by which calcium carbonate is dissolved by carbonic acid.

Dripstone and flowstone can be deposited only in caves that are at least partially filled with air and therefore lie at or above the water table. Yet many, perhaps most, caves are believed to have formed below the water table. This is suggested by their shapes and by the fact that some caves are lined with crystals, which can form only in an aqueous environment. How can we reconcile these apparently conflicting observations?

Figure 15.21 Dripstone Spectacular dripstone and flowstone formations ornament Lehman Caves in Great Basin National Park, Nevada.

The answer lies partly in the observation that both dissolution and deposition are known to take place in the zone of aeration. In some cases, the answer probably also lies in a change in the level of the water table. This change in

Figure 15.22 How Stalactites Form A drop of water collects at the end of a growing stalactite in Carlsbad Caverns, New Mexico. As the water loses carbon dioxide, a tiny amount of calcium carbonate precipitates from solution and is added to the end or sides of the stalactite.

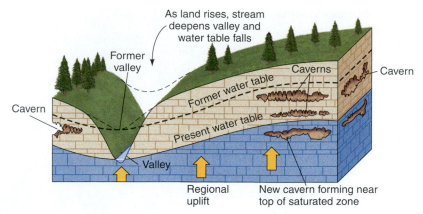

As land rises, stream deepens valley and water table falls

Former valley

Caverns

Cavern

Cavern

Former water table

Present water table

Valley

Regional uplift

New cavern forming near top of saturated zone

Figure 15.23 Formation of Caverns One possible history of a cavern containing carbonate deposits. The cavern formed in the saturated zone when the water table lay at a former, higher level. Uplift of the region caused streams to deepen their val-leys. The water table then lowered in response to valley deepening, leaving the cavern above the lowered, present water table (level 2). Stalactites and stalagmites could then form in the air-filled cavern.

level can result when uplift of the land causes a stream to cut downward into the landscape, thereby lowering the water table, or when a change of climate causes a lowering of the regional water table (Figure 15.23). Caves formed in the upper part of the saturated zone when the water table is high would shift into the zone of aeration as the water table falls. Dissolution could then give way to deposition of dripstone and flowstone.

SINKHOLES

In contrast to a cave, a **sinkhole** is a large dissolution cavity that is open to the sky. Some sinkholes are caves whose roofs have collapsed; others are formed at the surface, where rainwater is freshly charged with carbon dioxide and is most effective as a solvent. Many sinkholes located at the intersections of joints, where downward movement of water is most rapid, are funnel-shaped.

Sinkholes of the Yucatan Peninsula in Mexico, which are locally called *cenotes* (a word of Mayan origin), have high, vertical sides and contain water because their floors lie below the water table (Figure 15.24). The cenotes were the primary source of water for the ancient Maya and formerly supported a considerable population in Yucatan. A large cenote at the ruined city of Chichen Itza was sacred and dedicated to the rain gods. Remains of more than 40 human sacrifices, mostly young children, have been recovered from the cenote, together with huge quantities of jade, gold, and copper offerings.

In the carbonate landscape of Florida, new sinkholes are constantly forming (Figure 15.25). In one small area of about 25 km², more than 1000 collapses have occurred in recent years. In this case, lowering of the water table brought on by drought and excessive pumping of local wells has led to extensive collapse of cave roofs.

Some sinkholes form catastrophically. An account of one such event in rural Alabama describes how a resident was startled by a rumble that shook his house. He then distinctly heard the sound of trees snapping and breaking. A short time later, hunters walking through nearby woods discovered a huge 50 m deep sinkhole that measured 140 m long and 115 m wide.

KARST TOPOGRAPHY

In some regions of exceptionally soluble rocks, caves and sinkholes are so numerous that they form a peculiar topography characterized by many small, closed basins and a

Figure 15.24 Sacred Mayan Well The sacred well at Chichen Itza, a ruined Mayan city on the Yucatan Peninsula of Mexico. This *cenote*, formed in flat-lying limestone strata, contained a rich store of archeological treasures that were cast into the water with human sacrifices centuries ago.

Figure 15.25 Sinkhole in Florida Most of a city block in Winter Park, Florida disappeared into a widening crater as this sinkhole formed in underlying carbonate bedrock.

disrupted drainage pattern. Streams disappear into the ground and eventually reappear elsewhere as large springs. Such terrain is called **karst topography** after the classic karst regions of former Yugoslavia (extending from Slovenia to Montenegro), a remarkable limestone landscape of closely spaced sinkholes. Although most typical of carbonate landscapes, karst can also develop in areas underlain by gypsum and salt.

Several factors control the development of karst landscapes. First, the topography must produce a steep enough gradient to permit the flow of groundwater through soluble rock under the pull of gravity. In addition, precipitation must be adequate to supply the groundwater system, soil and plant cover must supply an adequate amount of carbon dioxide, and temperature must be high enough to promote dissolution. Although karst terrain is found throughout a wide range of latitudes and at varied altitudes, it often is best developed in moist temperate to tropical regions underlain by thick and widespread soluble rocks.

Several distinctive kinds of karst landscape are recognized. The most common is *sinkhole karst*, a landscape dotted with sinkholes of various sizes and shapes. Such landscapes are seen in southern Indiana, south-central Kentucky, central Tennessee, and Jamaica, among other places.

Cone karst and *tower karst* occur in thick, well-jointed limestone that separates into isolated blocks as it weathers. Cone karst consists of many closely spaced conical- or pinnacle-shaped hills separated by deep sinkholes (Figure 15.26). Tower karst, by contrast, consists of isolated tower-

like limestone hills separated by expanses of alluvium. If sediment being deposited in the bottom of sinkholes is not removed by the local drainage system, the depressions slowly fill, and areas between towers merge to form a flat alluvial surface. Cone and tower karst landscapes are found in parts of Mexico, Central America, and the Caribbean islands of Cuba and Puerto Rico, as well as in the South Pacific.

One of the most famous and distinctive of the world's tower karst regions lies near Guilin in southeastern China.

Figure 15.26 Astronomy and Karst Cone karst near Arecibo in northwestern Puerto Rico. The Arecibo radiotelescope (lower right) occupies a circular depression in the midst of a vast limestone landscape of closely spaced conical hills separated by deep sinkholes.

Figure 15.27   Tower Karst in China   Steep limestone pinnacles up to 200 m high, surrounded by flat expanses of alluvium, form a spectacular karst landscape around the Li River near Guilin, China.

There, vertical-sided peaks of limestone rise up to 200 m high. The dramatic landscape of this region has inspired both classical Chinese painters and present-day photographers (Figure 15.27).

Figure 15.28   Karst in Ireland   The Burren region in western Ireland is an extensive pavement karst developed on limestone. Crevices have developed by solution along prominent joints, giving the terrain a regular, geometric texture.

Pavement karst consists of broad areas of bare limestone in which joints and bedding planes have been etched and widened by dissolution, creating a distinctive land surface. Pavement karsts are especially common in high-latitude regions where continental glaciation has stripped away regolith and left carbonate bedrock exposed to weathering. These rocky, intricately etched landscapes, nearly devoid of vegetation, are found in such places as Spitzbergen, Greenland, and the Burren region of western Ireland (Figure 15.28).

Before you go on:

1. What factors affect rates of dissolution of carbonate rocks? How do these factors contribute to cave formation?

2. How does groundwater help convert sediment to sedimentary rock?

3. What are stalactites, stalagmites, and flowstone, and how do they form?

4. What features characterize karst landscapes? What do they tell us about the lithology of the region in which they form?

REVISITING PLATE TECTONICS AND THE EARTH SYSTEM

PLATE TECTONICS AND THE HIGH PLAINS AQUIFER

There are many aquifers in North America, but, the High Plains aquifer (p. 390), is so extensive and so important to so many people that it is unique.

The aquifer lies within the Ogallala Formation, of Tertiary age. It stretches from South Dakota to northern Texas, a region that is more than 1200 km long and 200 to 500 km wide (Figure 15.9). What possible connection could there be between this vast aquifer system and plate tectonics?

To answer the question we need to step back in geologic time. The ancient continent we call Pangaea began to break up in the early Jurassic Period, resulting in the formation of new oceans. The process was largely completed by the end of the Cretaceous Period. At the time of the continental breakup, an Andean-type subduction margin developed along the western edge of North America. The tectonic events caused many changes, but two are especially relevant to our story. First, a broad north–south lowland developed across the midcontinent region east of the present Rocky Mountains (which had not yet come into existence). Second, the new midocean ridges that formed as a result of continental rifting decreased the volume of the ocean basins and forced ocean water to flood low regions of the continent. The midcontinent lowland thus became a shallow sea. Over nearly 100 million years the sea entered and then withdrew from the lowland. During the maximum incursion, a Cretaceous seaway extended from the Gulf of Mexico to the Arctic Ocean.

Sediment deposited in the shallow seaway was sometimes marine, and sometimes terrestrial, but by the end of the Cretaceous the sea had withdrawn. The former seaway remained a lowland in which sediments accumulated, but now the sediments were terrestrial and deposited along streams and in lakes.

The land to the west of the Cretaceous seaway began to rise during the Tertiary Period. Although the causes are complex and poorly understood, probably they were related to plate tectonic processes along the western border of the continent. Terrestrial sediments from the western uplands were spread across the old seaway. In the Pliocene Epoch, extensive alluvial fans that headed in the western uplands coalesced to form a vast blanket of coarse, clastic sediment. That sediment, the Ogallala Formation, now constitutes the High Plains aquifer. The geologic history leading to the formation of the aquifer is long and complex, but plate tectonics was the driving force that led to the creation of this important source of underground water.

FROM RAINFALL TO MINERAL DEPOSITS

Karst, as we learned earlier in this chapter, is a term applied to landscapes formed as a result of the solution of soluble rocks by rainwater. Because rainwater is a weak acid, the rock most commonly involved in karstification is limestone. Drainage in karst regions is principally underground through caves and other solution-enlarged openings.

Modern karst landscapes are forming in many parts of the world—in the Yucatan Peninsula of Mexico and in northern Puerto Rico, for example—but in all cases modern karsts are merely solution features. With paleokarsts an entirely different picture emerges. Many paleokarsts hold pleasant surprises for miners.

Paleokarsts that are buried by younger sediments appear in the geologic record as karsts adjacent to an unconformity. Sediment above the unconformity serves as a seal to the porous openings in the paleokarst. As a result of subsequent geologic events, such as heated solutions expelled from compacting sedimentary basins, or vapors given off by cooling intrusives, mineralizing fluids may flow through the old paleokarst. When the solutions carry valuable mineral constituents, some of the dissolved matter may be deposited in the paleokarst openings and form valuable ore deposits. Zinc and lead deposits in Tennessee and Poland, fluorite in South Africa, antimony in China, bauxite in Jamaica, and uranium in Australia and Kyrgyzstan are all examples of paleokarst ores.

Formation of ore deposits in openings in limestone formed though the interaction of the atmosphere and hydrosphere with the lithosphere is an example of the Earth system at work.

What's Ahead?

For several million years, Earth has experienced a succession of glacial ages when much of the planet's fresh water was locked up in vast continental ice sheets and myriad smaller mountain glaciers. Although during the early years of the nineteenth century nobody knew of this, one of the great discoveries of late-nineteenth-century geology was the convincing evidence of these glacial events. Glaciated landscapes produced by former mountain glaciers are found on every continent, and deposits of ancient ice sheets cover vast regions of North America and Eurasia.

Studies have shown that modern glaciers are effective agents of erosion, surpassing streams as efficient sculptors of the land. Their ice-age predecessors not only eroded the landscape, but they caused sea level to fall, the crust to deform under their weight, streams to be diverted, and lakes to be impounded. In many countries of the Northern

Hemisphere, the modern landscape is a relict of the last glacial age.

In the next chapter we will see how glaciers develop, change in size, move across the land, and leave a geologic record of their passage. We'll examine the evidence for past glaciations and the causes of glacial–interglacial cycles. We also will ponder what may happen to glaciers during the present interval of rapidly warming climate and try to predict when the next ice age will occur.

CHAPTER SUMMARY

1. Groundwater is derived mostly from rainfall and occurs everywhere beneath the land surface.

2. The water table is the top of the saturated zone. In humid regions, the form of the water table is a subdued imitation of the overlying land surface.

3. Groundwater moves chiefly by percolation, at rates far slower than those of surface streams. In rock or sediment of constant permeability, the velocity of groundwater increases as the slope of the water table increases.

4. In moist regions, groundwater in recharge areas percolates downward under the pull of gravity. It moves away from hills toward valleys, where it may emerge to supply streams (groundwater discharge areas). In dry regions, the groundwater is principally recharged by water percolating downward beneath surface streams.

5. According to Darcy's Law, the discharge of water in a groundwater system is equal to the product of the cross-sectional area of flow, the coefficient of permeability, and the hydraulic gradient.

6. Springs often occur at places where the water table intersects the land surface.

7. Groundwater flows into most wells directly by gravity. Pumping of water from a well creates a cone of depression in the water table.

8. Major supplies of groundwater are found in aquifers, among the most productive of which are porous sand, gravel, and sandstone.

9. If the top of a well that penetrates an artesian aquifer lies below the altitude of the water table in the recharge area, hydrostatic pressure will allow water to rise in the well and flow out at the surface without pumping.

10. An unconfined aquifer is one that is not constrained above by an aquiclude, whereas a confined aquifer is bounded by aquicludes.

11. Excessive withdrawal of groundwater can lead to lowering of the water table and to land subsidence.

12. Water quality is influenced by the content of natural dissolved substances, seawater intrusion, and pollution by human and industrial wastes that percolate into groundwater reservoirs.

13. Hazardous (toxic and radioactive) wastes should be stored underground only if geologic conditions imply little or no change in groundwater systems over geologically long intervals of time.

14. Groundwater dissolves mineral matter from rock. It also deposits substances as cement between grains of sediment, thereby reducing porosity and converting the sediments to sedimentary rock.

15. In carbonate rocks, groundwater not only creates caves and sinkholes by dissolution but also deposits calcium carbonate as dripstone and flowstone.

16. Karst topography forms in areas of porous carbonate or other soluble rocks where the relief is great enough to permit gravitational flow of groundwater.

THE LANGUAGE OF GEOLOGY

aquiclude (p. 389)
aquifer (p. 390)
artesian aquifer (p. 393)

cone of depression (p. 390)
confined aquifer (p. 390)

Darcy's Law (p. 389)
discharge area (p. 387)
dripstone (p. 403)

flowstone (p. 403)

groundwater (p. 384)

hydraulic gradient (p. 389)

karst topography (p. 405)

percolation (p. 388)
permeability (p. 387)
porosity (p. 386)

recharge area (p. 387)
replacement (p. 402)

saturated zone (p. 385)
sinkhole (p. 404)
spring (p. 389)

unconfined aquifer (p. 390)

water table (p. 385)

zone of aeration (p. 385)

QUESTIONS FOR REVIEW

1. What is the ultimate source of groundwater?

2. Why does a thin zone immediately above the water table remain continuously moist?

3. Why do the flow paths of groundwater moving beneath a hill tend to turn upward toward a stream in an adjacent valley?

4. What variables determine how long it takes water to move from a recharge area to a discharge area?

5. Explain why it is possible to determine the discharge, velocity, or coefficient of permeability of groundwater passing through an aquifer provided two of these three factors are known.

6. What is the hydraulic gradient, and what importance does it have in determining the rate of flow of groundwater?

7. Why are sandstones generally better aquifers than siltstones or shales?

8. What features in igneous and metamorphic rocks promote the flow of groundwater through them?

9. How are springs related to the water table?

10. What causes a cone of depression to form around a producing well?

11. What causes water to rise to or above the ground surface in an artesian well?

12. Why is sand especially effective in purifying water flowing through it?

13. What is the origin of "hard" water in regions of carbonate bedrock?

14. Why do dripstone and flowstone formations not form in a cave that is in the saturated zone and therefore completely filled with water?

15. Why are karst pavements common in carbonate terrain at high latitudes?

16. A large area on a hillside has been suggested as a landfill site for garbage generated by a small nearby city. You are asked for a geologic appraisal of the site to determine if local subsurface water supplies might be affected. What geologic factors would you investigate and why?

17. What geologic and biologic factors are likely to disqualify a site from being selected for underground storage of high-level radioactive waste?

Click on *Presentation* and *Interactivity* in **The Water Cycle** module of your CD-ROM to further explore resources and activities presenting concepts from this chapter. Select *Assessment* in the same module to test your understanding of this chapter.

Rocky moraines mark the margins of coalescing tributary glaciers that merge to form a large valley glacier in Alaska's Glacier Bay National Park.

Glaciers and Glaciation

USGS research ship monitoring glacier retreat in Icy Bay, Alaska.

Modern Analogs of Ice-Age Glaciers

When the last continental ice sheet advanced from British Columbia into western Washington about 19,000 years ago, it moved forward at an average "glacial" pace of only about 130 meters per year. However, radiocarbon dating tells us that as the glacier began to retreat about 2000 years later, its average rate of recession was at least three times faster. How can such different rates of advance and retreat be explained? A possible answer can be sought by studying existing glaciers that are modern analogs of their ice-age predecessors.

In the summer of 1981, I was part of a small team studying recent glacier recession in Icy Bay, a large fjord system of southern Alaska at the foot of lofty Mount St. Elias. As we entered the bay aboard the U.S. Geological Survey's research vessel *Growler*, we encountered an array of icebergs that had calved (broken off) from five retreating tributary glaciers. When the Icy Bay glacier began to retreat at the end of the nineteenth century, the calving ice margin receded more and more rapidly up the 400-m-deep fjord. A regional study of Alaskan tidewater glaciers had

just shown that their retreat rate is correlated with the depth of water in which a fjord glacier terminates: the deeper the water, the faster is its rate of retreat. Like other large coastal glaciers, the measured retreat rate of the Icy Bay glacier also was closely related to water depth. Here was a likely explanation for the rapid retreat of the ice-age glacier in western Washington. As that ice margin retreated northward, it terminated in a series of large glacial lakes, as much as 400 m deep. By analogy with the modern Icy Bay glacier, the retreat rate of the calving ice margin in these deep-water lakes was many times the average rate of retreat on land, and led to rapid disintegration of the glacier. Like the Icy Bay glacier, its average rate of calving retreat was far greater than its earlier rate of advance.

Stephen C. Porter

KEY QUESTIONS

1. **How do glaciers form, and where are they found?**

2. **How do glaciers change in size?**

3. **What causes glaciers to move?**

4. **How do glaciers erode and create landforms?**

5. **How have past glacial ages affected Earth?**

6. **How can the onset and ending of glacial ages be explained?**

INTRODUCTION: THE EARTH'S CHANGING COVER OF SNOW AND ICE

Most people have never seen a glacier, except perhaps in a photograph. Most of us live in latitudes where glaciers cannot exist, except at very high altitudes. Nevertheless, a majority of the people in northern Europe, Canada, and the northern United States live on glacial deposits, left by the last of a long succession of continent-size ice sheets that occupied these regions during Earth's most recent glacial ages. Geologists studying these deposits and associated landforms, as well as oceanographers studying cores from the deep sea, know that we live in the midst of a glacial era that has lasted for more than 2 million years and likely will continue for many more millions of years. How do we know this? And how can we predict, with any confidence, what the future may hold in store?

In this chapter we will look closely at glaciers: why and how they form, how they move, and how they change in size. We will study the geologic evidence they produce in the form of sediments and landforms. We will examine evidence of past glacial ages and the effect of glaciers on rivers, sea level, and Earth's crust. Finally, we will consider the leading theories and hypotheses advanced to explain the causes of the glacial ages.

GLACIERS

At any place on the land where more snow accumulates than is melted during the course of a year, the snow will gradually grow thicker. As the snow piles up, the increasing weight of snow overlying the basal layers causes them to recrystallize, forming a solid mass of ice. This process is analogous to the way sedimentary rocks, buried under a deep pile of strata within Earth's crust, recrystallize to form metamorphic rocks. When the accumulating snow and ice become so thick that the pull of gravity causes the frozen mass to move, a glacier is born. Accordingly, we define a **glacier** as a permanent body of ice, consisting largely of recrystallized snow, that shows evidence of downslope or outward movement due to the pull of gravity.

Glaciers are found in regions where average temperature is so low that water can exist throughout the year in a frozen state. As we might expect, most glaciers are found in high latitudes, the coldest parts of our planet. However, because low temperatures also occur at high altitudes, many small glaciers exist in middle and low latitudes on high mountains.

Because glaciers vary considerably in their physical characteristics, we can distinguish several kinds based on their shape and size (Figure 16.1; Table 16.1), as well as on their internal temperature.

A.

B.

C.

D.

E.

F.

Figure 16.1 Kinds of Glaciers Examples of common glacier types, classified according to shape and size (Table 16.1) A. Small Cascade cirque glacier. B. Alaskan valley glacier. C. Alaskan fjord glacier. D. Alaskan piedmont glacier. E. South American mountain ice cap. F. Antarctic ice shelf. Ice sheets of continental size are shown in Figures 16.3 and 16.4.

TABLE 16.1   Principal Types of Glaciers, Classified According to Form

Glacier Type	Characteristics
Cirque glacier	Occupies a cirque on a mountainside.
Valley glacier	Flows from cirque(s) onto and along the floor of a valley.
Fjord glacier	Valley glacier that occupies a fjord. Base lies below sea level. May have a steep front that recedes rapidly as icebergs break off and float away.
Piedmont glacier	Broad lobe of ice that terminates on open slopes beyond a mountain front. Fed by one or more large valley glaciers.
Ice cap	Dome-shaped body of ice and snow that covers mountain highlands (or lower-lying lands at high latitudes) and displays generally radial outward flow.
Ice sheet	Continent-sized mass of ice that overwhelms nearly all land within its margins.
Ice shelf	Thick, slab-like glacier that floats on the sea and is fed by one or more glaciers on land. Commonly located in large embayments.

MOUNTAIN GLACIERS AND ICE CAPS

The shape and direction of movement of most mountain glaciers are determined by the surrounding bedrock topography. The smallest glacier occupies a **cirque**, a protected bowl-shaped depression on a mountainside, and is called a *cirque glacier* (Figure 16.1A). It typically is bounded upslope by a steep cliff, or *headwall*. A growing cirque glacier that spreads outward and downward along a valley will become a *valley glacier* (Figure 16.1B). Many of Earth's high mountain ranges (for example, the Alaska Range and the Himalaya) contain glacier systems that include valley glaciers tens of kilometers long (Figure 16.2).

Valley glaciers in some coastal mountain ranges at middle to high latitudes occupy deep glacier-carved valleys that are filled by an arm of the sea. Such a valley is a **fjord**, and a glacier that occupies it is a *fjord glacier* (Figure 16.1C).

A very large valley glacier may spread out onto gentle terrain beyond a mountain front where it becomes a *piedmont glacier* and forms a broad lobe of ice that resembles an inverted spoon (Figure 16.1D).

An *ice cap* covers a mountain highland or lower-lying land at high altitude and displays generally radial outward flow (Figure 16.1E).

ICE SHEETS AND ICE SHELVES

An *ice sheet* is the largest type of glacier on Earth. These continent-sized masses of ice overwhelm nearly all the land surface within their margins. Modern ice sheets,

Figure 16.2 Valley Glaciers A vertical satellite view of the valley-glacier complex that covers much of Denali National Park in south-central Alaska. Mount Denali (formerly McKinley), the highest peak in North America, lies near the center of the glacier-covered region.

which are found only on Greenland and Antarctica, include about 95 percent of the ice in existing glaciers. If all the ice in these vast ice sheets were to melt, their combined volume—close to 24 million km^3—would be sufficient to raise world sea level by nearly 66 m.

The Greenland Ice Sheet has about the same area as the United States west of the Rocky Mountains (Figure 16.3). It is so thick (about 3000 m) that its weight is enough to cause the land surface beneath much of it to be depressed below sea level.

Antarctica is covered by two large ice sheets that meet along the lofty Transantarctic Mountains (Figure 16.4). The East Antarctic Ice Sheet is the larger of the two and covers the continent of Antarctica. Because of its ice sheet, Antarctica has the highest average altitude and the lowest average temperature of all the continents. The smaller West Antarctic Ice Sheet overlies numerous islands of the Antarctic archipelago. Parts of it rest on land that rises above sea level, but extensive portions cover land lying below sea level.

Fed by one or more glaciers on land, an *ice shelf* is a thick, nearly flat sheet of floating ice that terminates seaward in a steep ice cliff as much as 50 m high (Figure

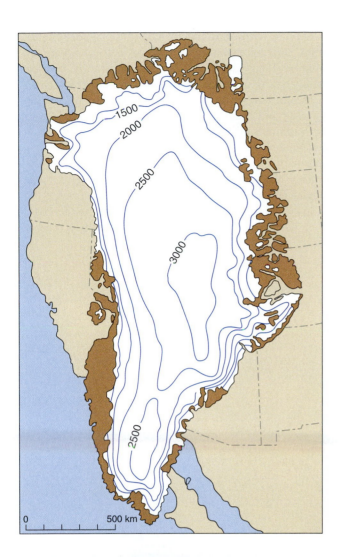

Figure 16.3   Greenland Ice Sheet   Map of Greenland superimposed on a map of western United States showing that the Greenland Ice Sheet covers an area approximating that of the United States west of the Rocky Mountains. Contours are the height of the ice surface above sea level, in meters. At the crest of the ice sheet, the ice is more than 3000 m above sea level.

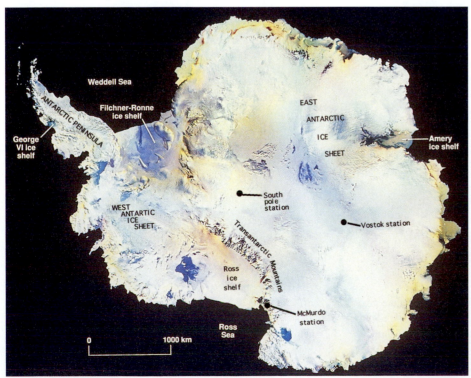

Figure 16.4   Antarctic Ice Sheets
Satellite view of Antarctica. The East Antarctic Ice Sheet overlies the continent of Antarctica, whereas the much smaller West Antarctic Ice Sheet overlies a volcanic island arc and surrounding seafloor. Three major ice shelves occupy large bays. The ice-covered regions of Antarctica nearly equal the combined area of Canada and the conterminous United States.

16.1F). Ice shelves are found at several places along the margins of the Antarctic ice sheets, where they are confined in large bays, and also in the Canadian Arctic islands. The largest Antarctic ice shelves extend hundreds of kilometers seaward from the coastline and reach a thickness of at least 1000 m near their landward margins.

TEMPERATE AND POLAR GLACIERS

Glaciers can be classified according to their internal temperature as well as their size and shape. Temperature is an important criterion for classification because it helps determine how glaciers move and how they shape the landscape.

Ice in a **temperate glacier** is at the **pressure melting point**, the temperature at which ice melts at a particular pressure (Figure 16.5A). Temperate glaciers are restricted mainly to low and middle latitudes under conditions where meltwater and ice exist together in equilibrium. At high latitudes and high altitudes, where the mean annual air temperature is below freezing, the temperature in a glacier remains below the pressure melting point and little or no seasonal melting occurs. Such a glacier, in which the ice remains below the pressure melting point, is called a **polar glacier** (Figure 16.5B).

If the temperature of snow crystals falling to the surface of a temperate glacier is below freezing, how can ice

A.

B.

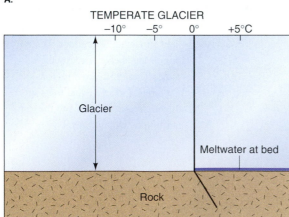

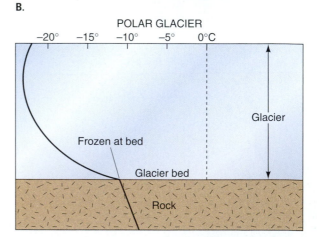

Figure 16.5 Temperate and Polar Glaciers Temperature profiles (red lines) through temperate and polar glaciers. A. Ice in a temperate glacier is at the pressure melting point from surface to bed. The terminus is rounded, as shown in this view of Pré de Bar Glacier in the Italian Alps, because melting occurs at the

surface. B. Ice in a polar glacier remains below freezing, and the ice is frozen to its bed. Subfreezing temperatures inhibit melting at the terminus, which forms a steep cliff of ice, as shown in this view of Commonwealth Glacier in Antarctica.

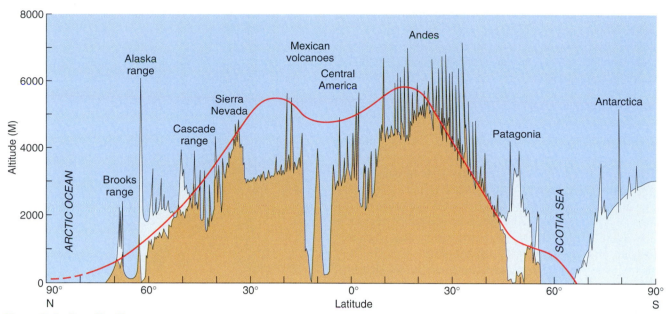

Figure 16.6 **Snowline Transect** Topographic transect along the cordillera of North and South America showing the average position of the snowline (dark blue line) and altitudinal distribution of glaciers (light blue).

throughout the glacier reach the pressure melting point? The answer lies in the seasonal fluctuation of air temperature and in what happens when water freezes to form ice. In summer, when air temperature rises above freezing, solar radiation melts the glacier's surface snow and ice. The meltwater percolates downward, where it encounters freezing temperatures and therefore freezes. When changing state from liquid to solid, each gram of water releases 335 J of heat. This released heat warms the surrounding ice and, together with heat flowing upward from the solid Earth beneath the glacier, keeps the temperature of the ice at the pressure melting point.

GLACIERS AND THE SNOWLINE

Glaciers can form only at or above the **snowline**, which is the lower limit of perennial snow. The snowline is sensitive to local climate, especially temperature and precipitation, and rises in altitude from near sea level in polar latitudes to as much as 6000 m in the tropics (Figure 16.6). It also rises inland from moist coastal regions toward the drier interiors of large islands and continents.

In Figure 16.6, the snowline is shown as a two-dimensional north-to-south profile. If we integrated similar profiles around the planet, the result would be an irregular, three-dimensional world-girdling surface, high near the equator and at or near sea level in the polar regions. Wherever this snowline surface intersects the land, a glacier potentially can form. Thus, glaciers are found not only at sea level in the polar regions, wherever land rises above the very low snowline, but also near the equator, where some high peaks in New Guinea, East Africa, and the Andes are high enough to rise above the snowline surface.

CONVERSION OF SNOW TO GLACIER ICE

Glacier ice is essentially a very low-temperature metamorphic rock that consists of interlocking crystals of the mineral ice. It owes its characteristics to deformation under the weight of overlying snow and ice.

Newly fallen snow is very porous and has a density less than a tenth that of water. The delicate points of snowflakes gradually disappear by evaporation. The resulting water vapor condenses, mainly in constricted places near the centers of snowflakes. In this way, the fragile snow crystals slowly become smaller and rounder (Figure 16.7), which

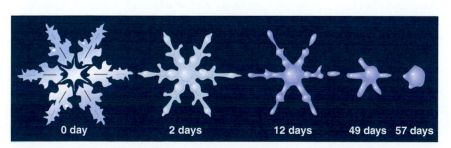

Figure 16.7 **From Snow to Old Snow** Conversion of a snowflake into a granule of old snow. The delicate points of a snowflake disappear as melting and evaporation occur. The resulting meltwater refreezes, and water vapor formed by evaporation at the margin condenses near the center of the granule, making it more compact.

allows the snow to settle and become more compact as the amount of pore space is reduced.

Snow that survives a year or more gradually increases in density until it is no longer permeable to air, at which point it becomes glacier ice. Although now a rock, this ice has a far lower melting temperature than any other naturally occurring rock, and its density of about 0.9 g/cm^3 means that it will float in water.

Further changes in ice take place as it becomes more deeply buried in a glacier. Figure 16.8 shows a core obtained by Russian glaciologists who drilled through the East Antarctic Ice Sheet at Vostok Station (Figure 16.4). As snowfall adds to the glacier's thickness, the increasing pressure causes initially small grains of glacier ice to grow in size until, near the base of the ice sheet, they reach a diameter of 1 cm or more. This increase in grain size is analogous to what happens in a fine-grained rock that is carried deep within Earth's crust over a long time; recrystallization causes large mineral grains to develop (Chapter 8).

WHY GLACIERS CHANGE IN SIZE

The pioneer Scottish glaciologist James D. Forbes visited the Alps in the mid-nineteenth century and carefully mapped the extent of long valley glaciers flowing from Mont Blanc, the highest peak in Europe. Since Forbes's day, these glaciers have shrunk greatly, exposing extensive areas of valley floor that only a century ago were buried beneath thick ice. Over this same period, changes in glacier size have been recorded worldwide. Like those in the Alps, most other glaciers have receded dramatically since the end of the last century, although some have remained relatively unchanged and a few have expanded. To understand why glaciers advance and retreat, and why different glaciers in any region may behave in different ways, we must examine how a glacier responds to a gain or loss of mass.

MASS BALANCE

The mass of a glacier is constantly changing as the weather varies from season to season and, on longer time scales, as local and global climates change. These ongoing environmental changes cause fluctuations in the amount of snow added to the glacier surface and in the amount of snow and ice lost by melting.

Think of a glacier as being like a checking account at the bank. The amount of money in the account at the end of the year is the difference between the amount of money added to the initial account value and the amount removed during the year. The glacier's account is measured in terms of the amount of snow added to the existing ice, mainly in the winter, and the amount of snow (and ice) lost, mainly during the summer. Additions to the glacier's account are collectively called **accumulation**, and losses are termed

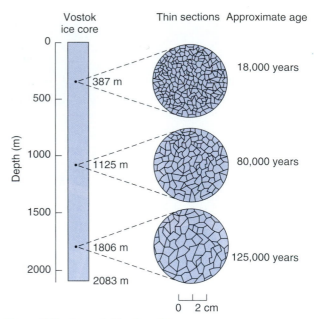

Figure 16.8 Recrystallization of Ice A deep ice core drilled at Vostok Station penetrated through the East Antarctic Ice Sheet to a depth of 2083 m. Thin-section samples of ice taken from different depths in the core show a progressive increase in the size of ice crystals, a result of slow recrystallization as the weight of overlying ice increases with time.

ablation. The total in the account at the end of a year—in other words, the difference between accumulation and ablation—is a measure of the glacier's **mass balance** (Figure 16.9). The account may have a surplus (a positive

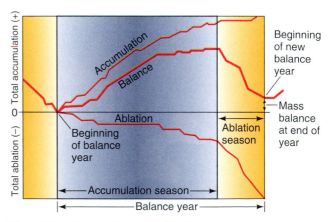

Figure 16.9 Mass Balance of a Glacier Graph showing how accumulation and ablation determine glacier mass balance (heavy line) over the course of a year. The balance curve, obtained by summing values of accumulation (positive values) and ablation (negative values), rises during the accumulation season as mass is added to the glacier, then falls during the ablation season as mass is lost. The mass balance at the end of the balance year reflects the difference between mass gain and mass loss.

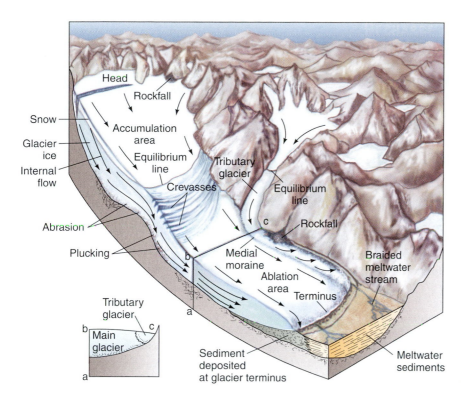

Figure 16.10   Features of a Valley Glacier
Main features of a valley glacier are shown in this oblique view, which includes a profile along the glacier's center line. Crevasses form where the glacier bed has a steep slope. Arrows show directions of ice flow. A band of rock debris marks the boundary between the main glacier and a tributary glacier that joins it from an adjacent valley.

balance) or a deficit (a negative balance), or it may be in exact balance (accumulation = ablation).

Two zones are generally visible on a glacier at the end of the summer ablation season (Figure 16.10). An upper zone, the *accumulation area*, is the part of the glacier covered by remnants of the previous winter's snowfall and is an area of net gain in mass. Below it lies the *ablation area*, a region of net loss where bare ice and old snow are exposed because the previous winter's snow cover has melted away.

The **equilibrium line** marks the boundary between the accumulation area and the ablation area (Figure 16.10). It lies at the level on the glacier where net mass loss equals net mass gain. The equilibrium line on a temperate glacier coincides with the local snowline, which marks the lower limit of fresh snow at the end of the ablation season. Being very sensitive to weather, the equilibrium line fluctuates in altitude from year to year and is higher in warm, dry years than in cold, wet years (Figure 16.11).

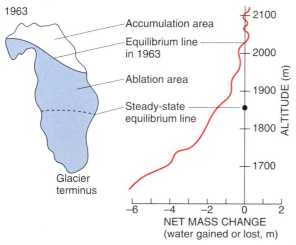

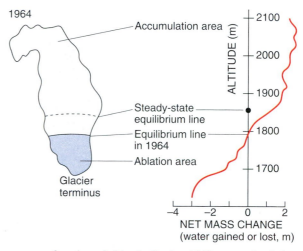

Figure 16.11   Differences in Mass Balance   Maps of South Cascade Glacier in Washington State at the end of the 1963 and the 1964 balance years showing the position of the equilibrium line relative to the position it would have under a balanced (steady-state) condition. The curves show values of mass bal-

ance as a function of altitude. During 1963, a negative balance year, the glacier lost mass and the equilibrium line was high (2025 m). In 1964, a positive balance year, the glacier gained mass and the equilibrium line was low (1800 m). Net change is measured in water lost or gained per unit area of the glacier.

FLUCTUATIONS OF THE GLACIER TERMINUS

If, over a period of years, a glacier's mass balance is positive more often than negative, the mass of the glacier increases. The front, or *terminus*, of the glacier is then likely to advance as the glacier grows. Conversely, a succession of years in which negative mass balance predominates will lead to retreat of the terminus. If no net change in mass occurs, the glacier is in a balanced state. If this condition persists, the terminus is likely to remain relatively stationary.

Response Lags. Advance or retreat of a glacier terminus does not necessarily give us an accurate picture of changing climate because a lag occurs between a change in accumulation due to a climate change and the response of the glacier terminus to that change. The lag reflects the time it takes for the effects of an increase or a decrease in accumulation above the equilibrium line to be transferred through the slowly moving ice to the glacier terminus. The length of the lag depends both on the size of a glacier and the way the ice flows; the lag will be longer for large glaciers than for small ones and longer for polar glaciers than for temperate ones. Temperate glaciers of modest size (like those in the European Alps) have response lags that range from several years to a decade or more. This lag time can explain why, in any area with glaciers of different sizes, some glaciers may be advancing while others are stationary or retreating.

Calving. During the last century and a half, many Alaskan fjord glaciers have receded at rates far in excess of typical glacier retreat rates on land. Their dramatic recession is due to **calving**, defined as the progressive breaking off of icebergs from the front of a glacier that terminates in deep water (Figure 16.1C). Although the base of a fjord glacier may lie far below sea level along much of its length, its terminus can remain stable as long as it is resting (or "grounded") against a shoal (Figure 16.12). However, if the terminus retreats off the shoal, water will replace the space that had been occupied by ice. With the glacier now terminating in water, conditions are right for calving. Because a fjord is a deep basin, the water becomes progressively deeper as the calving terminus continues to retreat upfjord. The deepening water leads to faster retreat. Studies of Alaskan tidewater glaciers have shown that retreat rate is related to water depth: the greater the depth of water at the glacier front, the faster the rate of calving. Once started, calving will continue rapidly and irreversibly until the glacier front recedes into water too shallow for much calving to occur, generally near the head of the fjord.

Icebergs produced by calving glaciers constitute an ever-present hazard to ships in subpolar seas. When the S.S. *Titanic* sank after striking a berg in the North Atlantic in 1912, the detection of approaching icebergs relied on sailors' vision. Today, with sophisticated electronic equipment, large bergs can generally be identified well before an encounter. Nevertheless, ice has a density of 0.9, so that 90 percent of an iceberg lies underwater, making it difficult to detect. In coastal Alaska, where calving glaciers are commonplace, icebergs pose a potential threat to huge oil tankers. For this reason, Columbia Glacier, which lies adjacent to the main shipping lanes from Valdez at the southern end of the Alaska Pipeline, is being closely monitored as its terminus pulls steadily back and multitudes of bergs are released.

Calving is the principal means of ablation for the large ice sheets of Antarctica. Large tabular icebergs breaking off the vast floating ice shelves that fill embayments between the East and West Antarctic ice sheets are many hundreds of meters thick. One exceptionally large berg about the size of Rhode Island drifted 2000 km in three years before it broke apart. Large bergs and countless smaller ones that enter the ocean every year constitute a vast potential resource of fresh water.

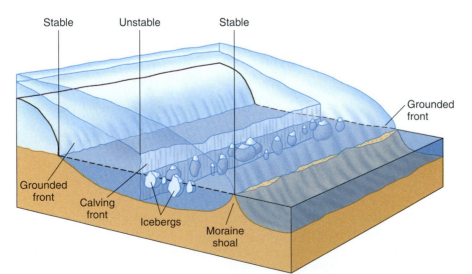

Figure 16.12 Calving Glacier Margin
The terminus of a fjord glacier remains stable if it is grounded against a moraine shoal, but if it retreats into deeper water, calving begins. The unstable terminus then retreats at a rate that increases with increasing water depth. Once it becomes grounded farther up the fjord, the terminus is stable once again.

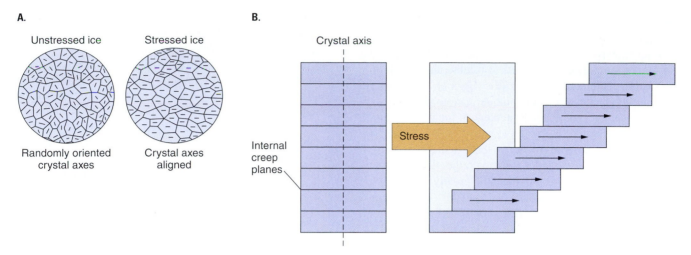

A.

Unstressed ice   Stressed ice

Randomly oriented
crystal axes

Crystal axes
aligned

B.

Crystal axis

Internal
creep
planes

Stress

Figure 16.13   Deforming Ice   Internal creep in ice crystals of a glacier. A. Randomly oriented crystals of ice in the upper layers of a glacier are transformed under stress so that their axes are aligned. B. When a stress is applied to an ice crystal, creep along internal planes results in slow deformation of the crystal.

HOW GLACIERS MOVE

We can easily prove to ourselves that glaciers move. One way is to visit a glacier near the end of the summer and carefully measure the position of a boulder lying on the ice surface with respect to some fixed point beyond the glacier margin. If we return a year later, we will find that the boulder has moved at least several meters in the downglacier direction. Actually, it is the ice that has moved, carrying the boulder along for the ride.

What causes a glacier to move may not be immediately obvious, but we can find clues by examining the ice and the terrain over which it lies. These clues tell us that ice moves in two ways: by internal flow and by sliding of the basal ice over underlying rock or sediment.

INTERNAL FLOW

When an accumulating mass of snow and ice on a mountainside reaches a critical thickness, the mass will begin to deform and flow downslope under the pull of gravity. The flow takes place mainly through movement within individual ice crystals, which are subjected to higher and higher stress as the weight of the overlying snow and ice increases. Under this stress, ice crystals are deformed by slow displacement (termed *creep*) along internal crystal planes in much the same way that cards in a deck of playing cards slide past one another if the deck is pushed from one end (Figure 16.13). As the compacted, frozen mass begins to move, stresses between adjacent ice crystals cause some to grow at the expense of others, and the resulting larger crystals end up with their internal planes being parallel. This alignment of crystals leads to increased efficiency of flow, for the internal creep planes of all crystals now are parallel.

In contrast to deeper parts of a glacier where the ice flows as a result of internal creep, the surface portion of a glacier has relatively little weight on it and is brittle. Where a glacier passes over an abrupt change in slope, such as a bedrock cliff, the surface ice cracks as tension pulls it apart. The cracks open up and form crevasses. A **crevasse** is a deep, gaping fissure in the upper surface of a glacier, generally less than 50 m deep (Figures 16.10 and 16.14). At depths greater than about 50 m, continuous flow of ice prevents crevasses from forming.

Figure 16.14   Crevasses   Climbers plot a route through a crevassed sector of a Canadian valley glacier. The crevasses are aligned perpendicular to the flow of the glacier as it flows over a steep portion of its bedrock valley.

BASAL SLIDING

Ice temperature is very important in controlling the way a glacier moves and its rate of movement. Meltwater at the base of a temperate glacier acts as a lubricant and permits the ice to slide across its bed (the rocks or sediments on which the glacier rests). In some temperate glaciers, such sliding accounts for up to 90 percent of the total observed movement (Figure 16.15). By contrast, polar glaciers are so cold that they are frozen to their bed. Their motion largely involves internal deformation rather than basal sliding, and so their rate of movement is greatly reduced.

ICE VELOCITIES AND DIRECTIONS OF FLOW

Measurements of the surface velocity across a valley glacier show that the uppermost ice in the central part of the glacier moves faster than ice at the sides, similar to the velocity distribution in a river (Figures 14.13 and 16.15). The reduced rates of flow toward the margins are due to frictional drag of the ice against the valley walls. A similar reduction in flow rate toward the bed is observed in a vertical profile of velocity (Figure 16.15).

Snow continues to pile up in the accumulation area each year, while melting removes snow and ice from the ablation area. The surface profile of a glacier does not change much, however, because ice is transferred from the accumulation area to the ablation area. In the accumulation area, the mass of accumulating snow and ice is pulled downward by gravity, and so the dominant flow direction is toward the glacier bed. However, the ice does not build up to ever-greater thickness because a downglacier component of flow is also present. Ice flowing downglacier replaces ice being lost from the glacier's surface in the ablation area, and so in this area the flow is upward toward the surface (Figure 16.10). Water molecules that fall as snowflakes on the glacier near its head therefore have a long path to follow before they emerge near the terminus. Those falling above but close to the equilibrium line, however, travel only a short distance through the glacier before reaching the surface again.

Even if the mass balance of a glacier is negative and the terminus is retreating, the downglacier flow of ice is maintained. Retreat does not mean that the ice-flow direction reverses; instead, it means that the rate of flow downglacier is insufficient to offset the loss of ice at the terminus.

In most glaciers, flow velocities range from only a few centimeters to a few meters a day, or about as fast as the rate at which groundwater percolates through crustal rocks. Hundreds or thousands of years have elapsed since ice now exposed at the terminus of a very long glacier fell as snow near the top of its accumulation area.

GLACIER SURGES

Although most glaciers slowly grow or shrink as the climate fluctuates, some glaciers experience episodes of very unusual behavior marked by rapid movement and dramatic changes in size and form. Such an event, called a **surge**, is unrelated, or only secondarily related, to a change in climate. When a surge occurs, a glacier seems to go berserk. Ice in the accumulation area begins to move rapidly downglacier, producing a chaos of crevasses and broken pinnacles of ice in the ablation area. Medial moraines, which are bands of rocky debris marking the boundaries between adjacent tributary glaciers (Figure 16.10), are deformed into intricate patterns (Figure 16.16). The termini of some glaciers have advanced up to several kilometers during surges. Rates of movement as great as 100 times those of nonsurging glaciers and reaching as much as 6 km a year have been measured.

The cause of surges is still imperfectly understood, but available evidence points to a reasonable hypothesis. We know that if water from melting ice is confined at the base of a glacier, the weight of the glacier will be supported by the water. Over a period of years, steadily increasing amounts of water trapped beneath the ice may lead to widespread separation of the glacier from its bed. The resulting effect is similar to the displacement of a beverage can on a sheet of glass in the experiment described in Chapter 13. According to this hypothesis, as the ice is floated off its bed, its forward mobility is greatly increased and it moves rapidly forward before the escape of water brings the surge to a halt.

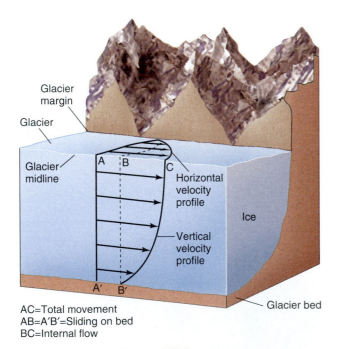

AC=Total movement
AB=A′B′=Sliding on bed
BC=Internal flow

Figure 16.15 Glacial Flow and Sliding   Three-dimensional view through half of a temperate glacier showing horizontal and vertical velocity profiles. A portion of the total observed movement is due to internal flow within the ice, whereas part is due to sliding of the glacier along its bed, lubricated by a film of meltwater.

GLACIAL EROSION AND SCULPTURE

In changing the surface of the land over which it moves, a glacier acts collectively like a plow, a file, and a sled. As a plow, it scrapes up weathered rock and soil and plucks out blocks of bedrock; as a file, it rasps away firm rock; and as a sled, it carries away the load of sediment acquired by plowing and filing, along with additional rock debris that falls onto the glacier from adjacent slopes.

SMALL-SCALE FEATURES OF GLACIAL EROSION

The base of a temperate glacier is studded with rock fragments of various sizes that are all carried along with the moving ice. Small fragments of rock embedded in the basal ice scrape away at the underlying bedrock and produce long, nearly parallel scratches called *glacial striations* (Figure 16.17). Larger rock fragments that the ice drags across the bedrock abrade *glacial grooves* aligned in

Figure 16.16 Surging Glaciers Surging tributary glaciers flowing from the mountains on the right of this view have deformed the medial moraines of Alaska's Yanert Glacier into a series of complex folds.

Before you go on:

1. What are the common glacier types?

2. What is the difference between a temperate glacier and a polar glacier?

3. How does a glacier's mass balance affect its size over time?

4. What are the two ways in which ice moves, and how do they differ?

GLACIATION

Glaciation, the modification of the land surface by the action of glacier ice, has occurred so recently over large areas of Europe, Asia, and North and South America that weathering, mass wasting, and erosion by running water have not yet had time to alter the landscape appreciably. Except for a cover of vegetation, the appearance of these glaciated landscapes has remained nearly unchanged since they emerged from beneath the ice. Like the geologic work of other surface processes, glaciation involves erosion and the transport and deposition of sediment.

Figure 16.17 Glacial Markings A deglaciated bedrock surface beyond Findelen Glacier in the Swiss Alps displays grooves and striations etched by rocky debris in the base of the moving glacier when it overlay this site. In the background rises the Matterhorn, a glacial horn sculpted by glaciers that surround its flanks.

Figure 16.18 Glacial Sculpturing Asymmetrical glacially sculptured rock surface beyond the terminus of Franz Josef Glacier in New Zealand's Southern Alps. The glacier flowed from right to left. Slopes facing toward the glacier are smooth and polished. Steep slopes facing downvalley result from the plucking of bedrock blocks by flowing ice.

Figure 16.19 Arêtes, Cirques, and Horns Sharp-crested arêtes flank the Western Cum, a deep cirque on the west side of Mount Everest in the central Himalaya. Sharp-crested peaks in the distance are horns that have resulted from headward growth of flanking cirque and valley glaciers.

the direction of glacier flow. Grains of silt in the basal ice act like fine sandpaper and polish the rock until it has a smooth, reflective surface.

Because striations and grooves are aligned with the direction of ice flow, geologists use these and other aligned erosional features to reconstruct the flow paths of former glaciers. Debris-laden ice striates and polishes the upglacier sides of small bedrock knobs or hills over which it moves, and plucks blocks of rock from the downglacier sides. The resulting asymmetrical landforms, smooth and gently sloping upglacier and steeper and angular downglacier, provide clear evidence of the direction in which the glacier was moving (Figure 16.18).

LANDFORMS OF GLACIATED MOUNTAINS

Skiers racing down the steep slopes at Alta, Mammoth, or Whistler, and rock climbers inching their way up the cliffs of Yosemite Valley, the granite spires of Mont Blanc, or the icy monoliths of the southern Andes owe a debt to the ancient glaciers that carved these mountain playgrounds. The scenic splendor of most of the world's high mountains is the direct result of glacial sculpturing that has produced a distinctive assemblage of alpine landforms.

Cirques. Cirques are among the most common and distinctive landforms of glaciated mountains. The characteristic bowl-like shape of a cirque is the combined result of frost-wedging, glacial plucking, and abrasion (Figure 16.19). Many cirques are bounded on their downvalley side by a bedrock threshold that impounds a small lake (a *tarn*).

A cirque probably begins to form beneath a large snowbank or snowfield at or just above the snowline. As meltwater infiltrates rock openings under the snow, it refreezes and expands, disrupting the rock and dislodging fragments. Small rock particles are carried away by snowmelt runoff during periods of thaw. A shallow depression in the land is

thereby created, and the depression gradually becomes larger and larger. As snow continues to accumulate in the deepening hollow, the snowbank thickens and becomes a glacier. Plucking then helps to enlarge the cirque still more, and abrasion at the bed further deepens it.

As cirques on opposite sides of a mountain grow larger and larger, their headwalls intersect to produce a sharp-crested ridge called an *arête*. Where three or more cirques have sculptured a mountain mass, the result can be a high, sharp-pointed peak (a *horn*), an example of which is the Matterhorn in the Swiss/Italian Alps (Figures 16.17 and 16.19).

Glacial Valleys. Valleys that were shaped by former glaciers differ from ordinary stream valleys in several ways. The chief characteristics of glacial valleys include a U-shaped cross profile and a floor that lies below the floors of smaller tributary valleys (Figure 16.20). Streams commonly descend as waterfalls, or cascades, as they flow from the tributary valleys into the main valley. The glaciated Cascade Range of western United States derives its name from such streams. The long profile of a glaciated valley floor may possess step-like irregularities and shallow basins. These usually are related to the spacing of joints in the rock, which influenced the ease of glacial plucking, or to changes in rock type along the valley. Finally, the valley typically heads in a cirque or a group of cirques.

Fjords. Fjords are common features along the mountainous, west-facing coasts of Norway, Alaska, British

Figure 16.20 Glacial Sculpture of Yosemite Repeated advances of valley glaciers carved the deep, U-shaped Yosemite Valley in California's Sierra Nevada. During each glaciation, the valley glacier was nourished by an extensive mountain ice cap that covered the undulating upland surface of the range, seen in the distance.

Columbia, Chile, and New Zealand, as well as in northern Canada (Figure 16.21). Fjords typically are shallow at their seaward end and deepen inland. This is a result of deep glacial erosion upstream of a glacier terminus at an earlier time, which created elongate basins in glacial valley floors, sometimes to depths of more than 300 m below modern sea level. Sognefjord in Norway reaches a depth of 1300 m, yet near its seaward end the water depth is only about 150 m.

Figure 16.21 Fjord Trekkers atop the Pulpit, a spectacular vantage point far above Lysefjord, can look far inland toward the source region of the glacier that carved this fjord, typical of many that indent the rocky western coast of Norway. If the fjord were drained of water, its cross-section would resemble that of Yosemite Valley (Figure 16.20).

LANDFORMS PRODUCED BY ICE CAPS AND ICE SHEETS

Although large ice caps and ice sheets produce many of the same landscape features that smaller glaciers do (for example, striations, moraines, fjords), they also generate some landforms not usually associated with small glaciers.

Abrasional Features. Landscapes that were shaped by overriding ice sheets display the same small-scale erosional features typical of valley glaciation. Striations have been especially helpful to geologists in reconstructing the flow lines of long-vanished ice sheets. Striations also demonstrate that basal ice in the thick central zones of former ice sheets was at the pressure melting point, a necessary condition for the sliding action required to produce these features. In peripheral zones, evidence of glacial erosion is sometimes less obvious, leading to the conclusion that the thinner ice there was cold and largely frozen to its bed.

Wherever ice sheets overwhelmed mountainous terrain, as in the high ranges of northwestern North America during the most recent glacial age, the highest evidence of glaciation can frequently be seen as the level where smooth, ice-abraded slopes pass abruptly upward into rugged, frost-shattered peaks and mountain crests that stood above the glacier surface. Some divides between adjacent drainages are broad and smooth and show evidence of glacial abrasion and plucking where they were overridden by ice.

Streamlined Forms. In many areas inside the limits of former ice sheets, the land surface has been molded into smooth, nearly parallel ridges, some of which are several kilometers long. Among the most distinctive of these landforms is the **drumlin**, a streamlined hill consisting largely of glacially deposited sediment that was elongated parallel to the direction of ice flow (Figure 16.22). Glacially molded drumlin-shaped hills of bedrock (called *rock drumlins*) also owe their shape to erosion by flowing ice. Drumlins and rock drumlins, like the streamlined bodies of supersonic airplanes that are designed to reduce air resistance, offered minimum resistance to glacier ice flowing over and around them.

TRANSPORT OF SEDIMENT BY GLACIERS

One way a glacier differs from a stream is the way in which it carries its load of rock particles. Unlike a stream, part of a glacier's coarse load can be carried at its sides as well as on its surface. A glacier can carry far larger pieces

Figure 16.22 Drumlins A field of drumlins in Dodge County, Wisconsin, each shaped like an inverted and elongated hull of a ship, are aligned parallel to the flow direction of the continental ice sheet that shaped them during the last glaciation.

of rock, and it can transport large and small pieces side by side without segregating them according to size and density into a bed load and a suspended load. Because of these differences, sediments deposited directly by a glacier are neither sorted nor stratified.

The load of a glacier typically is concentrated at its base and sides because these are the areas where glacier and bedrock are in contact and where abrasion and plucking are effective. In addition, much of the rock debris spread across the surface of valley glaciers arrived there by rockfalls from adjacent cliffs. If a rockfall reaches the accumulation area, the flow paths of the ice (Figure 16.10) will carry the debris downward through the glacier and then upward to the surface in the ablation area. If rocks fall onto the ablation area, the debris will remain at the surface and be carried along by the moving ice. Where two glaciers join, rocky debris at their margins merges to form a distinctive, dark-colored *medial moraine* (Figures 16.1B and 16.10).

Much of the load in the basal ice of a glacier consists of fine sand and silt grains informally called *rock flour*. These particles have sharp, angular surfaces that are produced by crushing and grinding.

GLACIAL DEPOSITS

Glaciers are efficient agents of erosion and transport. Many of the sediments and landforms they produce are distinctive, making it relatively easy for geologists to interpret the record of past glacier variations.

Sediments deposited by a glacier or by streams produced by melting glacier ice are collectively called **glacial drift**, or simply **drift**. The term dates from the early nineteenth century, when it was vaguely conjectured that all such deposits had been "drifted" to their resting places during the biblical flood of Noah or in some other ancient body of water. Glacial drift includes sediments associated both with moving ice and with stagnant ice. Several types of sediment are recognized. They form a gradational series ranging from nonsorted to sorted deposits.

ICE-LAID DEPOSITS

Sediments deposited directly from glacier ice range widely in character. Most are sufficiently unique that they enable geologists to identify glaciated landscapes and to map the extent of former glaciers.

Till and Erratics. At one end of the range is **till**, which is nonsorted drift deposited directly from ice. The term was used by Scottish farmers long before the origin of the sediment was understood. The rock particles in a body of till lie just as they were released from the ice (Figure 7.6). Most tills are a random mixture of rock fragments in which a matrix of fine-grained sediment surrounds larger stones of various sizes. The till matrix consists largely of sand and silt derived by abrasion of the glacier bed and from reworking of preexisting fine-grained sediments. Pebbles and larg-

Figure 16.23 Erratic Boulder Large erratic boulder of granite embedded in an end moraine of the last glaciation on Tierra del Fuego in southernmost Chile. The local bedrock is sedimentary. The nearest possible source area for the boulder lies in the high Cordillera Darwin to the south, on the opposite side of a deep fjord system.

er rock fragments in till often have smoothed and abraded surfaces, and some are striated. Both the stones and the coarser matrix grains in till tend to lie with their longest axis aligned with the direction of ice flow.

As we learned in Chapter 7, *tillite* is an ancient till that has been converted to rock. Tillites constitute a primary line of evidence for pre-Pleistocene glacial ages.

In many cases, not all the boulders and smaller rock fragments in a till are the same kind of rock as the underlying bedrock, indicating that these components of the till were carried to their present site from elsewhere. A glacially deposited rock or rock fragment with a lithology different from that of the underlying bedrock is an **erratic** (Latin for wanderer) (Figure 16.23). Some huge erratics weigh many tons and are found tens or even hundreds of kilometers from their sources. In areas of ice-sheet glaciation, erratics derived from distinctive bedrock sources may have a fanlike distribution, spreading out from the area of outcrop and reflecting the diverging pattern of ice flow.

Glacialmarine Drift. Closely resembling till, **glacialmarine drift** is sediment deposited on the seafloor from ice shelves or icebergs. As an iceberg or the base of an ice shelf slowly melts, the contained sediment is released and settles to the seafloor where it forms a nonsorted deposit. Individual stones released from icebergs that have drifted into the open sea fall to the seafloor and plunge into unconsolidated marine sediments. The impact causes any laminated structure in the uppermost sediment layers to be deformed. Such *dropstones* are also common in the sediments of lakes that are ponded along glacier margins.

Moraines. A moving glacier carries with it rock debris eroded from the land over which it is passing or dropped on

Figure 16.24 End Moraine
Tumbling glacier on Mount Robson in British Columbia has retreated upslope from a sharp-crested terminal moraine that it constructed during the nineteenth and twentieth centuries.

the glacier surface from adjacent cliffs. As the debris is transported past the equilibrium line and ablation reduces ice thickness, the debris begins to be deposited. Some of the basal debris is plastered directly onto the ground as till. Some also reaches the glacier margin where it is released by the melting ice and either accumulates there or is reworked by meltwater that transports it beyond the terminus.

An accumulation of till having a surface form that is unrelated to the underlying bedrock is called a **moraine**. If a till is widespread, has a relatively smooth-surface topography, and consists of gently undulating knolls and shallow, closed depressions, we call it *ground moraine*; commonly, it is a blanket of till 10 m or more thick that was deposited beneath a glacier. By contrast, an *end moraine* is a ridge-like accumulation of drift deposited along the margin of a glacier (Figure 16.24). An end moraine deposited at the glacier terminus is a *terminal moraine*, whereas a similar deposit along the side of a valley glacier is a *lateral moraine*. Normally, terminal and lateral moraines are segments of a single, continuous landform deposited below the equilibrium line.

End moraines can form in several ways: as sediment is bulldozed by a glacier advancing across the land, as loose surface debris on a glacier slides off and piles up along a glacier margin, or as debris melts out of ice and accumulates at the margin of a glacier. End moraines range in height from a few meters to hundreds of meters. The great thickness of some lateral moraines results from the repeated accretion of sediment from debris-covered glaciers during successive ice advances.

STRATIFIED DRIFT

In contrast to nonsorted till and glacialmarine drift, some glacial drift is both sorted and stratified. This kind of drift is not deposited directly by glacier ice but rather by melt-water flowing from the ice. **Stratified drift** ranges from coarse, very poorly sorted sandy gravels deposited by turbulent streams of meltwater to fine-grained, well-sorted silts and clays deposited in quiet water.

Outwash. Stratified sediment deposited by meltwater streams as they flow away from a glacier margin is called **outwash** (the sediment is washed "out" beyond the ice). Such streams typically have a braided pattern because of the large sediment load they are moving. If the streams are free to swing back and forth widely beyond the glacier terminus, they deposit outwash to form a broad *outwash plain*. Meltwater streams confined by valley walls build an outwash body called a *valley train* (Figures 16.10 and 16.25A).

Following glacier retreat, a stream's sediment load is reduced and the underloaded stream cuts down into its outwash deposits to produce *outwash terraces* (Figure 16.25B). A succession of terraces commonly is found in valleys that have experienced repeated glaciations (Figure 16.26). Generally, each prominent terrace can be traced upstream to an end moraine or to the limit of a former glacier.

Deposits Associated with Stagnant Ice. When ablation greatly reduces a glacier's thickness in its terminal zone, ice flow may virtually cease. Sediment carried by meltwater flowing over or beside such stagnant ice is deposited as stratified drift that slumps and collapses as the supporting ice slowly melts away. Such sediment, called **ice-contact stratified drift**, is recognized by (1) abrupt changes in grain size, (2) distorted, offset, and irregular stratification, and (3) extremely uneven surface form. Bodies of ice-contact stratified drift have many distinctive forms and are classified according to their shape (Figure 16.25B). Among the landforms most likely to be seen are *kames*, small hills of

A.

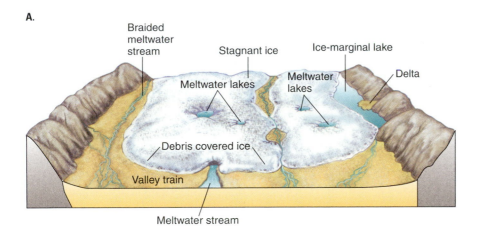

B.

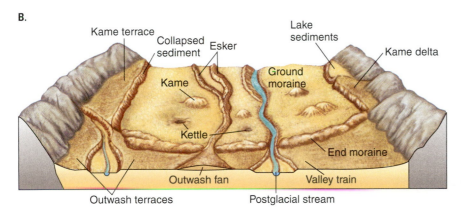

Figure 16.25 Landforms Related to Stagnant Ice Origin of ice-contact stratified drift and related landforms associated with stagnant ice. A. Ablating stagnant ice furnishes temporary retaining walls for bodies of stratified sediment deposited by meltwater streams and in meltwater lakes. B. As supporting ice melts, sediments slump, creating kettle-and-kame topography.

Figure 16.26 Outwash Terraces Flat-topped outwash terraces related to several phases of glaciation border meandering Cave Stream on South Island, New Zealand.

ice-contact stratified drift; *kettles*, basins in drift created by the melting away of a mass of underlying glacier ice; and *eskers*, long, sinuous ridges of sand and gravel deposited by a meltwater stream flowing under or within stagnant glacier ice. Extremely uneven terrain underlain by ice-contact stratified drift and marked by numerous kettles and kames is clear evidence of former stagnant-ice conditions.

> **Before you go on:**
> 1. What is meant by the term glaciation?
> 2. How do glaciers erode the land?
> 3. What landforms do glaciers produce?

PERIGLACIAL LANDSCAPES AND PERMAFROST

Land areas beyond the limit of glaciers where low temperature and frost action are important factors in determining landscape characteristics are called **periglacial** zones. Periglacial conditions are found over more than 25 percent of Earth's land areas, primarily in the circumpolar zones of each hemisphere and at high altitudes.

A common feature of periglacial regions is perennially frozen ground, also known as **permafrost**—sediment, soil, or even bedrock that remains continuously below freezing for an extended time (from two years to tens or hundreds

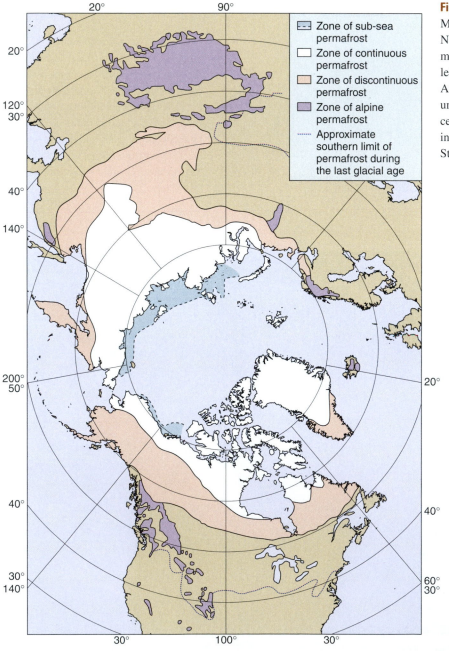

Figure 16.27   Distribution of Permafrost
Map showing extent of permafrost in the Northern Hemisphere. Continuous permafrost lies mainly north of the 60th parallel and is most widespread in Siberia and Arctic Canada. Extensive alpine permafrost underlies the high, cold plateau region of central Asia. Smaller, isolated bodies occur in the high mountains of the western United States and Canada.

Legend:
- Zone of sub-sea permafrost
- Zone of continuous permafrost
- Zone of discontinuous permafrost
- Zone of alpine permafrost
- Approximate southern limit of permafrost during the last glacial age

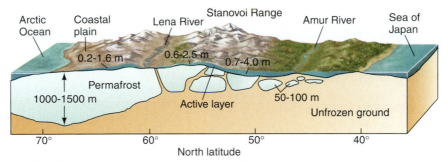

Figure 16.28 Permafrost in Siberia Diagrammatic transect across northeastern Siberia (vertical scale is greatly exaggerated) showing distribution and thickness of permafrost and thickness of the seasonally thawed (active) layer at the surface.

Thick, continuous permafrost under the Arctic coastal plain thins southward where it becomes discontinuous in response to warmer average annual temperature. The active layer increases in thickness southward due to warmer summer temperatures.

of thousands of years). The largest areas of permafrost occur in North America, northern Asia, and the high, cold Tibetan Plateau (Figure 16.27). It has also been found on many high mountain ranges, even including some lofty summits in tropical and subtropical latitudes. The southern limit of continuous permafrost in the Northern Hemisphere generally lies where the average annual air temperature is between -5 and -10°C (23 and 14°F).

Most permafrost is believed to have originated during either the last glacial age or earlier glacial ages. Remains of woolly mammoth and other extinct ice-age animals found well preserved in frozen ground indicate that permafrost existed at the time of their death.

The depth to which permafrost extends depends not only on the average air temperature but also on the rate at which heat flows upward from Earth's interior and on how long the ground has remained continuously frozen. The maximum reported depth of permafrost is about 1500 m (4900 ft) in Siberia (Figure 16.28). Thicknesses of about 1000 m (3300 ft) in the Canadian Arctic and at least 600 m (2000 ft) in northern Alaska have been measured. These areas of very thick permafrost all occur in high latitudes outside the limits of former ice sheets. The ice sheets would have insulated the ground surface and, where thick enough, caused ground temperatures beneath them to rise close to the pressure melting point of ice. On the other hand, open ground unprotected from subfreezing air temperatures by an overlying ice sheet could have become frozen to great depths during prolonged cold periods.

Permafrost presents a real challenge to people who live and work in permafrost regions (see *Understanding Our Environment*, Box 16.1, *Living with Permafrost*).

Before you go on:

1. What are periglacial zones, and where are they found?

2. What determines the extent and depth of permafrost?

THE GLACIAL AGES

As early as 1821, European scientists began to recognize features characteristic of glaciation in places far from any existing glaciers. They drew the then-remarkable conclusion that glaciers must once have covered extensive regions. The concept of a glacial age with widespread effects was first proposed in 1837 by Louis Agassiz, a Swiss scientist who achieved considerable fame through his hypothesis. Although at first many people regarded Agassiz's idea as outrageous, gradually, through the work of many geologists, the concept gained widespread acceptance. Today, the study of the glacial ages provides us with dramatic evidence of rapid global climatic changes on Earth and with clues about how natural physical and biological systems responded to these changes. It also gives us important information about how glaciers behave and helps us understand some of the basic physical processes of the crust and upper mantle.

ICE-AGE GLACIERS

Over tens of millions of years, the climate slowly grew cooler as Earth moved into a late Cenozoic glacial era. During the last few million years, the planet has experienced numerous glacial–interglacial cycles superimposed on the long-term cooling trend. At present, we find ourselves near a time of maximum warmth in such a cycle and likely poised to begin a slow decline into the next glacial age, which will culminate thousands of years in the future.

About 30,000 years ago, late in the Pleistocene Epoch, an extensive ice sheet that had formed over eastern Canada began to spread south toward the United States and west

UNDERSTANDING OUR ENVIRONMENT

LIVING WITH PERMAFROST

In permafrost terrain, a thin surface layer of ground that thaws in summer and refreezes in winter is known as the *active layer*. In summer this thawed layer tends to become very unstable. The permafrost beneath, however, is capable of supporting large loads without deforming. Many of the landscape features we associate with periglacial regions reflect movement of regolith within the active layer during annual freeze and thaw cycles.

Permafrost presents unique problems for people living on it. If a building were constructed directly on the surface,

Figure B16.1 Sinking Cabin on Thawing Permafrost
This cabin in central Alaska settled more than a meter in eight years as permafrost beneath its foundation thawed.

the warm temperature developed when the building was heated would be likely to thaw the underlying permafrost, making the ground unstable (Figure B16.1). Arctic inhabitants learned long ago that they must place the floors of their buildings above the land surface on pilings or open foundations so that cold air can circulate freely, thereby keeping the ground beneath the building frozen.

Wherever a continuous cover of low vegetation on a permafrost landscape is ruptured, melting can begin. As the permafrost melts, the ground collapses to form impermeable basins containing ponds and lakes. Thawing can also be caused by human activity, and the results can be environmentally disastrous. Large wheeled or tracked vehicles crossing the Arctic tundra can quickly rupture it. The water-filled linear depressions that result from thawing can remain as features of the landscape for many decades.

The discovery of a commercial oil field on the North Slope of Alaska in the 1960s generated the need to transport the oil southward by pipeline to an ice-free port. The company formed to construct the pipeline was faced with some unique problems. In order for the sticky oil to flow through a pipeline in the frigid Arctic environment, the oil had to be heated. However, an uninsulated, heated pipe in the frozen ground could melt the surrounding permafrost. Even if the pipe were insulated before placing it underground, the surface vegetation cover would be disrupted, likely leading to melting and instability. For these reasons, along much of its course, the pipeline was constructed in piers above the ground, thereby greatly reducing the possibility of ground collapse.

---

toward the Rocky Mountains. Simultaneously, another great ice sheet that originated in the highlands of Scandinavia spread southward across northwestern Europe and overwhelmed the landscape (Figure 16.29). Large ice sheets also grew over northern regions of North America and Eurasia, including some areas now submerged by shallow polar seas, and over the mountain ranges of western Canada. The ice sheets in Greenland and Antarctica expanded and advanced across areas of the surrounding continental shelves that were exposed by falling sea level. Glaciers also developed in the world's major mountain ranges, including the Alps, Andes, Himalaya, and Rockies, as well as in numerous smaller ranges and on isolated peaks scattered widely through all latitudes.

On a global scale, the areas of former glaciation add up to an impressive total of more than 44 million km^2, which is about 29 percent of Earth's present land area. Today, by comparison, only about 10 percent of the world's land area is covered with glacier ice; of this area, 84 percent lies in the Antarctic region.

DRAINAGE DIVERSIONS AND GLACIAL LAKES

The growth of ice sheets over the continents caused disruption of major stream systems. In North America, repeated encroachment of Pleistocene ice sheets displaced the Missouri and Ohio rivers into new courses beyond the ice margin. When glaciers blocked preglacial drainage paths, water was ponded to form ice-dammed lakes. Vast ice-marginal lakes that developed beyond the edges of the expanding ice sheet in eastern North America shifted in location and changed size as the glacier receded. Major disruption of drainage also produced large ice-marginal lakes in northern Asia as ice expanding southward in western Siberia blocked the courses of major north-flowing rivers.

LOWERING OF SEA LEVEL

Whenever large glaciers formed on the land, the moisture needed to produce and sustain them was derived primarily

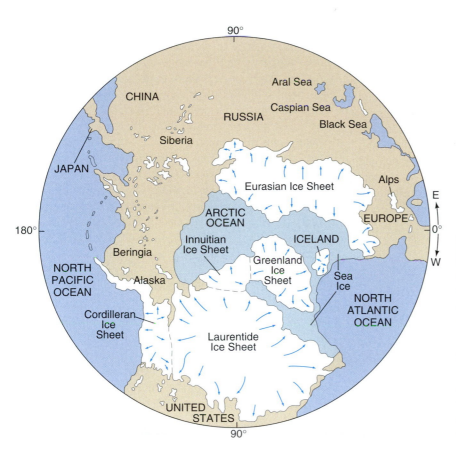

Figure 16.29 Northern Hemisphere Ice-Age Glaciers Areas of the Northern Hemisphere that were covered by glaciers during the last glacial age. Arrows show the general directions of ice flow. Coastlines are shown as they were at that time, when world sea level was at least 100 m lower than present. Sea ice covering the Arctic Ocean extended south into the North Atlantic. The extent of former glacier ice over shallow continental shelves of northern Eurasia, as well as in parts of northern North America, is uncertain.

from the oceans. As a result, sea level was lowered in proportion to the volume of ice on land. During the most recent glacial age, world sea level fell at least 100 m, thereby causing large expanses of the shallow continental shelves to emerge as dry land. At that time, the Atlantic coast of the United States south of New York lay as much as 150 km east of its present position. At the same time, lowering of sea level joined Britain to France where the English Channel now lies, and North America and Asia formed a continuous landmass across what is now the Bering Strait (Figure 16.29). These and other land connections allowed plants and animals, as well as humans, to pass freely between land areas that now are separated by ocean waters.

DEFORMATION OF EARTH'S CRUST

The weight of the massive ice sheets caused Earth's crust to subside beneath them, a process described in Chapter 2. Because ice (density 0.9 g/cm^3) is one-third as dense as average crustal rock (2.7 g/cm^3), an ice sheet 3 km thick could cause the crust to subside by as much as 1 km. The Hudson Bay region of Canada, which 20,000 years ago lay near the center of the vast Laurentide Ice Sheet (Figure 16.29), is still rising as the lithosphere and asthenosphere adjust to the removal of this ice load. Using glacial-geologic evidence, we can measure accurately the rates at

which the crustal rocks have risen over many thousands of years. Such measurements provide us with important information about how the lithosphere and asthenosphere behave when subjected to changing loads.

EARLIER GLACIATIONS

Until rather recently, it was thought that Earth had experienced four glacial ages during the Pleistocene Epoch. This assumption was based on studies of ice-sheet and mountain-glacier deposits and had its roots in early studies of the Alps, where geologists mapped moraines and outwash terraces and interpreted them as evidence of four Pleistocene glaciations. This traditional view had to be modified, however, when studies of deep-sea sediments disclosed a long succession of glaciations, the most recent of which was shown by radiocarbon dating to correspond to the youngest extensive glacial drift on land. Paleomagnetic dating of these marine sediments showed that during the last 800,000 years the length of each glacial–interglacial cycle averaged about 100,000 years. For the Pleistocene Epoch as a whole, more than 20 glacial ages are recorded, rather than the traditional four. Geologists now realize that the glacial record on land is incomplete and marked by numerous unconformities, whereas many areas of the deep sea contain a record of continuous sedimentation.

SEAFLOOR EVIDENCE

Deep-sea sediments provide some of the best evidence we have of the cyclic changes of climate from glacial to interglacial conditions. When we study the history of progressively earlier times by examining seafloor sediments that are sampled by drilling, we find that the biologic component of the sediments records repeated shifts in the composition of surface-water animal and plant populations—from warm interglacial forms to cold glacial forms and back to warm interglacial forms. The ratio of the amounts of the isotopes ^{18}O to ^{16}O in layers of calcareous ooze in these cores also fluctuates with a similar pattern. These

$^{18}O/^{16}O$ variations in Pleistocene marine sediments are thought primarily to represent changes in global ice volume. When water evaporates, water containing the light isotope ^{16}O evaporates more easily than water containing the heavier ^{18}O. Consequently, when water evaporates from the ocean and is precipitated on land to form glaciers, the ice is enriched in the lighter isotope relative to the ocean water that remains behind. As a result, Pleistocene glaciers contained more of the light isotope, while the oceans became enriched in the heavy isotope. Isotope curves derived from the seafloor sediments therefore provide a continuous reading of changing ice volume on the planet (Figure 16.30). Because glaciers wax and wane in response to climatic changes, the isotope curves also give a generalized view of global climatic change.

PRE-PLEISTOCENE GLACIATIONS

Ancient glaciations, identified mainly by tillites and striated rock surfaces, are known from older parts of the geologic column. The earliest recorded glaciation dates to about 2.3 billion years ago, in the early Proterozoic. Evidence of other glacial episodes has been found in rocks of late Proterozoic, early Paleozoic, and late Paleozoic age. During the late Paleozoic, 50 or more glaciations are believed to have occurred.

LITTLE ICE AGES

Old documents, lithographs, and paintings of Alpine valleys dating to the sixteenth and seventeenth centuries describe and depict glaciers that advanced over small villages and became larger than at any time in human memory (Figure 16.31). Similar glacier advances took place in other parts of the world and in many cases led to the greatest expansion of ice since the end of the last glacial age 10,000 years ago. During this recent interval of generally cool climate, which started in the mid-thirteenth century and lasted until the mid-nineteenth century, mountain glaciers expanded worldwide. This Little Ice Age, as it is commonly known, was similar to other brief episodes of glacier expansion that were superimposed on the much longer glacial–interglacial cycles.

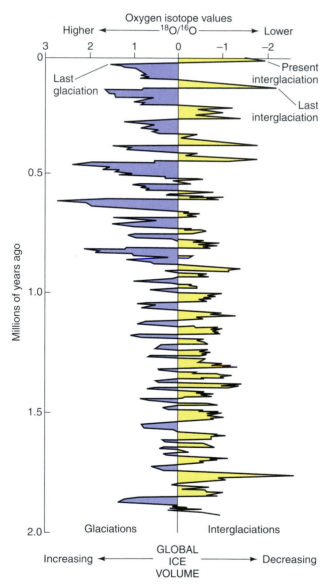

Figure 16.30 Isotopic Record of Ice Volume Average curve of oxygen-isotope variations in deep-sea cores representing global changes in ice volume during the last 2 million years. During most of the last million years, each glacial–interglacial cycle was about 100,000 years long; earlier cycles were about 40,000 years long.

Before you go on:

1. What evidence indicates that ice sheets once covered extensive areas?

2. How did the growth of the Pleistocene ice sheets over the continents affect stream systems? sea level? Earth's crust?

3. How and why have ideas changed about the number of glacial–interglacial cycles Earth has experienced?

4. When was the Little Ice Age and how did its climate differ from today's climate?

A.

B.

Figure 16.31 Historic Changes in Glaciers A. A lithograph made in the 1850s shows Rhone Glacier in the Swiss Alps close to its maximum extent during the Little Ice Age. The glacier terminates in a broad lobe that crosses the floor of the upper Rhone Valley. B. A photograph of the same area taken 110 years later shows the terminus of Rhone Glacier perched high up in the headward part of the valley. Extensive thinning and recession of this glacier and other glaciers worldwide marked the end of the Little Ice Age.

WHAT CAUSES GLACIAL AGES?

Ever since geologists first became convinced that Earth had experienced a succession of ice ages, the problem of their cause has been a subject of continuing debate and investigation. We are now much closer to resolving this problem than ever before, but many answers still elude us, for climatic change is a complex phenomenon that involves not only the atmosphere, but also the solid earth, the oceans, and the biosphere, as well as extraterrestrial factors. The ultimate solution to the problem of the ice ages, therefore, will involve the cooperative efforts of specialists from many different fields of science.

GLACIAL ERAS AND SHIFTING CONTINENTS

Several different episodes of glacial and interglacial ages, each episode lasting tens of millions of years, can be identified in the geologic record. What seems to be the only reasonable explanation for their pattern is suggested by slow but important geographic changes that affect the crust of the planet. These changes include (1) the movement of continents as they are carried along with shifting plates of lithosphere; (2) the large-scale uplift of continental crust where continents collide; (3) the creation of mountain chains where one plate overrides another; and (4) the opening or closing of ocean basins and seaways between moving landmasses.

The effect of such earth movements on climate is illustrated by the fact that low temperatures are found, and glaciers tend to form and persist, in two kinds of situations: (1) on landmasses at high latitudes and (2) at high altitudes. Furthermore, glaciers are particularly common in places where winds can supply abundant moisture evaporated from a nearby ocean. Today, 84 percent of Earth's glacier ice lies in Antarctica, where temperatures are constantly below freezing and the land is surrounded by ocean. The only glaciers found at or close to the equator lie at very high altitudes.

Abundant evidence leads us to conclude that the positions, shapes, and altitudes of landmasses have changed with time (Chapter 2), in the process altering the paths of ocean currents and atmospheric circulation. Where evidence of ancient ice-sheet glaciation is now found in low latitudes, we are led to infer that such lands were formerly located in higher latitudes where large glaciers could be sustained.

Evidence is clear that in the late Paleozoic Era, a continental ice sheet repeatedly covered much of ancient Gondwanaland (Figure 2.13) (see Revisiting Plate Tectonics: African Ice Sheets). The absence of widespread glacial deposits in rocks of Mesozoic age implies that during that era most of the world's landmasses had moved away from polar latitudes and that climates were mild. By the early Cenozoic, slowly shifting landmasses once again moved into polar latitudes, and tectonic movements were beginning to raise large areas of the western United States and central Asia to high altitudes. By middle Cenozoic time, Earth was again poised to enter another lengthy glacial era.

ICE AGES AND THE ASTRONOMICAL THEORY

As initially discovered through studies of glacial deposits and later verified by studies of deep-sea cores, glacial and inter-glacial ages have alternated for more than 2 million years (Figure 16.30). Determining their cause has long been a fun-damental challenge to the development of a comprehensive theory of climate. A preliminary answer was provided by Scottish geologist John Croll, in the mid-nineteenth century, and later elaborated by Milutin Milankovitch, a Serbian astronomer of the early twentieth century.

Croll and Milankovitch recognized that minor varia-tions in Earth's orbit around the Sun and in the tilt of Earth's axis cause slight but important variations in the amount of radiant energy reaching any given latitude on the planet's surface (see *The Science of Geology*, Box 16.2, *Understanding Milankovitch*). By reconstructing and dat-ing the history of climatic variations during the Quaternary Period, geologists have shown that fluctuations of climate on the time scale of glacial cycles correlate strongly with cyclical variations in Earth's tilt and orbital configuration. This persuasive evidence supports the theory that astro-nomical changes that determine the distribution of solar radiation reaching Earth's surface control the basic *timing* of the glacial–interglacial cycles.

ICE-CORE ARCHIVES OF CHANGING CLIMATE

In most temperate mountain glaciers of average size, the oldest ice is only a few hundred years old. By contrast, in the great polar ice sheets, and in some high-altitude glaciers at lower latitudes that have temperatures well below freez-ing, the oldest ice spans the last glacial age or even the last several glacial ages. For more than four decades, these gla-ciers have been the focus of a unique endeavor: obtaining from ice cores a high-resolution record of changing climate and atmospheric composition extending far into the past.

When snow accumulates on a glacier, it compacts and the air between snow crystals becomes trapped in bubbles as the snow slowly changes to glacier ice. The trapped air is a sample of the atmosphere over the glacier at the time the ice forms. By sampling the ice and analyzing its trapped gases and solid particles, an array of information can be obtained. Most of what we now know about the past composition of the atmosphere during glacial times has been obtained from such studies.

Measurements of the stable isotopes of oxygen in ice cores provide a measure of air temperature when the ice formed. In much the same way that annual rings enable us to date the age of trees, cyclic shifts in the isotopic signal, related to the annual solar cycle, permit ages within the upper part of ice cores to be dated to the nearest year. The

deeper part of ice cores loses the annual signal because of internal deformation and flow, and so the oldest ice is dated by modeling ice flow at the coring site. Among the remark-able discoveries is a high-amplitude, high-frequency varia-tion in oxygen isotopes in the Greenland Ice Sheet that characterized glacial times (Figure 16.32A, B). These sig-

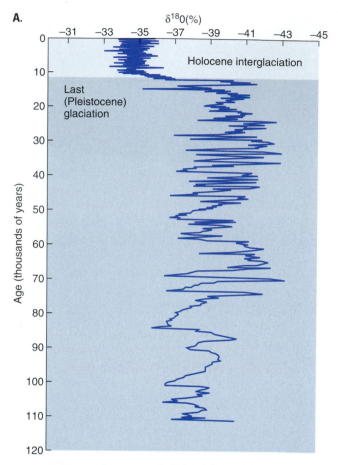

Figure 16.32 Temperature Variations Over Greenland A. Curve of oxygen-isotope variations in an ice core from the sum-mit of the Greenland Ice Sheet. The high variability of the signal indicates rapid changes in climate during the last glaciation that provide important clues about the causes of ice-age climate vari-ations. B. Drill site at top of the ice sheet.

nificant and rapid changes of temperature likely were related to oceanographic changes in the adjacent North Atlantic.

In addition to past temperatures, glaciologists can measure variations in atmospheric carbon dioxide and methane, and thereby show that major changes in the atmospheric content of these greenhouse gases occur during a glacial–interglacial cycle. Similar variations are seen in the content of microparticles in the ice (i.e., windblown dust) that demonstrate that the glacial atmosphere was very dusty compared to that of interglacial times. By analyzing the chemistry of the dust, source areas of the sediment can be determined, as well as major wind trajectories. Thus, we now know that most of the dust in Greenland ice cores originated in the desert basins of central Asia and that dust in the Antarctic Ice Sheet originated in Patagonia.

Glaciers suitable for obtaining ice core records occur not only in the two polar regions (Greenland and Antarctica), but also at the crests of many high plateaus and mountain ranges (for example, in Tibet, the high Andes, and the Himalaya). Thus, changes in atmospheric conditions through time can be reconstructed in a variety of geographic settings and then compared to identify both regional and global patterns.

ATMOSPHERIC COMPOSITION

Although orbital factors can explain the timing of the glacial–interglacial cycles, the variations in solar radiation reaching Earth's surface are too small to account for the average global temperature changes of 4 to 10°C that are implied by geologic and biologic evidence. We therefore must conclude that other factors are also involved. Somehow, the slight variations in radiant energy received from the Sun, caused by orbital changes, must be translated into a temperature change sufficiently large to generate and maintain the huge Pleistocene ice sheets. We are not yet certain how this was accomplished, but some of the factors involved are likely to be changes in the chemical composition and dustiness of the atmosphere, and changes in the reflectivity of Earth's surface.

Air bubbles in glacier ice of the Antarctic and Greenland ice sheets are samples of Earth's ancient atmosphere. Studies of the chemical composition of trapped air that dates back to the most recent ice age indicate that during glacial times the atmosphere contained less carbon dioxide and methane than it does today (Figure 16.33). These two gases are important "greenhouse" gases (Chapter 19). If their con-

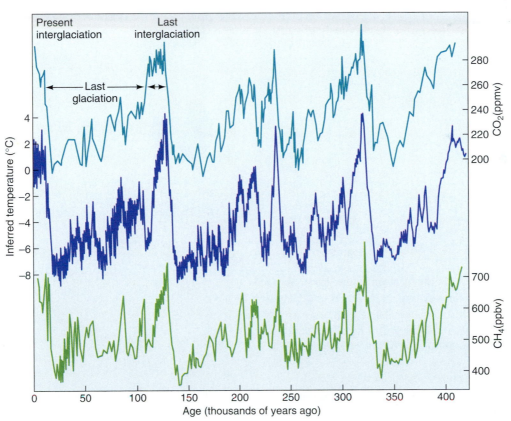

Figure 16.33 Changing Composition of the Atmosphere
Curves comparing changes in carbon dioxide and methane with temperature changes (based on oxygen-isotope values) in samples from a deep ice core drilled at Vostok Station, Antarctica. Concentrations of these greenhouse gases were high during the last interglaciation, just as they are now, but they were lower during glacial times. The curves are consistent with the hypothesis that these gases contributed to warm interglacial climates and cold glacial climates.

Box 16.2 THE SCIENCE OF GEOLOGY

UNDERSTANDING MILANKOVITCH

The concept that slow, cyclical changes in the path Earth takes around the Sun, and in the tilt of its axis, are related to changing climate on the scale of glacial and interglacial ages is generally referred to as the Milankovitch Theory. Three movements are involved:

First, the axis of rotation, which now points in the direction of the North Star, wobbles like a spinning top (Figure B16.2A). The wobbling movement causes Earth's axis to trace a cone in space, completing one full revolution every 26,000 years. If we were to look down on the surface from a point above Earth's axis, this movement would be clockwise. At the same time, the axis of Earth's elliptical orbit is also rotating, but much more slowly, in the opposite direc-

tion. These two motions together cause a progressive shift in the position of the four cardinal points of Earth's orbit (spring and autumn equinoxes and winter and summer solstices). As the equinoxes move slowly around the orbital path, a motion called *precession of the equinoxes*, they complete one full cycle in about 23,000 years.

Second, the *tilt* of the axis, which is now 23.5°, shifts over a range of about 3° during a span of about 41,000 years (Figure B16.2B).

Finally, the *eccentricity* of the orbit, which is a measure of the departure of the orbit from circularity, changes over periods of 100,000 and 400,000 years. About 50,000 years ago, the orbit was more circular (lower eccentricity) than it has been for the last 10,000 years (Fig B16.2C).

A. Precession of the equinoxes (period = 23,000 years)

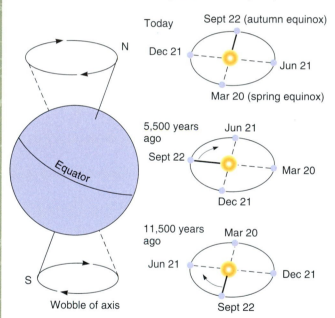

B. Tilt of the axis (period = 41,000 years)

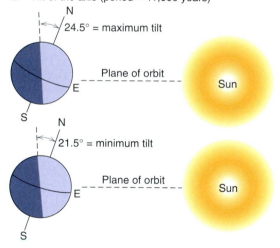

C. Eccentricity (dominant period =100,000 years)

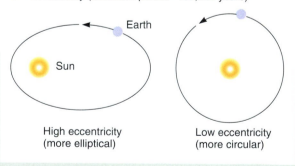

Figure B16.2 Geometry of the Earth's Orbit Geometry of Earth's orbit and axial tilt. A. *Precession*. Earth wobbles on its axis like a spinning top, making one revolution every 26,000 years. The axis of Earth's elliptical orbit also rotates, though more slowly, in the opposite direction. These motions together cause a progressive shift, or precession, of the spring and autumn equinoxes, with each cycle lasting about 23,000 years. B. *Tilt*. The tilt of Earth's axis, which now is about 23.5°, ranges from 21.5 to 24.5°, with each cycle lasting about 41,000 years. Increasing tilt means a greater difference, for each hemisphere, between the amount of solar radiation received in summer and that received in winter. C. *Eccentricity*.

Earth's orbit is an ellipse with the Sun at one focus. The shape of the orbit changes from almost circular (low eccentricity) to more elliptical (high eccentricity) over periods of 100,000 and 400,000 years. The higher the eccentricity, the greater is the seasonal variation in radiation received.

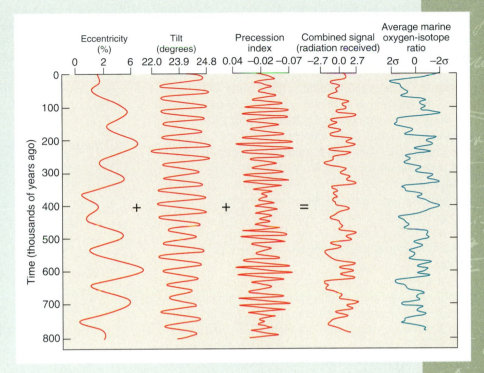

Figure B16.3 Orbital Changes Through Time Curves showing variations in orbital eccentricity, tilt, and precession index during the last 800,000 years. Summing the effects of these factors produces a combined signal that shows the amount of radiation received on Earth at a particular latitude through time.

The slow but predictable changes in orbital path and tilt cause long-term variations of as much as 10 percent in the amount of radiant energy that reaches any particular place on Earth's surface in a given season. Such variations are most pronounced at high latitudes. The effects of the three variables can be summed, and we can derive a combined signal that depicts the amount of radiation received at any specified latitude and at any specified time in the past (Figure B16.3). Because the astronomical variables are predictable, we can also use this theory to extrapolate the signal into the future and thereby forecast the timing of future glaciations and interglaciations.

centration in the atmosphere is high, they trap radiant energy emitted from Earth's surface that would otherwise escape to outer space. As a result, the lower atmosphere heats up and Earth's climate becomes warmer. If the concentration of these gases is low, as it was during glacial times, surface air temperatures are reduced. Calculations suggest that the low levels of these two important atmospheric gases during glacial times can account for nearly half of the total ice-age temperature lowering. Therefore, the greenhouse gases likely play a significant role in explaining the *magnitude* of past global temperature changes. Although we know that the percentages of these gases fell during glacial times, we do not yet know what caused them to drop.

Ice core studies have also shown that the amount of dust in the atmosphere was unusually high during glacial times. The fine dust was picked up by strong winds blowing across outwash deposits and dry desert basins. So much dust was delivered to the atmosphere that the sky must have appeared hazy much of the time. The fine atmospheric dust scattered incoming radiation back into space, thereby further cooling Earth's surface.

Whenever the world enters a glacial age, large areas of land are progressively covered by snow and glacier ice. The highly reflective surfaces of snow and ice scatter incoming radiation back into space, further cooling the lower atmosphere. Together with lower greenhouse gas concentrations and increased atmospheric dust, this additional cooling would favor the expansion of glaciers.

CHANGES IN OCEAN CIRCULATION

The circulation of ocean waters also plays an important role in global climate. As warm surface water moving northward into the North Atlantic evaporates, the remaining water becomes saltier and cooler. The resulting cold, salty water is dense and sinks deep within the ocean (Chapter 18). Heat released to the atmosphere as the water cools maintains a relatively mild climate in northwestern Europe. Consider, however, what would happen if this circulation ceased.

The rate of deep-ocean circulation is sensitive to surface salinity at sites where dense water forms. Studies have shown that during times of reduced salinity, the deep-ocean circulation is reduced. We therefore can postulate that as summer radiation decreased at the onset of a glaciation, driven by Earth's orbital changes, the high-latitude ocean and atmosphere cooled, thereby decreasing evaporation and leading to the expansion of sea ice. The resulting freshening of the high-latitude surface water would have halted the production of dense, salty water and shut down the vertical circulation system. Reduction of high-latitude evaporation, significantly reducing the release of heat to the atmosphere, would have maintained cold air masses moving eastward across the North Atlantic. Further cooled by an expanding sea-ice cover in the North Atlantic and by the growing ice sheets on the continents, the climate of Europe would have become increasingly colder, ultimately causing the formation of permafrost in a broad zone beyond the expanding Scandinavian sector of the Eurasian ice sheet.

Thus, a change in the ocean's circulation system could amplify the relatively small climatic effect attributable to astronomical changes. Furthermore, it may help explain why Earth's climate system appears to fluctuate between two relatively stable modes—one in which the ocean circulation system is operational (during interglaciations) and one in which it has shut down (during glaciations).

SOLAR VARIATIONS, VOLCANIC ACTIVITY, AND LITTLE ICE AGES

Climatic fluctuations lasting decades or centuries were responsible for the Little Ice Age and similar episodes of glacier expansion. However, such fluctuations are too brief to be caused either by movements of continents or variations in Earth's orbit; they therefore require us to seek other explanations for their cause. Two have received special attention.

One hypothesis regarding the cause of glacial events like the Little Ice Age is based on the concept that the energy output of the Sun fluctuates over time. The idea is appealing because it might explain climatic variations on several different time scales. Although correlations have been proposed between weather patterns and rhythmic fluctuations in the number of sunspots appearing on the surface of the Sun, as yet there has been no convincing demonstration that solar variations are responsible for climatic changes on the scale of the Little Ice Age.

Large explosive volcanic eruptions can eject huge quantities of ash into the atmosphere to create a veil of fine dust that circles the globe. Like other types of dust, the fine ash particles tend to scatter incoming solar radiation, resulting in a slight cooling at Earth's surface. The dust settles out rather quickly, generally within a few months to a year. Nevertheless, tiny droplets of sulfuric acid, produced by the interaction of volcanically emitted SO_2 gas with oxygen and water vapor, also scatter the Sun's rays, and such droplets remain in the upper atmosphere for several years.

After a large eruption, volcanic dust and gases in the atmosphere can lower average surface air temperature by 0.5 to 1.0 °C, sufficient to influence the mass balance of glaciers. A close association between intervals of glacier advance and periods of unusually strong volcanic activity during the last several centuries lends support to the hypothesis that volcanic emissions can produce detectable changes of climate on a decadal time scale.

Before you go on:

1. How do changes in Earth's orbit control the timing of glacial–interglacial cycles?

2. How do scientists use ice-core records to learn about climate changes? What do ice-core records tell us about Earth's atmosphere during glacial times?

3. What role does ocean circulation play in climate change?

4. What factors may control fluctuations of glaciers and climate on the scale of the Little Ice Age?

REVISITING PLATE TECTONICS AND THE EARTH SYSTEM

WHAT WILL HAPPEN TO GLACIERS IN A WARMER WORLD?

Global warming is much in the news these days. If it is real, and most knowledgeable scientists are convinced of it, what is the consequence of a warming Earth on glaciers? Does a warmer Earth mean that all glaciers will shrink, and many disappear? And what are the long-term prospects both for mountain glaciers and Earth's great ice sheets?

We learned earlier that a close relationship exists between glaciers and the snowline, and that glacier equilibrium lines fluctuate in altitude in response to climate change. If temperature exerts a strong control on glacier mass balance, then raising the temperature should lead to negative balances and glacier retreat. Recent observations clearly show that many of the world's glaciers are experiencing dramatic retreat as average world temperature rises at an unprecedented rate. For small glaciers in low- and middle-latitude mountain ranges, raising the snowline by 100 to 200 m will be their death knell, for all or most of these glaciers will then lie below the snowline. If the climate warms as much as 3.3°C by the end of the present century—the midpoint in the range of estimates by the Intergovernmental Panel for Climate Change (IPCC)—then the snowline could rise an average of 500 m or more (Figure 16.34). This would cause all but the largest glaciers in the Alps, Cascades, and Himalaya to disappear.

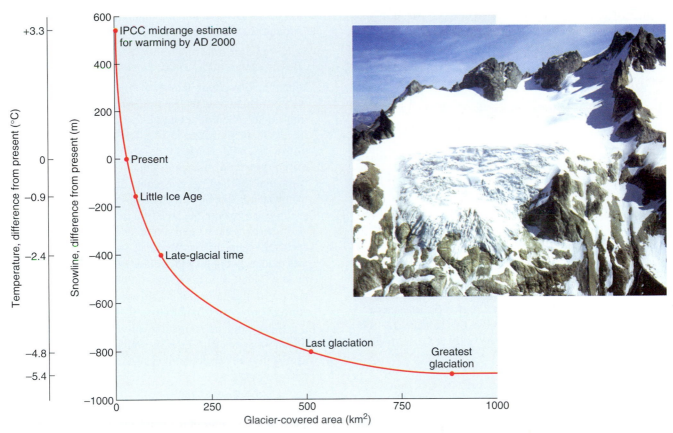

Figure 16.34   Shrinking Glaciers in a Warming Climate   Graph showing how glacier area changes as a function of snowline altitude for a basin in Washington's Cascade Range. During the last glaciation, about 500 km² of glacier ice existed in the basin, when the snowline was 800 m lower than today and average air temperature was about 5°C cooler. As the glaciation ended, the snowline rose and glacier cover diminished. Today the basin contains only a few small glaciers, like the cirque glacier shown in the photograph. An estimated rise of average temperature of 3.3°C by the end of the century would cause the snowline to rise another 500 m and lead to disappearance of most remaining glaciers. The small glacier in the photo would vanish with a snowline rise of only about 100 m.

There are greater uncertainties about ice sheets in a warming world. Those with marine calving margins will likely retreat in response to rising sea level, but warmer air temperatures also mean that more seawater can be evaporated and transferred to these glaciers as snow, adding to their volume. Does this mean that ice sheets will grow as mountain glaciers disappear? The linkages between climate, sea level, and glacier balance seem clear, yet we still know too little about the vast ice sheets to draw any convincing conclusions. One thing does seem very likely: if global warming is eventually halted by a shift away from fossil fuels as our primary energy source, Earth's climate should revert to a natural state. Then changes in Earth's orbit would propel the planet into the next full-scale glaciation—about 50,000 years from now.

AFRICAN ICE SHEETS

When we hear the name "Africa," what may come to mind is a varied continent of vast sandy deserts, tropical forests, expansive grasslands, great rift valleys, and abundant wildlife. Glaciers understandably are not included in a vision of the African landscape. The few glaciers to be found in Africa are restricted to the summit regions of a few high equatorial volcanoes, and they are rapidly disappearing. However, if we could travel back in time some 250–300 million years, our impression of Africa would be dramatically different. Then the landscape would have been a vast continental ice sheet, calving icebergs into the adjacent ocean. Evidence of this ice sheet is spread across parts of southern Africa, largely in the form of striated and grooved bedrock, tillite, and associated meltwater deposits (Figure 16.35). The stratigraphic record of glaciation is incomplete in Africa, but a contemporaneous record of sea-level fluctuations in central North America points to repeated marine invasions and sea-level fall that likely represent the buildup and decay of the continental glaciers. Between 40 and 50 events are recorded, implying a long series of glacial and interglacial ages during the Pennsylvanian and Permian periods.

Contemporaneous ice-sheet deposits in India, South America, Australia, and Antarctic indicate a former ice

Figure 16.35 Ancient Dwyka Tillite A geologist examines a large erratic boulder that is embedded in Dwyka tillite. The non-sorted tillite is exposed along a stream valley in South Africa.

sheet or ice sheets covering much of the ancient southern continent of Gondwanaland (Figure 2.13). Several of these modern continents lie in low latitudes, where continent-size ice sheets would be unlikely to form because of high temperatures. Reconstruction of the breakup of the ancient high-latitude continent and slow drift of several of its pieces to lower latitudes can explain the present distribution of the ancient glacial deposits. Africa is now moving slowly northeastward, making highly unlikely the return of an African ice sheet in the foreseeable geologic future.

What's Ahead?

When the first clear pictures of Mars were obtained by a camera aboard an orbiting satellite, it became immediately clear that wind likely was the most important agent that moves sediment on the surface of that dry, cold planet. On Earth, the wind is also an effective agent of transport, but because Earth is wet compared to Mars, wind is generally less effective than water. Exceptions include deserts and some sandy coasts where the lack of a dense vegetation cover permits the wind to erode and move sediment across the land surface. Wind, as we shall see, generates evidence that provides clues to Earth's climate history. Thick deposits of wind blown dust and dust contained in continental ice sheets tell us that the atmosphere was far more dusty in glacial times than it is today. Ancient formations of windblown sand, like those exposed at Zion National Park in Utah, tell us that dry, windy conditions also existed in the distant geologic past.

In the following chapter we will discuss the planetary wind system of Earth and see how wind acts as a geologic agent, eroding and depositing sediment. Because the effects of wind are most noticeable in desert environments, we will examine the distinctive sediments and desert landforms produced by wind, as well as the threat of desertification faced by people living in arid lands.

CHAPTER SUMMARY

1. Glaciers are permanent bodies of moving ice that consist largely of recrystallized snow.

2. Based on their geometry, we recognize cirque glaciers, valley glaciers, fjord glaciers, piedmont glaciers, ice caps, ice sheets, and ice shelves.

3. Ice in a temperate glacier is at the pressure melting point, and liquid water exists at the base of the glacier; in a polar glacier, ice is below the pressure melting point and is frozen to the rock on which it rests.

4. Glaciers can form only at or above the snowline, which is close to sea level in polar regions and rises to high altitudes in the tropics.

5. The mass balance of a glacier is measured in terms of accumulation and ablation. The equilibrium line separates the accumulation area from the ablation area and marks the level on the glacier where gain is balanced by loss.

6. Temperate glaciers move as a result of internal flow and basal sliding. In polar glaciers, which are frozen to their bed, motion is much slower and involves only internal flow. Surges involve extremely rapid flow, and probably are related to buildup of water at the base of a glacier.

7. Glaciers erode rock by plucking and abrasion. Rock debris, transported chiefly at the base and sides of a glacier, includes fragments of all sizes, from fine rock flour to large boulders.

8. Mountain glaciers erode stream valleys into U-shaped glacial valleys with cirques at their heads. Fjords are excavated far below sea level by glaciers in high-latitude coastal regions.

9. Glacial drift is sediment deposited by glaciers and glacial meltwater. Whereas till is deposited directly by glaciers, glacial-marine drift is deposited on the seafloor from floating glacier ice. Stratified drift includes outwash deposited by meltwater streams and ice-contact stratified drift deposited on or against stagnant ice.

10. Ground moraine is built up beneath a glacier, whereas end moraines (both terminal and lateral) form at a glacier margin.

11. Permafrost, a common feature of periglacial zones, is confined mainly to areas where average air temperature is less than –5°C. It reaches a maximum thickness of at least 1500 m and is believed to have formed during glacial ages in subfreezing landscapes not covered by continental ice sheets.

12. Permafrost can present unique engineering problems.

Thawing commences when the vegetation cover is broken, leading to collapse and extreme instability of the ground surface.

13. During glacial ages, huge ice sheets covered northern North America and Eurasia, causing the crust beneath the ice to subside and world sea level to fall.

14. Glacial ages have alternated with interglacial ages in which temperatures approximated those of today. Studies of marine cores indicate that more than 20 glacial–interglacial cycles occurred during the Pleistocene Epoch.

15. Glacial eras in Earth's history probably are related to the favorable positioning of continents and ocean basins, brought about by movements of lithospheric plates. The timing of glacial–interglacial cycles appears to be closely controlled by Earth's precession as well as by changes in the eccentricity of Earth's orbit and the tilt of the axis rotation, which affect the distribution of solar radiation received at the planet's surface.

16. Changes in the atmospheric concentration of carbon dioxide, methane, and dust may help explain the magnitude of global temperature lowering during glacial ages, while changes in ocean circulation may help explain the shifts between relatively stable glacial and interglacial modes of the climate system.

17. Climatic variations on the scale of centuries and decades have been ascribed to fluctuations in energy output from the Sun and/or to injections of volcanic dust and gases into the atmosphere.

THE LANGUAGE OF GEOLOGY

ablation (p. 418)
accumulation (p. 418)

calving (p. 420)
cirque (p. 414)
crevasse (p. 421)

drift (p. 427)
drumlin (p. 426)

equilibrium line (p. 419)
erratic (p. 427)

fjord (p. 414)

glacial drift (p. 427)
glacialmarine drift (p. 427)
glaciation (p. 423)
glacier (p. 412)

ice-contact stratified drift (p. 428)

mass balance (p. 418)
moraine (p. 428)

outwash (p. 428)

periglacial (p. 430)
permafrost (p. 430)
polar glacier (p. 416)
pressure melting point (p. 416)

snowline (p. 417)
stratified drift (p. 428)
surge (p. 422)

temperate glacier (p. 416)
till (p. 427)

QUESTIONS FOR REVIEW

1. What distinguishes temperate glaciers from polar glaciers?

2. What is the snowline and how are glaciers related to it?

3. Describe how snow is converted to glacier ice.

4. Why does the position of the equilibrium line provide a rough estimate of a glacier's mass balance?

5. Why is there a lag in time between a change of climate and the response of a glacier's terminus to the change?

6. In what ways does ice temperature influence the way a glacier moves?

7. Describe the unique motions of surging and calving glaciers.

8. Illustrate how small-scale and large-scale erosional features can be used to infer directions of flow of former glaciers.

9. Where and why would you expect to find permafrost at latitudes of less than 40°?

10. Describe what potential foundation problems a home builder might encounter in northern Alaska if the contractor were to clear the building site of vegetation and begin construction on the exposed ground surface.

11. How might you distinguish till from stratified drift in a roadside outcrop?

12. In what different ways are moraines built at a glacier margin?

13. Under what conditions do kettles and kames form?

14. What evidence obtained from deep-sea cores indicates that glacial–interglacial cycles have occurred repeatedly during the Pleistocene Epoch?

15. What natural factors explain the recurrence of glacial events on time scales of tens of thousands of years?

16. How might large volcanic eruptions influence climate and cause glaciers to grow or shrink?

Click on *Presentation* and *Interactivity* in the **Glaciers, Glaciations, and Ice Sheets** module of your CD-ROM to further explore resources and activities presenting concepts from this chapter. Select *Assessments* in the same module to test your understanding of this chapter.

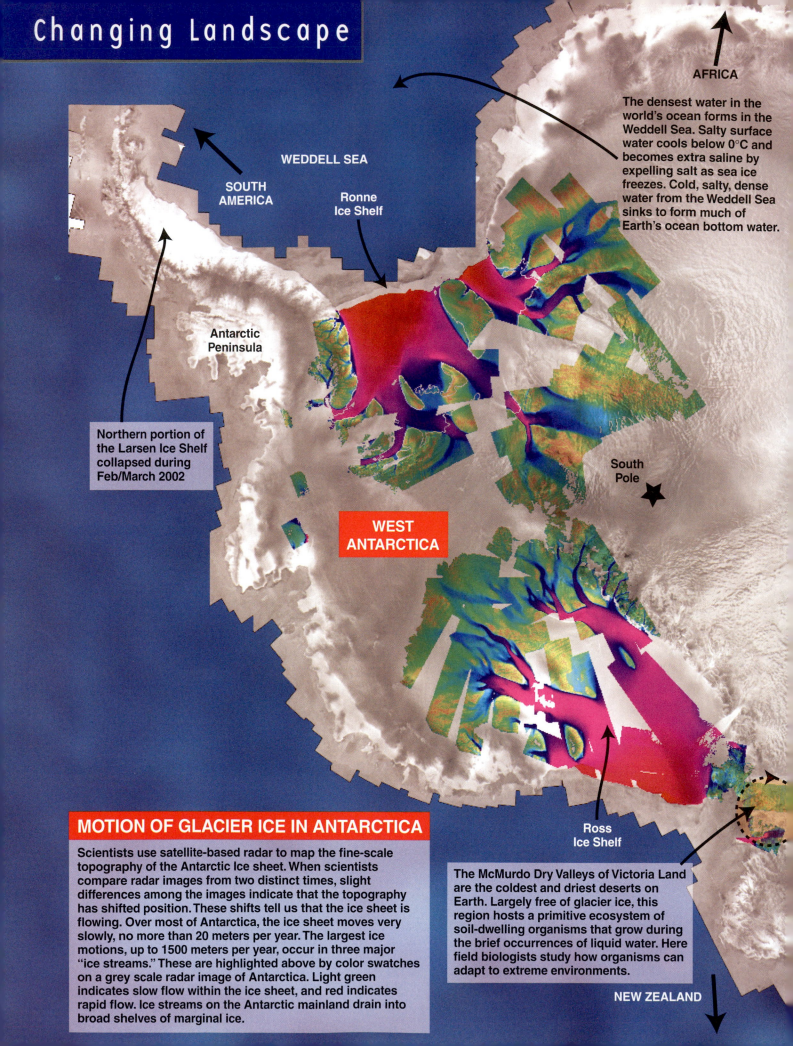

Changing Landscape

AFRICA

The densest water in the world's ocean forms in the Weddell Sea. Salty surface water cools below 0°C and becomes extra saline by expelling salt as sea ice freezes. Cold, salty, dense water from the Weddell Sea sinks to form much of Earth's ocean bottom water.

WEDDELL SEA

SOUTH AMERICA

Ronne Ice Shelf

Antarctic Peninsula

South Pole

Northern portion of the Larsen Ice Shelf collapsed during Feb/March 2002

WEST ANTARCTICA

Ross Ice Shelf

MOTION OF GLACIER ICE IN ANTARCTICA

Scientists use satellite-based radar to map the fine-scale topography of the Antarctic Ice sheet. When scientists compare radar images from two distinct times, slight differences among the images indicate that the topography has shifted position. These shifts tell us that the ice sheet is flowing. Over most of Antarctica, the ice sheet moves very slowly, no more than 20 meters per year. The largest ice motions, up to 1500 meters per year, occur in three major "ice streams." These are highlighted above by color swatches on a grey scale radar image of Antarctica. Light green indicates slow flow within the ice sheet, and red indicates rapid flow. Ice streams on the Antarctic mainland drain into broad shelves of marginal ice.

The McMurdo Dry Valleys of Victoria Land are the coldest and driest deserts on Earth. Largely free of glacier ice, this region hosts a primitive ecosystem of soil-dwelling organisms that grow during the brief occurrences of liquid water. Here field biologists study how organisms can adapt to extreme environments.

NEW ZEALAND

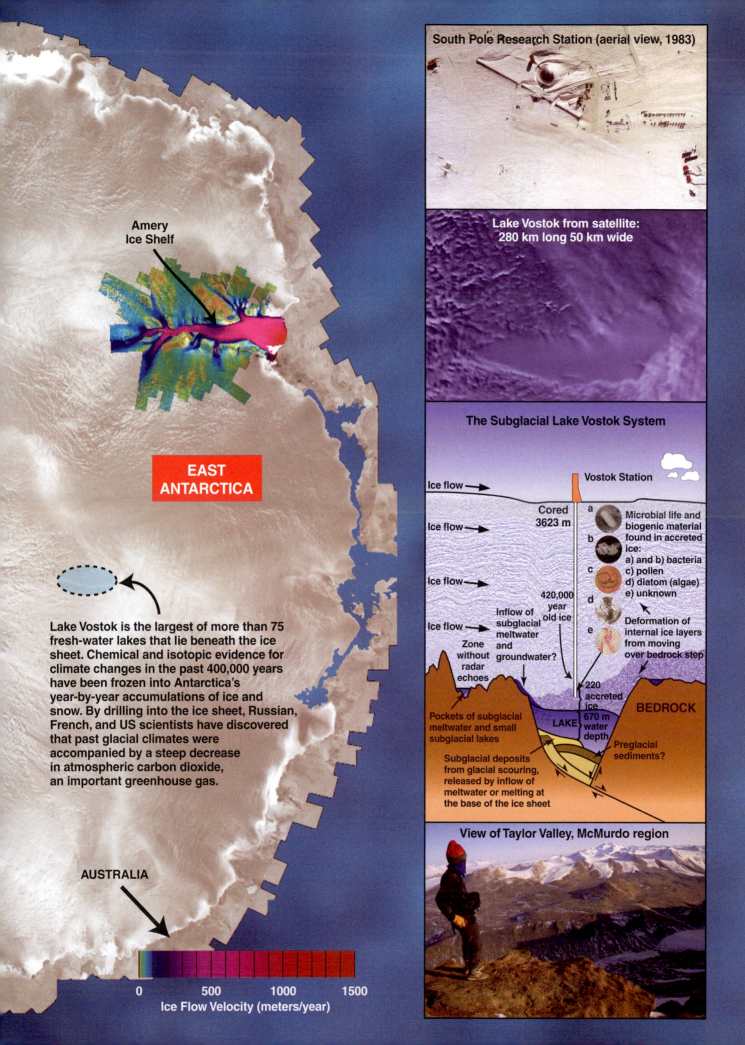

Amery
Ice Shelf

EAST
ANTARCTICA

Lake Vostok is the largest of more than 75 fresh-water lakes that lie beneath the ice sheet. Chemical and isotopic evidence for climate changes in the past 400,000 years have been frozen into Antarctica's year-by-year accumulations of ice and snow. By drilling into the ice sheet, Russian, French, and US scientists have discovered that past glacial climates were accompanied by a steep decrease in atmospheric carbon dioxide, an important greenhouse gas.

AUSTRALIA

0   500   1000   1500
Ice Flow Velocity (meters/year)

South Pole Research Station (aerial view, 1983)

Lake Vostok from satellite:
280 km long 50 km wide

The Subglacial Lake Vostok System

Ice flow

Ice flow

Ice flow

Ice flow

Vostok Station

Cored
3623 m

a
b
c
d
e

Microbial life and biogenic material found in accreted ice:
a) and b) bacteria
c) pollen
d) diatom (algae)
e) unknown

Deformation of internal ice layers from moving over bedrock step

420,000 year old ice

Inflow of subglacial meltwater and groundwater?

Zone without radar echoes

Pockets of subglacial meltwater and small subglacial lakes

Subglacial deposits from glacial scouring, released by inflow of meltwater or melting at the base of the ice sheet

220 accreted ice

670 m water depth

LAKE

BEDROCK

Preglacial sediments?

View of Taylor Valley, McMurdo region

Chapter 17

Death of a date palm grove. Advancing sand dunes are overwhelming a grove of date palms growing around an oasis near Chinguetti, Mauritania.

Atmosphere, Winds, and Deserts

Wind erosion of soil caused by farming, South Australia.

Living on the Edge of the Desert

I spent the first 12 years of my life living in the semidesert region of South Australia. Rainfall was low, bushfires a constant threat, and dust storms a menace to be

447

endured. Where I lived rainfall was sufficient in good years so that certain kinds of agriculture could be carried on. The bad years were terrible.

I remember an uncle who had a wheat farm in the region. He managed to survive reasonably well for a number of years, but a string of low-rain years in the 1930s finally forced him to pack up the family and leave his eroded and windblown farm. In retrospect, it is apparent that wheat farming should not have been attempted in a region with such an uncertain climatic environment.

The 1930s brought another Australian farming problem into focus. Sheep were put out to graze over huge areas of marginal lands. Not only did they eat down grasses close to the ground, their hard hooves broke up the surface. With reduced vegetation and a broken surface, there was little to stop wind erosion during dry years. Kangaroos and other native animals all have soft, padded feet and do not cause the same problems as sheep, so there is no doubt that the problem is a human one. Steps are now being taken to slow and possibly even reverse the damage of the past.

Brian J. Skinner

KEY QUESTIONS

1. **How are winds related to the global circulation of the atmosphere?**
2. **How does wind transport sand and dust?**
3. **How does wind create landforms, and what are some of the landforms?**
4. **Why are Earth's deserts where they are?**
5. **How have humans contributed to desertification?**

INTRODUCTION: WIND AS A GEOLOGIC AGENT

If we lived on Mars instead of Earth, a substantial percentage of this book would likely be devoted to the theme of wind action and deserts, for the entire Martian planet is an arid, windy, and dusty planet. When the *Mariner 9* spacecraft approached Mars in 1971, for instance, a dust storm of major proportions enveloped much of the planet and continued unabated for several months. Photographs taken both during this *Mariner* mission and during subsequent *Viking* missions revealed a planetary surface that has been extensively modified by wind action.

Wind is also an important agent of erosion and sediment transport on Earth, but its effects are visible mainly in desert regions, where few people live. Most of the world's population is concentrated in the relatively moist parts of the temperate and tropical latitudes, where a protective cover of vegetation makes wind an ineffective geologic agent. Nevertheless, even in these populated regions the occurrence of ancient sand dunes and widespread deposits of dust show us that wind has been important in shaping the landscape at times when the continents were drier and windier places than they are today.

Wind is the movement of air, principally horizontal movement, and is caused both as a consequence of heating by the Sun, and by rotation of Earth on its axis. In this chapter we first discuss the global wind system, then turn to all of the effects of the wind on the land surface—transport of dust particles, erosion, deposition of windborne sediment, and the formation of desert landscapes.

PLANETARY WIND SYSTEM

Winds are effective geologic agents in some regions but not in others. To help us understand why, we need to see how surface winds are related to the global circulation of the atmosphere.

CIRCULATION OF THE ATMOSPHERE

The atmosphere consists of a mixture of gases that together we call air. The atmosphere is always moving, a fact we are well aware of whenever we feel a gentle breeze or a strong wind blowing.

The basic reason the atmosphere is always in motion is that more of the Sun's heat is received per unit of land surface near the equator than near the poles. This unequal heating gives rise to convection currents. The heated air near the equator expands, becomes lighter, and rises. High up, it spreads outward in the direction of both poles. As the upper air travels both northward and southward, it gradually cools, becomes heavier, and sinks. On reaching Earth's surface, this cool, descending air flows back toward the equator, warms up, and rises, thereby completing a cycle of convection.

THE CORIOLIS EFFECT

If Earth did not rotate, the convection currents in the atmosphere would simply flow from the equator to the poles and back again. But of course Earth does rotate, and rotation complicates the convection currents in the atmosphere (and in the ocean).

The **Coriolis effect** is named for the nineteenth-century French mathematician, Gaspard-Gustave de Coriolis, who first analyzed it. To the observer on Earth, the Coriolis effect causes anything that moves freely with respect to the rotating Earth (such as a plane, a missile, or the wind) to veer off course. Imagine trying to throw a ball to a friend when you are both on spinning merry-go-rounds, and you have an idea of the complication caused by the effect. In the Northern Hemisphere the Coriolis effect causes a moving mass to veer to the right of the direction in which it is moving, and in the Southern Hemisphere to the left.

To understand how the Coriolis effect works, consider what happens to a small mass of air that flows northward from the equator. Because Earth and the atmosphere are rotating eastward, the northward-flowing air is also moving eastward. The eastward speed due to Earth's rotation is 1670 km/h at the equator, but at successively higher latitudes the eastward speed due to rotation is less and less, until at the north pole it is zero.

A northward mass of flowing air will keep the eastward velocity with which it started because it is not fixed to Earth's surface. Thus, as the northward-flowing mass of air moves farther and farther north, its eastward velocity becomes more and more rapid than the eastward velocity of Earth's surface immediately below. To an astronaut in a space vessel, the air mass would appear to flow in a straight line. An observer on the ground, however, would have an eastward velocity due to Earth's rotation. To this observer, the air mass would appear to have been deflected eastward, that is, to the right. The amount of the deflection is a function of the speed of the moving air mass and the latitude.

The Coriolis effect breaks up the simple flow of air between the equator and the poles into belts (Figure 17.1). The result, in both the Northern and Southern Hemispheres, is a large cell of circulating air lying between the equator and about 30° latitude. In these low-latitude cells (called *Hadley cells*), the prevailing winds are northeasterly in the Northern Hemisphere (that is, they flow *from* the northeast toward the southwest), while in the Southern Hemisphere

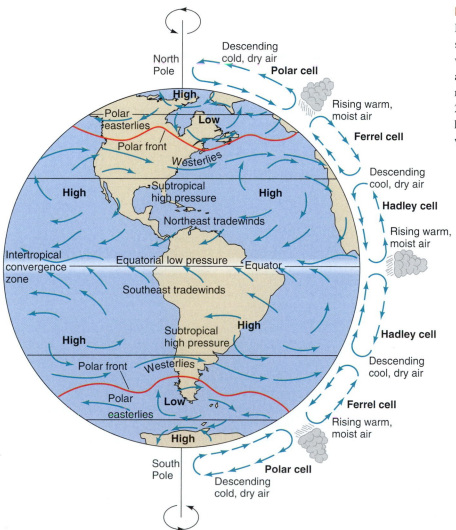

Figure 17.1 Global Wind Patterns
Earth's planetary wind system, shown schematically. Moist air, heated in the warm equatorial zone, rises convectively and forms clouds that produce abundant rain. Cool, dry air descending at latitudes 20–30° produces a belt of subtropical high pressure in which lie many of the world's great deserts.

they are southeasterly. These wind systems are called the *tradewinds*, for their direction and consistent flow carried trade ships across the tropical oceans at a time when winds were the chief source of sailing power.

In each hemisphere, a second cell of circulating air, called the *Ferrel cell*, lies poleward of the Hadley cell. In these second, middle-latitude cells, westerly winds prevail (blowing *from* the west). In these cells, cold upper air flowing toward the equator descends near 20–30° latitude in both hemispheres. At the same time, northward-moving surface air rises in higher latitudes, where it meets dense, cold air flowing from the polar regions.

A third cell of circulating air, a *polar cell*, lies over each polar region. In each cell, cold, dry, upper air descends near the pole and moves equatorward in a wind system called the polar easterlies. As this air slowly warms and encounters the belt of westerlies, it rises along the *polar front* and returns toward the pole.

The cold upper air converging where the low- and mid-latitude cells meet cannot hold as much moisture as warm air, so where this cold air descends, dry conditions are created at the land surface. As a result, much of the arid land in both the Northern and Southern hemispheres is centered between latitudes 20° and 30°. By contrast, abundant moisture in the warm air surrounding the equator condenses as the air rises and becomes cooler and denser, creating clouds that release their moisture as tropical rains.

CLIMATE

The global pattern of airflow ultimately controls the variety and pattern of Earth's climates. **Climate** is the average weather of a place, together with the degree of variability of that weather, over a period of years. It is measured by such factors as temperature, precipitation, cloudiness, and windiness.

In the preceding section, we described the global pattern of airflow created by temperature differences created by nonuniform heating of Earth's surface and the Coriolis effect. But these are not the only factors affecting global airflow and thus climate. The topography of the land—the distribution of oceans, continents, high mountains, and plateaus—also plays a role by causing local deviations of the surface winds. As a result, average temperature, precipitation, cloudiness, and windiness vary greatly from one place to another and give rise to distinct climatic regions. If Earth had no mountains and no oceans to affect the moving atmosphere, the major climatic zones would lie parallel to the equator.

Before you go on:

1. Why is the atmosphere always in motion?

2. How does the Coriolis effect influence the global wind circulation?

3. How does weather differ from climate?

MOVEMENT OF SEDIMENT BY WIND

We have all seen pictures of the tremendous destruction wreaked by hurricanes and typhoons when winds achieve speeds of at least 130 km/h and sometimes reach 300 km/h. The force of such a wind is so strong that trees are uprooted, houses are ripped apart, and large objects are thrown substantial distances. Fortunately, hurricane winds are the exception, but they provide a glimpse of the potential power of wind as a geologic agent. (See *The Science of Geology*, Box 17.1, *Measuring Earth's Hum*.)

In extraordinary wind storms, when wind speeds locally reach more than 300 km/h, coarse rock particles up to several centimeters in diameter can be lifted to heights of a meter or more. Pebbles swept aloft by exceptional winds have been found lodged in buildings, trees, and cracks in telephone poles. In most regions, however, wind speed rarely exceeds 50 km/h, a velocity described as the upper velocity range of a strong wind. In a strong wind, the largest particles of sediment that can be suspended in the air stream are grains of sand. Larger particles settle out too quickly to remain aloft. At lower wind speeds, sand moves along close to the ground surface, and only finer grains of dust move in suspension.

WINDBLOWN SAND

If a wind blows across a bed of sand, the grains begin to move when the wind speed reaches about 4.5 m/s (16 km/h). The resulting forward rolling motion of the sand is called *surface creep* (Figure 17.2). With increasing wind speed, turbulence near the ground lifts moving sand grains into the air, where they travel along arc-like paths, landing a short distance downwind. This is *saltation*, the same process we see in a stream, where grains of sand move close to the bottom, also following arc-like paths (see Chapter 14). Next we look more closely at saltation and at one of its effects, sand ripples.

SALTATION

Saltation accounts for at least three-quarters of the sand transport in areas covered by sand dunes. Measurements of the rate of sand movement in deserts of the Middle East indicate that sand movement increases rapidly with increasing wind speed. For example, a strong wind blowing at 58 km/h will move as much sediment in one day as a wind blowing at 29 km/h would move in three weeks.

If a wind is strong enough, it can start a grain rolling along the ground, where it may impact another grain and knock it into the air. As this second grain falls to the ground, it will impact other grains, some of which are thrown upward into the airstream. Within a very short time, the air close to the ground may contain a very large number of saltating sand grains, all moving along with the

THE SCIENCE OF GEOLOGY

MEASURING EARTH'S HUM

Sometimes Earth seems to clang like a bell. High-magnitude earthquakes, such as the great Chilean quake of 1960, cause the entire planet to vibrate for days; the oscillations move the ground up and down by as much as a centimeter. However, seismologists have recently identified a very different type of whole-earth vibration: a deep, soft hum that can be picked up only by the most sensitive seismographs.

At first, researchers speculated that the hum might result from the combined effects of all the earthquakes that occur throughout the world. But this would require a continual series of earthquakes with magnitudes of 5.8 or more. On the average such earthquakes occur every few days, yet Earth's low hum is constant, with periods ranging from three to eight minutes. All the world's known earthquakes, taken together, would not be enough to produce this result.

At the annual meeting of the American Geophysical Union in the fall of 1998, a group of seismologists led by Naoki Suda of Japan proposed another hypothesis: that

Earth hums along with the wind. They examined seismic records for periods of 50 to 80 days from seismically quiet sites in Europe, South Africa, and central Asia. After removing background noise, they found that at each location the strength of the hum increased and decreased at predictable times each day—it was strongest from noon to 8 P.M. and weakest from midnight to 6 A.M. The same pattern of activity can be seen in the world's thunderstorms. The researchers hypothesized that the seismic hum may be a result of turbulent winds striking Earth's surface.

Not everyone agrees with this hypothesis, and Suda's team has stated that it is preliminary. Other suggested explanations of Earth's hum include small undetected earthquakes, ocean currents, and the movement of tectonic plates. More daily seismic records from more sites will need to be analyzed before seismologists can reach a firm conclusion about why Earth hums. Research is continuing.

Source: Richard A. Kerr, "New Data Hint at Why Earth Hums and Mountains Rise," *Science,* January 15, 1999, pp. 320–321.

---

wind in arc-like paths somewhat similar to those of a Ping-Pong ball bouncing across a table (Figure 17.3). However, even in strong winds, saltating sand grains seldom rise far off the ground. We can see this from abrasion marks on utility poles and fence posts, which are sandblasted up to a height of about a meter.

SAND RIPPLES

Sheets of well-sorted sand that have accumulated on the land surface are inherently unstable, even under gentle winds. As the wind passes across such an accumulation, saltation moves the smaller, most easily transported grains. Sand grains too large to be moved are left behind. As the saltating finer grains impact the surface at some average distance downwind, additional fine particles are set in motion, and another accumulation of coarse grains develops as the fine sand moves onward.

By this process, the coarse grains form a series of small, linear ridges of sand called **sand ripples**. Sand ripples tend to be aligned in a regular pattern with their crests oriented

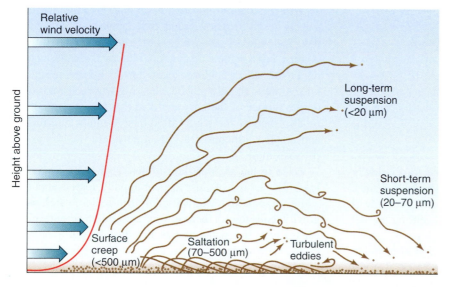

Figure 17.2 Suspension of Sediment
Under conditions of moderate wind, sand grains larger than 500 μm (0.5 mm) in diameter move by surface creep, while smaller grains (70–500 μm) saltate across the ground surface. Still finer particles (20–70 μm) are carried aloft in turbulent eddies and encounter faster-moving air that transports them downwind as they slowly settle to the ground. The finest dust (less than 20 μm) reaches greater heights and is swept along in suspension as long as the wind is blowing.

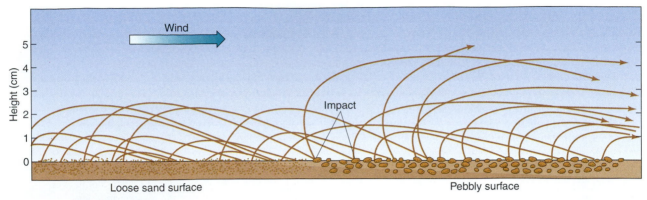

Figure 17.3 Saltation Strong wind causes movement of sand grains by saltation. Impacted grains bounce into the air and are carried along by the wind as gravity pulls them back to the land surface where they impact other particles, repeating the process.

Figure 17.4 Sand Ripples Sand ripples cross the surface of a desert sand sheet on the floor of Death Valley, California.

perpendicular to the wind direction (Figure 17.4). Under very strong winds, the ripples disappear because then all grains can be moved and sorting is less likely to occur.

WINDBLOWN DUST

Sand grains blown across the land surface travel slowly and are deposited quickly when wind velocity subsides. Fine particles of dust (silt- and clay-size sediment) travel faster, longer, and much farther before settling to the ground.

As you might guess, the dustiest places on Earth tend to coincide with some of the world's major desert regions.

Among the many types of terrain that can give rise to large quantities of dust, the following are especially important: dry lake and stream beds, alluvial fans, outwash plains of glacial streams, and regions underlain by deposits of wind-blown dust that have lost their vegetation cover as a result of either climatic change or human disturbance.

MOBILIZATION AND TRANSPORT OF DUST

Let's look more closely at how dust is transported. As a result of frictional drag, the velocity of moving air decreases sharply near the ground surface. Right at the surface lies a layer of relatively quiet air less than 0.5 mm thick, within which airflow is smooth and laminar rather than turbulent (Figure 17.5). Sand grains that protrude above this layer of quiet air can be swept aloft by rising turbulent eddies. Dust particles, however, are so small and often so closely packed that they form a very smooth surface that does not protrude above the quiet air. Even a strong wind blowing over such a surface may not disturb the dust. Mobilization of the fine sediment may require the impact of saltating sand grains or other physical disruption of the smooth surface.

We can see how such dust is set in motion by looking at a dusty desert road covered by dry, compact silt on a windy day. The wind blowing across the road generates little or no dust, but a vehicle driving over the road creates a choking cloud, which is blown away downwind before settling once more to the ground. The passing wheels have broken up the surface of the powdery dust that was too smooth to be disturbed by the wind, lifted dust particles into the air, and created turbulence in the atmosphere that keeps the particles aloft.

Once in the air, dust constitutes the wind's suspended load. The grains of dust are continually tossed about by eddies, like particles in a stream of turbulent water, while gravity tends to pull them toward the ground (Figure 17.2). Meanwhile, the wind carries the dust forward. In most cases suspended sediment is deposited fairly near its place of origin. However, strong winds associated with large

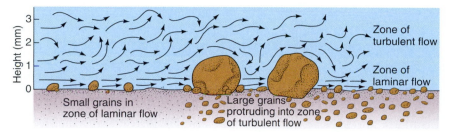

Figure 17.5   Laminar and Turbulent Flow   Particles of fine sand and silt at the ground surface lie within a zone of laminar airflow less than 0.5 mm thick where wind velocity is much lower than the turbulent flow. As a result, it is difficult for the wind to dis-lodge and erode these small grains. Larger grains protrude into a zone of faster-moving, turbulent air. The turbulence, which exerts a greater push on the top of the grains than does the laminar flow at their base, makes it easy to start the grains moving.

dust storms are known to carry very fine dust into the upper atmosphere, where it can be transported thousands of kilometers.

Dust storms are, in fact, the chief events leading to large-scale transport of dust. In a dust storm, the visibility at eye level is reduced to 1000 m or less by dust raised from the ground surface. Such storms are most frequent in the vast arid and semiarid regions of central Australia, western China, Russian Central Asia, the Middle East, and North Africa, as shown in Figure 17.6. In the United States, blowing dust is especially common in the southern Great Plains and in the desert regions of California and Arizona.

DEPOSITION OF DUST

Windborne dust can be deposited in several situations:

1. Wind velocity and air turbulence decrease so that particles can no longer remain in suspension.

2. The particles collide with rough or moist surfaces that trap them, or with surfaces having a weak electrical charge that attracts them.

3. The particles accumulate to form aggregates, which then settle out because of their greater mass.

4. The particles are washed out of the air by rain.

Vegetation acts as a trap for descending dust particles because wind velocity is reduced over vegetated land-scapes. Forest is more efficient at trapping dust than is low-lying vegetation because of the greater effect trees have on reducing wind velocity in the critical zone above the ground.

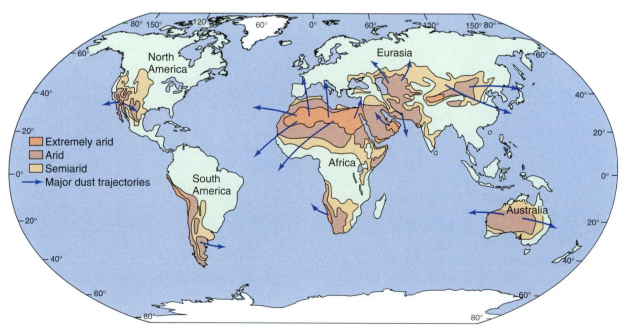

Figure 17.6   Dust Storms   Major dust storms are most frequent in arid and semiarid regions that are concentrated in the belts of subtropical high pressure north and south of the equatorial zone. Arrows show the most common trajectories of dust transported during major storms.

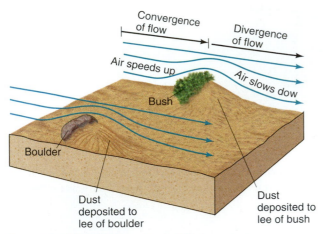

Figure 17.7 Effect of Obstacles on Wind Obstacles in the path of wind produce a convergence of air flow, causing the air to move faster. As the air passes across the obstacle, the flow paths diverge, leading to lower velocity. The lower velocity reduces the wind's carrying capacity, causing the air to drop some of its dust load, which accumulates on the lee side of the obstruction.

Deposition also occurs where a topographic obstacle causes a divergence of airflow, thereby leading to reduced wind velocity behind the obstruction (Figure 17.7). This explains why deposits of dust are generally thick on the *lee*, or downwind, side of obstacles (the side away from the wind) and either thin or absent on the *windward*, or upwind, side (the side from which the wind is blowing).

Coarse and medium-grained dust particles, which are carried at relatively low altitudes, are likely to settle out first. Finer particles are carried higher in the atmosphere and therefore may remain suspended for long periods.

DETRIMENTAL EFFECTS OF WINDBLOWN SEDIMENT

Wind action can create many problems for people. Each year the material losses can range into the billions of dollars, and the direct and indirect effects of wind can lead to a significant loss in human lives.

Blowing sand and dust can severely damage crops and other vegetation. Large wind storms are reported to have caused major damage to wheat fields, citrus orchards, and vineyards, as well as to root crops. In one severe California storm in 1977, cattle were asphyxiated by dense dust, and hair and skin were sandblasted from their hind quarters. A strong, dense dust storm can quickly turn a car's clear windshield into a sheet of frosted glass, and blowing sand can pit and strip away the shiny paint of a new car.

The engines of vehicles operating in dusty areas are particularly susceptible to damage. Dust can contaminate fuel, clog air filters, abrade cylinders, and cause electrical short circuits. During World War II, higher-than-average engine

failure among vehicles operating in desert areas was attributed to such abnormal cylinder wear. Dust similarly created a problem for military vehicles operating in the deserts of Saudi Arabia and Kuwait during the Gulf War of 1991.

Blowing dust can severely reduce visibility on roads and highways. Near Tucson, Arizona, clouds of dust caused so many accidents that a Dust Storm Alert system was introduced in 1976 to warn motorists of hazardous conditions.

Airborne dust also can pose a hazard to aircraft. In 1973 a Royal Jordanian Airlines plane crashed at an airport in northern Nigeria while flying through dense dust, killing 176 persons. An attempt to rescue U.S. hostages in Iran in 1979 was aborted after helicopter engine failure, attributed to dense airborne dust, caused a fatal crash in the desert staging area. Helicopter pilots operating in the Saudi Arabian desert in 1990–91 reported that dense dust raised while flying close to the ground produced a bright electrostatic glow around their rotor blades, thereby greatly reducing the effectiveness of light-sensitive night-vision goggles.

Inhalation of dust can lead to various medical problems. When too many fine mineral particles enter the lungs, tissue damage can cause emphysema. Quartz inhalation can lead to silicosis, a debilitating disease common among unprotected miners working in dusty conditions. In addition, disease-causing organisms may be carried in dust, where they may survive for long periods. Among the deadly germs that can be transported in windblown dust are anthrax and tetanus. In central China, a close correlation has been found between deaths due to cancer of the esophagus and the distribution of dust deposits, with the death rate increasing as the average grain size of the dust decreases. Outside China, the disease is mainly present in the dusty, arid regions of Iran, central Asia, Mongolia, and Siberia, suggesting that the cancer may somehow be related to persistent inhalation of fine dust.

WIND EROSION

As we might expect, wind is an important agent of erosion. It is particularly important wherever winds are strong and persistent and either the land is too dry to support vegetation or the influx of airborne sediment is so rapid that vegetation cannot gain a foothold and thereby stabilize the ground surface.

Flowing air erodes in two ways. The first, **deflation** (from the Latin word meaning "to blow away"), occurs when the wind picks up and carries away sand and dust. This process provides most of the wind's load. The second process, *abrasion*, results when wind-driven grains of sediment impact rock.

DEFLATION

Deflation on a large scale takes place only where little or no vegetation exists and where loose rock particles are fine

Figure 17.8   Deflation of Glacial Sediments   Active deflation of meltwater sediments downstream from Tasman Glacier in the Southern Alps of New Zealand produces clouds of dust. Loess is accumulating on vegetation-covered glacial deposits in the foreground.

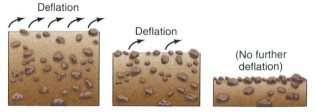

Figure 17.9   Deflation   Progressive deflation of a poorly sorted sediment leads to the development of a desert pavement.

Figure 17.10   Desert Pavement   A desert pavement on the floor of Searles Valley, California, consists of a layer of gravel, too coarse to be moved by the wind, that covers finer sediment and inhibits further deflation.

enough to be picked up by the wind. Areas of significant deflation are found mainly in deserts. Nondesert sites where deflation occurs include ocean beaches, the shores of large lakes, and the floodplains of large glacial streams (Figure 17.8). Of greatest economic importance is the deflation of bare plowed fields in farmland. Although deflation may occur seasonally when farmland is plowed, it is especially severe during times of drought, when no moisture is present to hold soil particles together.

Generally, deflation lowers the land surface slowly and irregularly, making measurements of wind erosion difficult. However, in the dry 1930s, deflation in parts of the western United States amounted to 1 m or more within only a few years. This is a tremendous rate compared with the long-term average rate of erosion for the region as a whole, which is only a few centimeters per thousand years.

DEFLATION HOLLOWS AND BASINS

Small saucer- or trough-shaped hollows and larger basins created by wind erosion are among the most conspicuous evidence of deflation. Tens of thousands of these basins occur in the semiarid Great Plains regions of North America from Canada to Texas. Most are less than 2 km long and only a meter or two deep. In wet years, the larger ones are carpeted with grass and may contain shallow lakes. However, in dry years soil moisture evaporates, grass dies away, and the wind deflates the bare soil.

Where sediments are particularly susceptible to erosion by wind, deflation basins can be excavated to depths of 50 m or more. The immense Qattara Depression in the desert

region of western Egypt, the floor of which lies more than 100 m below sea level, has been attributed to intense deflation.

In any basin, the depth to which deflation can reach is limited by the water table. As deflation lowers the land to the level at which the ground is saturated, the surface soil becomes moist, thereby encouraging the growth of vegetation that inhibits further deflation.

DESERT PAVEMENT

When sand and dust are either blown away from a deposit of alluvium or locally removed by sheet erosion (Chapter 14), stones too large to be moved become concentrated at the surface (Figure 17.9). Eventually, a continuous cover of stones forms a **desert pavement**, so called because the stones fit together almost like the cobbles in a cobblestone pavement (Figure 17.10). At this point deflation can no longer occur.

ABRASION

Abrasion occurs when rock is scoured by windborne grains of sediment. Results include distinctive rock shapes called ventifacts and landscape features called yardangs. Where bedrock and loose stones are abraded by wind-driven sand

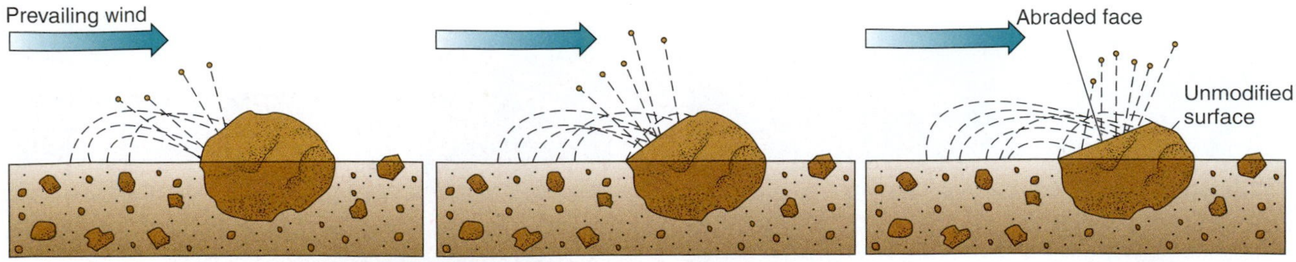

Prevailing wind

Abraded face

Unmodified surface

A.   **B.**   **C.**

Figure 17.11   Formation of Ventifacts   Stages in the formation of a ventifact. A. Saltating sand grains impact the upwind side of a stone that protrudes above a desert surface. B. As the sand grains chip away at the rock, a facet is cut across the upwind side of the stone. C. With continued sand blasting, the abraded face is reduced to a lower angle. Given time, the exposed face of the ventifact could be reduced to ground level.

Figure 17.12   Field of Ventifacts   Ventifacts litter the ground surface near Lake Vida in Victoria Valley, Antarctica. The most intensely abraded surfaces are inclined to the right, in the direction from which strong winds blow off the East Antarctic Ice Sheet. The larger ventifacts are about 50 cm high.

Figure 17.13   Yardang   A sharp-crested yardang carved from compact lake sediments rises above the floor of Rogers Lake playa in southeastern California. The crest of the yardang is oriented parallel to the prevailing wind direction.

and dust, they acquire a distinctive shape and surface polish. Any bedrock surface or stone that has been abraded and shaped by windblown sediment is a **ventifact** (Figure 17.11). A ventifact can be identified by its having at least one smooth, abraded surface that faces upwind (Figure 17.12). If erosion of surrounding sediment causes a stone to shift position, or if the wind direction changes, a new surface with a different orientation will be cut. Where this new abraded surface intersects the initial abraded surface, a sharp keel-like edge is produced. Because wind-abraded surfaces form facing upwind, ventifacts that have not been reoriented can be used to measure the prevailing wind direction.

Among the common landforms of some desert regions is an elongate, streamlined, wind-eroded ridge called a **yardang** (from the Turkic word *yar*, meaning *steep bank*). Typically, yardangs are sharp-crested and carved from hard, compacted sediments or from highly weathered crystalline rocks (Figure 17.13). Some have a shape that resembles an inverted ship's hull. Individual yardangs range up to a few tens of kilometers long and up to 100 m

high. Generally, these landforms occur in groups. Yardangs probably form initially by differential deflation along irregular depressions in the land surface that lie approximately parallel to the wind direction. Then they increase in size as the abrading action of windblown sand and dust further deepens and broadens the depressions, creating free-standing intervening ridges.

Before you go on:

1. How are sand ripples formed?
2. What is the difference between abrasion and deflation?
3. What are desert pavements and how do they form?

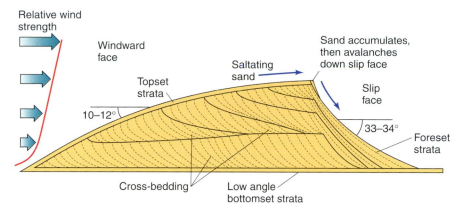

Figure 17.14 Anatomy of a Dune Cross section through a barchan dune showing typical gentle windward slope and steep slip face. Thin topset strata overlie sets of cross-bedded strata representing old slip faces. Sand grains saltating up the windward slope fall onto the top of the slip face where they accumulate before avalanching downward to produce foreset strata resting at the angle of repose.

EOLIAN DEPOSITS

Recall from Chapter 7 that sediments deposited by wind are known as eolian deposits. Although wind action is now chiefly confined to arid and semiarid lands, such deposits are found not only in these regions but also in regions that are moist and covered by vegetation. The explanation for this distribution lies partly in changing climatic regimes and partly in the fact that windblown dust often settles out far from its place of origin. Here we discuss the major kinds of eolian deposits: dunes, loess, dust in ocean sediments, glacial ice, and volcanic ash deposits.

DUNES

A **dune** is a hill or ridge of sand deposited by winds. Although little is known about how dunes initially form, it is likely that a dune develops where some minor surface irregularity or obstacle distorts the flow of air. Wind velocity within a meter or two of the ground varies with the slightest irregularity of the land surface. On encountering any small obstacle, wind sweeps over and around it but leaves a pocket of slower-moving air immediately downwind. In such a pocket of low wind velocity, sand grains moving with the wind drop out and begin to build a mound. The mound, in turn, influences the flow of air and may continue to grow until it forms a dune.

DUNE FORM AND SIZE

A typical isolated sand dune is asymmetrical. It generally has a gentle windward slope, and a steep leeward face that rests at the angle of repose, generally 33–34°.

Sand grains move up the windward slope by saltation to reach the crest of the dune. Grains making it past the crest generally fall onto the lee face near its top. The bulge thus created through grain-by-grain accumulation eventually reaches an unstable angle. The sand then avalanches (slips) downward, spreading the grains of the bulge down the lee face. For this reason, the lee face of an active dune is also known as the *slip face*. The continual avalanching of sand grains keeps the slip face at the angle of repose and pro-

duces cross strata much like the foreset layers in a delta (Chapter 14).

The angle of the windward slope of a dune varies with wind velocity and grain size but is always much less than that of the slip face. This asymmetry of form provides a means of telling the direction of the wind that shaped the dune, for the slip face always lies on the downwind side. Furthermore, cross strata within the dune represent former slip faces, and each stratum slopes in the direction toward which the wind was blowing when it was deposited.

Many dunes grow to heights of 30 to 100 m, and some massive desert dunes in the western Alashan Plain of China reach heights of 500 m or more (Figure 17.15). The height to which any dune can grow probably is determined by the maximum wind velocity, which increases above the land surface. At some height, the wind will reach a velocity great enough to carry the sand grains up into suspension

Figure 17.15 Giant Dune A huge sand dune in the Alashan Plain of western China towers over two horsemen, reaching a height of several hundred meters above the desert floor.

off the top of a dune as fast as they move up the windward slope by saltation. Thus, the dune can grow higher only as long as the rate at which sand is supplied to the crest exceeds the rate of removal by the wind.

DUNE TYPES

Five different dune types are shown in Table 17.1. Dune type is controlled by the amount of sand available, the variability of wind direction, and the amount of vegetation cover.

1. Where the amount of sand is limited and lack of moisture inhibits growth of vegetation, strong winds from one direction can build an isolated *barchan dune*, a crescent-shaped dune with horns that point downwind. They are the best-studied dune type.

2. Barchan dunes can migrate over great distances without much change in form. Where sand supply is greater, individual barchans may merge to form a *transverse dune* having a sinuous crest oriented perpendicular to the direction of the strongest wind.

3. A *linear dune* tends to be long and relatively straight and to occur in rather regularly spaced groups. Linear dunes form mainly in areas of limited sand supply and variable (often bidirectional) winds.

4. Wind blowing from all directions can produce a large, pyramidal *star dune* with sinuous radiating arms. Although they range in general from 50 to 150 m high, star dunes more than 300 m high are known.

TABLE 17.1 Principal Types of Dunes Based on Form

Dune Type	Definition and Occurrence	
Barchan dune	A crescent-shaped dune with horns pointing downwind; occurs on hard, flat desert floors in areas of constant wind direction and limited sand supply; height 1 m to more than 30 m	
Transverse dune	A dune forming an asymmetrical ridge transverse to dominant wind direction; occurs in areas with abundant sand; can form by merging of individual barchans	
Linear dune	A long, relatively straight, ridge-shaped dune in deserts with limited sand supply and variable (bidirectional) winds; slip faces change orientation as wind shifts direction	
Star dune	An isolated hill of sand having a base that resembles a star in plan; sinuous arms of dune converge to form central peak as high as 300 m; tends to remain fixed in place in areas where wind blows from all directions	
Parabolic dune	A dune shaped like a U or V, with open end facing upwind; trailing arms, generally stabilized by vegetation, also point upwind; common in coastal dune fields; some form by piling of sand along lee and lateral margins of deflated areas in older dunes	

Figure 17.16 Field of Barchans Sand dunes advance from right to left across irrigated fields in the Danakil Depression, Egypt.

5. A *parabolic dune* is shaped like a U or a V and has two trailing arms, generally vegetated, pointing upwind. These dunes are common in coastal dune fields, where relatively constant wind off the ocean creates a moist environment that allows vegetation to grow. Parabolic dunes almost always develop where local disturbance of the vegetation allows deflation to build an accumulation of sand that grows to large size, and they may develop multiple crests and slip faces.

DUNE MIGRATION

Transfer of sand from the windward to the lee side of an active dune causes the whole dune to migrate slowly downwind. Measurements of barchan dunes show rates of migration as great as 25 m/yr.

The migration of dunes, particularly along sandy coasts and across desert oases, has been known to bury houses and farmers' fields (Figure 17.16), fill in canals, and even threaten the existence of towns. In such places, sand encroachment is countered most effectively by planting vegetation that can survive in the very dry sandy soil of the dunes. Continuous plant cover inhibits dune migration for the same reason that it inhibits deflation: if the wind cannot move sand grains across it, a dune cannot migrate.

SAND SEAS

Some large deserts contain vast tracks of shifting sand known as **sand seas**. Among the best examples are those found in northern and western Africa, the vast desert region of the Arabian Peninsula, and the large deserts of western China (Figure 17.17). Sand seas contain a variety

of dune forms, ranging from low mounds of sand to barchans, transverse dunes, and star dunes. In a typical sand sea, huge dune complexes form a seemingly endless and monotonous landscape.

Figure 17.17 Sand Sea A sand sea stretches to the horizon in the center of China's vast Takla Makan desert, which lies thousands of kilometers from oceanic sources of precipitation.

LOESS

Although most regolith contains a small proportion of wind-laid dust, the dust is thoroughly mixed with other fine sediments, making it indistinguishable from them. However, in some regions wind-laid dust is so thick and uniform that it constitutes a distinctive deposit and may control the primary landscape characteristics. Known as **loess** (German for *loose*), this sediment is defined as wind-deposited dust consisting largely of silt but commonly accompanied by some fine sand and clay.

Loess is an important resource in countries where it is thick and widespread. Its importance lies in the productive soils developed on it. The rich agricultural lands of the upper Mississippi Valley, the Columbia Plateau of Washington State, the Loess Plateau region of central China, and much of eastern Europe are developed on rich loessial soils that provide food for millions.

Loess deposits can also provide shelter: throughout the loess region of central China, caves carved in loess make dry homes for thousands of families (Figure 17.18). Furthermore, paleosols interstratified in the Chinese loess constitute a primary material for brick making.

CHARACTERISTICS OF LOESS

Loess has two characteristics that indicate it was deposited by the wind rather than by streams, in marine water, or in lakes:

1. It forms a rather uniform blanket, mantling hills and valleys alike through a wide range of altitudes.

2. It contains fossils of land plants and air-breathing animals.

Loess typically is homogeneous, lacks stratification, and, where exposed, can stand at such a steep angle that it forms vertical cliffs, just as though it were firmly cemented rock. This last property results from the fine grain size of loess, for molecular attraction among the particles is strong enough to make them very cohesive.

A vertical face of loess will stand for a long time without collapsing, but if shaken or disturbed, it may suddenly fail. The loess region of central China is prone to violent earthquakes. Although major earthquakes are infrequent, when one does occur, the results can be disastrous. The most devastating earthquake on record struck the loess country of Shaanxi Province in 1556. Of the estimated 830,000 people killed, many likely were buried when violent shaking of the ground caused steep cliffs of loess to fail and loess caves to collapse.

ORIGIN OF LOESS

The distribution of loess shows that its principal sources are deserts and the floodplains of glacial meltwater streams. Consider two examples.

Figure 17.18 Loess Cave A cave excavated by hand in compact loess provides a roomy and comfortable home for a Chinese family in the loess region near Xian.

Loess of Central China The loess that covers some 800,000 km^2 in central China, and in places reaches a thickness of more than 300 m, was blown there from the floors of the great desert basins of central Asia. The source of this loess has generally been ascribed to weathering of rocks in the deserts of northern China and Mongolia. However, it is difficult to account for the volume of the deposits and the rather high sedimentation rate by weathering alone. At least part of the sediment may have come from the breakdown of rocks by frost action and glacial processes in the high glaciated mountains of inner Asia and the subsequent deflation of resulting fine particles that were transported by streams onto large alluvial fans and dry lake floors in adjacent desert basins.

Glacial Loess of North America and Europe Loess of glacial origin is widespread in the middle part of North America (especially Nebraska, South Dakota, Iowa, Missouri, and Illinois) and in east-central Europe (especially Austria, Hungary, the Czech Republic, and Slovakia). It has two distinctive features:

1. The shapes and compositions of its particles resemble those of the fine sediment produced by the grinding action of glaciers.

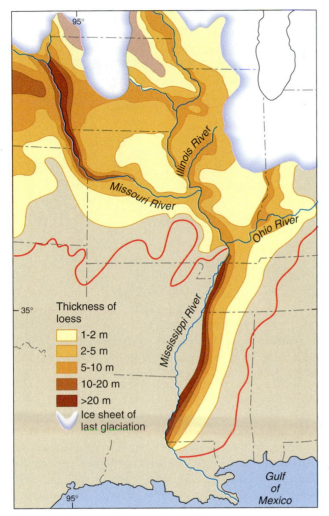

Figure 17.19 Loess in Central United States Loess deposits in the Mississippi Valley region of central United States. The loess is found largely beyond the limits of continental ice sheets that flowed into the northern United States during successive Pleistocene glaciations. The thickest loess lies adjacent to the courses of major streams that carried sediment-laden meltwater from the glacier margin. The loess thins downwind from the river source areas, showing that prevailing winds at times of loess deposition must have been predominantly from the west or northwest. Red line shows downwind limit of nearly continuous loess deposits.

2. Glacial loess is thickest downwind from former large braided meltwater streams, such as the Mississippi and Missouri rivers in North America and the Rhine and Danube rivers in Europe, which drained the North American and European ice sheets. Adjacent periglacial zones were cold and windswept. The floodplains of the constantly shifting meltwater streams were easily deflated by strong winds because they remained largely bare of vegetation. In the central United States, the coarsest fraction of windblown sediment settled out adjacent to the

source valleys, forming deposits 8 to 30 m thick (Figure 17.19). Downwind, with increasing distance from the sediment source, the loess decreases progressively in thickness and average grain size.

DUST IN OCEAN SEDIMENTS AND GLACIER ICE

Dust blown over the oceans falls out and settles to the seafloor, where it forms an important component of deep-sea sediments. Plume-like deposits of eolian dust, identified by the mineral content and chemical composition of deep-sea muds, trend eastward across the North Pacific from China, westward across the subtropical North Atlantic from Africa, and westward into the Indian Ocean from Australia (Figure 17.20). Fine particles of quartz deflated from Asian deserts have been found in soils of the Hawaiian Islands. Reddish dust from the Sahara has settled on the decks of ships in the Atlantic Ocean, is deposited on glaciers in the Alps, and has been identified in soils of the Caribbean islands.

Windblown dust is also found in cores drilled through polar ice sheets and low-latitude mountain glaciers. High rates of dust fall during the ice ages are correlated with dry conditions on the continents and strong winds that deflated fine sediment from desert basins and active outwash plains.

VOLCANIC ASH

Not all wind-transported sediment originates by deflation. Large quantities of tephra can be ejected into the atmosphere during explosive volcanic eruptions. Although deposits of volcanic ash may resemble loess, and even be interstratified with loess, a distinctive igneous mineralogy and tiny fragments of volcanic glass generally make tephra layers easy to recognize.

During an eruption, coarse and dense particles fall out quickly downwind from a vent, but small particles may be carried great distances. Fine ash that reaches the stratosphere may circle Earth many times before it finally settles to the ground. The particles that fall out during an eruption commonly form an elongate plume of sediment that decreases in particle size and thickness downwind from the source volcano (Figure 17.21). Such plume-like layers of prehistoric volcanic ash allow geologists to reconstruct the paths of winds that prevailed during ancient eruptions.

Before you go on:

1. Why are dunes asymmetrical in form?
2. What is loess? What is its principal source?

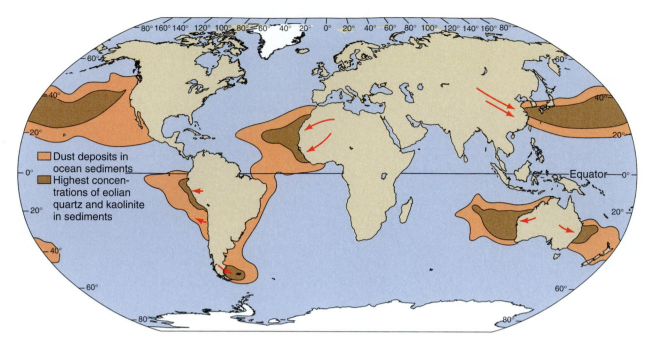

Figure 17.20 Windblown Dust on the Seafloor Fine grains of quartz and kaolinite identified in deep-sea cores define plumes of windblown dust derived from deserts in Asia, Australia, north Africa, and North and South America. The darker pattern within the plumes shows zones of highest concentration of these minerals, and therefore the primary paths of winds blowing from the desert regions. (Compare with Figure 17.6.)

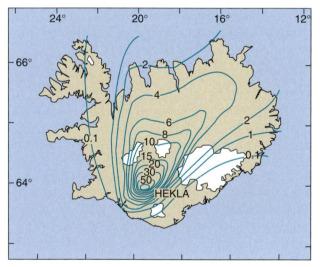

Figure 17.21 Windblown Tephra The distribution of a prehistoric tephra layer erupted from Hekla Volcano on Iceland about 4000 years ago is shown in this isopach map. The isopachs are lines of equal thickness of the deposit (in cm) and show that the tephra thins rapidly away from the volcano. The major plume of tephra extends to the north and northeast, indicating the direction of the prevailing winds at the time of the eruption.

DESERTS

The places where wind erosion, wind abrasion, and wind deposition are most obvious are deserts. Although the word *desert* literally means a deserted (relatively uninhabited) region, nearly devoid of vegetation, the modern development of artificial water supplies has changed the meaning of this word by making many desert regions livable and suitable for agriculture. As a result, the term **desert** is now generally used as a synonym for land where annual rainfall is less than 250 mm or in which the potential evaporation rate exceeds the precipitation rate, regardless of whether the land is "deserted." Aridity, then, is a chief characteristic of any desert.

TYPES AND ORIGINS OF DESERTS

Desert lands of various kinds total about 25 percent of the land area of the world outside the polar regions. In addition, a smaller though still large percentage of semiarid land exists in which the annual rainfall ranges between 250 and 500 mm. Together, these arid and semiarid regions form a distinctive pattern on the world map (Figure 17.22). They are not randomly scattered across the globe, but are related to Earth's geography and to atmospheric circulation.

In all, five types of desert are recognized: subtropical, continental, rainshadow, coastal, and polar (Table 17.2). When we compare Figure 17.22, showing the distribution of major deserts, with the general plan of atmospheric circulation shown in Figure 17.1, we can see how each type of desert relates to patterns of atmospheric circulation.

The most extensive deserts are the subtropical type, which are associated with the two circumglobal belts of dry, descending air centered between latitudes 20° and 30°.

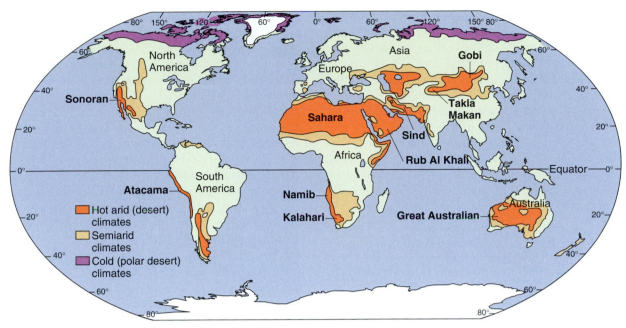

Figure 17.22 Deserts of the World Arid and semiarid climates of the world and the major deserts associated with them. Very dry areas of the polar regions include areas known as polar deserts.

Examples include the Sahara and Kalahari deserts of Africa, the Rub-al-Khali Desert of Saudi Arabia, and the Great Australian Desert.

A second type of desert (continental desert) is found in continental interiors, far from sources of moisture, where hot summers and cold winters prevail (a continental-type climate). The Gobi and Takla Makan deserts of central Asia fall into this category.

A third kind of desert is found where a mountain range creates a barrier to the flow of moist air, and produces a zone of low precipitation, called a *rainshadow*, on the lee side of mountains. As marine air moving onshore rises against the windward slope of a mountain range, it cools. Cooling lessens the amount of moisture the air can hold. As a result, most of the moisture is lost as precipitation falls on the windward slope. Air reaching the lee side of the mountain range now contains little moisture, resulting in a dry climate over the country beyond the mountains. The Cascade Range and Sierra Nevada of the western United States form such a barrier and are responsible for rainshadow deserts lying immediately east of these mountains.

Coastal deserts constitute a fourth category. They occur locally along the margins of continents, where cold, upwelling seawater cools maritime air flowing onshore. This decreases the air's ability to hold moisture. As the air encounters the warm land, it has limited moisture content. Because the air contains too little moisture to generate much precipitation, the coastal region remains a desert. In fact, coastal deserts of this type in Peru and southwestern Africa are among the driest places on Earth.

The four kinds of desert mentioned thus far are all hot deserts, where rainfall is low and summer temperatures are

TABLE 17.2 Main Types of Deserts and Their Origins

Desert Type	Origin	Examples
Subtropical	Centered in belts of descending, dry air about 20–30° north and south latitude	Sahara, Sind, Kalahari, Great Australian
Continental	In continental interiors, far from moisture sources	Gobi, Takla Makan
Rainshadow	To lee of mountain barriers that trap moist air flowing from ocean	Deserts to lee of Sierra Nevada, Cascades, and Andes Mountains
Coastal	Continental margins where cold, upwelling marine water cools maritime air flowing onshore	Coastal Peru and southwestern Africa
Polar	In regions where cold, dry air descends, creating very low precipitation	Northern Greenland, ice-free areas of Antarctica, Arctic Canada

high. In the fifth category are vast deserts of the polar regions, where precipitation is extremely low due to the sinking of cold, dry air. Cold deserts differ from hot deserts in one important respect: the surface of a polar desert, unlike those of warmer latitudes, is often underlain by abundant water, but nearly all in the form of ice. Even in midsummer, with the Sun above the horizon 24 hours a day, the air temperature may remain below freezing.

Polar deserts are found in northern Greenland, arctic Canada, and in the ice-free valleys of Antarctica. Such deserts are considered to be the closest earthly analogs to the surface of Mars, where temperatures also remain below freezing and the rarefied atmosphere is extremely dry.

DESERT CLIMATE

The arid climate of a hot desert results from the combination of high temperature, low precipitation, and high evaporation rate. The world's record temperature of 57.7°C (135.9°F) was measured in the Libyan desert of North Africa. In the Atacama Desert of northern Chile, intervals of a decade or more have passed without measurable rainfall. Temperature and evaporation are interrelated: the higher the temperature, the greater the rate of evaporation. In parts of the desert region of the southwestern United States, evaporation from lakes and reservoirs is 10 to 20 times more than the annual precipitation.

During daylight hours, the air over hot desert regions is heated, expands, and rises. Strong winds are produced as surface air moves in rapidly to take the place of the rising hot air. Thus, in addition to being arid, hot deserts can also be quite windy.

SURFACE PROCESSES AND LANDFORMS IN DESERTS

No major geologic process is restricted entirely to desert regions. Rather, the same processes operate with different intensities in moist and arid landscapes, but the most effective geologic agents in deserts are mechanical weathering, flash floods, and winds. We have already explored the importance of wind in shaping arid landscapes. Now we discuss surface sediments and landforms related to processes other than wind. In a desert, these sediments and landforms display some distinctive differences from those of the sediments and landforms of humid regions.

WEATHERING AND MASS WASTING IN DESERTS

In a moist region, regolith covers the ground almost universally. Usually, the regolith is comparatively fine textured because it contains clay, a product of chemical weathering. It is also covered with almost continuous vegetation. In contrast, the regolith in a desert is thinner, less

continuous, and coarser in texture. Much of it is the product of mechanical weathering. Although some chemical weathering does take place in deserts, its intensity is greatly diminished because of reduced soil moisture.

In both moist regions and deserts, regolith moves downslope mainly by creep. However, the results are different. That's because slope angles developed by downslope creep become adjusted to the average particle size of the regolith; the coarser the particles, the steeper the slope required to move them. In moist regions, fine-textured regolith moving downslope by creep tends to produce hillslopes that are smooth and rounded. In deserts, because particles created by mechanical weathering tend to be coarse, slopes are generally steeper and more angular.

Another characteristic of mechanical weathering is that mechanically weathered fragments of rock tend to break off along joints, leaving steep, rugged cliffs. Among the most distinctive landforms we can see in deserts is a *butte* (French for *knoll* or *small hill*, and pronounced *bewt*), which results from such breaks. Buttes may be isolated, steep-sided hills or precipitous pillars. They represent an erosional remnant carved from resistant, flat-lying rock units (Figure 17.23). A flat-topped *mesa* (Spanish for *table*) is a wider landform of the same origin.

In many desert areas, the light hues of recently deposited sediments contrast with the darker hues of older deposits. The darker color can generally be attributed to **desert varnish**, a thin, dark, shiny coating (commonly manganese oxide) formed on the surface of stones and rock outcrops in desert regions after long exposure (Figure

Figure 17.23   Butte in the Kalahari Desert   The Finger of God, a pillar-like butte of sandstone in the Kalahari Desert of southwestern Africa, rests precariously on a pyramid of erodible shale. Erosion has separated the butte from a sandstone-capped mesa in the distance.

Figure 17.24 Drawings in the Desert Varnish Aboriginal drawings have been etched in a dark coating of desert varnish on a rock outcrop at Newspaper Rock, Utah. Removal of the oxide coating has exposed unweathered rock of lighter color.

17.24). The presence of manganese in the varnish is thought to be due either to release of this element from desert dust (which settles on the ground and is weathered by summer rainstorms) or, alternatively, to the manganese-concentrating activity of microorganisms that live on rock surfaces.

DESERT STREAMS AND ASSOCIATED LANDFORMS

Contrary to popular belief and many Hollywood films, most deserts do not consist of endless expanses of sand dunes. Only a third of the Arabian Peninsula, the sandiest of all dry regions, and only a ninth of the Sahara are covered with sand.

What makes up the rest of a desert? Scattered oases—which mark places where the water table locally reaches the surface, allowing vegetation to develop—occupy little space. Much of the nonsandy area of deserts is land that is either crossed by systems of stream valleys or covered by alluvial fans and alluvial plains. Thus, in many deserts more geologic work apparently has been done by streams than by winds.

Most streams that flow into deserts from adjacent mountains never reach the sea, for they soon disappear as the water evaporates or soaks into the ground. Exceptions are long rivers like the Nile, which originates in the moist highlands of Ethiopia and East Africa and then flows across the arid expanses of the Sudan and Egypt. Such a river carries so much water that it keeps flowing to the ocean despite great evaporative loss where it crosses a desert.

FLASH FLOODS

The sparse vegetation cover in deserts presents no great impediment to surface runoff, which can readily erode loose, dry regolith. A major rainstorm is likely to be accompanied by a **flash flood**, a sudden, swift flood that can transport large quantities of sediment. The debris from such floods forms fans at the bases of mountain slopes and on the floors of wide valleys and basins.

Often streams in flood pass rapidly through desert canyons, where they erode preexisting alluvium and undercut valley sideslopes, causing the slopes to cave in. As a flood subsides, its load is deposited rapidly, creating a flat alluvial surface (Figure 17.25). The stratigraphy of such alluvial fills often discloses a complex history of cutting and filling.

Figure 17.25 Arroyo in Arizona A flash flood has just passed through this steep-walled arroyo on the Navajo reservation in northeastern Arizona. As the floodwater subsides, sediment is deposited across the flat alluvial floor of the canyon.

FANS AND BAJADAS

Alluvial fans develop under a wide range of climatic conditions, but they are especially common in arid and semiarid lands, where they typically are composed of both alluvium and debris-flow deposits. They are a characteristic landform of deserts and can be a major source of groundwater for irrigation. In some semiarid regions, entire cities have been built on alluvial fans or fan complexes (for example, San Bernardino, California, and Tehran, Iran). Alluvial fans in Iran, Afghanistan, and Pakistan are dotted with mounds of debris that mark the sites of deep artificial shafts passing downward from the surface to horizontal tunnel systems. The shafts were designed to collect water within the upper reaches of fans for use in surface irrigation. Some such systems date back a thousand years or more.

In desert basins of the southwestern United States, the Middle East, and central Asia, alluvial fans form a prominent part of the landscape. In these regions, the fans border highlands, with the top of each fan lying at the mouth of a mountain canyon. Where a mountain front is straight and its canyons are widely spaced, each fan will encompass an arc of about 180° (Figure 14.26). If canyons are closely spaced along the base of a mountain range, coalescing adjacent fans form a broad alluvial apron, or **bajada** (Spanish for *slope*), which has an undulating surface due to the convexities of the component fans (Figure 17.26).

DESERT LAKES AND PLAYAS

Runoff in arid regions is rarely abundant enough to sustain permanent lakes. Instead, the floor of a desert basin may contain a dry lakebed, called a **playa** (Spanish for *beach*) (Figure 17.26). Following a major rainstorm, runoff may be sufficient to form a temporary playa lake that will last up to several weeks. White or grayish salts at the dry surface of a playa, left by the repeated formation and evaporation of temporary lakes, can accumulate to thicknesses of tens of meters and constitute an important source of industrial chemicals.

PEDIMENTS

One of the most characteristic landforms of dry regions is the **pediment**, a broad, relatively flat surface, eroded across bedrock. The bedrock is thinly or discontinuously veneered with alluvium, which slopes away from the base of a highland (Figure 17.27). Although from a distance it may resemble a bajada, a pediment is a bedrock surface rather than a thick alluvial fill. Rock debris scattered over a pediment is carried by running water from adjacent mountains and is also derived by weathering of the pediment surface. Downslope, the scattered rock debris gradually forms a continuous cover of alluvium that merges with the thick alluvial fill of an adjacent valley.

The long profile of a pediment, like that of an alluvial fan, is concave upward, becoming progressively steeper toward a mountain front. Such a profile is typically associated with the work of running water (see Chapter 14). Faint, shallow channelways on pediment surfaces show that water is involved in their formation, and it is generally agreed that pediments are slopes across which sediment is transported by mass wasting and running water. Eyewitness accounts of lateral erosion by floodwaters associated with intense desert storms have led geologists to suggest that both these processes may be involved in pediment formation. However, the exact manner in which pediments form is still not firmly established.

As suggested, when the surface of a pediment rises toward a mountain, the surfaces meet at an abrupt angle. This suggests that desert mountain slopes do not become

Figure 17.26 Bajada in Death Valley
A vast salt-encrusted playa occupies the floor of Death Valley in California. The playa is bordered by a bajada, composed of coalescing alluvial fans constructed beyond the mouths of adjacent mountain valleys.

angle determined by the resistance of the bedrock and maintain that angle as they gradually retreat under the attack of weathering and mass wasting (Figure 17.28). In this way, as a mountain slope retreats, a pediment will increase in size by expanding at its upslope edge. The growth of the pediment, at the expense of the mountain, may continue until the entire mountain has been consumed.

INSELBERGS

Among the most impressive of Earth's landforms are steep-sided mountains, ridges, and isolated hills that rise abruptly from adjoining plains like rocky islands standing above the surface of a broad, flat sea. Ayers Rock in central Australia is a famous example (Figure 17.29). Called **inselbergs** (German for *island mountain*), these landforms appear in many environmental settings, ranging from coastal to interior and arid to humid. However, they are especially common and well developed in semiarid grasslands in the middle of tectonically stable continents. Numerous examples can be found in southern and central Africa, northwestern Brazil, and central Australia.

Field evidence suggests that inselbergs form in areas of relatively homogeneous, resistant rock (most commonly granite or gneiss but also sedimentary rocks such as conglomerate and sandstone) that are surrounded by rocks more susceptible to weathering. Differential weathering over long time intervals lowers adjacent terrains, leaving these resistant rock masses standing high.

Once formed, the bare rock hills tend to shed water, whereas surrounding debris-mantled plains absorb water, causing the underlying rocks to weather more rapidly. For this reason, inselbergs may remain as stable parts of a landscape and persist for tens of millions of years or more.

Figure 17.27 Pediment in Mojave A pediment in the Mojave Desert of southeastern California has left only a few residual hills near the crest of a former mountain ridge. The flat bedrock surface cut across crystalline rocks passes downslope beneath a thin cover of alluvium.

gentler with time, as slopes tend to do in moist regions, which are dominated by creep and chemical weathering. Instead, in arid lands the slopes apparently achieve an

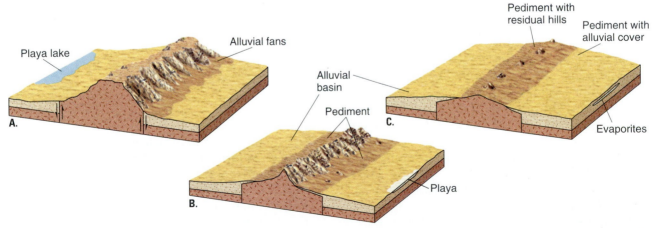

Figure 17.28 Formation of a Pediment Stages in the formation of a pediment. A. A mountain block, uplifted along bordering faults, is eroded by streams that contribute sediment to a growing alluvial basin fill. B. A pediment is cut across the margins of the uplifted block and grows headward into the mountains as sheetfloods and running water transport sediment across the planar rock surface toward the basin fill. C. Growing headward from both sides of the upland, the pediment slowly consumes the mountains, leaving only a few residual hills rising above the eroded bedrock surface.

Figure 17.29 Inselberg Ayers Rock, a massive inselberg, rises about 360 m above the surrounding flat plain in central Australia.

Some may even date back to the Mesozoic Era. If that is true, they have remained prominent landscape features since the time of the dinosaurs.

DESERTIFICATION

In the region south of the Sahara lies a belt of dry grassland known as the Sahel (Arabic for *border*). There the annual rainfall is normally only 100 to 300 mm, most of it falling during a single brief rainy season.

In the early 1970s, the drought-prone Sahel experienced the worst drought of the twentieth century (Figure 17.30). For several years in succession the annual rains failed to appear, causing adjacent desert to spread southward—according to one estimate as much as 150 km. The drought extended from the Atlantic to the Indian Ocean, a distance of 6000 km, and affected a population of at least 20 million people, many of them seminomadic herders of cattle, camels, sheep, and goats. The results of the drought were intensified by the fact that between about 1935 and 1970 the human population had doubled and the number of livestock had also increased dramatically. This increase in the number of people and animals led to severe overgrazing.

With the coming of the drought, the grass cover almost completely failed. About 40 percent of the cattle—a great many millions—died. Millions of people suffered from thirst and starvation, and many died as vast numbers migrated southward in search of food and water.

In the mid-1970s, the rains returned briefly. Then, in the 1980s, drought conditions resumed. Ethiopia and the Sudan were especially hard hit and experienced widespread famine. Mass starvation was alleviated only by worldwide relief efforts.

Figure 17.30 Drought in Mali Overgrazing during years of drought killed most of the vegetation around wells in the Azaouak Valley, Mali. Without vegetation, soil blows away and the desert advances.

Such invasion of desert into nondesert areas is referred to as **desertification**. The major symptoms are declining groundwater tables, increasing saltiness of water and topsoil, reduction in supplies of surface water, unnaturally high rates of soil erosion, and destruction of native vegetation. We can find evidence of natural desertification events in the geologic record. (See *Understanding Our Environment*, Box 17.2, *Tectonic Desertification*.) Regardless of natural climatic trends, however, there is increasing concern that human activities can in themselves promote widespread desertification (Figure 17.31).

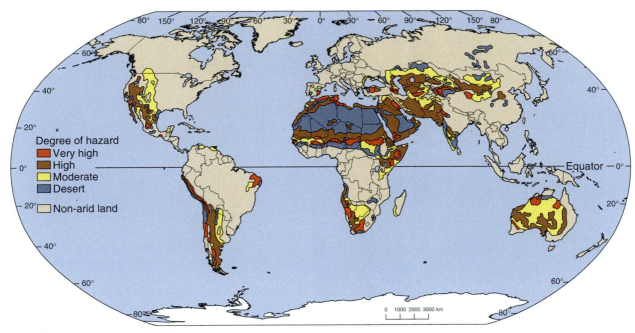

Figure 17.31 Possibility of Desertification Substantial portions of all the temperate-latitude continents face moderate to very hazard of high desertification.

DESERTIFICATION IN NORTH AMERICA

The impact of desertification on human life is less severe in North America than in more densely populated regions of the world. Nevertheless, desertification has important and far-reaching implications for the continent's food, water, and energy supplies, as well as its natural environment.

Nearly 37 percent of the dry regions of the continent have experienced "severe" desertification. In the southwestern United States, about 10 percent of the land area—approximately the same size as the original 13 states—has been affected by desertification over the past century. Large shifting sand dunes have formed, erosion has virtually denuded the landscape of vegetation, numerous gullies have developed, and salt crusts have accumulated on nearly impermeable irrigated soils. This desertification has been brought about primarily by overgrazing, excessive withdrawal of groundwater, and unsound water-use practices, combined with a population increase and expanded agricultural production.

COUNTERING DESERTIFICATION

How can the detrimental effects of desertification be halted? Can these effects be reversed? The answers lie largely in understanding the geologic principles discussed in this chapter and intelligently applying measures designed to reestablish a natural balance in the affected areas. Elimination of incentives to exploit arid lands beyond their natural capacity, coupled with long-range planning aimed

at minimizing the negative effects of human activity, should help in reaching the desired goal. Because arid lands of the western United States supply about 20 percent of the nation's total agricultural output, the long-term benefits could be substantial.

> **Before you go on:**
>
> 1. What are the chief factors affecting desert climate?
>
> 2. What are the origins of the five major types of deserts?
>
> 3. Why and how do desert landforms differ from moist climate landforms?

REVISITING PLATE TECTONICS AND THE EARTH SYSTEM

PLATE TECTONICS AND DESERTS OF THE PAST

Ancient desert sediments resemble those found in the deserts of today. Sandstones tend to be reddish in color, for example, caused mainly by thin coating of iron compounds such as limonite and hematite on the sand grains, and large dune structures are common.

In the central Asian deserts of today, such as the Gobi and Takla Makan deserts, there are large dune fields and the sands are reddish in color. The central Asian deserts are con-

UNDERSTANDING OUR ENVIRONMENT

TECTONIC DESERTIFICATION

Western China contains two of the world's major deserts. One, the Takla Makan, is a hot desert with vast regions of shifting sand dunes (Figure 17.17). The second encompasses the western Tibetan Plateau, a high, cold desert with sparse steppe vegetation

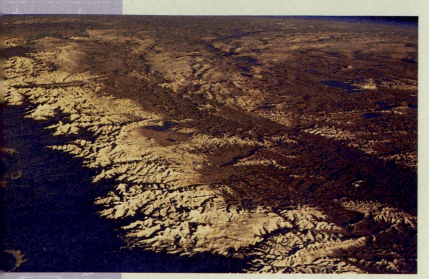

Figure B17.1 Himalayan Barrier The snow- and ice-capped Himalaya rising high above the dense green vegetated plains of India (lower left) keep moisture-bearing winds from reaching the high Tibetan Plateau, which receives less than 250 mm of annual precipitation.

underlain by perennially frozen ground (Figure B17.1). Both are relatively young, geologically speaking, and their origin is tied to the tectonics of converging lithospheric plates.

Evidence of the long-term natural desertification of western China has been assembled by Chinese paleontologists who have studied the fossil remains of plants and animals in Tertiary strata. The fossil sites lie scattered about the high Tibetan Plateau and adjacent mountain ranges, mostly at altitudes of 4000 to 6000 m where the landscape now consists of cold alpine steppe and cold desert. Fossil plants and pollen of Eocene and Oligocene age disclose a flora that consisted of evergreen broadleaf forests and included eucalyptus, magnolia, and fig. Today, these warmth-loving trees are found in moist, tropical environments at low altitudes. Remains of a giant rhinoceros, a larger relative of the rhinoceros that lives in tropical southeastern Asia today, have also been found in the Oligocene strata. From this evidence, we can infer that western China during the early Tertiary must have had a relatively low altitude (perhaps 500 to 1000 m) and a warm climate, and that there were no mountain barriers to block the passage of moist oceanic air from the south. The widespread distribution of *Hipparian* supports this view, for it implies that this primitive horse could migrate freely over the continent unimpeded by major mountain chains.

Changing plant and animal assemblages imply progressive uplift of the region during the Miocene and Pliocene, accompanied by cooling and drying of the climate. By the beginning of the Pleistocene, nearly 2 million years ago, the Tibetan Plateau had reached an altitude of 1000 to 2000 m, and alpine coniferous forests had replaced subtropical trees. During the last half of the Pleistocene, these forests gave way to dry alpine steppe as the rate of uplift acceler-

tinental deserts, resulting from the long distance that winds must flow over land without a source of moisture. Looking back in the geologic record we can find very similar evidence of reddish sandstones and large dune structures in rocks that formed during the Permian, Triassic, and Jurassic periods in northwestern Europe and North America. We hypothesize, on the basis of the Principle of Uniformitanianism, that desert conditions must have prevailed at the time the sediments were laid down. If the hypothesis has any validity, we should be able to find out why there should have been desert conditions. A look at a plate tectonic reconstruction for Permian and Triassic times provides the clue. The desert sediments formed when Pangea was at its maximum extent, and the places where the sediments were deposited were far from the oceans of the day. They are continental desert deposits.

Desert sediments are found in many parts of the world and in rocks of all ages. Deserts are natural phenomena that

have formed repeatedly during geologic ages. The rock record is not always sufficiently complete to allow us to say why a particular desert formed, but like the other natural record of climatic extremes, glacial deposits, they provide clear evidence of the climate at a given place and time.

AIR: A CONSEQUENCE OF THE EARTH SYSTEM

Is anything more important to the human race than air? But how well do we understand where air comes from and what maintains its composition?

Air is the colorless, odorless mixture of gases that surrounds one special planet, Earth. Because air pressure decreases with altitude, the amount of air per unit volume also decreases with altitude. In order to distinguish changes due to composition from those due to pressure, the composition of air is always discussed in terms of relative

ated, bringing the plateau and the adjacent Himalaya to high altitudes.

The progressive desertification of western China during the Tertiary coincides with the ongoing collision of India with Asia, enlarging of Asia, and resulting uplift of the Tibetan Plateau and associated mountain ranges (Figure B17.2). Prior to this collision, the area of Tibet and the Takla Makan desert stood at low altitude and lay close to an ancient seaway that separated India and Asia. As the two continents converged, the seaway narrowed and then disappeared, placing these future desert regions closer to the center of Asia and therefore in a more continental climatic environment. Uplift associated with continental collision further intensified the desertification process by raising Tibet into successively drier and colder climatic zones and simultaneously raising the lofty Himalaya across the path of the northward-flowing monsoonal air, thereby blocking off the primary source of precipitation. The cold desert of the lofty plateau and the adjacent hot Takla Makan therefore owe their existence both to their midcontinental position and the rainshadow formed by the high mountain barrier that separates the hot, humid plains of India from the frigid wastes of Tibet.

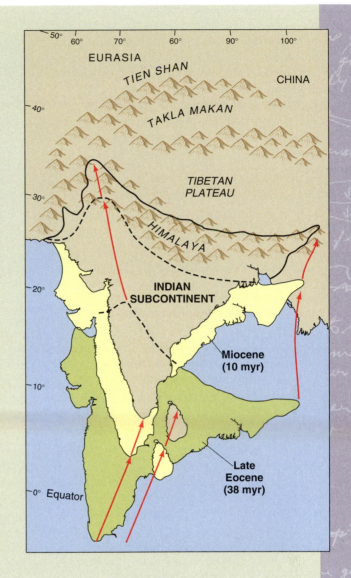

Figure B17.2 Asian Collision As India drifted northward during the Tertiary Period, the seaway separating it from Asia narrowed and disappeared. Successive positions of the continents during the late Eocene and the late Miocene are shown relative to the present continent of Asia. The collision of the two landmasses resulted in uplift of the Himalaya and associated high mountain ranges, and isolation of the Tibetan Plateau and Takla Makan Desert from moisture sources to the south. Arrows show movement of four different places with time. Times are in millions of years (myr).

rather than absolute amounts of the different constituents present. The relative composition, it turns out, varies from place to place at Earth's surface, even from time to time in the same place. The reason is water vapor, which varies considerably depending on the humidity. For this reason the relative composition of air is reported as dry air. Just three gases, nitrogen, oxygen, and argon, make up 99.96 percent of dry air (Figure 17.32). Eight gases make up the remaining 0.04 percent.

Initially all of the constituents in air came from inside Earth in the form of volcanic gases. But volcanic gases differ considerably from the gases in air, and the reason is the Earth system. Oxygen, for example, would not have been present in the original volcanic gases. Oxygen is present in the atmosphere because plants and other photosynthesizing organisms combine carbon dioxide with water to form carbohydrate compounds through photosynthesis. Oxygen is a

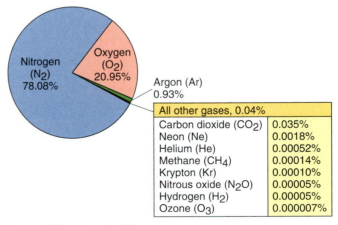

Figure 17.32 Gases in Air Composition of dry air in volume percent. Three gases—nitrogen, oxygen, and argon—make up 99.96 percent of the air.

waste product of photosynthesis that accumulates in the atmosphere. Because plants remove carbon dioxide from the atmosphere by photosynthesis, and aquatic creatures make shells from calcium carbonate, most of the carbon dioxide that was in the original atmosphere has been removed and is buried in sediments as organic matter and limestones.

All parts of the Earth system, the geosphere, hydrosphere, and the biosphere, play a role in maintaining the composition of air. If it were not so, we would be unable to live on Earth.

What's Ahead?

Many of the features discussed in this chapter have to do with a paucity of water. Wind erosion and wind transport of sediment, for example, are principally features of the low-rainfall regions of the world. But there is a large portion of the world where just the opposite is true, and the features we see are produced by an abundance of water. In the next chapter we turn to the main water reservoir of the world, the ocean.

CHAPTER SUMMARY

1. Unequal heating of Earth's surface by solar radiation sets up convective circulation in the atmosphere. The Coriolis effect breaks the equator-to-poleward atmospheric circulation in the Northern and Southern Hemispheres into cells, which contain prevailing lower-level winds called the tradewinds, westerlies, and polar easterlies.

2. Climate is the average weather conditions of a place or area over a period of years. Earth's climates are influenced by the distribution of land and oceans, and by surface topography.

3. Wind moves saltating sand grains close to the ground and suspended dust particles at higher levels. Sorting of sediment results.

4. Through deflation and abrasion, winds create deflation basins, desert pavement, ventifacts, and yardangs.

5. Dunes originate where obstacles distort the flow of air. Dunes have steep slip faces and gentler windward slopes. They migrate in the direction of windflow, forming cross strata that dip downwind.

6. Loess is deposited chiefly downwind from deserts and from the floodplains of glacial meltwater streams. Once deposited, it is stable and is little affected by further wind action.

7. Airborne tephra deposited during explosive volcanic eruptions decreases in thickness and grain size away from the source vent.

8. Hot deserts constitute about a quarter of the world's nonpolar land area and are regions of slight rainfall, high temperature, excessive evaporation, relatively strong winds, and sparse vegetation. Polar deserts occur at high latitudes where descending cold, dry air creates arid conditions.

9. Mechanical weathering, flash floods, and winds are especially effective geologic agents in deserts.

10. Fans, bajadas, and pediments are conspicuous features of many deserts. Pediments are probably shaped by running water and are eroded surfaces across which sediment is transported.

11. Inselbergs form in relatively homogeneous, resistant rocks and may remain as persistent landforms for millions of years.

12. Recurring natural droughts can lower the water table, cause high rates of soil erosion, and destroy vegetation, thereby leading to the invasion of deserts into nondesert areas. Overgrazing, excessive withdrawal of groundwater, and other human activities can promote desertification. Desertification can be halted or reversed by measures that restore the natural balance.

THE LANGUAGE OF GEOLOGY

bajada (p. 466)

climate (p. 450)
Coriolis effect (p. 449)

deflation (p. 454)
desertification (p. 468)
desert (p. 462)
desert pavement (p. 455)

desert varnish (p. 464)
dune (p. 457)

flash flood (p. 465)

inselberg (p. 467)

loess (p. 460)

pediment (p. 466)
playa (p. 466)

sand ripples (p. 451)
sand sea (p. 459)

ventifact (p. 456)

yardang (p. 456)

QUESTIONS FOR REVIEW

1. What atmospheric factors cause most of the world's large hot deserts to be concentrated in belts lying between 20° and 30° from the equator?

2. Explain what controls the depth to which deflation is effective in arid regions.

3. Why are the erosional effects of blowing sand generally confined to a zone extending only about a meter above the ground surface?

4. Explain the origin of the internal stratification of a sand dune. How do sand dunes migrate downwind?

5. How might you tell the former direction of the prevailing wind from the form and internal stratification of an ancient, inactive sand dune? from a ventifact? from a tephra deposit?

6. What measures could you recommend that a farmer in the southwestern United States take to halt the migration of sand dunes now threatening his agricultural fields?

7. How might you tell a deposit of loess from an alluvial silt having a similar range of particle sizes?

8. Why do hillslopes in arid landscapes tend to be steeper and sharper than those in humid landscapes?

9. What evidence can you cite that points to streams being effective agents of erosion and sediment transport in desert regions?

10. How would you tell a bajada from a pediment in the field? What process(es) are involved in the formation of each?

11. Why do playas often have a distinctive deposit of salts at their surface?

12. What factors influence the formation of inselbergs? Why are inselbergs likely to remain persistent features of a semiarid landscape?

13. What are some of the obvious symptoms of desertification? How might they be retarded or reversed by human intervention?

Click on *Presentation* and *Interactivity* in the **GeoHazards** module of your CD-ROM to further explore resources and activities presenting concepts from this chapter. Select *Assessment* in the same module to test your understanding of this chapter.

Chapter 18

Aerial view of Palau, a tropical island group in the western Pacific Ocean. Scattered islets in the lagoon are surrounded by a broad, shallow carbonate reef, beyond which water depths increase rapidly toward the floor of the deep ocean.

The Oceans and Their Margins

Lagoon margin, Rongelap Atoll, northern Marshall Islands.

Changing Sea Level and Atoll Evolution

On March 1, 1954, at 6:45 in the morning, an American nuclear device was detonated at Bikini Atoll. So powerful was the explosion that the resulting nuclear cloud rose

60 km above the atoll, far higher than anticipated. Instead of traveling northward, as expected, the cloud with its deadly rain of fallout traveled eastward across the inhabited atoll of Rongelap.

A decade later, I joined an Atomic Energy Commission expedition that was to monitor the lingering radiation levels in Rongelap's biota. As the only geologist, my research focused on the recent and long-term sea-level history of the atoll.

Rongelap atoll rises abruptly from the deep ocean floor and its geology was believed to record only glacial-age rise and fall of sea level, plus very slow long-term subsidence related to cooling of the Pacific Plate. The geologic evidence I recorded in my notebook affirmed this hypothesis. The highest deposits on the atoll's sandy islets rose only 5 m above sea level, and were related to recent large storms rather than to higher interglacial sea-level stands. Why was there no record of higher sea levels like that preserved on many volcanic islands in the Pacific?

The answer may lie in data showing that the maximum depths of Rongelap and 54 other atoll lagoons in this part of the Pacific average –52 m. This value is remarkably close to that of estimated average sea level during the last five glacial–interglacial cycles (ca. –55 m), based on isotopic analyses of deep-sea sediments. I reasoned that when world sea level fell during glacial times, exposed atoll reef tops were eroded down to the level of average glacial–interglacial sea level. As sea level rose during deglaciation, reef construction began anew and kept pace with the rising ocean surface. Although several factors contribute to the evolution of atoll lagoons, lagoon depth may be strongly influenced by long-term *average* sea level, and therefore by average glacial conditions.

Stephen C. Porter

INTRODUCTION: THE WORLD OCEAN

Seawater covers 70.8 percent of Earth's surface, and most of it is contained in three huge interconnected basins—the Pacific, Atlantic, and Indian oceans. All are connected with the Southern Ocean, a body of water south of 50° S latitude that completely encircles Antarctica. Collectively, these four vast interconnected bodies of water, together with a number of smaller ones, are often referred to as the *world ocean* (Figure 18.1).

In this chapter we begin with the properties of the world ocean, then discuss its circulation and the movements of tides and waves. Next we turn our attention to coastlines, where land and ocean meet, and will describe features of different kinds of coasts and their origin. Finally, we consider coastal hazards, including storms and landslides, and the protective measures we can take against the damage resulting from coastal erosion.

THE OCEANS' CHARACTERISTICS

Although the world ocean covers nearly 71 percent of Earth's surface, to most of us it is far less familiar than the land. What is this world-encircling body of water like? How does it affect Earth's human population, three-quarters of which live close to ocean margins? In this section we discuss several important properties of ocean water: its depth and volume, salinity, temperature and heat capacity, and stratification.

DEPTH AND VOLUME OF THE OCEANS

Over the past 70 years, the oceans have been criss-crossed many thousands of times by ships carrying acoustical instruments called echo sounders that measure ocean depth. As a result, the topography of the seafloor and the depth of the overlying water are known in considerable detail for all but the most remote parts of the ocean basins.

The greatest ocean depth yet measured (11,035 m) lies in the Mariana Trench near the island of Guam in the western Pacific. This is more than 2 km farther below sea level than Mount Everest rises above sea level. The average

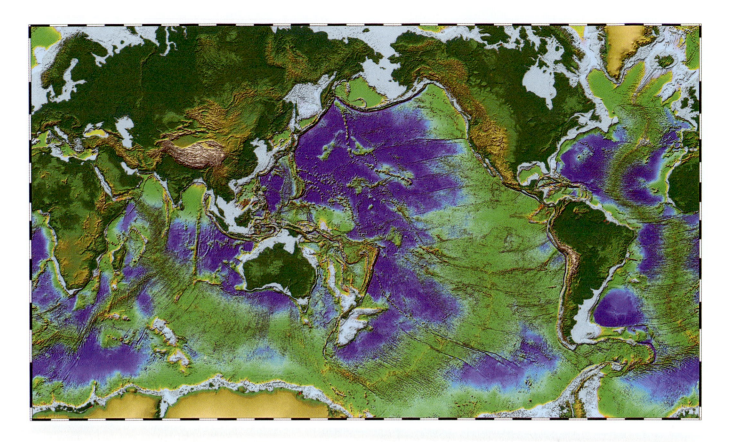

Figure 18.1 World Ocean Shaded relief map showing major and minor basins of the world ocean. The deepest basins are shown in purple and blue, shallower regions in shades of green and yellow, and the continental shelves in light blue.

depth of the oceans, however, is about 3.8 km, compared to an average height of the land of only 0.75 km.

If we measure the area of the oceans and calculate their average depth, we then can calculate that the present volume of seawater is about 1.35 billion cubic kilometers (Figure 18.2); more than half this volume resides in the Pacific Ocean. We say *present* volume because the amount of water in the oceans fluctuates somewhat over thousands of years with the growth and melting of continental glaciers (Chapter 16).

OCEAN SALINITY

About 3.5 percent of average seawater, by weight, consists of dissolved salts, enough to make the water undrinkable. It is enough, too, if these salts were precipitated, to form a layer about 56 m thick over the entire seafloor.

Salinity is the measure of the sea's saltiness, expressed in per mil (‰ = parts per thousand) rather than percent (% parts per hundred). The salinity of seawater normally ranges between 33 and 37‰. The principal elements that contribute to this salinity are sodium and chlorine. Not surprisingly, when seawater is evaporated, more than three-quarters of the dissolved matter is precipitated as common salt (NaCl). However, seawater contains most of the other natural elements as well, many of them in such low concentrations that they can be detected only by extremely sensitive analytical instruments.

More than 99.9 percent of the ocean's salinity reflects the presence of only eight ions: chloride, sodium, sulfate, magnesium, calcium, potassium, bicarbonate, and bromine. Where do these ions come from? The sources vary.

Cations are released by chemical weathering processes on land. As exposed crustal rocks interact with the atmosphere and the hydrosphere (rainwater), cations are leached out and become part of the dissolved load of streams. Each year streams carry 2.5 billion tons of dissolved substances to the sea.

The principal anions found in seawater are believed to have come from the mantle. Chemical analyses of gases released during volcanic eruptions show that the most important volatiles are water vapor (steam), carbon dioxide (CO_2), and the chloride (Cl^{1-}) and sulfate (SO_4^{2-}) anions. These two anions dissolve in atmospheric water vapor and return to Earth in precipitation, much of which falls directly into the ocean. Part of the remainder is carried to the sea dissolved in river waters. Volcanic gases are also released directly into the ocean from submarine eruptions.

Another source of ions is dust eroded from desert regions and blown out to sea. In addition, gaseous, liquid, and solid

A.

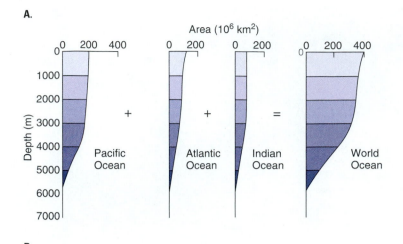

B.

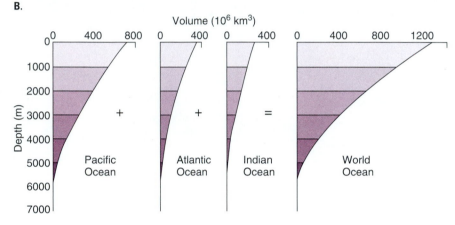

Figure 18.2 Area (A) and Volume (B) of the Oceans The Pacific represents nearly half the volume of the oceans, with the Atlantic and Indian oceans being comparable to each other in both size and volume. Although the deepest known place in the oceans lies more than 11,000 m below sea level, nearly all the water lies at a depth of less than 6000 m.

pollutants are either released by human activity directly into the oceans or carried there by streams or polluted air.

The quantity of dissolved ions added by rivers over the billions of years of Earth's history far exceeds the amount now dissolved in the sea. Why, then, doesn't the sea have a higher salinity? The reason is that chemical substances are being removed at the same time they are being added. Some elements, such as silicon, calcium, and phosphorus, are withdrawn from seawater by aquatic plants and animals to build their shells or skeletons. Other elements, such as potassium and sodium, are absorbed and removed by clay particles and other minerals as they slowly settle to the seafloor. Still others, such as copper and lead, are precipitated as sulfide minerals in claystones and mudstones rich in organic matter. Because these and other processes of extraction are essentially equal to the combined inputs, the composition of seawater remains virtually unchanged.

The most important factors affecting salinity of surface waters are:

• Evaporation (which removes water and leaves the remaining water saltier).

• Precipitation (which adds fresh water, thereby diluting the seawater and making it less salty).

• Inflow of fresh (river) water (which makes the seawater less salty).

• The freezing and melting of sea ice (when seawater freezes, salts are excluded from the ice, leaving the unfrozen seawater saltier).

As one might expect, salinity is high in the latitudes where Earth's great deserts lie, for in these zones evaporation exceeds precipitation, both on land and at sea (Figure 18.3A). In a restricted sea, like the Mediterranean, where there is little inflow of fresh water, surface salinity exceeds the normal range; in the Red Sea, which is surrounded by desert, salinity reaches 41 percent. Salinity is lower near the equator because precipitation is high, and cool water, which rises from the deep sea and sweeps westward in the tropical eastern Pacific and eastern Atlantic oceans, reduces evaporation. It also is low at high latitudes that are rainy and cool. Up to 100 km offshore from the mouths of large rivers, the surface ocean water can be fresh enough to drink.

TEMPERATURE AND HEAT CAPACITY OF THE OCEAN

An unsuspecting tourist from Florida who decides to take a swim on the northern coast of Britain quickly learns how varied the surface temperature of the ocean can be. A map of global summer sea-surface temperature displays a pronounced east–west zonation, with *isotherms* (lines con-

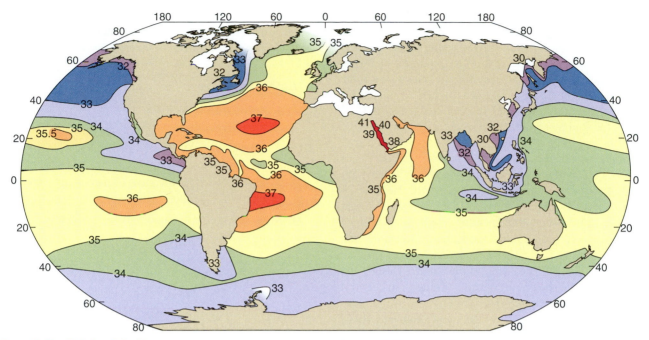

Figure 18.3A   Salinity of the Ocean   Average surface salinity of the oceans. High salinity values are found in tropical and subtropical waters where evaporation exceeds precipitation. The highest salinity has been measured in enclosed seas like the Persian Gulf, the Red Sea, and the Mediterranean Sea. Salinity values generally decrease poleward, both north and south of the equator, but low values also are found off the mouths of large rivers.

necting points of equal temperature) lying approximately parallel to the equator (Figure 18.3B). The warmest waters during August (>28°C) occur in a discontinuous belt between about 30° N and 10° S latitude in the zone where solar radiation reaching the surface is at a maximum. In winter, when the belt of maximum incoming solar radia-

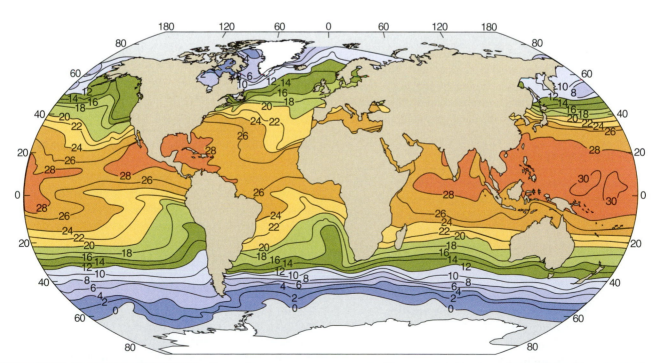

Figure 18.3B   Temperature of the Ocean   Sea-surface temperatures in the world ocean during August. The warmest temperatures (~28°C) are found in the tropical Indian and Pacific oceans. Temperatures decrease poleward from this zone, reaching values close to freezing in the north and south polar seas.

tion shifts southward, the belt of warm water also moves south until it is largely below the equator. Waters become progressively cooler both north and south of this belt and reach temperatures of less than 10°C poleward of 50° N and S latitude. The average surface temperature of the oceans is about 17°C, while the highest temperatures (>30°C) have been recorded in restricted tropical seas, such as the Red Sea and the Persian Gulf.

The ocean differs from the land in the amount of heat it can store. For a given amount of heat absorbed, water has a lower rise in temperature than nearly any other substance; that is, it has a high *heat capacity*. Because of water's ability to absorb and release large amounts of heat with very little change in temperature, both the total range and the seasonal changes in ocean temperatures are much less than what we find on land. The highest recorded land temperature is 58°C, measured in the Libyan Desert, and the lowest, measured at Vostok Station in central Antarctica, is –88°C; the range, therefore, is 146 C°. By contrast, the highest recorded ocean temperature is 36°C, measured in the Persian Gulf, and the coldest, measured in the polar seas, is –2°C, a range of only 38 C°.

Now let's consider the range of sea-surface temperatures. The annual range in sea-surface temperatures (difference between highest and lowest values) is 0–2 C° in the tropics, 5–8 C° in the middle latitudes, and 2–4 C° in the polar regions. Coastal inhabitants benefit from the mild climates resulting from this natural ocean thermostat. Along the Pacific coast of Washington and British Columbia, for example, winter air temperatures seldom drop to freezing, while east of the coastal mountain ranges they can plunge to –30°C or lower. In the interior of a continent, summer temperatures may exceed 40°C, whereas along the ocean margin they typically remain below 25°C.

Here, then, is a good example of the interaction of the hydrosphere, atmosphere, land surface, and biosphere. Ocean temperatures affect the climate, both over the ocean and over the land, and climate ultimately is a major factor in controlling the distribution of plants and animals.

VERTICAL STRATIFICATION

We have just been discussing temperature conditions of the surface water. It is important to realize, however, that temperature and other physical properties of seawater vary with depth.

To help understand why this is so, think about what happens when fresh river water meets salty ocean water at a coast. Being less dense, the fresh water flows over the denser saltwater, resulting in stratified water bodies. Such stratification is common in estuaries (Chapter 7).

The oceans also are vertically stratified, as a result of variations in the density of seawater. Seawater becomes denser as its temperature decreases and as its salinity increases. Gravity pulls dense water downward until it reaches a level where the surrounding water has the same density. These density-driven movements lead both to stratification of the oceans and to circulation in the deep ocean, subjects we turn to in the next section.

OCEAN CIRCULATION

When Christopher Columbus set sail from Spain in 1492 to cross the Atlantic Ocean in search of China, he took an indirect route. Instead of sailing due west, which would have made his voyage shorter, he took a longer route southwest toward the Canary islands, then west on a course that carried him to the Caribbean Islands where he first sighted land. In choosing this course, he was following the path not only of the prevailing winds but also of surface ocean currents. Instead of fighting the westerly winds and currents at 40° N latitude, he drifted with the Canary Current and North Equatorial Current, as the northeast tradewinds filled the sails of his three small ships.

In this section we examine surface ocean currents and what propels them. We also will look at the major water masses of the ocean and their role in generating deep currents and the global ocean circulation system.

SURFACE CURRENTS OF THE OPEN OCEAN

Surface ocean currents like those Columbus followed are broad, slow drifts of surface water set in motion by the prevailing surface winds. Air that flows across the sea drags the water slowly forward, creating a current of water as broad as the current of air but rarely more than 50 to 100 m deep. The ultimate source of this motion is the Sun, which heats Earth's surface unequally, thereby setting in motion the planetary wind system.

THE CORIOLIS EFFECT

The direction taken by ocean currents is also influenced by the Coriolis effect, the phenomenon by which all moving bodies veer to the right in the Northern Hemisphere and to the left in the Southern Hemisphere (Chapter 17). If a freely floating object on the ocean in the Northern Hemisphere moves away from the pole, its angular velocity about the pole will be slower than that of the water. This causes the object to lag behind the rotation, and so it will be deflected in a clockwise direction (to the right). If such an object were moving toward the pole, its angular velocity about the pole would be faster than the water, resulting in a counterclockwise deflection (again toward the right). Regardless of the direction of movement, an object in the Northern Hemisphere will be deflected to the right. In the Southern Hemisphere, the deflection is to the left, while at the equator the effect disappears. Although the Coriolis effect does not cause ocean currents, it deflects them once they are in motion.

CURRENT SYSTEMS

Low-latitude regions in the tradewind belts are dominated by the warm, westward-flowing North and South Equatorial currents (Figure 18.4). In their midst, and lying in the belt of light, variable winds (the doldrums), is the eastward-flowing Equatorial Countercurrent.

Each major current is part of a large subcircular current system called a **gyre**. Figure 18.4 shows Earth's five major ocean gyres—two each in the Pacific and Atlantic oceans and one in the Indian Ocean. Currents in the Northern Hemisphere gyres circulate in a clockwise direction; those in the Southern Hemisphere circulate counterclockwise.

The northern Indian Ocean has a unique circulation pattern: the direction of flow changes seasonally with the changing pattern of monsoonal air flow (see Chapter 17). During the summer, strong and persistent monsoon winds

blow the surface water eastward, but in winter, winds from Asia blow the water westward.

MAJOR WATER MASSES

In addition to wind-driven circulation of surface waters, ocean waters also circulate on a large scale within the deep ocean, driven by differences in water density.

The water of the oceans is organized into major *water masses*. These are bodies of seawater, each having a characteristic range of temperature and salinity. The identity and sources of these water masses have been determined by studying the salinity and temperature structure of the water column at many places. The water masses are stratified with respect to one another based on their relative densities. Cold water is denser than warm water, and salty water is denser than less-salty water. These density differences lead to large-scale circulation of water within the deep ocean, as water masses with different temperature, saltiness, and density move vertically or horizontally with respect to other masses of different density. The Atlantic Ocean provides a good example (Figure 18.5; Table 18.1).

THE GLOBAL OCEAN CONVEYOR SYSTEM

Dense, cold, and/or salty surface waters that flow toward adjacent warmer, less-salty waters will sink until they reach a level in which the water masses are of equal density. The resulting stratification of water masses is thus

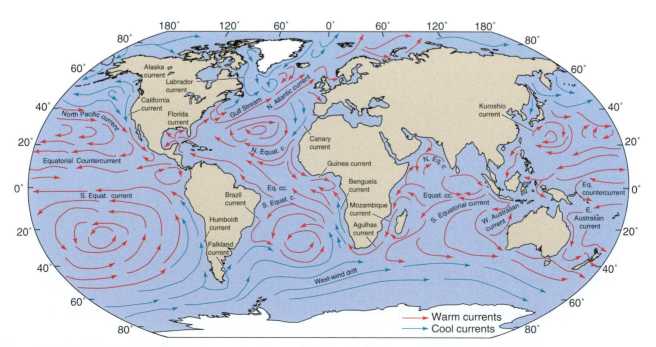

Figure 18.4 Ocean Currents Surface ocean currents form a distinctive pattern, curving to the right (clockwise) in the Northern Hemisphere and to the left (counterclockwise) in the Southern Hemisphere. The westward flow of tropical Atlantic and Pacific waters is interrupted by continents, which deflect the water poleward. The flow then turns away from the poles and becomes the eastward-moving currents that define the middle-latitude margins of the five great midocean gyres.

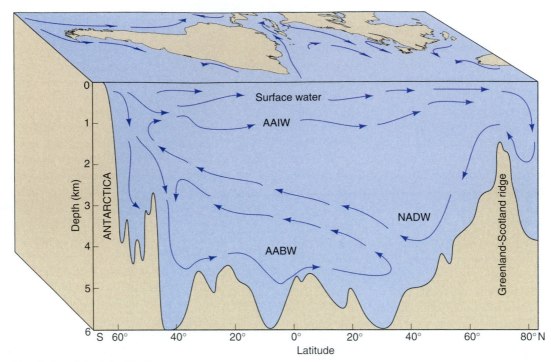

Figure 18.5 Circulation of the Atlantic Ocean Longitudinal transect along the western Atlantic Ocean showing water masses and general circulation pattern. North Atlantic Deep Water (NADW) originates near the surface in the North Atlantic as northward-flowing surface water cools, becomes increasingly saline, and plunges to depths of several km. As NADW moves into the South Atlantic, it rises over denser Antarctic Bottom Water (AABW). Antarctic Intermediate Water (AAIW) forms adjacent to the Antarctic continent and flows into the North Atlantic at a mean depth of about 1 km.

based on relative density. For example, dense, cold Antarctic Intermediate Water (AAIW; Table 18.1) spreads slowly northward after sinking downward beneath less-dense surface water (Figure 18.5). So, too, does the dense, salty water of the Mediterranean when it flows westward toward the open Atlantic.

The sinking dense water in the North Atlantic propels a global **thermohaline circulation** system, so called because

TABLE 18.1 Major Water Masses of the Atlantic Ocean

	Water Mass	Extent	Temperature (°C)	Salinity (per mil)	Source
Surface water	Central water mass Subarctic and Subantarctic water masses	35°N–35°S High latitudes	6–19	34–36.5	Regions of cool less-saline water
	Polar Surface water mass (Antarctic Circumpolar Current) (ACC)	Clockwise around Antarctica	0–2	34.6–34.7	
Intermediate water	Antarctic Intermediate Water (AAIW)	To 20°N	3–7	33.8–34.7	Cold Subantarctic surface water
	Saline Mediterranean Sea water	Beneath Antarctic Intermediate Water		37–38	Mediterranean Sea
Deep water	North Atlantic Deep Water (NADW)	North Atlantic to South Atlantic	2–4	34.8–35.1	Several sites in North Atlantic
	Antarctic Bottom Water (AABW)	To 30°N in Pacific	−0.4	34.7	Off Antarctica

it involves both the temperature (*thermo*) and salinity (*haline*) characteristics of the ocean waters. We can trace this circulation from the North Atlantic southward toward Antarctica and into the other ocean basins as well.

The largest mass of North Atlantic Deep Water (NADW; Table 18.1) forms in the Greenland and Norwegian seas, where relatively warm and salty surface water entering from the western North Atlantic cools, becomes denser, and sinks into a confined basin north of a submarine ridge connecting Scotland and Greenland (Figure 18.6A). The dense water then spills over low places along the ridge and plunges down into the deep ocean as NADW. Warm, salty surface and intermediate water is drawn toward the North Atlantic to take the place of the south-flowing deep water. Heat lost to the atmosphere by this warm surface water, together with heat from the warm Gulf Stream, maintains a relatively mild climate in northwestern Europe.

In the South Atlantic, south-flowing NADW enters the Antarctic Circumpolar Current, which travels clockwise around Antarctica. Surface and intermediate water flowing into the South Atlantic from the Pacific and from the Indian Ocean by way of the southern tip of Africa replenishes the NADW moving out of the basin (Figure 18.6B). Meanwhile, Antarctic Bottom Water plunging down to the ocean floor in the southernmost Atlantic moves northward, slowly rises, mixes with overlying NADW, and flows back toward Antarctica to join the Antarctic Circumpolar

Current (Figure 18.6B). This completes an important segment of the global system of ocean circulation. The Atlantic thermohaline circulation acts like a great conveyor belt, transporting low-density surface water northward and denser deep-ocean water southward.

Other circulation cells in the Pacific and Indian oceans are linked to the Atlantic circulation via the Antarctic Circumpolar Current. These, too, are driven by density contrasts related to temperature and salinity. Together, they move water along the globe-encircling ocean conveyor system, slowly replenishing the waters of the deep ocean (Figure 18.6B).

NADW is estimated to form at a rate of 15 to 20 million m^3/sec (equal to about 100 times the rate of outflow of the Amazon River). Antarctic Bottom Water forms at a rate of about 20 to 30 million m^3/s. Together, these water masses can replace all the deep water of the world ocean in about 1000 years.

Before you go on:

1. What is the Coriolis effect, and how does it affect ocean currents?

2. What properties differentiate the major water masses of the Atlantic Ocean?

3. What role does water density play in controlling global thermohaline circulation?

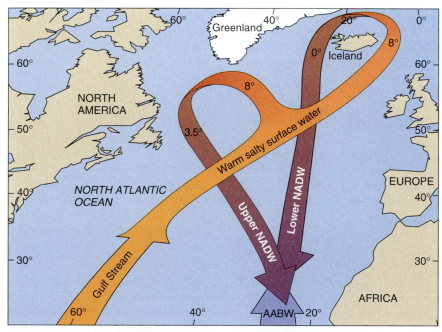

Figure 18.6A Deep Saline Water North Atlantic Deep Water (NADW) forms when the warm, salty water of the Gulf Stream cools, becomes increasingly saline due to evaporation, and plunges downward toward the ocean floor. The densest water then spills over the Greenland–Scotland ridge and flows south-

ward as lower NADW. Less-dense water forming between Greenland and North America moves south and east as upper NADW, overriding the lower, denser water. Because both water masses are less dense than northward-flowing Antarctic Bottom Water (AABW), they pass over it on their southward journey.

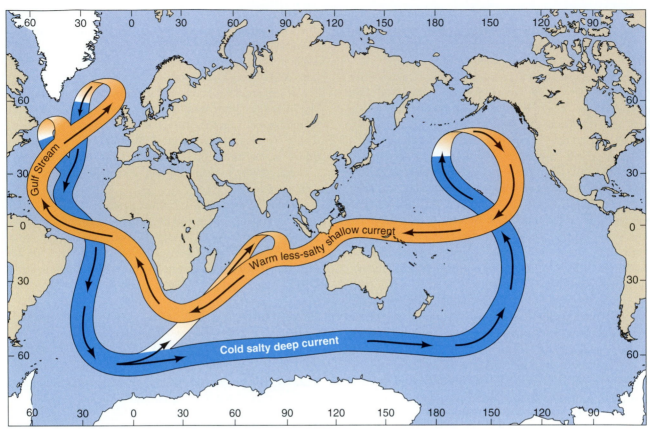

Figure 18.6B   Global Ocean Conveyor System   The major thermohaline circulation cells that make up the global ocean conveyor system are driven by exchange of heat and moisture between the atmosphere and ocean. Dense water forming at a number of sites in the North Atlantic spreads slowly along the ocean floor, eventually to enter both the Indian and Pacific oceans before slowly upwelling and entering shallower parts of the thermohaline circulation cells. Antarctic Bottom Water (AABW) forms adjacent to Antarctica and flows northward in fresher, colder circulation cells beneath warmer, more saline waters in the South Atlantic and South Pacific. It also flows along the Southern Ocean beneath the Antarctic Circumpolar Current to enter the southern Indian Ocean. Warm surface waters flowing into the western Atlantic and Pacific basins close the great global thermohaline cells.

OCEAN TIDES

Tides, the rhythmic, twice-daily rise and fall of ocean waters, are caused by the gravitational attraction between the Moon (and, to a lesser degree, the Sun) and Earth. A sailor in the open sea may not detect tidal motion. Near coasts, however, the effect of the tides is amplified and they become geologically important.

TIDE-RAISING FORCE

The Moon exerts a gravitational pull on the solid Earth. This pull is balanced by an equal but opposite *inertial force* created by Earth's rotation about the center of mass of the Earth–Moon system. The inertial force tends to maintain a body in uniform linear motion. At Earth's center, the gravitational and inertial forces are balanced. At Earth's surface the inertial force is everywhere the same (Figure 18.7). However, the magnitude of the Moon's gravitational attraction varies over Earth's nearly spherical surface, which creates a lack of balance between the two forces. A water particle in the ocean on the side facing the Moon is attracted more strongly by the Moon's gravitation than it would be if it were at Earth's center, which lies at a greater distance. Although the attractive force is small, liquid water is easily deformed, and so each water particle on this side of Earth is pulled toward a point located at the center of the Moon. This creates a bulge on the ocean surface.

We can think of this surface bulge of the ocean as resulting from a contest between the two forces. On the side nearest the Moon, gravitational attraction and inertial force combine, and the excess inertial force (called the *tide-raising force*) is directed toward the Moon. On the opposite side of Earth, however, the inertial force exceeds the Moon's gravitational attraction, and the tide-raising force is directed away from Earth (Figure 18.7). These unbalanced forces generate opposing *tidal bulges*.

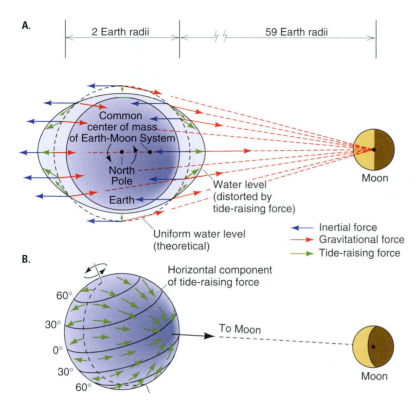

A.

2 Earth radii 59 Earth radii

Common center of mass of Earth-Moon System

North Pole

Earth

Water level (distorted by tide-raising force)

Uniform water level (theoretical)

Moon

→ Inertial force
→ Gravitational force
→ Tide-raising force

B.

Horizontal component of tide-raising force

60°
30°
0°
30°
60°

To Moon

Moon

Figure 18.7 Origin of the Tides A. Tide-raising forces are produced by the Moon's gravitational attraction and by inertial force. On the side toward the Moon, both forces combine to distort the water level from that of a sphere, raising a tidal bulge. On the opposite side of Earth, where inertial force is greater than the gravitational force of the Moon, the excess inertial force (called the tide-raising force) also creates a tidal bulge. B. The horizontal component of the tide-raising force is shown by arrows directed toward the point where a line connecting Earth and Moon intersects Earth's surface. This point shifts latitude with time as the relative position of Earth and Moon changes.

The tidal bulges created by the tide-raising force on opposite sides of Earth appear to move continually around Earth as it rotates. In fact, the bulges remain essentially stationary beneath the tide-producing body (the Moon) while Earth rotates. At most places on the ocean margins, two high tides and two low tides are observed each day as a coast passes through both tidal bulges. In effect, at every high tide, a mass of water runs into the coastline, where it piles up. This water then flows back to the ocean basin as the coastline passes beyond each tidal bulge.

Earth–Sun gravitational forces also affect the tides, sometimes opposing the Moon by pulling at a right angle and sometimes aiding by pulling in the same direction. Twice during each lunar month, Earth is directly aligned with the Sun and the Moon, whose gravitational effects are thereby reinforced, producing higher high tides and lower low tides (Figure 18.8). At positions halfway between these extremes, the gravitational pull of the Sun partially cancels that of the Moon, thus reducing the tidal range. However, the Sun is only 46 percent as effective as the Moon in producing tides, so the two tidal effects never entirely cancel each other.

Figure 18.8 Highest and Lowest Tides When Earth, Moon, and Sun are aligned (positions 1 and 3), tides of highest amplitude are observed. When the Moon and Sun are pulling at right angles to each other (positions 2 and 4), tides of lowest amplitude are experienced.

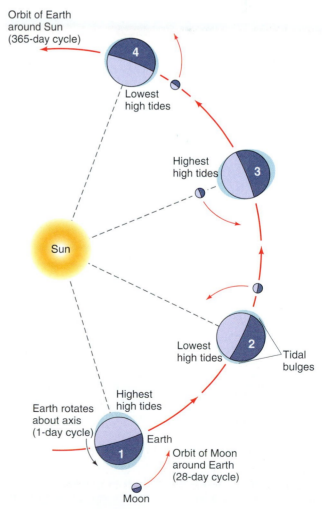

Orbit of Earth around Sun (365-day cycle)

Lowest high tides

Highest high tides

Sun

Lowest high tides

Tidal bulges

Earth rotates about axis (1-day cycle)

Highest high tides

Earth

Orbit of Moon around Earth (28-day cycle)

Moon

A.

B.

Figure 18.9 Big Tidal Range The tidal range in the Bay of Fundy, eastern Canada, is one of the largest in the world. A. Coastal harbor of Alma, New Brunswick at high tide. B. Same view at low tide.

In the open sea, the effect of the tides is small (~1 m), and along most coasts the tidal range commonly is no more than 2 m. However, in bays, straits, estuaries, and other narrow places along coasts, tidal fluctuations are amplified and may reach 16 m or more (Figure 18.9). Associated currents are often rapid and may approach 25 km/h. The incoming tide locally can create a wall of water a meter or more high (called a *tidal bore*) that moves up estuaries and

the lower reaches of streams. Fast-moving *tidal currents*, though restricted in extent, constitute a potential source of renewable energy that is still largely untapped.

TIDAL POWER

Energy obtained from the tides is renewable energy, for it never can be used up. However, harnessing tidal power for human use has seen only limited success. Water in a restricted bay, retained behind a dam at high tide, can drive a generator the same way that river water can. One important difference between hydroelectric power from rivers and that from tidal power is that rivers flow continuously, whereas tides can be exploited only twice a day; therefore, electrical supply is erratic.

Experts estimate that if all sites with suitable tidal ranges were developed to produce power, the total recoverable energy would be equivalent to only about a tenth that annually obtained from oil. Thus, although tidal power may prove important locally, it is unlikely ever to be a significant factor in global energy supply.

OCEAN WAVES

Whereas tides are generated by gravitational forces, ocean waves receive their energy from winds that blow across the water surface. The size of a wave depends on how fast, how far, and how long the wind blows. A gentle breeze blowing across a bay may ripple the water or form low waves less than a meter high. By contrast, storm waves produced by hurricane-force winds (>115 km/h) blowing for days across hundreds or thousands of kilometers of open water may become so high that they tower over ships unfortunate enough to be caught in them.

WAVE MOTION

How do waves travel through the ocean? Figure 18.10 shows the significant dimensions of a wave traveling in deep water, where it is unaffected by the bottom far below. In the figure, L represents **wavelength**, the distance between successive wave crests or troughs. Notice in the figure that each small parcel of water in the wave revolves in a loop, returning, as the wave passes, very nearly to its former position. This loop-like, or oscillating, motion of the water can be seen if droplets of dye are injected into a glass tank of water in which waves have been mechanically generated and the paths of the waves are then photographed with a video camera.

Because the waveform is created by a loop-like motion of water parcels, the diameters of the loops at the water surface exactly equal wave height (H in Figure 18.10). Downward from the surface, a progressive loss of energy occurs, resulting in a decrease in loop diameter. At a depth equal to half the wavelength ($L/2$ in Figure 18.10), the

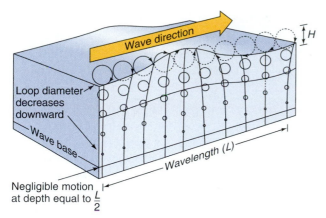

Figure 18.10 Motion of Water in a Wave Loop-like motion of water parcels in a wave in deep water. To trace the motion of a water parcel at the surface, follow the arrows in the largest loops from right to left. The resultant motion is the same as watching the wave crest travel from left to right. Parcels of water in smaller loops beneath the surface have corresponding positions, marked by nearly vertical lines. Dashed lines represent waveform and parcel positions one-eighth of a period later.

diameters of the loops have become so small that motion of the water is negligible.

WAVE BASE

The depth $L/2$, then, is the effective lower limit of wave motion (Figure 18.10). This depth is generally referred to as the **wave base**. In the Pacific Ocean, wavelengths as long as 600 m have been measured. For them, $L/2$ equals 300 m. However, the wavelengths of most ocean waves are far shorter than 600 m.

Landward of depth $L/2$, the circular motion of the lowest water parcels is influenced by the increasingly shallow

seafloor, which restricts movement in the vertical direction. As the water depth decreases, the orbits of the water parcels become flatter until the movement of water at the seafloor in the shallow water zone is limited to a back-and-forth motion (Figure 18.11).

BREAKING WAVES

As a wave nears the shore, it undergoes a rapid transformation. When the wave reaches depth $L/2$, its base encounters frictional resistance exerted by the seafloor. This resistance interferes with wave motion and distorts the wave's shape, causing the height to increase and the wavelength to decrease. Now the front of the wave is in shallower water than the rear part and is also steeper than the rear. Eventually, the front becomes too steep to support the advancing wave, and as the rear part continues to move forward, the wave collapses, or *breaks* (Figure 18.11).

The form of breaking waves differs from place to place. This is why surfers prefer some beaches to others. Board surfing originated in Hawaii, where the gently sloping seafloor immediately offshore allows approaching waves to increase in height gradually before rolling over and breaking. An expert surfer can catch a wave and ride its arcing front for a long distance before turning beneath the breaking crest and exiting to safety.

SURF

When a wave breaks, the motion of the water instantly becomes turbulent, like that of a swift river. Such "broken water," called **surf**, is wave activity between the line of breakers and the shore. Accordingly, this area is known as the *surf zone*.

In surf, each wave finally dashes against rock or rushes up a sloping beach until its energy is expended; then the water flows back toward the open sea. Water piled against

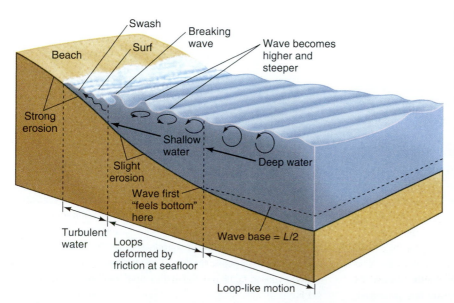

Figure 18.11 Waves Reach the Shore
Waves change form as they travel from deep water through shallow water to shore. In the process, the circular motion of water parcels in deep water changes to elliptical motion as the water becomes shallow and the wave encounters frictional resistance to forward movement. Vertical scale is exaggerated, as is the size of loops relative to the scale of waves. (Compare with Figure 18.10.)

Figure 18.12 Wave Refraction Waves arriving obliquely onshore along a coast near Oceanside, California, change orientation as they encounter the bottom and begin to slow down. As a result, each wave front is refracted so that it more closely parallels the bottom contours. The arriving waves develop a longshore current that moves from left to right in this view.

the shore returns seaward in an irregular and complex way, partly as a broad sheet along the bottom and partly in localized narrow channels as *rip currents*, which are responsible for dangerous undertows that can sweep unwary swimmers out to sea.

The geologic work of waves is mainly accomplished by the direct action of surf. Surf is a powerful erosional agent because it possesses most of the original energy of each wave that created it. This energy is quickly consumed in turbulence, in friction at the bottom, and in movement of sediment that is thrown violently into suspension from the bottom.

WAVE REFRACTION

A wave approaching a coast generally does not encounter the bottom simultaneously all along its length. As any segment of the wave touches the seafloor, that part slows down, the wavelength begins to decrease, and the wave height increases. Gradually, the trend of the wave becomes realigned to parallel the bottom contours. This process, known as **wave refraction**, causes a series of waves approaching the shoreline at an angle to change their direction of movement (Figure 18.12). For example, waves approaching the margin of a deep-water bay at an angle of 40 or 50° may, after refraction, reach the shore at an angle of 5° or less.

Wave refraction affects various sectors of a coastline differently. Waves passing over a submerged ridge off a rocky headland will be refracted and converge on the headland (Figure 18.13). The wave convergence, as well as the increased wave height that accompanies it, concentrates wave energy on the headland, which is vigorously eroded. Conversely, refraction of waves approaching a bay will make them diverge, diffusing their energy at the shore. The net tendency of these contrasting effects, in the course of time, is to make irregular coasts smoother and less indented.

Before you go on:

1. What roles does gravity play in controlling the ocean tides?

2. Why are there two high tides and two low tides each day?

3. How is the half wavelength of ocean waves related to the effective depth of wave erosion?

4. What is wave refraction, and how does it help control the shape of shorelines?

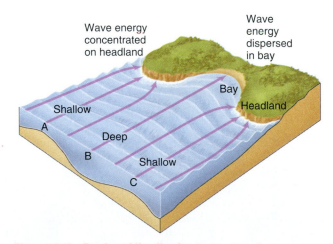

Figure 18.13 Erosion of Headlands Refraction of waves concentrates wave energy on headlands and disperses it along shores of bays. This oblique view shows how waves become progressively distorted as they approach the shore over a bottom that is deepest opposite the bay. The result is vigorous erosion on the exposed headland and sedimentation along the margin of the bay.

The world's ocean coastlines are dynamic zones of conflict where land and water meet. Erosional forces tear away at the land, and other forces move and deposit sediment, thereby adding to the land. Because the conflict is unending, few coasts achieve a condition of complete equilibrium.

EROSION BY WAVES

Most erosion along a seacoast is accomplished by waves moving onshore. Wave erosion takes place not only at sea level in the surf zone, but also below and above sea level.

EROSION BELOW SEA LEVEL

Ocean waves typically break at depths that range between wave height and 1.5 times wave height. Because waves are seldom more than 6 m high, the depth of vigorous erosion by surf should be limited to 6 m times 1.5, or 9 m below sea level. This theoretical limit is confirmed by observation of breakwaters and other coastal structures, which are only rarely affected by surf to depths of more than 7 m.

As we saw earlier, the lower limit of wave motion is half the wavelength of ocean waves ($L/2$ in Figure 18.10). This limit, then, is also the lower limit of erosion of the ocean floor by waves. At their seaward edge, the continental shelves have an average depth of about 200 m. Although wavelengths of most ocean waves are much less than 400 m, very large waves approaching or exceeding this size can erode even the outer part of the shelves.

ABRASION IN THE SURF ZONE

An important kind of erosion in the surf zone is the wearing down of rock by wave-transported rock particles. By continuous rubbing and grinding with these tools, the surf wears down and deepens the bottom and eats into the land, at the same time smoothing, rounding, and making smaller the tools themselves. Because surf is the active agent, this activity is limited to a depth of only a few meters below sea level. In effect, the surf is like an erosional knife edge or saw cutting horizontally into the land.

EROSION ABOVE SEA LEVEL

During great storms, surf can strike effective blows well above sea level. During one great storm that struck the coast of Scotland, a solid mass of stone, iron, and concrete weighing 1200 metric tons was ripped from the end of a breakwater and moved inshore. It was replaced with a block weighing more than 2300 metric tons, but five years later storm waves broke off and moved that block too. The pressures involved were about 27 metric tons/m².

Even waves having much smaller force can break loose and move blocks of bedrock from sea cliffs. Waves pounding against a cliff compress the air trapped in fissures. So great is the force of the compressed air that blocks of rock can be dislodged. Nearly all the energy expended by waves in coastal erosion is confined to a zone that lies between 10 m above and 10 m below mean sea level.

SEDIMENT TRANSPORT BY WAVES AND CURRENTS

Sediment is produced by waves pounding against a coast, or is brought to the sea by rivers, and is redistributed by currents. The currents build distinctive shoreline deposits or move sediment offshore onto the continental shelves.

LONGSHORE CURRENTS

Most waves reach the shore at an oblique angle (Figure 18.14). The path of an incoming wave can be resolved into two directional components, one oriented perpendicular to the shore and the other parallel to the shore. Whereas the perpendicular component produces the crashing surf, the parallel component sets up a **longshore current** within the surf zone that flows parallel to the shore (i.e., along the shore) (Figure 18.14). The direction of longshore currents may change seasonally if the prevailing wind direction changes, thereby causing changes in the direction of the arriving waves.

While surf erodes sediment at the shore, the longshore current moves the sediment along the coast. The greater the angle of waves to shore, the greater the rate of longshore movement.

BEACH DRIFT

On an exposed beach, incoming waves produce a second, irregular pattern of water movement directed along the shore. Because waves generally strike the beach at an

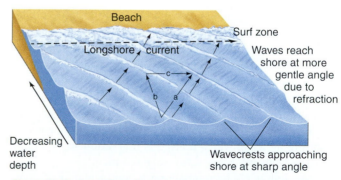

Figure 18.14 Longshore Current A longshore current develops offshore as waves approach a beach at an angle and are refracted. The line representing the front of each approaching wave (a) can be resolved into two components: the component oriented perpendicular to the shore (b) produces surf, whereas that oriented parallel to the shore (c) is responsible for the longshore current. Such a current can transport considerable amounts of sediment along the coast.

angle, the swash (uprushing water) of each wave travels obliquely up the beach before gravity pulls the water back directly down the slope of the beach. This zigzag movement of water carries sand and pebbles first up, then down the beach slope. The net effect of successive movements of this type is the progressive transport of sediment along the shore, a process known as **beach drift** (Figure 18.15). Marked pebbles have been observed to move along a beach at a rate of more than 800 m/day. When the volume of sand moved by beach drift is added to that moved by longshore currents, the total can be very large.

BEACH PLACERS

Gold, diamonds, and several other heavy minerals have been concentrated in beach sands by surf and longshore currents. Diamonds are being obtained in large quantities from gravelly beach placers, both above and below sea level, along a 350-km strip of coastal Namibia in southwestern Africa. Weathered from deposits in the interior, the diamonds were transported by the Orange River to the coast and were spread southward by longshore drift. Later, some beaches were raised tectonically, and others were drowned by rising sea level.

Although Nome, Alaska, has become famous as the finish line of the grueling Iditarod sled-dog race, the town initially attracted world attention because of its six beaches, four above sea level and two below, which are rich with placer gold. Ilmenite, a primary source of titanium, is highly concentrated (50 to 70%) along several beaches in India. Magnetite-rich sands occur on the coasts of Oregon, California, Brazil, and New Zealand, while chrome-rich sands are mined in Japan.

OFFSHORE TRANSPORT AND SORTING

Seaward of the surf zone, bottom sediment is moved by currents and by unusually large waves during storms, with net movement seaward. Each particle of sediment is picked up again and again, whenever the energy of waves or currents is great enough to move it. As the particle gets into ever-deeper water, it is picked up less and less frequently. With increasing depth, energy related to wave motion decreases. Therefore, far from shore only fine grains can be moved. As a result, the sediment becomes sorted according to diameter, from coarser in the surf zone to finer offshore. This subaqueous sorting is analogous to sorting by wind, which transports fine sediment to greater distances than it transports coarse sediment.

As sediments accumulate on a continental shelf, they normally grade seaward from sand into mud. This gradation is true not only of the particles eroded from the shore by surf but also of the particles contributed by rivers. For example, suspended sediment from the Columbia River in the northwestern United States moves obliquely across the continental shelf, where about 80 percent of it accumulates as a midshelf silt deposit (Figure 18.16). Of the remaining

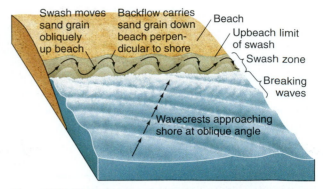

Figure 18.15 Beach Drift As surf rushes up a beach with each incoming wave, sand grains are picked up and carried shoreward. Arriving surf approaching the shore at an angle will travel obliquely up the beach. The return flow, pulled by gravity, flows back nearly perpendicular to the shoreline. A grain of sand will therefore move along a zigzag path as successive waves reach the shore. Net motion is down the beach and is called beach drift.

20 percent, about 13 percent is funneled down steep submarine canyons toward the continental rise, while the rest is deposited on the open continental slope.

> *Before you go on:*
>
> 1. How does marine erosion occur below sea level? above sea level?
>
> 2. How do longshore currents and beach drift move sediment along a shore?
>
> 3. Why are some precious metals and gemstones concentrated in beach sands?
>
> 4. Why does seafloor sediment become increasingly fine-grained with distance from the surf zone?

COASTAL DEPOSITS AND LANDFORMS

At a coast, waves that may have traveled unimpeded across thousands of kilometers of open ocean encounter an obstruction to further progress. They dash against firm rock, erode it, and move the eroded rock particles. Over the long term, the net effect on the coastline is substantial. Near the shoreline, waves and currents erode and redistribute sediments to produce a range of distinctive coastal deposits and landforms.

ELEMENTS OF THE SHORE PROFILE

To understand the changes along a coast, we must first examine the *shore profile*, a vertical section along a line drawn perpendicular to the shore. If we combine what we

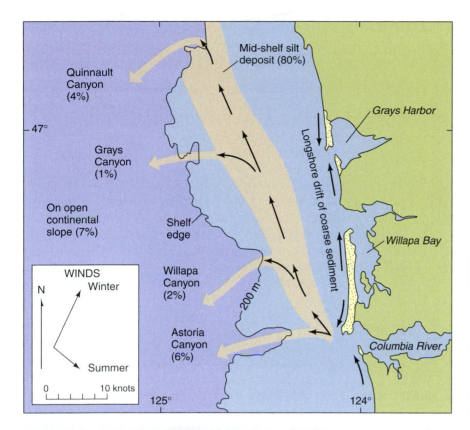

Figure 18.16 Sediment on the Continental Shelf Dispersal of sediment on the continental shelf off the state of Washington. Suspended sediment from the Columbia River moves northwest offshore, about 80 percent of it accumulating as a midshelf silt deposit. The remaining 20 percent either is deposited on the adjacent continental slope or moves down steep submarine canyons.

learn from such a profile with what we know about the forces that act parallel to the shore, we will have a three-dimensional picture of coastal activities. We look first at three important features of the shore profile: beaches, wave-cut cliffs, and wave-cut benches.

BEACHES

Beaches are characteristic features of many coasts, even those with steep, rocky cliffs. Along sandy coasts that lack cliffs, the beach constitutes the primary shore environment. A beach is regarded by most people as the sandy surface above the water along a shore. Actually, it is more than this. We define a **beach** as wave-washed sediment along a coast, including sediment in the surf zone. In this dynamic zone, sediment is continually in motion.

Part of the sediment of a beach may be derived from erosion of adjacent cliffs or cliffs elsewhere along the coast. However, along most coasts a much higher percentage of the sediment comes from alluvium brought to the shore by rivers.

On low, open shores an exposed beach typically has several distinct elements (Figure 18.17). The first is a rather gently sloping *foreshore*, a zone extending from the level of lowest tide to the average high-tide level. Here is found a *berm*, which is a nearly horizontal or landward-sloping bench formed of sediment deposited by waves. Beyond lies the *backshore*, a zone extending inland from the berm to the farthest point reached by surf. On some coasts, beach sand is blown inland by onshore winds to form belts of coastal dunes (Chapter 17).

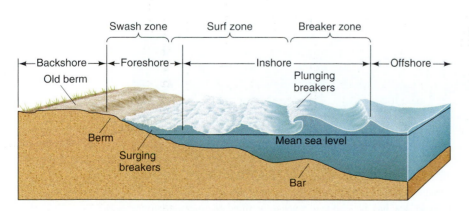

Figure 18.17 Beach Profile Section across a beach showing principal elements of the shore profile. Length of profile is about 75 m.

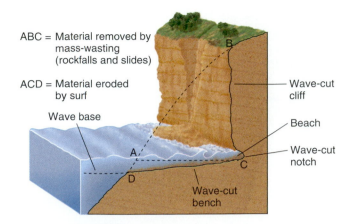

ABC = Material removed by mass-wasting (rockfalls and slides)

ACD = Material eroded by surf

Wave base

— Wave-cut cliff

— Beach

— Wave-cut notch

Wave-cut bench

Figure 18.18 Erosion of a Cliffed Coast Principal features of a shore profile along a cliffed coast. Notching of the cliff by surf action undermines the rock face, which then collapses and is reworked by surf. Note the large proportion of material removed by mass-wasting (ABC) relative to that eroded by surf (ACD).

ROCKY (CLIFFED) COASTS

Seen in profile, the usual elements of a cliffed coast (Figure 18.18) are a wave-cut cliff and wave-cut bench, both the work of erosion, and a beach, which is the result of deposition. A *wave-cut cliff* is a coastal cliff cut by surf. Acting like a horizontal saw, the surf cuts most actively at the base of the cliff. As the upper part of the cliff is undermined, it collapses and the resulting debris is redistributed by surf (Figure 13.26). An undercut cliff that has not yet collapsed may have a well-developed notch at its base.

Below a wave-cut cliff, you can often find a **wave-cut bench**, a platform cut across bedrock by surf. It slopes gently seaward and expands in the landward direction as the cliff retreats. Some benches are bare or partly bare, but most are thinly covered with sediment that is in transit from shore to deeper water. The shoreward parts of some benches are exposed at low tide (Figure 18.19). If the coast has been uplifted, a wave-cut bench and its sediment cover can be completely exposed.

Other erosional features associated with cliffed coasts are *sea caves*, *sea arches*, and *stacks* (Figure 18.20). Each is the result of differential erosion of rock as surf attacks a cliff, causing its gradual retreat.

FACTORS AFFECTING THE SHORE PROFILE

Through erosion and the creation, transport, and deposition of sediment, the form of a coast changes, often slowly but at some times very rapidly. At any given time, both the geometry of the shoreline and the shape of the shore

Figure 18.19 Wave-Cut Bench A nearly horizontal wave-cut bench has formed along the coast at Bolinas Point, California, as the surf, acting like an erosional saw, has cut into and beveled the tilted sedimentary strata.

Figure 18.20 Sea Stacks and Sea Arches Stack and sea arch along the French shore of the English Channel near Étretat carved in horizontally bedded white chalk. The surf first hollows out a sea cave in the most erodible part of the bedrock. A cave excavated completely through a headland is then transformed into a sea arch. An isolated remnant of the cliff stands as a stack on a wave-cut bench offshore.

profile represent a compromise among the constructive and destructive forces acting along the coast.

The compromise is reached in several ways. On a beach, for instance, a wave running up the beach as a thin sheet of water moves particles of sand and gravel upslope, while gravity pulls the particles back again (Figure 18.15). More energy is needed to move pebbles downslope than to move sand grains. Therefore, the pebbles moved shoreward by the wave remain until the slope becomes steep enough for the returning flow of water (the backwash) to carry them back. This partly explains why gravel beaches are generally steeper than beaches built of sand.

Another factor determining the beach profile is permeability. Some of the water rushing up a beach easily flows into the underlying porous beach sediment, thereby reducing the volume and transporting capability of the backwash.

During storms, the increased energy in the surf erodes the exposed part of a beach and makes it narrower. In calm weather, the exposed beach is likely to receive more sediment than it loses and therefore becomes wider. Storminess may be seasonal, resulting in seasonal changes in beach profiles. Along parts of the Pacific coast of the United States, winter storm surf tends to carry away fine sediment, and the remaining coarse fraction assumes a steep profile. In calm summer weather, fine sediment drifts in and the beach assumes a gentler profile.

MAJOR COASTAL DEPOSITS AND LANDFORMS

We have already seen some of the erosional effects of surf and its related shaping of the shore profile in the discussion of cliffed coasts. The deposits generated by surf and moved by currents are equally important. They are largely the result of longshore transport of sediment, and among the distinctive landforms they produce are marine deltas, spits and related features, beach ridges, barrier islands, and reefs.

MARINE DELTAS

In some places, constructional processes build out the coastline more rapidly than it can be destroyed by surf. The extent to which a marine delta projects seaward from the land is a compromise between the rate at which a river delivers sediment at its mouth and the ability of currents and waves to erode sediment along the delta front and move it elsewhere along the coast. The great size of the Mississippi Delta (Figure 18.21) testifies to the huge volume of sediment carried by the river and to the relative ineffectiveness of waves in destroying it. The Columbia River formerly transported a large load of sediment to the Pacific coast (much of the sediment is now trapped behind numerous hydroelectric and irrigation dams built along the river's course), and yet no delta has been constructed at its mouth. In this case, strong winter storms and persistent wave action erode sediment as quickly as it arrives at the coast.

Many deltas are compound features that have a long and complicated sedimentary history. The Mississippi Delta is really a complex of several coalescing subdeltas built successively over the last several thousand years (Figure 18.21). The present active margin of the delta has a very irregular front. The shores of the adjacent older subdeltas have been extensively modified by coastal erosion and by gradual submergence as the crust slowly subsides under the weight of the accumulating pile of sediment.

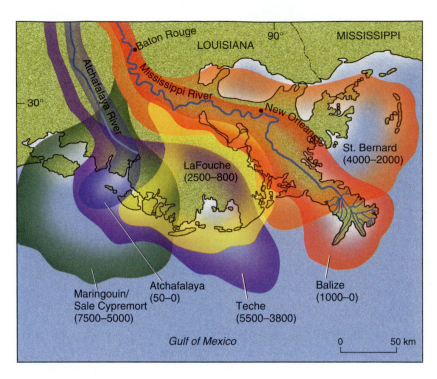

Figure 18.21   Mississippi Delta   The Mississippi River has built a series of overlapping subdeltas as it has continually dumped sediment into the Gulf of Mexico. The ages of subdeltas are given in radiocarbon years before the present.

SPITS AND RELATED FEATURES

Among other conspicuous landforms on many coasts is the **spit**, an elongated ridge of sand or gravel that projects from land and ends in open water (Figure 18.22). Most spits are merely continuations of beaches. Well-known examples are Sandy Hook on the southern side of the entrance to New York Harbor and Cape Cod, Massachusetts (Figure 18.23).

A spit is built of sediment moved by longshore drift and dropped at the mouth of a bay where the longshore current encounters deeper water and its velocity decreases. The free end curves landward in response to currents created by refraction as waves enter the bay.

A spit-like ridge of sand or gravel that connects an island to the mainland or to another island, called a *tombolo* (Figure 18.22), forms in much the same way as a spit does. The shape of a spit normally reflects the quantity and source of sediment and the prevailing transport direction. For example, coarse sediment carried to the Pacific Ocean by the Columbia River, together with sediment eroded by waves pounding against sea cliffs, is moved by longshore currents

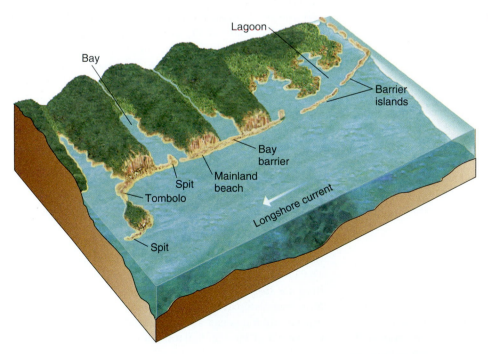

Figure 18.22   Shore Features
Some depositional shore features along a stretch of coast. Local direction of beach drift is toward the free end of the spits.

Figure 18.23   Cape Cod   The long, curved spit of Cape Cod, Massachusetts, has been built by longshore currents that rework glacial deposits forming the peninsula southeast of Cape Cod Bay (large bay to the left of the spit).

and forms large spits across major embayments (Figure 18.16). The spits are oriented in the directions of the prevailing summer (southward) and winter (northward) longshore currents, which are controlled by seasonal wind patterns.

Along an embayed coast with abundant sediment supply, a ridge of sand or gravel may be built across the mouth of a bay to form a *bay barrier* (Figure 18.22). A bay barrier develops as beach drift lengthens a spit across a bay in which tidal or river currents are too weak to scour away the spit as it is built.

BEACH RIDGES

Beaches on many sandy coasts lie seaward of, and nearly parallel to, a series of low, sandy *beach ridges*. Beach ridges are old berms, often built during major storms. Nearly all date to the last 5000 years, during which time sea level has been within several meters of its present position. Beach-ridge complexes in western Alaska, dated by prehistoric artifacts left by ancient maritime cultures, become progressively older inland from the present coast.

BARRIER ISLANDS

A **barrier island** (Figures 18.22 and 18.24A) is a long, narrow sandy island lying offshore and parallel to a coast. An elongate bay lying inshore from a barrier island or other low, enclosing strip of land (such as a coral reef) is

Figure 18.24   Barrier Islands   A. This sandy barrier island off the coast of Mississippi lies so close to sea level that waves can surge across its surface during large storms, eroding and redistributing the sediment. B. Cross section through Galveston Island, one of a series of barrier islands off the coast of Texas. Dashed lines show the former position of the seaward side of the island, based on radiocarbon dating. Since 3500 years ago, the island has grown southeastward toward the Gulf of Mexico.

A.

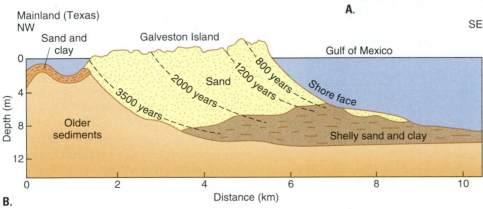

B.

called a *lagoon*. Barrier islands are found along most of the world's lowland coasts. Well-known examples are Coney Island and Jones Beach (New York City's coastal playground areas), and the long chain of islands centered at Cape Hatteras on the North Carolina coast. Padre Island, one of a succession of barrier islands paralleling the coast of Texas, is 130 km long.

Many barrier islands off the southeastern United States probably were built as the rate of sea-level rise at the end of the last glacial age began to slow during the middle Holocene (Figure 18.24B). Beach shells obtained at depths of 5 to 10 m from the basal deposits of some barrier islands have radiocarbon ages as old as 5000 years, an age that supports such a history. As the sea then rose more slowly across the very gentle continental shelf and the shoreline moved progressively inland, waves breaking in shallow water offshore eroded the bottom and piled up sand to form the long bars that ultimately became barrier islands.

A barrier island generally consists of one or more ridges of dune sand associated with successive shorelines that were occupied as the island formed. During great storms, surf washes across low places on the island and erodes it, cutting inlets that may remain permanently open. At such times, fine sediment is washed into the lagoon between barrier and mainland. In this way, the length and shape of barrier islands are always changing. Studies of the response of barrier islands to changing environmental conditions suggest that island development is closely related to sediment supply, the direction and intensity of waves and nearshore currents, the shape of the offshore profile, and the stability of sea level.

ORGANIC REEFS AND ATOLLS

Many of the world's tropical coastlines consist of organic reefs built by vast colonies of tiny organisms that secrete calcium carbonate to form corals. Such organisms require shallow, clear water in which the temperature remains above 18°C. Reefs therefore are built only at or close to sea level and are characteristic of low latitudes.

Three principal reef types are recognized. A **fringing reef** is either attached to or closely borders the adjacent land (Figure 18.25A). It therefore lacks an associated lagoon. Typically, a fringing reef has a flat upper surface as much as 1 km wide, and its seaward edge plunges steeply into deeper water.

A **barrier reef** is separated from the land by a lagoon that may be of considerable length and width (Figure 18.25B). Such a reef may surround an island or lie far off the coast of a continent, as in the case of the Great Barrier Reef off Queensland, Australia.

An **atoll**, a roughly circular coral reef enclosing a shallow lagoon (Figure 18.25C), is formed when a tropical volcanic island with a fringing reef slowly subsides (Figure 18.26A–C). The subsidence causes the reef to grow upward, for its organisms can survive only near sea level.

A.

B.

C.

Figure 18.25 Coral Reefs Chief kinds of tropical coral reefs. A. Fringing reef on the island of Oahu in the Hawaiian Islands. B. Barrier reef enclosing the island of Moorea in the Society Islands. A narrow lagoon separates the high island, which is the eroded remnant of a formerly active volcano, from a shallow reef. C. The reef of a small atoll in the Society Islands is surmounted by low, vegetated sandy islands that lie inside a line of breakers along the reef margin.

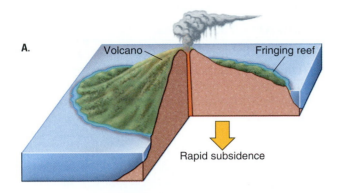

A.
Volcano Fringing reef

Rapid subsidence

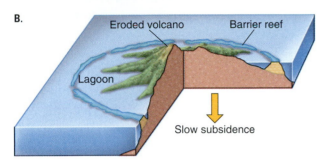

B.
Eroded volcano Barrier reef
Lagoon

Slow subsidence

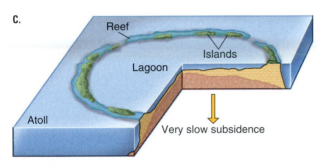

C.
Reef
Islands
Lagoon
Atoll Very slow subsidence

Figure 18.26 Origin of an Atoll Evolution of an atoll from a subsiding volcanic island. Rapid extrusion of lava to form an oceanic shield volcano causes the island to subside as the crust is loaded by the volcanic pile. A fringing reef grows upward, keeping pace with submergence, and becomes a barrier reef surrounding the eroding volcano. With continued subsidence and upward reef growth, the last remnants of the volcano disappear beneath sea level, and all that remains is an atoll reef surrounding a central lagoon.

As the island continues to subside, the area of exposed volcanic land becomes progressively smaller, and the fringing reef is transformed into an offshore barrier reef. Eventually, the volcanic island disappears beneath the sea, and the surrounding reef is thereby transformed into an atoll. Atolls generally lie in deep water in the open ocean and range in diameter from as little as 1 km to as much as 130 km. Their origin has been proved by drill holes that reach volcanic rock after penetrating thick sections of ancient reef rocks.

Before you go on:
1. What is the origin of sediment on beaches?
2. What distinctive erosional features are associated with rocky coasts? with sandy coasts?
3. Why do some large rivers have marine deltas at their mouths but others do not?
4. How could you prove that atolls are built atop submerged midocean volcanoes?

HOW COASTS EVOLVE

The world's coasts do not all fall into easily identifiable classes. Their variety is great because their configurations depend largely on the structure and erodibility of coastal rocks, on the active geologic processes at work, on the length of time over which these processes have operated, and on the history of world sea-level fluctuations.

TYPES OF COASTS

Some coasts—for instance, most of the Pacific coast of North America—are steep and rocky and consist of mountains or hills separated by deep valleys. Others, such as the Atlantic and Gulf coasts from New York City to Florida and westward into northern Mexico, border a broad coastal plain that slopes gently seaward and are festooned with barrier islands.

The Pacific and Atlantic/Gulf coasts represent two extremes, between which are many intermediate types. Each type owes its general character to its structural setting. The rugged and mountainous Pacific coast lies along the margin of the North American Plate, which is continually being deformed where it interacts with adjacent plates to the west. Uplifted and faulted marine terraces are common features along parts of this coast (for example, in southern Alaska, Oregon, and California) and along similar coasts that are emerging from the sea (Japan, New Zealand, and northern Chile are examples). By contrast, the eastern continental margin lies within the same lithospheric plate as the adjacent ocean floor and in a zone that is tectonically passive. The bedrock has low relief, and much of the coastal zone borders young sedimentary deposits of the Atlantic and Gulf coastal plains. Old shorelines inland from these coasts formed at times when glaciers were fewer, the ocean basins contained more water than they do now, and world sea level therefore was higher.

The coasts of New England, Sweden, and Finland are examples of low-relief landscapes that owe their special character to repeated ice-sheet glaciation and changing sea level. Each overriding ice sheet eroded bedrock in the coastal zone and depressed the land below sea level. With

Figure 18.27 Islands in the Adriatic Sea   Differential erosion of sedimentary rocks along the coast of Croatia on the Adriatic Sea. The tectonically deformed and eroded rocks, now partially submerged by the sea, form a series of linear islands along the trend of major structures.

retreat of the ice, both the land and world sea level rose (although not always at the same rate). The result is an embayed, rocky coastline that shows the effects of both differential glacial erosion and drowning of the land by the most recent sea-level rise.

Where rocks of different erodibilites are exposed along a coast, marine erosion typically will produce a shoreline that is strongly controlled by rock type and structure. Such control is especially impressive in folded sedimentary or metamorphic belts that have been partially submerged, as along the coasts of Norway, Ireland, and Croatia (Figure 18.27). It is also responsible for some of the world's deeply embayed fjord coasts, the pattern of which reflects deep glacial erosion along regional fracture systems and subsequent invasion by seawater as glaciers retreated to the heads of their fjords (Figure 16.21).

GEOGRAPHIC INFLUENCES ON COASTAL PROCESSES

The erosional and depositional processes we have discussed are not equally effective along all coasts. Coasts lying at latitudes between about 45 and 60° are subjected to higher-than-average storm waves generated by strong westerly winds, whereas subtropical east-facing coasts are subjected to infrequent but often disastrous hurricanes (called typhoons west of the 180th meridian). In the polar regions, sea ice becomes an effective agent of coastal erosion. These and other factors influence the amount of energy expended in erosion along the shore and, together with the structural and compositional properties of the exposed rocks, contribute to the variety of coastal landforms.

CHANGING SEA LEVEL

We have seen already that sea level fluctuates daily as a result of tidal forces. It also fluctuates over much longer time scales as a result of (1) changes in the volume of water in the oceans as continental glaciers wax and wane and (2) the motions of lithospheric plates that cause the volume of the ocean basins to change. Over the span of a human lifetime, these slow changes in sea level appear insignificant, but on geologic time scales they contribute importantly to the evolution of the world's coasts.

SUBMERGENCE: RELATIVE RISE OF SEA LEVEL

Whatever their nature, nearly all coasts have experienced **submergence**, a rise of water level relative to the land, owing to the worldwide rise of sea level that accompanied the most recent deglaciation. As water from melting glaciers returned to the oceans, rising sea level caused widespread submergence of coastal zones (Chapter 16; Figure 18.28). Most large estuaries, for example, are former river valleys that were drowned by this recent sea-level rise (see *Understanding Our Environment*, Box 18.1, *How to Modify an Estuary*).

Evidence of lower glacial-age sea levels is almost universally found seaward of the present coastlines and to depths of 100 m or more. Former beaches, sand dunes, and other coastal landscape features on the inner continental shelves mark shorelines built by the rising sea at the end of the glacial age (Figure 18.29) and later drowned.

EMERGENCE: RELATIVE FALL OF SEA LEVEL

Inland from the Atlantic coast of the United States, from Virginia to Florida, are many marine beaches, spits, and barriers. The highest reaches an altitude of more than 30 m. These landforms are related to a combination of broad up-arching of the crust, as well as submergence during past interglacial ages, when Earth's glacier cover was smaller than now, and sea level was therefore higher. The position

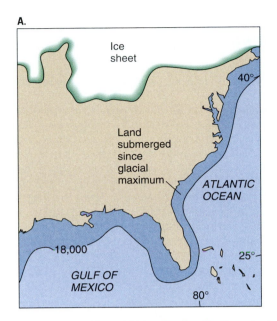

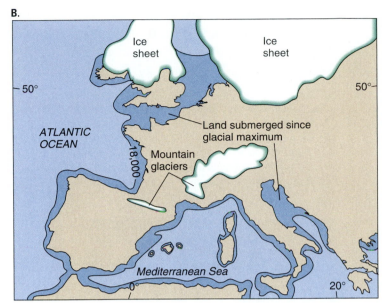

Figure 18.28 Effect of Rising Sea Level Coastal submergence of eastern North America and western Europe resulted when meltwater from wasting ice sheets returned to the ocean basins at the close of the last glaciation. A. Area of northeastern North America covered by glacier ice during the last glacial maximum, the approximate position of the shoreline at the glacial maximum (18,000 years ago), and coastal areas submerged by the postglacial rise of sea level. B. Areas covered by ice sheets in western Europe at the last glacial maximum and land areas that have been submerged during the postglacial rise of sea level.

of such features above present sea level points to **emergence**, a lowering of water level relative to the land.

SEA-LEVEL CYCLES

Varied evidence demonstrates that many coastal and offshore features date to times when relative sea level was either higher or lower than now. The youngest deposits along a coast often form a thin blanket over older, similar units that date to earlier episodes of submergence. Repeated emergence and submergence over many gla-

cial–interglacial cycles, each accompanied by erosion and redeposition of shoreline deposits, have resulted in complex coastal landform assemblages.

RELATIVE MOVEMENTS OF LAND AND SEA

The major rises and falls of sea level are global movements, affecting all parts of the world's oceans at the same time (Figure 18.29). By contrast, uplift and subsidence of the land, which cause emergence or submergence along a coast, are piecemeal movements, generally involving only parts of

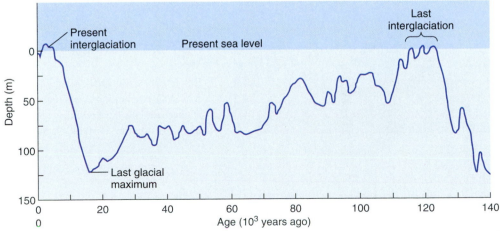

Figure 18.29 Global Sea-Level Variations Curve of sea-level variations spanning the last glacial–interglacial cycle. Prior to a recent high stand, global sea level had remained lower than at present since the last interglaciation, about 120,000 years ago.

Box 18-1

UNDERSTANDING OUR ENVIRONMENT

HOW TO MODIFY AN ESTUARY

San Francisco Bay is one of the world's best-known estuaries (Figure B18.1). When first discovered by Europeans, its shores and productive waters supported as many as 20,000 Native Americans. The population boom that followed the discovery of gold in California in 1848 and has continued unabated to the present day has led to major changes in the river systems that enter the bay

and has thereby has greatly affected the estuary environment.

The need for water is a continuing concern for Californians, who require it not only for human consumption and industry but also for support of huge agricultural enterprises. Diversion of water from natural streams has reduced fresh-water inflow to San Francisco Bay to far less than what it was in 1850; today, the flow is only about 30 percent of the post–1850 average. At the same time, increasing urbanization has led to a loss of 95 percent of the bordering tidal marshes as a result of filling and diking.

Hydraulic mining for gold in the foothills of the sierra Nevada following the initial discovery of gold produced vast quantities of fine sediment. The sediment choked streams, destroyed fish spawning grounds, obstructed navigation, and ultimately reached the bay, where it reduced the area and volume of the estuary and modified tidal circulation.

The unforeseen consequences of this human tampering with a natural stream and estuarine system have been many. Most commercial fisheries have disappeared. Reduced fresh-water flushing of the bay has concentrated agricultural, domestic, and industrial waste products in estuarine sediments, contaminating both them and the organisms that feed on them and raising increasing concerns about human health. Natural habitats of migrating birds have been extensively destroyed, with presumed major effects on bird populations.

Such changes are not unique to San Francisco Bay. Many other large estuaries, which are favored sites of urban development and industrialization, have similar problems. The Rhine River in the Netherlands, the Thames River in England, and the Susquehanna and Potomac rivers in the United States are examples. All are sensitive to human-induced changes and susceptible to steady deterioration. The best hope for reversing their decline is through an improved understanding of how human activities affect the physical, chemical, and biological processes in river-estuarine systems, and of the measures that must be taken to curtail potentially destructive actions.

Figure B18.1 Shrinking Size of San Francisco Bay
Vertical satellite image of the shrinking San Francisco Bay estuary. Filling and diking of tidal marshes to create farmland, evaporation ponds, and residential and industrial developments have reduced 2200 km^2 of marshland that existed before 1850 to less than 130 km^2 today.

landmasses. Nevertheless, such movements can cause rapid relative changes in sea level. Vertical tectonic movements at the boundary of converging lithospheric plates have uplifted beaches and tropical reefs to positions far above sea level

(Figure 18.30). Because movements of land and sea level may occur simultaneously, in either the same or opposite directions, unraveling the history of sea-level fluctuations along a coast can be a difficult and challenging exercise.

Figure 18.30 New Guinea Sea-Level Record Coastal emergence of eastern New Guinea. The emergent coast of the Huon Peninsula in eastern Papua New Guinea is flanked by a series of ancient coral reefs that form flat terrace-like benches parallel to the shoreline. Each reef formed at sea level and was subsequently uplifted along this active plate margin. The highest reefs lie several hundred meters above sea level and are hundreds of thousands of years old.

Before you go on:

1. What factors have controlled differences in the coastlines of New England and the Atlantic/Gulf coasts?

2. How does latitude affect processes of coastal erosion and deposition?

3. What causes changes in sea level on the scale of thousands and millions of years?

COASTAL HAZARDS

People who choose to live along a coast, like those who live on the floodplains of large rivers, are often exposed to natural hazards that can prove devastating. During infrequent large events, such as storms, the dynamic nature of the coastline environment becomes especially obvious.

In this section we will examine some of the primary hazards of the ocean's coastal zone and some of the ways that people have learned to minimize their impact.

STORMS

The approximate equilibrium among the forces that operate along coasts is occasionally interrupted by exceptional storms that erode cliffs and beaches at rates far greater than

the long-term average. During a single storm in 1944, for example, cliffs of compact sediment on Cape Cod retreated up to 5 m, which is more than 50 times the normal annual rate of retreat.

Such infrequent bursts of rapid erosion not only are important in the natural evolution of a coast but also can have a significant impact on coastal inhabitants. Thirty of the 50 states in the United States have coastlines on the Atlantic or Pacific oceans, the Gulf of Mexico, or the Great Lakes. These 30 states contain about 85 percent of the nation's population, and about half of these people live in the coastal zone. Clearly, large numbers of people and a great deal of property are at risk. Atlantic hurricanes that reach the eastern coast of the United States can be exceptionally devastating. The largest to strike the coast in the last decade have caused property damage running to tens of billions of dollars.

TSUNAMIS

A strong earthquake or other brief, large-scale disturbance of the ocean floor, such as a landslide or volcanic eruption, can generate a potentially dangerous *tsunami* (seismic sea wave; see Chapter 10). Its threat lies in the great speed at which it travels (as much as 950 km/h), its long wavelength (up to 200 km), its low observable height in the open ocean, and its ability to pile up rapidly to heights of 30 m or more as it moves into shallow water along an exposed coast.

The suddenness of a tsunami's arrival and consequent lack of warning time have resulted in numerous fatalities and considerable damage when tsunamis moved into populated areas. Hawaii is especially susceptible to dangerous tsunamis approaching from the numerous earthquake regions surrounding the Pacific basin (Figure 18.30 and Table 18.2). Today in Hawaii sirens and radio newscasts alert the population to arriving tsunamis, and people can

---

TABLE 18.2 Historic Tsunamis Striking Hawaii and Resulting in Moderate to Severe Damage[a]

Date	Location of Earthquake	Average Wave Speed (km/h)
November 7, 1837	South America	—
April 2, 1868	Hawaii	—
August 13, 1868	South America	—
July 25, 1869	South America (?)	—
May 10, 1877	South America	—
February 23, 1923	Kamchatka	695
April 1, 1946	Aleutian Islands	790
May 22, 1960	South America	710

[a] Data from G.A. Macdonald,, and A.T. Abbott, *Volcanoes in the Sea,* Honolulu, University of Hawaii Press, 1970, Table 16.

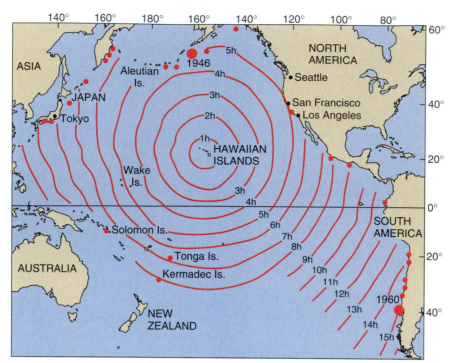

Figure 18.31   Tsunami Travel Times   Map showing the time required for a tsunami to reach the island of Oahu, Hawaii. Small red dots mark origins of historic tsunamis that struck Hawaii. Large dots mark places where the disastrous tsunamis of 1946 and 1960 originated (Table 18.2).

refer to maps printed in their telephone books that show the coastal zones at greatest risk (Figure 18.32); both measures have helped reduce the risk to humans.

LANDSLIDES

Cliffed shorelines are susceptible to frequent landsliding as erosion eats away at the base of a seacliff. Roads, buildings, and other structures built too close to such cliffs can become casualties when sliding occurs.

Sometimes landslides on cliffed shorelines give rise to giant waves that are even more destructive than the slides. We saw in Chapter 13 that large prehistoric submarine landslides on the submerged flanks of the Hawaiian Islands apparently produced huge waves that moved rapidly landward and transported beach gravels to altitudes of more than 200 m. A repetition of such an event in the now densely populated coastal zones of the islands would wreak great havoc and loss of life.

Very large waves have also been produced by massive coastal landslides at Lituya Bay, which lies along the Fairweather Fault on the southern coast of Alaska. During the last such event, in 1958, a 8.3-magnitude earthquake caused a coastal rock cliff to collapse into an arm of the bay. The massive landslide created a wave that rose more than 500 m against the opposite wall of the fjord. The wave swept away vegetation, producing a sharp trimline in the forest, and moved rapidly down the bay toward the sea. A boat anchored in the bay was caught up by the onrushing wave, carried over the spit at the bay mouth, and transported out into the open ocean.

PROTECTION AGAINST SHORELINE EROSION

Oceanfront land is often considered prime real estate. People who build houses and towns along seacoasts are not eager to lose their investment by the actions of an unruly ocean and therefore take steps to protect their land and property.

PROTECTION OF SEACLIFFS

A strip of shoreline that consists of easily erodible rock or sediment is not easy to protect. A cliff can be clad with an armor consisting of tightly packed boulders so large that they can withstand the onslaught of storm waves. It can also be defended by a strong seawall built parallel to the shore on foundations deep enough to prevent the undermining by surf during storms. Both structures offer cliffs some protection against ordinary storms, but both are expensive and may not be effective against large storms or hurricanes.

PROTECTION OF BEACHES

Because of their great recreational value, beaches in densely populated regions justify greater expense for mainte-

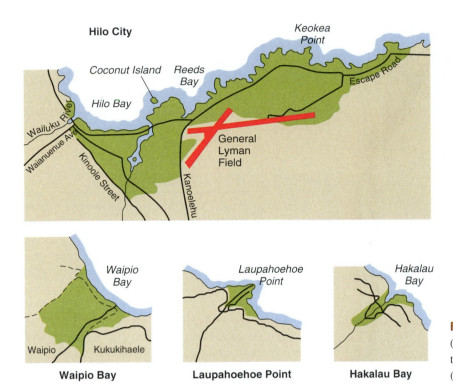

Figure 18.32 Tsunami Hazard Areas Areas (green) along coast of the island of Hawaii that are susceptible to inundation by tsunamis (from Hawaiian telephone book).

nance than most headlands do. A beach, however, presents a special sort of problem. As a result of beach drift, what happens on one part of a beach affects all other parts that lie in the downdrift direction. This is true because any beach is in a state of delicate balance between forces of erosion and deposition.

For example, a seawall, dock, or other structure built at the updrift end of a beach reduces the amount of sand available for beach drift. The surf becomes underloaded and makes good the loss by eroding sand from along the beach. Small beaches have been completely destroyed by this process in only a few years.

BREAKWATERS AND GROINS

A **breakwater** is an offshore barrier designed to protect a beach or boat anchorages from incoming waves. However, a breakwater upsets the natural balance of the adjacent beach, leading to shoreline changes. Breakwaters constructed along the shore at Tel Aviv in Israel (Figure 18.33) protect the beach from the onslaught of waves, but they

Figure 18.33 Breakwaters
Breakwaters constructed offshore from Tel Aviv, Israel, protect the beach zone from incoming waves. Wave refraction around the barriers has added sediment to the beach behind each breakwater, producing a scalloped coastline.

Figure 18.34 Groins Groins have been built perpendicular to the shoreline of Miami Beach, Florida, to prevent excessive loss of sand by longshore drift at this popular resort area.

have turned a straight coastline into a scalloped one. Arriving waves now are refracted around each breakwater. The currents created by the refracted waves transport sand toward the zone of protected water, causing the beach to expand seaward behind each barrier.

Beach erosion can be checked to some extent by building groins at short intervals along the beach. A **groin** is a low wall built out into the water at a right angle to the shoreline (Figure 18.34). A groin acts as a check on the rate of beach drift because it traps sand carried to it along the shore. However, erosion tends to occur on the downdrift side of a groin, where the beach has been deprived of its constant supply of sand. The net effect, therefore, is to protect one part of a beach at the expense of another part.

ARTIFICIAL NOURISHMENT

Another way of protecting an eroding beach is to haul in sand and pile it on the beach at the updrift end. Surf then erodes the pile and drifts the new sand down the length of the beach. Sand for artificially nourishing a beach must be continuously replenished, however. As can be imagined, the feeding of a beach, like the construction and maintenance of groins, can be expensive.

EFFECTS OF HUMAN INTERFERENCE

Many beaches around the world are deteriorating because of human interference. In southern California, for example, most of the sand on beaches is supplied not by erosion of wave-cut cliffs but by alluvium carried to the sea at times of flood. However, because buildings and other structures in stream valleys are vulnerable to floods, dams have been built across the stream courses to control flooding. Of course, the dams also trap the sand and gravel carried by the streams, thus preventing the sediment from reaching the sea. Halting the through-flow of sand has upset the natural balance among the factors involved in the longshore transport of sediment. The result has been significant erosion of some beaches.

A further dramatic example of human interference can be seen along the Russian coast of the Black Sea. Of the sand and pebbles that form the natural beaches there, 90 percent used to be supplied by rivers as they entered the sea. During the 1940s and 1950s, three things occurred: large resort developments were built at the beaches, large breakwaters were constructed so that two major harbors could be extended into the sea, and dams were built across some rivers inland from the coast. All this construction interfered with the steady state that had existed among the supply of sediment to the coast, longshore currents and beach drift, and deposition of sediment on beaches. By 1960, it was estimated that the combined area of all beaches along the coast had decreased by 50 percent. Then beachfront buildings began to sag or collapse as the surf ate away at their foundations. An ironic twist to the chain of events lies in the fact that large volumes of sand and gravel were removed from beaches and used as concrete aggregate, not only to construct the resort buildings but also to build the dams that cut off the supply of sediment to the coast.

Before you go on:

1. What role does population density play in assessing coastal hazards?

2. Why have tsunamis sometimes led to great losses of human life?

3. What measures can be used to reduce erosion along a sandy coast? a rocky coast?

4. What advice could you offer to people planning to build a house along a hurricane-prone section of coast?

REVISITING PLATE TECTONICS AND THE EARTH SYSTEM

OCEAN CIRCULATION AND THE CARBON CYCLE

More than one critical process within the global carbon cycle takes place in the ocean. Photosynthesizing marine organisms exchange dissolved CO_2 for dissolved O_2 in surface waters, and a wide variety of organisms draw bicarbonate anions out of seawater to form calcium carbonate shells. When these organisms die, their remains sink toward the seafloor. Tiny plankton have a large surface-to-volume ratio and tend to sink slowly, so their protoplasm often dissolves back into surrounding seawater. The remains and fecal matter of larger marine organisms are more likely to reach the seafloor.

Calcium carbonate accumulates on the seafloor if it is shallower than about 4 kilometers; in greater depths calcite tends to dissolve. As we discussed in Chapters 2 and 7, the carbonate deposit eventually lithifies to become limestone. Organic carbon from the protoplasm of dead organisms accumulates only if the seafloor is starved of dissolved O_2. The marine biosphere is an efficient consumer of edible muck. As long as animals can respire, scavenger organisms will eat the organic carbon.

Because O_2 enters seawater only at the ocean surface, the circulation of the ocean influences where organic carbon accumulates. At present, the thermohaline circulation system "ventilates" most of the deep ocean. Cold O_2-rich water sinks into the deep ocean from the surface waters of the North Atlantic and offshore Antarctica, spreads worldwide, and inhibits the accumulation of organic carbon in deep-ocean sediments. Organic-rich sediments can be found today where deep water is isolated from the global thermohaline circulation and has become anoxic, that is, starved of oxygen. One example is the Black Sea, which connects to the world ocean only through a shallow passage to the Mediterranean. Organic-rich sediments also accumulate within isolated depressions in the continental margin of southern California, where seafloor water stagnates beneath the coastal surface waters.

Along many continental shelves where marine life is abundant, the rain of organic matter is so large that scavengers cannot keep pace. They run out of O_2 before they run out of food. O_2-rich surface waters lie only a few hundred meters above the continental shelf and slope, but the buoyancy of the warmer surface waters prevents them from supplying O_2 to the colder water below. Most of today's offshore petroleum reserves were originally deposited in this manner, in the oxygen-starved sediments of continental margins.

SEDIMENTS AT THE BEGINNING AND END OF AN OCEAN'S LIFE CYCLE

The shifting arrangement of ocean basins during Earth's long history has strongly influenced the deposition of evaporites and organic carbon. Unusual depositional conditions are common when an ocean basin initially opens, and in its last stages of closure. When a continent rifts apart, the narrow, shallow ocean basin that forms often connects to the world ocean via a shallow channel. The Red Sea, for instance, connects to the Indian Ocean only via the Bab el Mandeb, a narrow passage between Yemen and Eritrea. When an ocean basin disappears between two colliding continents, shallow seas can be isolated. The Mediterranean Sea is a remnant of the Tethys Ocean, an ocean basin that largely disappeared in the convergence of India and Africa against Eurasia. The Mediterranean connects to the world ocean only via the Strait of Gibraltar between Spain and Morocco.

If evaporation dominates the regional climate, salinity increases in small semi-isolated ocean basins. Evaporite deposits can form if the connection to the world ocean is broken by tectonic activity or by a drop in sea level. Geologists have estimated that the Mediterranean would evaporate completely in only 1000 years if the Strait of Gibraltar were blocked. In fact, thick salt deposits beneath the Mediterranean seafloor tell us that it dried out as many as 40 times between 5 and 7 million years ago.

On the other hand, organic-carbon deposition can occur in a semi-isolated ocean basin if the climate is wet. The inflow of fresh river water may form a low-density layer above deeper marine water, which often lacks a connection to the deep water of the world ocean. Isolated from sources of dissolved O_2, the seafloor becomes anoxic, favoring the deposition of organic carbon. This situation occurred repeatedly in the young Atlantic Ocean as it opened during the Mesozoic Era. Numerous carbon-rich layers in limestones of the South Atlantic testify to frequent episodes of seafloor anoxia. Thin organic-rich layers, called *sapropels*, are widespread in Pleistocene sediments of the Mediterranean Sea. The sapropels likely indicate that transitions between glacial and interglacial climates on a worldwide scale were accompanied by episodes of wet climatic conditions in the Mediterranean region.

What's Ahead?

Prior to the last century, most people did not realize that Earth's climate changes, not only on the long time scale of the glacial ages, but also on much shorter time scales. We are now aware that during a human lifetime, the climate changes perceptibly, and this is especially evident in the present period of global warming.

In the next chapter we will examine a variety of evidence that leads us to recognize Earth's changing climate. We will discuss the possible causes of climate change on different time scales, how people are contributing to climate change, the potential impact of changing climate on the human population, and what we can say about the possible range of change in a warming world.

CHAPTER SUMMARY

1. Seawater covers nearly 71 percent of Earth's surface and is concentrated in the Pacific, Atlantic, and Indian oceans. Each is connected to the Southern Ocean, which encircles Antarctica.

2. Although the greatest ocean depth is more than 11 km, the average depth is 3.8 km. More than half the ocean water resides in the Pacific basin.

3. Almost all of the saltiness of seawater is due to eight ions that are released by chemical weathering of rocks on land and then transported by streams to the sea. Other sources of ions include airborne dust and human pollutants.

4. Sea-surface temperatures are strongly related to latitude, with the warmest temperatures measured in equatorial latitudes. Surface salinity is strongly latitude dependent and also related to both evaporation and precipitation.

5. Huge wind-driven surface ocean currents which, because of the Coriolis effect, circulate clockwise in the Northern Hemisphere and counterclockwise in the Southern Hemisphere, carry warm equatorial water toward the polar regions.

6. Sinking of cold and/or saline high-latitude surface waters leads to oceanwide thermohaline circulation. Operating like a great conveyor belt and driven by density contrasts, this global circulation system replenishes the deep water of the world ocean, replacing it in about 1000 years.

7. Twice-daily ocean tides, resulting from the gravitational attraction of the Moon and Sun, are produced as Earth's rotating surface passes through tidal bulges on opposite sides of the planet. Tidal currents move sediment in bays, straits, estuaries, and other restricted places along coasts.

8. The motion of wind-driven surface waves terminates downward at the wave base, a distance equal to half the wavelength. A wave breaks in shallowing water as interference with the bottom causes the wave to grow higher and steeper. Waves approaching shore are refracted as the wave base reaches the bottom, realigning the wave so that it reaches the shore at a gentler angle.

9. Longshore currents and beach drift can transport great quantities of sand along coasts.

10. On gentle, sandy coasts the beach typically consists of a foreshore, berm, and backshore. On rocky coasts the shore profile includes a wave-cut cliff, a wave-cut bench, and a beach.

11. Depositional shore features include beaches, marine deltas, spits, tombolos, bay barriers, and barrier islands. Barrier islands form offshore in areas where rising sea level causes a shoreline to advance across a gently sloping coastal plain.

12. Organic reefs form in tropical seas where water temperature averages at least 18°C. A fringing reef on a volcanic island can be transformed into a barrier reef as the island subsides, and eventually into an atoll.

13. The shape of coasts partly reflects the amount of energy available to erode and deposit sediment. Rock structure and degree of erodibility help dictate the form of rocky coasts.

14. Nearly all coasts have experienced recent submergence due to the postglacial rise of sea level. Some coasts have experienced more complicated histories of emergence and submergence due to crustal movements on which is superimposed the worldwide postglacial sea-level rise.

15. Infrequent but powerful storms, tsunamis, and large landslides can pose significant threats to people and structures in the coastal zone.

16. Erosional damage to a shore cliff can be minimized by a seawall or an armor of boulders. Beach erosion is a serious problem along many inhabited coasts, but beaches can be temporarily protected by breakwaters, groins, or artificial nourishment.

THE LANGUAGE OF GEOLOGY

atoll (p. 496)

barrier island (p. 495)
barrier reef (p. 496)
beach (p. 491)
beach drift (p. 490)
breakwater (p. 503)

emergence (of a coast) (p. 499)

fringing reef (p. 496)

groin (p. 504)
gyre (p. 481)

longshore current (p. 489)

salinity (p. 477)
spit (p. 494)

submergence (of a coast) (p. 498)
surf (p. 487)

thermohaline circulation (p. 482)

wave base (p. 487)
wave-cut bench (p. 492)
wavelength (p. 486)
wave refraction (p. 488)

QUESTIONS FOR REVIEW

1. If the quantity of dissolved ions carried to the oceans by streams throughout geologic history far exceeds the known quantity of these substances in modern seawater, why is seawater not far more salty than it is?

2. How and why are the temperature and salinity of surface ocean water related to latitude?

3. Explain why the oceans are vertically stratified with respect to water density.

4. Explain how the ocean's thermohaline circulation system operates and how it affects the climate of western Europe.

5. Why at any place on a coast are there two high tides and two low tides each day?

6. Describe the motion of a parcel of water below a passing wave in the open ocean, and explain how and why the motion changes as the wave moves into shallow water.

7. How is wave base related to wavelength?

8. What causes a wave to "break" near the shore?

9. What is the effective depth of erosion by surf, and what determines this depth?

10. How does wave refraction explain why rocky headlands are more vigorously eroded by surf than bays are?

11. What causes longshore currents to develop and, in some cases, to shift direction seasonally?

12. Why are the sediments in a beach generally well sorted and well stratified?

13. Describe how you think an atoll would look during the peak of a glacial age, and why.

14. Why might a ship in midocean not detect the passage of a tsunami, while one anchored in a constricted ocean-facing harbor would not miss its arrival?

15. What geologic features would you look for to determine whether a coastal region had experienced emergence or submergence in the recent geologic past?

16. Describe measures that can be taken to reduce the impact of erosion (a) along a cliffed coast and (b) along a sandy beach. What negative effects might the measures have, even though they reduce erosion?

17. How might the construction of a dam along the lower part of a large river affect the coast where the river enters the ocean?

Click on *Presentation* and *Interactivity* in the **GeoHazards** module of your **CD-ROM** to further explore resources and activities presenting concepts from this chapter. Select *Assessments* in the same module to test your understanding of this chapter.

Chapter 19

Catastrophic Ice Shelf Retreat on a Warming Earth. During the first months of 2002, the Larsen B ice shelf on the margin of the Antarctic Peninsula, larger than the state of Rhode Island and 200 meters thick, disintegrated into a vast number of floating icebergs.

Climate and Our Changing Planet

Thin layers of limestone and marl in Cretaceous sedimentary rock, Italy.

The Sensitivity of Earth's Climate to Small Changes

The sleepy town of Gubbio in central Italy is an unusual destination for a pilgrimage, but geologists are different. Near Gubbio a person can put his finger on a layer of clay that marks the demise of the dinosaurs. This clay layer is the weathered residue of debris that fell from the sky 65 million years ago after a meteor struck Earth. Most species of living organisms on Earth went extinct soon after, decimated by sudden environmental changes.

I made my own pilgrimage to Gubbio in 1992 to examine the geologic record of the Cretaceous, a period of warm climate (and dinosaurs) in Earth history. The surrounding hills expose thousands of thin layers of limestone and marl. Marl is a clay- and organic-rich calcareous sedimentary rock. The Gubbio sediments were deposited in the long-lost Tethys Ocean during the late Mesozoic and early Cenozoic eras. These sediments accumulated from a slow rain of dead marine organisms. When dissolved oxygen was present at the seafloor, scavenger organisms picked the skeletal material clean, leaving calcium carbonate behind to form limestone. A dark marl accumulated if seafloor oxygen was absent and organic carbon matter was preserved. All around Gubbio, the light–dark alternation of white limestone and gray-black marl told me that, for tens of millions of years,

seafloor chemistry had cycled between oxygen-poor and oxygen-rich conditions. From an Earth-system viewpoint, ocean chemistry connects with river runoff and ocean circulation, which in turn connect with rainfall and winds. The transitions between limestone and marl therefore indicated major transitions in past Earth climate.

The Cretaceous climate cycles are remarkable when compared to their underlying cause. Several researchers (including me) have analyzed the patterns of climate change recorded in these hills. We have found that they match the signature of small changes in Earth's orbit around the Sun, the so-called Milankovitch cycles (Chapter 16). These 20,000- to 100,000-year cycles cause small variations in the amount of sunlight received by Earth. Solar energy heats the atmosphere and drives climate processes. Milankovitch cycles alter patterns of solar heating by only a few percent, comparable to the extra heating from the recent increase in greenhouse gases. The rock record tells us that, when pushed, Earth's climate can jump.

Jeffrey Park

KEY QUESTIONS

1. **How does the carbon cycle work?**

2. **How does the greenhouse effect work?**

3. **Which greenhouse gases are increasing in abundance, and why?**

4. **What do historical data tell us about twentieth-century climate change?**

5. **How much global warming do computer climate models predict in the twenty-first century?**

6. **What can we learn about the causes and consequences of warm climates from Earth's history?**

7. **Will future climate change be slow or sudden?**

INTRODUCTION: THE CHANGING ATMOSPHERE

The climate on Earth has not always been like we find it today. We live on a dynamic planet, and conditions are always changing. Our own influence on Earth's climate, a new development in geologic history, is significant and growing. For some cities in arid regions, such as Los Angeles, California, the importation of water for drinking and landscaping has altered the local humidity. Smoke

from tens of millions of household fires in India has altered the amount of sunlight that reaches the surface in the region. On a global scale, our use of fossil fuels and our clearing of forest for farmland have increased the concentration of carbon dioxide, a key greenhouse gas, by more than 25 percent. A preponderance of scientific evidence argues that this CO_2 increase has already influenced weather throughout the world.

Earth's climate system consists of a number of interacting subsystems that involve the atmosphere, the hydrosphere, the solid Earth, and the biosphere (Figure 19.1). Climate operates on many time scales, and explanations of climatic change on a million-year time scale may have little or nothing to do with changes that take place on millennial or decadal time scales. So complicated are the interacting systems that only with the advent of powerful computers have we posed some of the basic questions about how the climate system works.

Geology plays an important role in this enterprise, for geologists have within their grasp a record of Earth's changing climates that extends into the remote past. Climate can be read from the stratigraphic record in many ways. For example, paleontologists infer past climates from the fossil plants and animals they find. Sedimentologists and stratigraphers can infer many things about past climates from the nature of the sediments they study: the present distribution of these sediments, their mineralogy, the varied facies represented, features that indicate the effects of transport and deposition, and the soils that represent former land surfaces. Isotope geologists can determine past surface temperatures from studies

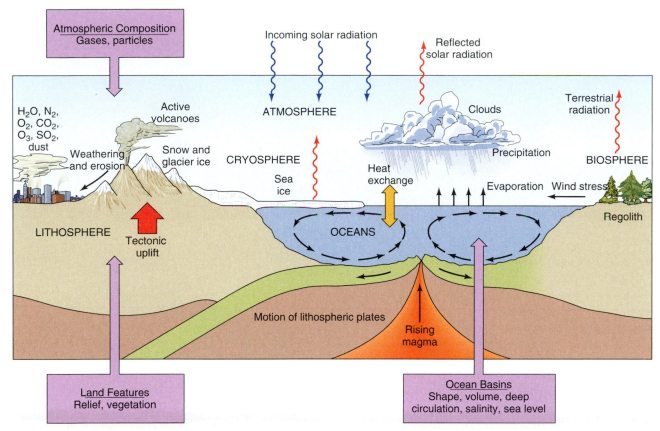

Figure 19.1 The Climate System Earth's climate system consists of four interacting subsystems: atmosphere, hydrosphere, lithosphere, and biosphere. Solar energy drives the system. Some of the incoming radiation is reflected back into space from clouds, atmospheric pollutants, ice, snow, and other reflective surfaces. Tectonic movements affect surface relief (which influences atmosphere circulation) and the geometry of continents and ocean basins (which controls ocean circulation and the location of ice sheets), while volcanic and industrial gases affect atmospheric composition.

of terrestrial and marine sediments and polar ice cores. By reconstructing past climates, geologists can determine the range of climate variability and use their data to test the accuracy of computer models.

One important reason for these studies is to learn how the climate system behaves, what controls it, and how it is likely to change in the future. We know it *will* change, for the geologic record of climatic change is abundantly clear. What we still lack is a clear view of *how* climate will change and at what rate. The answers are important not only scientifically, but socially and politically as well. By extracting and burning Earth's immense supply of fossil fuels, people have unwittingly begun a great geochemical "experiment" that is likely to have a significant impact on our planet and its inhabitants.

THE CARBON CYCLE

A basic chemical substance involved in the experiment is carbon, an element that is essential to all forms of life. We saw in Chapter 2 that carbon moves through the biosphere, the lithosphere, the hydrosphere, and the atmosphere in a major biogeochemical cycle driven ultimately by the subduction of oceanic lithosphere. To explain climate change, there are many other parts of the carbon cycle to consider.

Carbon occurs in five reservoirs: (1) in the atmosphere it occurs in carbon dioxide; (2) in the biosphere it occurs in organic compounds; (3) in the hydrosphere it occurs as dissolved carbon dioxide; (4) in the crust it occurs both in the calcium carbonate of limestone and in decaying and buried organic matter such as peat, coal, and petroleum; and (5) in the mantle some carbon has been present since Earth first formed, but subducting oceanic lithosphere supplies the portion that is vented as CO_2 in island-arc volcanism. Each reservoir is involved in the carbon cycle (Figure 19.2).

The key to the carbon cycle is the biosphere, where plants continually extract CO_2 from the atmosphere and then break the CO_2 down by the process of photosynthesis to form organic compounds. When plants die, the organic compounds decay by combining with oxygen from the atmosphere to form CO_2 again. The passage of material through the biosphere is so rapid that the entire content of CO_2 in the atmosphere cycles every 4.5 years.

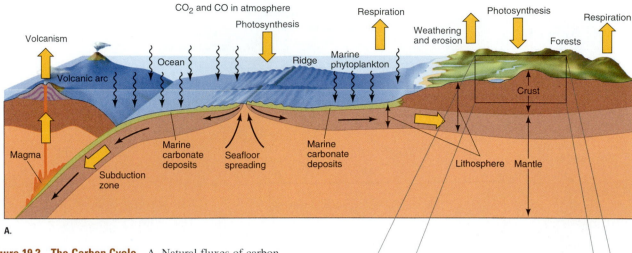

A.

Figure 19.2 The Carbon Cycle A. Natural fluxes of carbon between the atmosphere, hydrosphere, biosphere, crust, and mantle. Carbon enters the atmosphere through volcanism, weathering, biological respiration, and decay of organic matter in soils. Photosynthesis incorporates carbon in the biosphere, from which it can become part of the crust if buried with accumulating sediment. Subduction liberates carbon from oceanic crust into the mantle. B. Anthropogenic activities release carbon to the atmosphere through fossil-fuel use, combined with deforestation and burning of wood.

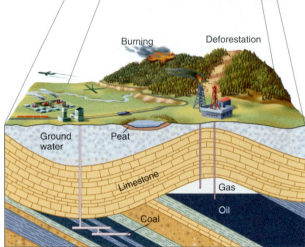

B.

The biospheric and atmospheric pathways interact with pathways in the hydrosphere, crust, and mantle, though on much longer time scales. Not all dead plant matter in the biosphere decays immediately back to CO_2. A small fraction is transported and redeposited as sediment; some is then buried and incorporated in sedimentary rock. The buried organic matter joins the slower-moving rock cycle and will reenter the atmosphere only when uplift and erosion have exposed the rock in which the organic matter is trapped.

Carbon dioxide from the atmosphere also is dissolved in the waters of the hydrosphere. There it is used by aquatic plants in the same way that land plants use CO_2 from the atmosphere. In addition, aquatic animals extract calcium and carbon dioxide from the water to make shells of $CaCO_3$. When the animals die, the shells accumulate on the seafloor, mixing with any $CaCO_3$ that may have been precipitated as chemical sediment. When compacted and cemented, the $CaCO_3$ forms limestone. In this way, too, some carbon joins the rock cycle. Eventually, the rock cycle may bring the limestone back to the surface where weathering and erosion will break it down; the calcium returns in solution to the ocean, and the carbon escapes as CO_2 to the atmosphere. Limestone that accompanies subducting lithosphere into the mantle eventually returns CO_2 to the atmosphere.

Next, we must consider what happens when human activities touch these carbon reservoirs and influence the fluxes between them. When we mine coal and extract oil

from the crust and then burn them, we convert organic matter to CO_2, thereby accelerating the carbon cycle. When we clear vast forests or allow deserts to expand when overgrazing kills off the vegetation cover, we are reducing the size of the biosphere reservoir. Although any individual action may appear insignificant, the combination of all human activities has had a measurable effect. Only the mantle reservoir of carbon seems to be outside our reach. Both the burning of fossil fuels and the clearing of land cause CO_2 to be released to the atmosphere at rates that are faster than the natural rate. Unless this CO_2 is either dissolved in the hydrosphere or buried rapidly in sediments, the CO_2 content of the atmosphere will increase. The best estimates suggest that only half of the CO_2 added by human activity to the atmosphere is currently being absorbed in this manner. Not surprisingly, CO_2 has been accumulating in the air around us.

THE GREENHOUSE EFFECT

The atmosphere is the engine that drives Earth's climate system, and the Sun provides the energy that allows the engine to work (Figure 19.3). Some solar radiation that

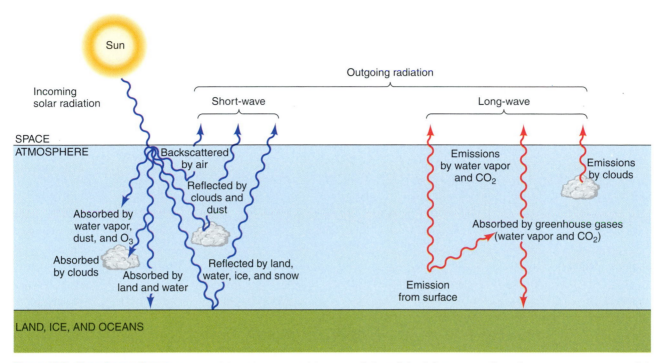

Figure 19.3   Greenhouse Gases   Some of the short-wave (visible) solar radiation reaching Earth is absorbed by the land and oceans, while reflects back into space off snow, ice, clouds, and atmospheric dust. Earth also radiates long-wave radiation (infrared) back into space. Greenhouse gases trap some of the outgoing long-wave radiation, causing the atmosphere to heat up. With increasing concentration of these trace gases, the air temperature of the lower atmosphere rises.

reaches the atmosphere reflects off clouds and dust and bounces back into space. Of the radiation that reaches the planet's surface, some is absorbed by the land and oceans, and some is reflected into space by water, snow, ice, and other reflective surfaces. This visible reflected solar radiation has a short wavelength. Earth also emits long-wave, infrared radiation. However, a portion of the outgoing long-wave radiation encounters gases in the atmosphere that have chemical properties that absorb it, preventing its energy from escaping. Instead, the energy is retained in the lower atmosphere, causing the temperature at Earth's surface to rise. The air temperature in a glass greenhouse is warmer than the air outside for the same reason: the glass acts in much the same way as the atmospheric gases. It allows visible (short-wavelength) sunlight in, but prevents the escape of infrared (long-wavelength) energy. Hence, we refer to this phenomenon as the **greenhouse effect**.

The greenhouse effect makes our planet habitable. If Earth lacked an atmosphere, its surface environment would be like that of the Moon, on the sunlit side of which temperatures are close to the boiling point of water, while on the dark side the temperature is far below freezing. The nearly airless surface of Mars is a frigid landscape whose closest earthly analogs are the frozen polar deserts. By contrast, Venus has a larger proportion of greenhouse gases in its atmosphere and is also closer to the Sun. The greenhouse effect on Venus generates surface temperatures hot enough to melt lead.

GREENHOUSE GASES

Greenhouse gases form a tiny fraction of Earth's atmosphere. Dry air consists mainly of three gases: nitrogen (79%), oxygen (20%), and argon (1%). However, water vapor is usually present in the Earth's atmosphere in concentrations of up to several percentage points and accounts for about 80 percent of the natural greenhouse effect. The remaining 20 percent is due to other gases present in very small amounts. Despite their very low concentrations, sometimes measurable only in parts per billion by volume (ppbv) of air, these *trace gases* contribute significantly to the greenhouse effect (Table 19.1).

Chief among the trace gases is carbon dioxide, which has an average concentration in surface air of about 370,000 ppbv, or 0.370 parts per thousand. Other significant gases, each a basic part of natural biogeochemical cycles and efficient in absorbing infrared radiation, are methane, nitrous oxide, and ozone. The commercially produced CFCs are notorious in ozone depletion (*Understanding Our Environment*, Box 19.1, *The Ozone Hole*), but are also an important group of greenhouse gases. Although these trace gases contribute a smaller greenhouse effect than that of water vapor, their concentration does not fluctuate with the weather. An increase in CO_2 concentration spreads globally and lasts a long time; an increase in water vapor concentration often is local or regional in extent and may disappear in the next rainstorm.

TABLE 19.1   Atmospheric Trace Gases Involved in the Greenhouse Effect

	Carbon Dioxide (CO_2)	Methane (CH_4)	Nitrous Oxide (N_2O)	Chlorofluoro-carbons (CFCs)	Tropospheric Ozone (O_3)	Water Vapor (H_2O)
Greenhouse role	Heating	Heating	Heating	Heating	Heating	Heats in air; cools in clouds
Effect on stratospheric ozone	Can increase or decrease	Can increase or decrease	Can increase or decrease	Decrease	None	Decrease
Principal anthropogenic sources	Fossil fuels; deforestation	Rice culture; cattle; fossil biomass burning	Fertilizer; land-use conversion	Refrigerants; aerosols; industrial processes	Hydrocarbons (with NO_x) biomass burning	Land conversion; irrigation
Principal natural sources	Balanced in nature	Wetlands	Soils; tropical forests	None	Hydrocarbons	Evapotranspiration
Atmospheric lifetime	50–200 yr	10 yr	150 yr	60–100 yr	Weeks to months	Days
Atmospheric concentration at surface (ppbv)	370,000	1720	310	CFC-11: 0.28 CFC-12: 0.48	20–40	3000–6000 in stratosphere
Preindustrial (1750–1800) concentration at surface (ppbv)	280,000	790	288	0	10	Unknown in stratosphere
Annual rate of increase	0.5%	1.1%	0.3%	5% prior to 1995 now near 0%	0.5—2.0%	Unknown
Relative contribution to the anthropogenic greenhouse effect	60%	15%	5%	12%	8%	Unknown

As discussed above and in Chapter 2, biochemical and geologic processes exert a strong influence on the CO_2 concentration in Earth's atmosphere, converting it into plant life and sedimentary rock. Our sister planet Venus has neither life, surface water, nor plate tectonics, and therefore CO_2 has accumulated in its atmosphere to become more than 100,000 times more abundant than on Earth. Some 96.5 percent of Venus's atmosphere is CO_2, and Venus's atmosphere is almost 100 times denser than Earth's atmosphere. The greenhouse effect on Venus therefore packs quite a wallop, leading to a surface temperature of almost 500°C.

TRENDS IN GREENHOUSE GAS CONCENTRATIONS

During recent decades, the greenhouse gases have received increasing scientific and public attention as it has become clear that the atmospheric concentration of each is rising.

CARBON DIOXIDE

In 1958, in conjunction with the International Geophysical Year, measurements of carbon dioxide concentration in the atmosphere were made near the top of Mauna Loa Volcano in Hawaii. This site was chosen because of its altitude and

remote location, far from sources of atmospheric pollution. The measurements are still being made, and they show two remarkable things. First, the amount of CO_2 fluctuates regularly with an annual rhythm (Figure 19.4). In effect, Earth's biosphere is breathing. During the growing season,

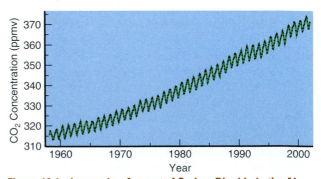

Figure 19.4   Increasing Amount of Carbon Dioxide in the Air Concentration of carbon dioxide in dry air, since 1958, measured at the Mauna Loa Observatory, Hawaii (given in ppmv [parts per million by volume] = ppbv/1000). Annual fluctuations reflect seasonal changes in biologic uptake of CO_2, while the long-term trend shows a persistent increase in the atmospheric concentration of this greenhouse gas.

Box 19.1

UNDERSTANDING OUR ENVIRONMENT

THE OZONE HOLE

Ozone, a pale-blue gas with a pungent odor, is present in the atmosphere in very small amounts (only 20–40 parts per billion by volume near the land surface). However, without this gas, life on Earth would be very different, for ozone provides living organisms with a protective shield against harmful ultraviolet radiation from the Sun. For humans, direct exposure to ultraviolet light will damage the immune system, produce cataracts, substantially increase the frequency of skin cancer, and cause genetic mutations. Maximum concentrations of ozone are found in the stratosphere between 25 and 35 km above Earth's surface, a region where ultraviolet radiation breaks down molecules of oxygen (O_2) into oxygen atoms that are then able to combine with other O_2 molecules to form molecules of ozone (O_3). The ozone, in turn, is broken down by ultraviolet radiation, thereby creating a balance among O, O_2, and O_3.

In 1985, British scientists reported a startling discovery: a vast ozone depletion, about the size of Canada, had developed temporarily in the stratosphere above the Antarctic region. By 1987, measurements showed that, for several weeks each austral spring, concentrations of this life-protecting gas over Antarctica had dropped more than 50 percent since 1979 and that between altitudes of 15 and 20 km the depletion had reached 95 percent. Studies have shown that the productivity of oceanic phytoplankton, which lie at the base of the marine food chain, has decreased 6 to 12 percent in the region of the ozone hole, where record amounts of ultraviolet radiation have been measured. Record-low ozone values have also been measured over Australia and New Zealand, and continuing surveys showed that ozone values at all latitudes south of 60° decreased by 5 percent or more after 1979. By early 1993, the globally averaged total ozone concentration, measured by satellite, had decreased to unprecedented low values.

What had happened to upset the natural atmospheric balance among the three gaseous forms of oxygen? A decade earlier, it was recognized that a group of synthetic industrial gases, the **chlorofluorocarbons (CFCs)**, were entering the lower atmosphere and spreading rapidly around the world. The CFCs ultimately rise into the upper atmosphere, where ultraviolet radiation breaks them down, releasing chlorine. It is chlorine that does the damage: the chlorine atoms destroy ozone in a catalytic reaction. Each chlorine atom is capable of destroying as many as 100,000 ozone molecules before other chemical reactions remove the chlorine from the atmosphere. The sunlight and very cold springtime temperatures (−80°C) in the upper atmosphere that are critical to the ozone destruction process are present in the south polar region, which is why the ozone hole is especially pronounced over Antarctica. In the Arctic, the period of the critical spring conditions is much shorter. With the documentation of ozone depletion in the upper atmosphere, scientists for the first time could show that human activity was having a detrimental global effect on one of Earth's natural systems.

The CFCs responsible for the ozone hole are used in the manufacture of plastic foams, as propellants in aerosol cans, and as refrigerants. In 1987, scientific concerns about ozone destruction led 49 industrial nations to ratify the Montreal Protocol on Substances That Deplete the Ozone Layer, an agreement calling for a reduction in CFC production to 50 percent of 1986 amounts by the end of the twentieth century. However, because even such cutbacks in CFC use would permit continued degradation of the ozone shield, in 1990 the United States and many other nations pledged to eliminate CFC production entirely by the year 2000.

Because substitute chemicals were available that posed a much smaller threat to stratospheric ozone, this pledge has been largely fulfilled.

CO_2 is absorbed by vegetation, and its atmospheric concentration falls. Then, during the winter dormant period, more CO_2 enters the atmosphere than is removed by vegetation, and its concentration rises. Second, the long-term trend is unmistakable and rising (Figure 19.4). Since 1958, the CO_2 has risen from 315,000 to 370,000 ppbv; moreover, the rise is not linear but exponential, like the money in a bank account that pays compound interest.

The rising curve of atmospheric CO_2 immediately raises two questions: Is the observed rise in CO_2 unusual? And how can it be explained? To answer the first question, we must turn to the geologic record. Ice cores obtained from the Antarctic and Greenland ice sheets contain samples of ancient atmosphere. These samples exist in tiny air bubbles that were trapped when snow falling on the glaciers slowly compressed and transformed into glacier ice (Figure 19.5;

Chapter 16). If the ice is melted in a laboratory and the air collected and analyzed, the amount of CO_2 and other trace gases in the ancient atmosphere can be measured. Ice from Vostok Station in Antarctica contains such a record extending back 400,000 years. The measurements show that during glacial ages, the atmosphere contained about 200,000 ppbv of CO_2, whereas during interglacial times the concentration rose to about 280,000 ppbv (Figure 16.33). Preindustrial levels of CO_2 were close to 280,000 ppbv as well, estimated both from glacial ice cores and air samples collected and stored by chemists at the dawn of the Industrial Revolution. The subsequent rapid increase to 366,000 ppbv by 1998 (Figure 19.4) is unprecedented in the ice core record and implies that something very unusual is taking place.

A possible explanation for the extraordinary recent rise in atmospheric CO_2 is immediately suggested by the rate

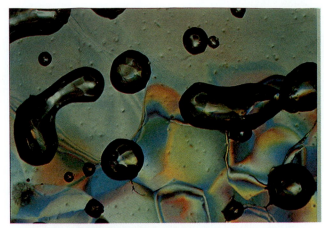

Figure 19.5 Samples of Ancient Air Air bubbles trapped in glacier ice. By melting the ice in a laboratory and collecting the gas, the content of CO_2 and other trace gases in these ancient samples of the atmosphere can be measured.

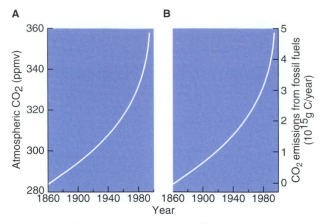

Figure 19.6 Carbon Dioxide and Fossil Fuels A. Since the beginning of the industrial revolution, the atmospheric concentration of CO_2 has risen at an increasing rate. B. The increase matches the growing rate at which CO_2 has been released through the burning of fossil fuels.

at which CO_2 has been added to the atmosphere since the beginning of the Industrial Revolution (Figure 19.6). The curve tracking the increase in CO_2 closely resembles the curve showing the increase in carbon released into the atmosphere by the burning of fossil fuels. No observed natural mechanism can explain such a rapid increase in CO_2 during this time. The inescapable conclusion is that anthropogenic (human-generated) burning of fossil fuels must be a primary factor in the observed increase in atmospheric CO_2. Additional contributing factors must be widespread deforestation, with its attendant burning and decay of cleared vegetation, and the use of wood as a primary fuel in many underdeveloped countries that have rapidly growing populations.

During the 1990s, the increase of atmospheric carbon dioxide has been irregular, and this has given scientists clues to the uptake of CO_2 by the biosphere. The largest yearly increases in CO_2 in the 1990s occurred in years with maximum El Niño conditions, which occur when the tropical Pacific Ocean is unusually warm. Warmer water holds less dissolved CO_2 (try warming a glass of soda!), so CO_2 is less likely to be absorbed by the ocean in an El Niño year. If the ocean fails to absorb its normal quota of CO_2, a larger fraction of human-generated CO_2 stays in the atmosphere. El Niño also induces unusual rainfall conditions over the entire globe, which may affect the uptake of CO_2 by land plants.

METHANE

Methane (CH_4) absorbs infrared radiation 25 times more effectively than CO_2, making it an important greenhouse gas despite its relatively low concentration (1720 ppbv). In prehistoric times, methane levels, like CO_2 levels, increased and decreased with the glacial/interglacial cycles (Figure 16.33). Starting in the late 1960s, when measurements of atmospheric methane began, the concentrations have increased at a rate of about 1 percent per year. Since 1984 the rate has decreased, possibly related to changes in

human-generated emissions. Methane levels for earlier times have been obtained from ice core studies that show an increase that parallels the rise in the human population. This relationship is not surprising, for much of the methane now entering the atmosphere is generated (1) by biological activity related to rice cultivation, (2) by leaks in domestic and industrial gaslines, and (3) as a byproduct of the digestive processes of domestic livestock, especially cattle. The global livestock population increased greatly in the twentieth century, and the total acreage under rice cultivation has increased more than 40 percent since 1950.

CHLOROFLOUROCARBONS (CFCS)

CFC-12, used mostly as a refrigerant, has 20,000 times the capacity of carbon dioxide to trap ultraviolet radiation, whereas CFC-11, which is widely used in making plastic foams and as an aerosol propellant, has 17,500 times the capacity. In the late twentieth century both compounds increased in the atmosphere at an annual rate of about 5 percent. As discussed in *Understanding Our Environment,* Box 19.1, *The Ozone Hole,* the increasing atmospheric concentration of CFCs has produced worldwide concern because the scientific consensus is that these gases destroy ozone in the upper atmosphere, thereby leading to the formation of the Antarctic ozone hole. As a result of an international treaty that limits the production of CFCs, the increase in CFC concentration has weakened and may reverse in coming decades. However, some of the chemicals used to replace CFCs are also greenhouse gases, and we can expect *their* concentrations to increase.

OZONE AND NITROUS OXIDE

Although ozone in the upper atmosphere is beneficial because it traps harmful infrared solar radiation, when this gas builds up in the troposphere (lower atmosphere) it contributes to the greenhouse effect. Both ozone and nitrous oxide, another greenhouse gas involved in biochemical cycles, are increasing annually at rates of 0.5 to 2 percent and 0.3 percent, respectively. Together, they account for about 13 percent of the anthropogenic greenhouse effect. Tropical forests are important in removing excess tropospheric ozone by processes related to photosynthesis. However, the wholesale destruction of these forests could lead to further concentration of this gas in the atmosphere. Nitrous oxide, released by microbial activity in soil, the burning of timber and fossil fuels, and the decay of agricultural residues, has a long lifetime in the atmosphere. Accordingly, atmospheric concentrations are likely to remain well above preindustrial levels even if emission rates stabilize.

Before you go on:

1. How does the yearly cycle of plant growth affect the carbon cycle? What evidence do we have?

2. How is the greenhouse effect similar to the effect of a blanket on a winter night? How do these effects differ?

3. Which greenhouse gases have increased most from human activities? Which have increased least?

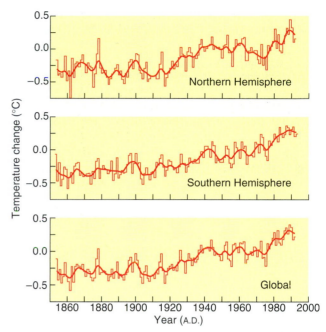

Figure 19.7 Historic Temperature Changes Hemispheric and global mean temperature changes since the mid-nineteenth century, incorporating both land and ocean data. Annual averages (thin line) and smoothed averages (bold line) are shown. In each of the curves, the long-term trend is a rise of temperature, with the total global rise for this interval being about 0.6°C.

GLOBAL WARMING

If the atmospheric concentration of the greenhouse gases is rising, what does this portend for future climate? Does it mean that Earth's surface temperature is warming, and, if so, by how much and at what rate? To try and answer these questions, we can first look at the historical record of climate and then see how forecasts of the future can be made using computer simulations of Earth's climate system.

HISTORICAL TEMPERATURE TRENDS

Correctly assessing recent global changes in temperature is a very complicated task. Few instrumental measurements were made before 1850, and vast regions have no weather data before World War II. The earliest records are from western Europe and eastern North America. Data for oceanic areas, which encompass 70 percent of the globe, are sparse, especially prior to 1945. Therefore, most "global" temperature curves are reconstructed primarily from land stations that are located mainly in the Northern Hemisphere. Numerous curves of average annual temperature variations since the mid- or late nineteenth century have been published. Although details differ, all show a long-term rise in temperature during the past century (Figure 19.7). The total temperature increase since the late nineteenth century is about 0.6°C. Because the interval of rising temperatures coincides with the time of rapidly increasing greenhouse gas emissions, it is tempting to assume that the two phenomena are causally related.

However, the temperature reconstructions prior to 1950 are based on relatively few data that are unequally distributed across the globe, and the upward trend in temperature is not as simple as the upward trends in CO_2 concentration and fossil-fuel consumption.

Some portion of the modest rise in global temperature during the past century could fall within the natural variability of Earth's climate system and would have occurred even if the greenhouse gas concentrations had not increased. Studies of historical data have demonstrated that some patterns of global temperature can fluctuate with durations of 3 to 17 years and make measurable contributions to the average global temperature. Indirect estimates of temperature can be obtained from the thicknesses of yearly growth rings in trees (on land) and coral (at sea), and annual mass balance of glaciers. These data demonstrate that some temperature patterns can vary on a 100-year time scale as well. Finally, very small fluctuations in the energy radiated by the Sun, which correlate with the varying number of sunspots observed by astronomers, may exert a significant influence on climate.

The historical temperature record by itself presents us with many possible causes for recent climate change. Just as with earthquake prediction (Chapter 10), it is risky to simply project historical climate trends into the future with no appreciation of the underlying processes. Fortunately, we understand how climate works much better than we understand how earthquakes work. We can explore the linkage between greenhouse gas emissions and climate change with computer models of the climate system.

CLIMATE MODELS

Global climate models are three-dimensional mathematical models of Earth's climate system and are an outgrowth of efforts to forecast the weather. The most sophisticated are **general circulation models (GCMs)** that attempt to link processes in the atmosphere, the hydrosphere, and the biosphere. The sheer complexity of the natural climate system means that such models, of necessity, are greatly simplified representations of the real world.

Many of the linkages and processes in the climate system are still poorly understood and therefore difficult to model. For instance, computer models do not yet adequately portray the dynamics of ocean circulation or cloud formation, two of the most important elements of the climate system. Some models try to mimic the biogeochemical processes that link climate to the biosphere, but scientists sometimes lack the global-scale observations that are necessary to test the model's accuracy. Despite their limitations, GCMs have been very successful in simulating the general character of present-day climates and have greatly improved weather forecasting. This success encourages us to use these models to gain a general picture of future climate change.

To run a modeling experiment that simulates the climate, scientists specify a set of **boundary conditions**. Boundary conditions are mathematical expressions of the physical state of Earth's climate system at the period of interest for the experiment. Thus, an experiment designed to simulate the present climate would prescribe as boundary conditions the solar radiation reaching Earth at a particular point in its orbit, the geographic distribution of land and ocean, the position and heights of mountains and plateaus, the concentrations of atmospheric trace gases, sea-surface temperatures, the limits of sea ice, the snow and ice cover on the land, the *albedo* (reflectivity) of land, ice, and water surfaces, and the effective soil moisture (the sum of water input and water loss).

The solution of the complex mathematical equations of a GCM requires considerable amounts of computer time. Even running on the fastest computers, climate simulations can calculate only broad averages of temperature, atmospheric pressure, rainfall, humidity, and so forth, usually at locations spaced by 200 km or more. Therefore, although these models can generate a reasonable picture of global and hemispheric climatic conditions, they are poor at resolving conditions at the scale of small countries, states, or counties.

MODEL ESTIMATES OF GREENHOUSE WARMING

Predictions of climatic change related to greenhouse warming are based mainly on the results of GCMs developed by research groups in various countries. The models differ in detail, as well as in the assumptions they employ. Nevertheless, they all predict that the anthropogenically generated greenhouse gases already in the atmosphere will lead to an average global temperature increase of at least 0.5 to 1.5°C. This prediction is consistent with the 0.6°C rise in temperature inferred from the instrumental record.

If fossil-fuel consumption continues to grow at an increasing rate, atmospheric CO_2 is projected to double by 2100, if not sooner. Climate models predict that if the greenhouse gases build up to a doubling of the preindustrial CO_2 concentration, then average global temperatures will rise between 1.5° and 4.5°C. This does not mean, of course, that the temperature will increase uniformly all over Earth. Instead, the projected temperature change varies geographically, with the greatest change occurring in the polar regions (Figure 19.8). Enhanced warming at high altitudes, such as in the state of Alaska in the United States, was indeed recorded in the late twentieth century.

The rate at which the projected warming will occur depends on a number of uncertainties: How rapidly will concentrations of the greenhouse gases increase? How rapidly will the oceans, a major reservoir of heat and a fundamental element in the climate system (Figure 19.1), respond to changing climate? How will changing climate affect ice sheets and cloud cover? What is the range of natural variations in the climate system on the century time scale?

The potential complexity is well illustrated by the response of clouds to a warming trend. If the temperature of the lower atmosphere increases, more water will evaporate from the oceans. The increased atmospheric moisture will create more clouds, but clouds reflect solar energy back into space. This will have a cooling effect on the surface air, thereby having an influence opposite that of the greenhouse effect. Clouds with different water-droplet sizes will reflect sunlight differently. It is not surprising that no computer model can yet match the complexity of natural cloud behavior.

Because of such uncertainties about the climate system, scientists tend to be cautious in their predictions. Nevertheless, there is a broad consensus among climate researchers that (1) human activities have led to increasing atmospheric concentrations of carbon dioxide and other trace gases that have enhanced the greenhouse effect; (2) global mean surface air temperature has increased by 0.6°C during the last 100 years, and this increase may be the direct result of the enhanced greenhouse effect; and (3) during the next few decades global average temperature will likely increase at about 0.3°C per decade on average, assuming emission rates continue to increase. This projected increase will lead to a global average temperature as much as 4.5°C warmer by the end of the next century.

If governmental controls lead to lower emission rates, the decadal rise in temperature may be only 0.1–0.2°C. Nevertheless, the temperature increase related to the continued release of greenhouse gases will be larger and more rapid than any experienced in human history. In effect, we may be about to experience a "super-interglacial" period warmer than any interglacial period of the past 2 million years (Figure 19.9).

A.

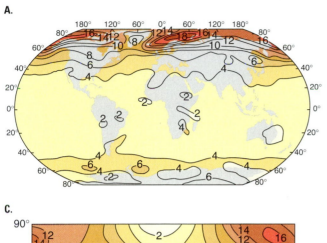

B.

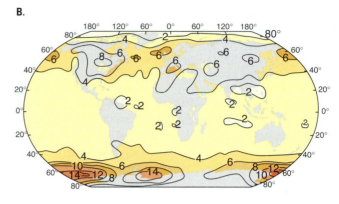

C.

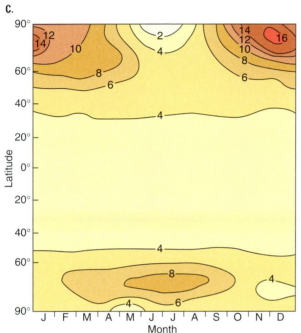

Figure 19.8 Temperature Rise if Carbon Dioxide Doubles A forecast of future changes in surface air temperature (in °C) that would result from an effective doubling of atmospheric CO_2 concentration relative to that of the present. A. Temperature increases for winter (December, January, February). For example, along the lines labeled 4, the projected temperature increase is everywhere 4°C. B. Temperature increases for summer (June, July, August). C. A latitudinal cross section showing changes in zonal average air temperature through the year. This graph is a summary of the map patterns shown in A and B, but includes the spring and autumn months as well. Greenhouse warming is greatest at high latitudes, where temperature increases as great as 16°C are forecast by the model for the Northern-Hemisphere winter.

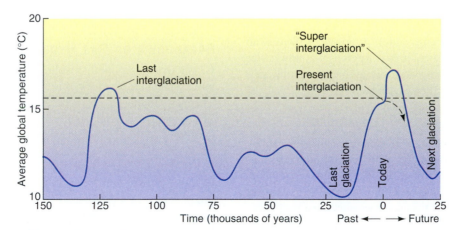

Figure 19.9 Temperature Changes Past and Future The course of average global temperature during the past 150,000 years and 25,000 years into the future. The natural course of climate (dashed arrow) would be declining temperatures leading to the next glacial maximum, about 23,000 years from now. With greenhouse warming, a continuing rise of temperature may lead to a "super-interglaciation" within the next several centuries. The temperature may then be warmer than during the last interglaciation and warmer than at any time in human history. The decline toward the next glaciation would thereby be delayed a millennium or more. The dashed black line just above 15°C marks the present average global temperature.

ENVIRONMENTAL EFFECTS OF GLOBAL WARMING

An increase in global surface air temperature by a few degrees does not sound like much. Surely, you say, we can put up with this rather insignificant change. However, if we stop and consider that the difference in average global temperature between the present and the coldest part of the last ice age was only about 5°C, we can begin to see how even a degree or two would have global repercussions.

Global warming is just one result of our great geochemical "experiment." There are many physical and biological side effects that are of considerable interest and concern.

GLOBAL PRECIPITATION CHANGES

A warmer atmosphere will lead to increased evaporation from oceans, lakes, and streams, and to greater precipitation. However, the distribution of the increased precipitation will be uneven. Climate models suggest that the equatorial regions will receive more rainfall, in part because warmer temperature will increase rates of evaporation over the tropical oceans and promote the formation of rain clouds. Some interior portions of large continents, which are distant from evaporation sources, will become both warmer and drier.

CHANGES IN VEGETATION

Shifts in precipitation patterns are likely to upset ecosystems, forcing them to adjust to new conditions. Forest boundaries may shift during coming centuries in response to altered temperature and precipitation regimes. Some prime midcontinental agricultural regions are likely to face increased drought and reduced soil moisture that will have a negative impact on crops. Higher-latitude regions with short, cool growing seasons may see increased agricultural production as summer temperatures increase.

INCREASED STORMINESS

The shift to a warmer, wetter atmosphere favors an increase in tropical storm activity. Regions that now suffer hurricanes and typhoons could see an increase in the size and frequency of these devastating storms.

MELTING (AND GROWING?) GLACIERS

Because warmer summers favor increased ablation, worldwide recession of low- and middle-latitude mountain glaciers is likely in a warmer world. On the other hand, warmer air in high latitudes could evaporate and transport more moisture from the oceans to ice sheets, causing them to grow larger.

REDUCTION OF SEA ICE

Enhanced heating in high northern latitudes (Figure 19.8) favors the shrinkage of sea ice. A reduction in polar sea ice, which has a high albedo, would reinforce the greenhouse effect by reducing the reflection of solar radiation.

Observations from submarines indicate that Arctic sea ice has already thinned by 40 percent since the mid-twentieth century. Models show much less heating in the high-latitude Southern Hemisphere, predicting little change in sea-ice cover there.

THAWING OF FROZEN GROUND

Rising summer air temperatures will begin to thaw vast regions of perennially frozen ground at high latitudes. The thawing will likely affect natural ecosystems as well as cities and engineering works built on frozen ground.

RISE OF SEA LEVEL

As the temperature of ocean water rises, its volume will expand, causing world sea level to rise. This rise in sea level, supplemented by meltwater from shrinking mountain glaciers, is likely to increase calving along the margins of tidewater glaciers and ice sheets, thereby leading to additional rise in sea level. The rising sea will inundate coastal regions where millions of people live and will make the tropical regions even more vulnerable to larger and more frequent cyclonic storms.

CHANGES IN THE HYDROLOGIC CYCLE

Shifting patterns of precipitation and warmer temperatures will likely lead to some significant local and regional changes in stream runoff and groundwater levels.

DECOMPOSITION OF SOIL ORGANIC MATTER

As the temperature rises, organic matter in soil will decompose more rapidly. Soil decomposition releases CO_2 to the atmosphere, reinforcing the greenhouse effect. If world temperature rises by 0.3°C per decade, during the next 60 years soils could add as much as 20 percent to the projected CO_2 release due to combustion of fossil fuels, assuming the present rate of fuel consumption continues.

BREAKDOWN OF GAS HYDRATES

Gas hydrates are ice-like solids in which gas molecules, mainly methane, are locked in the structure of solid H_2O. They are found in some ocean sediments and beneath frozen ground. By one estimate, gas hydrates worldwide may hold 10,000 billion metric tons of carbon, twice the carbon in all the coal, gas, and oil reserves on land. When gas hydrates break down, they release methane. Global warming at high latitudes will result in thawing of frozen ground that may well destabilize the hydrates there, release large volumes of methane, and thus amplify the greenhouse effect.

Before you go on:

1. What are the sources of uncertainty in estimating historical temperature trends?

2. What are the sources of uncertainty in estimating future temperature trends with computer simulations?

3. Which of the predicted environmental changes associated with greenhouse warming are likely to reinforce the warming trend?

THE PAST AS A KEY TO THE FUTURE

Although our present knowledge of how Earth works, coupled with computer modeling, enables us to make educated projections of surface environmental changes that will result from greenhouse warming, most scientists are reluctant to make firm forecasts. Instead, they hedge their bets with qualifying adjectives like "possible," "probable," and "uncertain." Their caution emphasizes the gap between what we know about Earth and what we would like to know and points to the many challenges that still face scientists who study climate change.

To increase the likelihood that our predictions about the changing global environment are correct, we can invoke "Ayer's Law." This useful tenet of geology states: "Anything that did happen, can happen." In other words, we can use the geologic record, which archives the history of natural environmental changes on Earth, as a key to understanding our future.

LESSONS FROM THE PAST

The most important lesson from the geologic record is that past climate has paralleled today's climate only occasionally. For most of the past 2 million years, Earth's climate was colder than at present. We live in an interglaciation within the late Cenozoic glacial ages that have dominated Earth's recent climate. Since well before the Pleistocene Epoch (Chapter 16), a large portion of the continental surface has been covered repeatedly by ice sheets. Our interglacial interval is now approximately 12,000 years old, close to the age where a return to glacial conditions might be expected.

Climate in the last million years has occasionally been warmer than today's climate. Of particular interest are the early Holocene, from 11,000 to about 6000 years ago, when average temperatures were 0.5 to 1°C warmer, and the warmest part of the last interglaciation, about 120,000 years ago, when global temperatures were 1 to 2°C higher. Geologists are examining the traces of past winds, rain, and ocean circulation for an indication of how our climate might be different in coming decades. For climates that match the projected temperature increase for a doubling of atmospheric CO_2, we must reach back in time before the Pleistocene ice ages, to the middle Pliocene, about 4.5 to 3

million years ago, when average temperatures may have been 3 to 4°C warmer than present.

Because human-induced climate change is occurring over the course of decades, rather than millions of years, geologic records of rapid environmental change are also important. We can use this information to help us anticipate the character of environmental changes that may happen on a warming Earth. One example is a rapid cooling and warming that occurred in the North Atlantic region at the end of the last glaciation. Study of these intervals enables us to see how plants and animals responded to climatic conditions that may have been broadly similar to those we may experience in the near future.

Still older intervals of unusually warm climate pose special problems for interpretation but have generated some imaginative solutions. They include the middle Cretaceous Period (about 100 million years ago) and the Eocene Epoch (about 40 to 50 million years ago).

WHY WAS THE MIDDLE CRETACEOUS CLIMATE SO WARM?

It's probably a good thing we did not live during the Middle Cretaceous Period. Not only was the world inhabited by huge carnivorous dinosaurs, but also the climate was one of the warmest in Earth's history. Evidence that the world was much warmer than today is compelling (Figure 19.10). Warm-water marine faunas were widespread, coral reefs grew 5 to 15° closer to the poles than they do now, and vegetation zones were displaced about 15° poleward of their present positions. Peat deposits

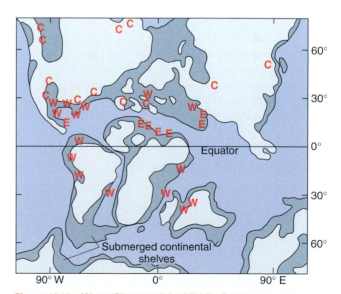

Figure 19.10 Warm Climate of the Middle Cretaceous During the Middle Cretaceous, sea level was 100 to 200 m higher than now and flooded large areas of the continents, producing shallow seas. Warm-water faunas (W) and evaporites (E) were present at low to middle latitudes, while coal deposits (C) developed in high-latitude regions, implying warmer year-round temperatures.

formed at high latitudes, and dinosaurs, which are generally thought to have preferred warm climates, ranged north of the Arctic circle. Sea level was 100 to 200 m higher, implying the absence of polar ice sheets, and isotopic measurements of deep-sea deposits indicate that intermediate and deep waters in the oceans were 15 to 20° warmer than now.

GCM simulations of the Middle Cretaceous world suggest that several factors were involved in producing such warm conditions: geography, ocean circulation, and atmospheric composition. The simulations show that the Middle Cretaceous arrangement of continents and oceans (Figure 19.10), which influenced ocean circulation and planetary albedo, could account for nearly 5°C of the warming; of this amount, about a third is attributable to the absence of polar ice sheets. However, geography alone is inadequate to explain warmer year-round temperatures at high latitudes. Could the poleward transfer of heat by ocean currents be the answer? The oceans now account for about a third of the present poleward heat transfer, but computer simulations suggest that even with the geography and ocean circulation rearranged as they were in the Middle Cretaceous, oceanic heat transfer cannot explain the greater high-latitude warmth. If the geologic data have been correctly interpreted, and the modeling results are reliable, some other factor must be involved. This factor appears to be CO_2, the major greenhouse trace gas.

Can an enhanced greenhouse effect be the key to explaining the exceptionally warm Middle Cretaceous climate? GCM experiments show that by rearranging the geography and also increasing carbon dioxide six to eight times above present concentrations, the warmer temperatures can be explained. Geochemical reconstructions of changing atmospheric CO_2 levels over the past 100 million years point to at least a tenfold increase in CO_2 during the Middle Cretaceous, leading to average temperatures as much as 8°C higher than now (Figure 19.11). Compared to the average conditions forecast for the twenty-first century (2–4°C hotter), the Middle Cretaceous climate must have been a scorcher!

MANTLE CONVECTION AND ATMOSPHERIC GREENHOUSE GASES

If CO_2 was an important factor in Middle Cretaceous warming, we still are faced with explaining how this gas increased so substantially. Unlike in the modern world, combustion of fossil fuels cannot provide the answer. The most likely source is volcanic activity, which today constitutes a major source of CO_2 entering the atmosphere.

Geologic evidence points to an unusually high rate of volcanic activity in the Middle Cretaceous. Rates of seafloor spreading were then about three times as great as now, implying increased rates of volcanism at midocean ridges. In addition, vast outpourings of lava created a succession of great undersea plateaus across the central Pacific Ocean between 135 and 115 million years ago, the

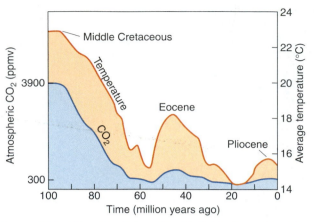

Figure 19.11 Changing Carbon Dioxide in the Atmosphere Over Geologic Time Geochemical reconstruction of changing atmospheric CO_2 concentration and average global temperature over the past 100 million years. High CO_2 values and temperatures in the Middle Cretaceous contrast with much lower values of the present. Other intervals of higher temperature and CO_2 occurred during the Eocene and the Middle Pliocene.

time of maximum Cretaceous warmth. One of these—the Ontong-Java plateau in the southwestern Pacific—has more than twice the area of Alaska and reaches a thickness of 40 km. Such a vast outpouring of lava likely released massive amounts of CO_2 from Earth's mantle. Could this gas emission have been sufficient to warm the climate to unprecedented levels? By one calculation, the eruptions could have released enough CO_2 to raise the atmospheric concentration to 20 times the preindustrial value, in the process raising average global temperature as much as 10°C. Other estimates range from 8 to 12 times the preindustrial value.

Recently, geologists have proposed that these vast lava outpourings are the result of a *superplume*. Mantle hot spots are thought to be due to rising plumes of hot rock (Chapter 2). Plumes like that responsible for the Hawaiian hot spot, which is about 200 to 300 km across near the surface, originate somewhere in the mesosphere, possibly as deep as the core–mantle boundary. A *superplume* would originate from a substantial overturn of mantle rock. Some computer simulations of mantle convection suggest that downgoing slabs of lithosphere can stall near 670 km depth, where a phase change from spinel to perovskite mineral structure marks the boundary between the upper and lower mantle (Chapter 10). At irregular intervals in the computer simulations, these stalled slabs fall together into the lower mantle in what is called a "flushing event." Hot lower mantle rises to replace the falling slabs, inducing a superplume. (See Figure 19.12.)

Some studies of seismic tomography (Chapter 10) support the flushing hypothesis, reporting zones of high wavespeed near 670 km depth that could be cool slabs that failed to penetrate into the lower mantle. There is also a

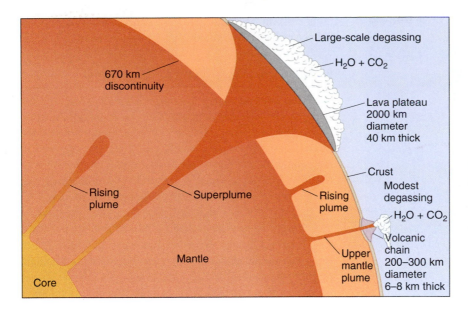

Figure 19.12 Superplumes Rise Through the Mantle Geologic data and mantle convection simulations suggest that rising superplumes from the core–mantle boundary may have built huge lava plateaus when they reached the lithosphere and given rise to large-scale degassing of CO_2 that greatly enhanced the greenhouse effect.

broad zone of low wavespeed at the bottom of the mantle that lies beneath the location of the presumed Cretaceous superplume. If the flushing hypothesis is correct, then plate tectonics cools the mantle in two styles. In "ordinary" times heat loss at spreading ridges couples with the downward plunge of cool lithosphere. In "extraordinary" times—when stalled slabs flush into the lower mantle—superplumes would allow heat to escape more efficiently from Earth's lower mantle and core. By this reasoning, the core and atmosphere would be linked dynamically, so that the warm Middle Cretaceous climate was a direct consequence of the cooling of Earth's deep interior.

EOCENE WARMTH AND LITHOSPHERE DEGASSING

By the end of the Cretaceous, world temperatures had fallen substantially below levels reached earlier in that period (Figure 19.11). A reversal in this trend brought temperatures to a new peak during the Early Eocene, about 50 million years ago. Varied evidence points to warmer conditions: alligator fossils on Ellesmere Island (at about 78° N latitude), tropical vegetation at up to 45° N latitude, tropical marine surface-water organisms at about 55° N in the Atlantic Ocean, and lateritic soils (indicating warm climate with seasonal rainfall) at up to 45° latitude in both hemispheres. Isotopic measurements of bottom-dwelling organisms show that the temperature of the deep ocean reached its highest Cenozoic value in the Early Eocene. The average atmospheric warming in the Eocene, relative to modern conditions, is estimated to be in the range of 1–4 C° and likely close to 2 C°.

Scientists speculate that two processes may have contributed to the higher Eocene temperatures: greater poleward transfer of heat by the oceans and an increased concentration of CO_2 in the atmosphere. Enhanced poleward

heat transfer could have prevented the formation of polar sea ice. Because ice is highly reflective relative to the open ocean, more solar radiation would be absorbed, thus warming the high latitudes.

Estimates of Cenozoic atmospheric CO_2 concentrations place the highest value—about two to six times the modern value—during the Eocene (Figure 19.11). The excess CO_2 originated partly from volcanism as in the Cretaceous, but other sources have been proposed. For instance, crustal metamorphism may have released a significant amount of greenhouse gas.

When carbonates and shales undergo regional metamorphism in orogenic belts, CO_2 is released as a byproduct of the metamorphic reactions. Low-grade (greenschist facies) metamorphic conditions may release a large proportion of this CO_2, because such conditions occur in the upper part of an orogen, where brittle fracturing and faulting provide escape pathways for gases. Studies show that regional metamorphism of the Himalaya (Figure 19.13) may have been contemporaneous with the Eocene warming. In addition, regional metamorphism at this time has been documented in the Mediterranean region (Greece, Turkey) and in the circum-Pacific region (New Caledonia, Japan, western North America). Although the interval of metamorphism in these regions is not yet tightly dated, calculations suggest that the quantity of CO_2 generated from these regions was sufficient by itself to explain its unusual buildup in the Eocene atmosphere.

An additional factor to consider is the removal of atmospheric CO_2 during weathering of orogenically uplifted silicate rocks. CO_2 is removed from the atmosphere when carbonic acid is produced (see Table 6.1). As the weak acid decomposes the rocks, bicarbonate released by the weathering reactions is carried by streams, in solution, to the sea and is there converted to calcium carbonate by marine organisms. When the organisms die, their skeletal remains become

Figure 19.13 Metamorphism and the Release of Carbon Dioxide Folded metasedimentary rocks from the Swat region of the northwestern Himalaya. Regional metamorphism of marine sedimentary rocks during collision of India with Asia is regarded as a possible source of carbon dioxide that contributed to the high concentration of this gas in the Eocene atmosphere.

stored in carbonate sediments at the seafloor. The net effect is to isolate in marine sediments much of the atmospheric CO_2 involved in weathering. It is difficult to estimate how much CO_2 was removed from the atmosphere in this way as a result of widespread mountain uplift, but it seems possible that the post-Eocene cooling trend that ultimately led to the glacial ages may be at least partly explained by high weath-

ering rates in orogenic mountain belts and the consequent storage of the carbon dioxide in ocean sediments.

MODELING PAST GLOBAL CHANGES

Paleoclimatic reconstructions offer a means of testing the accuracy of climate models. Our confidence in using GCMs for predicting the future will be strengthened if the models not only can accurately reproduce the present climate, but also can reproduce past climates. We can test, or "validate," these climate simulations by comparing the model results against independent geologic evidence.

The same GCMs that are used to model the present global weather are also used to model ancient climates. The main difference is in the boundary conditions. For example, one set of experiments has attempted to simulate Earth's climates at 3000-year intervals since the last glacial maximum about 18,000 years ago. The specified boundary conditions changed substantially over this time interval (Figure 19.14): solar radiation varied by as much as 8 percent as Earth went through one precessional cycle; the area of glacier ice shrank, and airborne dust decreased as the ice age came to an end; both sea-surface temperature and atmospheric CO_2 increased as the climate moved toward its present interglacial state. During each of the seven time periods modeled, the boundary conditions were different from those of the other periods. As a result, the successive simulations change as the model Earth passes from a glacial age to an interglacial age.

As an example of such a simulation, we can examine the results of an experiment focusing on the Eastern Hemisphere 9000 years ago (Figure 19.15A). The model "predicts" that 9000 years ago the climate in a broad belt across northern Africa and southern Asia experienced

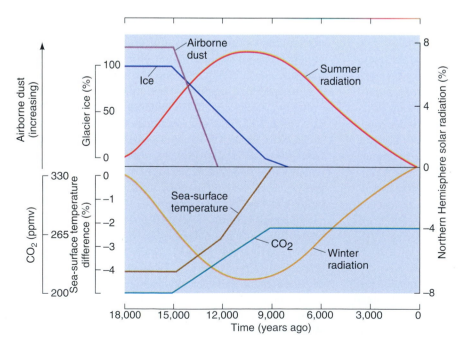

Figure 19.14 Global Climate Model Boundary conditions used in one set of GCM simulations of world climate between the last glacial maximum (18,000 years ago) and the present. Solar radiation values are based on Earth's known astronomical motions (Figure 16.3), while ice, dust, CO_2, and sea-surface temperature values are based on geologic data.

A.

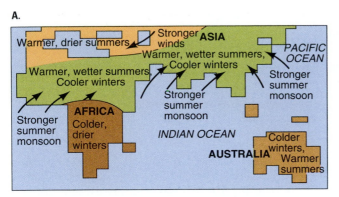

B.

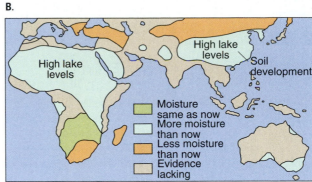

Figure 19.15   What the Climate Was Like Nine Millennia Ago
Climates of Africa and southern Asia during the early Holocene (9000 years ago). A. GCM simulation indicates that a belt crossing northern Africa and southeastern Asia had warmer, wetter summers and cooler winters 9000 years ago. These conditions resulted from a strengthening of summer monsoon winds, which brought precipitation to the continents from the warm tropical oceans. At that time, the Mediterranean region and Australia had warmer, drier summers, while southern Africa and Australia had colder winters. B. Geologic evidence generally bears out the GCM results: a belt running from northern Africa through the Middle East and China received more moisture 9000 years ago, as shown by high lake levels and paleosols that indicate increased soil moisture. Evidence also shows that southern Africa and Madagascar, as well as the northeastern Mediterranean region, were drier than now.

warmer and wetter summers and cooler winters; the Mediterranean region at that time had warmer, drier summers, southern Africa had colder, drier winters, and Australia had colder winters and warmer summers. The model also indicates that the increase in summer moisture was largely due to an increase in the strength of the summer monsoon in western Africa and southern Asia.

Figure 19.14 shows us that the boundary conditions for 9000 years ago were different from those of today mainly with respect to Northern Hemisphere solar radiation, which was about 8 percent greater in summer and 8 percent less in winter. This difference was a major factor in strengthening the summer monsoon.

To see how reasonable these model predictions are, we can examine geologic evidence of conditions about 9000 years ago (Figure 19.15B). The evidence points to a broad belt of land across northern Africa, the Middle East, and China that had greater effective moisture (i.e., precipitation minus evaporation) than now. In Africa and western China, evidence for this increased moisture consists of high water levels in closed-basin lakes and the widespread occurrence in China of an early Holocene soil that indicates an increase in monsoon precipitation. In this example, the results of the model simulation are generally consistent with the paleoclimatic data assembled by geologists.

Before you go on:

1. What evidence do we have that the Middle Cretaceous was unusually warm?

2. What factors contributed to the warmth of the Middle Cretaceous?

3. How could plate tectonics be responsible for both the warmth of the Eocene period and the cooling trend after the Eocene?

4. In what ways were climate conditions different 9000 years ago? Why did the climate boundary conditions change?

REVISITING PLATE TECTONICS AND THE EARTH SYSTEM

Plate tectonics has caused geologists to rethink the evidence for Earth's climate history. The presence of extensive coal deposits on Antarctica, for instance, implies a climate that supported abundant plant growth, because coal is composed primarily of plant fossils. Such a climate is hard to imagine near the South Pole. Plate tectonics allows us to contemplate the drift of Antarctica through more temperate latitudes during geologic history.

The processes that drive plate tectonics also influence climate, and this influence extends beyond the carbon cycle. Part of the warmth of the Cretaceous (about 100 million years ago) can be attributed to the presence of shallow seas over much of today's dry land (Figure 19.10). Sea surfaces absorb solar energy more efficiently than land surfaces do, and thus more sea surface encourages a warmer climate. But what caused the Middle Cretaceous oceans to spill over the land? Plate tectonics! Because seafloor spreading was faster in the Middle Cretaceous, the plate system had a large proportion of young, warm (and there-

BOX 19.2   THE SCIENCE OF GEOLOGY

THE YOUNGER DRYAS EVENT AND THE END OF THE LAST ICE AGE

At the end of the last glaciation about 11,000 to 10,000 radiocarbon years ago, the climate in the North Atlantic and adjacent lands experienced a rapid and remarkable change. For 2000 years, the climate had been warming, causing ice sheets in North America and Europe to retreat and allowing plants and animals to reoccupy the deglaciated landscape. Many mountain glaciers in Britain and Scandinavia had disappeared, and the southern limit of sea ice in the North Atlantic had shifted far north, close to its present limit. By all indications, the glacial age was drawing to a rapid close. Then, very abruptly, the climate cooled. Water temperatures in the North Atlantic fell as the southern limit of polar water shifted southward, nearly to its full-glacial extent. The retreating ice sheets halted, then readvanced, and mountain glaciers were reborn in former-ly ice-free cirques. Forests in northwestern Europe were rapidly replaced by low-growing herbaceous plants typical of full-glacial conditions. Among these plants was a distinctive flowering species, *Dryas octopetala*, now limited to polar latitudes and high altitudes. *Dryas* pollen is found abundantly in organic deposits dating to this interval and has provided the name used to identify this cold episode—the Younger Dryas.

Oxygen-isotope data obtained from Swiss lake sediments and a Greenland ice core show that the onset of the Younger Dryas episode was rapid. These records also indicate that the event terminated equally rapidly (Figure B19.1). In fact, the ice core indicates that the climate over Greenland warmed about 7°C in only 40 years, a rate that exceeds even the unusually rapid average global rate of temperature rise that climate models project for the coming century.

The effects of Younger Dryas cooling are most pronounced around the North Atlantic, and so the search for a cause has focused on this region. As pieces of the puzzle have been assembled, it has become clear that the solution likely lies in interactions of Earth's natural systems: the cryosphere, the oceans, the atmosphere, and the biosphere.

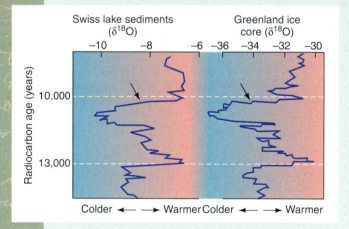

Figure B19.1   The Younger Dryas   Measurements of oxygen isotopes in the sediments of a Swiss lake and an ice core from the Greenland Ice Sheet show an abrupt and rapid change of climate at the end of the last glaciation (arrows). The curves, which can be viewed as recording changes in temperature, show a sudden shift to colder climate followed by an abrupt return to warmer postglacial climate. Detailed studies of the ice core indicate that, at the end of the Younger Dryas event, the average temperature in Greenland abruptly climbed about 7°C in only 40 years.

---

fore buoyant) oceanic lithosphere. As a consequence, Cretaceous mid-ocean ridges were wider than at present, decreasing the average depth of the ocean. If today's midocean ridges were wider, a substantial volume of water would be displaced, sea level would rise, and low-lying continental areas would become shallow seas. This is what happened during much of the Cretaceous.

Earth's climate is a complex part of the Earth system. Because CO_2 is an important greenhouse gas, climate is influenced profoundly by the biogeochemical cycle of carbon. Human activities altered the carbon cycle in the twentieth century, through fossil-fuel consumption and the conversion of forest into farmland. At present, we can conclude from the accumulation of scientific evidence and theory that it is very probable that the climate is warming and will continue to warm as we add greenhouse gases to the atmosphere. More specifically, there is a high probability that average global temperatures ultimately will increase by 2 to 4°C by the end of the 21st century, leading to widespread environmental changes. Food supply and human populations are delicately balanced in many parts of the world. Agriculture is sensitive to climate change, so we may look forward to some difficulties.

It is likely that our civilization will burn all accessible fossil fuels in the next few centuries, though energy conservation and the successful exploitation of offshore *gas hydrates* (p. 520) could extend this time considerably. In the simplest outcome, our greenhouse perturbation could last a

Glacial-geologic studies have shown that as the ice sheet over eastern North America retreated, vast meltwater lakes were ponded beyond the glacier margin. When the retreating ice uncovered a natural drainageway between these lakes and the North Atlantic, meltwater flowed rapidly into the ocean, where it formed a freshwater lid over the denser salty marine water. The cold surface meltwater reduced evaporation from the ocean surface, thereby shutting down the ocean's thermohaline circulation system (Chapter 18). Eastward-flowing air masses traveling across the colder North Atlantic cooled and moved across western Europe, bringing a return to frigid ice-age conditions (Figure B19.2). As the huge meltwater lakes drained and meltwater flow from the ice sheet eventually slowed, ocean circulation resumed and warmer climate returned to the North Atlantic region, heralding the rapid termination of the ice age.

As we learn more about this remarkable natural climatic event, and see how Earth's physical and biological systems responded to it, important insights will be gained that may help us anticipate future environmental changes linked to a rapidly warming world.

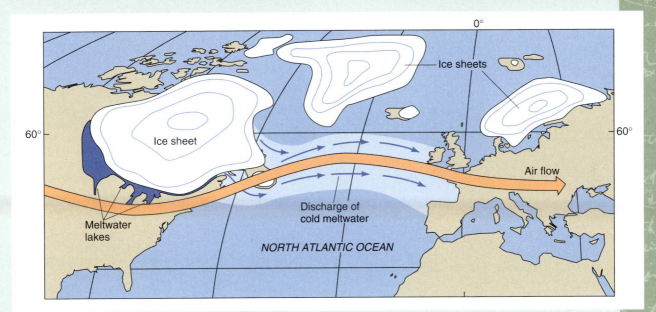

Figure B19.2   Causes of the Younger Dryas Event
Distribution of ice sheets in the North Atlantic region and North American ice-margin lakes during the Younger Dryas event. Rapid drainage of large volumes of meltwater into the western North Atlantic cooled the ocean surface and reduced its salinity, shutting down the thermohaline conveyor system. Air passing over the cold North Atlantic brought colder conditions to northwestern Europe that led to the growth of glaciers and a major change in vegetation communities.

thousand years or more, but ultimately the changing geometry of Earth's orbit would nudge the climate system back into the next glacial age (Figure 19.9). However, we do not yet understand the carbon cycle well enough to know the long-term future of greenhouse gases. Some scientists hypothesize that, if CO_2 concentration rises sufficiently, the next ice age may be aborted. Unfortunately, we do not know the rates of some critical processes. For instance, we know that plants can grow faster in an atmosphere with extra CO_2, but we do not know how rapidly, on a global scale, all or some excess CO_2 might be consumed by the biosphere. In the twentieth century the biosphere consumed roughly half of our CO_2 emissions, but the negative effects of future climate change (e.g., droughts and vegetation changes) are only now starting to emerge. There is another factor to consider: once fossil fuel is scarce, humans may turn to biofuels such as methane and alcohol, and consume biospheric carbon as fast as it can accumulate. Our descendants will have some interesting decisions to make.

Future climate changes are not guaranteed to be gradual. The geologic record of the Younger Dryas event argues that a sudden flood of glacial meltwater in the North Atlantic could shut down part of the ocean's thermohaline circulation. (See *The Science of Geology*, Box 19.2, *The Younger Dryas Event and the End of the Last Ice Age*.) Little more than 10,000 years ago, this caused a sudden transition to near-glacial conditions in Europe. In our time such a scenario could be caused by a rapid melting, or disintegration, of the Greenland ice sheet. It is surprising to associate global warming with a future transition to a *colder* climate, but the marked twentieth-century temperature increase in the high northern latitudes has caused some researchers to consider this scenario seriously. Such a climate change would severely weaken Europe's ability to feed its population.

What's Ahead?

We can reconstruct Earth's environmental conditions during the past 150 million years with greater confidence than for earlier times, because the former positions of most land masses can be determined by reversing the slow, steady spreading that occurs at today's midocean ridges. To estimate the locations of

continents before the Atlantic Ocean split the Americas from Europe and Africa, however, geologists need different observations, and must hypothesize the existence of now-vanished oceans. Traces of Earth's magnetic field are preserved in many ancient surface rocks, and these help determine past locations of continents. Geologists often find small slivers of former oceanic crust within the present continents and old, eroded mountain ranges that mark the remains of ancient subduction zones. Using present motions as a guide, geologists can glimpse how plates have formed and moved during most of Earth's history, over a span of nearly 4 billion years. The next chapter discusses how this is done.

CHAPTER SUMMARY

1. Earth's climate system involves the atmosphere, hydrosphere, lithosphere, and biosphere. Changes affecting it operate on time scales ranging from decades to millions of years.

2. Using information from the geologic record, geologists can measure the magnitude and geographic extent of past climatic changes, determine the range of climate variability on different time scales, and test the accuracy of computer models that simulate past climatic conditions.

3. The carbon cycle is among the most important of Earth's biogeochemical cycles. Carbon resides in the atmosphere, the biosphere, the hydrosphere, the crust and the mantle. It cycles through these reservoirs at different rates.

4. The anthropogenic extraction and burning of fossil fuels perturb the natural carbon cycle and have led to an increase in atmospheric CO_2 since the start of the Industrial Revolution.

5. The greenhouse effect, caused by the trapping of long-wave infrared radiation by water vapor and trace gases in the atmosphere, makes Earth a habitable planet.

6. The increase in atmospheric trace gases (CO_2, CH_4, O_3, N_2O, and the CFCs) due to human activities is projected to warm the lower atmosphere by 2° to 4°C by the end of the 21st century.

7. A 0.6°C increase in average global temperature since the mid-nineteenth century very likely represents the initial part of

this warming. The rate of warming is likely to reach 0.3°C per decade and may lead to a "super-interglaciation," making Earth warmer than at any time in human history.

8. Potential physical and biological consequences of global warming include worldwide changes in precipitation and vegetation patterns; increased storminess; melting of glaciers, sea ice, and frozen ground; worldwide rise of sea level; local and regional changes in the hydrologic cycle; and increased rates of organic decomposition in soils.

9. Evidence of past intervals of rapid environmental change in the geologic record and reconstructions of past warmer intervals can provide insights into physical and biological responses to global warming. Such reconstructions also permit evaluation of computer models that simulate climate.

10. Viewed from the geologic perspective, the enhanced-greenhouse interval will be a brief perturbation in Earth's climate history. On a human time scale, however, negative climate impacts can be expected to persist for many generations.

11. Synthetic chlorofluorocarbon (CFC) gases entering the upper atmosphere break down and release chlorine, which destroys the protective ozone layer. Discovery of a vast and recurring ozone hole over Antarctica has led to international efforts to eliminate CFC production by the end of the twentieth century.

THE LANGUAGE OF GEOLOGY

boundary conditions (p. 518)

chlorofluorocarbons (CFCs) (p. 515)

gas hydrates (p. 520)
general circulation model (p. 518)
greenhouse effect (p. 513)

QUESTIONS FOR REVIEW

1. Describe the carbon cycle and indicate why we regard it as one of the most important biogeochemical cycles.

2. In what ways can carbon be trapped on Earth and become

part of the rock cycle? How can such stored carbon once again find its way into the atmosphere?

3. If atmospheric CO_2 can be dissolved in streams, lakes,

groundwater, and the oceans, and also efficiently absorbed by vegetation, why is the burning of fossil fuels causing the CO_2 content of the atmosphere to increase?

4. How is Earth's atmosphere similar to a garden greenhouse, and why?

5. What are the anthropogenic sources of the principal greenhouse gases?

6. What geologic evidence indicates that the present concentrations of carbon dioxide and methane in the atmosphere are exceptional compared to those of the last several hundred thousand years?

7. What are the major boundary conditions that must be specified in GCM climate simulation experiments? How have these conditions changed since the maximum of the last glaciation?

8. Why is there a significant range in the predictions of the rise in average global temperature in the next century?

9. Give an example of an environmental effect arising from global warming that could enhance the greenhouse effect and lead to additional warming.

10. What factors are likely to cause world sea level to rise in a warming climate? In what ways is rising sea level likely to impact the human population?

11. Why is the geologic record important in helping predict the environmental effects of greenhouse warming? Give two examples.

12. Why does chlorine have such an adverse effect on the ozone layer, despite the fact that it is released into the atmosphere in very small amounts?

Click on *Presentation* and *Interactivity* in the **Glaciers, Glaciations, and Ice Sheets** module of your CD-ROM to further explore resources and activities presenting concepts from this chapter. Select *Assessment* in the same module to test your understanding of this chapter.

Chapter 20

Mistaya River, Alberta, Canada drains Peyto Lake (foreground). The lake is fed by meltwater from Peyto Glacier. The valley shape is evidence that Peyto Glacier was once more extensive. Simpson Pass Thrust fault runs down the center of the valley. Quartzites (right) have been thrust from left to right. Rocks on Mistaya Mountain (left) are Cambrian limestones.

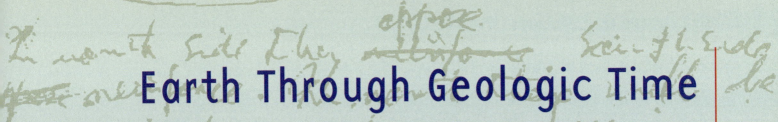

GPS receiver wrapped in plastic stands beside its antenna.

Watching a Continent Grow in Kamchatka

In 1998 I helped to install seismometers in Kamchatka, a rugged peninsula on the Pacific coast of Russia. Roughly as large as Arizona, Kamchatka is home to dozens of active volcanoes, most in the southern half of the peninsula above the subducting Pacific Plate. In northern Kamchatka subduction has stalled and the volcanoes are dormant. The boundary between active and inactive subduction lies near the tiny coastal settlement of Kruto-Berogovo, where we installed a seismometer in an underground vault behind a family home. In the photograph, the wooden shed above the vault sits beside a continuous GPS recorder and its antenna, which has been raised high enough to surmount the snows of the long Russian winter.

From the site, I could look east to a small mountain range on a large cape, known as Cape Kamchatka, that juts outward from Kruto-Berogovo into the Pacific Ocean. The rocks of the cape match the rocks of the Aleutian Islands, a volcanic arc that marks the boundary of the Pacific Plate from Alaska to Kamchatka. Geologists hypothesize that Cape Kamchatka was once an Aleutian island, but is now merging with the Eurasian mainland. Earthquakes and the movements on active thrust faults confirm this hypothesis. Two other capes south of Kruto-Berogovo also have rocks

of similar age and composition. Presumably they were once islands too, but their collisions with Kamchatka must have occurred long ago, because their thrust faults are inactive.

Throughout easternmost Russia, geologists have mapped continental fragments that resemble old pieces of island arc. Careful dating shows that the Eurasian continental margin has accumulated bit by bit from the flotsam of old plate boundaries. The process is still going on. I hypothesize that some 20 to 30 million years ago, the first Aleutian island collided with Kamchatka, splitting the subduction zone into north and south segments. The seismic data we collected shows that the slab beneath the north segment is missing—it must have broken off and sunk into the deep mantle when subduction stalled. GPS measurements from the other Aleutian Islands tell us that they continue to drift toward Kamchatka, and will eventually become part of Eurasia. The evidence is clear: continents can grow by adding volcanic arcs to their margins.

Jeffrey Park

KEY QUESTIONS

1. **Did the supercontinent Pangaea really exist?**

2. **Are Earth's poles fixed or do they wander?**

3. **How is it proved that seafloor spreads?**

4. **What features characterize ancient plate margins?**

5. **Were there supercontinents before Pangaea?**

6. **What are the key parts of a continent?**

INTRODUCTION: TRACKING PAST PLATE MOTIONS

Plate tectonics is the framework within which many of the topics of this book have been addressed. Plate tectonics was introduced in Chapter 1, then expanded and discussed in detail in Chapter 2. The evidence we cited in Chapter 2 in support of plate motions came from measurements made with the Global Positioning System (GPS). That evidence proves that Earth's surface is moving laterally today, and that large areas of Earth's surface are moving together in tandem, suggesting the drift of stiff surface plates. The boundaries of plates are where we find most earthquakes and most volcanic activity. By recycling the rocks, sediments, and volatile compounds of Earth's surface, plate tectonics plays a critical role in the Earth system.

Evidence of past plate motions is not obtained through GPS. Rather, evidence must be found and deciphered from rocks affected by past plate motions. This evidence includes great arc-shaped belts of metamorphic rocks formed by continental collisions, the eroded remains of island-arc volcanic rocks, and, most important, traces of Earth's past magnetic field preserved in old lava flows and certain sedimentary rocks. The evidence suggests that plates have been moving and changing Earth's surface for at least 2 billion years.

In this chapter we discuss plate tectonics of the past. We first turn to the evidence on which the continental drift hypothesis, the predecessor to plate tectonics theory, was proposed, then turn to magnetic evidence of plate motions, evidence of past plate boundaries, and finally the structure of continents and how they have grown as a byproduct of plate motions.

FORMER IDEAS ABOUT CONTINENTS

People wondered for a long time why continents have such irregular shapes and why ocean basins, mountain ranges, earthquake belts, and many other features occur where they do. When the first maps were made of the coastlines on either side of the Atlantic in the sixteenth century, it became apparent that the coasts were approximately parallel. People started to speculate why. They thought about a flood having cut an immense canyon—perhaps the great biblical flood. No realistic answers were forthcoming, but such speculations did get people thinking about why Earth is the way it is. Scientists eventually began to think that there might be a single, underlying cause for the whole array of Earth's major features. But what could that cause possibly be?

During the nineteenth century, the favored idea was that Earth was originally a molten mass that is cooling and contracting, with the crust being gradually compressed. The hypothesis was that mountain ranges full of folded strata are the places where past contraction has occurred, and seismic belts are places where contraction is happening at present. Contraction did explain some features, but it did not help with questions about the shapes and distribution of continents. Nor did it explain the great rift valleys and other features that are clearly caused by the crust having been stretched rather than compressed.

When scientists discovered at the beginning of the twentieth century that Earth's interior is kept hot by radioactive decay, some of them suggested that Earth might not be cooling but heating up (and therefore expanding). A much smaller Earth, they suggested, could once have been covered largely by continental crust. Heating would cause Earth to expand, and the continental crust would then crack into fragments. As expansion continued, the cracks would grow into ocean basins, and through the cracks basaltic magma would rise up from the mantle to build new oceanic crust. The theory of an expanding Earth does offer a plausible explanation for the approximately parallel coastlines of adjacent continents, but it does not easily account for mountain ranges formed by compression.

To get around the flaws in both the expansion theory and the contraction theory, geologists began to examine the effects of other forces on the crust. By the middle of the twentieth century, however, all reasonable suggestions concerning the shapes and positions of continents seemed to have been exhausted. The time was ripe for a totally new approach. The new approach turned out be to plate tectonics. When great slabs of lithosphere—called plates—slide sideways across the asthenosphere, some parts of the slabs can be in compression, others in tension (that is, being pulled apart). When a plate splits in two, the broken edges of continental crust match perfectly. The energy needed to move plates turns out to be Earth's internal heat energy, which causes great convective flows in the mantle.

Plate tectonics is the only theory ever developed that explains *all* of Earth's major features. A key proposal leading to the formulation of the theory was made early in the twentieth century, soon after the contraction theory collapsed. As we learned in Chapter 2, the German meteorologist Alfred Wegener proposed in 1912 that continents drift slowly across the surface of Earth, sometimes breaking into pieces and sometimes colliding with each other. Collisions formed supercontinents, and breakages formed smaller continents. According to Wegener, today's continents are the broken fragments of the most recent supercontinent.

PANGAEA

Wegener's theory of continental drift originated when he attempted, like many before him, to explain the striking match of the shorelines on the two sides of the Atlantic, especially along Africa and South America. Wegener suggested that the most recent supercontinent existed during the Permian period when all the world's landmasses were joined together in a single continent, which he dubbed Pangaea (pronounced *Pan-jée-ah*, meaning "all lands") (Figure 20.1A). The northern half of Pangaea is called Laurasia, the southern half Gondwanaland. Laurasia is a name derived from Laurentia, an old name for the Precambrian core of Canada, and from Eurasia, a combined term for Europe and Asia. Gondwanaland is a name derived from a distinctive group of plant fossils found in central India. The fossils are named for the land of the Gonds, a group of people living in central India. Similar fossils are found in Africa, Antarctica, Australia, and South America—this is one of the bits of evidence that suggest that India and today's Southern Hemisphere continents were once part of the same landmass. According to Wegener's hypothesis, Pangaea was somehow disrupted during the Mesozoic Era, and its fragments (the continents of today) slowly drifted to their present positions. Proponents of the theory likened the process to the breaking up of a sheet of ice that floats in a pond. The broken pieces, they argued, should all fit back together again, like pieces of a jigsaw puzzle. Figure 20.1A shows that a jigsaw reconstruction indeed works well.

One impressive line of evidence presented by Wegener that supports the former existence of Pangaea is that during the Late Carboniferous Period, about 300 million years ago, a continental ice sheet covered parts of South America, southern Africa, India, and southern Australia (Figure 20.1B). However, if 300 million years ago continents were in the positions they occupy today, an ice sheet would have had to cover all the southern oceans and in places would even have had to cross the equator. Such a huge ice sheet could mean only that the world climate was exceedingly cold. Yet if the climate were cold, why has no evidence of glaciation at that time ever been found in the Northern Hemisphere? In the Northern Hemisphere thick coal measures are evidence of warm, tropical climates. This dilemma is explained neatly by continental drift: 300 million years ago, the regions covered by ice lay in high, cold latitudes surrounding the south pole and North America and Eurasia were close to the equator (Figure 20.1A). No landmass covered the north pole, however, so there was no northern ice sheet. At that time, therefore, the Earth's climates need not have been greatly different from those of today. We will discuss other aspects of the Carboniferous glaciation later in this chapter.

Despite the impressive evidence supporting continental drift, many scientists remained unconvinced by Wegener's

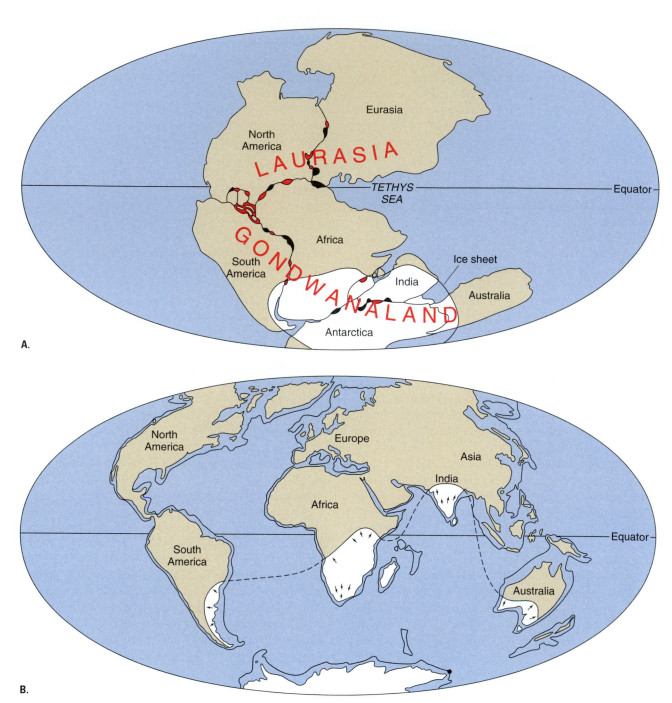

Figure 20.1 Pangaea The continents attained their present shapes when Pangaea began to break apart 200 million years ago. A. The shape of Pangaea, determined by fitting together pieces of continental crust along a contour line 2000 m below sea level. (This is the line halfway down the continental slope, along which continental crust meets oceanic crust.) In a few places in this drawing, some overlap (black) occurs; elsewhere, small gaps (red) are found. These are places where post-breakup events have modified the shapes of the continental margins. The white area is the region affected by continental glaciation 300 million years ago. B. Present continents and the 2000-m contour below sea level. The white areas are where evidence of the old ice sheets exists. Arrows show directions of movement of the former ice. The dashed line joining the glaciated regions indicates how large the ice sheet would have to have been if the continents were in their present positions at the time of glaciation.

ideas, largely because no one could explain how the solid rock of a continent could possibly overcome friction and slide across the oceanic crust. The process is like trying to slide two sheets of coarse sandpaper past each other.

APPARENT POLAR WANDERING

Wegener died in 1930, and although debate continued, its pace slowed down. A turning point came in the 1950s. From the mid-1950s to the mid-1960s, geophysicists made a number of remarkable discoveries. The first arose through studies of paleomagnetism. As we discussed in Chapter 11, certain igneous and sedimentary rocks can become weakly, but permanently magnetized and therefore preserve a fossil record of Earth's magnetic field at the time and place the rocks formed. Three essential bits of information are contained in that fossil magnetic record. The first is Earth's polarity—whether the magnetic field was normal or reversed at the time of rock formation. The second is the location of the magnetic poles at the time the rock formed. Just as a free-swinging magnet today will point toward today's magnetic poles, so too does paleomagnetism record the direction of the magnetic poles at the time of rock formation. The third piece of information, and the one that provides the data needed to say how far from the point of rock formation the magnetic poles lay, is the magnetic inclination, which is the angle with the horizontal assumed by a freely swinging bar magnet (Figure 20.2). Note in Figure 20.2 that the magnetic inclination varies regularly with latitude, from zero at the magnetic equator to 90° at the magnetic pole. The paleomagnetic inclination is therefore a record of the place between the pole and the equator (that is, the **magnetic latitude**) where the rock was formed.

In the 1950s, geophysicists studying paleomagnetic pole positions found evidence suggesting that magnetic poles had moved large distances over the globe. They referred to the strange plots of paleopole positions as *apparent polar wandering*. This evidence puzzled geophysicists because Earth's magnetic poles and the poles of rotation are close together. Determination of the magnetic latitude of any rock should therefore be a good indication of the geographic latitude at which the rock was formed. When it was discovered that the path of apparent polar wandering measured in North America differed from that in Europe (Figure 20.3), geophysicists were even more puzzled. Because no simple motion of the magnetic poles could explain all data, they concluded, somewhat reluctantly, that the continents and the magnetized rocks had moved instead. Paleomagnetic evidence revived Wegener's hypothesis of continental drift, but still did not provide scientists with a mechanism to explain how the movement occurred.

SEAFLOOR SPREADING

Help came from an unexpected quarter. All the early debate about continental drift, and even the data on apparent polar wandering, had centered on evidence drawn from the continental crust. But if continental crust moves, why shouldn't oceanic crust move too?

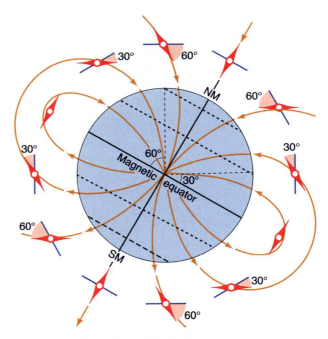

Figure 20.2 Magnetism and Latitude Change of magnetic inclination with latitude. The solid red diamonds show the magnetic inclinations of a free-swinging magnet. The solid blue line indicates a horizontal surface at each point.

In 1962, Harry Hess of Princeton University hypothesized that the topography of the seafloor could be explained if the seafloor moves sideways, away from the

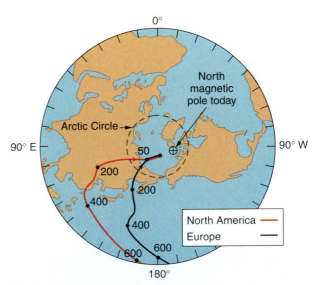

Figure 20.3 Apparent Polar Wandering Curves tracing the apparent path followed by the north magnetic pole through the past 600 million years. Numbers are millions of years before the present. The curve determined from paleomagnetic measurements in North America (red) differs from that determined from measurements made in Europe (black). Wide-ranging movement of the pole is unlikely; therefore, geologists conclude that it was the continents, not the pole, that moved.

oceanic ridges. His hypothesis came to be called the theory of **seafloor spreading** and was soon proved correct. Once again it was paleomagnetism that provided the proof.

Hess postulated that magma rose from Earth's interior and formed new oceanic crust along the midocean ridges. He could not explain what made the crust move away from the ridges, but he nevertheless proposed that it did and that as a consequence the oceanic crust far from any ridge was older than any crust nearer the ridge. Three geophysicists, Frederick Vine, Drummond Matthews, and Lawrence Morley, proposed a test of Hess's hypothesis that used the magnetism of the oceanic crust rather than its age. Radiometric dating of the ocean crust required dredging and drilling the seafloor for rock samples. The magnetism of the ocean crust could be determined with measurements at the sea surface, and is much easier to measure in detail over broad areas of the ocean.

When lava is extruded at any midocean ridge, the rock it forms becomes magnetized and acquires the magnetic polarity that exists at the time the lava cools. If new lava continuously generates new oceanic crust, and if the crust continuously moves away from the oceanic ridge, then this crust should contain a continuous record of the Earth's changing magnetic polarity. The oceanic crust is, in effect, a very slowly moving magnetic tape recorder. In fact, two oceanic tape recorders commence at each midocean ridge,

one on each side of the ridge. Successive strips of oceanic crust are magnetized with normal and reversed polarity (Figure 20.4), so it is a straightforward matter to match the sort of magnetic pattern observed in Figure 20.4 with a record of magnetic polarity reversals, such as that shown in Figure 11.19. The magnetic striping allowed the age of any place on the seafloor to be determined.

Vine, Matthews, and Morley based their test of Hess's seafloor spreading hypothesis on existing surveys of seafloor magnetism, which had indeed showed stripes of normal and reversed polarity parallel to midocean ridges. Once the ages of magnetic polarity reversals were determined from rock samples, magnetic striping also provided a means of estimating the speed with which the seafloor had moved. Wider stripes implied faster spreading. In some places, such movement was found to be remarkably fast: as high as 10 cm/yr.

MAGNETIC RECORDS AND PLATE VELOCITIES

The most recent magnetic reversal occurred 780,000 years ago (Figure 11.19), and can be found at midocean ridges worldwide. The oldest reversals so far found in oceanic crust date back to the middle Jurassic, about 175 million years ago. From the symmetrical spacing of magnetic time lines on the two sides of nearly all midocean ridges (Figure 20.5), it appears that both plates move away from a spread-

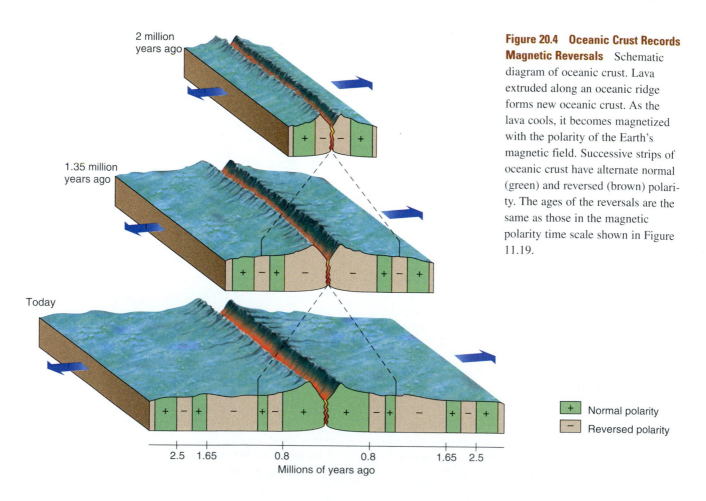

Figure 20.4 Oceanic Crust Records Magnetic Reversals Schematic diagram of oceanic crust. Lava extruded along an oceanic ridge forms new oceanic crust. As the lava cools, it becomes magnetized with the polarity of the Earth's magnetic field. Successive strips of oceanic crust have alternate normal (green) and reversed (brown) polarity. The ages of the reversals are the same as those in the magnetic polarity time scale shown in Figure 11.19.

2 million years ago

1.35 million years ago

Today

☐ Normal polarity
☐ Reversed polarity

2.5 1.65   0.8   0.8   1.65 2.5
Millions of years ago

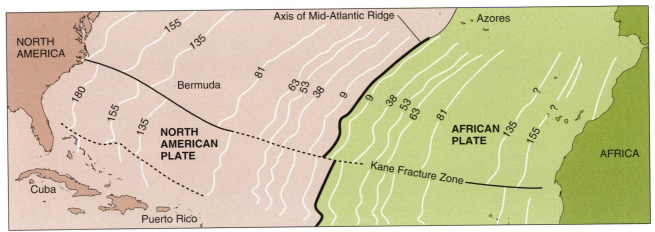

Figure 20.5 Age of North Atlantic Age of the ocean floor in the central North Atlantic, deduced from magnetic striping. Numbers give ages in millions of years before the present. The Kane Fracture Zone, which is the extension of a transform fault, continues across the Atlantic and causes consistent displacement of the age contours.

ing center at equal rates. Appearances can be deceiving, however, because the spreading center can drift as well. All that can be deduced from magnetic time lines is the *relative velocity* of two plates. An answer to the question of *absolute velocities* requires more information, either from GPS measurements (for present-day plate motions) or hot spot tracks (for past plate motions).

VARIATIONS IN PLATE VELOCITIES

Some plates move much faster than others (Figure 20.6). The differences in speed appear to be related to the amount of continental lithosphere in a plate. Plates with only oceanic lithosphere tend to have high relative velocities. This is the case for the Pacific and Nazca plates. Plates with lots of thick continental lithosphere, such as the African, North American, and Eurasian plates, have low relative velocities.

A second reason plate velocities vary has to do with the geometry of motion on a sphere. One might think, intuitively, that all points on a plate move with the same velocity relative to Earth's deep interior, but that is incorrect. Our intuition would be correct if plates of lithosphere were flat and moved over a flat asthenosphere, like plywood floating on water. However, plates of lithosphere are pieces

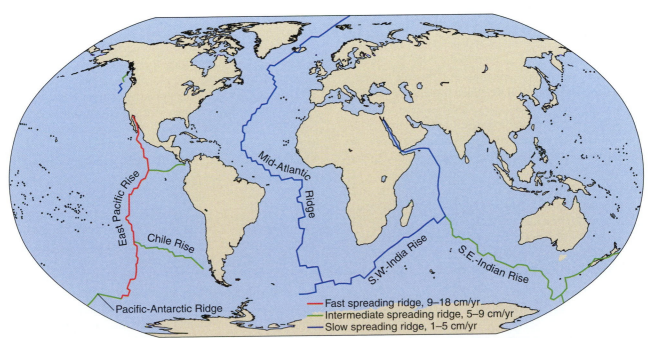

Figure 20.6 Spreading Rates Spreading rates of principal midoceanic ridges. Fast spreading rates mean plates move away from each other between 9 and 18 cm/yr. Intermediate rates are 5 to 9 cm/yr; slow rates are 1 to 5 cm/yr.

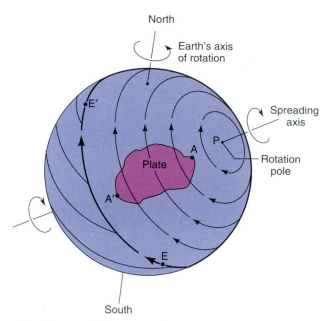

Figure 20.7   Movement on a Sphere   Movement of a curved plate on a sphere. The movement of each plate of lithosphere on Earth's surface can be described as a rotation about the plate's own spreading axis. Point P has zero velocity because it is the fixed point around which rotation occurs. Point A´, at the edge of the plate closest to the equator of the spreading axis, EE´, has a high velocity. Point A, closest to the pole of the spreading axis, has a low velocity.

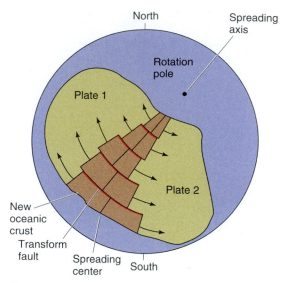

Figure 20.8   Geometry of Plate Motions   Relationship between spreading axis, oceanic ridge, and transform faults in two adjacent plates. Plates 1 and 2 have a common spreading center (black) displaced by transform faults (red). Each segment of the oceanic ridge lies on a line of longitude that passes through the rotation pole. Each transform fault lies on a line of latitude with respect to the rotation pole. The width of new oceanic crust increases away from the rotation pole.

of a shell on a spherical Earth; they are curved, not flat. In the geometry of a sphere, any movement on the surface is a rotation about an axis of the sphere. A consequence of such rotation is that different parts of a rigid plate move with different velocities relative to the underlying material, as shown in Figure 20.7.

Plate A in Figure 20.7 moves independently of Earth's rotation and instead rotates about an axis of its own. In the figure, point P, where the spreading axis reaches the surface, is called the **rotation pole.** One consequence of different plate velocities is that the width of new oceanic crust bordering a spreading center increases with distance from the rotation pole (Figure 20.8). A second consequence is that the projection of a spreading center passes through the rotation pole. Such a projection is analogous to a line of longitude. A third consequence is that each transform fault lies on a line analogous to a line of latitude around the rotation pole. Geologists use these geometric properties to determine the variations in past plate motions from the clues recorded in the magnetic stripes, transform faults, and fracture zones of the seafloor.

RELICT PLATE BOUNDARIES IN THE GEOLOGIC RECORD

Seafloor magnetic stripes help us reconstruct plate motion only as far back in time as the Jurassic, some 175 million

years ago. All large expanses of older oceanic crust have been subducted back into the mantle at convergent plate boundaries. The paleomagnetism of continental rock can be used to follow plate motion further back in time, but without the breadth and continuity of the seafloor data. Continental lava flows occur at infrequent intervals in the geologic record, and are found at scattered locations. Apparent polar wander paths for different locations within a single continent often become separate as one goes farther back in time, indicating that today's continents were assembled from many distinct plates or plate fragments. Small fragments of continental crust that have drifted as a single unit in Earth history are called *terranes*. Indeed, geologists often look for rock formations that they can identify as belonging originally to plate boundaries, but which now form sutures between terranes. We describe several of these below.

OPHIOLITES

As we learned in Chapter 4, the igneous rock formed at a spreading center, known as *midocean ridge basalt* (MORB), has a distinctive chemistry that geologists can look for in other places on Earth. When MORB is found on land, it usually lies within a body of rock that resembles a fragment of oceanic crust that was caught up in a continental collision. The minerals that characterize basalt, if buried deep within the collision zone, transform into an assemblage dominated by a distinctive green fibrous mineral called *serpentine* (Figure 20.9). As first mentioned in

Figure 20.9 Serpentine Reveals an Old Collision Zone
Serpentine is formed by alteration of olivine and pyroxene present in the parent rock. This specimen is from an ophiolite in Marin County, California. The green color is typical of the mineral serpentine, and the shattered nature of the rock reflects the grinding and crushing of rock caught in a subduction zone. The photographed area is about 30 cm across.

Chapter 4, these serpentine-dominated fragments of oceanic crust found on continents are called ophiolites from the Greek word for serpent, *ophis*.

Ophiolites tend to be quite similar wherever they are found, and their structure matches well the crustal structure expected at a midocean ridge. Figure 4.23 is an idealized cross-section of an ophiolite. At the top is a thin veneer of sediment that was deposited on the ocean floor. Beneath the sediment is a layer of pillowed basalt—pillows indicate that the basaltic lava erupted under water. Still deeper are sills of gabbro, the plutonic equivalent of basalt. Cutting through the gabbro sills and basalt pillows are numerous vertical dikes of gabbro. The basalts and

gabbro are chemically similar to MORB and clearly formed from the same partial melting event. The gabbro dikes are probably the former conduits that fed the eruptions that formed the basalt pillows.

Many ophiolites also contain the apparent source rocks of the basalts and gabbros. Beneath the gabbro sills there is often a layer of *peridotite*, the dense ultramafic rock that forms most of the upper mantle. The contact between gabbro and peridotite is interpreted to be the former Moho at the base of what was formerly oceanic crust. The presence of an ophiolite at the boundary between two continental terranes is strong evidence that an ocean once lay between them.

SUBDUCTION MÉLANGE AND BLUESCHISTS

Many features on Earth's surface occur as a result of deformation along convergent margins. A distinctive feature of some margins is the development of a **mélange**, a chaotic mixture of broken, jumbled, and thrust-faulted rock. Once a subduction zone forms and a seafloor trench is created, sediment accumulates in the trench. A sinking plate drags the sedimentary rock formed from this accumulated sediment downward beneath the overriding plate. Sedimentary rock has a low density. As a result, it is buoyant and cannot be dragged down very far. Caught between the overriding plate and the sinking plate, the sediment becomes shattered, crushed, sheared, and thrust-faulted to form a mélange (Figure 20.10). As the mélange thickens, it undergoes metamorphism. The cold sedimentary rock can be dragged down so rapidly that it remains cooler than adjacent rock at the same depth. The kind of metamorphism that is common in many mélange zones, therefore, is that which occurs along curve C of Figure 8.16, a high-pressure, low-temperature metamorphism distinguished by blueschists and eclogites. The blue color of a blueschist comes from a bluish amphibole called glaucophane. The presence of blueschists clearly indicates a former subduction zone.

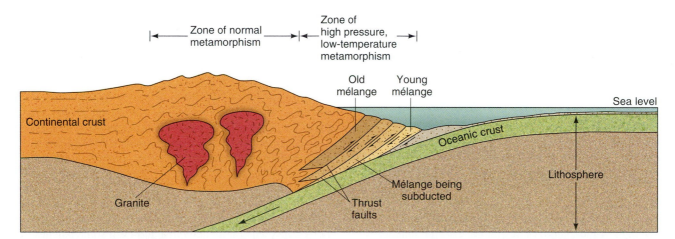

Figure 20.10 Mélange of Broken Rock A mélange is formed when young sediment in a trench is smashed by moving lithosphere and dragged downward in slices bounded by thrust faults. As successive slices are dragged down, older mélange, closer to the overriding plate, is pushed back up. The process is like lifting a deck of cards by adding new cards at the base of the deck.

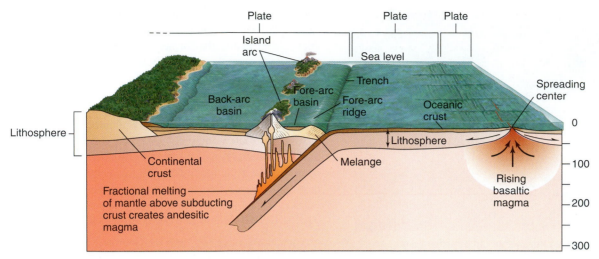

Figure 20.11   Features of a Convergent Plate Margin   Structure of a tectonic plate at a convergent margin. Lithosphere is capped by oceanic crust formed by basaltic magma rising from the asthenosphere. Moving laterally, the lithosphere accumulates a thin layer of marine sediment and eventually starts sinking into the asthenosphere. Along the line of subduction, an oceanic trench is formed, and sediment deposited in the trench, plus sediment on the moving plate, is compressed and deformed to create a mélange. The sinking oceanic crust eventually reaches the temperature where wet partial melting commences and forms andesitic magma, which then rises to form an island arc of stratovolcanoes on the adjacent plate. Behind the island arc, tensional forces lead to the development of a back-arc basin.

BACK-ARC BASINS AND PLATE EXTENSION

When the sinking rate of a subducting plate is faster than the forward motion of the overriding plate, the margin of the overriding plate can be subjected to tensional (pulling) stress. The leading edge of the overriding plate must remain in contact with the subduction edge or else a huge void will open. Instead, the overriding plate stretches at a rate equal to the difference in velocities between the two plates. If the overriding plate is continental, this process most commonly starts with a thinning of the crust as the plate edge stretches. If the mantle beneath is hot enough to melt partially, this process of continental extension may involve the accumulation of a layer of basaltic magma in the lower crust. Extended continental margins are found at several locations along the Pacific coast of Eurasia, for instance, in the lithosphere that underlies the Okhotsk Sea and the Sea of Japan.

If the overriding plate is oceanic, or if the extension of a continental margin has progressed to an extreme state, an arc-shaped basin forms behind and parallel to the magmatic arc of the subduction zone (Figures 20.11 and 20.12). Basaltic magma may rise into such a **back-arc basin** at a newly formed spreading center, and new oceanic crust may form. The largest current example of back-arc spreading is the Philippine Sea Plate in the western Pacific Ocean. This oceanic plate has formed as a result of the oldest, coolest lithosphere of the Pacific Plate sinking into the mantle.

Continental extension and back-arc spreading seem, somewhat surprisingly, to be important features of some continental collisions. The Mediterranean Sea, where the African and Eurasian plates converge, is underlain partially by extended continental crust and even some patches of recently formed oceanic lithosphere. The Mediterranean example has inspired geologists to reinterpret the geology of some ancient plate collisions. For example, roughly 400

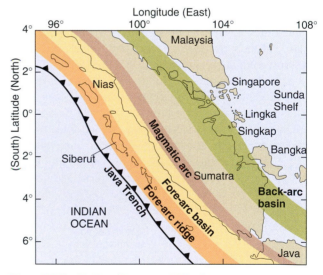

Figure 20.12   Modern Convergent Margin   Map of portion of Sumatra showing the positions of the major topographic features in a present-day convergent plate boundary. Compare with Figure 20.11. The fore-arc ridge is often underlain by mélange. The fore-arc and back-arc basins both tend to be filled with sediment derived from the magmatic arc.

million years ago the supercontinent of Pangaea formed with the closure of the Iapetus Ocean along a long suture zone that, in North America, follows the present-day Appalachian mountain range. Canadian geologists now recognize a variety of unusual terranes in Quebec, Nova Scotia, and New Brunswick that suggest a complex collision process in which both marginal basins and mountain ranges formed.

Before you go on:

1. How was paleomagnetic evidence used to prove that the seafloor moves away from a midocean ridge?

2. Why does the velocity differ from place to place on a tectonic plate?

3. What evidence can be used to identify former convergent plate boundaries?

SUPERCONTINENTS AND VANISHED OCEANS

The life cycle of an oceanic plate is the principal story of plate tectonics. It is not the only story. As an oceanic plate drifts slowly from a spreading center rift to a subduction zone, pieces of continental lithosphere drift with it. Although continental lithosphere is relatively cool compared to oceanic lithosphere, the average composition of continental lithosphere is quartz-rich compared to olivine- and pyroxene-rich oceanic lithosphere. As a result, continental lithosphere is always less dense than oceanic lithosphere and is not subducted.

The structure of continental lithosphere remains something of a puzzle. Seismic evidence suggests that a root of mantle rock is attached to the base of old, cool continental crust. Many geologists hypothesize that continental buoyancy can be enhanced by the detachment and loss of this dense mantle root during the final stages of a continental collision. We encountered this hypothesis in Chapter 12 as one mechanism for tectonic uplift. However, there is evidence from seismic waves for the presence of former slabs "frozen" into the mantle beneath some ancient continental sutures. If a slab survives long after a continental collision is complete, we must conclude that the mantle root of the continent has not been lost, at least at the sutures where an ancient slab can be detected.

The unceasing drift of the plates tends to sweep together patches of lithosphere that are too buoyant to subduct back into the deeper mantle. Modern-day examples for such patches of buoyant lithosphere within an oceanic plate include island arcs and the earliest (and most intense) volcanism associated with a mantle plume. Welded together into small continental fragments, these larger patches of

lithosphere have, over Earth history, repeatedly sutured together to form supercontinents. If you refer back to Figure 2.18, note the broad expanses of high-standing Earth topography compared to the irregular patches of high topography on our sister planet Venus. The high-standing topography on Venus probably corresponds to thicker crust that, like Earth's continental crust, is more evolved in its chemical content and mineral assemblage. Because plate tectonics does not appear to operate on Venus, it is likely that each patch of its high-standing topography has remained where it was first formed.

Although plate movement tends to bunch continental fragments together, the heat of the mantle beneath still must escape Earth's interior. A supercontinent impedes heat flow from the deep mantle to the surface, effectively forming a layer of insulation. Convective motion within Earth's mantle will accumulate heat and cause thermal buoyancy at the base of the continental lithosphere. The heat has nowhere to go but up, so the supercontinent lithosphere eventually warms, softens, and begins to rift. The opening of the Atlantic Ocean was heralded by the eruption of large basalt flows at points on what would later become the Atlantic coastline. In some cases, these eruptions persisted at a much reduced level of magmatic output, and became hot spots. One example is the Parana flood basalts in Brazil, which now connect with a hot spot near the southern Mid-Atlantic Ridge via a hot spot track that is traced by a ridge in the seafloor.

We have good evidence for two supercontinents in the past: Pangaea, the Phanerozoic-aged supercontinent proposed by Wegener, and Rodinia, an earlier Proterozoic landmass. Pangaea was formed roughly 350 million years ago with the closure of the Iapetus Ocean and broke apart with the first opening of the Atlantic Ocean nearly 200 million years ago. Rodinia formed roughly 1100 million years ago and split apart roughly 750 million years ago, but we are less certain of these dates than we are of the Pangaea dates.

THE FORMATION OF PANGAEA

The history of the formation of Pangaea has been deciphered from a combination of paleomagnetic data, studies of old regions of continental collision, and indicators of environmental change recorded in sedimentary rocks. Continental fragments that are sutured together to form a supercontinent will share a common path of apparent polar wander. The sutures between continental fragments can be found in the many mountain ranges and belts of ophiolites that formed in the late Paleozoic Era, roughly between 450 and 350 million years ago. Examples include the Appalachian mountains in eastern North America, the Atlas mountains in Morocco, the Ural mountains in Russia that divide Europe from Asia, and the Hercynian mountains in Europe. Each of these mountain ranges is heavily eroded now, in some locations exposing rocks that once experienced the pressures and temperatures of metamorphism that are characteristic of the lower continental crust.

The collisions that formed Pangaea happened in stages as fragment after fragment accreted. Extensive volcanism and the formation of large volumes of granitic magma accompanied many of the accretions. The uplift of the Appalachians in present-day New England, for example, continued for tens of millions of years after the Iapetus Ocean had closed, indicating that the collision zone was hot and magma was generated for a long time.

As Pangaea formed, sediment sequences on continental shelves around the world record evidence that global sea-level fell, draining shallow continental seas and exposing large areas of continental shelves. Geologists interpret the sea level drop as evidence for a general slowdown in plate movement and, particularly, a slowdown in the rate of formation of young, buoyant oceanic lithosphere at midocean ridges. This meant that the oceanic lithosphere at the time of Pangaea was, on average, older, colder, and denser than oceanic lithosphere had been before Pangaea. Shorelines would regress outward onto the continental shelves as seawater receded to fill a deeper world ocean. The coastline of Pangaea was quite irregular. A large notch in Pangaea at equatorial latitudes is known as the Tethys Sea (Fig 20.1A). This ocean persisted, in one form or another, for some 300 million years, and a fragment still survives in the form of the Mediterranean Sea.

Pangaea lasted as a supercontinent for at least 150 million years, but eventually, starting in the late Triassic, thermal stresses caused breakup to start. New oceans formed—the Atlantic and Southern oceans both came into being, and movement of fragments of continental crust—today's continents—commenced. The world is still in a phase of continental dispersion, but sometime in the future, perhaps 150 million years, a new supercontinent will form.

RODINIA: A SUPERCONTINENT OF THE LATE PROTEROZOIC

There is convincing evidence that a supercontinent existed in the late Proterozoic Eon, about 1100 million years ago. The evidence is in the paleomagnetic record, the evidence of now deeply eroded belts of regionally metamorphosed rocks, and other evidence such as glacially transported sediments in places now far from the poles. Given the name Rodinia, the late Proterozoic supercontinent continues to be the subject of much study. The challenge is to see back through geologic time to the formation of a supercontinent that was completed 1100 million years ago. Confusing data have led to many different attempts to reconstruct the old supercontinent. One such attempt, shown in Figure 20.13, puts Laurentia in the center, while Antarctica, Australia, and most of the other bits of continent are arrayed around Laurentia. The possibility of other ancient supercontinents prior to Rodinia remains controversial and the subject of much research.

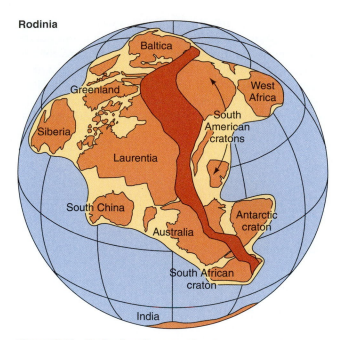

Figure 20.13   An Ancient Supercontinent   A proposed arrangement of continental fragments to form Rodinia, a Proterozoic landmass that spanned much of Earth one billion years ago. The red strip marks the location of an orogen that once formed a common suture of several cratons, but that now is dispersed among Earth's present-day continents.

ICE AGES IN EARTH'S HISTORY

Large continental ice sheets have existed several times during Earth's history. Evidence of their former existence includes features such as polished and ice-scratched pavements, and tillites, as discussed in Chapter 16.

Continental glaciers were prevalent in the Carboniferous Period, 286 to 360 million years ago, and also in the Ordovician and Silurian periods, roughly 450–425 million years ago. Both glacial periods occurred at times when large continental landmasses lay over Earth's south pole, providing a platform for the accumulation of vast quantities of snow and ice at high latitudes (Figure 20.14). Continental placement, however, was probably not the only factor leading to an ice sheet. The amount of carbon dioxide in the atmosphere, and therefore the carbon cycle, must also have played a role.

For the Carboniferous Period, geologists have sufficient evidence to estimate the competing factors in the carbon cycle. The glaciations occurred following the assembly of Pangaea, at a time when plate motion was slow. Slow plate motion means that spreading-ridge volcanism, and thus the emission of CO_2 from volcanoes, was slower than usual. At the same time, the Carboniferous takes its name from extensive deposits of coal. Plant fossils in the coal tell us that vast low-lying swamps full of huge trees and ferns dominated the biosphere during the Carboniferous, indicating vigorous consumption of CO_2 by photosynthesis, and rapid burial of carbon in the coal.

A combination of rapid carbon burial and reduced volcanic activity meant lower concentrations of CO_2 in the atmosphere, reducing the greenhouse effect and encouraging glaciation. Earth was not a frozen wasteland for millions of years, however. Carboniferous coal deposits occur within repetitive layered sedimentary sequences of marine and nonmarine sediments called **cyclothems**. The repetition of marine and nonmarine sediments indicates repeated transgressions and regressions by the sea. Geologists have connected the timing of these cyclothems to the longer Milankovitch orbital cycles of 100,000 years or more, indicating that glacial and interglacial climate conditions may have alternated in a regular pattern.

We do not yet know all the details of the way Earth's climate cycled between glacial and interglacial conditions in the Carboniferous. We can hypothesize that coastal coal swamps would have been sensitive to changes in sea level that accompany the growth and shrinkage of ice sheets. We can be sure that carbon burial in coal swamps would have reduced the CO_2 level of the atmosphere and thereby reduced the greenhouse effect, but new continental glaciers would have stolen water from the oceans, causing sea level to drop. If coastal coal swamps drained as the sea level fell, carbon consumption would decrease and atmospheric CO_2 levels could return to normal levels. An increase in greenhouse warming would melt the glaciers and raise sea level, rejuvenating the coal swamps and repeating the glacial cycle. Many details of this hypothesis remain to be tested, but it seems likely that this, or some other feedback effect within the Earth system, can best explain the glacial cycles of the Carboniferous.

Prior to the Phanerozoic Eon, there is evidence for continental glaciation near the end of the Proterozoic Eon, between 800 and 600 million years ago. Paleomagnetic data for this time span indicate that there were no large continental landmasses near either of Earth's poles. Surprisingly, much of the evidence of continental glaciation in the late Proterozoic is in rock sequences for which paleomagnetic data indicate a location near the paleomagnetic equator rather than the paleomagnetic poles. If the data are correct we have to conclude that extensive low-latitude glaciation occurred during a climatic extreme and the whole Earth must have experienced glaciation.

Extensive low-latitude glaciation did not occur at any time during the Phanerozoic, so data suggesting such an extreme state of our planet's environment pose an interesting challenge. The hypothesis that continental glaciation existed at low latitudes, possibly even at the equator, is colloquially called the Snowball Earth hypothesis. As we write this book, the hypothesis is being hotly debated and intensively researched. If the Snowball Earth hypothesis survives testing and becomes an accepted theory, geologists will have to explain how such extreme changes in the Earth system could have occurred and persisted for as long

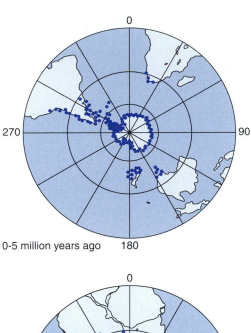

0-5 million years ago   180

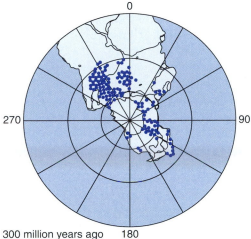

300 million years ago   180

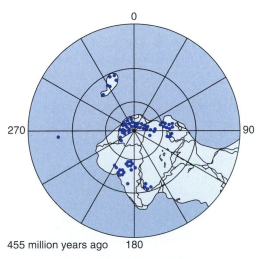

455 million years ago   180

Figure 20.14 Phanerozoic Glaciations Three major periods of high-latitude glaciation have been recognized in the Phanerozoic: Pleistocene (0 to 5 million years ago), Carboniferous (roughly 300 million years ago), and Ordovician (roughly 455 million years ago). The three panels show the arrangement of Earth's continents in the Southern Hemisphere. Blue dots mark the sites of glacial deposits from the relevant period.

as 100 million years. Although most fossils such as shells, corals, and bones are absent in Proterozoic rocks, the chemistry of sediments indicates that single-celled life forms flourished during much of the eon. Thus, the biosphere probably had a significant effect on the Precambrian carbon cycle, but its possible role in the Snowball Earth is highly speculative at this time.

> **Before you go on:**
>
> 1. Why do supercontinents eventually break apart?
> 2. Why did the Carboniferous ice sheets wax and wane?
> 3. What evidence leads to the hypothesis of a Snowball Earth?

REGIONAL STRUCTURES OF CONTINENTS

In a sense, continental crust is simply a passenger being rafted on large plates of lithosphere. But it is a passenger that is buffeted, stretched, fractured, and altered by the ride. Someone once characterized continental crust as the product of bump-and-grind tectonics. Each bump between two crust fragments forms a mountain belt, each grind a strike-slip fault, each stretch a rift valley. Scars left in the continental crust by bump-and-grind tectonics are evidence of former plate edges and plate motions. That this evidence exists is fortunate because the most ancient oceanic crust known to exist in the ocean dates only from the mid-Jurassic Period. Indeed, the only direct evidence concerning geologic events more ancient than the mid-Jurassic comes from the continental crust. In order to discuss how the continental crust has been affected by plate tectonics, it is helpful to look first at the large-scale structure of continents.

CRATONS

On the scale of a continent, two kinds of structural units can be distinguished within the continental crust. The first is a core of very ancient rock called a **craton** (Figure 20.15). The term is applied to any portion of the Earth's crust that has attained tectonic and isostatic stability. Rocks within cratons may be deformed, but the deformation is invariably ancient.

OROGENS

Draped around cratons are the second kind of crustal building unit, **orogens**, which are elongate regions of crust that have been intensely folded and faulted during continental collisions. Crust in an orogen is commonly thicker than crust in a craton, and many orogens—even some very old ones—have not yet attained isostatic equilibrium. Orogens are the eroded roots of ancient mountain ranges that formed

as a result of collisions between the cratons. Orogens differ from each other in age, history, size, and details of origin; however, all were once mountainous terrains, and all are younger than the cratons they surround. Only the youngest orogens are mountainous today; ancient orogens, now deeply eroded, reveal their history through the kinds of rock they contain and the kind of deformation present.

CONTINENTAL SHIELDS

An assemblage of cratons and ancient orogens that has reached isostatic equilibrium is called a **continental shield**. That portion of a continental shield that is covered by a thin layer of little-deformed sediments is called a *stable platform*. North America has a huge continental shield at its core, and around the shield are four younger orogens (Figure 20.15). Because the North American shield crops out in Canada (especially Ontario and Quebec), but is mostly covered by flat-lying sedimentary rocks in the United States, geologists often refer to it as the Canadian Shield.

Through careful mapping and radiometric dating, geologists have identified several ancient cratons and orogens in the Canadian Shield (Figure 20.15). Within the cratons, all rocks are older than 2 billion years. Such rocks can be observed in many places in eastern Canada, but within the United States cratonic rocks crop out only in a small region around Lake Superior. Nevertheless, by drilling through the cover of sedimentary rocks on the stable platform, geologists have discovered that cratons and orogens similar to those that surface in eastern Canada also lie below much of the central United States and part of western Canada.

The small cratons within the Canadian Shield shown in Figure 20.15 were probably minicontinents during the Archean Eon and early part of the Proterozoic Eon. By about 1.6 billion years ago, these minicontinents had become welded together to form the assemblage of cratons and ancient orogens we see in North America today. Each time two cratonic fragments collided, an orogen was formed between them. The existence of ancient collision belts—orogens—is the best evidence available to support the idea that plate tectonics operated at least as far back as 2 billion years ago.

> **Before you go on:**
>
> 1. What are the main structural units of a continent?
> 2. Where could you drill and find cratonic rocks in the United States?

CONTINENTAL MARGINS

The fragmentation, drift, and welding together of pieces of continental crust are inevitable consequences of plate tectonics. Various combinations of these processes are

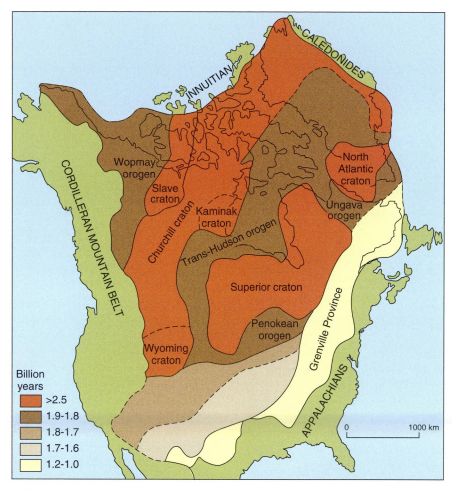

Figure 20.15 Cratons and Orogens The North American cratons and associated orogens. The Caledonide, Appalachian, Cordilleran, and Innuitian orogens are each younger than 600 million years. The assemblage of ancient cratons and orogens that is surrounded by the four young orogens is the Canadian Shield.

responsible for the five types of continental margins we know of today: passive, convergent, collision, transform fault, and accreted terrane. Before we discuss additional details of plate tectonics, it will be helpful to review briefly the features associated with each of the continental margins.

PASSIVE CONTINENTAL MARGINS

A **passive continental margin** is one that occurs in the stable interior of a plate. The Atlantic Ocean margins of the Americas, Africa, and Europe are passive continental margins. The eastern coast of North America, for example, is in the stable interior of the North American Plate, far from the plate margins. Passive continental margins develop when a new ocean basin forms by the rifting of continental crust, as illustrated in Figure 20.16. This process can be seen in the Red Sea, which is a young ocean with an active spreading center running down its axis (Figure 20.17). New, passive continental margins have formed along both edges of the Red Sea.

Passive continental margins are places where a great thickness of sediment accumulates. The kind of sediment deposited is distinctive, and the Red Sea provides an example. Deposition commenced with clastic, nonmarine sediments followed by evaporites and then clastic marine shales. The sequence apparently arises in the following manner. Basaltic magma, associated with formation of the new spreading edge that splits the continent, heats and expands the lithosphere so that a plateau forms with an elevation of as much as 2.5 km above sea level. Tensional forces cause normal faults and form a rift so that there is a pronounced topographic relief between the plateau and the floor of the rift. The earliest rifting of the Red Sea must have looked very much the way the African Rift Valley looks today, although the sense of movement in the African Rift Valley is more complicated than simply extension, as discussed in *The Science of Geology*, Box 20.1, *Watching a Continent Splinter*. Before the rift floor sank low enough for seawater to enter, clastic nonmarine sediments, such as conglomerates and sandstones, were shed from the steep valley walls and accumulated in the rift. Associated with

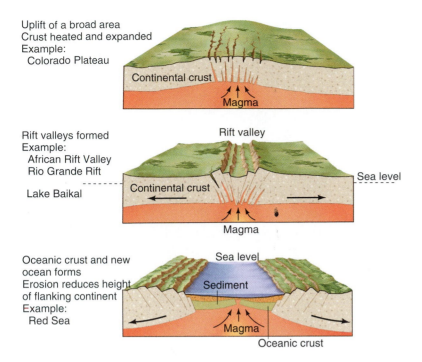

Uplift of a broad area
Crust heated and expanded
Example:
  Colorado Plateau

Continental crust

Magma

Rift valleys formed
Example:
  African Rift Valley
  Rio Grande Rift

Lake Baikal

Rift valley

Sea level

Continental crust

Magma

Oceanic crust and new
ocean forms
Erosion reduces height
of flanking continent
Example:
  Red Sea

Sea level

Sediment

Magma

Oceanic crust

Continental crust, thinned by erosion, cools,
contracts, and sinks beneath the sea
Example:
  Atlantic Ocean

Sediment   Oceanic crust

Magma

Figure 20.16   Rifting of Continental Crust
The rifting of continental crust to form a
new ocean basin bounded by passive continental margins. The rifting can cease at
any stage. It is not necessarily correct to
conclude that the African Rift Valley, for
example, will open to form a new ocean.

these sediments are basaltic lavas, dikes, and sills, all
formed by magma rising up the normal faults. As the rift
widened, a point was reached where seawater entered. The
early flow was apparently restricted, and the water was
shallow, resembling a shallow lake more than an ocean.
The rate of evaporation would have been high, and as a
result strata of evaporite salts were laid down on top of the
clastic nonmarine sediments. Finally, as rifting continued
and the depth of the seawater increased, normal clastic
marine sediments were deposited. This is the stage the Red
Sea is in today. Eventually, as further rifting exposes new
oceanic crust, the Red Sea will evolve into a younger version of the Atlantic Ocean.

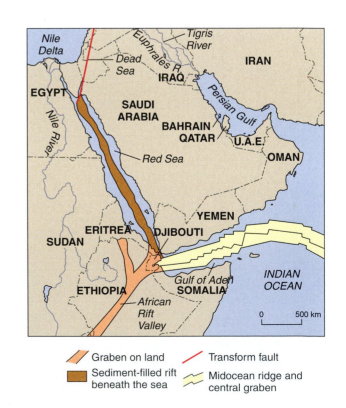

Figure 20.17   Triple Junctions   Three spreading centers meet at
a triple junction. Two, the Gulf of Aden and the Red Sea, are
actively spreading, and there are passive continental margins
along the adjacent coastlines. The African Rift, however,
appears to be a failing rift that will not develop into an open
ocean.

Box 20.1 THE SCIENCE OF GEOLOGY

WATCHING A CONTINENT SPLINTER

Travelers in East Africa are awed by the Rift Valley. There, two sections of Earth's crust are pulling apart, creating an enormous gash—more than 1800 meters deep in some places—that runs lengthwise from the Red Sea at the top of Africa to Mozambique in the south (Figure B20.1). For years geoscientists wondered why the rift disappears south of Mozambique. They now have an answer.

The Rift Valley marks the boundary between two tectonic plates: the Nubian Plate to the west and the Somalian Plate to the east. Such rifts are usually linked to other plate boundaries, but researchers were unable to trace the East African rift beyond the point where it disappeared in Mozambique. Unlike conditions along the rift to the north, there are no earthquakes or other signs of disturbance to the south. How, then, do these plates connect to the rest of Earth's tectonic system?

Two researchers, Dezhi Chu of Exxon Production Research Company and Richard G. Gordon of Rice University, solved the mystery by studying rocks of the ocean floor south of Mozambique. In the seafloor between Africa and Antarctica there is a rift system that produces new igneous rock. The rift pushes the African plates apart but it also pushes them away from Antarctica. By using Antarctica as a reference point, the researchers were able to plot the motion of the Nubian and Somalian plates. They found that the two plates do not pull apart in a straight line; instead, they rotate around a pivot point east of South Africa. To the north, in the region of the East Africa rift, they separate by 6 millimeters per year. To the south of the pivot point, they move toward each other at a rate of 2 millimeters per year, so slowly that they don't produce many large earthquakes.

Armed with a new understanding of the motion, or *kinematics*, of these two plates, geoscientists are in a better position to figure out how other tectonic plates interact. In so doing, they will create a fuller picture of Earth's overall plate-tectonic scheme.

Figure B20.1   A Split in a Continent   The African Rift Valley seen from the air. The view is looking north in northern Kenya. The cliff is a fault escarpment. In the distance is a small volcanic cone.

Notice in Figure 20.17 that the Gulf of Aden, the Red Sea, and the northern end of the African Rift Valley meet at angles of 120°. Such a meeting point formed by three spreading edges is called **plate triple junction**. Two of the edges, the Gulf of Aden and the Red Sea, are active and still spreading. The third, the African Rift Valley, is spreading so slowly that possibly it will not evolve into an ocean. If a new ocean does not form, what will remain on the African continent is a long, narrow sequence of grabens filled primarily with nonmarine sediment. The formation of three-armed rifts with one of the arms not developing into an ocean is apparently a characteristic feature of passive continental margins. This can be seen from Figure 20.18, which shows the reassembled positions of the continents flanking the Atlantic Ocean prior to breakup. Note that some of the world's largest rivers flow down valleys formed by failed rifts associated with the opening of the Atlantic Ocean.

CONTINENTAL CONVERGENT MARGINS

At a **continental convergent margin**, the edge of a continent coincides with a convergent plate margin in which oceanic lithosphere is being subducted beneath continental lithosphere. The Andean coast of South America is an example. On this coast, the Nazca Plate (capped by oceanic crust) is being subducted beneath the South American Plate (capped by continental crust). Wet partial melting in the mantle activated by water released by the subducted Nazca Plate produced the andesitic magma that formed the Andes (a **continental volcanic arc**).

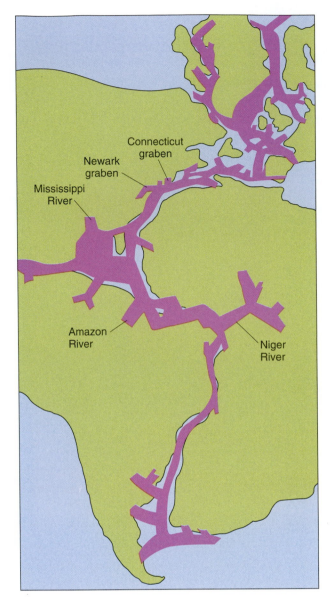

Figure 20.18   Rifts Around the Atlantic   Map of a closed Atlantic Ocean showing the rifts formed when Pangaea was split by a spreading center. The rifts on today's continents are now filled with sediment. Some of them serve as the channelways for large rivers.

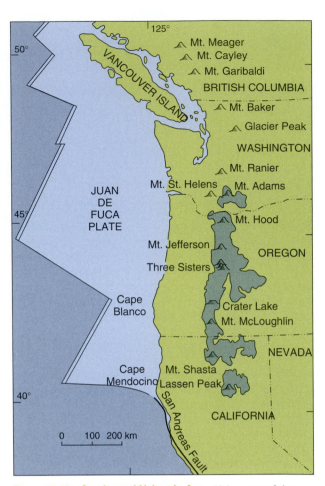

Figure 20.19   Continental Volcanic Arc   Volcanoes of the Cascade Range, a continental volcanic arc. Each volcano has been active during the last 2 million years. Magma to form the volcanoes comes from wet partial melting above the oceanic crust of the Juan de Fuca Plate as it subducts beneath the North American Plate.

Subduction produces intense deformation of a continental margin (together with characteristic magmatic activity and a distinctive style of metamorphism and deformation of sediments deposited in the trench). Sediment subjected to deformation in such a setting forms a mélange. The tectonic setting in which sediments are subjected to high-pressure, low-temperature metamorphism is a mélange at a subduction zone. Adjacent and parallel to the belt of mélange, the crust beneath the continental volcanic arc is thickened and also metamorphosed, but here the metamorphism is regional metamorphism of the kind that occurs along curve B in Figure 8.16. One distinctive

feature of a continental convergent margin, therefore, is a pair of parallel metamorphic belts.

The most distinctive feature of a continental convergent margin is the continental volcanic arc. Modern examples are the chains of volcanoes in the Andes and the Cascade Range (Figure 20.19). Where the volcanoes have been eroded and the deeper parts of the underlying magmatic arc exposed, granitic batholiths can be observed. They are remnants of the magma chambers that once fed stratovolcanoes far above. The strings of huge elongate batholiths that run from southern California to northern British Columbia (Figure 4.14) provide a striking example.

CONTINENTAL COLLISION MARGINS

At a **continental collision margin**, the edges of two continents, each on a different plate, come into collision (Figure 20.20). A modern example of a continental collision margin

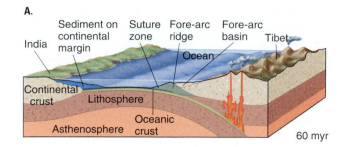

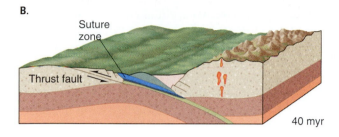

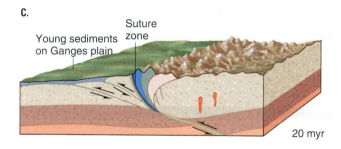

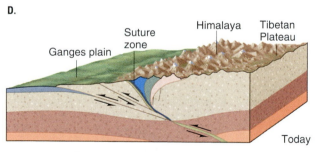

Figure 20.20 Collision of India with Asia Collision between two fragments of continental crust shown schematically for the collision between India and Tibet. Sediment is folded and faulted, and the lithosphere is thickened. The downward-moving plate of lithosphere capped by oceanic crust breaks off and continues to sink, but the edge of the remaining segment of plate, which is capped by buoyant continental crust, is partly thrust under the edge of the overriding plate, causing further elevation of the collision zone.

margin, the two continental fragments must eventually collide. The collision sweeps up and deforms any sediment that accumulated along the margins of both continents and forms a mountain system characterized by intense folding and thrust faulting.

All modern continental collision margins are young orogens that are fold-and-thrust mountain systems. Occurring in great arc-shaped systems a few hundred kilometers wide, fold-and-thrust mountain systems commonly reach several thousand kilometers in length. Within a fold-and-thrust mountain system, strata are compressed, faulted, folded, and crumpled, commonly in an exceedingly complex manner. Metamorphism and igneous activity are always present. Examples are widespread: the Alps, the Himalaya, and the Carpathians are all young fold-and-thrust mountain systems formed during the Mesozoic and Cenozoic eras. The Appalachians and the Urals are older, Paleozoic-aged examples.

All fold-and-thrust mountain systems develop from piles of sedimentary strata, commonly 15,000 m or more thick. The strata are predominantly marine. Some strata, as in the Appalachians, accumulated mainly in shallow water, while others, as in the Alps, accumulated mainly in deep water. Today we recognize that sediments in fold-and-thrust mountains accumulated along passive continental margins, such as the modern Atlantic margin of North America. Before the theory of plate tectonics, however, the problem of huge catchment sites for sediment was a puzzle. The American geologist J. D. Dana coined the term *geosyncline* to describe what he perceived to be a great trough that received thick deposits of sediment during slow subsidence through long geologic periods. The site of sediment deposition is not a trough, however. Rather, it is the passive margin of a continent where oceanic and continental crust is joined. It is here that sediment derived by weathering and erosion of the adjacent continent accumulates. Shallow-water sediment forms on the continental shelf, and deep-water sediment accumulates on the continental slope and rise.

A passive continental margin eventually becomes a continental convergent margin when old oceanic lithosphere fractures close to the join between oceanic and continental crust and subduction commences. The Atlantic margin of North America will probably become a continental collision margin at some time in the future. As discussed earlier in this chapter, the collision line between the two masses of deformed sediment caught up in a continental collision is commonly marked by ophiolites, the remnants of ancient seafloor.

One distinctive feature of mountain systems formed by collision is that an ocean disappears and a new mountain system lies in the interior of a major landmass. Modern examples are the previously mentioned Himalayan mountain chain, formed by the collision of India with Asia, and the Alps, which were formed by the collision of Africa and Europe starting in early Mesozoic time. The Ural Mountains in Russia and Kazakhstan and the Appalachians

is the line of collision between the Australian–Indian Plate and the Eurasian Plate. India, on the Australian–Indian Plate, and Asia, on the Eurasian Plate, have collided and the Himalayan mountain chain is the result.

When continental crust is carried on a plate of lithosphere being subducted beneath a continental convergent

in eastern North America were both formed by Paleozoic collisions, and both mountain systems were originally in the interior of Pangaea. The Urals are still in the center of a great landmass—they separate Europe from Asia—but the Appalachians are again near a continental margin following the breakup of Pangaea. Orogens are such distinctive and important features of Earth's crust that we will return to them later in the chapter and discuss the features of several mountain systems.

TRANSFORM FAULT MARGINS

A **transform fault continental margin** occurs when the margin of a continent coincides with a transform fault boundary of a plate. The most striking example of a modern transform fault boundary is the western margin of North America, from the Gulf of California to San Francisco, where it is bounded by the San Andreas Fault.

The San Andreas Fault apparently arose when the westward-moving North American continent overrode part of the East Pacific Rise, as shown in Figure 20.21. The San Andreas is the transform fault that connects the two remaining segments of the old spreading center.

ACCRETED TERRANE MARGINS

An **accreted terrane continental margin** is a former convergent or transform fault margin that has been further modified by the addition of rafted-in, exotic fragments of crust such as island arcs. They are the most complex of the five kinds of continental margin. The northwestern margin

of North America, from central California to Alaska, is an example of an accreted terrane margin.

Plate motion can raft fragments of crust tremendous distances. Eventually, any fragment not consumed by subduction is added (accreted) to a larger continental mass. Some of the fragments are island arcs formed by subduction of oceanic crust beneath oceanic crust. Other fragments form when they are sliced off the margin of a large continent by a transform fault, much as the San Andreas Fault is slicing a fragment off North America today. Other combinations of volcanism, rifting, faulting, and subduction can also form fragments of crust that are too buoyant to be subducted. In the western Pacific Ocean, there are many such small fragments of continental crust; examples include the island of Taiwan, the Philippine islands, and the many islands of Indonesia. Each fragment, called a *terrane*, is a geologic entity characterized by a distinctive stratigraphic sequence and structural history. The ultimate fate of all terranes is to be accreted to a larger continental mass. An accreted terrane is always fault-bounded and differs so markedly in its geologic features from adjacent terranes that many geologists still use the term *suspect terranes*. The term dates from the time when the terranes had been recognized as being unusual but their origin had not yet been deciphered.

The western margin of North America, where a large number of suspect terranes have now been recognized, is the most carefully studied example of a young, accreted terrane margin (Figure 20.22). Using a combination of lithologic and paleontologic studies, structural analysis, and paleomagnetism, geologists have recently identified

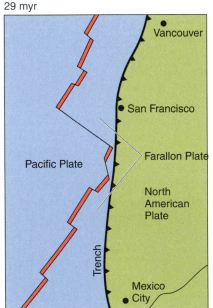
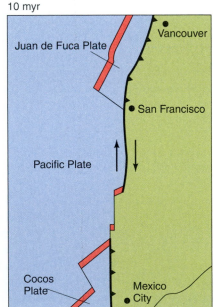
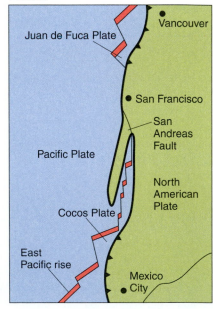

Figure 20.21 How the San Andreas Fault Began Origin of the San Andreas Fault. Twenty-nine million years ago, the edge of North America overrode a portion of the Farallon Plate, creating two smaller plates in the process, the Cocos Plate and the Juan de Fuca Plate. The San Andreas Fault is the transform fault that connects the remaining pieces of the severed spreading ridge.

the sources of many of the terranes, all of which have been accreted since the early Mesozoic, about 200 million years ago. One of the terranes is an ancient seamount; another is an older limestone platform formed on some other continental margin; still others are fragments of island arcs and even fragments of old metamorphic rocks. Paleomagnetic studies suggest that some terranes have moved 5000 km or more and that, once accreted, the process of movement did not necessarily cease. Later motion along transform faults caused still further reorganizations.

The recognition of accreted terrane margins of continents is a relatively new discovery. In a sense, it is a second stage of complexity in the plate-tectonic revolution. A great deal still remains to be discovered concerning terranes. How to recognize a terrane, how to work out where it came from, and how it was moved are all challenges facing geologists. Although it is a new and exciting concept, accreted terrane continental margins seem to be yet another distinctive piece of supporting evidence for ancient plate motions.

> **Before you go on:**
>
> 1. How does a passive continental margin form, and what sequence of sediments is deposited there?
>
> 2. Where are two places on Earth where continental convergent plate margins currently exist?
>
> 3. What sequence of events leads to a continental collision plate margin?
>
> 4. How does an accreted terrane continental margin occur?

MOUNTAIN BUILDING

Today's fold-and-thrust mountain ranges are the orogens that formed during the last few hundred million years. They are such distinctive and impressive features of the Earth's crust that we will briefly describe some examples.

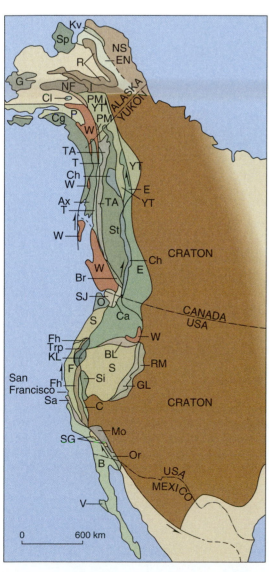

Figure 20.22 Terranes of Western North America   The western margin of North America is a complex jumble of terranes accreted during the Mesozoic and Cenozoic eras. Some terranes, such as Wrangellia (W), were fragmented during accretion and now occur in several different places.

Symbol	Name
Ax	Alexander
B	Baja
BL	Blue Mountains
Br	Bridge River
C	Calaveras
Cg	Chugach
Ch	Cache Creek
Cl	Chulitna
E	Eastern assemblages
EN	Endicott
F	Franciscan and Great Valley
Fh	Foothills Belt
GL	Golconda
I	Innoko
KL	Klamath Mountains
Kv	Kagvik
Mo	Mohave
NF	Nixon Fork
NS	North Slope
O	Olympic
P	Peninsular
PM	Pingston and McKinley
R	Ruby
RM	Roberts Mountains
S	Siletzia
Sa	Salinian
SG	San Gabriel
Si	Northern Sierra
SJ	San Juan
Sp	Seward Peninsula
St	Stikine
T	Taku
TA	Tracy Arm
Trp	Western Triassic and Paleozoic of Klamath Mountains
V	Vizcaino
W	Wrangellia
YT	Yukon-Tanana

THE APPALACHIANS

The Appalachians are a Paleozoic fold-and-thrust mountain system, 2500 km long, that borders the eastern and southeastern coasts of North America (Figure 20.23) and that continues offshore as eroded remnants beneath the sediment of the modern continental shelf. The sedimentary strata in the ranges contain mud cracks, ripple marks, fossils of shallow-water organisms, and, in places, freshwater materials such as coal. Evidence is strong for deposition of sediment on the continental shelf of an old, passive continental margin. The sedimentary strata thicken from west to east, and are underlain by a basement of metamorphic and igneous rocks.

Most but not all of the old strata of the Appalachians have now been deformed. Today, if we approach the central Appalachians from western New York or western Pennsylvania, we see first the sedimentary rocks occurring as essentially flat-lying, undisturbed strata. Continuing eastward, we notice that the same strata thicken and become gently folded and thrust-faulted. Many of Pennsylvania's oil pools were found in these gently folded strata. In eastern Pennsylvania, in the region known as the Valley and Ridge Province, the strata have been bent into broad anticlines and synclines (Figure 20.24A).

Approaching the Appalachians of Tennessee and the Carolinas from the west, we see a different style of deformation. Here, thrust faults predominate (Figure 20.25). Huge, thin slices of sedimentary strata were pushed westward, each successive slice riding upward and over earlier slices. To the west the strata are nearly flat lying, but as we move further east the strata dip more steeply to the east. The surface along which movement occurred is known as a **detachment surface**, and the slice that moved is commonly referred to by its French name, **décollement**. A distinctive feature of a décollement is that the style of deformation above the detachment surface is usually different from that below. That is, the brittle sedimentary strata above the detachment surface have been fractured, moved along thrust faults, and stacked like a series of thin cards, while the older rocks below tend to have resisted faulting and large-scale translation and to have been deformed by ductile deformation.

Proceeding east, toward the region from which the thrust slices came, we find the core of the Appalachians. Here the ancient basement rocks and the deep-water sediments that were deposited on the old continental rise have been pushed upward and can be examined. The strata are increasingly metamorphosed, and deformation becomes more intense as the line of collision is approached. Folds become isoclinal and then recumbent, and faulting is prevalent. In places, fragments of the old basement have been thrust up over younger sedimentary strata. Finally, we reach a region where intense metamorphism has occurred and where granite batholiths have been emplaced.

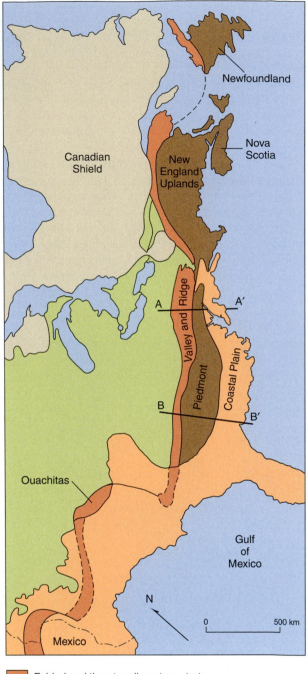

	Folded and thrust sedimentary strata
	Metamorphic and igneous rocks
	Mesozoic and Cenozoic rocks of the Coastal Plain

Figure 20.23 The Appalachians The Appalachian Mountain System runs from Newfoundland to the Mexican border. Parts of the eastern and southern margins of the system are covered by younger sediments of the coastal plain. Figures 20.24 and 20.25 are cross sections drawn approximately along the lines A–A′ and B–B′, respectively.

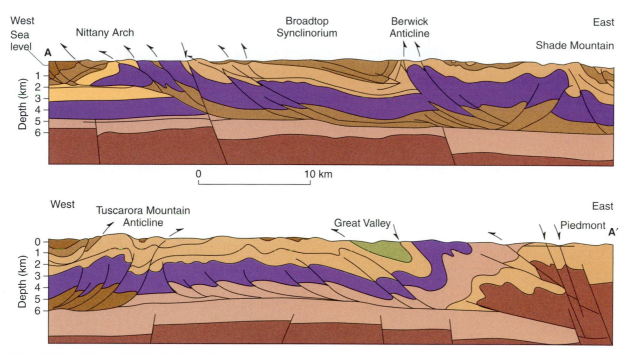

Figure 20.24   Valley and Ridge in Pennsylvania   Section through the Valley and Ridge Province of the Appalachians in Pennsylvania, along the A–A´ line in Figure 20.23. The structure includes both folding and thrusting. The prominent stratum blue is a limestone of middle Cambrian to early Ordovician age. (Note that A–A´ has been cut in two to fit on the page so that the left-hand edge of the bottom section joins the right-hand edge of the top section.)

The Appalachians have a complex history that started more than 600 million years ago, as demonstrated in Figure 20.26. Notice that three collision events are postulated. The section drawn in Figure 20.25 is in the southeastern United States, where the rocks that form today's continental margin are thought by some geologists to be an accreted terrane of what was once a fragment of Africa.

THE ALPS

We naturally ask how well the Appalachian picture can be applied to other fold-and-thrust mountains. The answer is that similar features are found in all of them. The Alps and associated mountain ranges in southern Europe (Figure 20.27) were formed later than the Appalachians, during the

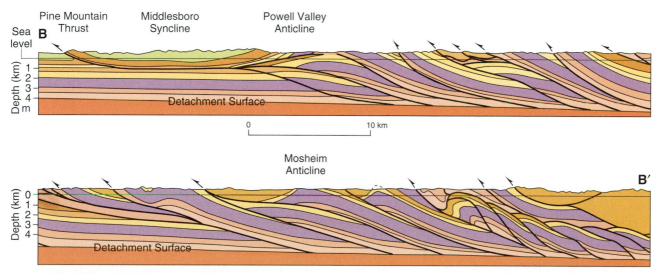

Figure 20.25   Valley and Ridge in Tennessee   Section through the Valley and Ridge Province of the Appalachians in Tennessee and North Carolina, along the line B–B´ in Figure 20.23 (again cut in two to fit on the page). Compare with Figure 20.24. Development of décollement by thrust faulting is predominant in the southern Appalachians.

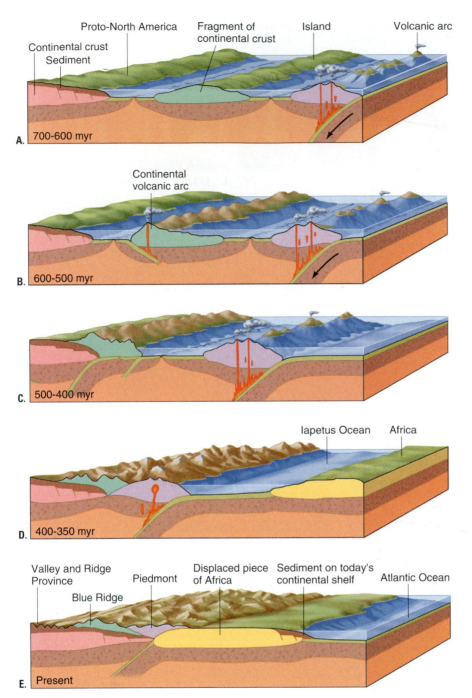

A. 700-600 myr
- Continental crust
- Sediment
- Proto-North America
- Fragment of continental crust
- Island
- Volcanic arc

B. 600-500 myr
- Continental volcanic arc

C. 500-400 myr

D. 400-350 myr
- Iapetus Ocean
- Africa

E. Present
- Valley and Ridge Province
- Blue Ridge
- Piedmont
- Displaced piece of Africa
- Sediment on today's continental shelf
- Atlantic Ocean

Figure 20.26   How the Southern Appalachians Were Formed   A suggested sequence of events explaining the evolution of the southern Appalachians in terms of plate tectonics. A sequence of subduction, collision, thrusting, and accretion produced the present-day structure. Iapetus is the posthumous name of the ocean that disappeared about 350 million years ago when Africa collided with North America. Iapetus was one of the minor Greek gods and father of Atlas and Prometheus.

Mesozoic and Cenozoic eras, as a consequence of a collision between the European and African plates. Nevertheless, the two systems have many features in common. For instance, the Jura Mountains, which mark the northwestern edge of the Alps, have the same folded form and origin as the Valley and Ridge Province (Figure 20.28A). Also, the Jura Mountains were formed from shallow-water sediments deposited on an ancient continental shelf. In the high Alps, which correspond to the now deeply eroded Appalachians in Connecticut, Vermont, Virginia, and Maryland, thrusting appears to have developed on a much grander scale than in the Appalachians.

The high Alps are composed of deeper-water marine strata that were deposited on an ancient continental slope and continental rise. Just as the collision that formed the Appalachians left a bit of Africa behind (Figure 20.26), so the Alpine collision left some of Africa behind.

THE CANADIAN ROCKIES

The Canadian Rocky Mountains are a magnificent fold-and-thrust system that can also be compared with the Appalachians. A section at about the latitude of Calgary, Alberta, reveals all the features described for the

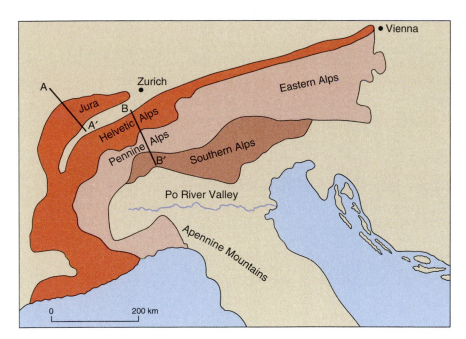

Figure 20.27 The Alps Major units of the Alps in Switzerland and Austria. The mountains were formed as a result of compressive forces operating in a south-eastern-to-northwestern direction. Figures 20.28A and B are along the lines A–A´ and B–B´, respectively.

Appalachians (Figure 20.29). A central zone has been intensely metamorphosed. In it, parts of the older basement rocks have been thrust upward, and folding and thrust faulting are evident on the margins. The thrust sheets in the Canadian Rockies moved eastward away from the core zone (Figure 20.30). Each sedimentary unit becomes thinner from west to east, indicating that the eastern portion formed from shallow-water sediments, while the core zone coincides with the thickest section of sediments, which were deposited in deep water.

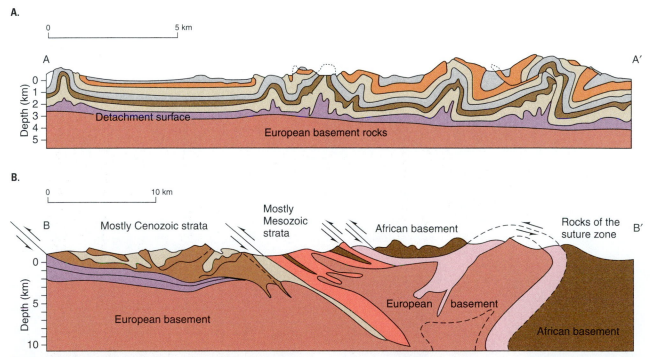

Figure 20.28 Anatomy of the Jura and the Alps Two sections through portions of the Alps. A. Section through the Jura Mountains along line A–A´ in Figure 20.27. The weak sedimentary rocks were deformed by slippage along the detachment surface. B. Section in central Switzerland along the line B–B´ in Figure 20.27. Strata have moved northward along great thrust faults that later were themselves folded. Part of the old African basement has been thrust over the European strata as a result of the collision.

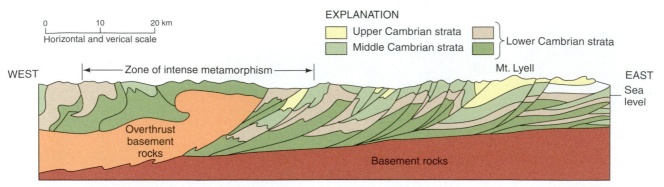

Figure 20.29 Anatomy of the Rockies in Canada Section through the Canadian Rocky Mountains at about the latitude of Calgary, Alberta. The zone of intense metamorphism coincides with the region of maximum uplift and maximum deformation. Farther east, where strata become progressively thinner, the pile has been greatly thickened by movement along thrust faults. The sense of movement is such that each fault block has moved toward the east, riding over the block beside it.

Figure 20.30 Thrust Sheets of the Rockies The Canadian Rockies. This early-morning photograph is looking west in the area of Banff National Park, Alberta. The ranges are the upthrust edges of thrust sheets. The sense of movement on the thrusts is toward the viewer.

REVISITING PLATE TECTONICS AND THE EARTH SYSTEM

A PLATE COLLISION AND THE MONSOONS

For more than 2000 years, Arab traders sailed to India during the hot summer months, a time when India and Southeast Asia are drenched by rain. They sailed back home again during the pleasant, dry, winter months. They did so because summer winds blow from the west while winter winds blow from the east. The Arabic name for this seasonal reversal of wind and weather is *mausim*, and from it we derive our word *monsoon*. The monsoon winds of India and Southeast Asia are a consequence of the collision between the Eurasian Plate and the Australian–Indian Plate. The collision has pushed India under the edge of the Eurasian Plate and the uplift of the Himalaya and the Tibetan Plateau is the result.

Computer modeling has shown that atmospheric circulation in the absence of the Himalaya and Tibetan Plateau would be very different from what it is today (Figure 20.31): the modern monsoon climate would not exist. The high mountains and plateau now divert the normal flow of westerly winds and are the primary factor in the present monsoon circulation. Summer heating of the high plateau causes warm air to rise, and this creates a low-pressure region that draws moist air inland from the adjacent ocean to produce the summer monsoon rainy season—the Arabs sailed to India. In winter, the snow-covered surface of the plateau reflects solar radiation and cools down, creating a region of high pressure from which cold, dry air flows outward and downward—the Arabs sailed home.

A. Before uplift

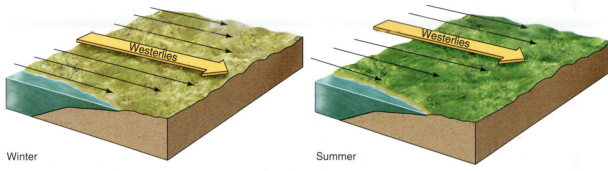

Winter   Summer

B. Post-uplift

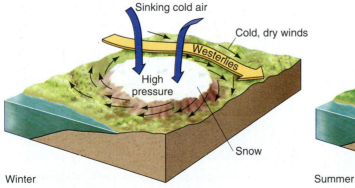

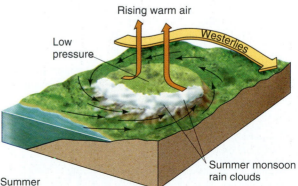

Winter   Summer

Figure 20.31 Development of a Monsoon Climate System A. Before uplift, westerly winds pass directly across a relatively flat landscape, the axis of flow shifting seasonally north and south. B. As a large plateau is uplifted, the westerly flow is diverted around the upland, a winter high-pressure system develops over the high snow-covered plateau, and cold, dry winds blow clockwise off the upland region. In summer, as the plateau surface warms, the warm air rises, drawing moist marine air inland. The summer monsoon clouds move inland toward the plateau margin and lose their moisture as heavy rainfall.

The history of uplift and erosion of the Himalaya and Tibetan Plateau provides a striking example of the Earth system at work, because it shows how closely land, ocean, and atmospheric records are linked. Chinese geologists have discovered that these high lands are rather recent topographic features. They base their conclusion in part on evidence of plant fossils of the Pliocene Epoch (5.3 to 1.6 million years ago), collected at altitudes of 4000 to 6000 m, that include many subtropical forms that today exist only at altitudes below 2000 m. Additional evidence is provided by studies of changing rates of *denudation*, the continually destructive effects of weathering, mass wasting, and erosion discussed in Chapter 12.

For the Himalaya and Tibetan Plateau most of the sediment ultimately is deposited in deltas, as a blanket across the continental shelves, or in vast submarine sediment fans. The volume of sediment deposited during a specific time interval can be estimated using drill-core and seismic records of the seafloor. Calculating the equivalent rock volume of this sediment and averaging it over the area from which it was derived provides both an average denudation rate and indications of any rate changes that may have occurred.

Sediment cores from the northern Indian Ocean show low rates of sedimentation, implying reduced erosion rates and low-altitude source areas, until about 10 million years ago when sediment accumulation rates rose sharply as a result of rapid uplift. Two peaks in sediment supply (9 to 6 and 4 to 2 million years ago), seen both in the deep-sea cores and in alluvium on land, are evidence of major intervals of Himalayan uplift.

There is suggestive evidence that monsoon winds started only during the second major phase of uplift. About 8 million years ago, fine dust, eroded by strong winds from desert basins north of the Tibetan Plateau, began accumulating in central China. The dust also was carried far out into the Pacific Ocean where it settled to form deep-sea clay. An increase in the rate of dust accumulation occurred between 3.5 and 2.5 million years ago. The onset of this vigorous wind erosion seems to have coincided with the beginning of the East Asian monsoon climate system. The possibility that the uplift of the Himalaya and the Tibetan Plateau may have had a wider, even global, influence on the climate is a matter of current research.

What's Ahead?

We have reached the end of discussion about the way Earth works, but there is a very important economic aspect that still needs to be addressed—mineral and energy resources. All mineral and most energy resources have been formed as a result of the geologic processes discussed throughout the book. We turn next, therefore, to a discussion of resources.

CHAPTER SUMMARY

1. The lithosphere is broken into six large and many small plates, each about 100 km thick and each slowly moving over the top of the weak asthenosphere beneath it.

2. Three kinds of margins are possible between plates. Divergent margins (spreading centers) are those where new lithosphere forms; plates move away from them. Convergent margins (subduction zones) are lines along which plates compress against each other and along which lithosphere capped by oceanic crust is subducted back into the mantle. Transform fault margins are lines where two plates slide past each other.

3. Plate movement can be described in terms of rotation across the surface of a sphere. Each plate rotates around a spreading axis. The spreading axis does not necessarily coincide with Earth's axis of rotation.

4. Because plate movement is a rotation, the velocity varies from place to place on the plate.

5. Each segment of oceanic ridge that marks a divergent margin between two plates lies on a line of longitude passing through the rotation pole. Each transform fault margin between two plates lies on a line of latitude of the rotation pole.

6. Two major structural units can be discerned in the continental crust. Cratons are ancient portions of the crust that are tectonically and isostatically stable. Separating and surrounding the cratons are orogens of highly deformed rock, marking the site of mountain ranges.

7. An assemblage of cratons and deeply eroded orogens that forms the core of a continent is a continental shield.

8. There are five kinds of continental margins: passive, convergent, collision, transform fault, and accreted terrane.

9. Passive margins develop by rifting of the continental crust. The Red Sea is an example of a young rift; the Atlantic Ocean is a mature rift.

10. A characteristic sequence of sediments forms along a passive continental margin, starting with clastic nonmarine sediments, followed by marine evaporites, and then marine clastic sediments.

11. Continental convergent margins are the locale of paired metamorphic belts, chains of stratovolcanoes (magmatic arc), and linear belts of granitic batholiths.

12. Collision margins are the locations of fold-and-thrust mountain systems.

13. Transform fault margins occur where the edge of a continent coincides with the transform fault boundary of a plate.

14. Accreted terrane margins arise from the addition of blocks of crust brought in by subduction and transform fault motions.

THE LANGUAGE OF GEOLOGY

accreted terrane continental margin (p. 550)

back-arc basin (p. 540)

continental collision margin (p. 548)
continental convergent margin (p. 547)
continental shield (p. 547)
continental volcanic arc (p. 547)
craton (p. 544)

cyclothem (p. 543)

décollement (p. 552)
detachment surface (p. 552)

magnetic latitude (p. 535)
mélange (p. 539)

orogen (p. 544)

passive continental margin (p. 545)
plate triple junction (p. 547)

rotation pole (p. 538)

seafloor spreading (p. 536)

transform fault continental margin (p. 550)

QUESTIONS FOR REVIEW

1. Long ago it was suggested that the location and shape of mountain ranges and other major topographic features were the result of contraction as Earth cooled. Why is that explanation incorrect?

2. Explain how the apparent wandering of magnetic poles throughout geologic history can be used to help prove continental drift.

3. What are the main features of seafloor spreading? What critical test proved that the seafloor does move?

4. How are the velocities of tectonic plates in the geologic past determined? Do magnetic time lines on plates provide relative or absolute plate velocities?

5. What is a rotation pole and how does plate velocity depend on the position of a plate relative to the rotation pole?

6. What is a mélange and what kind of metamorphism is associated with mélanges? Where might mélanges be forming today?

7. What are cratons and how do they differ from orogens? Name three orogens in North America that are less than a billion years old.

8. Name three fold-and-thrust mountain ranges.

9. What evidence indicates that plate tectonics has been operating for at least the last 2 billion years of Earth history?

10. Name the five kinds of continental margins and describe how they form.

11. Describe the sequence of events that leads to the opening of a new ocean basin flanked by two passive continental margins.

12. How does an accreted terrane margin form? Name a continental margin that was modified by terrane accretion.

Click on *Presentation* and *Interactivity* in **The Rock Record and Geologic Time** module of your CD-ROM to further explore resources and activities presenting concepts from this chapter. Select *Assessment* in the same module to test your understanding of this chapter.

Hope Springs Eternal. Prospecting for gold, French Guyana, South America. Prospectors dig out sample of soil in order to test for the presence of gold. Man on left is holding prospector's pan used in the test.

Resources of Minerals and Energy

The incredibly rich, gold-bearing conglomerates of the Witwatersrand, South Africa.

The Golden Puzzle of the Witwatersrand

Hundreds of thousands of gold mines have been worked since humans first started to prize gold at least 20,000 years ago. Because gold is inert to air and rainwater and people don't throw it away after use, most of the gold ever mined is still in use. Some has been recycled dozens of times. If piled together, all of the gold mined through human history would make a pile about 55 feet on an edge.

A 55-foot cube is a lot of gold, but there is a great puzzle concerning it—about 40 percent of all the gold has come from one location, a sedimentary basin in South Africa called the Witwatersrand, that is between 2.7 and 3.0 billion years old. I have been privileged to visit and examine these extraordinary deposits on several occasions, and I have concluded, in concert with many others, that the deposits are fossil placers, which means that the gold must

have been transported by streams, together with the other constituents of the sandstones and conglomerates in which the gold is now found.

But if the gold came in as sedimentary particles, where did it come from? No evidence of the primary mineralization has been found, despite extensive prospecting. Why and how so much of the world's gold came to be concentrated in one small area remains a puzzle, and how all evidence of primary deposits, which must have been truly huge, can have disappeared, is a truly golden puzzle.

Brian J. Skinner

KEY QUESTIONS

1. **How important are mineral and energy resources to society?**
2. **How are mineral deposits formed?**
3. **How much energy do we use today?**
4. **Are energy resources large enough for future needs?**
5. **How did fossil fuels form?**
6. **What are possible energy sources besides fossil fuels?**

INTRODUCTION: NATURAL RESOURCES AND HUMAN HISTORY

Civilization and natural resources—the former would not have been possible without the latter. Millions of years ago our ancestors started using special kinds of resources—rocks and minerals—when they picked up suitably shaped stones and used them as hunting aids. After a while they discovered that flint, chert, obsidian, and other tough stones were best for knives and spear points. Because the most desired stones could be found in only a few restricted places, trading started. Next our ancestors started gathering and trading salt. Originally, eating the meat brought home by hunters satisfied dietary needs for salt. When farming started, however, diets became cereal-based, and extra salt was needed. We don't know when or where the mining of salt started, but long before recorded history salt routes criss-crossed the globe.

Metals were first used more than 20,000 years ago. Copper and gold are both found as native metals, and these were the earliest metals to be used. But native copper is rare, and so eventually other sources of copper were needed. By 6000 years ago our ancestors had learned how to extract

copper from certain minerals by the process called smelting. Before another thousand years had passed, they had discovered how to smelt minerals of lead, tin, zinc, silver, and other metals. The technique of mixing metals to make alloys came next; bronze (copper and tin) and pewter (tin, lead, and copper) came into use. The smelting of iron is more difficult than the smelting of copper, and so development of an iron industry came much later—about 3300 years ago.

The first people to use oil instead of wood for fuel were the Babylonians, about 4500 years ago. The Babylonians lived in what is now Iraq, and they used oil from natural seeps in the valleys of the Tigris and Euphrates rivers. The first people to mine and use coal were the Chinese, about 3100 years ago. At about the same time, the Chinese drilled the first wells for natural gas—some nearly 100 m deep.

By the time that first the Greek and then the Roman empires came into existence about 2500 years ago, our ancestors had come to depend on a very wide range of resources—not just metals and fuels, but also processed materials such as cements, plasters, glasses, and porcelains. The list of materials we mine, process, and use has grown steadily larger ever since. Today we have industrial uses for almost all of the naturally occurring chemical elements, and more than 200 kinds of minerals are mined and used.

In many of the previous chapters, we mentioned that geologic processes such as weathering, sedimentation, and volcanism can, under suitable conditions, form mineral and energy deposits. The "suitable conditions" are not common, however, and for this reason mineral and energy resources are limited in quantity and hard to find. No geologic challenge is more difficult or more rewarding than the search for, and discovery of, new resources of minerals and energy. In this chapter we focus on how and where the different kinds of mineral and energy deposits form.

MINERAL RESOURCES

Can you imagine a world without machines? Our modern world with its 6.0 billion inhabitants couldn't operate without them. Machines are used to produce our food, to make our clothes, to transport us around, and to help us communicate. The metals needed to build machines and the fuels needed to run them are dug from Earth. Geologic processes form deposits of metallic minerals and fuels. Geologic processes, therefore, influence the daily lives of each and every one of us.

We turn first to the mineral substances that provide materials from which machines and myriad other necessary things can be made. The number and diversity of such substances are so great that to make a simple classification covering all of them is almost impossible. Nearly every kind of rock and mineral can be used for something.

SUPPLIES OF MINERALS

Many industrialized nations are rich in many kinds of **mineral deposits** (any volume of rock containing an enrichment of one or more minerals), which they are exploiting vigorously. Yet no nation is entirely self-sufficient in mineral supplies, and so each must trade with other nations to fulfill its needs (Figure 21.1).

Mineral resources have three distinctive aspects. First, occurrences of usable minerals are limited in abundance and distinctly localized at places within Earth's crust. This is the main reason no nation is self-sufficient in mineral supplies. Because usable minerals are localized, they must be searched out, and the search ranges over the entire globe. The special branch of geology concerned with discovering new supplies of usable minerals is called, appropriately, **exploration geology**.

Second, the quantity of a given material available in any one country is rarely known with accuracy, and the likelihood that new deposits will be discovered is difficult to assess. As a result, production over a period of years can

be difficult to predict. Thus, a country that today can supply its need for a given mineral substance may face a future in which it will become an importing nation. Britain, for example, once supplied most of its own mineral needs but can no longer do so today. A little more than a century ago, Britain was a great mining nation, producing and exporting such materials as tin, copper, tungsten, lead, and iron. Today, the known deposits have been worked out.

Third, unlike plants and animals, which are harvested yearly or seasonally and then replenished, deposits of minerals are depleted by mining and eventually exhausted. This disadvantage can be offset only by finding new occurrences or by using the same material repeatedly—that is, by recycling and making use of scrap.

The peculiarities of the mineral industry place a premium on the skills of exploration geologists and engineers, who play the essential roles in finding and mining mineral deposits. The finding is accomplished through application of the basic geologic principles set forth in this book. Much ingenuity has been expended in bringing the production of minerals to its present state. Because known deposits are being rapidly exploited while demands for minerals continue to grow, we can be sure that even more ingenuity will be needed in the future.

ORE

Minerals for industry are sought in deposits from which the desired substances can be recovered least expensively. The more concentrated the desired minerals, the more valuable the deposit. In some deposits the desired minerals are so highly concentrated that even very rare substances such as gold and platinum can be seen with the naked eye. For every desired mineral substance, a *grade* (level of concentration) exists, below which the deposit cannot be worked economically (Figure 21.2). To distinguish

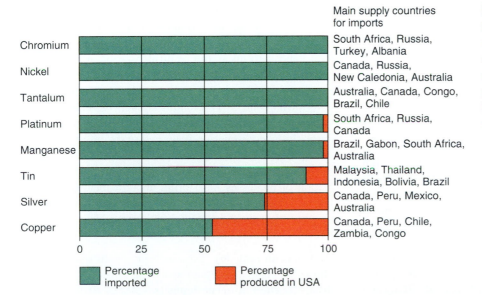

Figure 21.1 Commodities That Must Be Imported Selected mineral substances for which U.S. consumption exceeds production. The difference must be supplied by imports. Data are plotted for 1989, but the percentages have changed little over the past 12 years.

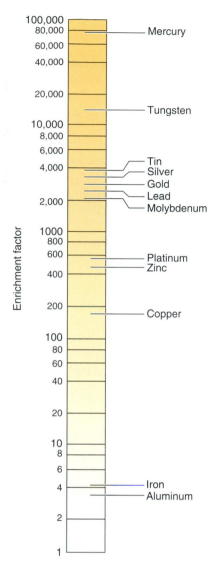

Figure 21.2 Enrichment Needed to Form an Ore Before a mineral deposit can be worked profitably, the percentage of valuable metal in the deposit must be greatly enriched above its average percentage in Earth's crust. The enrichment is greatest for metals that are least abundant in the crust, such as gold and mercury. As mining and mineral processing become more efficient and less expensive, it is possible to work leaner ore, and enrichment factors decline. Note that the scale is a magnitude (logarithmic) scale, in which the major divisions increase by multiples of 10.

between profitable and unprofitable mineral deposits, the word **ore** is used, meaning an aggregate of minerals from which one or more minerals can be extracted profitably. It is not always possible to say exactly what the grade must be, or how much of a given mineral must be present, in order to constitute an ore. Two deposits may have the same grade and be the same size, but only one of them is ore. There could be many reasons for the difference; for example, the uneconomical deposit could be too deeply buried or located in so remote an area that the costs of mining and

transport would be high enough to make the final product noncompetitive with the same product from other deposits. Furthermore, as costs and market prices fluctuate, a particular aggregate of minerals may be an ore at one time but not another.

The ore challenge is twofold: (1) to find the ores (which all together underlie an infinitesimally small proportion of Earth's land area), and (2) to mine the ores and get rid of the contaminants as cheaply as possible. Both steps are technical problems; engineers have been so successful in solving them that some deposits now considered ore are only one-sixth as rich as were the lowest-grade ores 100 years ago. Copper provides an interesting example. The lowest-grade ores ever mined—about 0.5 percent copper—were worked at a time of high metal prices, in the 1970s. In 2002, as this book is being revised, the lowest grade of mineable copper ore is closer to 1 percent copper. The reason for this is that overproduction of copper around the world coincided with an economic recession and produced a glut of newly mined copper. This, in turn, drove the price of copper down and led to the closing of many mines, in particular those mines with the lowest grades.

Gangue

As we learned in Chapter 3, sphalerite, galena, and chalcopyrite are ore minerals from which zinc, lead, and copper respectively can be extracted. But ore minerals rarely, if ever, occur alone. Rather, they are mixed with other, nonvaluable minerals, collectively termed **gangue** (pronounced *gang*). Familiar minerals that commonly occur as gangue are quartz, feldspar, mica, calcite, and dolomite.

ORIGIN OF MINERAL DEPOSITS

All ores are mineral deposits because each of them is a local enrichment of one or more minerals or mineraloids. The reverse is not true, however. Not all mineral deposits are ores. "Ore" is an economic term, whereas "mineral deposit" is a geologic term. How, where, and why a mineral deposit forms is the result of one or more geologic processes. Whether a given mineral deposit is an ore is determined by how much we human beings are prepared to pay for its content. Fascinating though the economics of ores and mining is, the topic cannot be explored in this book. Instead, discussion is limited to the origin of mineral deposits without regard to questions of economics.

In order for a deposit to form, some process or combination of processes must bring about a localized enrichment of one or more minerals. A convenient way to classify mineral deposits is through the principal concentrating process. Minerals become concentrated in five ways:

1. Concentration by hot, aqueous solutions flowing through fractures and pore spaces in crustal rock to form **hydrothermal mineral deposits** (Chapter 8).

2. Concentration by magmatic processes within a body of igneous rock to form **magmatic mineral deposits** (Chapter 4).

3. Concentration by precipitation from lake water or seawater to form **sedimentary mineral deposits** (Chapter 7).

4. Concentration by flowing surface water in streams or along the shore, to form **placers** (Chapter 14).

5. Concentration by weathering processes to form **residual mineral deposits** (Chapter 6).

HYDROTHERMAL MINERAL DEPOSITS

Many of the most famous mines in the world contain ores that were formed when their ore minerals were deposited from hydrothermal solutions. However, the origins of hydrothermal solutions are often difficult to decipher. Some solutions originate when water dissolved in magma is released as the magma rises and cools. Other solutions are formed from rainwater or seawater that circulates deep in the crust.

An example of the way a hydrothermal solution can form from deeply circulating seawater is shown in Figures 21.3 and 21.4. Because the heat source for seawater hydrothermal solutions of the kind illustrated in the figures is midocean ridge volcanism, and because the ore minerals are sulfides, mineral deposits formed from such solutions are called **volcanogenic massive sulfide deposits**.

The ore–mineral constituents in volcanogenic massive sulfide deposits are derived from the igneous rocks of the oceanic crust. Heated seawater reacts with the rocks it is in contact with, causing changes in both mineral composition and solution composition. For example, feldspars are changed to clay and epidote, and pyroxenes are changed to chlorite. As the minerals are transformed, trace metals such as copper and zinc, present by atomic substitution, are released and become concentrated in the slowly evolving hydrothermal solution.

Hydrothermal solutions having similar compositions can form in many different ways. Which ore constituents are carried in solution depends on the kinds of rocks involved in the formation of the solution. For example, copper and zinc are present in pyroxenes by atomic substitution, so that the pyroxene-rich rocks of the oceanic crust yield solutions charged with copper and zinc. As a result, volcanogenic massive sulfide deposits are rich in copper and zinc.

The most important question concerning hydrothermal solutions is not *where* the water and dissolved mineral constituents came from, but rather what made the solutions precipitate their soluble mineral load and form a mineral deposit?

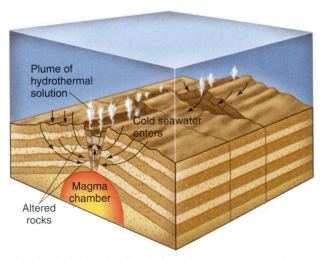

Figure 21.3 Submarine Hot Spring Seawater penetrates volcanic rocks on the seafloor. Heated by magma chamber and by the hot rocks that surround the magma chamber, seawater becomes a hydrothermal solution, produces metasomatic alterations, and rises at the midocean ridge as a seafloor hot spring.

Figure 21.4 Hydrothermal Vent Hydrothermal solution forming an ore deposit on the seafloor. A so-called black smoker photographed at a depth of 2500 m below sea level on the East Pacific Rise at 21° N latitude. The "smoker" has a temperature of 320°C. The rising hydrothermal fluid is actually clear; the black color is due to fine particles of iron sulfide and other minerals precipitated from solution as the plume is cooled by contact with cold seawater. The chimney-like structure is composed of pyrite, chalcopyrite, and other ore minerals deposited by the hydrothermal solution.

CAUSES OF PRECIPITATION

When a hydrothermal solution moves slowly upward, as with groundwater percolating through an aquifer, the solution cools very slowly. If dissolved minerals were precipitated from such a slow-moving solution, they would be spread over a large volume of rock and would not be sufficiently concentrated to form an ore. But when a solution flows rapidly, as in an open fracture, or through a mass of shattered rock, or through a layer of porous tephra where flow is less restricted, cooling can be sudden and can occur over short distances. Rapid precipitation and a concentrated mineral deposit are the result. Other effects—such as boiling, a rapid decrease in pressure, composition changes of the solution caused by reactions with adjacent rock, and cooling as a result of mixing with seawater—can also cause rapid precipitation and form concentrated deposits. When valuable minerals are present, an ore can be the result.

EXAMPLES OF PRECIPITATION

Veins form when hydrothermal solutions deposit minerals in open fractures, and many such veins are found in regions of volcanic activity (Figure 21.5). The famous gold deposits at Cripple Creek, Colorado, were formed in fractures associated with a small caldera, and the huge tin and silver deposits in Bolivia are in fractures that are localized in and around stratovolcanoes. In each case, the fractures formed as a result of volcanic activity, and the magma chambers that fed the volcanoes served as the sources of the hydrothermal solutions that rose up and formed the mineralized veins.

Figure 21.5 Mining a Rich Vein A rich vein in Potosí, Bolivia, containing chalcopyrite, sphalerite, and galena cutting andesite. The andesite has been altered metasomatically by the hydrothermal solution that deposited the ore minerals.

A cooling granitic stock or batholith is a source of heat just as the magma chamber beneath a volcano is—and it can also be a source of hydrothermal solutions. Such solutions move outward from a cooling stock and will flow through any fracture or channel, metasomatically altering the surrounding rock in the process and commonly depositing valuable minerals. Many famous ore bodies are associated with intrusive igneous rocks. The tin deposits of Cornwall, England, and the copper deposits at Butte, Montana, Bingham, Utah, and Bisbee, Arizona, are examples. For examples of hydrothermal deposits forming today, see *Understanding Our Environment*, Box 21.1, *Hydrothermal Mineral Deposits Forming Today*.

MAGMATIC MINERAL DEPOSITS

The processes of partial melting and fractional crystallization discussed in Chapter 4 are two ways of separating some minerals from others. Fractional crystallization, in particular, can lead to the creation of valuable mineral deposits. The processes involved are entirely magmatic, and so such deposits are referred to as magmatic mineral deposits.

Pegmatites formed by fractional crystallization of granitic magma (Chapter 5) commonly contain rich concentrations of such elements as lithium, beryllium, cesium, and niobium. Much of the world's lithium is mined from pegmatites such as those at King's Mountain, North Carolina, and Bikita in Zimbabwe. The great Tanco pegmatite in Manitoba, Canada, produces much of the world's cesium, and pegmatites in many countries yield beryl, one of the main ore minerals of beryllium.

Crystal settling, another process of fractional crystallization, can also form valuable mineral deposits. The process is especially important in low-viscosity basaltic magma. When a large chamber of basaltic magma crystallizes, one of the first minerals to form is chromite, the main ore mineral of chromium. As shown in Figure 4.21B, settling of the dense chromite crystals to the bottom of the magma chamber can produce almost-pure layers of chromite. The world's principal ores of chromite, in the Bushveld Igneous Complex in South Africa and the Great Dike of Zimbabwe, were both formed as a result of crystal settling.

SEDIMENTARY MINERAL DEPOSITS

The term *sedimentary mineral deposit* is applied to any local concentration of minerals formed through processes of sedimentation. Any process of sedimentation can form localized concentrations of minerals, but it has become common practice to restrict use of the term *sedimentary* to those mineral deposits formed through precipitation of substances carried in solution.

EVAPORITE DEPOSITS

The most direct way in which sedimentary mineral deposits form is by evaporation of lake water or seawater. The layers of salts that precipitate as a consequence of evaporation are called evaporite deposits (Chapter 7).

Examples of salts that precipitate from lake waters of suitable composition are sodium carbonate (Na_2CO_3), sodium sulfate (Na_2SO_4), and borax ($Na_2B_4O_7 \cdot 10H_2O$) (Figure 7.11B). Huge evaporite deposits of sodium carbonate were laid down in the Green River basin of Wyoming during the Eocene Epoch. This is the same lake in which the rich Green River oil shales were deposited (Figure 7.29). Borax and other boron-containing minerals are mined from evaporite lake deposits in Death Valley and Searles and Borax lakes, all in California, and in Argentina, Bolivia, Turkey, and China.

Much more common and important than lake water evaporites are the marine evaporites formed by evaporation of seawater. The most important salts that precipitate from seawater are gypsum ($CaSO_4 \cdot 2H_2O$), halite ($NaCl$), and carnallite ($KCl \cdot MgCl_2 \cdot 6H_2O$). Low-grade metamorphism of marine evaporite deposits causes another important mineral, sylvite (KCl), to form from carnallite. Marine evaporite deposits are widespread; in North America, for example, strata of marine evaporites underlie as much as 30 percent of the land area (Figure 21.6). Most of the salt that we use, plus the gypsum used for plaster and the potassium used in plant fertilizers, is recovered from marine evaporites.

IRON DEPOSITS

Sedimentary deposits of iron minerals are widespread, but the amount of iron in average seawater is so small that such deposits cannot have formed from seawater that is the same as today's seawater.

All sedimentary iron deposits are tiny by comparison with the class of deposits characterized by the *Lake Superior–type iron deposits*. These remarkable deposits, mined principally in Michigan and Minnesota, were long the mainstay of the U.S. steel industry but are declining in importance today as imported ores replace them. The deposits are of early Proterozoic age (about 2 billion years or older) and are found in sedimentary basins on every craton, particularly in Labrador, Venezuela, Brazil, Russia, India, South Africa, and Australia. Every aspect of the Lake Superior–type deposits indicates chemical precipitation; they are chemical sediments, as discussed in Chapter 7. The deposits are interbedded layers of chert and several different kinds of iron minerals. Because the deposits are

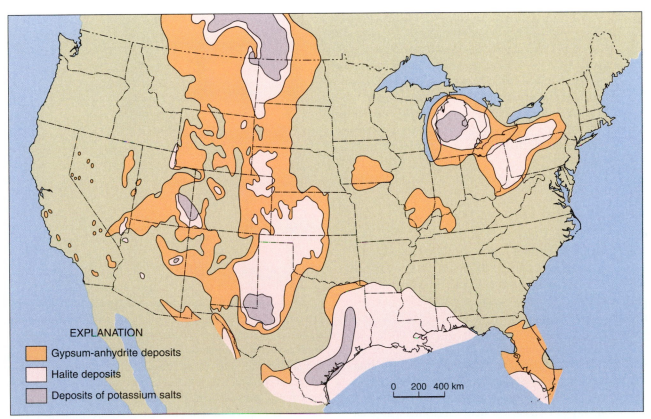

EXPLANATION

- Gypsum-anhydrite deposits
- Halite deposits
- Deposits of potassium salts

0   200   400 km

Figure 21.6 Evaporite Deposits Portions of the United States known to be underlain by marine evaporite deposits. The areas underlain by gypsum and anhydrite do not contain halite. The areas underlain by halite are also underlain by gypsum and anhydrite. The areas underlain by potassium salts are also underlain by halite and by gypsum and anhydrite.

Box 21.1

UNDERSTANDING OUR ENVIRONMENT

HYDROTHERMAL MINERAL DEPOSITS FORMING TODAY

Three extraordinary discoveries over a 15-year period have changed our thinking about mineral deposits. The first discovery, in 1962, was accidental. Until that year no one was sure where to look for modern hydrothermal solutions or even how to recognize one when it was found. Drillers seeking oil and gas in the Imperial Valley of southern California were astonished when they struck a 320°C brine at a depth of 1.5 km. As the brine flowed upward, it cooled and precipitated minerals it had been carrying in solution. Over three months, the well deposited 8 tons of siliceous scale containing 20 percent copper and 8 percent silver by weight. The drillers had found a hydrothermal solution that could, under suitable flow conditions, form a rich mineral deposit.

The Imperial Valley is a sediment-filled graben covering the join between the Pacific and North American plates, where the East Pacific Rise passes under North America (Figure B21.1). Magmatism is the source of heat for the brine solution discovered in 1962. These brines provided the first unambiguous evidence that hydrothermal solutions can leach metals from ordinary sediments.

Before geologists had a chance to fully absorb the significance of the Imperial Valley discovery, a second remarkable find was announced. In 1964, oceanographers

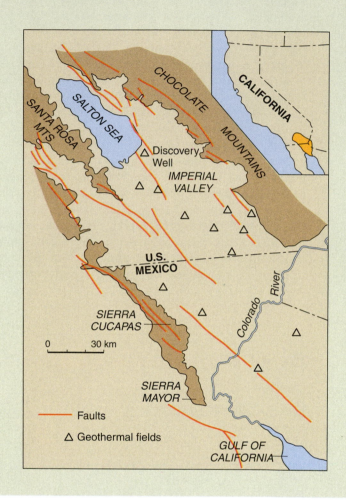

Figure B21.1   Present-day Hydrothermal Solution   The Imperial Valley graben (also known as the Salton Trough) is bounded by the Chocolate Mountains on the east and the Santa Rosa Mountains on the west. Hydrothermal solutions were discovered in a well drilled on the southern end of the Salton Sea. Places where geothermal activity is known, and where other hydrothermal solutions may be present at depth, are marked with triangles.

so large, it is inferred that the iron and silica must have been transported in surface water to the ocean, but the cause of precipitation remains uncertain. Many experts suspect that Lake Superior–type deposits may be ancient evaporites that formed from seawater of a different composition than today's seawater.

Lake Superior–type iron deposits are not ores. The grades of the deposits range from 15 to 30 percent Fe by weight, and the deposits are so fine grained that the iron minerals cannot be easily separated from the gangue. Two additional processes can form ore. First, leaching of silica during weathering can lead to secondary enrichment and can produce ores containing as much as 66 percent Fe. Compare Figure 7.25 of a Lake Superior–type iron deposit in the Hamersley Range, Western Australia, with Figure 21.7, a

sample of ore developed by secondary enrichment in the Hamersley Range. The rocks in Figure 7.25 contain about 25 percent Fe, whereas those in Figure 21.7 have had most of the silica leached out and contain about 60 percent Fe.

The second way a Lake Superior–type iron deposit can become an ore is through metamorphism. Two changes occur as a result of metamorphism. First, grain sizes increase so that separating ore minerals from the gangue becomes easier and cheaper. Second, new mineral assemblages form, and iron silicate and iron carbonate minerals originally present can be replaced by magnetite or hematite, both of which are desirable ore minerals. The grade is not increased by metamorphism. It is the changes in grain size and mineralogy that transform the sedimentary rock into an ore. Iron ores formed as a result of meta-

discovered a series of hot, dense brine pools at the bottom of the Red Sea. The brines are trapped in the graben formed by the spreading center between the Arabian and African plates (Figure B21.2), and they are so much saltier, and therefore more dense, than seawater that they remain ponded in the graben even though they are as hot as 60°C. Many such brine pools have now been discovered.

The Red Sea brines rise up the normal faults associated with the central rift of a spreading center and, like the Imperial Valley brines, have evolved to their present compositions through reactions with the enclosing rocks. The Red Sea brine discovery was surprising, but even more surprising was the discovery that sediments at the bottom of the pools contained ore minerals such as chalcopyrite, galena, and sphalerite. In other words, the oceanographers had discovered modern stratabound mineral deposits in the process of formation.

The third remarkable discovery was really a series of discoveries that commenced in 1978. Scientists using deep-diving submarines made a series of dives on the East Pacific Rise at 21° N latitude. To their amazement, they found 300°C hot springs emerging from the seafloor 2500 m below sea level. Around the hot springs lay a blanket of sulfide minerals. The submariners watched a modern volcanogenic sulfide deposit forming before their eyes.

Each of the discovery sites—Imperial Valley, Red Sea, and 21° N—is on a spreading center, and so there is no doubt that the deposits are forming as a result of plate tectonics. Soon the hunt was on to see if seafloor deposits could be found above subduction zones. In 1989, a joint German-Japanese oceanographic expedition to the western Pacific discovered the first modern subduction-related deposits. No longer are geologists limited to speculating about how certain mineral deposits *might* have formed. Today we can study them as they grow.

Figure B21.2 Ore Deposit Forming on the Seafloor Topography of the Red Sea graben near the Atlantis II brine pool. Hot, dense brines rise up normal faults, pond on the floor of the graben, and form stratabound deposits rich in copper and zinc.

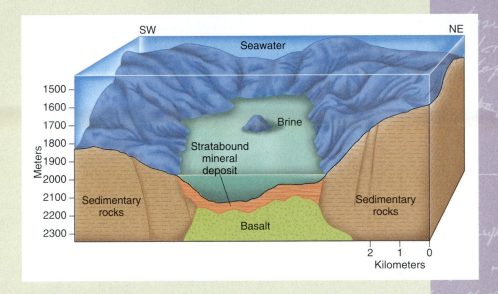

morphism are called *taconites*, and they are now the main kind of ore mined in the Lake Superior region.

Figure 21.7 Iron Ore Leaching of silica from an iron-rich sedimentary rock leads to the formation of a secondarily enriched mass of limonite and hematite. This outcrop, in the Hamersley Range of Western Australia, developed by secondary enrichment from the kind of iron-rich sedimentary rock in Figure 7.25. Remnants of the original silica band can be seen to the right of the hammer.

STRATABOUND DEPOSITS

Some of the world's most important ores of lead, zinc, and copper occur in sedimentary rocks. The ore minerals—galena, sphalerite, chalcopyrite, and pyrite—occur in such regular, fine layers that they look like sediments (Figure 21.8). The sulfide mineral layers are enclosed by and par-

Figure 21.8   A Stratabound Ore   Stratabound ore of lead and zinc from Kimberley, British Columbia. The layers of pyrite (yellow), sphalerite (brown), and galena (gray) are parallel to the layering of the sedimentary rock in which they occur. The specimen is 4 cm across.

allel to the sedimentary strata in which they occur, and for this reason such deposits are called *stratabound mineral deposits*. They look like sediments but are not, strictly speaking, sediments. It is more correct to consider most stratabound deposits as being diagenetic in origin.

Stratabound deposits form when a hydrothermal solution invades and reacts with a muddy sediment. Reactions between sediment grains and the solution cause deposition of the ore minerals. Deposition commonly occurs before the sediment has become a sedimentary rock.

The famous copper deposits of Zambia, in central Africa, are stratabound ores, as are the great Kupferschiefer deposits of Germany and Poland. The world's largest and richest lead and zinc deposits, at Broken Hill and Mount Isa in Australia and at Kimberley in British Columbia, are also stratabound ores.

PLACERS

The way in which a mineral with a high specific gravity can become concentrated by flowing water was discussed in Chapters 14 and 18. Deposits of minerals having high specific gravities are *placers*. The most important minerals concentrated in placers are gold, platinum, cassiterite (SnO_2), and diamond. Typical locations of placers are illustrated in Figure 21.9.

Gold is the most valuable mineral recovered from plac-

ers; more than half of the gold recovered throughout all of human history has come from placers. This is the result of the huge gold production from South Africa, which has come from placers.

The South African gold deposits are really fossil placers, and they have many unusual features. Most placers are found in stream gravels that are geologically young. The South African fossil placers are a series of gold-bearing conglomerates (Figure 21.10) that were laid down 2.7 billion years ago as gravels in the shallow marginal waters of a marine basin. Associated with the gold are grains of pyrite and uranium minerals. As far as size and richness are concerned, nothing like the deposits in the Witwatersrand basin has been discovered anywhere else. Nor has the original source of all the placer gold been discovered, and so it is not possible to say why so much of the world's mineable gold should be concentrated in this one sedimentary basin.

Mining in the Witwatersrand basin has reached a depth of 3600 m (11,800 ft). This is the deepest mining in the world, and there are plans to continue mining to depths as great as 4500 m. Despite such ambitious plans, the heyday of gold mining in South Africa has passed because the deposits are running out of ore.

Through the middle years of the 1980s, the price of gold fluctuated between about $14 and $16 a gram. This led to a boom in gold prospecting and to the discovery of a large number of new ore deposits in the United States, Canada, Australia, the Pacific islands, and elsewhere. Most of the new discoveries are hydrothermal deposits. Despite all the new discoveries, South Africa with its huge fossil placers continues to dominate the world's gold production. In 2001, South Africa still supplied about 25 percent of all the gold produced in the world, but the production rate is dropping slowly. By contrast, the production of gold in the United States is rising as a result of discoveries in Nevada.

RESIDUAL MINERAL DEPOSITS

Weathering occurs because newly exposed rock is not chemically stable when it is in contact with rainwater and the atmosphere. Chemical weathering, in particular, leads to mineral concentration through the removal of soluble materials in solution and the concentration of a less soluble residue. A common example of a deposit formed through residual concentration is laterite (Figure 21.11), and the most important kind of laterite is bauxite (Chapter 6).

Bauxites are the source of the world's aluminum. They are widespread in the world, but they are concentrated in the tropics because that is where lateritic weathering occurs. Where bauxites are found in present-day temperate conditions, such as France, China, Hungary, and Arkansas, it is clear that the climate was tropical when the bauxites formed.

All bauxites are vulnerable to erosion. They are not found in glaciated regions, for example, because overriding glaciers scrape off the soft surface materials. The vul-

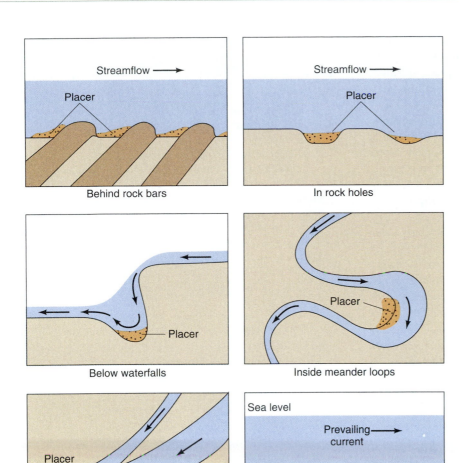

Figure 21.9   Placer Deposits   Placers occur where barriers allow flowing water to carry away the suspended load of lightweight particles while trapping particles carried in the bed load. Placers can form whenever water moves but are most commonly associated with streams of longshore currents.

Figure 21.10   Paleoplacer   Gold is recovered from ancient fossil placer deposits of the Witwatersrand, South Africa. The gold is found at the base of conglomerate layers interbedded with finer-grained sandstone, here seen in weathered outcrop at the site where gold was first discovered in 1886.

Figure 21.11   Laterite   Red, iron-rich laterite near Khao Yai, Thailand. Originally a conglomerate, the rock is now a mixture of hematite and clay as a result of extreme tropical leaching.

nerability of bauxites means that most deposits are geologically young. More than 90 percent of all known deposits formed during the last 60 million years, and all of the very large deposits formed less than 25 million years ago.

When a preexisting mineral deposit is chemically weathered, the results can sometimes be a spectacular upgrading in the metal content. We have already mentioned the enrichment in Lake Superior–type iron formations as a result of the leaching of silica, but other kinds of deposits can also be enriched by the leaching process, which is known as **secondary enrichment**.

Many of the world's manganese deposits have been formed by secondary enrichment of low-grade primary deposits, particularly in tropical regions. Nickel is always present in gabbros and peridotites in trace amounts. When such rocks are subjected to chemical weathering, the pyroxene, olivine, and plagioclase feldspar break down, silica and other constituents are removed in solution, and in the limonite residue there is a mixture of nickel hydroxide minerals called garnierite. One of the largest nickel deposits ever found, in New Caledonia, was formed in this manner.

Secondary enrichment is striking in many hydrothermal deposits containing sulfide minerals. Striking examples are found in the great deposits of the arid southwestern United States and desert regions of northern Chile. The primary minerals are pyrite (FeS_2) and chalcopyrite ($CuFeS_2$). These minerals are oxidized by rainfall and the atmosphere to form a copper-bearing solution of sulfuric acid and a residue of limonite. The acid solution percolates down and reacts with unoxidized pyrite and chalcopyrite by dissolving iron and precipitating copper to form chalcocite (CuS_2). The chalcocite blanket lies below the weathered residue and above the leaner primary materials below.

USEFUL MINERAL SUBSTANCES

It is convenient for purposes of discussion to group mineral products on the basis of the way they are used rather than the way they occur. Excluding substances used for energy, there are two broad groups: (1) minerals from which metals such as iron, copper, and gold can be recovered, and (2) nonmetallic minerals, such as salt, gypsum, and clay, used not for the metals they contain but for their properties as chemical compounds. The nonmetallic substances can be further subdivided on the basis of more specialized uses (Table 21.1).

GEOCHEMICALLY ABUNDANT METALS

Metals can be usefully subdivided on the basis of their average percentage in the crust. Those present in such abundance that they make up 0.1 percent by weight or

TABLE 21.1 Principal Mineral Substances, Grouped According to Use

Metals
  Geochemically abundant metals
    Iron, aluminum, magnesium, manganese, titanium
  Geochemically scarce metals
    Copper, lead, zinc, nickel, chromium, gold, silver, tin, tungsten, mercury, molybdenum, uranium, platinum, palladium, and many others
Nonmetallic substances
  Used for chemicals
    Sodium chloride (halite), sodium carbonate, sulfur, borax, fluorite
  Used for fertilizers
    Calcium phosphate (apatite), potassium chloride (sylvite), sulfur, calcium carbonate (limestone), sodium nitrate
  Used for building
    Gypsum (for plaster), limestone, clay (for brick and tile), asbestos, sand gravel, crushed rock of various kinds, shale (for cement)
  Used for ceramics and abrasives
    Ceramics: Clay, feldspar, quartz
    Abrasives: Diamond, garnet, corundum, pumice, quartz

more of the crust are considered to be **geochemically abundant**. These metals are iron, aluminum, manganese, magnesium, and titanium.

Geochemically abundant metals require comparatively small enrichment factors to form large deposits. The minerals that are concentrated in deposits of the geochemically abundant metals tend to be oxides and hydroxides. The most important kinds of deposits are residual, sedimentary, and magmatic deposits.

GEOCHEMICALLY SCARCE METALS

Metals that make up less than 0.1 percent by weight of the crust are said to be **geochemically scarce**.

With the exception of copper, zinc, and chromium, minerals of the geochemically scarce metals are not found in common rocks. However, chemical analysis of any rock will reveal that even though the minerals are absent, geochemically scarce metals are certainly present. Further research will reveal that the scarce metals are present exclusively as a result of atomic substitution (Chapter 3). Atoms of the scarce metals (such as nickel, cobalt, and copper) can readily substitute for more common atoms (such as magnesium and calcium). In order for a mineral deposit to form, therefore, some gathering and concentrating agent, such as a hydrothermal solution, must react with the rock-forming minerals and leach the scarce metals from them. The solution must then transport the metals in solution and deposit them as separate minerals in a localized place. With such a complicated chain of events, it is hardly surprising that deposits of geochemically scarce metals are rarer and much smaller than deposits of the abundant metals.

Most ore minerals of the scarce metals are sulfides; a few, such as the ore minerals of tin and tungsten, are oxides. In the case of gold, platinum, palladium, and a few less common elements, the metal itself is the most important mineral. Most scarce metal deposits form as hydrothermal or magmatic mineral deposits. In the case of gold and platinum, placer concentration is also important.

> **Before you go on:**
>
> 1. What is the difference between a mineral deposit and an ore deposit?
> 2. What are the four main families of mineral deposits?
> 3. What role does plate tectonics play in the location of mineral deposits?
> 4. Which metals are geologically abundant, and which are geologically scarce?

ENERGY RESOURCES

A healthy, hard-working person can produce just enough muscle energy to keep a single 75-watt lightbulb burning for 8 hours a day. It costs about 10 cents to purchase the same amount of energy from the local electrical utility. Viewed strictly as machines, humans aren't worth much. By comparison, the amount of mechanical and electrical energy used each 8-hour working day in North America could keep 400 75-watt bulbs burning for every person living there.

To see where all this energy is used, it is necessary to sum up all the energy employed to grow and transport food, make clothes, cut lumber for new homes, light streets, heat and cool office buildings, and do a myriad of other things. The uses can be grouped into three categories: transportation, home and commerce, and industry (meaning all manufacturing and raw material processing plus the growing of foodstuffs). The present-day uses and sources of energy in the United States are summarized in Figure 21.12.

How much energy do all the people of the world use? The total is enormous. The energy drawn annually from major fuels—coal, oil, and natural gas—plus that from nuclear power plants is 3×10^{20} J. Nobody keeps accurate accounts of all the wood and animal dung burned in the cooking fires of Africa and Asia, but the amount has been estimated to be so large that when it is added to the 3×10^{20} J figure, the world's total energy consumption rises to about 3.3×10^{20} J annually. This is equivalent to the burning of about 2 metric tons of coal or 9 barrels of oil for

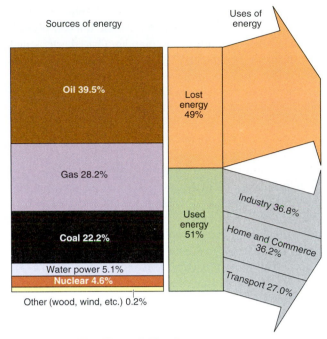

Figure 21.12 How Energy Is Used Uses and sources of energy in the United States. Lost energy arises both from inefficiencies of use and from the fact that the laws of thermodynamics impose a limit to the efficiency of any engine and therefore a limit on the fraction of available energy that can be usefully employed.

every living man, woman, and child each year! Energy consumption around the world is very uneven, however. In less developed countries such as India and Tanzania, energy use is equivalent to burning only 4 or 5 barrels of oil per person per year, whereas in a developed country such as the United States, energy use is equivalent to burning more than 50 barrels of oil per person per year.

SUPPLIES OF ENERGY

The chief sources of the energy consumed in highly industrialized nations are few: the fossil fuels (coal, oil, natural gas), hydroelectric and nuclear power, wood, wind, and a very small amount of muscle energy. As recently as a century ago, wood was an important fuel in industrial societies, but now it is used mainly for space heating in some dwellings.

FOSSIL FUELS

The term *fossil fuel* refers to the remains of plants and animals trapped in sediment that can be used for fuel. The kind of sediment, the kind of organic matter trapped, and the changes in the organic matter that take place as a result of burial and diagenesis, determine the kind of fossil fuel that forms.

In the ocean, microscopic photosynthetic phytoplankton and bacteria are the principal sources of trapped organic matter. Shales do most of the trapping. Once bacteria and phytoplankton are trapped in the shale, the organic compounds they contain—*proteins, lipids,* and *carbohydrates*—become part of the shale, and it is these compounds that are transformed (mainly by heat) to oil and gas.

On land, it is trees, bushes, and grasses that contribute most of the trapped organic matter; these large land plants are rich in resins, waxes, and lignins, which tend to remain solid and form coal rather than oil or natural gas.

In many marine and lake shales, burial temperatures never reach the levels at which the original organic molecules are converted to the organic molecules found in oil and natural gas. Instead, an alteration process occurs in which wax-like substances containing large molecules are formed. This material, which remains solid, is called *kerogen*, and it is the substance in so-called *oil shale*. Kerogen can be converted to oil and gas by mining the shale and heating it in a retort.

COAL

The combustible sedimentary rock we call coal is the most abundant fossil fuel. Most of the coal mined is eventually either burned under boilers to make steam for electric generators, or else converted to coke, an essential ingredient in the smelting of iron ore and the making of steel. In addition to its use as a fuel, coal is a raw material for nylon and many other plastics, plus a multitude of other organic chemicals.

The conditions under which organic matter accumulates in swamps as peat, then during burial and diagenesis is converted to coal, are discussed in Chapter 7. Coalification involves the loss of volatile materials such as H_2O, CO_2, and CH_4 (methane). As the volatiles escape, the remaining coal is increasingly enriched with carbon. Through coalification, peat is converted successively to *lignite* (one type of coal), *subbituminous coal*, and *bituminous coal* (Figure 21.13). These coals are sedimentary rocks. However, *anthracite*, a still later phase in the coalification process, is a metamorphic rock. Because of its low volatile content, anthracite is hard to ignite, but once alight it burns with almost no smoke. In contrast, lignite is rich in volatiles, burns smokily, and ignites so easily that it is dangerously subject to spontaneous ignition.

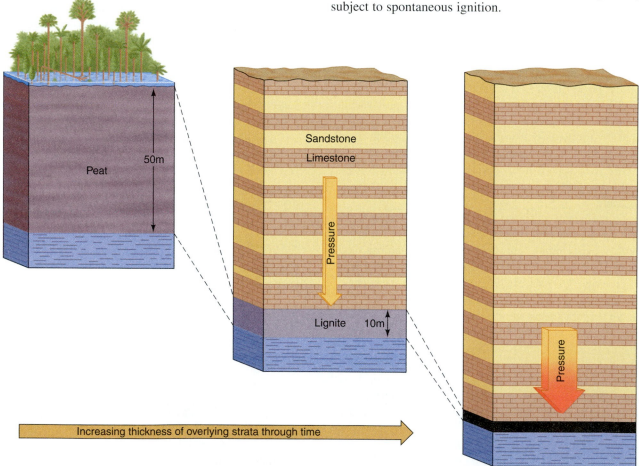

Figure 21.13 From Peat to Coal Plant matter in peat (a biogenic sediment) is converted to coal (a biogenic sedimentary rock) by decomposition and increasing pressure as overlying sediments increase in thickness. By the time a layer of peat 50 feet thick has been converted to bituminous coal, its thickness has been reduced by 90 percent. In the process, the proportion of carbon has increased from 60 to 80 percent.

In regions where metamorphism has been intense, coal has been changed so thoroughly that it has been converted to graphite, in which all volatiles have been lost. Graphite will not burn in an ordinary fire.

Occurrence of Coal

A coal seam is a flat, lens-shaped body having the same surface area as the swamp in which it originally accumulated. Most coal seams tend to occur in groups. In western Pennsylvania, for example, 60 seams of bituminous coal are found. This clustering indicates that the coal must have formed in a slowly subsiding site of sedimentation.

Coal swamps seem to have formed in many sedimentary environments, of which two types predominate. One consists of slowly subsiding basins in continental interiors and the swampy margins of shallow inland seas formed at times of high sea level. This is the home environment of the bituminous and subbituminous coal seams in Utah, Montana, Wyoming, and the Dakotas. The second sedimentary environment consists of continental margins with wide continental shelves (that is, continental margins in plate interiors) that were flooded at times of high sea level. This is the environment of the bituminous coals of the Appalachian region.

Coal-Forming Periods

Although peat can form under even subarctic conditions, it is clear that the luxuriant plant growth needed to form thick and extensive coal seams develops most readily in a tropical or semitropical climate. The Great Dismal Swamp in Virginia and North Carolina is one of the largest modern peat swamps. It contains an average thickness of 2 m of peat. However, unless this swamp lasts millions of years, even that dense growth is insufficient to produce a coal seam as thick as some of the seams in Pennsylvania.

Peat formation has been widespread and more or less continuous from the time land plants first appeared about 450 million years ago, during the Silurian Period. The size of peat swamps has varied greatly, however, and so, as a consequence, has the amount of coal formed. By far the greatest period of coal swamp formation occurred during the Carboniferous and Permian periods, when Pangaea existed. The great coal beds of Europe and the eastern United States formed at this time, when the plants of coal swamps were giant ferns and scale trees (gymnosperms). The second great period of coal deposition peaked during the Cretaceous Period but commenced in the early Jurassic and continued until the mid-Tertiary. The plants of the coal swamps during this period were flowering plants (angiosperms) much like flowering plants today.

Petroleum: Oil and Natural Gas

As was mentioned in the introductory essay to this chapter, rock oil is one of the earliest resources our ancestors learned to use. However, the major use of oil really started about 1847 when a merchant in Pittsburgh, Pennsylvania, started bottling and selling rock oil from natural seeps to be used as a lubricant. Five years later, in 1852, a Canadian chemist discovered that heating and distillation of rock oil yielded kerosene, a liquid that could be used in lamps. This discovery spelled doom for candles and whale-oil lamps. Wells were soon being dug by hand near Oil Springs, Ontario, in order to produce oil. In Romania in 1856, using the same hand-digging process, workers were producing 2000 barrels a year (a barrel is equivalent to 42 U.S. gallons). In 1859, the first oil well was drilled in Titusville, Pennsylvania. On August 27, 1859, at a depth of 21.2 m, oil-bearing strata were encountered and up to 35 barrels of oil a day were pumped out. Oil was soon discovered in West Virginia (1860), Colorado (1862), Texas (1866), California (1875), and many other places.

The earliest known use of natural gas was about 3000 years ago in China, where gas seeping out of the ground was collected and transmitted through bamboo pipes to be ignited and used to evaporate saltwater in order to recover salt. It wasn't long before the Chinese were drilling wells to increase the flow of gas. Modern use of gas started in the early seventeenth century in Europe, where gas made from wood and coal was used for illumination. Commercial gas companies were founded as early as 1812 in London and 1816 in Baltimore. The stage was set for the exploitation of an accidental discovery at Fredonia, New York, in 1821. A water well drilled in that year produced not only water but also bubbles of a mysterious gas. The gas was accidentally ignited and produced such a spectacular flame that a new well was drilled at the same site and wooden pipes were installed to carry the gas to a nearby hotel, where 66 gaslights were installed. By 1872, natural gas was being piped as far as 40 km from its source.

Origin of Petroleum

Petroleum is a term used for both oil and natural gas. Petroleum is a product of the decomposition of organic matter trapped in sediment; once formed, the petroleum migrates through aquifers and becomes trapped in reservoirs.

The migration of petroleum deserves further discussion. The sediment in which organic matter is accumulating today is rich in clay minerals, whereas most of the strata that constitute oil or gas pools are sandstones (consisting of quartz grains), limestones and dolostones (consisting of carbonate minerals), and much-fractured rock of other kinds. Long ago, geologists realized that oil and gas form in one kind of material (shale) and at some later time migrate to a more porous rock (sandstone or limestone).

Petroleum migration is analogous to groundwater migration. When oil and gas are squeezed out of the shale in which they originated and enter a body of sandstone or limestone, they can migrate more easily than before because most sandstones and limestones are both more porous and more permeable than any shale. The force of molecular attraction between oil and quartz or carbonate minerals is weaker than that between water and quartz or carbonate minerals. Hence, because oil and water do not mix, water remains fastened to the quartz or carbonate grains while oil occupies the central parts of the larger openings in the porous sandstone or limestone. Because it is lighter than water, the oil tends to glide upward past the carbonate- and quartz-held water. In this way it becomes segregated from the water; when it encounters a trap, it can form a pool.

Most of the petroleum that forms in sediments does not find a suitable trap, and eventually makes its way, along with groundwater, to the surface. It is estimated that no more than 0.1 percent of all the organic matter originally buried in a sediment is eventually trapped in an oil or a gas pool. It is not surprising, therefore, that the highest ratio of oil and gas pools to volume of sediment is found in rock no older than 2.5 million years, and that nearly 60 percent of all the oil and gas discovered so far has been found in strata of Cenozoic age (Figure 21.14). This does not mean that older rocks produced less petroleum. It simply means that oil in older rocks has had a longer time in which to escape.

DISTRIBUTION OF OIL

Petroleum deposits, like coal deposits, are frequent but are distributed unevenly. The reasons for uneven petroleum distribution are not as obvious as they are with coal. Suitable source sediments for petroleum are very widespread and seem as likely to form in subarctic waters as in tropical regions. The critical controls seem to be an anoxic (lack of oxygen) bottom water so that organic matter gets buried instead of oxidizing, a supply of heat to effect the conversion of solid organic matter to oil and gas, and the formation of a suitable trap before the petroleum has leaked away.

Solid organic matter is converted to oil and gas within a specific range of depth and temperature defined by the geothermal gradients shown in Figure 21.15. If a thermal gradient is too low (less than 1.8°C/100 m), conversion does not occur to either oil or gas. If the gradient is above 5.5°C/100 m, conversion to gas starts at such shallow depths that little trapping occurs. Once oil and gas have been formed, they will accumulate in pools only if suitable traps are present. Most oil and gas pools are found beneath anticlines; the timing of the folding event is therefore a critical part of the trapping process. If folding occurs after petroleum has formed and migrated, pools will not form. The great oil pools in the Middle East arose through the fortunate coincidence of the right thermal gradient and the development of anticlinal traps during the collision of Europe and Asia with Africa.

How much oil is there in the world? This is an extremely controversial question. Approximately 800 billion barrels have already been pumped out of the ground and about 30 billion more are pumped each year that passes. A lot of additional oil has been located by drilling and is still waiting to be pumped. Probably a great deal more oil remains

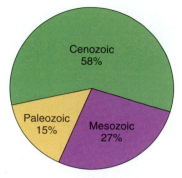

Figure 21.14 Age of Strata and Yield of Oil Percentage of world's total oil production from strata of different ages.

to be found by drilling. Unlike coal, for which the volume of strata in a basin of sediment can be accurately estimated even before drilling, the volume of undiscovered oil can only be surmised. The way estimates are made is to use the accumulated experience of a century of drilling. Knowing how much oil has been found in an intensively drilled area, such as eastern Texas, experts make estimates of probable oil volumes in other regions where rock types and structures are similar to what we find in eastern Texas. Using this approach, and considering all the sedimentary basins of the world (Figure 21.16), experts estimate that about 2100 billion barrels of oil will eventually be discovered.

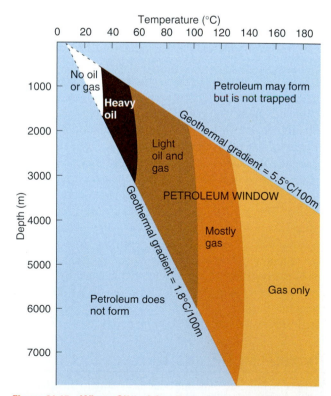

Figure 21.15 Where Oil and Gas Form Regions of depth and temperature within which oil and gas are generated and trapped.

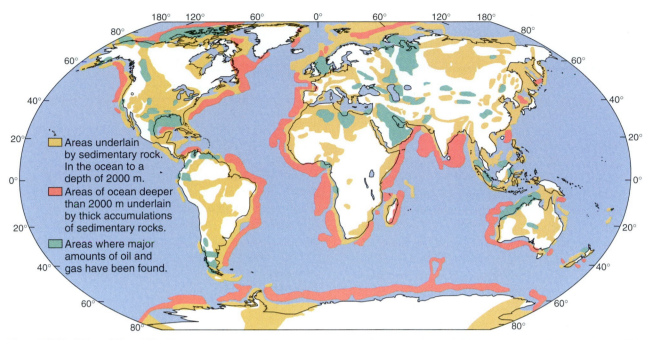

Figure 21.16   Where Oil and Gas Occur   Areas underlain by sedimentary rock and regions where large accumulations of oil and gas have been located. Where the ocean is deeper than 2000 m, sedimentary rock has yet to be tested for its oil and gas potential.

TARS

Oil that is exceedingly viscous will not flow easily and cannot be pumped. Colloquially called **tar**, heavy, viscous oil acts as a cementing agent between mineral grains in an oil pool. The tar can be recovered only if the sandstone is mined and heated enough to make the tar flow. The resulting tar must then be processed to recover the valuable gasoline fraction. The cost of mining and treating "tar sands," as heavy, viscous oil deposits are called, is high, but it is technologically possible and someday tar sands may be an important source of fuel. The largest known occurrence of tar sands is in Alberta, Canada, where the *Athabasca Tar Sand* covers an area of 5000 km^2 and reaches a thickness of 60 m (Figure 21.17). Similar deposits almost as large are known in Venezuela and in Russia.

Figure 21.17   Tar Sands   Athabasca tar sand being mined in Alberta, Canada. After mining, the tar-cemented sands are heated in order to soften and remove the tar prior to processing.

OIL SHALE

Another potential source of petroleum is kerogen in shale. If the kerogen is heated, it breaks down and forms liquid and gaseous hydrocarbons similar to those in oil and gas. All shales contain some kerogen, but to be considered a resource for energy, the kerogen must yield more energy than is required to mine and heat it. Only those shales that yield 40 or more liters of distillate per ton can be considered because the energy needed to process and mine a ton of shale is equivalent to that created by burning 40 liters of oil.

The world's largest deposit of rich oil shale is in the United States. During the Eocene Epoch, many large, shallow lakes existed in basins in Colorado, Wyoming, and Utah; in three of them a series of rich organic sediments were deposited that are now the Green River Oil Shales (Figure 7.29). The richest shales were deposited in the lake in Colorado now called the Piceance basin. These shales are capable of producing as much as 240 liters of oil per ton. Scientists of the U.S. Geological Survey estimate that, in the Green River Oil Shales alone, oil-shale resources capable of producing 50 liters or more of oil per ton of shale can ultimately yield about 2000 billion barrels of oil.

Rich deposits of oil shale in other parts of the world have not been adequately explored, but there is a huge deposit in Brazil called the Irati Shale. Another very large deposit is known in Queensland, Australia, and others have been reported in such widely dispersed places as South Africa and China. Although oil shales have been mined and processed in an experimental fashion in the United States, the only countries where extensive commercial production has been tried are Estonia and China. Production expenses today make exploitation of oil shales in all coun-

tries unattractive by comparison with oil and gas. Many experts believe, however, that large-scale mining and processing of oil shale will eventually occur.

HOW MUCH FOSSIL FUEL?

Are supplies of fossil fuels adequate to meet future demands? If we use a barrel of oil as our unit of measurement, we can compare quantities of all fossil fuels directly. Approximately 0.22 ton of coal produces the same amount of heat energy as one barrel of oil. Thus, the world's recoverable coal reserves of $13,800 \times 10^9$ tons are equivalent to about 63,000 billion barrels of oil.

Considering the approximate world-use rate of oil (30 billion barrels a year), and comparing the estimated recoverable amounts of fossil fuels (Table 21.2), we see that only coal seems to have the capacity to meet long-term demands.

OTHER SOURCES OF ENERGY

Three sources of energy other than fossil fuels have already been developed to some extent: Earth's plant life (so-called biomass energy), hydroelectric energy, and nuclear energy. Five others—the Sun's heat, winds, waves, tides, and Earth's internal heat (geothermal power)—have been tested and developed on a limited basis, but none has yet been shown to be developable on a large scale. Many experts believe geothermal heat is the energy source most likely to be further exploited in the near future.

BIOMASS ENERGY

Scientists working for the United Nations estimate that wood and animal dung used for cooking and heating fires now amount to energy production of 3×10^{19} J annually. This is approximately 10 percent of the world's total energy use. The greatest use of wood as a fuel occurs in developing countries, where the cost of fossil fuel is very high in relation to income.

Measurements made on living plant matter indicate that new plant growth on land equals 1.5×10^{11} metric tons of dry plant matter each year. If all of this were burned, or used in some other way as a biomass energy source, it would produce almost nine times more energy than the world uses each year. Obviously, this is a ridiculous suggestion because in order to do so all the forests would have to be destroyed, plants could not be eaten, and agricultural soils would be devastated. Nevertheless, controlled harvesting of fuel plants could probably increase the fraction of the biomass now used for fuel without serious disruption to forests or to food supplies. In several parts of the world, such as Brazil, China, and the United States, experiments are already under way to develop this energy source.

TABLE 21.2 Amounts of Fossil Fuels Possibly Recoverable Worldwide (Unit of Comparison is a Barrel of Oil)*

Fossil Fuel	Total Amount in Ground (billions of barrels)	Amount Possibly Recoverable (billions of barrels)
Coal	About 100,000	63,000[a]
Oil and gas (flowing)	1500–3000	1500–3000
Trapped oil in pumped-out pools	1500–3000	0–?
Viscous oil (tar sands)	3000–6000	500–?
Oil shale	Total unknown: much greater than coal	1000–?

*A barrel is equal to 42 U.S. gallons and is the volume generally used when commercial production of oil is discussed.
[a] 0.22 ton of coal = 1 barrel of oil.

HYDROELECTRIC POWER

Hydroelectric power is recovered from the potential energy of stream water as it flows to the sea. As discussed in Chapter 14, in order to convert the power of flowing water to electricity efficiently, it is necessary to dam streams. Unfortunately, reservoirs behind dams fill with silt, and so even though water-power is continuous, dams and reservoirs have limited lifetimes.

Water-power has been used in small ways for thousands of years, but only in the twentieth century did it begin to be used to any significant extent for generating electricity. All the water flowing in the streams of the world has a total recoverable energy estimated at 9.2×10^{19} J/yr, an amount equivalent to burning 15 billion barrels of oil per year. Thus, even if all the possible hydropower in the world were developed, we could satisfy only about one-third of the present world energy needs. We have to conclude that, for those fortunate countries with large rivers and suitable dam sites, hydropower is very important, but for most countries hydropower holds limited potential for development.

NUCLEAR ENERGY

Nuclear energy is the heat energy produced during controlled transformation of suitable radioactive isotopes (a process called **fission**). Three of the radioactive atoms that keep the Earth hot by spontaneous decay—^{238}U, ^{235}U, and ^{232}Th—can be mined and used to obtain nuclear energy. Fission is accomplished by bombarding the radioactive atoms with neutrons, thus accelerating the rate of decay and the release of heat energy. The device in which this operation is carried out is called a **pile**.

When ^{235}U fissions, it not only releases heat and forms new elements but also ejects some neutrons from its nucleus. These neutrons can then be used to induce more ^{235}U atoms to fission, and a continuous chain reaction occurs. The function of a pile is to control the flux of neutrons so that the rate of fission can be controlled. When a chain reaction proceeds without control, an atomic explosion occurs. Controlled fission, therefore, is the method used by nuclear power plants, and a tremendous amount of energy can be obtained in the process. The fissioning of one gram of ^{235}U produces as much heat as the burning of 13.7 barrels of oil. Unfortunately, however, ^{235}U is the only natural radioactive isotope that will maintain a chain reaction, and it is the least abundant of the three radioactive isotopes that are mined for nuclear energy. Only one atom of each 138.8 atoms of uranium in nature is ^{235}U. The remaining atoms are ^{238}U, which will not sustain a chain reaction. However, if ^{238}U is placed in a pile with ^{235}U that is undergoing a chain reaction, some of the neutrons will bombard the ^{238}U and convert it to plutonium-239 (^{239}Pu). This new isotope can, under suitable conditions, sustain a chain reaction of its own. The pile in which the conversion

of ^{238}U takes place is called a **breeder reactor**. The same kind of device can be used to convert ^{232}Th into ^{233}U, which also will sustain a chain reaction. Unfortunately, breeder reactors and nuclear power plants based on them are more complex and less safe than ^{235}U plants, so present nuclear power plants use ^{235}U.

Already there are more than 300 piles in nuclear power plants operating around the world. They utilize the heat energy from fission to produce steam that drives turbines and generates electricity. Approximately 8 percent of the world's electrical power is derived from nuclear power plants. In France, more than half of all the electrical power comes from nuclear plants; the fraction is rising sharply in some other European countries and in Japan, too. The reason for the increase is obvious. Japan and many European countries do not have adequate supplies of fossil fuels in order to be self-sufficient.

Many problems are associated with nuclear energy. The isotopes used in power plants are the same isotopes used in atomic weapons, and so a security problem exists. The possibility of a power plant failing in some unexpected way creates a safety problem. The dreadful Chernobyl disaster in 1986 in Ukraine, a member country of the former USSR, is an example of such an event. Finally, the problem of safe burial of dangerous radioactive waste matter must be faced. Some of the waste matter will retain dangerous levels of radioactivity for thousands of years, an issue that is discussed in *Understanding Our Environment*, Box 21.2, *The Waste Disposal Problem: Geology and Politics*.

GEOTHERMAL POWER

Geothermal power, as Earth's internal heat flux is called, has been used for more than 50 years in New Zealand, Italy, and Iceland and more recently in other parts of the world, including the United States. How this is done was discussed in Box 5.2 on geothermal energy in Chapter 5.

Most of the world's geothermal steam reservoirs are close to plate margins because plate margins are where most recent volcanic activity has occurred. A depth of 3 km seems to be a rough lower limit for big geothermal steam and hot-water pools. It is estimated that the world's geothermal reservoirs could yield about 8×10^{19} J—equivalent to burning 13 billion barrels of oil. This estimate incorporates the observation that, in New Zealand and Italy, only about 1 percent of the energy in a geothermal reservoir is recoverable. If the recovery efficiency were to rise, the estimate of recoverable geothermal resources would also rise. But even if the efficiency rose to 50 percent, geothermal power, like hydropower, could satisfy only a small part of human energy needs. For this reason, as we discussed in Box 5.2 on geothermal power in Chapter 5, a good deal of attention is being given to creating artificial geothermal steam fields. So far, experiments have been only partially successful.

BOX 21.2 — UNDERSTANDING OUR ENVIRONMENT

THE WASTE DISPOSAL PROBLEM: GEOLOGY AND POLITICS

Waste materials containing substances that are hazardous to human health are produced in ever-increasing amounts in industrial societies. The major kinds of waste are poisonous metals like mercury and lead, organic chemicals from industrial processes, and radioactive elements generated in nuclear reactors. When toxic wastes are disposed of, it is necessary to ensure that if these substances are accidentally released into the environment they will not become concentrated in unsafe levels in food, drinking water, or other materials used in ordinary life.

The problem of controlling release is especially difficult in the case of radioactive waste. Radioactive waste contains one or more chemical elements whose atoms are continually emitting radiation in the form of small atomic particles and γ-rays. In disposing of this material, the goal is not to eliminate the potential exposure to radiation completely but to make certain that amounts of radioactive elements escaping into the environment are so small that the general background level of radiation is not raised appreciably.

If radioactive waste, or any other kind of toxic waste, is to be disposed of by burial, a major worry is the effect of underground water. Rocks below the land surface are saturated with water, and water is the principal means by which toxic substances might escape into surface environments. Dissolved poisonous substances might be carried to the surface in springs or seepages, or through wells drilled by farmers.

Some low-level waste (which contains small amounts of dangerous materials) can be safely disposed of by shallow landfill, a method that is commonly used to dispose of nontoxic industrial waste. The problems posed by high-level wastes are much more difficult. Concentrations of toxic substances in these materials are so great that even small quantities escaping into the environment can be a menace to living creatures. The most abundant type of high-level waste consists of spent fuel rods from nuclear reactors. For the most part these are stored underwater at reactor sites, where the radiation they produce is absorbed by the water and the metal-containing basins. Fuel rods can also be kept in dry storage, surrounded by metal or some other solid material that will prevent the escape of radiation.

Why, then, should we worry about the disposal of high-level waste? It is not harming the environment, so why not just leave it where it is? The difficulty is that constant vigilance is needed to make sure nothing goes wrong. Water can evaporate, pumps malfunction, metals crack or corrode. If any of these should happen, someone has to be present to prevent waste from leaking into the environment. This job would not be so difficult were it not for the times involved. Radioactive elements in waste will eventually decay to nonradioactive isotopes, but some remain dangerously active for a very long time. Is there any place where these wastes can be stored with a guarantee that none will leak out in 24,000 years, or in 100,000 years?

It seems likely that sites can be found deep in solid rock where waste could be placed with a very small chance of any appreciable amount escaping, even after a thousand centuries. Such a site must be a few hundred meters underground and consist of rock that is solid, durable, and free of large cracks. The site must also be geologically stable—a place where erosion is slow, earthquakes are infrequent, and there is no sign of recent volcanic activity. But the chief worry will still be groundwater. Is the water in the rock noncorrosive and slow moving? Does it travel a long way before it emerges at the surface? These are standard geologic questions, and with a little investigation a geologist can find answers to them.

Most geologists agree that the best way to accomplish disposal is to sink a shaft a few hundred meters deep at a carefully selected site, excavate tunnels from the base of the shaft, and drill holes in the tunnel floors into which the waste containers would be placed. The holes and the tunnels would then be filled with clay so that the waste would be completely isolated even if the containers eventually corroded. In the United States a site of this kind is being prepared in an area of southern Nevada known as Yucca Mountain. However, no high-level waste has yet been disposed of there.

What has gone wrong? The difficulty lies in the very long times for which geologists must give assurance that no significant amount of radioactive material will escape Earth's surface. How sure is it that waste buried at the Yucca Mountain site would remain in place for 100,000 years? Very sure, say geologists, but they cannot provide an absolute guarantee. For people living in the area, the lack of such a guarantee makes them suspicious and inclined to oppose the disposal project with all the political means at their command.

Thus, an environmental problem, seemingly solved from a geologic standpoint, has become a political problem that remains unsolved.

ENERGY FROM WINDS, WAVES, TIDES, AND SUNLIGHT

The most obvious source of energy is the Sun. The amount of energy reaching Earth each year from the Sun is approximately 4×10^{24} J—that is, 10,000 times more than we humans use. We already put some of the Sun's energy to work in greenhouses and in solar homes, but the amount so used is tiny. The major challenge is to convert solar energy directly to electricity (Figure 21.18). Devices that effect such a conversion, called photovoltaic devices, have been invented. So far their costs are too high and efficiencies too

Figure 21.18 Capturing Solar Energy
Photovoltaic panels convert the energy
of sunlight to electricity at a power-
generating site in southern California.

low for most uses, although they are already widely used in small calculators, radios, and other devices that use very little power.

Winds and waves are both secondary expressions of solar energy. Winds, in particular, have been used as an energy source for thousands of years through sails on ships and windmills. Today, huge farms of windmills are being erected in suitably windy places (Figure 21.19). Although there are problems and high costs with windmills, it seems very likely that by the year 2010 or sooner, windmills will be cost-competitive with coal-burning electrical power plants. Unfortunately, much of the wind energy is in very-high-altitude winds. Steady surface winds have only about 10 percent of the energy the human race now uses. As with hydro- and geothermal power, therefore, wind power may become locally significant but cannot satisfy global needs.

Waves, which arise from winds blowing over the ocean, contain an enormous amount of energy. We can see how powerful waves are along any coastline during a storm. Wave power has been used to ring bells and blow whistles as navigational aids for centuries, but so far no one has discovered how to tap wave energy on a large scale. Devices that have been designed to do this tend to fail because of corrosion or storm damage.

Tides arise from the gravitational forces exerted on Earth by its Moon and the Sun. If a dam is put across the mouth of a bay so that water can be trapped at high tide, the outward flowing water at low tide can drive a turbine. Unfortunately, the process is inefficient, and few places around the world have tides high enough to make tidal energy feasible.

It is clear that numerous sources of energy exist and that more energy is available than we can use. What is not yet clear is when, or even whether, we will be clever enough to learn how to tap the different energy sources in ways that are not hazardous and that do not disrupt the environment.

Figure 21.19 Wind Energy A field of windmills near Palm Springs, California. The windmills generate electricity using the kinetic energy of wind.

Before you go on:
1. What is the explanation for the two great coal-forming periods in geologic history?
2. Some shales that are rich in organic matter do not form petroleum. Why?
3. What steps are necessary for the formation of an oil pool?
4. If we stop using fossil fuels, what other sources of energy could we turn to?

REVISITING PLATE TECTONICS AND THE EARTH SYSTEM

PLATE TECTONICS AND MINERAL DEPOSITS

Many kinds of mineral deposits tend to occur in groups and to form what exploration geologists call **metallogenic provinces**. These are defined as limited regions of the crust within which mineral deposits occur in unusually large numbers. A striking example is the metallogenic province shown in Figure 8.19, which runs along the western side of the Americas. Within the province is the world's greatest concentration of large hydrothermal copper deposits. These deposits are associated with intrusive igneous rocks that are invariably porphyritic (Chapter 5), and they are therefore called **porphyry copper deposits**. The intrusive igneous rocks, and therefore the deposits themselves, were formed as a consequence of subduction because they are in, or adjacent to, old stratovolcanoes.

Magmatic, hydrothermal, and stratabound deposits all form near present or past plate boundaries (Figure 21.20). This is hardly surprising as the deposits are related directly or indirectly to igneous activity, and most igneous activity we now know is related to plate tectonics.

MINERAL PRODUCTION AND THE EARTH SYSTEM

In the first chapter we mentioned three ways that we humans are influencing the Earth system—by changing the composition of the atmosphere through the burning of fossil fuels (global warming), by releasing gases to the atmosphere, causing ozone to be destroyed and thereby affecting the shielding effects of the atmosphere, and by the building of huge reservoirs. Throughout the book we have brought other human-caused effects to your attention. Now we draw your attention to the issue of what might be called conspicuous consumption.

Mining companies, cement makers, oil companies, and all other organizations that dig or pump material from the ground report the extent of their operations to the governments of the countries where they work. As a result we know reasonably accurately how much is produced and used each year. By adding up all of the crushed rock, cement, sand and gravel, fertilizer, oil, gas, coal, metals, and other commodities we learn that the amount used in North America is about 20 tons per person per year.

The figure for the world as a whole is not quite as certain as for North America, but is about 9 tons per person per year. There are about 6 billion people, which means that about 54 billion tons of material is dug up and used each year. This does not include all of the soil and rock moved around in order to build roads, and to prepare home sites and for other kinds of site preparation. Use means that the material is dug up, processed, moved away from its original site, and placed elsewhere. By doing so we humans have become a potent geologic agent. Compare the estimated human figure of 54 billion tons a year with the estimate of all the sediment transported to the oceans each year—25 billion tons. We do not, as yet, understand all of the changes our activities are causing, but it is quite clear that each and everyone of us must be changing the environment simply by going about our daily activities and using the materials made available for us.

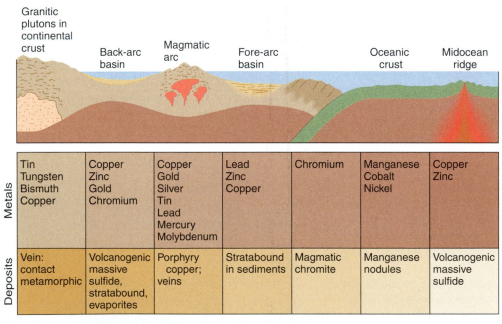

	Back-arc basin	Magmatic arc	Fore-arc basin		Oceanic crust	Midocean ridge
Metals	Tin Tungsten Bismuth Copper	Copper Zinc Gold Chromium	Copper Gold Silver Tin Lead Mercury Molybdenum	Lead Zinc Copper	Chromium	Manganese Cobalt Nickel	Copper Zinc
Deposits	Vein: contact metamorphic	Volcanogenic massive sulfide, stratabound, evaporites	Porphyry copper; veins	Stratabound in sediments	Magmatic chromite	Manganese nodules	Volcanogenic massive sulfide

Figure 21.20 Plate Tectonics and Mineral Deposits Locations of certain kinds of mineral deposits in terms of plate structures.

What's Ahead?

You have just completed the last chapter in the book. We hope that you have enjoyed the course and the text. What lies ahead is your use of the understanding you have gained of the world around you and the natural processes that combine to create the environment in which you live.

CHAPTER SUMMARY

1. When a mineral deposit can be worked profitably, it is called an ore. The waste material mixed with ore minerals is gangue.

2. Mineral deposits form when minerals become concentrated in one of five ways: (1) precipitation from hydrothermal solutions to form hydrothermal mineral deposits; (2) concentration through crystallization to form magmatic mineral deposits; (3) concentration from lake water or seawater to form sedimentary mineral deposits; (4) concentration in flowing water to form placers; and (5) concentration through weathering to form residual deposits.

3. Hydrothermal solutions are brines, and they can either be given off by cooling magma or else form when either groundwater or seawater penetrates the crust, becomes heated, and reacts with the enclosing rocks.

4. Hydrothermal mineral deposits form when hydrothermal solutions deposit dissolved minerals because of cooling, boiling, pressure drop, mixing with colder water, or through chemical reactions with enclosing rocks.

5. Chromite, the main ore mineral of chromium, is an important mineral concentrated by fractional crystallization.

6. Sedimentary mineral deposits are varied. The largest, and most important, are evaporites. Marine evaporite deposits supply most of the world's gypsum, halite, and potassium minerals.

7. Gold, platinum, cassiterite, diamonds, and other minerals are commonly found mechanically concentrated in placers.

8. Bauxite, the main ore of aluminum, is the most important kind of residual mineral deposit. Bauxite forms as a result of tropical weathering.

9. The distribution of many kinds of mineral deposits is controlled by plate tectonics because most magmas and most sedimentary basins are where they are because of plate tectonics.

10. Geochemically abundant metals, which make up 0.1 percent or more of the crust, tend to form residual, sedimentary, or magmatic mineral deposits. The amounts available for exploitation in such deposits are enormous.

11. Deposits of geochemically scarce metals, present in the crust in amounts less than 0.1 percent, form mainly as hydrothermal, magmatic, and placer deposits. Amounts of scarce metals available for exploitation are limited and geographically restricted.

12. Nonmetallic substances are used mainly as chemicals, fertilizers, building materials, and ceramics and abrasives.

13. Coal originated as plant matter in ancient swamps and is both abundant and widely distributed.

14. Oil and gas originated as organic matter trapped in shales and decomposed chemically due to heat and pressure following burial. Later, these fluids moved through reservoir rocks and were caught in geologic traps to form pools.

15. When heated, part of the solid organic matter found in shale—called kerogen—will convert to oil and gas. Oil from shales is the world's largest resource of fossil fuel. Unfortunately, most shale contains so little kerogen that more oil must be burned to heat the shale than is produced by the conversion process.

16. Biomass energy, mainly the burning of wood and animal dung, accounts for about 9 percent of the world's energy.

17. Hydropower, the energy of flowing streams, is exploited at many places around the world, but only locally does it have the potential to be a major energy source.

18. Nuclear energy, derived from atomic nuclei of radioactive isotopes, chiefly uranium, accounts for 7.6 percent of the world's electrical power but presents problems of plant safety and waste disposal.

19. Minor amounts of energy are derived from the Sun's heat, tides, waves, winds, and geothermal power.

THE LANGUAGE OF GEOLOGY

breeder reactor (p. 579)

exploration geology (p. 563)

fission (p. 579)

gangue (p. 564)
geochemically abundant metals (p. 572)

geochemically scarce metals (p. 572)
geothermal power (p. 579)

hydrothermal mineral deposit (p. 564)

magmatic mineral deposit (p. 564)
metallogenic province (p. 582)
mineral deposit (p. 563)

ore (p. 564)

pile (p. 579)
placer (p. 565)
porphyry copper deposit (p. 582)

residual mineral deposit (p. 565)

secondary enrichment (p. 572)
sedimentary mineral deposit (p. 565)

tar (p. 577)

volcanogenic massive sulfide deposit
(p. 565)

QUESTIONS FOR REVIEW

1. What are mineral deposits? Describe five ways by which a mineral deposit can form.

2. If there are any mineral deposits in the area where you live or study, what kind of deposits are they and how did they form?

3. What factors determine whether a mineral deposit is ore?

4. How do hydrothermal solutions form and how do they form mineral deposits?

5. Briefly describe the formation of three kinds of sedimentary mineral deposits.

6. What factors control the concentration of minerals in placers? Name four minerals mined from placers.

7. Compare and contrast mineral deposits of the geochemically abundant and geochemically scarce metals.

8. What is a fossil fuel? Name four kinds of fossil fuel.

9. During what two periods in Earth's history was most coal formed? Explain why coal formed, when and where it did, and why the coal is now found where it is.

10. What kind of rocks serve as source rocks for petroleum? In what kinds of rocks does petroleum tend to be trapped? Why?

11. Oil drillers find more petroleum per unit volume of rock in Cenozoic rocks than in Paleozoic rocks of the same kind. Explain.

12. Oil shales are rich in organic matter. Explain why such shales have not served as source rock for petroleum.

13. Discuss the relative amounts of energy available from the different fossil fuels. What is your opinion about how fossil fuels will be used in the future?

14. What is nuclear energy? How is it used to make electricity, and what possible dangers are there in developing it?

15. What limitations does the development of hydroelectric power present? of wave and wind power?

Appendix A

Units and Their Conversions

ABOUT SI UNITS

Regardless of the field of specialization, all scientists use the same units and scales of measurement. They do so to avoid confusion and the possibility that mistakes can creep in when data are converted from one system of units, or one scale, to another. By international agreement the SI units are used by all, and they are the units used in this text. SI is the abbreviation of Système International d'Unités (in English, the International System of Units).

Some of the SI units are likely to be familiar, some unfamiliar. The SI unit of length is the meter (m), of area the square meter (m^2), and of volume the cubic meter (m^3). The SI unit of mass is the kilogram (kg), and of time the second (s). The other SI units used in this book can be defined in terms of these basic units. Three important ones are:

1. The newton (N), a unit of force defined as that force needed to accelerate a mass of 1 kg by 1 m/s^2; hence 1 N = 1 kg·m/s^2. (The period between kg and m indicates multiplication.)

2. The joule (J), a unit of energy or work, defined as the work done when a force of 1 newton is displaced a distance of 1 meter; hence 1 J = 1 N·m. One important form of energy as far as the Earth is concerned is heat. The outward flow of the Earth's internal heat is measured in terms of the number of joules flowing outward from each square centimeter each second; thus, the unit of heat flow is $J/cm^2/s$.

3. The pascal (Pa), a unit of pressure defined as a force of 1 newton applied across an area of 1 square meter; hence 1 Pa = 1 N/m^2. The pascal is a numerically small unit. Atmospheric pressure, for example (15 lb/in^2), is 101,300 Pa. Pressure within the Earth reaches millions or billions of pascals. For convenience, earth scientists sometimes use 1 million pascals (megapascal, or MPa) as a unit.

Temperature is a measure of the internal kinetic energy (expressed as movement) of the atoms and molecules in a body. In the SI system, temperature is measured on the Kelvin scale (K). The temperature intervals on the Kelvin scale are arbitrary, and they are the same as the intervals on the more familiar Celsius scale (°C). The difference between the two scales is that the Celsius scale selects 100°C as the temperature at which water boils at sea level, and 0°C as the freezing temperature of water at sea level. Zero degrees Kelvin, on the other hand, is absolute zero, the temperature at which all atomic and molecular motions cease. Thus, 0°C is equal to 273.15 K, and 100°C is 373.15 K. The tempera-

tures of processes on and within the Earth tend to be at or above 273.15 K. Despite the inconsistency, earth scientists still use the Celsius scale when geological processes are discussed.

Appendix A provides a table of conversion from older units to Standard International (SI) units.

PREFIXES FOR MULTIPLES AND SUBMULTIPLES

When very large or very small numbers have to be expressed, a standard set of prefixes is used in conjunction with the SI units. Some prefixes are probably already familiar; an example is the centimeter (which is one hundredth of a meter, or 10^{-2} m). The standard prefixes are

tera	$1,000,000,000,000 =$	10^{12}
giga	$1,000,000,000 =$	10^9
mega	$1,000,000 =$	10^6
kilo	$1,000 =$	10^3
hecto	$100 =$	10^2
deka	$10 =$	10
deci	$0.1 =$	10^{-1}
centi	$0.01 =$	10^{-2}
milli	$0.001 =$	10^{-3}
micro	$0.000001 =$	10^{-6}
nano	$0.000000001 =$	10^{-9}
pico	$0.000000000001 =$	10^{-12}

One measure used commonly in geology is the nanometer (nm), a unit by which the sizes of atoms are measured; 1 nanometer is equal to 10^{-9} meter.

COMMONLY USED UNITS OF MEASURE

LENGTH

METRIC MEASURE

1 kilometer (km)	= 1000 meters (m)
1 meter (m)	= 100 centimeters (cm)
1 centimeter (cm)	= 10 millimeters (mm)
1 millimeter (mm)	= 1000 micrometers (μm)
	(formerly called microns)
1 micrometer (mm)	= 0.001 millimeter (mm)
1 angstrom (Å)	= 10^{-8} centimeters (cm)

NONMETRIC MEASURE

1 mile (mi)	=	5280 feet (ft) = 1760 yards (yd)
1 yard (yd)	=	3 feet (ft)
1 fathom (fath)	=	6 feet (ft)

CONVERSIONS

1 kilometer (km)	=	0.6214 mile (mi)
1 meter (m)	=	1.094 yards (yd) = 3.281 feet (ft)
1 centimeter (cm)	=	0.3937 inch (in)
1 millimeter (mm)	=	0.0394 inch (in)
1 mile (mi)	=	1.609 kilometers (km)
1 yard (yd)	=	0.9144 meter (m)
1 foot (ft)	=	0.3048 meter (m)
1 inch (in)	=	2.54 centimeters (cm)
1 inch (in)	=	25.4 millimeters (mm)
1 fathom (fath)	=	1.8288 meters (m)

AREA

METRIC MEASURE

1 square kilometer (km^2)	=	1,000,000 square meters (m^2)
	=	100 hectares (ha)
1 square meter (m^2)	=	10,000 square centimeters (cm^2)
1 hectare (ha)	=	10,000 square meters (m^2)

NONMETRIC MEASURE

1 square mile (mi^2)	=	640 acres (ac)
1 acre (ac)	=	4840 square yards (yd^2)
1 square foot (ft^2)	=	144 square inches (in^2)

CONVERSIONS

1 square kilometer (km^2)	=	0.386 square mile (mi^2)
1 hectare (ha)	=	2.471 acres (ac)
1 square meter (m^2)	=	1.196 square yards (yd^2)
	=	10.764 square feet (ft^2)
1 square centimeter (cm^2)	=	0.155 square inch (in^2)
1 square mile (mi^2)	=	2.59 square kilometers (km^2)
1 acre (ac)	=	0.4047 hectare (ha)
1 square yard (yd^2)	=	0.836 square meter (m^2)
1 square foot (ft^2)	=	0.0929 square meter (m^2)
1 square inch (in^2)	=	6.4516 square centimeter (cm^2)

VOLUME

METRIC MEASURE

1 cubic meter (m^3)	=	1,000,000 cubic centimeters (cm^3)
1 liter (l)	=	1000 milliliters (ml)
	=	0.001 cubic meter (m^3)
1 centiliter (cl)	=	10 milliliters (ml)

1 milliliter (ml)	=	1 cubic centimeter (cm^2)

NONMETRIC MEASURE

1 cubic yard (yd^3)	=	27 cubic feet (ft^3)
1 cubic foot (ft^3)	=	1728 cubic inches (in^3)
1 barrel (oil) (bbl)	=	42 gallons (U.S.) (gal)

CONVERSIONS

1 cubic kilometer (km^3)	=	0.24 cubic miles (mi^3)
1 cubic meter (m^3)	=	264.2 gallons (U.S.) (gal)
	=	35.314 cubic feet (ft^3)
1 liter (l)	=	1.057 quarts (U.S.) (qt)
	=	33.815 ounces (U.S. fluid) (fl. oz.)
1 cubic centimeter (cm^3)	=	0.0610 cubic inch (in^3)
1 cubic mile (mi^3)	=	4.168 cubic kilometers (km^3)
1 acre-foot (ac-ft)	=	1233.46 cubic meters (m^3)
1 cubic yard (yd^3)	=	0.7646 cubic meter (m^3)
1 cubic foot (ft^3)	=	0.0283 cubic meter (m^3)
1 cubic inch (in^3)	=	16.39 cubic centimeters (cm^3)
1 gallon (gal)	=	3.784 liters (l)

MASS

METRIC MEASURE

1000 kilograms (kg)	=	1 metric ton (also called a tonne) (m.t)
1 kilogram (kg)	=	1000 grams (g)

NONMETRIC MEASURE

1 short ton (sh.t)	=	2000 pounds (lb)
1 long ton (l.t)	=	2240 pounds (lb)
1 pound (avoirdupois) (lb.)	=	16 ounces (avoirdupois) (oz) = 7000 grains (gr)
1 ounce (avoirdupois) (oz)	=	437.5 grains (gr)
1 pound (Troy) (Tr. lb)	=	12 ounces (Troy) (Tr. oz)
1 ounce (Troy) (Tr. oz)	=	20 pennyweight (dwt)

CONVERSIONS

1 metric ton (m.t)	=	2205 pounds (avoirdupois) (lb)
1 kilogram (kg)	=	2.205 pounds (avoirdupois) (lb)
1 gram (g)	=	0.03527 ounce (avoirdupois) (oz)
	=	0.03215 ounce (Troy) (Tr. oz)
	=	15,432 grains (gr)
1 pound (lb)	=	0.4536 kilogram (kg)
1 ounce (avoirdupois) (oz)	=	28.35 grams (g)
1 ounce (avoirdupois) (oz)	=	1.097 ounces (Troy) (Tr. oz)

PRESSURE

1 pascal (Pa)	= 1 newton/square meter (N/m^2)
1 kilogram/square centimeter (kg/cm^2)	= 0.96784 atmosphere (atm) = 14.2233 pounds/square inch (lb/in2) = 0.098067 bar
1 bar	= 0.98692 atmosphere (atm) = 105 pascals (Pa) = 1.02 kilograms/square centimeter (kg/cm^2)

ENERGY AND POWER

ENERGY

1 joule (J)	= 1 newton meter (N.m)
	= 2.390×10^{-1} calorie (cal)
	= 9.47×10^{-4} British thermal unit (Btu)
	= 2.78×10^{-7} kilowatt-hour (kWh)
1 calorie (cal)	= 4.184 joule (J)
	= 3.968×10^{-3} British thermal unit (Btu)
	= 1.16×10^{-6} kilowatt-hour (kWh)
1 British thermal unit (Btu)	= 1055.87 joules (J)
	= 252.19 calories (cal)
	= 2.928×10^{-4} kilowatt-hour (kWh)

1 kilowatt hour	= 3.6×10^6 joules (J)
	= 8.60×10^5 calories (cal)
	= 3.41×10^2 British thermal units (Btu)

POWER (ENERGY PER UNIT TIME)

1 watt (W)	= 1 joule per second (J/s)
	= 3.4129 Btu/h
	= 1.341×10^{-3} horsepower (hp)
	= 14.34 calories per minute (cal/min)
1 horsepower (hp)	= 7.46×10^3 watts (W)

TEMPERATURE

To change from Fahrenheit (F) to Celsius (C)

$$°C = \frac{(°F - 32°)}{1.8}$$

To change from Celsius (C) to Fahrenheit (F)

$$°F = (°C \times 1.8) + 32°$$

To change from Celsius (C) to Kelvin (K)

$$K = °C + 273.15$$

To change from Fahrenheit (F) to Kelvin (K)

$$K = \frac{(°F - 32)}{1.8} + 273.15$$

Appendix B
Tables of the Chemical Elements and Naturally Occurring Isotopes

TABLE B.1 Alphabetical List of the Elements

Element	Symbol	Atomic Number	Crustal Abundance, Weight Percent	Element	Symbol	Atomic Number	Crustal Abundance, Weight Percent
Actinium	Ac	89	Human-made	Iridium	Ir	77	0.00000002
Aluminum	Al	13	8.00	Iron	Fe	26	5.80
Americium	Am	95	Human-made	Krypton	Kr	36	Not known
Antimony	Sb	51	0.00002	Lanthanum	La	57	0.0050
Argon	Ar	18	Not known	Lawrencium	Lw	103	Human-made
Arsenic	As	33	0.00020	Lead	Pb	82	0.0010
Astatine	At	85	Human-made	Lithium	Li	3	0.0020
Barium	Ba	56	0.0380	Lutetium	Lu	71	0.000080
Berkelium	Bk	97	Human-made	Magnesium	Mg	12	2.77
Beryllium	Be	4	0.00020	Manganese	Mn	25	0.100
Bismuth	Bi	83	0.0000004	Meitnerium	Mt	109	Human-made
Boron	B	5	0.0007	Mendelevium	Md	101	Human-made
Bromine	Br	35	0.00040	Mercury	Hg	80	0.000002
Cadmium	Cd	48	0.000018	Molybdenum	Mo	42	0.00012
Calcium	Ca	20	5.06	Neodymium	Nd	60	0.0044
Californium	Cf	98	Human-made	Neon	Ne	10	Not known
Carbon[a]	C	6	0.02	Neptunium	Np	93	Human-made
Cerium	Ce	58	0.0083	Nickel	Ni	28	0.0072
Cesium	Cs	55	0.00016	Nielsbohrium	Ns	107	Human-made
Chlorine	Cl	17	0.0190	Niobium	Nb	41	0.0020
Chromium	Cr	24	0.0096	Nitrogen	N	7	0.0020
Cobalt	Co	27	0.0028	Nobelium	No	102	Human-made
Copper	Cu	29	0.0058	Osmium	Os	76	0.00000002
Curium	Cm	96	Human-made	Oxygen[b]	O	8	45.2
Dysprosium	Dy	66	0.00085	Palladium	Pd	46	0.0000003
Einsteinium	Es	99	Human-made	Phosphorus	P	15	0.1010
Erbium	Er	68	0.00036	Platinum	Pt	78	0.0000005
Europium	Eu	63	0.00022	Plutonium	Pu	94	Human-made
Fermium	Fm	100	Human-made	Polonium	Po	84	Footnote[d]
Fluorine	F	9	0.0460	Potassium	K	19	1.68
Francium	Fr	87	Human-made	Praseodymium	Pr	59	0.0013
Gadolinium	Gd	64	0.00063	Promethium	Pm	61	Human-made
Gallium	Ga	31	0.0017	Protactinium	Pa	91	Footnote[d]
Germanium	Ge	32	0.00013	Radium	Ra	88	Footnote[d]
Gold	Au	79	0.0000002	Radon	Rn	86	Footnote[d]
Hafnium	Hf	72	0.004	Rhenium	Re	75	0.00000004
Hahnium	Ha	105	Human-made	Rhodium[c]	Rh	45	0.00000001
Hassium	Hs	108	Human-made	Rubidium	Rb	37	0.0070
Helium	He	2	Not known	Ruthenium[c]	Ru	44	0.00000001
Holmium	Ho	67	0.00016	Samarium	Sm	62	0.00077
Hydrogen[b]	H	1	0.14	Scandium	Sc	21	0.0022
Indium	In	49	0.00002	Seaborgium	Sg	106	Human-made
Iodine	I	53	0.00005	Selenium	Se	34	0.000005

A4

TABLE B.1   Alphabetical List of the Elements, *continued*

Element	Symbol	Atomic Number	Crustal Abundance, Weight Percent	Element	Symbol	Atomic Number	Crustal Abundance, Weight Percent
Silicon	Si	14	27.20	Thulium	Tm	69	0.000052
Silver	Ag	47	0.000008	Tin	Sn	50	0.00015
Sodium	Na	11	2.32	Titanium	Ti	22	0.86
Strontium	Sr	38	0.0450	Tungsten	W	74	0.00010
Sulfur	S	16	0.030	Uranium	U	92	0.00016
Tantalum	Ta	73	0.00024	Vanadium	V	23	0.0170
Technetium	Tc	43	Human-made	Xenon	Xe	54	Not known
Tellurium[c]	Te	52	0.000001	Ytterbium	Yb	70	0.00034
Terbium	Tb	65	0.00010	Yttrium	Y	39	0.0035
Thallium	Tl	81	0.000047	Zinc	Zn	30	0.0082
Thorium	Th	90	0.00058	Zirconium	Zr	40	0.0140

Source: After K. K. Turekian, 1969.

[a] Estimate from S. R. Taylor (1964).

[b] Analyses of crustal rocks do not usually include separate determinations for hydrogen and oxygen. Both combine in essentially constant proportions with other elements, so abundances can be calculated.

[c] Estimates are uncertain and have a very low reliability.

[d] Elements formed by decay of uranium and thorium. The daughter products are radioactive with such short half-lives that crustal accumulations are too low to be measured accurately.

TABLE B.2   Naturally Occurring Elements Listed in Order of Atomic Numbers, Together with the Naturally Occurring Isotopes of Each Element, Listed in Order of Mass Numbers

Atomic Number[a]	Name	Symbol	Mass Numbers[b] of Natural Isotopes	Atomic Number[a]	Name	Symbol	Mass Numbers[b] of Natural Isotopes
1	Hydrogen	H	1, 2, [3][c]	27	Cobalt	Co	59
2	Helium	He	3, 4	28	Nickel	Ni	58, 60, 61, 62, 64
3	Lithium	Li	6, 7	29	Copper	Cu	63, 65
4	Beryllium	Be	9, [10]	30	Zinc	Zn	64, 66, 67, 68, 70
5	Boron	B	10, 11	31	Gallium	Ga	69, 71
6	Carbon	C	12, 13, [14]	32	Germanium	Ge	70, 72, 73, 74, 76
7	Nitrogen	N	14, 15	33	Arsenic	As	75
8	Oxygen	O	16, 17, 18	34	Selenium	Se	74, 76, 77, 80, 82
9	Fluorine	F	19	35	Bromine	Br	79, 81
10	Neon	Ne	20, 21, 22	36	Krypton	Kr	78, 80, 82, 83, 84, 86
11	Sodium	Na	23	37	Rubidium	Rb	85, [87]
12	Magnesium	Mg	24, 25, 26	38	Strontium	Sr	84, 86, 87, 88
13	Aluminum	Al	27	39	Yttrium	Y	89
14	Silicon	Si	28, 29 30	40	Zirconium	Zr	90, 91, 92, 94, 96
15	Phosphorus	P	31	41	Niobium	Nb	93
16	Sulfur	S	32, 33, 34, 36	42	Molybdenum	Mo	92, 94, 95, 96, 97, 98, 100
17	Chlorine	Cl	35, 37				
18	Argon	Ar	36, 38, 40	44	Ruthenium	Ru	96, 98, 99, 100, 101, 102, 104
19	Potassium	K	39, [40], 41				
20	Calcium	Ca	40, 42, 43, 44, 46, [48]	45	Rhodium	Rh	103
21	Scandium	Sc	45	46	Palladium	Pd	102, 104, 105, 106, 108, 110
22	Titanium	Ti	46, 47, 48, 49 50				
23	Vanadium	V	[50], 51	47	Silver	Ag	107, 109
24	Chromium	Cr	50, 52, 53, 54	48	Cadmium	Cd	106, 108, 110, 111, 112, 113, 114, 116
25	Manganese	Mn	55				
26	Iron	Fe	54, 56, 57, 58	49	Indium	In	113, [115]

TABLE B.2   Naturally Occurring Elements Listed in Order of Atomic Numbers, Together with the Naturally Occurring Isotopes of Each Element, Listed in Order of Mass Numbers, *continued*

Atomic Number[a]	Name	Symbol	Mass Numbers[b] of Natural Isotopes	Atomic Number[a]	Name	Symbol	Mass Numbers[b] of Natural Isotopes
50	Tin	Sn	112, 114, 115, 116, 117, 118, 119, 120, 122, 124	68	Erbium	Er	162, 166, 167, 168, 170
51	Antimony	Sb	121, 123	69	Thulium	Tm	169
52	Tellurium	Te	120, 122, 123, 124, 125, 126, 128, 130	70	Ytterbium	Yb	168, 170, 171, 172, 173, 174, 176
53	Iodine	I	127	71	Lutetium	Lu	175, [176]
54	Xenon	Xe	124, 126, 128, 129, 130, 131, 132, 134, 136	72	Hafnium	Hf	174, 176, 177, 178, 179, 180
55	Cesium	Cs	133	73	Tantalum	Ta	180, 181
56	Barium	Ba	130, 132, 134, 135, 137, 138	74	Tungsten	W	180, 182, 183, 184, 186
57	Lanthanum	La	[138], 139	75	Rhenium	Re	185, [187]
58	Cerium	Ce	136, 138, 140, [142]	76	Osmium	Os	184, 186, 187, 188, 189, 190, 192
59	Praseodymium	Pr	141	77	Iridium	Ir	191, 193
60	Neodymium	Nd	142, 143, [144], 145, 146, 148, 150	78	Platinum	Pt	190, 192, 195, 196, 198
62	Samarium	Sm	144, [147], [148], [149], 150, 152, 154	79	Gold	Au	197
63	Europium	Eu	151, 153	80	Mercury	Hg	196, 198, 199, 200, 201, 202, 204
64	Gadolinium	Gd	[152], 154, 155, 156, 157, 158, 160	81	Thallium	Tl	203, 205
65	Terbium	Tb	159	82	Lead	Pb	204, 206, 207, 208
66	Dysprosium	Dy	156, 158, 160, 161, 162, 163, 164	83	Bismuth	Bi	209
67	Holmium	Ho	165	84	Polonium	Po	[210]
				86	Radon	Rn	[222]
				88	Radium	Ra	[226]
				90	Thorium	Th	[232]
				91	Protactinium	Pa	[231]
				92	Uranium	U	[234], [235], [238]

[a] Atomic number = number of protons.

[b] Mass number = protons + neutrons.

[c] ☐ indicates isotope is radioactive.

Appendix C

Tables of the Properties of Common Minerals

TABLE C.1 Properties of the Common Minerals with Metallic Luster

Mineral	Chemical Composition	Form and Habit	Cleavage	Hardness	Specific Gravity	Other Properties	Most Distinctive Properties
Bornite	Cu_5FeS_4	Massive. Crystals very rare.	None. Uneven fracture.	3	5	Brownish bronze on fresh surface. Tarnishes purple, blue, and black. Grayish-black streak.	Color, streak.
Chalcocite	Cu_2S	Massive. Crystals very rare.	None. Conchoidal fracture.	2.5	5.7	Steel-gray to black. Dark gray streak.	Streak.
Chalcopyrite	$CuFeS_2$	Massive or granular.	None. Uneven fracture.	3.5–4	4.2	Golden yellow to brassy yellow. Dark green to black streak.	Streak. Hardness distinguishes from pyrite.
Chromite	$FeCr_2O_4$	Massive or granular.	None. Uneven fracture.	5.5	4.6	Iron black to brownish black. Dark brown streak.	Streak and lack of magnetism distinguishes from ilmenite and magnetite.
Copper	Cu	Massive, twisted leaves and wires.	None. Can be cut with a knife.	2.5–3	9	Copper color but commonly stained green.	Color, specific gravity, malleable.
Galena	PbS	Cubic crystals, coarse or fine-grained granular masses.	Perfect in three directions at right angles.	2.5	7.6	Lead-gray color. Gray to gray-black streak.	Cleavage and streak.
Gold	Au	Small irregular grains.	None. Malleable.	2.5	19.3	Gold color. Can be flattened without breakage.	Color, specific gravity, malleability.
Hematite	Fe_2O_3	Massive, granular, micaceous.	Uneven fracture.	5–6	5	Reddish-brown, gray to black. Reddish-brown streak.	Streak, hardness.
Ilmenite	$FeTiO_3$	Massive or irregular grains.	Uneven fracture.	5.5–6	4.7	Iron-black. Brown-reddish streak differing from hematite.	Streak distinguishes hematite. Lack of magnetism distinguishes magnetite.
Limonite (Goethite is most common.)	A complex mixture of minerals, mainly hydrous oxides.	Massive, coatings, botryoidal crusts, earthy masses.	None.	1–5.5	3.5–4	Yellow, brown, black, yellowish-brown streak.	Streak.
Magnetite	Fe_3O_4	Massive, granular. Crystals have octahedral shape.	None. Uneven fracture.	5.5–6.5	5	Black. Black streak. Strongly attracted to a magnet.	Streak, magnetism.

TABLE C.1 Properties of the Common Minerals with Metallic Luster, *continued*

Mineral	Chemical Composition	Form and Habit	Cleavage	Hardness	Specific Gravity	Other Properties	Most Distinctive Properties
Pyrite ("Fool's gold")	FeS_2	Cubic crystals with striated faces. Massive.	None. Uneven fracture.	6–6.5	5.2	Pale brass-yellow, darker if tarnished. Greenish-black streak.	Streak. Hardness distinguishes from chalcopyrite. Not malleable, which distinguishes from gold.
Pyrolusite	MnO_2	Crystals rare. Massive, coatings on fracture surfaces.	Crystals have a perfect cleavage. Massive, breaks unevenly.	2–6.5	5	Dark gray, black or bluish black. Black streak.	Color, streak.
Pyrrhotite	FeS	Crystals rare. Massive or granular.	None. Conchoidal fracture.	4	4.6	Brownish-bronze. Black streak. Magnetic.	Color and hardness distinguish from pyrite, magnetism from chalcopyrite.
Rutile	TiO_2	Slender, prismatic crystals or granular masses.	Good in one direction. Conchoidal fracture in others.	6–6.5	4.2	Reddish-brown (common), black (rare). Brownish streak. Adamantine luster.	Luster, habit, hardness.
Sphalerite (zinc blende)	ZnS	Fine to coarse granular masses. Tetrahedron shaped crystals.	Perfect in six directions.	3.5–4	4	Yellowish-brown to black. White to yellowish-brown streak. Resinous luster.	Cleavage, hardness, luster.
Uraninite	UO_2 to U_3O_8	Massive, with botryoidal forms. Rare crystals with cubic shapes.	None. Uneven fracture.	5–6	6.5–10	Black to dark brown. Streak black to dark brown. Dull luster.	Luster and specific gravity distinguish from magnetite. Streak distinguishes from ilmenite and hematite.

TABLE C.2 Properties of Rock-Forming Minerals with Nonmetallic Luster

Mineral	Chemical Composition	Form and Habit	Cleavage	Hardness	Specific Gravity	Other Properties	Most Distinctive Properties
Amphiboles (A complex family of minerals, *Hornblende* is most common.)	$X_2Y_5Si_8O_{22}$ $(OH)_2$ where X = Ca, Na; Y = Mg, Fe, Al.	Long, six-sided crystals; also fibers and irregular grains.	Two; intersecting at 56° and 124°.	5–6	2.9–3.8	Common in metamorphic and igneous rocks. Hornblende is dark green to black; actinolite, green; tremolite, white.	Cleavage, habit.
Andalusite	Al_2SiO_5	Long crystals, often square in cross-section.	Weak, parallel to length of crystal.	7.5	3.2	Found in metamorphic rocks. Often flesh-colored.	Hardness, form.
Anhydrite	$CaSO_4$	Crystals are rare. Irregular grains or fibers.	Three, at right angles.	3	2.9	Alters to gypsum. Pearly luster, white or colorless.	Cleavage, hardness.
Apatite	$Ca_5(PO_4)_3$ (F_3,OH, Cl)	Granular masses. Perfect six-sided crystals.	Poor. One direction.	5	3.2	Green, brown, blue, or white. Common in many kinds of rocks in small amounts.	Hardness, form.
Aragonite	$CaCO_3$	Massive, or slender, needle-like crystals.	Poor. Two directions.	3.5	2.9	Colorless or white. Effervesces with dilute HCl.	Effervescence with acid. Poor cleavage distinguishes from calcite.

TABLE C.2   Properties of Rock-Forming Minerals with Nonmetallic Luster, *continued*

Mineral	Chemical Composition	Form and Habit	Cleavage	Hardness	Specific Gravity	Other Properties	Most Distinctive Properties
Asbestos			See Serpentine				
Augite			See Pyroxene				
Biotite			See Mica				
Calcite	$CaCO_3$	Tapering crystals and granular masses.	Three perfect oblique angles to give a rhomb-shaped fragment.	3	2.7	Colorless or white. Effervesces with dilute HCl.	Cleavage, effervescence with acid.
Chlorite	$(Mg, Fe)_5$ $(Al, Fe)_2$ $Si_3O_{10}(OH)_8$	Flaky masses of minute scales.	One perfect; parallel to flakes.	2–2.5	2.6–2.9	Common in metamorphic rocks. Light to dark green. Greasy luster.	Cleavage—flakes not elastic, distinguishes from mica. Color.
Dolomite	$CaMg(CO_3)_2$	Crystals with rhomb-shaped faces. Granular masses.	Perfect in three directions as in calcite.	3.5	2.8	White or gray. Does not effervesce in cold, dilute HCl unless powdered. Pearly luster.	Cleavage. Lack of effervescence with acid.
Epidote	Complex Ca, Fe and Al.	Small elongate crystals. Fibrous.	One perfect, one poor.	6–7	3.4	Yellowish-green to dark green. Common in metamorphic rocks.	Habit, color. Hardness distinguishes from chlorite.
Feldspars: *Potassium* feldspar *(orthoclase)* is a common variety	$KAlSi_3O_8$	Prism-shaped crystals, granular	Two perfect, at right angles.	6	2.6	Common mineral. Flesh-colored, pink, white, or gray.	Color, cleavage.
Plagioclase	$NaAlSi_3O_8$ (albite) and $CaAl_2Si_2O_8$ (anorthite) and all compositions between.	Irregular grains, cleavable masses. Tabular crystals.	Two perfect, not quite at right angles.	6–6.5	2.6–2.7	White to dark gray. Cleavage planes may show fine parallel striations.	Cleavage. Striations on cleavage planes will distinguish from potassium feldspar.
Fluorite	CaF_2	Cubic crystals, granular masses.	Perfect in four directions.	4	3.2	Colorless, bluish green. Always an accessory mineral.	Hardness, cleavage does not effervesce with acid.
Garnets	$X_3Y_2(SiO_4)_3$; X = Ca, Mg, Fe, Mn; Y = Al, Fe, Ti, Cr.	Perfect crystals with 12 or 24 sides. Granular masses.	None. Uneven fracture.	6.5–7.5	3.5–4.3	Common in metamorphic rocks. Red, brown, yellowish-green, black.	Crystals, hardness, no cleavage.
Graphite	C	Scaly masses.	One, perfect. Forms slippery flakes.	1–2	2.2	Metamorphic rocks. Black with metallic to dull luster.	Cleavage, color. Marks paper.
Gypsum	$CaSO_4 \cdot 2H_2O$	Elongate or tabular crystals. Fibrous and earthy masses.	One, perfect. Flakes bend but are not elastic.	2	2.3	Vitreous to pearly luster. Colorless.	Hardness, cleavage.
Halite	NaCl	Cubic crystals.	Perfect to give cubes.	2.5	2.2	Tastes salty. Colorless, blue.	Taste, cleavage.
Hornblende			See Amphibole				
Kaolinite	$Al_2Si_2O_5$ $(OH)_4$	Soft, earthy masses. Submicroscopic crystals.	One, perfect.	2–2.5	2.6	White, yellowish. Plastic when wet; emits clayey odor. Dull luster.	Feel, plasticity, odor.
Kyanite	Al_2SiO_5	Bladed crystals.	One perfect. One imperfect.	4.5 parallel to blade, 7 across blade.	3.6	Blue, white, gray. Common in metamorphic rocks.	Variable hardness, distinguishes from sillimanite. Color.

TABLE C.2   Properties of Rock-Forming Minerals with Nonmetallic Luster, *continued*

Mineral	Chemical Composition	Form and Habit	Cleavage	Hardness	Specific Gravity	Other Properties	Most Distinctive Properties
Mica: *Biotite*	$K(Mg, Fe)_3$-$AlSi_3O_{10}$-$(OH)_2$	Irregular masses of flakes.	One, perfect.	2.5–3	2.8–3.2	Common in igneous and metamorphic rocks. Black, brown, dark green.	Cleavage, color. Flakes are elastic.
Muscovite	$KAl_3Si_3O_{10}$ $(OH)_2$	Thin flakes.	One, perfect.	2–2.5	2.7	Common in igneous and metamorphic rocks. Colorless, pale green or brown.	Cleavage, color. Flakes are elastic.
Olivine	$(Mg, Fe)_2SiO_4$	Small grains, granular masses.	None. Conchoidal fracture.	6.5–7	3.2–4.3	Igneous rocks. Olive green to yellowish-green.	Color, fracture, habit.
Orthoclase			See Feldspar				
Plagioclase			See Feldspar				
Pyroxene (A complex family of minerals. *Augite* is most common.)	$XY(SiO_3)_2$ $X = Y = Ca, Mg,$ Fe	8-sided stubby crystals. Granular masses.	Two, perfect, nearly at right angles.	5–6	3.2–3.9	Igneous and metamorphic rocks. Augite, dark green to black; other varieties white to green.	Cleavage.
Quartz	SiO_2	6-sided crystals, granular masses.	None. Conchoidal fractures.	7	2.6	Colorless, white, gray, but may have any color, depending on impurities. Vitreous to greasy luster.	Form, fracture, striations across crystal faces at right angles to long dimension.
Serpentine (Fibrous variety is *asbestos*.)	$Mg_3Si_2O_5$ $(OH)_4$	Platy or fibrous.	One, perfect.	2.5–5	2.2–2.6	Light to dark green. Smooth, greasy feel.	Habit, hardness.
Sillimanite	Al_2SiO_5	Long, needle-like crystals, fibers.	Breaks irregularly, except in fibrous variety.	6–7	3.2	White, gray. Metamorphic rocks.	Hardness distinguishes from kyanite. Habit.
Talc	$Mg_3Si_4O_{10}$ $(OH)_2$	Small scales, compact masses.	One, perfect.	1	2.6–2.8	Feels slippery. Pearly luster. White to greenish.	Hardness, luster, feel cleavage.
Tourmaline	Complex silicate of B, Al, Na, Ca, Fe, Li and Mg.	Elongate crystals, commonly with triangular cross section.	None.	7–7.5	3–3.3	Black, brown, red, pink, green, blue, and yellow. An accessory mineral in many rocks.	Habit.
Wollastonite	$CaSiO_3$	Fibrous or bladed aggregates of crystals.	Two, perfect.	4.5–5	2.8–2.9	Colorless, white, yellowish. Metamorphic rocks. Soluble in HCl.	Habit. Solubility in HCl and hardness distinguish amphiboles, kyanite, sillimanite.

TABLE C.3 Properties of Some Common Gemstones

Mineral	Chemical Composition	Form and Habit	Cleavage	Hardness	Specific Gravity	Other Properties	Most Distinctive Properties
Beryl: *Aquamarine* (blue) *Emerald* (green) *Golden beryl* (golden-yellow)	$Be_3Al_2Si_6O_{18}$	Six-sided, elongate crystals common.	Weak.	7.5–8	2.75	Bluish green, green, yellow, white, colorless. Common in pegmatites.	Form. Distinguished from apatite by its hardness.
Corundum: *Ruby* (red) *Sapphire* (blue)	Al2o3	Six-sided, barrel-shaped crystals.	None, but breaks easily across its crystal.	9	4	Brown, pink, red, blue, colorless. Common in metamorphic rocks. Star sapphire is opalescent with a six-sided light spot showing.	Hardness.
Diamond	C	Octahedron-shaped crystals.	Perfect, parallel to faces of octahedron.	10	3.5	Colorless, yellow; rarely red, orange, green, blue, or black.	Hardness, cleavage.
Garnet: *Almandite* (red) *Grossularite* (green, cinnamon-brown) *Andradite* (variety *demantoid* is green)			A rock-forming mineral — see Table C.2.				
Opal (A mineraloid)	$SiO_2 \cdot nH_2O$	Massive, thin coating. Amorphous.	None. Conchoidal fracture.	5–6	2–2.2	Colorless, white, yellow, red, brown, green, gray, opalescent.	Hardness, color, form.

Quartz:
(1) Coarse crystals
    Amethyst (violet)
    Cairngorm (brown)
    Citrine (yellow)
    Rock crystal (colorless)
    Rose quartz (pink)
(2) Fine-grained
    Agate (banded, many colors)
    Chalcedony (brown, gray)
    Heliotrope (green)
    Jasper (red)
			A rock-forming mineral — see Table C.2.				

Topaz	$Al_2SiO_4(OH, F)_2$	Prism-shaped crystals, granular masses.	One, perfect.	8	3.5	Colorless, yellow, blue, brown.	Hardness, form, color.
Tourmaline			A rock-forming mineral — see Table C.2				
Zircon	$ZrSiO_4$	Four-sided elongate crystals, square in cross section.	None.	7.5	4.7	Brown, red, green, blue, black.	Habit, hardness.

Glossary

Some definitions are not included in the glossary; *units of measurement* can be found in Appendix A, *chemical elements* are listed in Appendix B, and *mineral names* are given in Appendix C.

A horizon. A soil horizon that either underlies the O horizon or is the uppermost horizon. Generally dark colored and characterized by an accumulation of organic matter.

Aa. A rubbly, rough-looking form of lava, usually basaltic in composition.

Ablation. The loss of mass from a glacier.

Ablation area. A region of net loss on a glacier characterized by a surface of bare ice and old snow from which the last winter's snowcover has melted away.

Absolute velocity (of a plate). The velocity of a plate of lithosphere measured against a fixed, external frame of reference. Compare with *relative velocity*.

Abyssal plain. A large flat area of the deep seafloor having slopes less than about 1 m/km, and ranging in depth below sea level from 3 to 6 km.

Abyssal seafloor. Broad flat expanses of seafloor, 3 to 6 km deep. Young oceanic plate grows dense and subsides as it cools. Abyssal floors form after an oceanic plate has cooled enough for heat rising from the deep mantle to balance heat loss through the seafloor.

Accreted terrane. Block of crust moved laterally by strike-slip faulting or by a combination of strike-slip faulting and subduction, then accreted to a larger mass of continental crust. Also called a *suspect terrane*.

Accreted terrane continental margin. Continental margin modified by the addition of island arcs and other rafted-in, exotic fragments of crust.

Accretion. The process by which solid bodies gather together to form a planet or a continent.

Accumulation. The addition of mass to a glacier.

Accumulation area. An upper zone on a glacier, covered by remnants of the previous winter's snowfall and representing an area of net gain in mass.

Active margin. A continental coastline that is coincident with a convergent plate boundary.

Active zone. The margin of a tectonic plate where deformation is occurring.

Adiabatic expansion. The tendency of a substance to decrease in temperature, without loss of heat to its surroundings, as it expands in response to decreasing pressure. Conversely, a substance tends to experience an increase in temperature, without a gain of heat from its surroundings, as it contracts in response to increasing pressure.

Aftershocks. Earthquakes that occur after a large earthquake on the same fault, or a nearby fault.

Agglomerate. The pyroclastic rock consisting of bomb-sized tephra, that is, tephra in which the average particle diameter is greater than 64 mm.

Aggradation. Depositional upbuilding, as by a stream.

Albedo. The reflectivity of the surface of a planet.

Alluvial fan. A fan-shaped body of alluvium typically built where a stream leaves a steep mountain valley.

Alluvium. Sediment deposited by streams in nonmarine environments.

Alpha particle (α-particle). An atomic particle expelled from an atomic nucleus during certain radioactive transformations, equivalent to an $_2^4$He nucleus stripped of its electrons.

Amorphous. A term applied to solids that lack internal atomic order.

Amphibolite. A metamorphic rock of intermediate grade, generally coarse-grained, containing abundant amphibole.

Amygdule. A vesicle filled by secondary minerals such as calcite and quartz deposited by groundwater.

Andesite. A fine-grained igneous rock with the composition of a diorite.

Andesite line. A line on a map that roughly surrounds the Pacific Ocean basin and inside of which andesite is not found.

Andesite magma. One of the three common magma types; a magma with an SiO_2 content of about 60 percent by weight.

Angle of repose. The steepest angle, measured from the horizontal, at which rock debris remains stable.

Angular unconformity. An unconformity marked by angular discordance between older and younger rocks.

Anhydrous. A term applied to a substance that is H_2O free. Opposite of hydrous.

Anion. An ion with a negative electrical charge.

Anorthosite. A coarse-grained igneous rock consisting largely of plagioclase.

Antecedent stream. A stream that has maintained its course across an area of the crust that was raised across its path by folding or faulting.

Anthracite. A metamorphic rock derived from coal by heat and pressure.

Anticline. An upfold in the form of an arch.

Aphanite. An igneous rock in which the constituent mineral grains are so small they can only be seen clearly by using some kind of magnification.

Apparent polar wandering. The apparent motions of the magnetic poles derived from measurements of pole positions using paleomagnetism.

Aquiclude. A body of impermeable or distinctly less permeable rock adjacent to an *aquifer*.

Aquifer. A body of permeable rock or regolith saturated with water and through which groundwater moves.

Archean Eon. The eon that follows the Hadean Eon.

Arête. A jagged, knife-edged ridge crest created where glaciers have eroded back into a ridge.

Arkose (-arkosic sandstone). A sandstone in which feldspar is a major mineral component.

Artesian aquifer. An aquifer in which water is under hydraulic pressure.

Artesian spring. A natural spring that draws its supply of water from an artesian aquifer.

Artesian system. An inclined aquifer that permits water confined in it to rise to the surface in a well or along a fissure.

Artesian well. A well in which water rises above the aquifer.

Ash. Tephra in which particles have an average diameter less than 2 mm. Also called *volcanic ash*.

Ash tuff. Pyroclastic rock in which the tephra particles are less than 2 mm in diameter.

Asphalt. See *tar*.

Asteroids. Irregularly shaped rocky bodies that have orbits lying between the orbits of Mars and Jupiter.

Asthenosphere. The layer of the mantle where rocks are relatively ductile and are easily deformed. It lies at a depth of 100 to 350 km below the surface.

Asymmetric fold. A fold in which one limb dips more steeply than the other.

Atmosphere. The mixture of gases, predominantly nitrogen, oxygen, carbon dioxide, and water vapor, that surrounds Earth.

Atoll. A coral reef, often roughly circular in plan, that encloses a shallow lagoon.

Atom. The smallest individual particle that retains the distinctive properties of a given chemical element.

Atomic number. The number of protons in the nucleus of an atom.

Atomic substitution. See *ionic substitution.*

Auxiliary plane. A surface that lies at a right angle to the fault plane of an earthquake. The sliding of rock on either the fault plane or the auxiliary plane of an earthquake would produce identical first-motions of P waves.

Axial plane. An imaginary plane that divides a fold as symmetrically as possible, and that passes through the axis.

Axis (of a fold). The median line between the limbs, along the crest of an anticline or the trough of a syncline.

B horizon. A soil horizon generally lying below an A horizon, usually brownish or reddish in color, and commonly enriched in clay and iron oxides.

Back-arc basin. An arc-shaped basin formed by crustal thinning behind a magmatic arc.

Backshore. A zone extending inland from a berm to the farthest point reached by waves.

Backwash. The seaward return of water down a beach following the swash of a wave.

Badlands. A hilly landscape generated largely by the rapid erosion of near-surface sediments and sedimentary rock by rainwater runoff.

Bajada. A broad alluvial apron composed of coalescing adjacent fans.

Bar. An accumulation of alluvium formed in a channel where a decrease in stream velocity causes deposition.

Barchan dune. A crescent-shaped sand dune with horns pointing downwind.

Barrier island. A long island built of sand, lying offshore and parallel to the coast.

Barrier reef. A reef separated from the land by a lagoon.

Basalt. A fine-grained igneous rock with the composition of a gabbro.

Basaltic magma. One of the three common types of magma; contains about 50 percent SiO_2 by weight.

Base level. The limiting level below which a stream cannot erode the land.

Batholith. The largest kind of pluton. A very large, igneous body of irregular shape that cuts across the layering of the rock it intrudes.

Bauxite. An aluminous laterite formed by tropical weathering. The preferred ore of aluminum.

Bay. A wide, open, curving indentation or inlet of a sea or lake into an adjacent land mass.

Bay barrier. A ridge of sand or gravel that completely blocks the mouth of a bay.

Beach. Wave-washed sediment along a coast, extending throughout the surf zone.

Beach drift. The irregular movement of particles along a beach as they travel obliquely up the slope of a beach with the swash and directly down this slope by the backwash.

Beach ridge. A low ridge of sand parallel to and on the landward side of a beach.

Bed. The smallest formal unit of a body of sediment or sedimentary rock.

Bedding. The layered arrangement of strata in a body of sediment or sedimentary rock.

Bedding plane. The top or bottom of a bed.

Bed load. Coarse particles that move along the bottom of a stream channel.

Bedrock. The continuous mass of solid rock that makes up the crust.

Benioff strain seismograph. See *strain seismograph.*

Benioff zone. A narrow, well-defined zone of deep earthquake foci beneath a seafloor trench.

Berm. A nearly horizontal or landward-sloping bench formed of sediment deposited by waves.

Beta particle. (β-particle). An electron expelled from an atomic nucleus during certain radioactive transformations.

Biogenic rock. Rock formed by lithification of biogenic sediment.

Biogenic sediment. Sediment composed mainly of fossil remains.

Biogeochemical cycles. Natural cycles describing the movements and interactions through Earth's spheres of the chemicals essential to life.

Biomass energy. The energy obtained through burning plant matter.

Biosphere. The totality of Earth's organisms and, in addition, organic matter that has not yet been completely decomposed.

Bituminous coal. The highest grade of coal.

Blueschist. A metamorphic rock formed under conditions of high pressure and low temperature containing blue-colored amphiboles.

Body waves. Seismic waves that travel outward from an earthquake focus and pass deeply through Earth's interior.

Bolide. An impacting body, either a meteorite, an asteroid, or a comet.

Bombs. Tephra particles having average diameters greater than 64 mm.

Bonding. The electrostatic forces that hold atoms together to form compounds by sharing and transfer of electrons. See also *covalent bonding, ionic bonding, metallic bonding* and *van der Waals bond.*

Bottomset layer. A gently sloping, fine, thin part of each layer in a delta.

Boundary conditions. A mathematical expression of the physical state of Earth's climate system at a particular time or location within a climate-simulation experiment.

Bowen's reaction series. A schematic description of the order in which different minerals crystallize during the cooling and progressive crystallization of a magma. See also *continuous* and *discontinuous reaction series.*

Braid delta. A delta composed of coarse-grained sediment built by a braided stream into a standing body of water.

Braided stream. A channel system consisting of a tangled network of two or more smaller branching and reuniting channels that are separated by islands or bars.

Breaker. An oversteepened wave that collapses in a mass of turbulent water against a shore or reef.

Breakwater. An offshore barrier built to protect a beach or anchorage from incoming waves.

Breccia. A coarse-grained rock composed of cemented angular fragments.

Breeder reactor. A nuclear reactor in which nonfissionable isotopes such as ^{238}U are converted to fissionable isotopes.

Brittle deformation. Any response of a substance to stress that involves internal breakage and the formation of fractures.

Brittle fracture. Rupture of a solid body that is stressed beyond its elastic limit.

Burial metamorphism. Metamorphism caused solely by the burial of sedimentary or pyroclastic rocks.

Butte. Isolated, often flat-topped, steep-sided, desert hill. Smaller than a *mesa.*

C horizon. The deepest soil horizon, lying beneath the A horizon and/or B horizon of a soil profile; often yellowish-brown in color and consisting of weathered parent rock or sediment.

Calcareous ooze. A deep-sea pelagic sediment composed largely of calcareous skeletal remains. See also *deep-sea ooze.*

Caldera. A roughly circular, steep-walled volcanic basin several kilometers or more in diameter.

Caliche. A solid, almost impervious layer of whitish calcium carbonate in a soil profile.

Calving. The progressive breaking off of icebergs from a glacier that terminates in deep water.

Capillary attraction. The adhesive force between a liquid, such as water, and a solid.

Carbon cycle. The continuous transfer of the element carbon between different parts of the Earth system, mediated by photosynthesis, rainfall, rock weathering, secretion of calcium carbonate, sedimentation, subduction and volcanism.

Carbonate shelf. A shallow marine shelf where sedimentation is dominated by carbonate-secreting organisms.

Carbonic acid. A weak acid resulting from the solution of small quantities of carbon dioxide in rain or groundwater.

Cataclastic metamorphism. Metamorphism that involves change of texture caused by mechanical effects such as crushing and shearing, but no change in mineral assemblage.

Catastrophism. The concept that all of Earth's major features, such as mountains, valleys, and oceans, have been produced by a few great catastrophic events.

Cation. A positive ion.

Cave. A natural underground opening, generally connected to the surface and large enough for a person to enter.

Cavern. A large cave or system of interconnected cave chambers.

Celsius scale. A temperature scale in which the boiling point of water is 100° and the freezing point is 0°.

Cementation. The diagenetic process by which clastic sediments are converted to rock through deposition or precipitation of minerals in the spaces between the grains.

Cenozoic era. The youngest era of the Phanerozoic Eon.

Central rift valley. A long, narrow valley at the crest of a midocean ridge.

Chalk. Rock composed of compacted carbonate shells of minute floating organisms.

Channel. The passageway in which a stream flows.

Chemical differentiation by partial melting. The process of forming magma by the incomplete melting of rock.

Chemical elements. The most fundamental substances into which matter can be separated by chemical means.

Chemical sediment. Sediment formed by precipitation of minerals from solutions in water.

Chemical weathering. The decomposition of rocks through chemical reactions such as hydration and oxidation.

Chert. A hard, very compact sedimentary rock composed almost entirely of very fine-grained, interlocking quartz crystals that occurs as extensive continuous layers (bedded chert) or as nodules in carbonate rocks.

Chloroflurocarbons. Synthetic industrial gases that destroy ozone in the upper atmosphere and contribute to the greenhouse effect. Also called CFCs.

Chrons. See *magnetic chrons*.

Cirque. A bowl-shaped hollow on a mountainside, open downstream, bounded upstream by a steep slope (headwall), and excavated mainly by frost wedging and by glacial abrasion and plucking.

Cirque glacier. A glacier that occupies a bowl-shaped hollow on the side of a mountain.

Clast. Any individual particle of clastic sediment.

Clastic sediment. See *detritus*.

Cleavage. The tendency of a mineral to break in preferred directions along bright, reflective plane surfaces.

Climate. The average weather conditions of a place or area over a period of years.

Closed system. Any system with a boundary that allows the passage in or out of energy but not of matter.

Closure temperature. Threshold temperature for either (1) the escape of radioactive product elements, such as *radiogenic helium*, from a rock, or (2) the annealing of *fission tracks* caused by the spontaneous fission of uranium.

Coal. A black, combustible, sedimentary or metamorphic rock consisting chiefly of decomposed plant matter and containing more than 50 percent organic matter.

Coalification. The stages by which plant matter is converted first to peat, then lignite, subbituminous coal, and bituminous coal.

Col. A gap or pass in a mountain crest where the headwalls of two cirques intersect.

Collision zone. A convergent plate margin where two plates collide.

Collisional uplift. The raising of topography as a byproduct of convergence of lithospheric plates.

Colluvium. Loose, incoherent deposits on or at the base of slopes and moving mainly by creep.

Column. A stalactite joined with a stalagmite, forming a connection between the floor and roof of a cave.

Columnar joints. Joints that split igneous rocks into long prisms or columns.

Comet. A small celestial body that circles the Sun with a highly elliptical orbit.

Compaction. A decrease in porosity and bulk of a body of sediment as additional sediment is deposited above it, or due to pressures resulting from deformation.

Complex ion. A strongly bonded pair of ions that act in the same way as a single ion, forming compounds by bonding with other elements.

Composition (of a mineral). The proportions of the various chemical elements in a mineral.

Compound. A combination of atoms of different elements bonded together.

Compressional stress. Differential stress that squeezes and compresses a body.

Compressional waves. See *P waves*.

Conchoidal fracture. Breakage resulting in smooth, curved surfaces.

Concretion. A hard, localized body, having distinct boundaries, enclosed in sedimentary rock, and consisting of a substance precipitated from solution, commonly around a nucleus.

Conduction. The means by which heat is transmitted through solids without deforming the solid.

Cone of depression. A conical depression in the water table immediately surrounding a well.

Confined aquifer. An aquifer bounded by aquicludes.

Confining stress. See *uniform stress*.

Conformity. Strata that have been deposited layer upon layer without interruption.

Conglomerate. A sedimentary rock composed of clasts of rounded gravel set in a finer-grained matrix.

Consequent stream. A stream whose pattern is determined solely by the direction of slope of the land.

Contact metamorphism. (also called *thermal metamorphism*). Metamorphism adjacent to an intrusive igneous rock.

Continental collision margin. A plate margin along which two continental masses collide.

Continental convergent margin. The margin of a continent that is adjacent to a subduction zone.

Continental crust. The part of Earth's crust that comprises the continents, which has an average thickness of 45 km.

Continental divide. A line that separates streams flowing towards opposite sides of a continent, usually into different oceans.

Continental drift. The slow, lateral movements of continents across Earth's surface.

Continental rise. A region of gently changing slope where the floor of the ocean basin meets the margin of a continent.

Continental shelf. A submerged platform of variable width that forms a fringe around a continent.

Continental shield. An assemblage of cratons and orogens that has reached isostatic equilibrium.

Continental slope. A pronounced slope beyond the seaward margin of the continental shelf.

Continental volcanic arc. An arcuate chain of andesitic volcanoes on the continental crust formed as a result of subduction.

Continuous GPS measurement. Measurement of the slow motion of fixed structures on Earth by continuous reference to orbiting satellites.

Continuous reaction series. The continuous change of mineral composition, through ionic substitution, as a magma crystallizes See also *discontinuous reaction series*.

Convection. The process by which hot, less dense materials rise upward, being replaced by cold, more dense, downward-flowing material to create a convection current.

Convection current. The flow of material as a result of convection.

Convergent margin. The zone where plates meet as they move toward each other. See also *subduction zone*.

Coquina. A limestone composed solely or chiefly of loosely aggregated shells and shell fragments.

Core. The spherical mass, largely metallic iron, at the center of Earth.

Coriolis effect. An effect that causes any body that moves freely with respect to the rotating solid Earth to veer toward the right in the northern hemisphere and toward the left in the southern hemisphere, regardless of the initial direction of the moving body.

Coronae. Ring-like structural features on a planetary surface. Many coronae are found on Venus.

Correlation. Determination of the equivalence in time-stratigraphic units of the succession of strata in two or more different places.

Correlation of strata. Determination in time stratigraphic age of the succession of strata found in two or more different areas.

Cosmogenic nuclides. Radioactive atoms generated by high-energy subatomic particles emitted by the Sun, via a collision with a nonradioactive atom on Earth.

Covalent bonding. The force between two atoms that have filled their energy-level shells by sharing one or more electrons.

Crater. A funnel-shaped depression, opening upward, at the top of a volcano from which gases, fragments of rock, and lava are ejected.

Craton. A core of ancient rock in the continental crust that has attained tectonic and isostatic stability.

Creep. The imperceptibly slow downslope movement of regolith.

Creep of glacier ice. Slow deformation of glacier ice, with movement occurring along the internal planes of ice crystals.

Crevasse. A deep, gaping fissure in the upper surface of a glacier.

Cross bedding. Beds that are inclined with respect to a thicker stratum within which they occur.

Cross section. See *geologic cross section*.

Crust. The outermost and thinnest of Earth's compositional layers, which consists of rocky matter that is less dense than the rocks of the mantle below.

Cryosphere. The portion of the hydrosphere that is ice, snow, and frozen ground.

Crystal. A solid compound composed of ordered, three-dimensional arrays of atoms or ions chemically bonded together and displaying crystal form.

Crystal faces. The planar surfaces that bound a crystal.

Crystal form. The geometric arrangement of crystal faces.

Crystalline. See *crystal structure*.

Crystal settling. The process of fractional crystallization by which dense minerals sink and form segregated layers of one or more minerals in a magma chamber.

Crystal structure. The geometric pattern that atoms assume in a solid. Any solid that has a crystal structure is said to be *crystalline*.

Curie point. A temperature above which permanent magnetism is not possible.

Cyclothems. A repetition of marine and nonmarine sediments that indicates repeated transgressions and regressions by the sea.

Dacite. A fine-grained igneous rock with the composition of a granodiorite.

Darcy's law. The relationship between discharge, coefficient of permeability, and hydraulic gradient in percolating groundwater.

Daughter product. (-daughter). The product arising from radioactive decay Compare *parent*.

Debris avalanche. A granular flow of regolith moving at high velocity ($\geq$10 m/s).

Debris fall. The relatively free fall or collapse of regolith from a steep cliff or slope.

Debris flow. The downslope movement of a mass of unconsolidated regolith more than half of which is coarser than sand.

Debris slide. The slow to rapid downslope movement of regolith across an inclined surface. Compare *rockslide*.

Décollement. A body of rock above the detachment surface of a thrust fault.

Decomposition. (of rocks). Chemical weathering.

Deep-sea fan. Huge fan-shaped body of sediment at the base of the continental slope that spreads downward and outward to the deep seafloor.

Deep-sea ooze. A muddy marine sediment composed mainly of the remains of microscopic marine organisms. See also *calcareous ooze and siliceous ooze*.

Deflation. The picking up and removal of loose particles by wind.

Dehydrate. The loss of water.

Delta. A body of sediment deposited by a stream where it flows into standing water.

Density. The average mass per unit volume.

Denudation. The sum of the weathering, mass-wasting, and erosional processes that result in the progressive lowering of Earth's surface.

Desert. Arid land, whether "deserted" or not, in which annual rainfall is less than 250 mm (10 in) or in which the evaporation rate exceeds the precipitation rate.

Desertification. The invasion of desert into nondesert areas.

Desert pavement. A surface layer of coarse particles concentrated chiefly by deflation.

Desert varnish. A thin, dark, shiny coating consisting mainly of manganese and iron oxides, formed on the surfaces of stones and rock outcrops in desert regions after long exposure.

Detachment surface. The surface along which a large-scale thrust or normal fault moves.

Detrital sediment. The accumulated particles of broken rock and skeletal remains of dead organisms (also called *clastic sediment*).

Detritus. (also called *clastic sediment* and *detrital sediment*). The loose fragmented debris produced by the mechanical breakdown of older rocks.

Diagenesis. Chemical, physical, and biological changes that affect sediment after its initial deposition and during and after its slow transformation into sedimentary rock.

Diatomite. A sedimentary rock formed by lithification of *siliceous ooze*.

Diatremes. Breccia-filled volcanic pipes formed by volcanic explosions.

Differential stress. Stress in a solid that is not equal in all directions.

Differential weathering. Weathering that occurs at different rates or intensity as a result of variations in the composition and structure of rocks.

Dike. Tabular, parallel-sided sheets of intrusive igneous rock that cut across the layering of the intruded rock.

Diorite. A coarse-grained igneous rock consisting mainly of plagioclase and ferromagnesian minerals. Quartz is sparse or absent.

Dip. The angle in degrees between a horizontal plane and an inclined plane, measured down from horizontal in a plane perpendicular to the strike.

Dip-slip fault. A normal or reverse fault on which the only component of movement lies in a plane normal to the strike of the fault surface.

Discharge. The quantity of water that passes a given point in a stream channel per unit time.

Discharge area. Area where subsurface water is discharged to streams or to bodies of surface water.

Disconformity. An irregular surface of erosion between parallel strata.

Discontinuous reaction series. The discontinuous sequence of reactions by which early formed minerals in a crystallizing magma react with residual liquid to form new minerals. See also *continuous reaction series*.

Dispersion. Waves of different wavelengths traveling at different velocities.

Dissolution. The chemical weathering process whereby minerals and rock material pass directly into solution.

Dissolved load. Matter dissolved in stream water.

Divergent margin (of a plate). A fracture in the lithosphere where two plates move apart. Also called a *spreading center*.

Divide. The line that separates adjacent drainage basins.

Dolostone. A sedimentary rock composed chiefly of the mineral dolomite.

Drainage basin. The total area that contributes water to a stream.

Drift. See *glacial drift*.

Dripstone. A deposit chemically precipitated by dripping water in an air-filled cavity.

Dropstone. A stone released from a melting iceberg that plunges into unconsolidated sediment on the seafloor or a lake bottom.

Drumlin. A streamlined hill consisting of glacially deposited sediment and elongated parallel with the direction of ice flow.

Ductile deformation. The irreversible deformation induced in a solid that is stressed beyond its elastic limit but before rupture occurs.

Dune. A mound or ridge of sand deposited by wind.

E horizon. Soil horizon, sometimes present below the A horizon, that is grayish or whitish in color.

Earth system. The assemblage of all of Earth's parts and the relationships between them.

Earthflow. A granular flow of regolith with velocities ranging from 10^{-5} to 10^{-1} m/s.

Earthquake focus. The point of the first release of energy associated with an earthquake.

Earthquake forecasting. Estimating the long-term likelihood (decades to centuries) of a damaging earthquake in a particular region, in order to develop cost-effective building codes and emergency-response plans.

Earthquake magnitude. See *Richter magnitude scale*.

Earthquake precursors. Phenomena that occur in the hours, days, months and years prior to an earthquake, and which have the potential for predicting future earthquakes.

Earthquake prediction. Estimating the short-term likelihood (days to years) of a damaging earthquake in a particular region, in order to warn local authorities and thereby minimize casualties and property damage.

Earth's gravity. An inward-acting force with which Earth tends to pull all objects toward its center.

Eccentricity (of Earth's orbit). The degree to which the shape of Earth's orbit departs from perfect circularity.

Ecliptic. Plane of Earth's orbit around the Sun.

Eclogite. A metamorphic rock containing garnet and jadeitic pyroxene.

Ejecta blanket. Layer of broken rock surrounding an impact crater.

Elastic deformation. The reversible or non-permanent deformation that occurs when an elastic solid is stretched and squeezed and the force is then removed.

Elastic limit. The limiting stress beyond which a body suffers irreversible deformation.

Elastic rebound theory. The theory that earthquakes result from the release of stored elastic energy by slippage on faults.

Electrons. Negatively charged atomic particles.

Element (chemical). The most fundamental substances into which matter can be separated by chemical means.

Emergence of a coast. An increase in the area of land exposed above sea level resulting from uplift of the land and/or fall of sea level.

End moraine. A ridgelike accumulation of drift deposited along the margin of a glacier.

Energy-level shell. The specific energy level of electrons as they orbit the nucleus of an atom.

Eolian. Pertaining to the wind, especially erosional and depositional processes, as well as landforms and sediments resulting from wind action.

Eon. The largest interval of geologic time. We are now in the fourth eon.

Epicenter. That point on Earth's surface that lies vertically above the focus of an earthquake.

Epidote amphibolite. A metamorphic rock containing both amphibole and epidote as major constituents.

Epoch. The time during which a geologic series accumulates.

Equilibrium line. A line that marks the level on a glacier where net mass loss equals net gain.

Era. The primary time division of eons.

Erosion. The complex group of related processes by which rock is broken down physically and chemically and the products are moved.

Erratic. A glacially deposited rock fragment whose composition differs from that of the bedrock beneath it.

Eruption column. A mixture of ash and hot gases that rises upward as a column above an erupting volcano.

Esker. A long narrow ridge, often sinuous, composed of stratified drift.

Estuary. A semi-enclosed body of coastal water within which seawater is diluted with fresh water.

Evaporite. Sedimentary rock composed chiefly of minerals precipitated from a saline solution through evaporation.

Evaporite deposits. Layers of salts that precipitate as a consequence of evaporation.

Exfoliation. The spalling off of successive shells, like the "skins" of an onion, around a solid rock core.

Exhumation. The gradual exposure of sub-surface rocks by stripping off (eroding) the surface layers.

Exploration geology. The special branch of geology concerned with discovering new supplies of usable minerals.

Exposure (also called an *outcrop*). A place where rock or sediment is exposed at Earth's surface.

Extensional uplift. The raising of topography as a byproduct of divergent stretching within Earth's crust, often associated with normal faulting.

External processes. All the activities involved in erosion and in the transport and deposition of the eroded materials.

Extraterrestrial material. Material originating outside Earth.

Extrusive igneous rock. Rock formed by the solidification of magma poured out onto Earth's surface.

Facies. A distinctive group of characteristics, within a rock unit, that differs as a group from those elsewhere in the same unit. See also *sedimentary facies* and *metamorphic facies*.

Fan. See *alluvial fan*.

Fan delta. A gravel-rich delta formed where an alluvial fan builds outward into a standing body of water.

Fault. A fracture in a rock along which movement occurs.

Fault breccia. Crushed and broken rock adjacent to a fault.

Fault plane. A surface on which rock masses slip past each other during an earthquake.

Ferromagnesian minerals. The common rock-forming minerals that contain iron and/or magnesium as essential constituents.

Fiord. See *fjord*.

Fission. Radioactive transformation by nuclear fragmentation.

Fission tracks. Damage to the atomic lattice structure of a mineral, caused by the passage of charged nuclear fragments emitted by the spontaneous fission of uranium.

Fission-track dating. Determining the time a rock has been accumulating fission tracks from radioactive decay. Fission tracks anneal when a rock is warmer than its *closure temperature*, so this technique measures the time that a rock has been close to Earth's surface, and therefore relatively cool.

Fissure eruption. Extrusion of lava or pyroclasts and associated gases along an extended fracture.

Fjord. A deep, glacially carved valley submerged by the sea. Also spelled *fiord*.

Fjord glacier. A glacier that occupies a fjord.

Flash flood. A local and sudden flood of water through a stream channel, generally of relatively great volume and short duration.

Flood. A discharge great enough to cause a stream to overflow its banks.

Floodplain. The part of any stream valley that is inundated during floods.

Flowstone. A deposit chemically precipitated from flowing water in the open air or in an air-filled cavity.

Fluvial. Of, or pertaining to, streams or rivers, especially erosional and depositional processes of streams and the sediments and landforms resulting from them.

Focal sphere. Graphical representation of the pattern of P-wave first motion that radiates from an earthquake. The sphere is divided into four quadrants by the fault and auxiliary planes of an earthquake.

Foliation. The planar texture of mineral grains, principally micas, produced by metamorphism.

Fold. An individual bend or warp in layered rock.

Folding. The bending of rocks or sediments.

Footwall block of a fault. The block of rock below an inclined fault.

Fore-arc basin. A basin parallel to a deep-sea trench and separated from the trench by a *fore-arc ridge*.

Fore-arc ridge. See *fore-arc basin*.

Foreset layer. The coarse, thick, steeply sloping part of each layer in a delta.

Foreshocks. Small earthquakes that precede a larger earthquake on the same fault, or a nearby fault.

Foreshore. A zone extending from the level of lowest tide to the average high-tide level.

Formation. A body of rock distinctive enough on the basis of physical properties to constitute a basic unit for geologic mapping. The basic unit of rock stratigraphy.

Fossil. The naturally preserved remains or traces of an animal or a plant.

Fossil fuel. Remains of plants and animals trapped in sediment that may be used for fuel.

Fracture. Irreversible deformation of a rock in which the limits of both elastic and ductile deformation have been exceeded; breakage.

Fringing reef. A coral reef attached to or bordering the adjacent land.

Frost heaving. The lifting of regolith by the freezing of water contained within it.

Frost wedging. The formation of ice in a confined opening within rock, thereby causing the rock to be forced apart.

Fumarole. A volcanic vent that emits only gases.

Gabbro. A coarse-grained igneous rock in which olivine and pyroxene are the predominant minerals and plagioclase is the feldspar present. Quartz is absent.

Gamma rays (γ-rays). Very short wavelength electromagnetic radiation given off by an atomic nucleus during certain radioactive transformations.

Gangue. The nonvaluable minerals of an ore.

Garnet peridotite. A coarse-grained igneous rock consisting largely of olivine, garnet, and pyroxene.

Gas hydrates. Ice-like solids in which gas molecules, mainly methane, are locked in the structure of solid H_2O. They are found in some ocean sediments and beneath frozen ground.

Gelifluction. Downslope movement of the thawed surface layer of regolith in a region of perennially frozen ground.

General Circulation Model (GCM). A mathematical model used to simulate present and past climate on Earth.

Geochemically abundant elements. Those chemical elements that individually comprise 0.1 percent or more by weight of the crust.

Geochemically scarce elements. Those chemical elements that individually comprise less than 0.1 percent by weight of the crust.

Geodesy. The science of measuring Earth's shape and its gravity, including small changes in position induced by deformation of Earth's interior.

Geographic cycle. A hypothesis of landscape evolution proposed by W. M. Davis in which an uplifted land area passes through sequential stages of development as it erodes down to a surface of low relief.

Geologic column. A composite diagram combining in chronological order the succession of known strata, fitted together on the basis of their fossils or other evidence of relative or actual age.

Geologic cross section. A diagram showing the arrangement of rocks in a vertical plane.

Geologic map. A map that shows the distribution, at the surface, of rocks of various kinds or of various ages.

Geologic time scale. A sequential arrangement of geologic time units.

Geologists. Scientists who study Earth.

Geology. The science of our planet.

Geomorphology. The study of Earth's landforms.

Geosphere. The solid portion of Earth.

Geosyncline. A great trough that has received thick deposits of sediment during its slow subsidence through long geologic periods.

Geothermal gradient. The rate of increase of temperature downward in Earth.

Geothermal power. Heat energy drawn from Earth's internal heat.

Geyser. A thermal spring equipped with a system of plumbing and heating that causes intermittent eruptions of water and steam.

Glacial drift. Sediment deposited directly by glaciers or indirectly by meltwater in streams, in lakes, and in the sea. Also called *drift*.

Glacial grooves. See *glacial striations*.

Glacialmarine drift (also referred to as glacial-marine sediment). Terrigenous sediment dropped onto the seafloor from floating ice shelves or from icebergs.

Glacial striations. Subparallel scratches inscribed on a clast or a bedrock surface by rock debris embedded in the base of a glacier. Wider and deeper markings on bedrock are *glacial grooves*.

Glaciation. The modification of the land surface by the action of glacier ice.

Glacier. A permanent body of ice, consisting largely of recrystallized snow, that shows evidence of downslope or outward movement, due to the stress of its own weight.

Global change. Changes produced in the environment, especially the atmosphere, as a result of human activities.

Global positioning system (GPS). A network of satellites whose orbits form a reference frame for precise determination of position on Earth.

Global warming. Rising temperature of the atmosphere as a result of rising levels of greenhouse gases.

Gneiss. A high-grade metamorphic rock, always coarse-grained and foliated, with marked compositional layering but with imperfect cleavage.

Gondwanaland. The southern half of Pangaea, consisting of present-day Australia, India, Madagascar, Africa, Antarctica, and South America.

Graben (also called a *rift*). A trenchlike structure bounded by parallel normal faults. See also *half-graben*.

Grade. A term for the level of concentration of a metal in an ore. Usually expressed as a percentage.

Graded bed. A layer in which the size of the sedimentary particles grades upward from coarse to finer.

Graded stream. A stream in which the slope has become so adjusted, under conditions of available discharge and prevailing channel characteristics, that the steam is just able to transport the sediment load available to it.

Gradient (topographic). A measure of the vertical drop over a given horizontal distance.

Grain flow. Mass-wasting of dry or nearly dry granular sediment with air filling the pore space.

Granite. A coarse-grained igneous rock containing quartz and feldspar, with potassium feldspar being more abundant than plagioclase.

Granitic. Any coarse-grained igneous or metamorphic rock having a texture and composition resembling that of a granite.

Granodiorite. A coarse-grained igneous rock resembling a granite, in which plagioclase is more abundant than potassium feldspar.

Granular flow. A type of flow in which the weight of the flowing mass is supported by grain-to-grain contact or repeated collision between grains.

Granular texture. The interlocking arrangements of mineral grains in granitic rocks.

Granulite. A high-grade metamorphic rock, usually coarse-grained and indistinctly foliated, containing pyroxenes as a major mineral.

Gravimeter (also called a *gravity meter*). A sensitive device for measuring the pull of gravity at any locality.

Gravity anomalies. Variations in the pull of gravity after correction for latitude and altitude.

Gravity meter. See *gravimeter*.

Greenhouse effect. The property of Earth's atmosphere by which long-wavelength heat radiation from Earth's surface is trapped or reflected back by the atmosphere.

Greenhouse gases. The gases in the atmosphere, mainly H_2O, CO_2, CFCs, and CH_4, that cause the greenhouse effect.

Greenschist. A low-grade metamorphic rock rich in chlorite.

Greywacke. See *lithic sandstone*.

Groin. A low wall, built on a beach, that crosses the shoreline at a right angle.

Groundmass. The fine-grained matrix of a porphyry.

Ground moraine. Widespread drift with a relatively smooth surface topography consisting of gently undulating knolls and shallow closed depressions.

Groundwater. All the water contained in the spaces within bedrock and regolith.

Growth habit. A characteristic growth form of a mineral.

Gyre. A large subcircular current system.

Hadean eon. The oldest eon.

Half-graben. A trenchlike structure formed when the hanging-wall block moves downward on a curved fault surface. See *graben*.

Half-life. The time required for radioactive decay to reduce the number of parent atoms by one-half.

Hand specimen. A rock sample of convenient size to hold in the hand for study.

Hanging valley. A glacial valley whose mouth is at a relatively high level on the steep side of a larger glaciated valley.

Hanging-wall block (of a fault). The block of rock above an inclined fault.

Hardness. Relative resistance of a mineral to scratching.

Hard water. Water containing an unusually high amount of calcium carbonate.

Headwall. The steep cliff that bounds the upslope side of a cirque.

Heat. The energy a body has due to the motions of its atoms.

Heat conduction. The transport of heat through a substance. On a microscopic scale the heat energy moves via the exchange of kinetic energy between atoms and molecules.

Heat energy. The energy of a hot body.

Heat flow. The outward flow of heat from Earth's interior.

High grade metamorphism. Metamorphism under conditions of high temperature and high pressure.

Highlands. See *lunar highlands*.

Hinge fault. A fault on which displacement dies out perceptibly along strike and ends at a definite point.

Historical geology. Study of the chronology of Earth's past events, both physical and biological.

Homogeneous stress. See *uniform stress*.

Horn. A sharp-pointed peak bounded by the intersecting walls of three or more cirques.

Hornfels. A hard, fine-grained rock developed during contact metamorphism of a shale.

Horst. An elevated elongate block of crust bounded by parallel normal faults. See also *graben*.

Hot spot. A fixed point on the Earth's surface defined by long-lived volcanism.

Hot spot hypothesis. A scientific hypothesis that mantle hot spots form a stable reference frame for the determination of plate motions over geologic time.

Humus. The decomposed residue of plant and animal tissues.

Hurricane. A tropical cyclonic storm having winds that exceed 120 km/h. See also *typhoon*.

Hydration. The incorporation of water into a crystal structure.

Hydraulic gradient. The slope of the water table.

Hydrocarbons. Any organic compound (gaseous, liquid, or solid) consisting wholly of carbon and hydrogen.

Hydroelectric power. Energy recovered from the potential energy of rivers as they flow downward to the sea.

Hydrologic cycle. The day-to-day and longterm cyclic changes in the hydrosphere.

Hydrolysis. A chemical reaction in which the H^+ or OH^- ions of water replace ions of a mineral.

Hydrosphere. The totality of Earth's water, including the oceans, lakes, streams, water underground, and all the snow and ice, including glaciers.

Hydrothermal mineral deposit. Any local concentration of minerals formed by deposition from a hydrothermal solution.

Hydrothermal solutions. Hot brines either given off by cooling magmas, or produced by reactions between hot rock and circulating water, that concentrate minerals in solutions.

Hydrous. A term applied to substances that contain H_2O or (OH).

Hypothesis. An unproved explanation for the way things happen.

Iapetus. Name given to the ocean that disappeared when North America and Europe collided during the Paleozoic Era.

Ice cap. A dome-shaped body of ice and snow that covers a mountain highland, or lower-lying land at high latitude, and that displays generally radial outward flow.

Ice-contact stratified drift. Stratified sediment deposited in contact with supporting glacier ice.

Ice field. A broad, nearly level area of glacier ice in a mountainous region consisting of many interconnected mountain glaciers.

Ice sheet. A continent-sized mass of ice that overwhelms nearly all the land surface within its margin.

Ice shelf. Thick glacier ice, connected to glaciers on land, that floats on the sea and commonly is located in large coastal embayments at high latitudes.

Igneous rock. Rock formed by the cooling and consolidation of magma.

Ignimbrite. The poorly sorted mass of tephra deposited by a pyroclastic flow. See also *welded tuff*.

Impact cratering. The process by which a planetary surface is deformed as a result of a transfer of energy from a bolide to the planetary surface.

Index fossil. A fossil that can be used to identify and date the strata in which it is found and is useful for local correlation of rock units.

Index mineral. A mineral whose first appearance marks the outer limits of a specific zone of metamorphism.

Inertia. The resistance a large mass has to sudden movement.

Inertial seismograph. A device for measuring earthquake waves based on inertia of a mass suspended on a sensitive spring.

Inner core. The central, solid portion of Earth's core.

Inorganic compound. Chemical compounds that do not consist largely of the elements carbon and hydrogen.

Inselberg. Steep-sided mountain, ridge, or isolated hill rising abruptly from adjoining monotonously flat plains.

Intergranular fluids. The fluids, both liquid and gas, that fill the tiny pore spaces in a rock.

Intermediate grade of metamorphism. Metamorphism under conditions of intermediate pressures and temperatures.

Internal processes. All activities involved in movement or chemical and physical change of rocks in Earth's interior.

Intrusive igneous rocks. Any igneous rock formed by solidification of magma below Earth's surface.

Ion. An atom that has excess positive or negative charges caused by electron transfer.

Ionic bonding. The electrostatic attraction between negatively and positively charged ions.

Ionic radius. The distance from the center of the nucleus to the outermost shell of the orbiting electrons.

Ionic substitution. (also called *atomic substitution*). The substitution of one ion for another in a random fashion throughout a crystal structure.

Island arc. An arcuate chain of andesitic stratovolcanoes sitting on oceanic crust, parallel to a seafloor trench, and separated from it by a distance of 100 to 400 km.

Isoclinal fold. A fold in which both limbs are parallel.

Isograd. A line on a map connecting points of first occurrence of a given mineral in metamorphic rocks.

Isolated system. Any system that has a boundary that prevents the passage in or out of energy and matter.

Isostasy, Principle of. The ideal property of flotational balance among segments of the lithosphere.

Isostatic uplift. The raising of topography as a byproduct of a gravitational balance, for instance, by the emplacement of warm buoyant rock or magma beneath a landscape, or by the removal of surface rocks by erosion.

Isotope. Atoms of an element having the same atomic number but differing mass numbers.

Joints. Fractures in a rock on which no observable movement has occurred.

Jovian planets. Giant planets in the outer regions of the solar system that are characterized by great masses, low densities, and thick atmospheres consisting primarily of hydrogen and helium.

K horizon. A horizon, present in some arid-zone soils beneath the B horizon, that is impregnated with calcium carbonate.

Kame. A short, steep-sided knoll of stratified drift.

Kame terrace. A terrace of ice-contact stratified drift along a valley side.

Karst topography. An assemblage of topographic forms resulting from dissolution of carbonate bedrock and consisting primarily of closely spaced sinkholes.

Kerogen. Insoluble, waxlike organic matter found in sedimentary rocks, especially shales.

Kettle. A basin within a body of drift created by melting out of a mass of underlying ice.

Key bed. A thin and generally widespread bed with sedimentary characteristics so distinctive that it can be easily recognized but not confused with any other bed.

Kimberlite pipes. Narrow, pipelike masses of igneous rocks, sometimes containing diamonds that intrude the crust but originate deep in the mantle.

Kinetic energy. The energy possessed by a body associated with its movement.

Laccolith. A lenticular pluton intruded parallel to the layering of the surrounding rock, above which the layers of the invaded country rock have been bent upward to form a dome.

Lacustrine. Pertaining to, produced by, or formed in a lake.

Lagoon. A bay inshore from an enclosing reef or island paralleling a coast.

Lake Superior-type iron deposit. Iron-rich chemical-sedimentary rocks in which chert and iron-rich layers are interbedded on a fine scale. All known deposits are early Proterozoic in age, or older.

Laminar flow. A pattern of flow in which fluid particles move in parallel layers.

Landslide. Any perceptible downslope movement of a mass of bedrock or regolith, or a mixture of the two.

Lapilli. Tephra with particles having an average diameter between 2 and 64 mm.

Lapilli tuff. Pyroclastic rock in which the average diameter of tephra particles ranges between 2 and 64 mm.

Lateral moraine. An end moraine built along the side of a valley glacier.

Laterite. A hardened soil horizon characterized by extreme weathering that has led to concentration of secondary oxides of iron and aluminum.

Latitude. Part of a grid used for describing positions on Earth's surface, consisting of parallel circles called *parallels of latitude*.

Laurasia. The northern half of Pangaea, consisting of present-day Asia, Europe, and North America.

Lava. Magma that reaches Earth's surface through a volcanic vent.

Lava dome. A dome-shaped mass of sticky, gas-poor lava erupted from a volcanic vent following a major eruption.

Law (scientific). A statement that some aspect of nature is always observed to happen in the same way and that no deviations have ever been seen.

Law of faunal succession. Fossil faunas and floras succeed one another in a definite, recognizable order.

Law of original horizontality. See *original horizontality*.

Leaching. The continued removal, by water solutions, of soluble matter from bedrock or regolith.

Left-lateral fault. A strike-slip fault in which relative motion is such that to an observer looking directly at the fault, the motion of the block on the opposite side of the fault is to the left. A *right-lateral fault* has right-handed movement.

Levee. See *natural levee*.

Lignite. A low-grade coal with a calorific value between that of peat and bituminous coal.

Limbs of a fold. The sides of a fold.

Limestone. A sedimentary rock consisting chiefly of calcium carbonate, mainly in the form of the mineral calcite.

Linear dune. A long, straight, ridge-shaped dune paralleling the wind direction.

Liquefaction. The rapid fluidization of sediment as a result of an abrupt shock such as earthquakes.

Lithic sandstone. A dark-colored sandstone containing quartz, feldspar, and a large amount of tiny rock fragments. Also called *greywacke*.

Lithification. The process that converts a sediment into a sedimentary rock.

Lithology. The systematic description of rocks in terms of mineral assemblage and texture.

Lithosphere. The outer 100 km of solid Earth, where rocks are harder and more rigid than those in the plastic asthenosphere.

Little Ice Age. The interval of generally cool climate between the middle thirteenth and middle nineteenth centuries, during which mountain glaciers expanded worldwide.

Load. The material that is moved or carried by a natural transporting agent, such as a stream, the wind, a glacier, or waves, tides, and currents.

Local base level. Any base level, other than sea level, below which a stream controlled by that base level cannot erode the land.

Loess. Wind-deposited silt, sometimes accompanied by some clay and fine sand.

Long profile. A line drawn along the surface of a stream from its source to its mouth.

Longitude. Part of a grid used for describing positions on Earth's surface, consisting of half circles joining the poles. The half circles are called *meridians*.

Longshore current. A current, within the surf zone, that flows parallel to the coast.

Low grade metamorphism. Metamorphism under conditions of low temperature and low pressure.

Low-velocity zone. A region in the mantle, approximately between a depth of 100 and 350 km, where seismic-wave velocities decrease.

Lunar highlands. Mountainous regions on the Moon, believed to consist of anorthosite and gabbro.

Luster. The quality and intensity of light reflected from a mineral.

Magma. Molten rock, together with any suspended mineral grains and dissolved gases, that forms when temperatures rise and melting occurs in the mantle or crust.

Magma ocean. Molten outer layer of Earth's moon initially formed as a result of intense rain of bolides.

Magmatic arc. An arcuate chain of magmatic activity lying above a subduction zone, parallel to and separated from the seafloor trench by 100 to 400 km.

Magmatic differentiation by fractional crystallization. Compositional changes that occur in magmas by the separation of early formed minerals from residual liquids.

Magmatic differentiation by partial melting. The process of forming magmas with differing compositions by the incomplete melting of rocks.

Magmatic mineral deposit. Any local concentration of minerals formed by magmatic processes in an igneous rock.

Magnetic chrons. Periods of predominantly normal polarity (as at present), or predominantly reversed polarity.

Magnetic declination. The clockwise angle from true north assumed by a magnetic needle.

Magnetic field. Magnetic lines of force surrounding Earth.

Magnetic inclination. The angle with the horizontal assumed by a freely swinging bar magnet.

Magnetic latitude. The latitude of a place on the Earth with respect to the magnetic poles.

Magnitude (of an earthquake). See *Richter magnitude scale*.

Mantle. The thick shell of dense, rocky matter that surrounds the core.

Mantle convection. Convection of Earth's mantle, driven both by the thermal buoyancy of hot rock rising from the depth and by sinking of cooler, denser lithosphere.

Marble. A metamorphic rock derived from limestone and consisting largely of calcite.

Mare. Dark-colored lowland region of the Moon underlain by basalt.

Mare basins. Huge circular impact structures on the Moon that were later filled by basaltic flows.

Maria. Plural of *mare*.

Mass balance (of a glacier). The sum of the accumulation and ablation on a glacier during a year.

Mass number. The sum of the protons and neurons in the nucleus of an atom.

Mass-wasting. The movement of regolith downslope by gravity without the aid of a transporting medium.

M-discontinuity. See *Mohorovičić discontinuity*.

Meander. A looplike bend of a stream channel.

Mechanical deformation. The changes in texture of a rock due to grinding, crushing and development of foliation during metamorphism.

Megascopic. Features that can be seen by the unaided eye, or by the eye assisted by a simple lens that magnifies up to 10 times.

Mélange. A chaotic mixture of broken, jumbled, and thrust-faulted rock above a subduction zone.

Mercalli Scale. See *Modified Mercalli Scale*.

Mesa. An isolated, flat-topped, steep-sided desert landform larger than a *butte*.

Mesosphere. The region between the base of the asthenosphere and the core-mantle boundary.

Mesozoic Era. The middle era of the Phanerozoic Eon.

Metallic bonding. A form of covalent bonding between atoms in which electron sharing occurs with inner energy-level shells rather than the outermost shells.

Metallogenic provinces. Limited regions of the crust within which mineral deposits are unusually abundant.

Metamorphic aureole. A shell of metamorphic rock, produced by contact metamorphism, surrounding an igneous intrusion.

Metamorphic facies. Contrasting assemblages of minerals that reach equilibrium during metamorphism within a specific range of physical conditions belonging to the same metamorphic facies.

Metamorphic rock. Rock whose original compounds or textures, or both, have been transformed to new compounds and new textures by reactions in the solid state as a result of high temperature, high pressure, or both.

Metamorphic zone. The region on a map between isograds.

Metamorphism. All changes in mineral assemblage and rock texture, or both, that take place in sedimentary and igneous rocks in the solid state within the Earth's crust as a result of changes in temperature and pressure.

Metasomatism. The process by which rocks have their composition distinctly altered by the addition or removal of materials in solution.

Meteorites. Small stony or metallic objects from interplanetary space that impact a planetary surface

Microscopic. Those features of rocks that require high magnification in order to be viewed.

Midocean ridge. Continuous rocky ridges on the ocean floor, many hundreds to a few thousand kilometers wide with a relief of more than 0.6 km. Also called *oceanic ridge* and *oceanic rise*.

Migmatite. A composite rock containing both igneous and metamorphic portions.

Mineral. Any naturally formed, crystalline solid with a definite chemical composition and a characteristic crystal structure.

Mineral assemblage. The variety and abundance of minerals present in a rock.

Mineral deposit. Any volume of rock containing an enrichment of one or more minerals.

Mineral group. A mineral that displays extensive ionic substitution without changing the cation-anion ratio.

Mineralogy. The special branch of geology that deals with the classification and properties of minerals.

Mineraloid. A naturally occurring mineral-like solid that lacks either a crystal structure or a definite composition, or both.

Modified Mercalli Scale. A scale used to compare earthquakes based on the intensity of damage caused by the quake.

Moho. See *Mohorovičić discontinuity*.

Mohorovičić discontinuity (also called *M-discontinuity* and *Moho*). The seismic discontinuity that marks the base of the crust.

Moh's relative hardness scale. A scale of relative mineral hardness determined by scratching, divided into 10 steps, each marked by a common mineral.

Molecule. The smallest unit that retains the distinctive properties of a compound.

Monocline. A local steepening in an otherwise uniformly dipping pile of strata.

Moraine. An accumulation of drift deposited beneath or at the margin of a glacier and having a surface form that is unrelated to the underlying bedrock.

MORB. Acronym for midocean ridge basalt. The magma that rises at seafloor spreading centers and solidifies to form oceanic crust.

Mountain chain. A large scale, elongate geologic feature consisting of numerous ranges or systems, regardless of similarity in form or equivalence in ages.

Mountain range. An elongate series of mountains forming a single geologic feature.

Mountain system. A group of ranges similar in general form, structure and alignment, and presumably owing their origin to the same general causes.

Mudcracks. Cracks caused by shrinkage of wet mud as its surface dries.

Mudflow. A flowing mass of predominantly fine-grained rock debris that generally has a high enough water content to make it highly fluid; a rapidly moving type of *debris flow*.

Mudstone. A clastic sedimentary rock composed of mineral fragments finer than those in a siltstone.

Natural gas. A gaseous component of petroleum; chiefly methane.

Natural levee. A broad, low ridge of fine alluvium built along the side of a stream channel by water that spreads out of the channel during floods.

Neutron. An electrically neutral particle with a mass 1833 times greater than that of the electron.

Nonconformity. Stratified rocks that unconformably overlie igneous or metamorphic rocks.

Normal fault. A fault, generally steeply inclined, along which the hanging-wall block has moved relatively downward.

Nuclear energy. The heat energy produced during controlled fission or fusion of atoms.

Nucleus (of an atom). The assemblage of protons and neutrons in the core of an atom.

O horizon. An accumulated layer of humus that is the uppermost horizon in many soil profiles.

Oblique-slip fault. A fault on which movement includes both horizontal and vertical components.

Obsidian. An extrusive igneous rock that is wholly or largely glass.

Ocean trenches. Linear seafloor depressions at plate boundaries where an oceanic plate sinks into Earth's asthenosphere beneath a neighboring plate.

Oceanic crust. The crust beneath the oceans.

Oceanic ridge. See *midocean ridge.*

Oceanic rise. See *midocean ridge.*

Oil. The liquid component of petroleum.

Oilfield. A group of oil pools, usually of similar type, or a single pool in an isolated position.

Oil pool. An underground accumulation of oil and gas in a reservoir limited by geologic barriers.

Oil shale. A shale containing waxlike substances that will break down to liquid and gaseous hydrocarbons when heated.

Oolitic limestone. A sedimentary rock composed of accumulations of tiny, round, calcareous bodies called *oolites.*

Open fold. A fold in which the two limbs dip gently and equally, and away from the axis.

Open system. Any system that has a boundary that allows the passage in or out of both energy and matter.

Ophiolites. Fragments of oceanic crust found on continents.

Ore. An aggregate of minerals from which one or more minerals can be extracted profitably.

Organic compound. Chemical compounds made from carbon and hydrogen, with or without other elements such as nitrogen and oxygen.

Original horizontality (law of). Waterlaid sediments are deposited in strata that are horizontal, or nearly horizontal, and parallel, or nearly parallel, to the Earth's surface.

Orogenic belts. See *orogens.*

Orogens. Elongate regions of the crust that have been intensively folded, faulted, and thickened as a result of continental collisions.

Orogeny. The process by which large regions of the crust are deformed and uplifted to form mountains.

Outcrop. See *exposure.*

Outcrop area. The area on a geological map shown as occupied by a particular rock unit.

Outer core. The outer portion of Earth's core, which is molten.

Outwash. Stratified drift deposited by meltwater streams.

Outwash plain. A body of outwash that forms a broad plain.

Outwash terrace. A terrace formed by dissection of an outwash plain or valley train.

Overland flow. The movement of runoff in broad sheets or groups of small, interconnecting rills.

Overturned fold. A fold in which the strata in one limb have been tilted beyond vertical.

Oxbow lake. A crescent-shaped, shallow lake occupying the abandoned channel of a meandering stream.

Oxidation. A process in which a chemical element loses electrons.

Oxidizing environment. A sedimentary environment in which oxygen is present and organic remains are readily converted by oxidation into carbon dioxide and water.

Pahoehoe. A smooth, ropy-surfaced lava flow, usually basaltic in composition.

Paleomagnetism. Remanent magnetism in ancient rock recording the direction of the magnetic poles at some time in the past.

Paleosol. A soil that formed at the ground surface and subsequently was buried and preserved.

Paleozoic Era. The oldest era of the Phanerozoic Eon.

Pangaea. The name given to a supercontinent that formed by collision of all the continental crust during the late Paleozoic.

Parabolic dune. A sand dune of U-shape with open end of the U facing upwind.

Parallel of latitude. See *latitude.*

Parallel strata. Strata whose individual layers are parallel.

Parent (radioactive). An atomic nucleus undergoing radioactive decay. Compare *daughter product.*

Parent material (of a soil). The regolith from which a soil develops.

Passive continental margin. A continental margin in a plate interior.

Peat. An unconsolidated deposit of plant remains that is the first stage in the conversion of plant matter to coal.

Pediment. A sloping surface cut across bedrock and thinly or discontinuously veneered with alluvium that slopes away from the base of a highland in an arid or semiarid environment.

Pegmatite. An exceptionally coarse-grained intrusive igneous rock, commonly granitic in composition and texture.

Pelagic sediment. Sediment consisting of the remains of marine organisms living in the open ocean.

Perched water body. A water body perched atop an aquiclude that lies above the main water table.

Percolation. The movement of groundwater in the saturated zone.

Periodotite. A coarse-grained igneous rock consisting largely of olivine, with or without pyroxene.

Periglacial. A land area beyond the limit of glaciers where low temperature and frost action are important factors in determining landscape characteristics.

Period. The time during which a geologic system accumulated.

Permafrost. Sediment, soil, or bedrock that remains continuously at a temperature below 0°C for an extended time.

Permeability. A measure of how easily a solid allows a fluid to pass through it.

Petroleum. Gaseous, liquid, and semi-solid substances occurring naturally and consisting chiefly of chemical compounds of carbon and hydrogen.

Petrology. The special branch of geology that deals with the occurrence, origin, and history of rocks.

Phanerite. An igneous rock in which the constituent mineral grains are readily visible to the unaided eye.

Phanerozoic. The eon that follows the Proterozoic Eon.

Phase transition. Atomic repacking caused by changes in pressure and temperature.

Phenocrysts. The isolated large mineral grains in a porphyry.

Photosynthesis. The process by which plants combine water and carbon dioxide to make carbohydrates and oxygen.

Phyllite. A well-foliated metamorphic rock in which the component platy minerals are just visible.

Physical geology. The study of the processes that operate at or beneath the surface of Earth, and the materials on which those processes operate.

Physical weathering. Disintegration of rocks by mechanical processes, such as frost-wedging.

Piedmont glacier. A broad glacier that terminates on a piedmont slope beyond confining mountain valleys and is fed by one or more large valley glaciers.

Pile. A device in which nuclear fission can be controlled.

Pillow basalt. Discontinuous, pillowshaped masses of basalt, ranging in size from a few centimeters to a meter or more in greatest dimension.

Placer. A deposit of heavy minerals concentrated mechanically.

Plane of the ecliptic. See *ecliptic.*

Planet. A large celestial body that revolves around the Sun in an elliptical orbit.

Planetary accretion. The process by which bits of condensed solid matter were gathered to form the planets.

Planetary nebula. A flattened, rotating disk of gas surrounding a proto-sun.

Planetology. A comparative study of the Earth with the Moon and with the other planets.

Plateau basalt. Flat plains of lava formed as a result of a fissure eruption of basalt.

Plate friction. Frictional resistance to lateral drift encountered by lithosphere at its bottom boundary with the underlying asthenosphere.

Plate tectonics. The special branch of tectonics that deals with the processes by which the lithosphere is moved laterally over the asthenosphere.

Plate tectonics theory. Grand unifying theory of geology saying that Earth's outermost 100 km "eggshell" (the lithosphere) is cracked, *composed of about a dozen pieces* ("plates"), which float on the hot, deformable asthenosphere, moving ponderously in various directions, spreading apart, slowly colliding, and grinding past one another.

Plate triple junction. Junction between three plate spreading edges. The angle between any two edges is approximately 120°.

Playa. A dry lake bed in a desert basin.

Plunge (of a fold). The angle between a fold axis and the horizontal.

Plunging fold. A fold with an inclined axis.

Pluton. Any body of intrusive igneous rock, regardless of shape or size.

Point bar. An arcuate deposit of sand or gravel along the inside of the bend of a meander loop.

Polar (cold) glacier. A glacier in which the ice is below the pressure melting point throughout, and the ice is frozen to its bed.

Polar easterlies. Globe-encircling belts of easterly winds in the high latitudes of both hemispheres.

Polar front. The region where equatorward-moving polar easterlies meet poleward-moving westerlies.

Polarity reversals. Changes of the Earth's magnetic field to the opposite polarity.

Polymerization. The process of linking silicate tetrahedra into large anion groups.

Polymorph. A compound that occurs in more than one crystal structure.

Pores (-pore space). The innumerable tiny openings in rock and regolith that can be filled by water or other fluids.

Porosity. The proportion (in percent) of the total volume of a given body of bedrock or regolith that consists of pore spaces.

Porphyry. Any igneous rock consisting of coarse mineral grains scattered through a mixture of fine material grains.

Porphyry copper deposit. A class of hydrothermal mineral deposit associated with intrusions of porphyritic igneous rocks.

Potential energy. Stored energy.

Precession of the equinoxes. A progressive change in the orientation of Earth's tilted rotation axis.

Pressure melting point. The temperature at which ice can melt at a given pressure.

Primary waves. See *P waves*.

Principle. A statement that some aspect of nature is always observed to happen in the same way and that no deviations have ever been observed.

Principle of stratigraphic superposition. See *stratigraphic superposition*.

Principle of Uniformitarianism. The same external and internal processes we recognize in action today have been operating unchanged, though at different rates, throughout most of Earth's history.

Progradation. The outward extension of a shoreline into the sea or lake due to sedimentation.

Prograde metamorphic effects. The metamorphic changes that occur while temperatures and pressures are rising.

Proterozoic Eon. The eon that follows the Archean Eon.

Proton. A positively charged particle with a mass 1832 times greater than the mass of an electron.

Pumice. A natural glassy froth made by gases escaping through a viscous magma.

P waves. Seismic body waves transmitted by alternating pulses of compression and expansion. *P* waves pass through solids, liquids, and gases.

Pyroclast. A fragment of rock ejected during a volcanic eruption.

Pyroclastic flow. A hot, highly mobile flow of tephra that rushes down the flank of a volcano during an eruption.

Pyroclastic rocks. Rocks formed from pyroclasts.

Pyrometer. An optical device for measuring temperature.

Quartzite. A metamorphic rock consisting largely of quartz, and derived from a sandstone.

Radiation. Transmission of heat energy through the passage of electromagnetic waves.

Radioactive decay. Natural process of transformation by which the nucleus of one kind of atom transforms to the nucleus of another kind of atom.

Radioactivity. The process by which isotopes of one element transform spontaneously to other isotopes of the same or different elements.

Radiogenic helium. Helium gas created as a byproduct of the alpha-particle decay of radioactive isotopes of uranium, thorium and many other elements.

Radiometric age. The length of time a mineral has contained its built-in radioactivity clock.

Radiometric dating. Determination of radiometric age through the use of radioactive isotopes.

Rainshadow. A dry region on the downwind side of a mountain range where precipitation is noticeably less than on the windward side.

Rayleigh number. A ratio of physical factors that govern the transport of heat in a system. Factors that tend to favor convection are divided by factors that discourage convection.

Recharge. The addition of water to the saturated zone of a groundwater system.

Recharge area. Area where water is added to the saturated zone.

Recrystallization. The formation of new crystalline minerals within a rock.

Recumbent fold. A fold in which the axial plane is horizontal.

Reducing environment. An environment in which oxygen is lacking and organic matter does not decay, but instead is slowly transformed into solid carbon.

Red giant. A large, cool star with a high luminosity and a low surface temperature (about 2500 K), which is largely convective and has fusion reactions going on in shells.

Reef. A generally ridgelike structure composed chiefly of the calcareous remains of sedentary marine organisms (e.g., corals, algae).

Reef limestone. A carbonate sedimentary rock formed of fossil reef organisms.

Reflection. The bouncing of a wave off the surface between two media.

Refraction. The change in direction when a wave passes from one medium to another.

Regional metamorphism. Metamorphism affecting large volumes of crust and involving both mechanical and chemical changes.

Regolith. The irregular blanket of loose, noncemented rock particles that covers Earth's surface.

Relative velocity (of a plate). The apparent velocity of one plate relative to another.

Relief. The range in altitude of a land surface.

Replacement. The process by which a fluid dissolves matter already present and at the same time deposits from solution an equal volume of a different substance.

Reservoir rock. A permeable body of rock in which petroleum accumulates.

Residual mineral deposit. Any local concentration of minerals formed as a result of weathering.

Resurgent dome. The uplifting of the collapsed floor of a caldera to form a structural dome.

Retrograde metamorphic effects. Metamorphic changes that occur as temperature and pressure are declining.

Reverse fault. A fault, generally steeply inclined, along which the hanging-wall block has moved relatively upward.

Rhyolite. A fine-grained extrusive igneous rock with the composition of a granite.

Rhyolite magma. One of the three common magma types. A magma with an SiO_2 content of about 70 percent by weight.

Richter magnitude scale. A scale, based on the recorded amplitudes of seismic waves, for comparing the amounts of energy released by earthquakes.

Ridge push. The tendency for young oceanic lithosphere to slide laterally down the slope of a midocean ridge under the influence of gravity, thereby pushing the rest of the plate away from the ridge.

Rift. See *graben* and *half-graben*.)

Right-lateral fault. See *left-lateral fault*.

Rind. See *weathering rind*.

Rip current. A high-velocity current flowing seaward from the shore as part of the backwash from a wave.

Ripple mark. One of a series of small, fairly regular, subparallel ridges preserved in rock and representing a former rippled sedimentary surface.

Rock. Any naturally formed, nonliving, firm, and coherent aggregate mass of mineral matter that constitutes part of a planet.

Rock cleavage (also called *slaty cleavage*). The property by which a rock breaks into platelike fragments along flat planes.

Rock cycle. The cyclic movement of rock material, in the course of which rock is created, destroyed, and altered through the operation of internal and external Earth processes.

Rockfall. The free falling of detached bodies of bedrock from a cliff or steep slope.

Rock flour. Fine rock particles produced by glacial crushing and grinding.

Rock glacier. A lobe of ice-cemented rock debris that moves slowly downslope in a manner similar to glaciers.

Rockslide. The sudden and rapid downslope movement of detached masses of bedrock across an inclined surface.

Rock-stratigraphic unit. Any distinctive rock unit that can be distinguished from other strata on the basis of composition and physical properties.

Roof rock. A rock, such as shale, that is impermeable and caps a petroleum reservoir.

Rotation pole. Another term for a *spreading pole*, a point on Earth's surface about which a plate rotates.

Runoff. The fraction of precipitation that flows over the land surface.

Salinity. The measure of the sea's saltiness; expressed in parts per thousand (per mil.).

Saltation. The progressive forward movement of a sediment particle in a series of short intermittent jumps along arcing paths.

Sand ripples. A series of small and rather regular ridges on the surface of a body of sand, such as a dune.

Sand sea. Vast tract of shifting sand.

Sandstone. A medium-grained clastic sedimentary rock composed chiefly of sand-sized grains.

Saturated zone. The groundwater zone in which all openings are filled with water.

Scale (of a map). The proportion between a unit of distance on a map and the unit it represents on the Earth's surface.

Schist. A well-foliated metamorphic rock in which the component platy minerals are clearly visible.

Schistosity. The parallel arrangement of coarse grains of the sheet-structure minerals, like mica and chlorite formed during metamorphism under conditions of differential stress.

Scientific method. The use of evidence that can be seen and tested by anyone who has the means to do so, consisting often of observation, formation of a hypothesis, testing of that hypothesis and forma-

tion of a theory, formation of a law, and continued reexamination.

Sea arch. An opening through a headland, generally produced by wave erosion, that forms a bridge of rock over water.

Sea cave. A cave at the base of a seacliff produced by wave erosion.

Seafloor spreading (theory of). A theory proposed during the early 1960s in which lateral movement of the oceanic crust away from midocean ridges was postulated.

Seamount. An isolated submerged volcanic mountain standing more than 1000 m above the seafloor.

Secondary enrichment. The process by which a sulfide mineral deposit is chemically weathered and enriched in its metal content as a result.

Secondary mineral. A mineral formed later than the rock enclosing it, usually at the expense of an earlier formed primary mineral.

Secondary waves. See *S waves*.

Sediment. Regolith that has been transported by any of the external processes.

Sediment drifts. Huge bodies of sediment, up to hundreds of kilometers long, deposited and shaped by deep ocean currents along a continental margin.

Sediment flows. Mass wasting of mixtures of sediment, water, and air.

Sedimentary facies. A distinctive group of characteristics within a sedimentary unit that differs, as a group, from those elsewhere in the same unit.

Sedimentary mineral deposit. Any local concentration of minerals formed through processes of sedimentation.

Sedimentary rock. Any rock formed by chemical precipitation or by sedimentation and cementation of mineral grains transported to a site of deposition by water, wind, ice, or gravity.

Seismic belts. Large tracts of the Earth's surface that are subject to frequent earthquake shocks.

Seismic moment. Common measure of earthquake size, related to the mechanical energy released by fault motion.

Seismic sea waves (also called *tsunami*). Long-wavelength ocean waves produced by sudden movement of the seafloor following an earthquake. Incorrectly called tidal waves.

Seismic tomography. A way of revealing inhomogeneities in the mantle by measuring slight differences in the arrival times of seismic waves.

Seismic waves. Elastic disturbances spreading outward from an earthquake focus.

Seismograph. See *seismometer*.

Seismology. The study of earthquakes.

Seismometer. Instrument that measures and records vibrational motions caused by earthquakes, explosions and similar disturbances. *Seismograph* is an equivalent, but older, term for such an instrument.

Serpentinite. A rock composed largely of the mineral serpentine.

Setting time. The moment a mineral starts accumulating a daughter product produced by radioactive decay.

Shale. A fine-grained, clastic sedimentary rock.

Shard. See *volcanic shard*.

Shear strength. The internal resistance of a body to shape distortion.

Shear stress (on a free-standing body). The force acting on a body that causes slippage or translation.

Shear waves. See *S waves*.

Sheet erosion. The erosion performed by overland flow.

Shield volcano. A volcano that emits fluid lava and builds up a broad dome-shaped edifice with a surface slope of only a few degrees.

Shore profile. A vertical section along a line perpendicular to a shore.

Silicate (*-silicate mineral*). A mineral that contains the silicate anion.

Silicate anion. A complex ion $(SiO_4)^{-4}$, that is present in all silicate minerals.

Silicate mineral. See *silicate*.

Siliceous ooze. Any pelagic deep-sea sediment of which at least 30 percent consists of siliceous skeletal remains. See also *deep-sea ooze*.

Sill. Tabular, parallel-sided sheets of intrusive igneous rock that are parallel to the layering of the intruded rock.

Siltstone. A sedimentary rock composed mainly of silt-sized mineral fragments.

Sinkhole. A large solution cavity open to the sky.

Slab friction. Frictional resistance encountered by lithosphere as it sinks into Earth's interior.

Slab pull. The tendency for sinking lithosphere at a convergent plate boundary to pull the rest of the plate laterally into the convergent zone.

Slate. A low-grade metamorphic rock with a pronounced slaty cleavage.

Slaty cleavage. See *rock-cleavage*.

Slickensides. Striated or highly polished surfaces on hard rocks abraded by movement along a fault.

Slip face. The straight, lee slope of a dune.

Slump. A type of slope failure in which a downward and outward rotational movement of rock or regolith occurs along a concave-up slip surface.

Slurry flow. A moving mass of sediment that is saturated with water trapped among the grains and transported with the flowing mass.

Snowline. The lower limit of perennial snow.

Soft water. Groundwater that contains little dissolved matter and no appreciable calcium.

Soil. The part of the regolith that can support rooted plants.

Soil horizons. The subhorizontal weathered zones formed as a soil develops.

Soil profile. A vertical section through a soil that displays its component horizons.

Solar system. The Sun and the group of objects in orbit around it.

Sole marks. Irregularities formed by currents together with tracks and other markings preserved on the bedding plane of sandstone or siltstone.

Solifluction. The very slow downslope movement of waterlogged soil and surficial debris.

Sorting. A measure of the range of particle size of sediment.

Source-rock. A sedimentary rock containing organic matter that is a source of petroleum.

Spall. Flakey material broken away from the surface of a rock exposed to intense heat.

Spatter cone. A cone-shaped pile of bits of lava surrounding a volcanic vent.

Spatter rampart. A linear pile of bits of lava erupted along a fissure.

Specific gravity. A number stating the ratio of the weight of a substance to the weight of an equal volume of pure water. A dimensionless number numerically equal to the density.

Spheroidal weathering. The successive loosening of concentric shells of decayed rock from a solid rock mass as a result of chemical weathering.

Spit. An elongate ridge of sand or gravel that projects from land and ends in open water.

Spreading axis. The axis of rotation of a plate of lithosphere.

Spreading center (also called a *divergent margin*). The new, growing edge of a plate. Coincident with a midocean ridge.

Spreading pole. The point where a spreading axis reaches the Earth's surface.

Spring. A flow of groundwater emerging naturally at the ground surface.

Stable platform. That portion of a craton that is covered by a layer of little-deformed sediments.

Stable zone. The interior part of a tectonic plate.

Stack. An isolated rocky island or steep rock mass near a cliffy shore, detached from a headland by wave erosion.

Stalactite. An iciclelike form of dripstone and flowstone, hanging from cave floors.

Stalagmite. An "icicle" of dripstone and flowstone projecting upward from cave floors.

Star dune. An isolated hill of sand having a base that resembles a star in plan.

Steady state. A condition in which the rate of arrival of material or energy equals the rate of escape.

Steady-state landscape. A landscape that maintains a constant altitude and topographic relief while still undergoing uplift, exhumation and denudation.

Stock. A small, irregular body of intrusive igneous rock, smaller than a batholith, that cuts across the layering of the intruded rock.

Stoping. The process by which a rising body of magma wedges off fragments of overlying rock that then sink through the magma chamber.

Strain. The measure of the changes in length, volume, and shape in a stressed material.

Strain rate. The rate at which a rock is forced to change its shape or volume.

Strain seismograph. A device for recording earthquake waves based on the flexure of a long, rigid rod.

Strata. See *stratum.*

Stratabound mineral deposits. Ores of lead, zinc, copper, and other metals enclosed in sedimentary rocks in such a way that they closely resemble primary sediments.

Stratification. The layered arrangement of sediments, sedimentary rocks, or extrusive igneous rocks.

Stratified drift. Glacial drift that is both sorted and stratified.

Stratigraphic superposition (principle of). In a sequence of strata, not later overturned, the order in which they were deposited is from bottom to top.

Stratigraphy. The study of strata.

Stratovolcano. Volcano that emits both tephra and viscous lava, and that builds up steep conical mounds.

Stratum (plural = *strata*). A distinct layer of sediment that accumulated at Earth's surface.

Streak. A thin layer of powdered mineral made by rubbing a specimen on a nonglazed porcelain plate.

Stream. A body of water that carries detrital particles and dissolved substances and flows down a slope in a definite channel.

Streamflow. The flow of surface water in a well-defined channel.

Stress. The magnitude and direction of a deforming force exerted on a surface.

Striations. See *glacial striations.*

Strike. The compass direction of a horizontal line that marks the intersection of an inclined plane with Earth's surface.

Strike-slip fault. A fault on which displacement has been horizontal and parallel to the strike of the fault.

Structural geology. The branch of geology devoted to the study of rock deformation.

Structure (of minerals). See *crystal structure.*

Subatomic particles. The small particles that combine to form an atom—electrons, protons, and neutrons.

Subduction. The process by which old, cold lithosphere sinks into the asthenosphere.

Subduction zone (also called a *convergent margin*). The linear zone along which a plate of lithosphere sinks down into the asthenosphere.

Submarine canyon. A steep-sided valley on the continental shelf or slope resembling a river-cut canyon on land.

Submergence (of a coast). A rise of water level relative to the land so that areas formerly dry are inundated.

Subsequent stream. A stream whose course has become adjusted so that it occupies belts of weak rock or other geologic structures.

Superposed stream. A stream that was let down, or superposed, from overlying strata onto buried bedrock having composition or structure unlike that of the covering strata.

Surf. Wave activity between the line of breakers and the shore.

Surface waves. Seismic waves that are guided by the Earth's surface and do not pass through the body of the Earth.

Surge. An unusually rapid movement of a glacier marked by dramatic changes in glacier flow and form.

Suspect terrane. See *accreted terrane.*

Suspended load. Fine particles suspended in a stream.

Swash. The surge of water up a beach caused by waves moving against a coast.

S waves. Seismic body waves transmitted by an alternating series of sideways (shear) movements in a solid. *S* waves cause a change of shape and cannot be transmitted through liquids and gases.

Symmetrical fold. A fold in which both limbs dip equally away from the axial plane.

Syncline. A downfold with a troughlike form.

Synthetic seismograms. Predictions of seismic motions caused by earthquakes, calculated by computer programs that incorporate estimates of fault plane orientation and the stiffness of subsurface rock.

System. The primary unit in a time-stratigraphic sequence of rocks.

Taconite. Iron ore found as a result of metamorphism of Lake Superior-type iron deposits.

Talus. The apron of rock waste sloping outward from the cliff that supplies it.

Tar (also called *asphalt*). An oil that is viscous and so thick it will not flow.

Tarn. A small, generally deep mountain lake occupying a cirque.

Tectonic cycle. A model that describes the movement and interactions of lithospheric plates, the internal processes that drive plate motion, and the types of rock and rock formations that develop as a result of tectonic movement and interactions.

Tectonics. The study of movement and deformation of the lithosphere.

Temperate (warm) glacier. A glacier in which the ice is at the pressure-melting point and water and ice coexist in equilibrium.

Tensional stress. The differential stress on a body that causes stretching and elongation.

Tephra. A loose assemblage of pyroclasts.

Tephra cone. A cone-shaped pile of tephra deposited around a volcanic vent.

Terminal moraine. An end moraine deposited at the front of a glacier.

Terminus. The outer, lower margin of a glacier.

Terrace. An abandoned floodplain formed when a stream flowed at a level above the level of its present channel and floodplain.

Terrane. A large piece of crust with a distinctive geological character.

Terrestrial planets. The innermost planets of the solar system (Mercury, Venus, Earth, and Mars), which have high densities and rocky compositions.

Tethys. The name of a narrow sea separating Gondwanaland from Laurasia.

Texture. The overall appearance that a rock has because of the size, shape, and arrangement of its constituent mineral grains.

Theory. A hypothesis that has been examined and found to withstand numerous tests.

Thermal expansion. The tendency of a substance to expand as its temperature rises, and to contract as it cools.

Thermal metamorphism. See *contact metamorphism.*

Thermal plume. A vertically rising mass of heated rock in the mantle.

Thermal spring. A natural spring that emits hot water.

Thermohaline circulation. Global patterns of water circulation propelled by the sinking of dense cold and salty water.

Thin section. A thin slice of rock glued to a glass slide and used for microscopic examination.

Threshold effects. Geologic processes that operate only when one or more critical conditions have been satisfied. For instance, rapid landscape erosion may occur once rainfall and river runoff exceed a critical value.

Thrust faults (also called *thrusts*). Low-angle reverse faults with dips less than 45°.

Tidal bore. A large, turbulent, wall-like wave of water caused by the meeting of two tides or by the rush of tide up a narrowing inlet, river, estuary, or bay.

Tidal bulge. A bulge in bodies of marine and fresh water, produced by the gravitational attraction of the Moon and Sun, that moves around the Earth as it rotates.

Tide. The twice-daily rise and fall of the ocean surface resulting from the gravitational attraction of the Moon and Sun.

Till. A nonsorted sediment deposited directly from glacier ice.

Tillite. A nonsorted sedimentary rock of glacial origin (i.e., a lithified till).

Tilt (of axis). The angle of the Earth's rotational axis with respect to the plane of the Earth's orbit.

Time-stratigraphic unit. All the rocks or sediments that formed during a specific interval of geologic time. Compare *rock-stratigraphic unit.*

Titius–Bode rule. The distance to each planet is approximately twice as far as the next inner one, measured from the Sun.

Tombolo. A ridge of sand or gravel that connects an island to the mainland or to another island.

Topography. The relief and form of the land.

Topset layer. A layer of stream sediment that overlies the foreset layers in a delta.

Tradewind. A globe-encircling belt of winds in the low latitudes. They blow from the northeast in the northern hemisphere and the southeast in the southern hemisphere.

Transform. The junction point where one of the major deformation features—a midocean ridge, a seafloor trench, or a strike-slip fault—meets another.

Transform faults. The special class of strike-slip fault that links major structural features.

Transform fault continental margin. The margin of a continent that coincides with a transform fault.

Transform fault margin. A fracture in the lithosphere along which two plates slide past each other.

Transverse dune. A sand dune forming a wavelike ridge transverse to wind direction.

Trap. A reservoir rock plus a roof rock that serve to accumulate petroleum.

Trenches. Long, narrow, very deep, and arcuate basins in the seafloor.

Tributary. A stream that joins a larger stream.

Triple junction. See *plate triple junction.*

Tsunami. See *seismic sea waves.*

Tuff. A pyroclastic rock consisting of ash- or lapilli-sized tephra, hence *ash tuff* and *lapilli tuff.*

Turbidite. A graded layer of sediment deposited by a turbidity current.

Turbidity current. A gravity-driven current consisting of a dilute mixture of sediment and water having a density greater than the surrounding water.

Turbulent flow. A pattern of flow in which particles of fluid move in swirls and eddies.

Typhoon. A term used in the western Pacific Ocean for a tropical cyclonic storm. See *hurricane.*

Unconfined aquifer. An aquifer with an upper surface that coincides with the water table.

Unconformity. A substantial break or gap in a stratigraphic sequence that marks the absence of part of the rock record.

Unconformity-bounded sequence. A grouping of strata that is bounded at its base and top by unconformities of regional or interregional extent.

Uniform layer. A layer of sediment or sedimentary rock that consists of particles of about the same diameter.

Uniform stress. Stress that is equal in all directions. Also called *confining stress* or *homogeneous stress.*

Uniformitarianism. See *Principle of Uniformitarianism.*

Unsaturated zone (zone of aeration). The groundwater zone in which open spaces in regolith or bedrock are filled mainly with air.

Uplift. The raising of topography by geologic processes.

Valley glacier. A glacier that flows from a cirque or cirques onto and along the floor of a valley.

Valley train. A body of outwash that partly fills a valley.

van der Waals bond. A weak electrostatic attraction that arises because certain ions and atoms are distorted from a spherical shape.

Varve. A pair of sedimentary layers deposited during the seasonal cycle of a single year.

Velocity-weakening behavior. A tendency for friction on a fault surface to decrease as slippage begins. Thought to be a key factor in earthquakes.

Ventifact. Any bedrock surface or stone that has been abraded and shaped by wind-blown sediment.

Vesicle. A small opening, in extrusive igneous rock, made by escaping gas originally held in solution under high pressure while the parent magma was underground.

Viscosity. The internal property of a substance that offers resistance to flow.

Volcanic ash. See *ash.*

Volcanic breccia. A breccia formed as a result of explosive volcanic activity.

Volcanic neck. The approximately cylindrical conduit of igneous rock forming the feeder pipe of a volcanic vent that has been stripped of its surrounding rock by erosion.

Volcanic sediment (in the ocean). Sediment from submarine volcanoes, together with ash from oceanic and nonoceanic volcanic eruptions.

Volcanic shard. A particle of ash-sized, glassy tephra.

Volcano. The vent from which igneous matter, solid rock, debris, and gases are erupted.

Volcanogenic massive sulfide deposit. A mineral deposit formed by deposition of sulfide materials from a submarine hot spring.

Wall-rock alteration. Changes produced in the mineral assemblages of rocks lining the flow channel of a hydrothermal solution.

Water quality. The fitness of water for human use, as affected by physical, chemical, and biological factors.

Water table. The upper surface of the saturated zone of groundwater.

Wave (in a body of water). An oscillatory movement of water characterized by alternate rise and fall of the surface.

Wave base. The effective lower limit of wave motion, which is half of the wavelength.

Wave-cut bench. A bench or platform cut across bedrock by surf.

Wave-cut cliff. A coastal cliff cut by surf.

Wavelength. The distance between the crests or troughs of adjacent waves.

Wave refraction. The process by which the direction of a series of waves, moving into shallow water at an angle to the shoreline, is changed.

Weathering. The chemical alteration and mechanical breakdown of rock materials during exposure to air, moisture, and organic matter.

Weathering rind. A discolored rim of weathered rock surrounding an unweathered core.

Welded tuff (also called *ignimbrite*). Pyroclastic rocks, the glassy fragments of which were plastic and so hot when deposited that they fused to form a glassy rock.

Well. An excavation in the ground designed to tap a supply of underground liquid, especially water or petroleum.

Westerlies. Globe encircling belts of winds centered at about 45° latitude in both hemispheres.

Windchill factor. The heat loss from exposed skin as a result of the combined effects of low temperature and wind speed.

Xenoliths. Fragments of country rock still enclosed in a magmatic body when it solidifies.

Yardang. An elongate, streamlined, wind-eroded ridge.

Zone of aeration. See *unsaturated zone*.

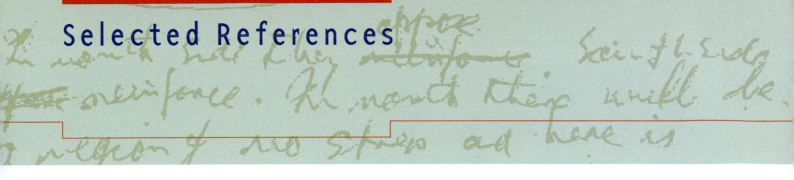

Selected References

Chapter 1 A First Look at Planet Earth

Alvarez, W., and Asaro, F., 1990. What caused the mass extinction? An extraterrestrial impact: *Scientific American*, October, pp. 78–84.

Berner, E.K. and Berner, R.A., 1996. *The Global Environment: Water, Air and Geochemical Cycles.* Prentice-Hall, Upper Saddle River, NJ.

Erwin, D.H. 1996, The mother of mass extinctions: *Scientific American*, July, pp. 72–78.

Grieve,R.A.F., 1990. Impact cratering on the Earth: *Scientific American*. v. 262, no. 4, pp.66–73

National Academy of Sciences, 1989. *On being a scientist.* Washington, DC, National Academy Press.

National Research Council, 2001. *Basic Research Opportunities in Earth Science.* Washington, DC, National Academy Press.

Skinner, B.J., Porter, S.C., and Botkin, D.B., 1999. *The Blue Planet: An Introduction to Earth System Science.* New York, John Wiley and Sons.

Stanley, S.M., 1996. *Children of the Ice Age; How a Global Catastrophe Allowed Humans to Evolve.* New York, Harmony Books.

Chapter 2 Global Tectonics

Bercovici, D., 2003. The generation of plate tectonics from mantle convection: *Earth and Planetary Science Letters* (Frontiers), v. 205, pp. 107–121.

Cattermole, P.J., 1994. *Venus: The Geological Story.* Baltimore, Johns Hopkins Press.

Conrad, C. P, and Lithgow-Bertelloni, C., 2002. How mantle slabs drive plate tectonics: *Science*, v. 298, pp. 207–209.

International GPS service (IGS): http://igscb.jpl.nasa.gov.

Kearey, P., and Vine, F.J. 1990, *Global tectonics.* Oxford, Blackwell Scientific.

Koch, D., 1997, A spreading drop plume model for Venus: *Journal of Geophysical Research*, v. 99, pp. 2035–2052.

Kogan-M.-G; Steblov-G.-M.; King-R.-W; Herring-T.-A; Frolov-D.-I, Egorov-S.-

G-; Levin-V.-Y., Lerner-Lam, A.; Jones-A., 2000. Geodetic constraints on the rigidity and relative motion of Eurasia and North America: *Geophysical Research Letters*, v. 27; pp. 2041–2044.

Macdonald, K.C., and Fox, P.J., 1990, The mid-ocean ridge: *Scientific American*, v. 262, pp. 72–79.

National Research Council, 1995. *The Global Positioning System: A Shared National Asset.* Washington, DC, National Academy Press.

Rappaport, N. J., Konopliv, A. S., Kucinskas, A. B., and Ford, P. G., 1999. An improved 360 degree and order model of Venus topography: *Icarus*, v. 139, pp. 19–31.

Schubert, G., Turcotte, D.L., and Olson, P., 2001. *Mantle Convection in the Earth and Planets.* Cambridge, Cambridge University Press.

Scripps Orbit and Permanent Array Center (SOPAC): http://sopac.ucsd.edu.

Silver-P-G and Holt-W-E, 2002. The mantle flow field beneath western North America: *Science*, v. 295; pp. 1054–1058.

Solomon, S.C., and Toomey, D.R., 1992, The structure of mid-ocean ridges: *Annual Review of Earth and Planetary Sciences*, v. 20, pp. 329–364.

Chapter 3 Atoms and Minerals

Brady, J., and Holum, J., 1996, *Chemistry: The Study of Matter and Its Changes.* New York, John Wiley and Sons.

Dietrich, R.V., and Skinner, B.J., 1990, *Gems, Granites, and Gravels.* Cambridge: Cambridge University Press.

Dietrich, R.V., and Skinner, B.J., 1979, *Rocks and Rock Minerals.* New York: John Wiley and Sons.

Klein, C., and Hurlburt, C.S. Jr., 2002, *Manual of Mineralogy*, 22nd ed. New York, John Wiley and Sons.

Skinner, B.J., 1976, A second Iron Age ahead?: *American Scientist*, v. 64, pp. 258–269.

Chapter 4 Igneous Rocks

Blatt, H., and Tracy, R.J., 1996. *Petrology: Igneous, Sedimentary, and Metamorphic*, 2nd ed. New York, W.H. Freeman.

Faure, G. 1998. *Principles and Applications of Geochemistry: A Comprehensive Textbook for Geology Students.* Upper Saddle River, N.J., Prentice Hall.

Kelemen, P.B., Hirth-G; Shimizu-N; Spiegelman-M; Dick-Henry-J-B, 1999. A review of melt migration processes in the adiabatically upwelling mantle beneath oceanic spreading ridges. In *Mid-Ocean Ridges; Dynamics of Processes Associated with Creation of New Ocean Crust*, pp. 67–102, Cambridge, Cambridge University Press.

Philpotts, A.R., 1990, *Principles of Igneous and Metamorphic Petrology.* Englewood Cliffs, NJ, Prentice-Hall.

Plank, T., and Langmuir, C. H., 1993. Tracing trace elements from sediment input to volcanic output at subduction zones: *Nature*, v. 362, pp. 739–743.

Chapter 5 Volcanism

Decker, R., and Decker, B., 1989, *Volcanoes.* San Francisco, W.H. Freeman.

Foxworthy, B.L., and Hill, M., 1982. Volcanic eruption of 1980 at Mount St. Helens. The first 100 days: Professional Paper no. 1249, U.S. Geological Survey.

Haymon, R.M., and Macdonald, K.C., 1985, The geology of deep-sea hot springs: *American Scientist*, v. 73, pp. 441–449.

National Research Council, 2000. Review of the U.S. Geological Survey's Volcano Hazards Program, Washington, DC, National Academy Press.

Rona, P.A., 1986, Mineral deposits from seafloor hot springs: *Scientific American*, v. 254, January, pp. 84–92.

Simkin, T., and Fiske, R.S., 1983. *Krakatau, 1883: The Volcanic Eruption and Its Effects.* Washington, DC, Smithsonian Institute Press.

Chapter 6 Weathering and Soils

Birkeland, P.W., 1999. *Soils and Geomorphology.* New York, Oxford University Press.

Brown, L.R., and Wolf, E.C., 1984. Soil erosion: Quiet crisis in the world economy: Worldwatch Institute, *Worldwatch Paper 60*.

Carroll, D., 1970. *Rock Weathering.* New York, Plenum.

Nahon, D.B., 1991. *Introduction to the Petrology of Soils and Chemical Weathering*: New York, John Wiley and Sons.

Retallack, G., 1989. *Soils of the Past, an Introduction to Paleopedology*: New York, HarperCollins Academic.

Robinson, D.A., and Williams, R.B.G., eds., 1994. *Rock Weathering and Landform Evolution*. New York.

Chapter 7 Sedimentation
Blatt, H., 1992. *Sedimentary Petrology*, 2nd ed. New York, W.H. Freeman.

Blatt, H., and Tracy, R.J., 1996. *Petrology: Igneous, Sedimentary, and Metamorphic*, 2nd ed. New York, W.H. Freeman.

McLane, M., 1995. *Sedimentology*. New York, Oxford University Press.

Siever, R., 1988. *Sand*. New York, Scientific American Library.

Reineck,H.H., and Singh, I.B., 1980. *Depositional Sedimentary Environments*, 2nd ed. New York, Springer-Verlag.

Chapter 8 New Rocks from Old
Blatt, H., and Tracy, R.J., 1996, *Petrology: Igneous, Sedimentary, and Metamorphic*, 2nd ed. New York, W.H. Freeman.

Dietrich, R.V., and Skinner, B.J., 1979. *Rocks and Rock Minerals*: New York, John Wiley and Sons.

Philpotts, A.R., 1990. *Principles of Igneous and Metamorphic Petrology*. Englewood Cliffs, NJ, Prentice-Hall.

Yardley, B.W.D., 1989. *An Introduction to Metamorphic Petrology*. London, Longman Scientific.

Yardley, B.W.D., MacKenzie, W.S., and Guilford, C., 1990. *Atlas of Metamorphic Rocks and Their Textures*. New York, John Wiley and Sons.

Chapter 9 Deformation of Rock
Davis, G.H., and Reynolds, S.J., 1996. *Structural Geology of Rocks and Regions*. New York, John Wiley and Sons.

Suppe, J., 1985. *Principles of Structural Geology*. Englewood Cliffs, NJ, Prentice-Hall.

Twiss, R.J., and Moores, E.M., 1992. *Structural Geology*. New York, W.H. Freeman.

Chapter 10 Earthquakes
Atwater, B.F., 1987. Evidence for great Holocene earthquakes along the outer coast of Washington State: *Science*, v. 236, pp. 942–944.

Bolt, B.A., 1993. *Earthquakes*, 3rd ed. New York, W.H. Freeman.

Bolton, H., and Masters, G., 2001. Travel times of P and S from global digital seismic networks; implications for the relative variation of P and S velocity in the mantle: *Journal of Geophysical Research*, v. 106, pp. 13527–13540.

Incorporated Research Institutions for Seismology: http://www.iris.edu.

Johnston, A.C., and Kanter, L.R., 1990. Earthquakes in stable continental crust: *Scientific American*, v. 262, no. 3, pp. 68–75.

National Research Council, 1994. *Practical Lessons from the Loma Prieta Earthquake*. Washington, DC, National Academy Press.

National Research Council, 1996. Review of Recommendations for Probabilistic Seismic Hazard Analysis: Guidance on Uncertainty and Use of Experts: Washington, DC, National Academy Press.

Obara, K., 2002. Nonvolcanic deep tremor associated with subduction in Southwest Japan: *Science*, v. 296, pp. 1679–1681.

Satake-K.; Shimazaki-K.; Tsuji-Y.; Ueda-K., 1996. Time and size of a giant earthquake in Cascadia inferred from Japanese tsunami records of January 1700: *Nature*, v. 369, pp. 246–249.

Scholz, C. H., 1998. Earthquakes and friction laws: *Nature*, v. 391, pp. 37–42.

Stein, R. S., 1999. The role of stress transfer in earthquake occurrence: *Nature*, v. 402, pp. 605–609.

U.S. Geological Survey Earthquake Hazards Program http://earthquake.usgs.gov/.

Wysession, M., 1995. The inner workings of the Earth: *American Scientist*, v. 83, pp. 134–146.

Yeats, R. S., Sieh, K., and Allen, C. R., 1997. *The Geology of Earthquakes*, New York, Oxford University Press.

Chapter 11 Relative Time
Dalrymple, G.B., 1991. *The Age of the Earth*. Stanford, Stanford University Press.

Faure, G., 1986. *Principles of Isotope Geology*, 2nd ed. New York, John Wiley and Sons.

Palmer, A.R., 1984. Decade of North American Geology Geologic Time Scale. Boulder, CO, Geological Society of America, Map and Chart Series MC50.

Stanley, S.M., 1999. *Earth System History*. New York, W. H. Freeman.

Tarling, D.H., 1983. *Paleomagnetism*. London, Chapman and Hall.

Chapter 12 Changing Face of the Land
Balestrieri, M.L., Bernet, M., Brandon, M.T., Picotti, V., Reiners, P., and Zattin, M., 2003. Pliocene and Pleistocene exhumation and uplift of two key areas of the Northern Apennines: *Quaternary International*, v. 101–102, pp. 67–73.

Bull, W.B., and Brandon, M.T., 1998. Lichen dating of earthquake-generated regional rockfall events, Southern Alps, New Zealand: *Geological Society of America Bulletin*, v. 110, pp. 60–84.

Burbank, D. and Anderson, R., 2000. *Tectonic Geomorphology*. Malden, MA. Blackwell.

Chorley, R.J., Schumm, S.A., and Sugden, D.E., 1995. *Geomorphology*. New York, Methuen.

Farabegoli, E., and Agostini, C., 2000. Identification of calanco, a badland landform in the northern Apennines, Italy: *Earth Surface Processes and Landforms*, v. 25, pp. 307–318.

Molnar, P. England, P., and Martinod, J., 1993. Mantle dynamics, uplift of the Tibetan Plateau, and the Indian monsoon: *Reviews of Geophysics*, v. 31, pp. 357–396.

Pinter, N., Brandon, M.T., 1997. How erosion builds mountains: *Scientific American*, v. 276, pp. 74–79.

Reiners, P.W., 2002. (U-Th)/He chronometry experiences a renaissance: *Eos*, v. 83, pp. 21–27.

Ruddiman, W.F., ed., 1997. *Tectonic Uplift and Climate Change*. New York, Plenum.

Ruddiman, W.F., and Kutzbach, J.E., 1991. Plateau uplift and climatic change: *Scientific American*, v. 264, no. 4, pp. 66–75.

Chapter 13 Mass Wasting
Costa, J.E., and Wieczorek, G.F., 1987. Debris flows/avalanches:process, recognition, and mitigation: Boulder, CO, Geological Society of America, *Reviews in Engineering Geology VII*.

National Research Council, 1987. *Confronting natural disasters. An International Decade for Natural Hazard Reduction*. Washington, DC, National Academy Press.

Parsons, A.J., 1988. *Hillslope Form*. London, Routledge.

Porter, S.C., and Orombelli, G., 1981. Alpine rockfall hazards: *American Scientist*, v. 69, pp. 67–75.

Chapter 14 Streams and Drainage
Dunne, Thomas, and Leopold, L.B., 1978. *Water in Environmental Planning*. San Francisco, W.H. Freeman.

McCullough, D.G., 1968. *The Johnstown Flood*. New York, Simon and Schuster.

National Research Council, 1996. *Alluvial Fan Flooding*. Washington, DC, National Academy Press.

Bridge, John, 2003. *Rivers and floodplains; Forms, Processes,m and Sedimentary Record*. Malden, MA, Blackwell.

Chapter 15   Groundwater

Back, W., Rosenshein, J.S., and Seaber, P.R., eds, 1988. *Hydrology. The Geology of North America, vol. O-2*. Boulder, CO, Geological Society of America.

Dolan, R., and Goodell, H.G., 1986. Sinking cities: *American Scientist*, v. 74, pp. 38–47.

Jennings, J.N., 1983. Karst landforms: *American Scientist*, v. 71, pp. 578–586.

Palmer, A.N., 1991. Origin and morphology of limestone caves: *Geological Society of America Bulletin*, v. 103, pp. 1–21.

Trudgill, S., 1985. *Limestone Geomorphology*. White Plains, NY, Longman.

Chapter 16   Glaciers

Alley, R. B., 2000. *Two-Mile Time Machine: Ice Cores, Abrupt Climate Change, and Our Future*. Princeton, Princeton University Press.

Bell-R-E; Studinger-M; Tikku-A-A; Clarke-G-K-C; Gutner-M-M; Meertens-C, 2002. Origin and fate of Lake Vostok water frozen to the base of the East Antarctic ice sheet: *Nature*, v. 416, pp. 307–310.

Benn, D.I., and Evans, D.J.A., 1998. *Glaciers and Glaciation*. New York, Arnold.

Broecker, W.S., and Denton, G.H., 1990. What drives glacial cycles?: *Scientific American*, v. 262, pp. 48–56.

Dawson, A.G., 1992. *Ice-Age Earth*. New York, Routledge.

Imbrie, J., and Imbrie, K.P., 1979. *Ice Ages: Solving the Mystery*. Short Hills, NJ, Enslow.

Post, A. and LaChapelle, E. R. 2000. *Glacier Ice*. Seattle, University of Washington Press.

Sharp, R.P., 1988. *Living Ice. Understanding Glaciers and Glaciation*. Cambridge, Cambridge University Press.

Swiss National Tourist Office, 1981. *Switzerland and Her Glaciers*. Berne, Kummerly, and Frey.

Chapter 17   Atmosphere

Greeley, R., and Iverson, J., 1985. *Wind as a Geological Process*. Cambridge, Cambridge University Press.

Pye, K., 1987, *Aeolian Dust and Dust Deposits*. New York, Academic Press.

Pye, K., and Tsoar, H., 1990. *Aeolian Sand and Sand Dunes*. New York, HarperCollins Academic.

Walker, A.S., 1982. Deserts of China: *American Scientist*, v. 70, pp. 366–376.

Chapter 18   Oceans

Carter, R.W.G., and Woodroffe, C.D., eds., 1994. *Coastal Evolution*. Cambridge, Cambridge University Press.

Davis, R. A., Jr., and Fitzgerald, D., 2003. *Beaches and Coasts*. Malden, MA. Blackwell.

Duxbury, A.C., and Duxbury, A.B., 1994. *An Introduction to the World's Oceans*, 4th ed. Dubuque, Iowa, Wm. C. Brown.

Hardisty, J., 1990. *Beaches, form and process*. New York, HarperCollins Academic, 352 pp.

National Research Council, 1997. *The Global Ocean Observing System: Users, Benefits, and Priorities*. Washington, DC, National Academy Press.

National Research Council, 2000. *Illuminating the Hidden Planet: The Future of Seafloor Observatory Science*. Washington, DC, National Academy Press.

Sheridan and J.A. Grow, eds., The geology of North America, v. I-2, The Atlantic Continental Margin: Boulder, CO, Geological Society of America, pp. 549–556.

Chapter 19   Climate/Changing Planet

Houghton, J. T., Ding , Y., Griggs, D. J., Noguer, M., van der Linden, P. J., Dai, X., Maskell, K., and Johnson, C. A., (eds.) 2001. *Climate Change 2001: The Scientific Basis: Contribution of Working Group I to the Third Assessment Report of the Intergovernmental Panel on Climate Change*. Cambridge, Cambridge University Press.

Jones, P.D., and Wigley, T.M.L., 1990. Global warming trends: *Scientific American*, v. 263, pp. 84–91.

Mungall, C., and McLaren, D.J., eds., 1990. *Planet under stress: The Challenge of Global Change*: New York, Oxford University Press.

National Research Council, 1994. *Solar Influences on Global Change*. Washington, DC, National Academy Press.

National Research Council, 1996. *Natural Climate Variability on Decade-to-Century Time Scales*. Washington, DC, National Academy Press.

National Research Council, 2000. *Reconciling Observations of Global Temperature Change*. Washington, DC, National Academy Press.

National Research Council, 2001. *Climate Change Science: An Analysis of Some Key Questions*. Washington, DC, National Academy Press.

Robert, N., ed., 1994. *The Changing Global Environment*. Oxford, Blackwell.

Schneider, S.H., 1987. Climate modeling: *Scientific American*, v. 256, pp. 72–80.

Silver, C.S., and DeFries, R.S., 1990. *One Earth, one future: Our Changing Global Environment*. Washington, DC, National Academy Press.

Toon, O.B., and Turco, R.P., 1991. Polar stratospheric clouds and ozone depletion: *Scientific American*, v. 264, pp. 68–74.

Trenberth, K.E., ed., 1992. *Climate System Modeling*. Cambridge, Cambridge University Press.

White, R.M., 1990. The great climate debate: *Scientific American*, v. 263, pp. 36–44.

Chapter 20   Earth Through Time

Berner, R.A., 1997. The rise of plants and their effect on weathering and atmospheric CO_2: *Science*, v. 276, pp. 544–546.

Berner, R.A., 1999. A new look at the long-term carbon cycle: *GSA Today*, v. 9, pp. 1–6.

Condie, K. C., 1997. *Plate Tectonics and Crustal Evolution*, 4th ed. Oxford, Butterworth-Heinemann.

Cox, A., and Hart, R.B., 1986. *Plate tectonics; How it works*. Palo Alto, CA, Blackwell.

Dalziel, I.W.D., 1995. Earth before Pangaea: *Scientific American*, January, pp. 58–63.

Evans, D.A.D., 2000. Stratigraphic, geochronological, and paleomagnetic constraints upon the Neoproterozoic climatic paradox: *American Journal of Science*, v. 300, pp. 347–433.

Hoffman, P.F., 1988. United plates of America, the birth of a craton: Early Proterozoic assembly and growth of Laurentia: *Annual Reviews of Earth and Planetary Sciences*, v. 16, pp. 543–603.

National Research Council, 2001. *Review of EarthScope Integrated Science*. Washington, DC, National Academy Press

Park, J., Levin, V., Brandon, M., Lees, J., Peyton, V., Gordeev, E., Ozerov, A., 2002. A dangling slab, amplified arc volcanism, mantle flow, and seismic anisotropy in the Kamchatka plate corner. In S. Stein, J. Freymuller, Plate Boundary Zones, *American Geophysical Union Geodynamics Series* v. 30, pp. 295–324.

Chapter 21   Resources

Craig, J.R., Vaughan, D.J., and Skinner, B.J., 1988. *Resources of the Earth*. Englewood Cliffs, NJ, Prentice-Hall.

Davis, G.R., 1990. Energy for the planet Earth: *Scientific American*, v. 263, no. 3, pp. 54–62.

Fulkerson, W., Judkins, R.R., and Sanghvi, M. K., 1990. Energy from fossil fuels: *Scientific American*, v. 263, no. 3, pp. 128–135.

Häfele, W., 1990. Energy from nuclear power: *Scientific American*, v. 263, no. 3, pp. 136–144.

Hodges, C.A., 1995. Mineral resources, environmental issues, and land uses: *Science*, v. 268, pp. 1305–1312.

National Research Council, 2002. *Evolutionary and Revolutionary Technologies for Mining*. Washington, DC, National Academy Press.

Sawkins, F.J., 1990. *Metal Deposits in Relation to Plate Tectonics*, 2nd ed. Berlin, Springer-Verlag.

Skinner, B.J., 1986. Earth Resources, 3rd ed. Englewood Cliffs, NJ, Prentice-Hall.

Weinberg, Carl J., and Williams, R.H., 1990. Energy from the Sun: *Scientific American*, v. 263, no. 3, pp. 146–155.

Index

A

AABW, *see* Antarctic Bottom Water
Aa (lava), 125
Ablation, 418, 419
Ablation area, 419
Abrasion, 454–456, 489
Absolute plate motion, 54
Absolute plate velocity, 537
Abyssal floor, 37
ACC, *see* Antarctic Circumpolar Current
Accreted terrane continental margin, 550–551
Accumulation, 418
Accumulation area, 419
Active margin, 37
Adiabatic expansion, 40
Aeration, zone of, 385, 388
African ice sheet, 441–442
African Rift Valley, 546, 547
Aftershock, 265
Agassiz, Louis, 431
Agate, 88
Age. *See also* Dating
    of Earth, 12, 27, 297, 300
    numerical, 278
    radiometric, 278, 291
    relative, 278
Agglomerates, 104
Agricultural waste, 399
Agriculture, dust damage to, 454
A horizon, 161
Air, as consequence of Earth system, 470–471
Alfisols, 162
Alloy, 562
Alluvial fan, 372, 373, 466
Alluvium, 366
Alpha decay, 289
Alps, 553, 554
Aluminum, 16, 17
Amino acid, 24
Amorphous solids, 74
Amphibole, 86, 102
Amphibolite, 213
Amphibolite facies, 221
Amu Dar'ya River, 7
Amygdules, 126
Anaerobic diagenesis, 190
Andesite, 103, 104
Andesite Line, 110
Andesitic magma, 110, 111, 113, 123
Andisols, 162
Angle of repose, 332
Angular unconformity, 280, 281
Anhydrite, 90
Animals, burrowing, 159
Anion, 71
Antarctica, 4
Antarctic Bottom Water (AABW), 482–484

Antarctic Circumpolar Current (ACC), 482–484
Antarctic Ice Sheet, 415
Antecedent stream, 378
Anthracite, 574
Anticline, 240
Apatite, 89
Aphanite, 101
Appalachians, 552–554
Apparent polar wandering, 535
Aquiclude, 389
Aquifer, 390–394
Aragonite, 89
Aral Sea, 7
Archean eon, 287
Area, measures of, A2
Arête, 424
Aridsols, 163
Arkose, 191
Arkosic sandstone, 191
Arroyo, 465
Artesian aquifer, 393
Artesian spring, 393
Artesian system, 392–393
Artesian well, 393
Artificial nourishment of beaches, 504
Asbestos, 91
Ash, volcanic, 126–127, 461, 462
Ash-flow, 136–137
Asian collision, 471
Aspidin, Joseph, 215
Assemblage, mineral, 93
Asthenosphere, 17, 18, 269
Astronomical theory, 436
Aswan Dam, 379
Athabasca Tar Sand, 577
Atlantic margin, 198
Atlantic Ocean:
    birth of, 46–48
    major water masses, 482t
    mass wasting in, 342
Atmosphere:
    chemical composition, 470–472
    circulation of, 448
    definition, 21
    formation, 17
    and glacial cycles, 437, 439
    greenhouse gases, 513–514
    ozone, 516–517
    wind, 447–450
Atoll, 475–476, 496–497
Atom, 68–70
Atomic number, 70
Australia:
    human effects on environment, 447–448
    weathering in, 169–170
Auxiliary plane, 261
Avalanche, debris, 338, 339, 339t

Axis (of fold), 240
Ayer's Law, 521

B

Back-arc basin, 540–541
Backshore, 491
Bacon, Francis, 7
Badland, 307
Bajada, 466
Banded iron deposits, 192
Barchan dune, 458
Barrier island, 495–496
Barrier reef, 496
Barton, Paul, 226
Basalt, 104
    metamorphism of, 212–213
    Midocean ridge basalt, 117, 538
    pillow, 136
    plateau, 136
    progressive metamorphism, 213t
Basaltic magma, 110, 111, 113, 123
Base level, 360–361
Batholith, 108, 109, 113
Bathymetry, 34
Bauxite, 570, 572
Bay barrier, 495
Beach(es), 491
    effect of human activity on, 504
    protection against shoreline erosion, 502, 503
Beach drift, 489, 490
Beach placer, 490
Beach ridge, 495
Beach sediment, 185, 491
Bed, 175
Bedding:
    cross, 180
    definition, 174–175
    graded, 180
Bedding plane:
    definition, 175
    environmental clues on, 196
Bed load, 366
Bench, wave-cut, 492
Benioff zone, 51, 260, 271
Berm, 491
Beta decay, 289
Beta particle, 289
Beverage-can experiment, 328
B horizon, 161–162
Bikini Atoll, 475–476
Bioclastic sediment, 182
Biogenic sediment, 175, 182
Biogenic sedimentary rock, 193–195
    coal, 194–195
    dolostone, 194
    limestone, 193–194
    oil shale, 195

I1

Biogeochemical cycle, 23–25
Biomass energy, 578
Biosphere, 21
Bituminous coal, 194, 574, 575
Blueschist, 221, 539
Body wave, 250, 252
Bonds, electrical, 71–73
Bonding, 72–73, 82–83, 83t
Bottomset layer, 374
Boundary conditions, 518
Bowen, N. L., 114
Braid delta, 374
Braided stream, 364, 365
Branner, J. C., 159
Breakage surface, 79
Breaking waves, 487
Breakwater, 503, 504
Breccia, 191
Bretz, J. Harlen, 360
Brittle deformation, 38
Brittle—ductile transition, 231
Brooks Range, 174
Buoyancy, 16–17, 34–35
Burial metamorphism, 217
Butte, 464

C
Calcareous ooze, 182
Calcite, 89, 90
Calderas, 133, 134
California gold rush, 366
Calving, 420
Cambrian period, 287–288
Canadian Rockies, 554–556
Canadian Shield, 544
Capillary attraction, 328, 385–386
Carbonates, 83, 89
Carbonate caves/caverns, 402–403
Carbonate sedimentation, 186, 199
Carbonate shelves, 186
Carbon cycle:
    and climate change, 511–512
    and Earth system, 33
    and ocean circulation, 505
    and plate tectonics, 319–320
Carbon dating, 293–294
Carbon dioxide (CO2), 318–321
    and global climate change, 199
    and greenhouse gas concentrations,
        514–516
    and oceanic lithosphere, 48, 319–320
    and uplift, 320–321
    and Venus's atmosphere, 514
    and volcanism, 49, 319
Carbonic acid, 152–153
Cassiterite, 92
Cataclastic metamorphism, 215–216
Catastrophic floods, 359, 360
Catastrophism, 8–12
Cation, 71
Caves and caverns, 402–404
Celsius, conversion to/from
    Fahrenheit/Kelvin, A2
Cement, 191, 215
Cementation, 190, 402
Cenote, 404
Cenozoic era, 287

Centi- (prefix), A1
Central-vent volcanoes, 129–136
CFCs, see Chlorofluorocarbons
Chains, 86
Chalcedony, 88
Chalcopyrite, 90
Chalk, 194
Channels, 362–365
    definition, 354
    patterns, 355–356
Channeled Scabland, 360
Chemical alteration, 190
Chemical cementation, 402
Chemical differentiation by partial melting,
    111
Chemical equilibrium, 114
Chemical recrystallization, 214, 215
Chemical sediment, 175, 181
Chemical sedimentary rock, 192–193
    banded iron deposits, 192
    chert, 193
    evaporite deposits, 192
    phosphorous deposits, 192–193
Chemical weathering, 148–149, 152–157, 153t
Chemistry:
    groundwater, 396, 398–399
    mineral, 69–74
Chert, 193
China:
    loess of, 460
    monsoons, 147–148
    tectonic desertification, 470–471
Chlorites, 86
Chlorofluorocarbons (CFCs), 513, 515, 516
C horizon, 162
Chu, Dezhi, 547
Circulation, ocean, 480–484
Circum-Pacific belt, 259
Cirque, 414, 424
Cirque glaciers, 414, 424
Clastic sediment, 175–181
    cross bedding, 180
    definition, 175
    graded bedding, 180
    mineral composition, 180–181
    nonsorted, 178–179
    particle shape, 179
    particle size, classification by, 175–176,
        176t
    rhythmic layering, 179
    sorting, 176, 178
Clastic sedimentary rock, 191–192, 195
Clays, 86
Cleavage, 79–80, 210, 211
Cliff, wave-cut, 492
Climate:
    and denudation, 309
    desert, 464
    and tectonism, 244
    and weathering, 158–159
    and wind, 450
Climate change, 508–528. See also Global
    warming
    and carbon cycle, 511–512
    and carbon dioxide (CO2), 199
    and glaciation, 436–437
    historical record, 521–525

    and plate tectonics, 525–528
    Younger Dryas event, 526–527
Climate system, 511
Closed system, 19, 20
Closure temperature, 314, 316
CO2, see Carbon dioxide
Coal, 194–195, 574–575
Coal swamps, 575
Coasts:
    erosion, protection against, 502–504
    evolution, 497–501
    hazards, 501–502
    rocky (cliffed), 492
Coastal desert, 463
Coastal erosion/sediment transport, 489–491
Coastal landforms/deposits, 489–504
    atolls, 475–476, 496–497
    barrier islands, 495–496
    beaches, 491
    beach ridges, 495
    coastal erosion/sediment transport,
        489–491
    erosion, protection against, 502–504
    evolution, 497–501
    human activity's effect on, 504
    marine deltas, 493–494
    reefs, 496–497
    rocky (cliffed) coasts, 492
    spits, 494, 495
Coefficient of permeability, 389
Collisional uplift, 306
Colluvium, 336
Color:
    mineral, 80
    rock, 196–197
Columbus, Christopher, 480
Columns (stalactite/stalagmite), 403
Compaction, 190
Complex ions, 73
Compounds, chemical, 71
Compressional stress, 227
Compressional wave, 250, 252
Concrete, 215
Conduction, 38, 39
Cone karst, 405
Cone of depression, 389
Confined aquifer, 390
Confining stress, 227, 229
Conformity, 280
Conglomerate, 191, 192
Consequent stream, 378
Contact metamorphism, 216–217
Contessa Valley (Italy), 11
Continents, 532–533, 544. See also
    Supercontinents
Continental collision margin, 548–550
Continental collision zone, 198
Continental convergent margins, 547, 548
Continental crust, 17
Continental desert, 463
Continental divides, 310–311
Continental drift, 33–34, 533–534. See also
    Plate motion; Plate tectonics
Continental lithosphere, 18
Continental margins, 198, 544–551
Continental rise, 37
Continental shelf, 36

Continental shelf depositional environments, 185–186
Continental shield, 544
Continental slope, 36
Continental slope/rise depositional environments, 186–189
   deep-sea fans, 187–188
   sediment drifts, 188–189
   turbidity currents, 186–187
Continental volcanic arc, 49–50, 547, 548
Continuous reaction series, 114
Convection, 38-42
Convergent margin/collision zone, 43, 51–52
Convergent margin/subduction zone, 42–43, 48–51
Conversions:
   Fahrenheit—Celsius—Kelvin, A2
   metric—nonmetric, A2
Copper, 562, 582
Coquina, 194
Coral reefs, 496, 497
Core (of Earth), 18
   definition, 17
   inner/outer, 18
   seismic waves, 267, 269
   wave shadows, 267
Coriolis, Gaspard-Gustave de, 449
Coriolis effect, 448–450, 480–481
Coronae, 57
Correlation (of strata), 283–285
Cosmogenic nuclides, 314, 316–317
Covalent bonding, 72, 83t
Craters, 11
   volcanic, 132–133
Cratons, 544
Creep, 335–337, 337t, 340, 421
Crevasse, 421
Croll, John, 436
Cross bedding, 180
Crust, 18
   definition, 17
   glacial deformation of, 433
   Mohorovičić discontinuity, 266–267
   seismic waves, 266–267
   thickness/composition of, 267
Crystal, 77, 150, 151
Crystal faces, 77
Crystal form, 77–78
Crystalline solids, 74
Crystal settling, 116
Crystal structures, 74–79
Currents, ocean, 480–484, 489
Current systems, 481
Cycles:
   biogeochemical, 23–25
   carbon, 33, 319–320, 505, 511–512
   geochemical, 271–272
   geographic, 311–312
   geologic, 10, 21–27
   glacial, 437, 439
   hydrologic, 23, 28, 520
   Milankovitch, 510, 543
   nitrogen, 24–25
   open system interactions, 21–23
   rock, see Rock cycle
   sea-level, 499
   tectonic, 23, 26, 28

Cyclotherms, 543

D
Dacites, 103
Dams, 361
Dana, James Dwight, 55
Darcy, Henri, 389
Darcy's Law, 389, 390
Darwin, Charles, 159
Dating:
   carbon (radiocarbon), 293–294
   potassium-argon, 292–293, 295
   radiometric, 12, 289–297, 293t, 536
Daughter (radioactive decay product), 289
Davis, William Morris, 312
Deadman Wash (Wupatki National Monument, Arizona), 8
Debris avalanche, 338, 339
Debris fall, 331
Debris flow, 333
Debris slide, 331
Decay (of buildings and monuments), 154
Deci- (prefix), A1
Décollement, 552
Decompression melting, 117
Deep-sea depositional environments, 189–190
Deep-sea fans, 187–188
Deep-sea ooze, 182, 183, 189
Deflation, 454–455
Deflation hollows/basins, 455
Deformation, 38, 225–244
   abrupt movement, 231
   by bending (folds), 240–244
   ductile, 38, 227–229
   elastic, 38, 227
   evidence of, 232–244
   faults, 233–240
   fracture, 233–240
   and geologic time, 229, 230
   gradual movement, 231, 232
   mechanical, 214, 215
   ocean floor, 434
   oil pools, 241
   stages of, 227–228
   strike/dip, 232
Dehydration, 155
Deka- (prefix), A1
Delta, 341–342, 372–375, 493–494
Deltaic sediments, 185
Density of minerals, 81, 82
Denudation, 304, 309–310, 558
Denudation rate, calculation of, 314–317
Depositional environments, 183
   continental slope/rise, 186–189
   deep-sea, 189–190
   marine, 185–190
   nonmarine, 184–185
   shoreline/continental shelf, 185–186
Desert, 462–472
   definition, 462
   desertification, 468–470
   landforms, 465–468
   and plate tectonics, 469–471
   surface processes, 464–465
   types/origins, 462–463, 463t
Desertification, 468–471
Desert lake, 466

Desert pavement, 455
Desert soils, 163
Desert streams, 465–468
Desert varnish, 464–465
Detachment surface, 552
Diagenesis, 190–191
Diapir, 57, 58
Diatremes, 134, 135
Dietz, Robert, 68
Differential stress, 38, 205, 207, 227
Differential weathering, 158
Dikes, 107, 108
Diorite, 103
Dip, 232
Discharge, 354, 356
Discharge area, 387
Disconformity, 280, 281
Discontinuous reaction series, 114, 115
Dispersion, 253
Dissolution, 153, 401–402
Dissolved load, 366, 368–369
Distorted nonconformity, 281
Distributaries, 375
Divergent margin, 42, 46–48
Divide, 375
Dolomite, 89, 90
Dolostone, 194
Downslope movement, 327
Drainage, evolution of, 376–377
Drainage basin, 375
Drainage systems, 375–377. See also Streams
Drift, 427–430
   glacialmarine, 427
   stratified, 428–430
Dripstone, 403
Dropstone, 427
Drumlins, 426
Dryas octopetala, 526
Ductile deformation, 38, 227–229
Ductile rock, 45
Dunes, 457–459
   definition, 457
   form/size, 457–458
   migration, 459
   sand seas, 459
   types, 458, 459
   and Uniformitarianism, 9–10
Dust, 91
   damage caused by, 454
   on Mars, 448
   in ocean sediments/glacier ice, 461, 462
   windblown, 452–454
Dust storms, 448, 453

E
Early Eocene, 523–524
Early Holocene, 521
Earth:
   age of, 12, 27, 297
   equatorial bulge, 34
   hum of, 451
   internal phenomena, 37–41
   internal structure, 16–19
   layers, 17–18
   orbit, 438–439
   rotation on axis, change of, 27
   surface features, 34–37

Earthflow, 337–338
Earth history. *See also* Geologic time
    environmental clues in sedimentary
        rocks, 195–198
    isotopes, 197
    magnetic properties, 197
Earth probes, seismic waves as, 266–270
Earthquakes, 246–272. *See also* Seismic waves
    damage from, 257–259
    disasters, 256–257, 257t
    epicenter, 249, 254, 255
    first-motion studies, 260–261
    focus, 249–250
    forecasting of, 262
    forecasting/prediction, 261–266
    hazards, 255–259
    magnitude/frequency of, 251, 254–255,
        259t
    and plate tectonics, 270–272
    precursors to, 263–264
    prediction of, 262
    seismometers, 248–249
    study of, 248–255
    in subduction zones, 50–51
    world distribution of, 259–260
Earthquake displacements, uplift rate calcu-
        lation and, 313
Earthquake focus, 249
Earthquake waves, 17–18
Earth system, 20–21
    air as consequence of, 470–471
    definition, 20
Eclogite facies, 221
Ecological disasters, *see* Human activity
Elastic deformation, 38, 227
Elastic rebound theory, 262–264
Electricity generation, 578
Electron capture, 289
Elements, 68–70
    definition, 68
    periodic table, 73, 74t
    tables of, A4–A6
Emergence, 498, 499
End moraines, 428
Energy:
    geothermal, 138, 139
    heat, 16
    impact, 16
    kinetic, 16
    measures of, A3
    and sedimentation, 198
Energy-level shells, 70–71
Energy reserves, 578
Energy resources, 573–581
    biomass energy, 578
    fossil fuels, 573–578
    geothermal power, 579
    hydroelectric power, 579
    nuclear energy, 579
    solar energy, 580–581
    tidal energy, 581
    waste disposal, 580
    wave energy, 581
    wind energy, 581
Entisols, 162, 163
Environmental clues, minerals as source of,
    92

Eocene, 523–524
Eolian deposits, 457–461
    dunes, 457–459
    loess, 460–461
Eolian sediments, 185
Eons, 286–287
Epicenter, 249, 254, 255
Epochs, 287–288, 287t
Equatorial bulge, 34
Equilibrium line, 419
Eras, 287, 287t
Erosion, 366
    above sea level, 489
    below sea level, 489
    of cliffed coast, 492
    coastal, protection against, 502–504
    definition, 21
    glacial, 423–426
    headlands, 488
    Hutton's observations on, 9
    by running water, 366
    soil, 166–168
    by waves, 489
    wind as cause of, 454–456
Erosion surface, 10
Erratics, 427
Eruptions:
    central, 129–136
    explosive, 126–128
    fissure, 136–137
    lateral, 128–129
    nonexplosive, 125, 126
    Plinian, 127
    tephra, 140
    undersea, 140
Eruption column, 127, 128
Estuarine sediments, 185
Estuary, 185, 500
Europa (Jovian moon), 54
Europe, loess of, 460–461
Evaporite, 186, 192, 567
Evolution, landscape, *see* Landscape evolu-
        tion
Exfoliation, 156
Exhumation, 304
Exhumation rate, calculation of, 314–317
Exploration geology, 563
Explosive eruptions, 126–128
Extensional uplift, 308
Extinction of species, 11
Extrusive igneous rock, 101, 103, 104

F
Facies, 182–184
Fahrenheit, conversion to/from
        Celsius/Kelvin, A2
Failure, 328
Fans, 466
Fan delta, 374
Fault, 45, 46, 231, 233–240
    and folds, 242
    normal, 45, 46, 235–236
    reverse, 236
    strike-slip, 237–239
    thrust, 236, 237
    transform, 237–239
Fault behavior, 265–266

Fault breccia, 240
Fault interaction, 265
Fault plane, 261
Feldspar, 88, 89, 102
Ferrel cell, 450
Fire, rock alteration and, 151–152
Fission (nuclear), 578
Fission tracks, 314, 316, 317
Fissure eruptions, 136–137
Fjords, 424, 425
Fjord glaciers, 414, 424, 425
Flash floods, 465–466
Flint, 88
Floods, 357–360, 359t
Flood-frequency curve, 359
Floodplain, 371
Flood prediction, 358, 359
Floridan aquifer, 393, 394
Flowstone, 403
Fluids:
    and fault behavior, 266
    and metamorphism, 208–209
Focal sphere, 261
Focus, earthquake, 249
Folds, 240–244
Footprints, 196
Footwall block, 234
Foreset layer, 374
Foreshocks, 264–265
Foreshore, 491
Formation, rock-stratigraphic record and, 282
Fossils:
    clastic sedimentary rock, 191
    environmental clues from, 196
Fossil fuels, 573–578
    and climate change, 516
    coal, 574–575
    petroleum, 575–578
    recoverable quantities, worldwide, 578t
Fossil magnetic record, 535
Fossil placers, 562, 570
Fractional crystallization, magmatic differ-
        entiation by, 114
Fracture, 227–229
    deformation by, 233–240
    ductile deformation vs., 228–229
Friction, plate tectonics and, 60
Fringing reef, 496
Frost heaving, 340
Frost wedging, 151
Frozen ground, thawing due to global
        warming, 520

G
Gabbro, 103, 539
Galena, 90
Gamma ray emission, 289, 290
Gangue, 564
Garnet peridotites, 111
Garnets, 85–86
Gases, in magma, 123, 137
Gas hydrates, 520, 526
GCMs, *see* General circulation models
Gelifluction, 340
Gemstones, A11
General circulation models (GCMs), 518,
        522, 524–525

Geochemical cycles, 271–272
Geochemically abundant metals, 572
Geochemically scarce metals, 572–573
Geodesy, 42
Geographic cycle, 311–312
Geologic column, 285–288
Geologic cycles, 10, 21–27
Geologic maps, 232, 233
Geologic processes:
    human effect on, 4–5, 7
    rate of, 8–12, 26–27
Geologic record, global climate change and, 521–525
Geologic time, 12, 277–300
    and correlation of rock units, 283–285
    early measurement attempts, 288–289
    geologic column, 285–288
    intervals in, 283
    magnetic polarity time scale, 297–299
    numerical measurement of, 288–297
    and plate tectonics, 300
    radioactivity as measure of, 289–297
    and rock deformation, 229, 230
    role of, in metamorphism, 209
    stratigraphic classification, 282–283
    stratigraphy, 278–282
    and weathering, 159–160
Geologic time scale, 286–288
    definition, 286
    eons, 286–287
    epochs, 287–288
    eras, 287
    periods, 287–288
Geologist, 5
Geology:
    definition, 5
    historical vs. physical, 7–8
Geomorphologists, 312
Geomorphology, 304. See also Landscape evolution
Geosphere, 21
Geothermal energy, 138, 139
Geothermal gradients, 37, 38
Geothermal power, 579
Geyser, 137
Giga- (prefix), A1
Glacial ages, 431–440
    causes of, 435–440
    earlier glaciations, evidence of, 433–434
    Ice Ages, 431–433
    Little Ice Ages, 434, 440
Glacial cycles, 437, 439
Glacial drift, see Drift
Glacial grooves, 423–424
Glacial loess, 460–461
Glacialmarine drift, 427
Glacial sediments, 185
Glacial striations, 423–424
Glacial valleys, 424, 425
Glaciation, 423–430
    definition of, 423
    drift resulting from, 427–430
    erosion and sculpturing, 423–426
    sediment, transport of, 426–427
Glaciers, 412–423, 414t. See also Glaciation
    and African ice sheets, 441–442
    basal sliding in, 422

changes due to global warming, 520
changes in size of, 418–420
cirque, 413, 414, 424
definition of, 412
fjord, 413, 414, 424, 425
and global warming, 440–441
ice caps, 413, 414, 426
ice sheets, 414–415
ice shelves, 415–416
internal flow in, 421
mass balance in, 418–419
mountain, 414
movement of, 421–423
periglacial zones, 430–431
piedmont, 413, 414
snow, conversion to ice in, 417–418
and snowline, 417
surges, glacier, 422–423
temperate vs. polar, 416–417
terminus fluctuations in, 420
types of, 412–414
valley, 413, 414, 424
Glacier ice, dust in, 461, 462
Global ocean conveyor system, 481–484
Global Positioning System (GPS), 41–42, 263
Global warming (global change), 517–521
    climate models, 518, 519
    environmental effects of, 519, 520
    fossil fuels, 516
    glaciation, 440–441
    greenhouse effect, 512–517
    greenhouse warming, model estimates of, 518
    historical temperature trends, 517–518
    human effect on, 27, 510, 515–518, 526–528
    and limestone/carbon dioxide, 199
Gneiss, 207, 212
Gold, 561–562, 570
Gondwanaland, 47, 378, 442, 533
Gordon, Richard G., 547
GPS, see Global Positioning System
Graben, 235
Graded bedding, 180
Grade (of ore), 563
Gradient, geothermal, 37, 38
Gradient, stream, 356
Gradualism, 9
Grain flow, 338
Grain size, 176t
Granite, 103
Granitic batholiths, 113
Granitic rocks, 103
Granodiorite, 103
Granular flow, 332, 335–339
Granulite, 213
Graphite, 73, 575
Gravitational attraction, planetary accretion and, 15
Gravity:
    and mass wasting, 327
    perpendicular vs. tangential, 327
Graywacke, 191
Greasy luster, 80
Greenhouse effect, 512–517
    effect on Earth's life forms, 33
    gases involved in, 514t

and Venus, 61
Greenhouse gases, 199, 514t
    concentration trends, 514–517
    and mantle convection, 522, 523
    as portion of atmosphere, 513–514
Greenhouse warming, model estimates of, 518
Greenland Ice Sheet, 415
Greenschist, 212, 213, 221
Groin, 504
Grooves, glacial, 423–424
Ground moraines, 428
Groundwater, 383–407
    and agricultural poisons, 399
    aquifers, 390–394
    chemistry, 396, 398–399
    contamination, 396–399
    definition, 384
    depth of, 385
    geologic activity of, 400–406
    and hazardous waste, 399, 400
    mining of, 394–395
    motion, 386–389
    origin of, 384–385
    and plate tectonics, 407
    quality/contamination, 396–400
    recharge/discharge, 387–389
    seawater contamination, 396–397
    selenium contamination, 398–399
    sewage pollution, 396
    springs, 389
    and toxic wastes, 397, 399
    water table, 385–386
    wells, 389, 390
Growth habit, 78
Growth surface, 79
Gypsum, 90
Gyre, 481

H
Hadean eon, 286
Hadley cells, 449
Half–graben, 235
Half–life, 290–291
Hall, James, 112
Halley, Edmund, 288
Hanging–wall block, 234
Hardness of minerals, 81
Hard water, 396
Hawaiian Islands:
    mass wasting in, 342, 343
    and plate motion, 55–56
    tsunami, 501–502, 501t
Hazardous waste, underground storage of, 399, 400
Headlands, 488
Heat:
    internal, 198
    latent, 76
Heat capacity, 480
Heat energy, Earth's internal structure and, 16
Hecto- (prefix), A1
Helium, radiogenic, 314, 316
Hematite, 92
Hess, Harry, 535, 536
High-grade metamorphism, 205, 210

High Plains aquifer, 390–392
Historical geology, 7–8
Historical record, 521–525
Historical temperature trends, 517–518
Histosols, 162
Holocene, 521
Horizons, soil, 160–162
Horizontality, *see* Law of Original
    Horizontality
Horns, 424
Hornblende, 86
Hornfels, 216
Horst, 235
Hot spots, 54–59
Hot spot hypothesis, 55
Hot spot track, 55
Hubbert, M. King, 328
Human activity:
    and Aral Sea, 7
    and Australia, 447–448
    and beaches, 504
    and Black Sea coast, 504
    and climate change, 27, 510, 515–518,
        526–528
    and coastal landforms, 504
    and denudation, 306, 307, 315, 317
    and desertification, 448, 469
    and geologic processes, 4–5, 27–28
    and Nile River, 379
    and ozone layer, 27, 515
    and San Francisco Bay, 500
    and soil erosion, 166–167
Hutton, James:
    and geologic time measurement, 278
    and gradualism hypothesis, 9
    and igneous rock, 102
    and rock cycle, 92
    and sandstone, 10
    and Uniformitarianism, 26
Hydration, 155
Hydraulic conductivity, 389
Hydraulic gradient, 389
Hydroelectric power, 361–362, 579
Hydrograph, 358
Hydrologic cycle:
    change in, from global warming, 520
    definition, 23
    interconnectivity of, with other cycles,
      28
Hydrolysis, 153
Hydrosphere, 21
Hydrothermal solutions, 221
Hydrothermal vent, 565
Hypotheses, 8

I

Ice Ages, 431–433, 542–543
Icebergs, 420
Ice caps, 414, 426
Ice–contact stratified drift, 428–430
Ice sheets, 414–415, 426, 441–442
Ice shelves, 413, 415–416, 508
Igneous rock, 92, 100–119, 105t
    definition, 23
    extrusive, 101, 103, 104
    intrusive, 101, 103
    and life on earth, 117–118
    and magma, 110–116

    mineral assemblage in, 102–104
    mineral grains in, 101–102
    and plate tectonics, 117, 118
    and plutons, 106–109
    pyroclasts, 104
    tephra, 104
    texture in, 101–102
    tuffs, 104, 106
Ignimbrite, 128
Imperial Valley, 568
Inceptisols, 162, 163
Index fossil, 285
Index minerals, 217–218
Inertia, 248–249
Inertial force, 484
Inertial seismometers, 249
Inner core, 18, 269
Inselberg, 467, 468
Intergranular fluid, 208
Internal heat, sedimentation and, 198
Internal structure of Earth, 16–19
Intrusive igneous rock, 101, 103
Io (moon of Jupiter), 14, 15
Ions:
    complex, 73
    definition, 71
    and ocean salinity, 477, 478
Ion electrical charge, 75–77
Ionic bonding, 72, 83t
Ionic radius, 74, 75
Ionic substitution, 74–77
Iridium, 10, 11
Iron, 17
Iron deposits, 567–569
Iron meteorites, 18, 267
Ironstone, 164
Irrigation, 7
Isograd, 218
Isolated system, 19
Isolated tetrahedra, 85–86
Isostasy, 34–35
Isostatic feedbacks, uplift and, 321–322
Isostatic uplift, 306, 308
Isotherms, 478–479
Isotopes, 289
    definition, 70
    environmental clues from, 197
    half-life of, 290–291
    in radiometric dating, 293t

J

Jasper, 88
Joints, 149–150, 233
Joly, John, 288
Jovian Planets, 12–15
Jupiter, 14, 15, 54

K

Kamchatka, 531–532
Kames, 428, 429
Kanamori, Hiroo, 251
Kaolinite, 153
K/Ar dating, *see* Potassium-argon dating
Karst, 404–407
Karst topography, 404–406
Kelvin, conversion to/from
    Celsius/Fahrenheit, A2
Kelvin, Lord William Thompson, 288–289

Kerogen, 574
Kettles, 428–430
Key bed, 285
K horizon, 162
Kilauea volcano (Hawaii), 99–100, 121–122
Kilo- (prefix), A1
Kimberlite pipes, 267, 268
Kinematics, 547
Kinetic energy, Earth's internal structure
    and, 16

L

Laccoliths, 108
Lagoon, 496
Lahar, 139, 140
Lake sediments, 184–185
Lake Superior-type iron deposits, 567–569
Laminar flow, 366, 453
Land-derived sediment, 189–190
Landscapes:
    ancient, 317, 318
    steady-state, 312
Landscape evolution, 303–322
    ancient landscapes of low relief, 317,
      318
    causes of, 303–311
    continental divides, 310–311
    denudation, 309–310
    exhumation/denudation rates, 314–317
    hypothetical models for, 311–313
    and plate tectonics, 318–322
    and streams, 378–379
    uplift, 305–309, 313–314
Landslides, 328
    cliffed shoreline, 502
    fatalities, 345t
    and plate tectonics, 347–349
    submarine, 342, 343
Land surface, subsidence of, 395
Larsen B ice shelf, 508
Latent heat, 76
Lateral moraines, 428
Laterites, 164, 169–170, 571
Lateritic crust, 164
Laurasia, 533
Laurentia, 542
Lava, 123–125. *See also* Magma
Lava domes, 133
Lava fountains, 125, 126
Lava tubes, 125
Laws, scientific, 8
Law of Original Horizontality, 8, 278
Layered rocks, *see* Stratigraphy
Leaching, 154
Leaning Tower of Pisa, 395
Left-lateral strike-slip fault, 237
Length, measures of, A1–A2
Levee, natural, 371
Life on the Mississippi (Twain), 364
Lignite, 194, 574
Limbs, 240
Limestone, 193–194
    and global climate change, 199
    metamorphism of, 213, 214
Linear dune, 458
Liquefaction, 338
Lithic particles, 191
Lithic sandstone, 191

Lithification, 190
Lithographic limestone, 194
Lithology, 309
Lithosphere, 17
    brittle—ductile properties of, 230–231
    definition, 18
    tectonic cycle, 26
Lithosphere degassing, 523–524
Little Ice Ages, 434, 440
Load, 354, 356, 366–370
Loess, 460–461
Loma Prieta earthquake (1989), 246–248, 264
Long profile, 356
Longshore current, 489
Love waves, 250
Low-grade metamorphism, 205, 210
Low-velocity zone, 269
Luster, 80
Lyell, Charles, 288

M
Magellan satellite, 53, 54, 57
Magma, 110–116, 122–129
    andesitic, 110, 111, 113
    basaltic, 110, 111, 113
    composition of, 123
    and divergent margins, 46
    explosive eruptions, 126–128
    nonexplosive eruptions, 125, 126
    rhyolitic, 110, 113
    solidification of, 114–116
    tectonic cycle, 26
    temperature of, 123, 124
    viscosity of, 124–125
Magmatic differentiation by fractional crystallization, 114
Magnetic chrons, 299
Magnetic latitude, 535
Magnetic polarity time scale, 297–299
Magnetic properties, environmental clues from, 197
Magnetic records, 535–538
Magnetic reversal, 536
Magnetite, 92
Major plutons, 108–109
Man and Nature (Marsh), 306
Manganese, 572
Mantle, 18
    convection, 37–42
    definition, 17
    layers of, 269–270
    low-velocity zone, 269
    seismic waves, 267
Mantle convection, 522, 523
Mantle outgassing, 266
Mantle plumes, 54–57
Mantle plume hypothesis, 55
Marble, 213, 214
Margins, 36-37
    continental, 198, 544–551
    continental convergent, 547, 548
    divergent, 42, 46–48
    passive, 36, 545–547
    plate, 42–52
    rift, 378
    transform fault, 43, 52, 550
Marine deltas, 341–342, 493–494
Marine depositional environments, 185–190

Marine evaporite basins, 186
Mariner 9 spacecraft, 448
Marine skeletons, 199
Marl, 509, 510
Mars:
    dust storms on, 448
    greenhouse effect, lack of, 513
    volcanoes/volcanism, 57–59
Marsh, Charles Perkins, 306, 307
Mass:
    measures of, A2–A3
    planetary, 12
Mass balance, 418–419
Mass extinctions, 11
Mass number, 70, 289
Mass wasting, 325–349
    n cold climates, 340
    definition, 326
    in deserts, 464
    events causing, 343–345
    and gravity, 327
    hazards to life/property, 345–347
    and plate tectonics, 347–349
    processes, 329–339
    rock debris, downslope movement of, 326–329
    sediment flows, 332–339
    slope failures, 329–332
    under water, 340–343
Matrix, 191
Matter, states of, 74, 76
Matthews, Drummond, 536
Mawson, Sir Douglas, 3–4
M-discontinuity, see Mohorovičić discontinuity
Meander, 363
Mechanical deformation, 214
Medial moraines, 427
Mega- (prefix), A1
Megascopic features, 93
Megathrust, 262
Mélange, 539
Melting (of rock), 111–113
Melting point, pressure, 416
Mendeleev, Dmitri, 73
Mercalli Scale, 258–259
Mesa, 464
Mesosphere, 17, 18
Mesozoic era, 287
Metals, 83, 90–92
Metallic bonding, 72–73, 83t
Metallic luster, 80
Metallogenic provinces, 582
Metamorphic aureole, 216
Metamorphic dehydration, 266
Metamorphic facies, 218–219, 219t
Metamorphic rock, 93, 203–204, 211–214.
    See also Metamorphism
    amphibolite, 213
    conventions for naming, 212
    definition, 25
    gneiss, 212
    granulite, 213
    greenschist, 212, 213
    marble, 213, 214
    phyllite, 211
    quartzite, 213, 214
    schist, 212

    slate, 211
    tectonic history, 206
    types of, 211–214
Metamorphic zones, 218
Metamorphism, 203–222
    of basalt, 212–213
    burial, 217
    cataclastic, 215–216
    contact, 216–217
    definition, 204
    factors in, 205–209
    high-grade, 205
    of limestone/sandstone, 213, 214
    limits of, 209–210
    low-grade, 205
    metamorphic facies, 218–219
    metasomatism, 217, 219–221
    and plate tectonics, 220–222
    prograde, 206
    progressive, 213t
    regional, 217–218
    and rock cycle, 221–222
    rocks' response to temperature/pressure in, 210–212
    types of, 214–218
Metasomatism, 217, 219–221
Meteor Carter (Arizona):
    catastrophism, 11
    planetary accretion, 16
Meteorites:
    catastrophism, 10–11
    impact energy, 16
    iron, 18, 267
    planetary accretion, 16
Methane, climate change and, 516
Metric measures, A1–A2
Mica, 102
Micro- (prefix), A1
Microscopic features, 93
Middle Cretaceous Period, 521–524
Midocean ridge basalt (MORB), 117, 538
Midocean ridges, 46
Migmatites, 209
Milankovitch, Milutin, 436, 438
Milankovitch cycles, 510, 543
Milli- (prefix), A1
Mines, abandoned, 82
Minerals, 68–95, 572t
    carbonates, 83, 89
    and chemical weathering, 158, 158t
    composition of, 69–74
    crystal structure of, 74–77
    definition, 68–69
    environmental clues provided by, 92
    hardness of, 81
    index, 217–218
    key characteristics, 69
    ore minerals, 83, 90–92
    oxides, 83, 92
    phosphates, 83, 89–90
    and plate tectonics, 93–95
    properties of, 77–83, A7–A11
    rocks as mixtures of, 92–93
    rocks vs., 69
    silicates, 83–89
    sulfates, 83, 90
    sulfides, 83, 90
Mineral assemblage, 93, 210, 212

Mineral chemistry, 69–74
    bonds, 71–73
    complex ions, 73
    compounds, 71
    energy-level shells, 70–71
    ions, 71, 73
    periodic table, 73, 74
Mineral deposits:
    hydrothermal, 565–566, 568–569
    magmatic, 116, 565, 566
    and plate tectonics, 582
    residual, 565, 571, 572
    sedimentary, 565–570
Mineraloids, 69
Mineral resources, 561–582
    energy resources, 573–581
    fossil fuels, 573–578
    gangue, 564
    ore, 563–564
    placers, 570, 571
    and plate tectonics, 582
    useful substances, 572–573, 572t
Minor plutons, 107–108
Mississippi Delta, 493, 494
Modified Mercalli Scale, 258–259
Mohorovičić, Andrija, 266
Mohorovičić discontinuity (Moho), 266–267
Mohs relative hardness scale, 81, 81t
Molecule, 71–72
Mollisols, 163
Monocline, 240
Monsoons, 147–148
Montreal Protocol on Substances that
    Deplete the Ozone Layer, 515
Moon, 484, 485, 513
Moraine, 427–428
MORB, *see* Midocean ridge basalt
Morley, Lawrence, 536
Motion of plates, 54–59
Mountains:
    Alps, 553, 554
    Appalachians, 552–554
    Canadian Rockies, 554–556
    creation of, 549–556
    and global weathering rates, 168–169
Mount St. Helens, 128–131, 133, 180
Mudflow, 139, 333–335
Mudstone, 192

N

NADW, *see* North Atlantic Deep Water
Nanometer, A1
Nano- (prefix), A1
Natural gas, 562
Natural resources, 561–582
Nile River, 379
Nitrogen cycle, 24–25
Nitrous oxide, 516–517
Noble gases, 73
Nonconformity, 280
Nonexplosive eruptions, 125, 126
Nonmarine depositional environments, 184–185
Nonmetric measures, A1–A2
Nonsorted sediment, 178–179
Normal earthquake, 45
Normal faults, 45, 46, 235–236
North America, 460–461, 469
North Atlantic Deep Water (NADW), 482, 483

Nuclear energy, 579
Nuclear testing, 475–476
Nugget, 368
Numerical age, 278

O

Observation, 8
Obsidian, 102
Oceans, 475–490, 505. *See also* Coastal
    landforms/deposits
    changes in circulation of, 439–440
    characteristics, 476–480
    circulation, 439–440, 480–484
    coastal erosion/sediment transport,
      489–491
    depth of, 476, 477
    and Earth's surface, 35–37
    life cycle, 505
    and plate tectonics, 505
    salinity, 477–479
    sediment, 505
    submarine volcanism in, 141–142
    temperature, 478–480
    tides, 484–486
    waves, 486–488
Ocean floor:
    glacial deformation of, 434
    and plate tectonics, 117
    spreading of, 535, 536
    topography of, 53–54
    windblown dust on, 462
Oceanic crust, 17, 535–536
Oceanic lithosphere, 18
Ocean sediments, 461, 462
Offshore sediments, 185–186
O horizon, 160
Oil, *see* Petroleum
Oil pools, 241
Oil reserves, 576, 578
Oil shale, 195, 574, 578
Old Red Sandstone, 10
Olivines, 76–77, 85–86, 102
Oolitic limestone, 194
Ooze, deep-sea, *see* Deep-sea ooze
Open fold, 240
Open system(s):
    definition, 19, 20
    interactions among Earth's, 21-27
Ophiolites, 117, 538-539
Oracoke Island (North Carolina), 196
Orbit, Earth's, 438–439
Orbitals, 70
Ore, 563–564
Ore minerals, 83, 90–93
Organic compounds, 574
Original Horizontality, Law of, 8, 278
Orogen, 306, 307, 544
Outer core:
    definition, 18
    wave shadows, 267
Outwash, 428
Outwash plain, 428
Outwash terraces, 428
Overland flow, 356
Oxbow lake, 364
Oxidation, 154–155
Oxides, 83, 92
Oxisols, 163

Oxygen, biogeochemical cycle and, 24
Ozone (in lower atmosphere), 516–517
Ozone layer, 27, 515, 516

P

Pahoehoe, 124–125
Paleokarsts, 407
Paleomagnetism, 535
Paleoplacer, 571
Paleosols, 165–166
Paleozoic era, 287
Pangaea, 46, 533–534
    formation of, 541–542
    Wegener's continental drift theory, 533–534
Parabolic dune, 458, 459
Parent material (soil), 160
Parent (radioactive), 289
Particle shape in clastic sediment, 179
Passive continental margins, 545–547
Passive margins, 36, 545–547
Pavement karst, 406
Peat, 190, 191, 574, 575
Peat swamps, 575
Pediment, 466–467
Pegmatite, 101
Perched water body, 390
Percolation, 388
Peridotite, 103, 111, 539
Periglacial zones, 430–431
Period, 283, 287–288, 287t
Periodic table, 73, 74t
Permafrost, 430–432
Permeability, 387
Perovskite, 270
Perrault, Pierre, 385
Petrified wood, 402
Petrogenetic grid, 219
Petroleum, 575–578
    distribution of, 576
    oil reserves, 576, 578
    oil shale, 195, 574, 578
    origins of, 575–576
    tar, 577
Phanerite, 101
Phanerozoic eon, 287
Phenocrysts, 101
Phosphates, 83, 89–90
Phosphorous deposits, 192–193
Photosynthesis, 24
Phyllite, 211
Physical geology, 7–8
Physical weathering, 149–152
Pico- (prefix), A1
Piedmont glaciers, 414
Pile (nuclear), 578
Pillow basalts, 136
Placer, 367, 490, 565
Placers, 570, 571
Placer deposits, 366–368, 571
Plain, outwash, 428
Planets:
    classification of, 12
    Jovian, 12–15
    Jupiter, 14, 15
    Mars, 57–59, 448, 513
    temperature gradient, 15
    terrestrial, 12–16
    Venus, 33, 53–54, 60, 513, 514

Planetary accretion, 15–16
Plants, rock alteration by, 152
Plates:
  and mantle convection, 41–42
  motion of, 54–59
  velocity of, 536–538
Plateau basalts, 136
Plate boundaries, 538–541
Plate friction, 60
Plate margins, 42–52
  convergent margin/collision zone, 43, 51–52
  convergent margin/subduction zone, 42–43, 48–51
  divergent margin, 42, 46–48
  transform fault margin, 43, 52
Plate motion, 54–59
  absolute vs. relative, 54
  and mantle plumes, 54, 55
  tracking of past motion, 532–541
Plate tectonics, 18–19, 26, 32–34
  and carbon cycle, 319–320
  causes of, 59–61
  and climate, 244
  and climate change, 525–528
  CO_2 levels, uplift and, 320–321
  and continental drift hypothesis, 34
  and deserts, 469–471
  and earthquakes, 270–272
  and geologic time, 300
  and groundwater, 407
  isostatic feedbacks, uplift and, 321–322
  and landscape evolution, 318–322
  and mass wasting, 347–349
  and metamorphism, 220–222
  and minerals, 93–95
  and monsoons, 557–558
  and oceans, 505
  and ore minerals, 93
  and plate collision, 557–558
  and sedimentation, 198–199
  and streams, 377–379
  and strike-slip faults, 244
  tectonic cycle, 23, 26, 28
  theory of, 19
  Uniformitarianism, 10
  and volcanoes, 140–141
  and weathering, 168–170
Plate triple junction, 546, 547
Plate velocity:
  and magnetic records, 536–538
  relative vs. absolute, 537
  variations in, 537, 538
Playa, 466
Plinian eruption, 127
Plume volcanism, 57–59
Plunging fold, 240
Plutons, 106–109
  definition of, 107
  major, 108–109
  minor, 107–108
Point bar, 364
Polar cell, 450
Polar desert, 463, 464
Polar front, 450
Polar glaciers, 416–417
Polarity-reversal time scale, 298–299
Polar soils, 162

Polar wandering, 535
Pollution, 154, 396–400. See also Human activity
Polymerization, 84, 85
Polymorphs, 79
Polymorphic transition, 270
Poorly sorted sediment, 176, 178
Pores, 386
Porosity, 386–387
Porphyry, 101, 102
Porphyry copper deposits, 582
Portland cement, 215
Positron emission, 289
Potassium-argon dating, 292–293, 295
Power, measures of, A3
Precession, 438
Precipitation, 566
  changes due to global warming, 520
  and mass wasting, 344
Pressure melting point, 416
Principle of Isostasy, see Isostasy
Principle of stratigraphic superposition, 279
Principle of Uniformitarianism, see Uniformitarianism
Principle (scientific), 8
Prograde metamorphism, 206, 209
Progressive metamorphism, 213t
Proterozoic eon, 287
Pumice, 126
P waves (primary waves), 252, 254
Pyrite, 90
Pyroclastic flow, 128, 138
Pyroclastic rocks, 104
Pyroclasts, 104
Pyroxenes, 86, 102
Pyrrhotite, 90

Q
Quartz, 88, 102
Quartzite, 213, 214

R
Radioactive dating, see Radiometric dating
Radioactive decay, 289–290
  and Earth's internal structure, 16
  and exhumation rate calculation, 314, 316–317
Radioactive waste, underground storage of, 399, 400, 580
Radioactivity, 289–297
  and Earth's age, 27, 297
  and geologic column, 294, 296–297
  natural radioactivity, 289
  radiocarbon dating, 293–294
  rates of decay, 290–293
Radiocarbon dating, 293–294
Radiogenic helium, 314, 316
Radiometric age, 278, 291
Radiometric dating, 12, 289–297, 293t
  and geologic column, 294, 296–297
  isotopes used in, 293t
  of ocean crust, 536
  potassium-argon dating, 292–293, 295
  radiocarbon dating, 293–294
Raindrop impressions, 196
Rainshadow desert, 463
Rainwater, 266, 384–385
Rayleigh, Lord, 39

Rayleigh number, 39
Rayleigh waves, 250
Recharge area, 387
Recrystallization, 190
Recurrence interval, 359
Red Sea, 569
Reefs, 496–497
Reef limestone, 194
Refraction, 488
Regional metamorphism, 217–218
Regolith, 21
  creep, 337t
  environmental clues from, 92
Relative age, 278
Relative plate motion, 54
Relative plate velocity, 537
Relict plate boundaries, 538–541
Relief:
  definition, 305
  and denudation, 309, 310
  and plate tectonics, 349
Replacement, 402
Reservoirs, 21–23
Reservoir rock, 241
Resinous luster, 80
Resources of minerals/energy, 561–582
Resurgent domes, 134
Retrograde metamorphism, 209
Reverse faults, 236
Rhone Glacier, 435
Rhyodacites, 103
Rhyolites, 103
Rhyolitic magma, 110, 113, 116, 123
Rhythmic layering, 179
Richter, Charles, 251
Richter magnitude scale, 251, 254–255, 258–259
Ridge push, 60
Rift, 235
Rift margins, 378
Rift Valley, 546, 547
Right-lateral strike-slip fault, 237
Rip currents, 488
Ripple marks, 196
Rivers, see Streams
River terraces, 313–314
Rock(s):
  adiabatic expansion of, 40–41
  biogenic sedimentary, 193–195
  chemical sedimentary, 192–193
  clastic sedimentary, 191–192, 195
  definition, 23, 68
  deformation of, 38, 225–244
  ductile, 45
  extrusive igneous, 101, 103, 104
  granitic, 103
  igneous, 23, 92, 100–119, 105t
  intrusive igneous, 101, 103
  layered, see Stratigraphy
  magnetism in, 297, 298
  metamorphic, 25, 93, 203–204, 206, 211–214
  minerals vs., 69
  as mixtures of minerals, 92–93
  pyroclastic, 104
  sedimentary, 23, 25, 92, 173–198
  viscosity of, 38
  weathering effect on, 147–170

Rock cycle, 23, 25–26
    color, environmental clues from, 196–197
    definition, 23
    and Earth's age, 300
    Hutton and, 92
    interconnectivity with other cycles, 28
    and metamorphism, 221–222
    origins of, 300
    and rock types, 93–95
    sediment in, 174
Rock debris, downslope movement of, 326–329
Rock drumlins, 426
Rockfall, 329–331
Rock flour, 427
Rock glacier, 340
Rock record, see Geologic time
Rockslide, 331, 332
Rock-stratigraphic record, 282
Rock-stratigraphic unit, 282
Rocky (cliffed) coasts, 492
Rodinia, 542
Roof rock, 241
Rotation pole, 538
Roundness, sphericity vs., 179
Rubey, William, 328
Runoff, 356
Rupture front, 249
Rutile, 92

S
Salinity, of ocean, 477–479
Salt, 7, 181, 562
Saltation, 366, 450, 451
Salt domes, 192
San Andreas Fault, 52, 265
    strike-slip fault, 244
    and transform fault continental margins, 550
Sand, windblown, 450–452
Sand dunes, see Dunes
Sand ripples, 451, 452
Sand seas, 459
Sandstone:
    clastic sedimentary rock, 191, 192
    metamorphism of, 213, 214
    Uniformitarianism, 10
San Francisco Bay, 500
Saturated zone, 385, 388
Schist, 212
Schistosity, 210
Schmidt, Jack, 6
Scientific method, 8, 9
Sculpturing, glacial, 423–426
Sea arches, 492, 493
Sea caves, 492
Seacliffs, protection of, 502
Seafloor, see Ocean floor
Seafloor spreading, 536
Sea ice, reduction due to global warming, 520
Sea level:
    and coastal evolution, 498–501
    lowering of, 432–433
    rise due to global warming, 520
Seamounts, 55
Sea stacks, 492, 493

Seawater. See also Oceans
    contamination of groundwater by, 396–397
    mineral composition, 92
    role of, at spreading centers, 48
    and submarine volcanism, 141–142
    tectonic cycle, 26
Secondary enrichment, 572
Sediment, 23, 173–190
    beach, 185, 491
    bioclastic, 182
    biogenic, 175, 182
    carbonate, 186, 199
    chemical, 175, 181
    clastic, 175–181, 192
    deep-sea fans, 187–188
    deep-sea ooze, 189
    definition, 23
    deltaic, 185
    eolian (windblown), 185
    estuarine, 185
    glacial, 185
    glaciation and transport of, 426–427
    grain size, 176t
    lake, 184–185
    land-derived, 189–190
    movement by wind, 450–454
    nonsorted, 178–179
    ocean, 461, 462, 505
    and ocean life cycle, 505
    offshore, 185–186
    offshore transport/sorting of, 489–491
    poorly sorted, 176, 178
    in rock cycle, 174
    sorting, 176, 178
    stream, 184
    transport/deposition, 174, 489–491
    turbidite, 186–188
    wave transport, 489
    wind transport, 450–454
Sedimentary depositional environments, see Depositional environments
Sedimentary facies, 183–184
Sedimentary rock, 92, 173–198
    banded iron deposits, 192
    biogenic, 193–195
    chalk, 194
    chemical, 192–193
    chert, 193
    clastic, 191–192, 195
    coal, 194–195
    coquina, 194
    definition, 23, 25
    dolostone, 194
    environmental clues in, 195–198
    evaporite deposits, 192
    limestone, 193–194
    lithographic limestone, 194
    oil shale, 195
    oolitic limestone, 194
    phosphorous deposits, 192–193
    reef limestone, 194
Sedimentation, 174–175, 198–199
Sediment drifts, 188–189
Sediment flows, 332–339
Sediment yields, 315, 369, 370
Seismic discontinuity zones, 270

Seismic gaps, 263
Seismic moment, 251
Seismic sea waves, 258. See also Tsunami
Seismic shadow zones, 267, 268
Seismic tomography, 270, 271
Seismic waves, 38, 250, 252–254, 266–270
Seismology, 248. See also Earthquakes
Seismometers, 248–249
Selenium, 398–399
Serpentine/serpentine group, 86, 88, 91, 117, 538–539
Sewage, pollution of groundwater by, 396
Shadow zones, 267, 268
Shale, 574
    clastic sedimentary rock, 192
    oil shale, 195
    progressive metamorphism, 213t
Shear stress, 227
Sheets, 86, 88
Sheet erosion, 366
Shield volcanoes, 129, 130
Shocks, mass wasting and, 343
Shoreline depositional environments, 185–186
Shore profile, 490–493
Siberia, 16
Siccar Point (Scotland), 10, 278
Silica, 125
Silicate anion, 84
Silicates, 83–89
Siliceous ooze, 182
Silicon, 16, 17
Silicosis, 454
Sills, 107–108
Siltstone, 191, 192
Sinkhole, 404
Sinkhole karst, 405
SI units, A1
Skinner, H. Catherine, 278
Slab friction, 60
Slab pull, 60
Slate, 211
Slaty cleavage, 210
Slickensides, 239
Slip face, 457
Slope angle, 327
Slope failures, 329–332
Slope failures (submarine), 345
Slope modification, 343–344
Slump, 329, 331
Slurry flow, 332–335
Smeaton, John, 215
Smith, William, 283–284
Snow, 417–418
Snowball Earth hypothesis, 543–544
Snowline, 417
SO_2 (sulfur dioxide), 15
Soft water, 396
Soil:
    classification, 160–164, 163t
    erosion, 166–168
    formation, rate of, 164–165
    and global warming, 520
    and monsoon history, 147–148
    origin, 160
    paleosoil, 165–166
Soil decomposition, 520

Soil horizons, 160–162
Soil profile, 160
Solar energy, 198, 580–581
Solar system, 12–16. *See also* Planets
Sole marks, 196
Solifluction, 333
Sorting:
    offshore, 490, 491
    sediment, 176, 178
Source rock, 241
Spall, 151, 152
Spatter cone, 125
Specific gravity, 82
Sphalerite, 90
Sphericity, roundness vs., 179
Spheroidal weathering, 156
Spinels, 270
Spits, 494, 495
Spodosols, 162
Spreading centers, 48
Springs, 389, 390
Stable platform, 544
Stalactites, 403
Stalagmites, 403
Star dune, 458
States of matter, 74, 76
Steady-state landscapes, 312
Steno, Nicolaus, 77
Stocks, 109
Stoping, 109
Storms, 501, 520
Strain, 227
Strain rate, 230
Strata, *see* Stratum
Stratabound deposits, 569–570
Stratification, 174–175
Stratification (oceanic), 480
Stratified drift, 428–430
Stratigraphic classification, 282–283
Stratigraphic record, gaps in, 280–282
Stratigraphic superposition, 279–280
Stratigraphy, 278–282
    definition, 278
    gaps in stratigraphic record, 280–282
    laws of, 278–280
Stratovolcanoes, 130, 132
Stratum(strata):
    definition, 175
    relative ages of, 279–280
    warped, 313
Streak, 80–81
Streams, 353–379
    channel patterns, 362–365
    channels, 355–356
    definition, 354
    deposits, 370–375
    drainage systems, 375–377
    erosion, 366
    load, 366–370
    and plate tectonics, 377–379
    streamflow dynamics, 356–362
Stream capture, 377
Streamflow, 356–362
Stream order, 375–376
Stream sediments, 184
Strength regions, 18
Stress, 38, 227

Striations, glacial, 423–424
Strike, 232
Strike-slip earthquakes, 45
Strike-slip faults, 45, 46, 237–239
    definition, 45
    and plate tectonics, 244
San Andreas Fault, 244
Structural geology, 232
Subbituminous coal, 574, 575
Subduction, 41, 49
Subduction mélange, 539
Subduction zones:
    convergent margin/subduction zone,
        42–43, 48–51
    earthquakes in, 50–51
    volcanoes above, 49–50
Submarine hot spring, 565
Submarine landslides, 342, 343
Submarine slope failures, 345
Submergence, 498, 499
Subsequent stream, 378
Subsidence of land surface, 395
Subtropical desert, 462–463
Sulfates, 83, 90
Sulfides, 83, 90
Sulfur, 15
Sulfur dioxide (SO2), 15
Sun, 13–14, 485
Supercontinents, 541–544
    Pangaea, 533–534, 541–542
    Rodinia, 542
Supernova, 13
Superplume, 522, 523
Superposed stream, 378
Surf, 487, 488
Surface creep, 450
Surface currents, 480–481
Surface waves, 250, 252–253
Surf zone, 487, 489
Surges, 422–423
Suspect terranes, 550
Suspended load, 366, 368
S waves (secondary waves), 252, 254
Syncline, 240
Synthetic seismograms, 251
Syr Dar'ya River, 7
System(s), 19–21
    artesian, 392–393
    climate, 511
    closed, 19, 20
    current, 481
    definition, 19
    drainage, 375–377. *See also* Streams
    Earth, 20–21, 470–471
    global ocean conveyor, 481–484
    isolated, 19
    open, 19–27
    as unit of time-stratigraphic record, 282,
        283
    wind (planetary), 447–450

T

Taconites, 569
Talus, 332
Tarns, 424
Tectonics. *See also* Plate tectonics
    definition, 26

of Europa, 54
of Venus, 53
Tectonic cycle, 26
    definition, 23
    interconnectivity with other cycles, 28
Tectonic desertification, 470–471
Temperate glaciers, 416–417
Temperate-latitude soils, 162–163
Temperature:
    measures of, A3
    ocean, 478–480
    and rock deformation, 229
    trends, historical, 517–518
Temperature gradient, 15
Tensional stress, 227
Tephra, 104, 138, 139
Tephra cones, 130
Tephra eruption, 140
Tephra fall, 127, 128
Tera- (prefix), A1
Terminal moraines, 428
Terminus, 420
Terrace(s), 371–372
    outwash, 428
Terrane, 550–551
Terrestrial planets, 13–16
    accretion, 15–16
    chemical composition, 14, 15
    definition, 12
Testability, 8
Tetrahedra, 85–87
Texture, rock:
    definition, 93
    stress and, 205, 207, 208, 210
Theory, 8
Theory of the Earth, with Proofs and
    Illustrations (James Hutton), 9
Thermal expansion, 39
Thermohaline circulation, 482, 483
Threshold effect, 312, 313
Thrust earthquake, 45
Thrust faults, 45, 46, 236, 237
Tidal bore, 486
Tidal bulges, 484, 485
Tidal energy, 581
Tidal power, 486
Tidal range, 486
Tides, 484–486
Tide-raising force, 484, 485
Till, 427
Tillites, 427, 442
Tilt (of Earth's axis), 438
Time, *see* Geologic time
Time scale, geologic, *see* Geologic time
    scale
Time-stratigraphic record, 282, 283
Time-stratigraphic unit, 282
Tombolo, 494, 495
Topography, 34, 243, 244
Topset layer, 374
Tower karst, 405, 406
Toxic groundwater, 398–399
Toxic waste, 397, 399
Trace gases, greenhouse effect and, 513,
    514
Tradewinds, 450
Trails, 196

Transform faults, 45, 237–239
Transform fault continental margin, 550
Transform fault margin, 43, 52, 550
Transport/deposition of sediment, 174, 489–491
Transporting medium, 208
Transverse dune, 458
Trench, 37
Tributary, 357
Triple junction, 546, 547
Tropical soils, 163, 164
Tsunami, 140, 258, 501–502, 501t
Tuffs, 104, 106
Tunguska explosion, 16
Turbidite, 186–188
Turbidity currents, 186–187
Turbulent flow, 366, 453
Twain, Mark, 364

U

Ultisols, 163
Unconfined aquifer, 390
Unconformity, 280–282
Undercutting, 344
Uniformitarianism, 9–10
  geologic processes, rates of, 26–27
  Principle of, 9–10
Uniform stress, 38, 205, 207, 227
Unsaturated zone, 385
Uplift, 304, 313–314
  and CO2 levels, 320–321
  collisional, 306
  extensional, 308
  factors controlling, 305–309
  isostatic, 306
  and isostatic feedbacks, 321–322
Uplift rate, calculation of, 313–314
Uranite, 92
Uranium, 578
Uzbekistan, 7

V

Valleys, glacial, 424, 425
Valley glaciers, 414, 419, 424
Valley train, 428, 429
Van der Waals bonding, 73, 83t
Varve, 179
Vegetation, changes due to global warming, 520
Vein, 208–209
Velocity-weakening behavior, 265
Ventifact, 456
Venus:
  geology of, 33
  greenhouse effect, 513, 514

and plate tectonics, 60
topography of, 53–54
Vertical stratification, 480
Vertisols, 163
Vesicles, 126
Vesicular texture, 126
Vine, Frederick, 536
Viscosity, 39
  of magma, 124
  of rocks, 38
Vitreous luster, 80
Volcanic arc, 49–50
Volcanic ash, 126–127, 461, 462
Volcanic crater, 132–133
Volcanic disasters, 140t
Volcanic island arc, 49
Volcanic neck, 108
Volcanic pipe, 108
Volcanoes/volcanism, 129–142, 140t. See also Eruptions; Magma
  above subduction zones, 49–50
  central eruptions, 129–136
  and climate of Middle Cretaceous, 522
  and Earth's internal structure, 17
  eruptions, 125–137, 140
  fissure eruptions, 136–137
  Hawaiian submarine landslides, 342, 343
  and hot spots, 57
  on Mars, 57–59
  and mass wasting, 344, 345
  and plate tectonics, 140–141
  posteruption effects, 137
  and seawater, 141–142
  shield, 129, 130
  stratovolcanoes, 130, 132
  tectonic cycle, 26
  tephra cones, 130
  on Venus, 57, 58
Volcanogenic massive sulfide deposits, 565
Volume, measures of, A2
Von Laue, Max, 77, 78
Voyager 2, 14

W

Warped strata, uplift rate calculation and, 313
Waste disposal, 580
Water. See also Groundwater; Oceans; Seawater; Streams
  and mass wasting, 328–329
  and metamorphism limits, 209–210
  and metasomatism, 219, 220
  and plate tectonics, 60–61
  rainwater, 266, 384–385
  states of, 76

Waterflow, velocity of, 356
Water masses, 481, 482
Water table, 385–386
  definition, 385
  lowering of, 394–395
Waves (ocean), 486–489
Wave base, 487
Wave-cut bench, 492
Wave-cut cliff, 492
Wave energy, 581
Wavelength, 486
Wave motion, 486–488
Wave refraction, 488
Wave shadows, 267, 268
Weak fault behavior, 265–266
Weathering, 147–170
  chemical, 148–149, 152–157, 153t
  in deserts, 464–465
  factors affecting, 157–160
  physical, 149–152
  and plate tectonics, 168–170
  soil as product of, 160–166
  soil erosion, 166–168
Weathering rind, 161
Wedging, 151, 152
Wegener, Alfred:
  continental drift hypothesis, 33–34, 41, 59, 533
  and Pangaea, 46, 533–534, 541
Welded tuff, 106
Wells, 389–391
Well-sorted sediment, 176, 178
Wilson, J. Tuzo, 55, 237
Wind, 447–462. See also Desert
  dust in ocean sediments/glacier ice, 461, 462
  eolian deposits, 457–461
  erosion caused by, 454–456
  as geologic agent, 447
  and global atmosphere circulation, 447–450
movement of sediment by, 450–454
Wind energy, 581
Wind system, planetary, 447–450
Witwatersrand basin (South Africa), 562, 570
World ocean, 476, 477

Y

Yangtze River, 362
Yardang, 456
Younger Dryas event, 526–527

Z

Zeolites, 217
Zone of aeration, 385, 388

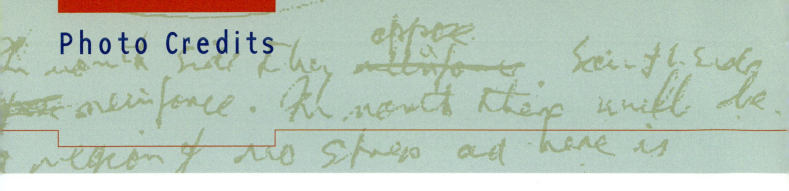

Photo Credits

Chapter 1
Chapter Opener 1: © Anne Heimann/Corbis Stock Market. Chapter Opener 1 (inset): Brian J. Skinner. Fig. 1.1a: © Robert Holland/DRK. Photo. 1.1b: Lunar & Planetary Institute. Fig. 1.1c: © K & M Krafft/ Explorer/Photo Researchers. Fig. 1.1d: © Catherine Ursillo/Photo Researchers. Fig. 1.1e: David Austen/Stock, Boston. Fig. 1.1f: © Catherine Ursillo/Photo Researchers. Fig. 1.2: Tom & Susan Bean, Inc. Fig. 1.3a: © John S. Shelton. Fig. 1.3b: Francois Gohier/Ardea London. Fig. 1.4: Dr. K. Roy Gill. Fig. 1.5: W. Alvarez/Photo Researchers. Fig. 1.6: Michael Collier. Fig. 1.8: Comstock Images. Fig. 1.11: Space Photography Lab of Arizona. Fig. 1.12: Breck Kent. Fig. 1.17: Tom Bean/DRK. Box 1.1: David Turnley/Corbis Images. Box 1.4: Phil Degginger/Bruce Coleman, Inc.

Chapter 2
Chapter Opener 2: Michael Collier/DRK Photo. Chapter Opener 2 (inset): Jeffrey Park. Fig. 2.9 a,b,c: Jeffrey Park. Fig. 2.14: Ric Ergenbright. Fig. 2.15, 2.18, 2.19, 2.21-2.23: Jeffrey Park. Fig. 2.23 (inset): NASA Media Services. Fig. 2.24: USGS/Tom Stack & Associates. Fig. 2.25 & 2.26: Jeffrey Park. Changing Landscape: Map on Pages 64 & 65: Jeffrey Park. Page 64 (top): Morton Beebe/Corbis Images. Page 64 (center): Paul A. Soders/Corbis Images. Page 65 (top): Corbis Images. Page 65 (center): Kevin Schafer/Corbis Images. Page 65 (bottom): Corbis Images.

Chapter 3
Chapter Opener 3(a): E.R. Degginger/Photo Researchers. Chapter Opener 3(b): Breck Kent. Chapter Opener 3 (inset): Brian J. Skinner. Fig. 3.7a: Michael Hochella. Fig. 3.9: William Sacco. Fig. 3.10: Brian J. Skinner. Fig. 3.11: Breck Kent. Fig. 3.12 & 3.13a: William Sacco. Fig. 3.14: Breck Kent/Animals Animals. Fig. 3.15a: William Sacco. Fig. 3.15b: Breck Kent. Fig. 3.16: William Sacco. Fig. 3.17: The Natural History Museum. Fig. 3.18: William Sacco. Fig. 3.23a: Brian J. Skinner. Fig. 3.23b: Boltin Picture Library. Fig. 3.23c-g: Breck Kent. Fig. 3.25: Brian J. Skinner. Fig. 3.26: William Sacco. Fig. 3.27 & 3.28: Brian J. Skinner. Fig. 3.29: William Sacco. Fig. 3.30a: Breck P. Kent/Animals Animals/Earth Scenes. Fig. 3.30b&c: The Natural History Museum. Fig. 3.32: Brian J. Skinner. Box

3.3: © John Cancalosi/DRK Photo. Box 3.4: Photo Index.

Chapter 4
Chapter Opener 4: Breck Kent. Chapter Opener 4 (inset): Brian J. Skinner. Fig. 4.1a: Breck Kent. Fig. 4.1b: William Sacco. Fig. 4.2a: William E. Ferguson. Fig. 4.2b: Breck Kent. Fig. 4.3: Martin Miller. Fig. 4.4: Tony Waltham Geophotos. Fig. 4.5 & 4.7: Brian J. Skinner. Fig. 4.8a: William E. Ferguson. Fig. 4.8b: J.P. Lockwood//USGS. Fig. 4.8c: Steven L. Nelson/Alaska Stock Images. Fig. 4.9a: William E. Ferguson. Fig. 4.9b: Martin Miller. Fig. 4.10a: Ron Sanford/f/STOP Pictures. Fig. 4.12a: John Eastcott/Yva Momtiuk/DRK Photo. Fig. 4.12b: William E. Ferguson. Fig. 4.13: Breck Kent. Fig. 4.15: Fred Hirschmann. Fig. 4.19 & 4.21b: Brian J. Skinner.

Chapter 5
Chapter Opener 5: Visuals Unlimited. Chapter Opener 5 (inset): Stephen Porter. Fig. 5.2: G. Brad Lewis/Liaison Agency, Inc./Getty Images. Fig. 5.3 & 5.5: J.D. Griggs/USGS. Fig. 5.4: D.A. Swanson/Corbis Images. Fig. 5.6: G. R. Roberts. Fig. 5.7: A.J. Copley/Visuals Unlimited. Fig. 5.8: Roger Werth/Woodfin Camp & Associates. Fig. 5.9: ©AP/Wide World Photos. Fig. 5.11: Stephen Porter. Fig. 5.12a: Krafft Explorer/Photo Researchers. Fig. 5.12b: Tom Bean/DRK Photo. Fig. 5.14: Steve Vidler/Leo de Wys, Inc. Fig. 5.13: Stephen Porter. Fig. 5.15: Krafft-Explorer/Photo Researchers. Fig. 5.16: Lyn Topinka//USGS. Fig. 5.17: Rich Buzzelli/Tom Stack & Associates. Fig. 5.18: Brian J. Skinner. Fig. 5.20: Philip Richardson, Gallo Images/Corbis Images. Fig. 5.21: Stephen Porter. Fig. 5.22: Courtesy NOAA. Fig. 5.23: William E. Ferguson. Fig. 5.24: Stephen Porter. Fig. 5.25: Brian J. Skinner. Fig. 5.26: David Hiser/Picture Network International, Ltd. Box 5.1a: © Jeffery Hutcherson/DRK Photo. Box 5.1b: Fred Hirschmann. Changing Landscape: Page 144 (top left): Paul A. Soders/Corbis Images. Page 144 (top right): Corbis Images. Page 144 (bottom left): Jeffrey Park. Page 144 (bottom right): AFP/Corbis Images. Page 145: AFP/Corbis Images.

Chapter 6
Chapter Opener 6: Joseph Sohm/Photo Researchers. Chapter Opener 6 (inset): Stephen Porter. Fig. 6.1: Stephen Porter. Fig. 6.2: Jeff Gnass. Fig. 6.3: Tony Waltham

Geophotos. Fig. 6.4: Bill Hatcher. Fig. 6.5 & 6.6: Stephen Porter. Fig. 6.7: Stan Osolinski/Oxford Scientific Films Ltd. Fig. 6.8: Kenneth W. Fink/Ardea London. Fig. 6.9 (top): Stephen Porter. Fig. 6.9 (bottom): Telegraph Colour Library /Taxi/Getty Images. Fig. 6.10: © Natural History Photographic Agency. Fig. 6.11: Stephen Porter. Fig. 6.13: G. R. Roberts. Fig. 6.17-6.19, 6.21: Stephen Porter. Fig. 6.22: Frans Lanting/Minden Pictures, Inc. Fig. 6.23: Claus Meyer/Black Star. Fig. 6.24: ChromoSohn Inc./Corbis Images. Fig. 6.25: Stone/Getty Images. Fig. 6.26: Gordon Wiltsie/AlpenImages Ltd. Fig. 6.27: Brian J. Skinner. Box 6.1: J.A. Wilkinson/Valan Photos. Box 6.2: Stephen Porter.

Chapter 7
Chapter Opener 7 & inset: Stephen Porter. Fig. 7.1: Brian J. Skinner. Fig. 7.2, 7.4-7.7: Stephen Porter. Fig. 7.8: David Muench Photography. Fig. 7.9 & 7.10: Stephen Porter. Fig. 7.11a: Bruce Jackson/Jeff Gnass Photography. Fig. 7.11b: Martin Miller. Fig. 7.12: Kevin and Betty Collins/Visuals Unlimited. Fig. 7.13a&b: Part (a) photo by Patrizia Liveri & Part (b) photo by Henry Elderfield, "Carbonate Mysteries," Science 31 May 2002, Vol 296, pg 1618. Fig. 7.16, 7.19, 7.22: Stephen Porter. Fig. 7.23: G. Whitely/Photo Researchers. Fig. 7.24a: Stephen Porter. Fig. 7.24b- f: Stephen Porter. Fig. 7.25: Brian J. Skinner. Fig. 7.26a: Stephen Porter. Fig. 7.26b: Color-Pic, Inc. Fig. 7.27: Stephen Porter. Fig. 7.28: Kristin Finnegan/Stone/Getty Images. Fig. 7.29: William E. Ferguson. Fig. 7.30a: Stephen Porter. Fig. 7.30b: Stephen Trimble/DRK Photo. Fig. 7.31a: Gordon Wiltsie/Bruce Coleman, Inc. Fig. 7.31b: Stephen Porter. Fig. 7.32: Tom & Susan Bean, Inc.

Chapter 8
Chapter Opener 8: Brian J. Skinner. Chapter Opener 8 (inset): Breck Kent. Fig. 8.2 & 8.3: Brian Skinner. Fig. 8.4: William E. Ferguson. Fig. 8.5 & 8.6: Brian J. Skinner. Fig. 8.7: William E. Ferguson. Fig. 8.8 & 8.10: Brian J. Skinner. Fig. 8.11a: William E. Ferguson. Fig. 8.11b: Craig Johnson. Fig. 8.12a: Photo Researchers. Fig. 8.12b: Craig Johnson. Fig. 8.13, 8.17, and Box 8.1: Brian J. Skinner. Box 8.3: H.P. Merten/Corbis Stock Market.

Chapter 9
Chapter Opener 9: Francois Gohier/Photo